Remote Sensing Handbook, Volume VI (Six Volume Set)

Volume VI of the Six Volume *Remote Sensing Handbook*, Second Edition, is focused on the use of remote sensing in the assessment and monitoring of droughts, dry lands, biomass burning, disasters such as volcanoes and fires, and urban studies and nightlights. It discusses land degradation assessment and monitoring, greenhouse gas (GHG) emissions, and pollution from nightlights in megacities. Chapters include remote sensing of agricultural droughts, including US drought monitoring, dryland studies, coal fires, biomass burning and GHG emissions, volcanoes, humanitarian disasters, smart cities, and night lights mapping. This thoroughly revised and updated volume draws on the expertise of a diverse array of leading international authorities in remote sensing and provides an essential resource for researchers at all levels interested in using remote sensing. It integrates discussions of remote sensing principles, data, methods, development, applications, and scientific and social context.

FEATURES

- Provides the most up-to-date comprehensive coverage of remote sensing science for droughts, disasters, and GHG emissions.
- Discusses and analyzes data from old and new generations of sensors.
- Highlights remote sensing of agricultural droughts, humanitarian and natural disasters, and GHG emissions from coal and stubble burning.
- Includes numerous case studies on advances and applications at local, regional, and global scales.
- Introduces advanced methods in remote sensing, such as machine learning, cloud computing, and AI.
- Highlights scientific achievements over the last decade and provides guidance for future developments.

This volume is an excellent resource for the entire remote sensing and GIS community. Academics, researchers, undergraduate and graduate students, as well as practitioners, decision makers, and policymakers, will benefit from the expertise of the professionals featured in this book and their extensive knowledge of new and emerging trends.

Remote Sensing Handbook, Volume VI (Six Volume Set)

Droughts, Disasters, Pollution, and Urban Mapping

Second Edition

Edited by Prasad S. Thenkabail, PhD

CRC Press is an imprint of the
Taylor & Francis Group, an **informa** business

Designed cover image: © Prasad S. Thenkabail

Second edition published 2025
by CRC Press
2385 NW Executive Center Drive, Suite 320, Boca Raton FL 33431

and by CRC Press
4 Park Square, Milton Park, Abingdon, Oxon, OX14 4RN

CRC Press is an imprint of Taylor & Francis Group, LLC

© 2025 selection and editorial matter, Prasad S. Thenkabail; individual chapters, the contributors

First edition published by CRC Press 2016

Reasonable efforts have been made to publish reliable data and information, but the author and publisher cannot assume responsibility for the validity of all materials or the consequences of their use. The authors and publishers have attempted to trace the copyright holders of all material reproduced in this publication and apologize to copyright holders if permission to publish in this form has not been obtained. If any copyright material has not been acknowledged please write and let us know so we may rectify in any future reprint.

Except as permitted under U.S. Copyright Law, no part of this book may be reprinted, reproduced, transmitted, or utilized in any form by any electronic, mechanical, or other means, now known or hereafter invented, including photocopying, microfilming, and recording, or in any information storage or retrieval system, without written permission from the publishers.

For permission to photocopy or use material electronically from this work, access www.copyright.com or contact the Copyright Clearance Center, Inc. (CCC), 222 Rosewood Drive, Danvers, MA 01923, 978-750-8400. For works that are not available on CCC please contact mpkbookspermissions@tandf.co.uk

Trademark notice: Product or corporate names may be trademarks or registered trademarks and are used only for identification and explanation without intent to infringe.

Library of Congress Cataloging-in-Publication Data
Names: Thenkabail, Prasad Srinivasa, 1958- editor.
Title: Remote sensing handbook / edited by Prasad S. Thenkabail ; foreword by Compton J. Tucker.
Description: Second edition. | Boca Raton, FL : CRC Press, 2025. | Includes bibliographical references and index. |
  Contents: v. 1. Remotely sensed data characterization, classification, and accuracies — v. 2. Image processing, change detection, GIS and spatial data analysis — v. 3. Agriculture, food security, rangelands, vegetation, phenology, and soils — v. 4. Forests, biodiversity, ecology, LULC, and carbon — v. 5. Water, hydrology, floods, snow and ice, wetlands, and water productivity — v. 6. Droughts, disasters, pollution, and urban mapping.
Identifiers: LCCN 2024029377 (print) | LCCN 2024029378 (ebook) | ISBN 9781032890951 (hbk ; v. 1) |
  ISBN 9781032890968 (pbk ; v. 1) | ISBN 9781032890975 (hbk ; v. 2) | ISBN 9781032890982 (pbk ; v. 2) |
  ISBN 9781032891019 (hbk ; v. 3) | ISBN 9781032891026 (pbk ; v. 3) | ISBN 9781032891033 (hbk ; v. 4) |
  ISBN 9781032891040 (pbk ; v. 4) | ISBN 9781032891453 (hbk ; v. 5) | ISBN 9781032891477 (pbk ; v. 5) |
  ISBN 9781032891484 (hbk ; v. 6) | ISBN 9781032891507 (pbk ; v. 6)
Subjects: LCSH: Remote sensing—Handbooks, manuals, etc.
Classification: LCC G70.4 .R4573 2025 (print) | LCC G70.4 (ebook) | DDC 621.36/780285—dc23/eng/20240722
LC record available at https://lccn.loc.gov/2024029377
LC ebook record available at https://lccn.loc.gov/2024029378

ISBN: 978-1-032-89148-4 (hbk)
ISBN: 978-1-032-89150-7 (pbk)
ISBN: 978-1-003-54141-7 (ebk)

DOI: 10.1201/9781003541417

Typeset in Times
by Apex CoVantage, LLC

Contents

Foreword by Compton J. Tucker xiii
Preface xxi
About the Editor xxix
List of Contributors xxxiii
Acknowledgments xxxvii

PART I   Droughts and Drylands

Chapter 1   Agricultural Drought Monitoring Using Space-Derived Biophysical Products: A Global Perspective 3

Felix Kogan and Wei Guo

     1.1   Introduction 3
     1.2   Drought: Unusual Weather Disaster 4
     1.3   How to Measure Drought 4
          1.3.1   Traditional Approach 4
          1.3.2   Remote Sensing Approach 4
     1.4   Vegetation Health Interpretation 5
     1.5   Unusual Droughts of the 21st Century 6
     1.6   Drought Products 7
     1.7   Droughts in a Warmer World 11
     1.8   Conclusion 12

Chapter 2   Agricultural Drought Monitoring Using Space-Derived Vegetation and Biophysical Products: A Global Perspective 16

F. Rembold, M. Meroni, O. Rojas, C. Atzberger, F. Ham, and Erwann Fillol

     2.1   Introduction 17
          2.1.1   What Is Drought? 17
          2.1.2   How Does Drought Affect Agricultural Production and Food Security? 17
          2.1.3   What Can Remote Sensing Do? 18
     2.2   Operational Methods and Techniques 20
          2.2.1   NDVI-Based Vegetation Anomaly Indicators Used Mainly for Drought or Crop Monitoring Bulletins 21
          2.2.2   Agricultural Stress Index System Approach Targeting Agriculture 24
          2.2.3   Action Contre la Faim Approach Targeting Pastures 24
          2.2.4   JRC Approach for the Early Detection of Biomass Production Deficit Hot Spots 28
     2.3   Other Recent Methodological Approaches 29
     2.4   Discussion and Future Developments 32
     2.5   Conclusions 33

Chapter 3 Remote Sensing of Drought: Satellite-Based Monitoring Tools for the United States 38

Brian D. Wardlow, Martha A. Anderson, Tsegaye Tadesse, Mark S. Svoboda, Brian Fuchs, Chris R. Hain, Wade T. Crow, and Matt Rodell

    3.1   Introduction 39
           3.1.1   Drought Definitions and Monitoring 40
           3.1.2   Traditional, Station-Based Drought Monitoring Approaches 40
           3.1.3   US Drought Monitor 42
           3.1.4   Traditional Remote Sensing of Drought 44
           3.1.5   New Era of Remote Sensing Observations 45
    3.2   New Drought Monitoring Tools 45
           3.2.1   Vegetation Drought Response Index (VegDRI) 45
           3.2.2   The Forest Drought Index (ForDRI) 50
           3.2.3   Evaporative Stress Index (ESI) 57
           3.2.4   Microwave-Based Surface Soil Moisture Retrievals 63
           3.2.5   Satellite Gravimetry–Based Soil Moisture and Groundwater 68
    3.3   Concluding Thoughts 74

Chapter 4 Regional Drought Monitoring Based on Multi-Sensor Remote Sensing 87

Jinyoung Rhee, Jungho Im, and Seonyoung Park

    4.1   Introduction 88
    4.2   Recent Trends of Remote Sensing–Based Drought Monitoring 90
           4.2.1   Customized Drought Monitoring Approaches 90
           4.2.2   Development of New Remote Sensing–Based Drought Indices 91
           4.2.3   Data Assimilation for Drought Monitoring and Prediction 92
           4.2.4   Machine Learning for Drought Monitoring and Prediction 94
           4.2.5   Issues of Multi-Sensor Data Combination 94
    4.3   Development of the SDCI 95
           4.3.1   Drought in Humid Regions 95
           4.3.2   Introduction of SDCI, Solely Based on Remote Sensing Data 95
           4.3.3   Validation of the SDCI 95
           4.3.4   Development of the Advanced SDCI 98
    4.4   Case Study: Linearly Combined SDCI in the Korean Peninsula 99
    4.5   Case Study: Nonlinearly Combined SDCI in the USA 101
    4.6   Conclusions 104
    4.7   Acknowledgments 105

Chapter 5 Land Degradation Assessment and Monitoring of Drylands 109

Marion Stellmes, Ruth Sonnenschein, Achim Röder, Thomas Udelhoven, Gabriel del Barrio, and Joachim Hill

    5.1   Introduction 110
           5.1.1   Drylands 110
           5.1.2   Land Use in Dryland Areas 111
           5.1.3   Land Degradation and Desertification 112
           5.1.4   Scientific Perception of Land Degradation 113

Contents

5.2	Remote Sensing of Dryland Degradation Processes	113
	5.2.1 Suitable Remote Sensing Indicators for Dryland Observation	114
	5.2.2 Earth Observation Platforms Used in Dryland Observation	115
	5.2.3 Time Series Analysis Techniques	118
5.3	Assessing Land Condition	119
	5.3.1 Assessment of Land Condition Related to the Biological Productivity of Ecosystems	119
	5.3.2 Assessment of Land Condition Including Climate and Its Variability	120
5.4	Monitoring of Land Use/Land Cover Changes to Assess Land Degradation Processes	122
	5.4.1 Local-Scale Studies to Detect Land Degradation–Related Modifications	124
	5.4.2 Regional- to Global-Scale Studies to Detect Land Degradation–Related Modifications	125
5.5	Integrated Concepts to Assess Land Degradation	128
	5.5.1 Integrated Studies at Local Scale	128
	5.5.2 Integrated Studies at Regional to Global Scale	130
5.6	Uncertainties and Limits	132
	5.6.1 Uncertainties Regarding the Definition of Land Degradation and Its Derivation	132
	5.6.2 Uncertainties Regarding Remote Sensing Data	133
5.7	Summary and Conclusions	137

PART II Disasters

Chapter 6 Disasters: Risk Assessment, Management, and Post-disaster Studies Using Remote Sensing 153

Norman Kerle

6.1	Introduction	154
6.2	From Hazards to Disaster Risk—Terms and Concepts	155
6.3	Domain Developments and Selected Technical Advances	156
	6.3.1 Early Disaster Mapping with Remote Sensing	156
	6.3.2 Dawn of the Satellite Era	157
	6.3.3 Disaster Risk as a Multi-Faceted Spatial Phenomenon, and the Role of Operational Remote Sensing	159
6.4	Trends and Developments	176
	6.4.1 New Platforms	176
	6.4.2 New Sensors	179
	6.4.3 Better Data Analysis Methods	185
	6.4.4 Organization	185
6.5	Gaps and Limitations	186
	6.5.1 The Military Side	186
	6.5.2 Methodological Gaps in DRM	187
	6.5.3 Lack of Standards and Suitable Legislation	187
6.6	Summary	188

Chapter 7 Humanitarian Emergencies: Causes, Traits, and Impacts as Observed by Remote Sensing 199

Stefan Lang, Petra Füreder, Olaf Kranz, Brittany Card, Shadrock Roberts, and Andreas Papp

- 7.1 Introduction 200
  - 7.1.1 Humanitarian Disasters: A Particular Case? 200
  - 7.1.2 Forced Migrations and Regional Conflicts 200
  - 7.1.3 The Role of Satellite Remote Sensing in Humanitarian Action 202
  - 7.1.4 Information Needs for Humanitarian Action 207
- 7.2 Crisis-Related Earth-Observable Indicators 208
  - 7.2.1 "Early Warning" 208
  - 7.2.2 Crisis Monitoring 209
  - 7.2.3 Mid- to Long-Term Impact 210
- 7.3 (Satellite) Earth Observation Capacities 211
  - 7.3.1 Usage of EO Data—Chances and Challenges in the Humanitarian Development Peace (HDP) Nexus 211
  - 7.3.2 Different Tasks—Different Sensors 212
- 7.4 Image Analysis Techniques 218
  - 7.4.1 General Workflow—Example Population Monitoring 218
  - 7.4.2 Visual Image Interpretation 220
  - 7.4.3 Automated Feature Extraction and Image Classification ... 221
  - 7.4.4 Spatial Analysis and Modeling 222
  - 7.4.5 Validation and Ground Reference Information 223
- 7.5 Case Studies 224
  - 7.5.1 Monitoring Population Dynamics in the Refugee Camp Dagahaley, Kenya 224
  - 7.5.2 Evolution and Impact of the IDP Camp Zam Zam, Sudan at Local and Regional Scale 226
  - 7.5.3 Indications of Destructions—The Case of Abyei, Disputed Border Region between Sudan and South Sudan 227
  - 7.5.4 Logging and Mining Activities in Relation to the Conflict Situation in the Democratic Republic of the Congo (DRC) 232
- 7.6 Conclusion 235

PART III Volcanoes

Chapter 8 Remote Sensing of Volcanoes 243

Robert Wright

- 8.1 Introduction 243
- 8.2 What Do Volcanoes Do That We Might Be Interested in Measuring? 244
- 8.3 Why Use Satellite Remote Sensing to Study Active Volcanism? 246
- 8.4 Remote Sensing of Volcano Deformation 247
  - 8.4.1 Quantifying the Surface Deformation Field 248
  - 8.4.2 Quantifying Subsurface Magma Bodies 252
  - 8.4.3 Quantifying Volcano Topography 254
- 8.5 Remote Sensing of Volcanic Degassing 254
  - 8.5.1 Quantifying Volcanic Sulfur Dioxide Emissions in the Ultraviolet 255

Contents

	8.5.2	Quantifying Volcanic Sulfur Dioxide Emissions in the Thermal Infrared	258
	8.5.3	Other Gases	261
8.6	Remote Sensing of Geothermal and Hydrothermal Activity	261
8.7	Remote Sensing of Active Lavas—Effusive Eruptions	265
	8.7.1	Detecting the Thermal Signature of Erupting Volcanoes	265
	8.7.2	Quantifying the Thermo-Physical Characteristics of Active Lava Bodies	269
8.8	Remote Sensing of Volcanic Ash Clouds—Explosive Eruptions	270
8.9	Conclusion	275

PART IV Fires

Chapter 9 Satellite-Derived Nitrogen Dioxide Variations from Biomass Burning in a Subtropical Evergreen Forest, Northeast India 287

Krishna Prasad Vadrevu and Kristofer Lasko

9.1	Introduction	287
9.2	Study Area	289
9.3	Datasets	291
	9.3.1	Active Fires and Fire Radiative Power (FRP)	291
	9.3.2	OMI-NO_2	291
	9.3.3	SCIAMACHY-NO_2	292
	9.3.4	MODIS Aerosol Optical Depth (AOD) Variations	293
9.4	Methods	293
	9.4.1	Descriptive Statistics	293
	9.4.2	Time-Series Regression	293
9.5	Results	294
	9.5.1	Fires in Northeast India	294
	9.5.2	NO_2 Temporal and Seasonal Variations	294
	9.5.3	Correlations and Time-Series Regression	295
9.6	Discussion	301
9.7	Conclusions	302
9.8	Acknowledgments	303

Chapter 10 Remote Sensing–Based Mapping and Monitoring of Coal Fires 309

Anupma Prakash, Claudia Kuenzer, Santosh K. Panda, Anushree Badola, and Christine F. Waigl

10.1	Introduction	309
	10.1.1	Terminology	309
	10.1.2	Causes	310
	10.1.3	Hazards	313
	10.1.4	Attributes of Surface and Subsurface Coal Fires	314
10.2	Remote Sensing of Coal Fires	314
	10.2.1	History and Recent Evolution	314
	10.2.2	Spatial and Temporal Resolution	315
	10.2.3	Spectral Resolution	316

	10.3	Methods 318
		10.3.1 Atmospheric Correction and Land Surface Emissivity Variations 319
		10.3.2 Fire Temperature Estimation 319
		10.3.3 Thresholding to Delineate Fire Area 320
		10.3.4 Multi-Source Data for Improved Coal Fire Mapping 321
		10.3.5 Time Series Analysis for Fire Monitoring 321
	10.4	Selected Results 322
	10.5	Discussion 325
		10.5.1 Coal Fire Research Is Local in Nature 325
		10.5.2 Coal Fires Are Dynamic and Baseline Data Does Not Exist 326
		10.5.3 Timing of Data Acquisition Is Important 326
		10.5.4 Spaceborne TIR Imagery Is Coarse and Discontinuous ...326
	10.6	Conclusions 327
	10.7	Acknowledgments 328

PART V Urban

Chapter 11 Urban Growth and Climatic Mapping of Mega cities: Multi-Sensor Approach 337

Hasi Bagan, Chaomin Chen, and Yoshiki Yamagata

	11.1	Introduction 337
	11.2	Grid Cell Process 339
	11.3	Multi-Sensor Approach for Urban Mapping 339
		11.3.1 Spatial-Temporal Changes of Land Cover in Tokyo 339
		11.3.2 Relationship between Land Cover Changes and Population Census in Tokyo 346
		11.3.3 Relationship among DMSP, Urban Area, and Population Census in Tokyo 349
		11.3.4 Relationship between Local Climate Zones and Land Surface Temperatures in Tokyo 23-Ku 351
		11.3.5 Spatial-Temporal Analyses of Local Climate Zones in Shanghai 356
	11.4	Conclusions 357

Chapter 12 High-Resolution Remote Sensing and Visibility Analysis Method for Smart Environment Design 362

Yoshiki Yamagata, Daisuke Murakami, Hajime Seya, and Takahiro Yoshida

	12.1	Introduction 362
		12.1.1 Remote Sensing and Urban Analysis 362
		12.1.2 Remote Sensing and Smart Environment 363
	12.2	Visibility Analysis 364
		12.2.1 Classical Visibility Evaluation Approaches 364
		12.2.2 Viewshed and Isovist Analyses 365
		12.2.3 Three-Dimensional (3D) View Indexes 365
		12.2.4 Hedonic Analysis of View 367

		12.3	Application: Three-Dimensional (3D) View Analysis of Yokohama City, Japan.. 369
			12.3.1 Outline... 369
			12.3.2 Three-Dimensional (3D) View Evaluation....................................... 370
			12.3.3 Hedonic Analysis Results 373
		12.4	Concluding Remarks.. 374

PART VI   Nightlights

Chapter 13 Nighttime Light Remote Sensing—Monitoring Human Societies from Outer Space 381

Qingling Zhang, Noam Levin, Christos Chalkias, Husi Letu, and Di Liu

	13.1	Introduction.. 381
		13.1.1 The Rationale That Underlies Nighttime Light (NTL) Remote Sensing... 383
		13.1.2 History of Nighttime Light Remote Sensing 385
		13.1.3 The DMSP/OLS Data Archive... 388
		13.1.4 The VIIRS/DNB Data Archive... 391
	13.2	Major Applications of Nighttime Lights... 391
		13.2.1 Urban Extent and Socioeconomic Variables................................... 392
		13.2.2 Nighttime Light Pollution (NLP)... 392
	13.3	Study of Urban Dynamics with NTL Time Series 396
		13.3.1 Mapping Urban Extent Dynamics 396
		13.3.2 Detecting Socioeconomic Changes.. 397
		13.3.3 Tracking Social Events with High Temporal Frequency NTL Time Series 399
	13.4	Challenges in Remote Sensing of Night Lights 399
		13.4.1 Correcting Saturation in DMSP/OLS NTL Imagery 401
		13.4.2 Intercalibrating the DMSP/OLS NTL Time Series 402
		13.4.3 Intercalibrating DMSP/OLS and VIIRS/DNB for Generating Continuous Time Series 404
		13.4.4 Quantifying the Impacts of the Transitions to LED Lighting on Light Pollution... 404
	13.5	Outlook—Fine Resolution Nighttime Light Remote Sensing 405
		13.5.1 Applications of Fine Spatial Resolution Night Lights 405
		13.5.2 Calibrating with Ground Measurements.. 406
	13.6	Summary and Future Directions 408

PART VII   Summary and Synthesis for Volume VI

Chapter 14 Summary Chapter for Remote Sensing Handbook, Volume VI: Droughts, Disasters, Pollution, and Urban Mapping 425

Prasad S. Thenkabail

	14.1	Agricultural Droughts Using Vegetation Health Methods 426
	14.2	Agricultural Drought Monitoring Using Biophysical Parameters............... 427

14.3	Drought Monitoring Advances and Toolkits Using Remote Sensing	430
14.4	New Remote Sensing–Based Drought Indices	432
14.5	Remote Sensing of Drylands	436
14.6	Disaster Risk Management through Remote Sensing	437
14.7	Humanitarian Disasters and Remote Sensing	440
14.8	Remote Sensing of Volcanoes	440
14.9	Biomass Burning and Greenhouse Gas Emissions Studied Using Remote Sensing	442
14.10	Coal Fires Studies Using Remote Sensing	444
14.11	Urban Remote Sensing	446
14.12	Remote Sensing in Design of Smart Cities	448
14.13	Nighttime Light Remote Sensing	450
14.14	Acknowledgments	452

Index 461

Foreword

Satellite remote sensing has progressed tremendously since the first Landsat was launched on June 23, 1972. Since the 1970s, satellite remote sensing and associated airborne and *in situ* measurements have resulted in geophysical observations for understanding our planet through time. These observations have also led to improvements in numerical simulation models of the coupled atmosphere-land-ocean systems at increasing accuracies and predictive capabilities. This was made possible by data assimilation of satellite geophysical variables into simulation models, to update model variables with more current information. The same observations document the Earth's climate and have driven consensus that *Homo sapiens* are changing our climate through greenhouse gas emissions.

These accomplishments are the work of many scientists from a host of countries and a dedicated cadre of engineers who build and operate the instruments and satellites that collect geophysical observation data from satellites, all working toward the goal of improving our understanding of the Earth. This edition of the *Remote Sensing Handbook* (Second Edition, Volumes I–VI) is a compendium of information for many research areas of the Earth System that have contributed to our substantial progress since the 1970s. The remote sensing community is now using multiple sources of satellite and *in situ* data to advance our studies of Planet Earth. In the following paragraphs, I will illustrate how valuable and pivotal satellite remote sensing has been in climate system study since the 1970s. The chapters in the *Remote Sensing Handbook* provide other specific studies on land, water, and other applications using Earth observation data of the past 60+ years.

The Landsat system of Earth-observing satellites led the way in pioneering sustained observations of our planet. From 1972 to the present, at least one and frequently two Landsat satellites have been in operation (Wulder et al. 2022; Irons et al. 2012). Starting with the launch of the first NOAA-NASA Polar Orbiting Environmental Satellites NOAA-6 in 1978, improved imaging of land, clouds, and oceans and atmospheric soundings of temperature were accomplished. The NOAA system of polar-orbiting meteorological satellites has continued uninterrupted since that time, providing vital observations for numerical weather prediction. These same satellites are also responsible for the remarkable records of sea surface temperature and land vegetation index from the Advanced Very High-Resolution Radiometers (AVHRR) that now span more than 46 years as of 2024, although no one anticipated valuable climate records from these instruments before the launch of NOAA-6 in 1978 (Cracknell 2001). AVHRR instruments are expected to remain in operation on the European MetOps satellites into 2026 and possibly beyond.

The successes of data from the AVHRR led to the MODerate resolution Imaging Spectrometer (MODIS) instruments on NASA's Earth Observing System of satellite platforms that improved substantially upon the AVHRR. The first of the EOS platforms, Terra, was launched in 2000, and the second of these platforms, Aqua, was launched in 2002. Both of these platforms are nearing their operational end of life and many of the climate data records from MODIS will be continued with the Visible Infrared Imaging Suite (VIIRS) instrument on the Joint Polar Satellite System (JPSS) meteorological satellites of NOAA. The first of these missions, the NPOES Preparation Project, was launched in 2012 with the first VIIRS instrument that is operating currently along with similar instruments on JPSS-1 (launched in 2017) and JPSS-2 (launched in 2022). However, unlike the morning/afternoon overpasses of MODIS, the VIIRS instruments are all in an afternoon overpass orbit. One of the strengths of the MODIS observations was morning and afternoon data from identical instruments.

Continuity of observations is crucial for advancing our understanding of the Earth's climate system. Many scientists feel the crucial climate observations provided by remote sensing satellites are among the most important satellite measurements because they contribute to documenting the current state of our climate and how it is evolving. These key satellite observations of our climate are second in importance only to the polar orbiting and geostationary satellites needed for numerical weather prediction that provide natural disaster alerts.

The current state of the art for remote sensing is to combine different satellite observations in a complementary fashion for what is being studied. Climate study is an example of using disparate observations from multiple satellites coupled with *in situ* data to determine if climate change is occurring, where it is occurring, and to identify the various component processes responsible.

1. **Planet warming quantified by satellite radar altimetry:** Remotely sensed climate observations provide the data to understand our planet and identify what forces our climate. The primary sea level climate observation come from radar altimetry that started in late 1992 with Topex-Poseidon and has been continued by Jason-1, Jason-2, Jason-3, and Sentinel-6 to provide an uninterrupted record of global sea level. Changes in global sea level provide unequivocal evidence that our planet is warming, cooling, or staying at the same temperature. Radar altimetry from 1992 to date has shown global sea level increases of ~3.5 mm/yr, hence our planet is warming (Figure 0.1). Sea level rise has two components, ocean thermal expansion and ice melt from the ice sheets of Greenland and Antarctica, and to a lesser extent for glacier concentrations in places like the Gulf of Alaska and Patagonia. The combination of GRACE and GRACE Follow-On gravity measurements quantifies the ice mass losses of Greenland and Antarctica to a high degree of accuracy. Combining the gravity data with the flotilla of almost 4,000 Argo floats provides the temperature data with the depth necessary to quantify ocean temperatures and isolate the thermal component of sea level rise.
2. **Our Sun is remarkably stable in total solar irradiance**. Observations of total solar irradiance have been made from satellites since 1979 and show that total solar irradiance has varied only ±1 part in 500 over the past 35 years, establishing that our Sun is not to blame for global warming (Figure 0.2).

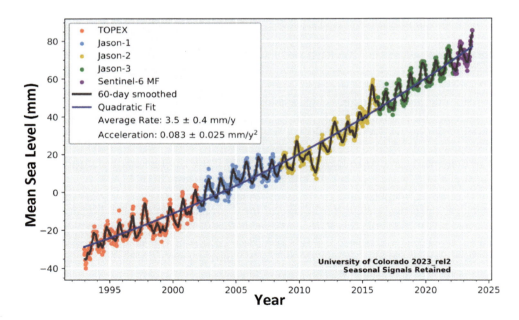

FIGURE 0.1 Seasonal sea level from five satellite radar altimeters from late 1992 to the present. Sea level is the unequivocal indicator of the Earth's climate—when sea level rises, the planet is warming; when sea level falls, the planet is cooling (Nerem et al. 2018 updated to 2023; https://sealevel.colorado.edu/data/total-sea-level-change).

Foreword xv

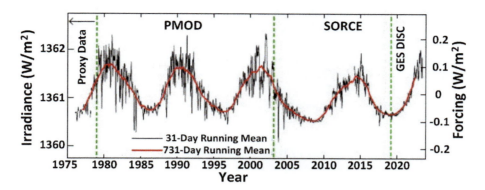

FIGURE 0.2 The Sun is not to blame for global warming, based on total solar irradiance observations from satellites. The few Watts per m² solar irradiance variations covary with the sunspot cycle. The luminosity of the Sun varies 0.2% over the course of the 11-year solar and sunspot cycle. The SORCE TSI data set continues these important observations with improved accuracy on the order of ±0.035 (Kopp et al. 2024) and from https://lasp.colorado.edu/sorce/data/tsi-data/.

3. **Determining ice sheet contributions to sea level rise**. Since 2002 gravity observations from the Gravity Recovery and Climate Experiment Satellite or GRACE mission and the GRACE Follow-On mission have been measured. GRACE data quantify ice mass changes from the Antarctic and Greenland ice sheets that constitute 98% of the ice mass on land (Luthcke et al. 2013). GRACE data are truly remarkable—their retrieval of variations in the Earth's gravity field is quantitatively and directly linked to mass variations. With GRACE data we are able for the first time to determine the mass balance with time of the Antarctic and Greenland ice sheets and concentrations of glaciers on land. GRACE data show sea level rise is 60% explained by ice sheet mass loss (Figure 0.3). GRACE data have many other uses, such as changes in ground water storage. See: http://www.csr.utexas.edu/grace/.
4. **Forty percent sea level rise explained by thermal expansion in the planet's oceans measured by in situ ~3700 Argo drifting floats**. The other contributor to sea level rise is the thermal expansion or "steric" component of our planet's oceans. To document this necessitates using diving and drifting floats or buoys in the Argo network to record temperature with depth (Roemmich and The Argo Steering Team 2009 and Figure 0.4). Argo floats are deployed from ships, they then submerge and descend slowly to 1000 m depth, recording temperature, pressure, and salinity as they descend. At 1000 m depth, they drift for ten days, continuing their measurements of temperature and salinity. After ten days, they slowly descend to 3000 m and then ascend to the surface, all the time recording their measurements. At the surface, each float transmits all the data collected on the most recent excursion to a geostationary satellite and then descend again to repeat this process.

Argo temperature data show that 40% of sea level rise results from warming and thermal expansion of our oceans. Combining radar altimeter data, GRACE and GRACE Follow-On data, and Argo data provide confirmation of sea level rise and show what is responsible for it and in what proportions. With total solar irradiance being near-constant, what is driving global warming can be determined. Analysis of surface *in situ* air temperature coupled with lower tropospheric air temperature and stratospheric temperature data from remote sensing infrared and microwave sounders show the surface and near-surface is warming while the stratosphere is cooling. This is an unequivocal confirmation that greenhouse gases are warming the planet.

Combining sea level radar altimetry, GRACE and GRACE Follow-On gravity data to quantify ice sheet mass losses, and Argo floats to measure ocean temperatures with depth enables the reconciliation of sea level increases with the mass loss of ice sheets and ocean thermal expansion. The ice and steric expansion explains 95% of sea level rise (Figure 0.5).

5. **The global carbon cycle**. Many scientists are actively working to study the Earth's carbon cycle, and there are several chapters in this *Remote Sensing Handbook* (Volumes I–VI) on various components under study.

Carbon cycles through reservoirs on the Earth's surface in plants and soils, exists in the atmosphere as gases such as carbon dioxide (CO_2), and exists in ocean water in phytoplankton and in marine sediments. CO_2 is released to the atmosphere from the combustion of fossil fuels, by land cover changes on the Earth's surface, by the respiration of green plants, and by the decomposition of carbon in dead vegetation and in soils, including carbon in permafrost.

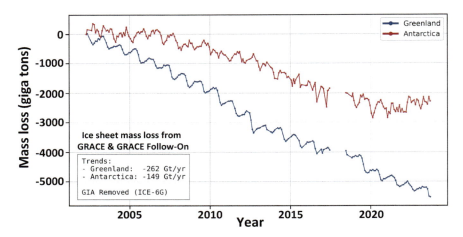

FIGURE 0.3 Sixty percent sea level rise explained by mass balance of melting of ice measured by GRACE and GRACE Follow-On satellites. Ice mass variations from 2003 to 2023 for the Antarctic and Greenland ice sheets using gravity data (Croteau et al. 2021 updated to 2023). The Antarctic and Greenland ice sheets constitute 98% of the Earth's land ice.

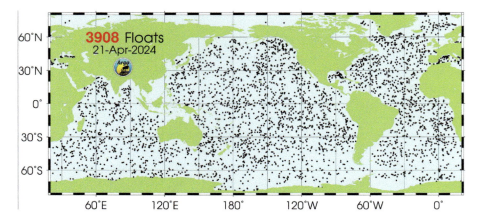

FIGURE 0.4 Forty percent sea level rise explained by thermal expansion in the planet's oceans measured *in situ* by ~3908 drifting floats that were in operation on March 25, 2024. These floats provide the data needed to document thermal expansion of the oceans (Roemmich and The Argo Steering Team 2009, updated to 2024 and http://www.argo.ucsd.edu/).

Foreword   xvii

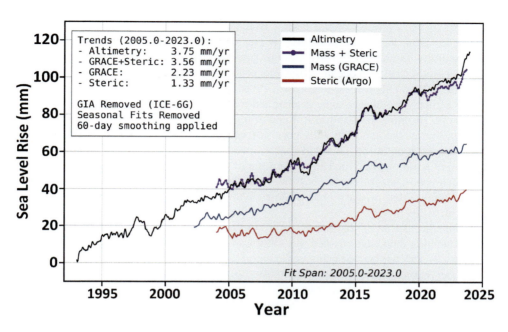

FIGURE 0.5 Sea level rise with the gravity ice mass loss and the Argo thermal expansion quantities added to the plot of global mean sea level. The GRACE and GRACE Follow-On ice sheet gravity term and Argo thermal expansion terms together explain 95% of sea level rise (Croteau et al. 2021 updated to 2023).

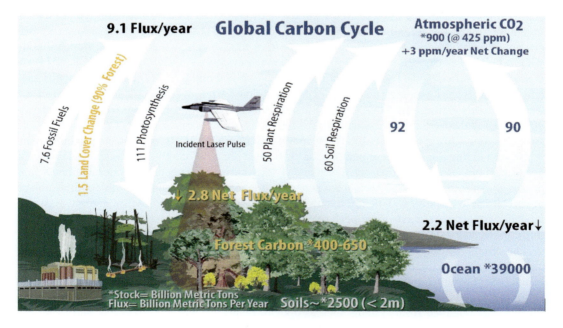

FIGURE 0.6 Global carbon cycle measurements from a multitude of satellite sensors. A representation of the global carbon cycle showing our best estimates of carbon fluxes and carbon reservoirs as of 2024. A series of satellite observations are needed simultaneously to understand the carbon cycle and its role in the Earth's climate system (Ciais, et al. 2014 updated to 2023). The major unknowns in the global carbon cycle are fluxes between different reservoirs, oceanic gross primary production, carbon in soils, and the carbon in woody vegetation.

Land gross primary production has been a MODIS product that is extended into the VIIRS era (Running et al. 2004; Román et al. 2024). MODIS data also provide burned area and CO_2 emissions from wildfire (Giglio et al. 2016). Oceanic gross primary production will be provided by the Plankton, Aerosol, Cloud, and ocean Ecosystem, or PACE, satellite that was launched in early 2024 (Gorman et al. 2019). This complements the GPP land portion of the carbon cycle and will enable global gross primary production to be determined by MODIS-VIIRS and PACE.

Furthermore, Harmonized Landsat-8, Landsat-9, and Sentinel-2 30 m data (HLS) provide multispectral time-series data at 30 m with a revisit frequency of three days at the equator (Crawford et al. 2023; Masek et al. 2018). This will enable time-series improvements in spatial detail to 30 m from the 250 m scale of MODIS. The revisit time of Sentinel-2 with 10 m data is five days at the equator, which is a major improvement from 30 m. Multispectral time-series observations are the basis for providing gross primary production estimates on land that is also used for food security (Claverie et al. 2018).

Refinements in satellite multispectral spatial resolution to the 50 cm to 3–4 m scale provided by commercial satellite data have enabled tree carbon to be determined from large areas of trees outside of forests. NASA has started using commercial satellite data to complement MODIS, Landsat, and other observations. One of the uses for Planet 3–4 m and Maxar < 1 m data have been for mapping trees outside of forests (Brandt et al. 2020; Reiner et al. 2023; Tucker et al. 2023). Tucker et al. (2023) mapped 10 billion trees at the 50 cm scale over 10 million km^2 and converted them into carbon at the tree level with allometry. The value of Planet and Maxar (formerly Digital Globe) data allows carbon studies to be extended into areas with discrete trees, and Huang et al. 2024 has successfully mapped one tree species across the entire Sahelian and Sudanian Zones of Africa.

The height of trees is an important measurement to determine their carbon content. For areas of contiguous tree crowns, GEDI and ICESat laser altimetry (Magruder et al. 2024) coupled with Landsat and Sentinel-2 observations, enable improved estimates of carbon in these forests (Claverie et al. 2018).

The key to closing several uncertainties in the carbon cycle is to quantify fluxes among the various components. Passive CO_2 retrieval methods from the Greenhouse gases Observing SATellite (GOSAT) (Noël et al. 2021) and the Orbiting Carbon Observatory-2 (OCO-2) (Jacobs et al. 2024) are inadequate to provide this. Passive methods are not possible at night, in all seasons, and require specific Sun-target-sensor viewing perspectives and conditions. A recent development of the Aerosol and Carbon dioxide Detection LiDAR (ACDL) instrument (Dai et al. 2023) by our Chinese colleagues offers a ten-fold coverage improvement in CO_2 retrievals over those provided by OCO-2 and 20-fold coverage improvement over GOSAT. The reported uncertainty of ACDL is in the order of ±0.6 ppm.

Understanding the carbon cycle requires a "full court press" of satellite and *in situ* observations, because all of these observations must be made at the same time. Many of these measurements have been made over the past 30–40 years, but new measurements are needed to quantify carbon storage in vegetation, to quantify CO_2 fluxes, to quantify land respiration, and to improve numerical carbon models. Similar work needs to be performed for the role of clouds and aerosols in climate and to improve our understanding of the global hydrological cycle.

The remote sensing community has made tremendous progress over the last six decades, as captured in various chapters of the *Remote Sensing Handbook* (Second Edition, Volumes I–VI). Handbook chapters provide a comprehensive understanding of land and water studies through detailed methods, approaches, algorithms, syntheses, and key references. Every type of remote sensing data obtained from systems such as optical, radar, LiDAR, hyperspectral, and hyperspatial are presented and discussed in different chapters. Chapters in this volume addressing remote sensing data characteristics, within and between sensor calibrations, classification methods, and accuracies taking a wide array of remote sensing data from a wide array of platforms over the last five decades. Volume I also brings in new remote sensing technologies, such as radio occultation and reflectometry from the global navigation satellite system or GPS satellites, crowdsourcing, drones, cloud computing, artificial intelligence, machine learning, hyperspectral, radar, and remote sensing law. The chapters in the *Remote Sensing Handbook* are written by leading remote sensing scientists of the world and ably edited by Dr. Prasad S. Thenkabail, Senior Scientist (ST), at US Geological Survey (USGS) in Flagstaff, Arizona. The importance and the value of the *Remote Sensing Handbook* is clearly demonstrated by the need for a second edition. The *Remote Sensing Handbook* (First Edition, Volumes I–III) was published in 2014, and now after 10 years the *Remote Sensing Handbook* (Second Edition, Volumes I–VI) with 91 chapters and nearly 3500 pages will be published. It is certainly monumental work in remote sensing science, and for this I want to compliment Dr. Prasad Thenkabail. Remote sensing is now important to a large number of scientific disciplines beyond our community, and I recommend the *Remote Sensing Handbook* (Second Edition, Volumes I–VI) to not only remote sensors but to the entire scientific community.

We can look forward in the coming decades to improving our quantitative understanding of the global carbon cycle, understanding the interaction of clouds and aerosols in our radiation budget, and understanding the global hydrological cycle.

by Compton J. Tucker
Satellite Remote Sensing Beyond 2025
NASA/Goddard Space Flight Center
Earth Science Division
Greenbelt, Maryland 20771 USA

REFERENCES

Brandt, M., Tucker, C.J., Kariryaa, A., et al. 2020. An unexpectedly large count of trees in the West African Sahara and Sahel. *Nature* 587: 78–82. http://doi.org/10.1038/s41586-020-2824-5

Ciais, P., et al. 2014. Current systematic carbon-cycle observations and the need for implementing a policy-relevant carbon observing system. *Biogeosciences* 11(13): 3547–3602.

Claverie, M., Ju, J, Masek, J.G., Dungan, J.L., Vermote, E.F, Roger, J.-C., Skakun, S.V., et al. 2018. The Harmonized Landsat and Sentinel-2 surface reflectance data set. *Remote Sensing of Environment* 219: 145–161. http://doi.org/10.1016/j.rse.2018.09.002

Cracknell, A. 2001. The exciting and totally unanticipated success of the AVHRR in applications for which it was never intended. *Advances in Space Research* 28: 233–240. http://doi.org/10.1016/S0273-1177(01)00349-0

Crawford, C.J., Roy, D.P., Arab, S., Barnes, C., Vermote, E., Hulley, G., et al. 2023. The 50-year Landsat collection 2 archive. *Science of Remote Sensing* 8(2023): 100103. ISSN 2666–0172. https://doi.org/10.1016/j.srs.2023.100103. https://www.sciencedirect.com/science/article/pii/S2666017223000287

Croteau, M.J., Sabaka, T.J., Loomis, B.D. 2021. GRACE fast mascons from spherical harmonics and a regularization design trade study. *Journal of Geophysical Research—Solid Earth* 126: e2021JB022113. https://doi.org/10.1029/2021JB022113.10.1029/2021JB022113

Dai, G., Wu, S., Long, W., Liu, J., Xie, Y., Sun, K., Meng, F., Song, X., Huang, Z., Chen, W. 2023. Aerosols and clouds data processing and optical properties retrieval algorithms for the spaceborne ACDL/DQ-1. *EGUsphere* [preprint]. https://doi.org/10.5194/egusphere-2023-2182.

Giglio, L., Schroeder, W., Justice, C.O. 2016. The collection 6 MODIS active fire detection algorithm and fire products. *Remote Sensing of Environment* 178: 31–41. http://doi.org/10.1016/j.rse.2016.02.054

Gorman, E.T., Kubalak, D.A., Patel, D., Dress, A., Mott, D.B., Meister, G., Werdell, P.J. 2019. The NASA Plankton, Aerosol, Cloud, ocean Ecosystem (PACE) mission: An emerging era of global, hyperspectral Earth system remote sensing. *Sensors, Systems, and Next Generation Satellites* XXIII: 11151. http://doi.org/10.1117/12.2537146

Huang, K. et al. 2024. Mapping every adult baobab (Adansonia digitata L.) across the Sahel to uncover the co-existence with rural livelihoods. *Nature Ecology and Evolution*. http://doi.org/10.21203/rs.3.rs-3243009/v1

Irons, J.R., Dwyer, J.L., Barsi, J.A. 2012. The next Landsat satellite: The Landsat data continuity mission. *Remote Sensing of Environment* 122: 11–21. http://doi.org/10.1016/j.rse.2011.08.026

Jacobs, N., et al. 2024. The importance of digital elevation model accuracy in X_{CO_2} retrievals: Improving the Orbiting Carbon Observatory-2 Atmospheric Carbon Observations from Space version 11 retrieval product. *Atmospheric Measurement Techniques* 17(5): 1375–1401. http://doi.org/10.5194/amt-17-1375-2024

Kopp, G., Nèmec, N.E., Shapiro, A. 2024. Correlations between total and spectral solar irradiance variations. *Astrophysical Journal* 964(1). http://doi.org/10.3847/1538-4357/ad24e5

Luthcke, S.B., Sabaka, T.J., Loomis, B.D., Arendt, A.A., McCarthy, J.J., Camp, J. 2013. Antarctica, Greenland, and Gulf of Alaska land-ice evolution from an iterated GRACE global mascon solution. *Journal of Glaciology* 59(216). http://doi.org/10.3189/2013JoG12J147.

Magruder, L.A., Farrell, S.L., Neuenschwander, A., Duncanson, L., Csatho, B., Kacimi, S., et al. 2024. Monitoring Earth's climate variables with satellite laser altimetry. *Nature Reviews of Earth and Environment* 5(2):120–136. http://doi.org/10.1038/s43017-023-00508-8

Masek, J., Ju, J., Roger, J.-C., Skakun, S., Claverie, M., Dungan, J. 2018. Harmonized Landsat/Sentinel-2 products for land monitoring. *IGARSS 2018—2018 IEEE International Geoscience and Remote Sensing Symposium*, Valencia, Spain, 2018, pp. 8163–8165. http://doi.org/10.1109/IGARSS.2018.8517760

Nerem, R.S., Beckley, B.D., Fasullo, J.T., Mitchum, G.T. 2018. Climate-change–driven accelerated sea-level rise detected in the altimeter era. *Proceeding of the National Academy of Sciences* 115(9): 2022–2025. http://doi.org/10.1073/pnas.1717312115

Noël, S., et al. 2021. XCO_2 retrieval for GOSAT and GOSAT-2 based on the FOCAL algorithm. *Atmospheric Measurement Techniques* 14(5): 3837–3869. http://doi.org/10.5194/amt-14-3837-2021

Reiner, F., et al. 2023. More than one quarter of Africa's tree cover is found outside areas previously classified as forest. *Nature Communications*. http://doi.org/10.1038/s41467-023-37880-4

Roemmich, D., The Argo Steering Team. 2009. Argo—The challenge of continuing 10 years of progress. *Oceanography* 22(3): 46–55.

Román, M., et al. 2024. Continuity between NASA MODIS collection 6.1 and VIIRS collection 2 land products. *Remote Sensing of Environment* 302. http://doi.org/10.1016/j.rse.2023.113963

Running, S.W., Nemani, R.R., Heinsch, F.A., Zhao, M.S., Reeves, M., Hashimoto, H. 2004. A continuous satellite-derived measure of global terrestrial primary production. *Bioscience* 54(6): 547–560. http://doi.org/10.1641/0006-3568

Tucker, C., Brandt, M., Hiernaux, P., Kariryaa, A., et al. 2023. Sub-continental-scale carbon stocks of individual trees in African drylands. *Nature* 615: 80–86. http://doi.org/10.1038/s41586-022-05653-6

Wulder, M.A., Roy, D.P., Radeloff, V.C., Loveland, T.R., Anderson, M.C., Johnson, D.M., et al. 2022. Fifty years of Landsat science and impacts. *Remote Sensing of Environment* 280(2022): 113195. ISSN 0034-4257. https://doi.org/10.1016/j.rse.2022.113195. https://www.sciencedirect.com/science/article/pii/S0034425722003054

Preface

REMOTE SENSING HANDBOOK
SECOND EDITION, SIX VOLUMES
VOLUME VI
DROUGHTS, DISASTERS, POLLUTION, AND URBAN MAPPING
ADVANCES OF THE LAST 60 YEARS AND A VISION FOR THE FUTURE

The overarching goal of this six-volume, 91-chapter, about 3500-page, *Remote Sensing Handbook* (Second Edition, Vols. I–VI) was to capture and provide the most comprehensive state of the art of remote sensing science and technology development and advancement in the last 60+ years, by clearly demonstrating the: (1) scientific advances, (2) methodological advances, and (3) societal benefits achieved during this period, as well as to provide a vision of what is to come in the years ahead. The book volumes are, to date and to my best knowledge, the most comprehensive documentation of the scientific and methodological advances that have taken place in understanding remote sensing data, methods, and a wide array of applications. Written by 300+ leading global experts in the area, each chapter: (1) focuses on a specific topic (e.g., data, methods, and a specific set of applications), (2) reviews existing state-of-the-art knowledge, (3) highlights the advances made, and (4) provides guidance for areas requiring future development. Chapters in the book cover a wide array of subject matters of remote sensing applications. The *Remote Sensing Handbook* (Second Edition, Vols. I–VI) is planned as reference material for a broad spectrum of remote sensing scientists to understand the fundamentals as well as the latest advances, including a wide array of applications such as for land and water resource practitioners, natural and environmental practitioners, professors, students, and decision-makers.

Special features of the six-volume *Remote Sensing Handbook* (Second Edition) include:

1. Participation of an outstanding group of remote sensing experts, an unparalleled team of writers for such a book project.
2. Exhaustive coverage of a wide array of remote sensing science: data, methods, and applications.
3. Each chapter is led by a luminary, and most chapters are written by writing teams, which further enriched the chapters.
4. Broadening the scope of the book to make it ideal for expert practitioners as well as students.
5. Global team of writers, global geographic coverage of study areas, wide array of satellites and sensors.
6. Plenty of color illustrations.

Chapters in the book have covered remote sensing:

State of the art on satellites, sensors, science, technology, and applications
Methods and techniques
Wide array of applications, such as land and water applications, natural resources management, and environmental issues
Scientific achievements and advancements over the last 60+ years
Societal benefits
Knowledge gaps
Future possibilities in the 21st century

Great advances have taken place over the last 60+ years in the study of Planet Earth from remote sensing, especially using data gathered from the multitude of Earth Observation (EO) satellites launched by various governments as well as private entities. A large part of the initial remote sensing technology was developed and tested during the two world wars. In the 1950s remote sensing slowly began its foray into civilian applications. But, during the years of the Cold War civilian and military remote sensing applications increased swiftly. But it was also an age when remote sensing was the domain of very few top experts, often having multiple skills in engineering, science, and computer technology. From the 1960s onwards, there have been many governmental agencies that have initiated civilian remote sensing. The National Aeronautics and Space Administration (NASA) of the USA has been at the forefront of many of these efforts. Others who have provided leadership in civilian remote sensing include, but are not limited to, the European Space Agency (ESA) of the European Union, the Indian Space Research Organization (ISRO), The Centre national d'études spatiales (CNES) of France, The Canadian Space Agency (CSA), the Japan Aerospace Exploration Agency (JAXA), the German Aerospace Center (DLR), the China National Space Administration (CNSA), the United Kingdom Space Agency (UKSA), and the Instituto Nacional de Pesquisas Espaciais (INPE) of Brazil. Many private entities, such as Planet Labs PBC, have launched and operate satellites. These government and private agencies and enterprises have launched, and continue to launch and operate, a wide array of satellites and sensors that capture data of Planet Earth in various regions of the electromagnetic spectrum and in various spatial, radiometric, and temporal resolutions, routinely and repeatedly. However, the real thrust for remote sensing advancement came during the last decade of the 20th century and the beginning of the 21st century. These initiatives included the launch of a series of new-generation EO satellites to gather data more frequently and routinely, the release of pathfinder datasets, web-enabling of the data for free by many agencies (e.g., the USGS release of the entire Landsat archives as well as real-time acquisitions of the world for free by making them web accessible), and providing processed data ready to users (e.g., the Harmonized Landsat and Sentinel-2 or HLS data, surface reflectance products of MODIS). Other efforts, like Google Earth, made remote sensing more popular and brought in a new platform for easy visualization and navigation of remote sensing data. Advances in computer hardware and software made it possible to handle big data. Crowdsourcing, web access, cloud computing such as in thenGoogle Earth Engine (GEE) platform, machine learning, deep learning, coding, artificial intelligence, mobile apps, and mobile platforms (e.g., drones) added new dimensions to how remote sensing data is used. Integration with global positioning systems (GPS) and global navigation satellite system (GNSS), and inclusion of digital secondary data (e.g., digital elevation, precipitation, temperature) in analysis has made remote sensing much more powerful. Collectively, these initiatives provided a new vision in making remote sensing data more popular, widely understood, and increasingly used for diverse applications, hitherto considered difficult. The availability of free archival data when combined with more recent acquisitions has also enabled quantitative studies of change over space and time. The *Remote Sensing Handbook* (Vols. I–VI) is targeted to capture these vast advances in data, methods, and applications, so a remote sensing student, scientist, or professional practitioner will have the most comprehensive, all-encompassing reference material in one place.

Modern-day remote sensing technology, science, and applications are growing exponentially. This growth is a result of a combination of factors that include: (1) advances and innovations in data capture, access, processing, computing, and delivery (e.g., big data analytics, harmonized and normalized data, inter-sensor relationships, web enabling of data, cloud computing such as in the Google Earth Engine (GEE), crowdsourcing, mobile apps, machine learning, deep learning, coding in Python and Java Script, and artificial intelligence); (2) increasing number of satellites and sensors gathering data of the planet, repeatedly and routinely, in various portions of the electromagnetic spectrum as well as in an array of spatial, radiometric, and temporal resolutions; (3) efforts at integrating data from multiple satellites and sensors (e.g., Sentinels with Landsat); (4) advances in data normalization, standardization, and harmonization (e.g., delivery of data in surface reflectance, inter-sensor calibration); (5) methods and techniques for handling very large

Preface   xxiii

data volumes (e.g., global mosaics); (6) quantum leap in computer hardware and software capabilities (e.g., ability to process several terabytes of data) (7) innovation in methods, approaches, and techniques leading to sophisticated algorithms (e.g., spectral matching techniques, neural network perceptron); and (8) development of new spectral indices to quantify and study specific land and water parameters (e.g., hyperspectral vegetation indices or HVIs). As a result of these all-round developments, remote sensing science is today very mature and is widely used in virtually every discipline of Earth Sciences for quantifying, mapping, modeling, and monitoring our planet. Such rapid advances are captured in a number of remote sensing and Earth Science journals. However, students, scientists, and practitioners of remote sensing science and applications have significant difficulty in gathering a complete understanding of various developments and advances that have taken place because of their vastness spread across the last 60+ years. Thereby, the chapters in the *Remote Sensing Handbook* are designed to give a whole picture of the scientific and technological advances of the last 60+ years.

Today the science, art, and technology of remote sensing is truly ubiquitous and increasingly part of everyone's everyday life, often without even the user knowing it. Whether looking at your own home or farm (e.g., following figure), helping you navigate when you drive, visualizing a phenomenon occurring in a distant part of the world (e.g., following figure), monitoring events such as droughts and floods, reporting weather, detecting and monitoring troop movements or nuclear sites, studying deforestation, assessing biomass carbon, addressing disasters like earthquakes or

FIGURE 0.7 Google Earth can be used to seamlessly navigate and precisely locate any place on Earth, often with very high spatial resolution data (VHRI; sub-meter to 5 m) from satellites such as IKONOS, QuickBird, and GeoEye (Note: the image is from one of the VHRI). Here, the Editor-in-Chief (EiC) (Thankabail) of this *Remote Sensing Handbook* (Volumes I–VI) located his village home and surroundings that have the land cover such as secondary rainforests, lowland paddy farms, areca nut plantations, coconut plantations, minor roads, walking routes, open grazing lands, and minor streams (typically, first and second order) (Note: Land cover is based on ground knowledge of the EiC). The first primary school attended by the EiC is located precisely. Precise coordinates (1345 39.22 Northern latitude, 75 06 56.03 Eastern longitude) of Thankabail's village house and the date of image acquisition (March 1, 2014). Google Earth Images are used for visualization as well as for numerous science applications, such as accuracy assessment, reconnaissance, determining land cover, and establishing land use for various ground surveys.

tsunamis, and a host of other applications (e.g., precision farming, crop productivity, water productivity, deforestation, desertification, water resources management), remote sensing plays a key role. Already, many new innovations are taking place. Companies such as Planet Labs PBC and Skybox are capturing very high spatial resolution imagery and even videos from space using large numbers of microsatellite (CubeSat) constellations. Planet Labs also will soon launch hyperspectral satellites called Tanager. There are others (e.g., Pixxel, India) who have launched and continue to launch constellations of hyperspectral or other sensors. China is constantly putting a wide array of satellites into orbit. Just as the smartphone and social media connected the world, remote sensing is making the world our backyard (e.g., following figure). No place goes un-observed, and no event gets reported without an image. True liberation for any technology and science comes when it is widely used by common people who often have no idea on how it all comes together but understand the information provided intuitively. That is already happening (e.g., how we use smartphones is significantly driven by satellite data–driven maps and GPS-driven locations). These developments make it clear that not only do we need to understand the state of the art, but we must also have a vision of where the future of remote sensing is headed. Thereby, in a nutshell, the goal of the *Remote Sensing Handbook* (Vols. I–VI) is to cover the developments and advancement of six distinct eras (listed next) in terms of data characterization and processing as well as myriad land and water applications:

Pre-civilian remote sensing era of the pre-1950s: World War I and II when remote sensing was a military tool.

Technology demonstration era of the 1950s and 1960s: Sputnik-I and NOAA AVHRR era of the 1950s and 1960s.

Landsat era of the 1970s: when the first truly operational land remote sensing satellite (Earth Resources Technology Satellite, or ERTS, later re-named Landsat) was launched and operated.

Earth observation era of the 1980s and 1990s: when a number of space agencies began launching and operating satellites (e.g., Landsat 4,5 by the USA, SPOT-1,2 by France; IRS-1a,1b by India).

Earth observation and new millennium era of the 2000s: when data dissemination to users became as important as launching, operating, and capturing data (e.g., MODIS Terra/Aqua, Landsat-8, Resourcesat).

Twenty-first-century era starting in the 2010s: when new-generation micro/nano satellites or CubeSats (e.g., Planet Labs PBC, Skybox), hyperspectral satellite sensors (e.g., Tanager-1, DESIS, PRISMA, EnMAP, upcoming NASA SBG) add to the increasing constellation of multi-agency sensors (e.g., Sentinels, Landsat-8,9, upcoming Landsat-Next).

Motivation to take up editing the six-volume *Remote Sensing Handbook* (second edition) wasn't easy. It is a daunting work and requires extraordinary commitment over two to three years. After repeated requests from Ms. Irma Shagla-Britton, Manager and Leader for Remote Sensing and GIS books of Taylor & Francis Group/CRC Press, and considerable thought, I finally agreed to take the challenge in 2022. Having earlier edited the three-volume *Remote Sensing Handbook*, published in 2015, I was pleased that the books were of considerable demand for a second edition. This was enough motivation. Further, I wanted to do something significant at this stage of my career that would make a considerable contribution to the global remote sensing community. When I edited the first edition during 2012–2014, I was still recovering from colon cancer surgery and chemotherapy. But this second edition is a celebration of my complete recovery from the dreaded disease. I have not only fully recovered, but I have never felt so completely full of health and vigor. This naturally gave me the energy and enthusiasm required to back my motivation to edit this monumental six-volume *Remote Sensing Handbook*. At least for me this is the *magnum opus* that I feel proud to have accomplished and feel confident of the immense value for students, scientists, and professional practitioners of remote sensing who are interested in a standard reference on the subject. They will find in this six-volume *Remote Sensing Handbook*: "Complete and comprehensive coverage of state-of-the-art remote sensing, capturing the advances that have taken place over last 60+ years, which will set the stage for a vision for the future."

Above all, I am indebted to some 300+ authors and co-authors of the chapters who have spent so much of their creative energy to work on the chapters, deliver them in time, and patiently address all edits and comments. These are amongst the very best remote sensing scientists from around the world. Extremely busy people, making time for the book project and making outstanding contributions. I went back to everyone who contributed to the *Remote Sensing Handbook* (First Edition, three volumes) published in 2014 and requested them to revise their chapters. Most of the lead authors of the chapters agreed to revise, which was reassuring. However, some were not available, some retired, and some declined for other reasons. In such cases I adopted two strategies: (1) invite a few new chapter authors to make up for this gap and (2) update the chapters myself in other cases. I am convinced this strategy worked very well to ensure to capture the latest information and maintain the integrity of every chapter. What was also important was to ensure the latest advances in remote sensing science were adequately covered. Authors of the chapters amazed me with their commitment and attention to detail. First, the quality of each of the chapters was of the highest standards. Second, with very few exceptions, chapters were delivered on time. Third, edited chapters were revised thoroughly and returned on time. Fourth, all my requests on various formatting and quality enhancements were addressed. My heartfelt gratitude to these great authors for their dedication to quality science. It has been my great honor and privilege to work with these dedicated legends. Indeed, I call them my "heroes" in a true sense. These are highly accomplished renowned pioneering scientists of highest merit in remote sensing science, and I am ever grateful to have their time, effort, enthusiasm, and outstanding intellectual contributions. I am indebted to their kindness and generosity. In the end we had 300+ authors writing 91 chapters.

Overall, the *Remote Sensing Handbook* (Vols. I–VI) took about two years, from the time book chapters and authors were identified to the final publication of the book. The six volumes of the *Remote Sensing Handbook* were designed in such a way that a reader could have all six volumes as a standard reference or have individual volumes to study specific subject areas. The six volumes are:

Remote Sensing Handbook, Second Edition, Vol. I
Volume 1: *Sensors, Data Normalization, Harmonization, Cloud Computing, and Accuracies—9781032890951*

Remote Sensing Handbook, Second Edition, Vol. II
Volume 2: *Image Processing, Change Detection, GIS, and Spatial Data Analysis—9781032890975*

Remote Sensing Handbook; Second Edition, Vol. III
Volume 3: *Agriculture, Food Security, Rangelands, Vegetation, Phenology, and Soils—9781032891019*

Remote Sensing Handbook; Second Edition, Vol. IV
Volume 4: *Forests, Biodiversity, Ecology, LULC, and Carbon—9781032891033*

Remote Sensing Handbook; Second Edition, Vol. V
Volume 5: *Water Resources: Hydrology, Floods, Snow and Ice, Wetlands, and Water Productivity—9781032891453*

Remote Sensing Handbook; Second Edition, Vol. VI
Volume 6: *Droughts, Disasters, Pollution, and Urban Mapping—9781032891484*

There are 18, 17, 17, 12, 13, and 14 chapters, respectively, in the six volumes.

A wide array of topics are covered in the six volumes.

The topics covered in Volume I include: (1) satellites and sensors; (2) global navigation satellite systems (GNSS); (3) remote sensing fundamentals; (4) data normalization, harmonization, and standardization; (5) vegetation indices and their within- and across-sensor calibration; (6) crowdsourcing; (7) cloud computing; (8) Google Earth Engine–supported remote sensing; (9) accuracy assessments; and (10) remote sensing law.

The topics covered in Volume II include: (1) digital image processing fundamentals and advances; (2) digital image classifications for applications such as urban, land use, and land cover; (3) hyperspectral image processing methods and approaches; (4) thermal infrared image processing principles and practices; (5) image segmentation; (6) object-oriented image analysis (OBIA), including geospatial data integration techniques in OBIA; (7) image segmentation in specific applications like land use and land cover; (8) LiDAR digital image processing; (9) change detection; and (10) integrating geographic information systems (GIS) with remote sensing in geoprocessing workflows, democratization of GIS data and tools, fronters of GIScience, and GIS and remote sensing policies.

The topics covered in Volume III include: (1) vegetation and biomass, (2) agricultural croplands, (3) rangelands, (4) phenology and food security, and (5) soils.

The topics covered in Volume IV include: (1) forests, (2) biodiversity, (3) ecology, (4) land use/land cover, and (5) carbon. Under each of thesebroad topics, there are one or more chapters.

Volume V has anfocus on hydrology, water resources, ice, wetlands, and crop water productivity. The chapters are broadly classified into: (1) geomorphology, (2) hydrology and water resources, (3) floods, (4) wetlands, (5) crop water use and crop water productivity, and (6) snow and ice.

Volume VI has a focus on water resources, disasters, and urban remote sensing. The chapters are broadly classified into: (1) droughts and drylands, (2) disasters, (3) Volcanoes, (4) fires, and (5) nightlights.

There are many ways to use the *Remote Sensing Handbook* (Second Edition, six volumes). A lot of thought went into organizing the volumes and chapters. So, you will see a "flow" from chapter to chapter and from volume to volume. As you read through the chapters, you will see how they are inter-connected and how reading all of them provides you with greater in-depth understanding. You will also realize, as someone deeply interested in one of the topics, you will have greater interest in one volume. Having all six volumes as reference material is ideal for any remote sensing expert, practitioner, or student. However, you can also refer to individual volumes based on your interest. We have also made great attempts to ensure chapters are self-contained. That way, you can focus on a chapter and read it through, without having to be overly dependent on other chapters. Taking this perspective, there is slight (~5–10%) material that maybe repeated across chapters. This is done deliberately. For example, when you are reading a chapter on LiDAR or radar, you don't want to go all the way back to another chapter to understand the characteristics of these data. Similarly, certain indices (e.g., vegetation condition index or VCI, temperature condition index or TCI) that are defined in one chapter (e.g., on drought) may be repeated in another chapter (also on drought). Such minor overlaps are helpful to the reader to avoid having to go back to another chapter to understand a phenomenon, an index, or a characteristic of a sensor. However, if you want a lot of details of these sensors, indices, or phenomena, then you will have to read the appropriate chapter where there is in-depth coverage of the topic.

Each volume has a summary chapter (the last chapter of each volume). The summary chapter can be read two ways: (1) either as the last chapter to recapture the main points of each of the chapters and/or (2) or as an initial overview to get the feeling for what is in the volume before diving in to read each chapter in detail. I suggest the readers do it both ways: read it first before reading the chapters in detail to gather an idea of what to expect in each chapter, and then read it at the end to recapture what is being read in each of the chapters.

It has been a great honor as well as a humbling experience to edit the *Remote Sensing Handbook* (Vols. I–VI). I truly enjoyed the effort, albeit I felt overwhelmed at times with never ending work. What an honor to work with luminaries in your field of expertise. I learned a lot from them and am very grateful for their support, encouragement, and deep insights. Also, it has been a pleasure

Preface

working with the outstanding professionals of Taylor & Francis Group/CRC Press. There is no greater joy than being immersed in the pursuit of excellence, knowledge gain, and knowledge capture. At the same time, I am happy it is over. If there will be a third edition a decade or so from now, it will be taken up by someone else (individually or as a team) and certainly not me!

I expect the book to be a standard reference of immense value to any student, scientist, professional, and practical practitioners of remote sensing. Any book that has the privilege of 300+ truly outstanding and dedicated remote sensing scientists ought to be a *magnum opus* deserving to be a standard reference on the subject.

Dr. Prasad S. Thenkabail, PhD
Editor-in-Chief (EiC)
Remote Sensing Handbook (Second Edition, Vols. I–VI)

Volume 1: Sensors, Data Normalization, Harmonization, Cloud Computing, and Accuracies
Volume 2: Image Processing, Change Detection, GIS, and Spatial Data Analysis
Volume 3: Agriculture, Food Security, Rangelands, Vegetation, Phenology, and Soils
Volume 4: Forests, Biodiversity, Ecology, LULC, and Carbon
Volume 5: Water Resources: Hydrology, Floods, Snow and Ice, Wetlands, and Water Productivity
Volume 6: Droughts, Disasters, Pollution, and Urban Mapping

About the Editor

Dr. Prasad S. Thenkabail, PhD, is a Senior Scientist with the US Geological Survey (USGS), specializing in remote sensing science for agriculture, water, and food security. He is a world-recognized expert in remote sensing science with multiple major contributions in the field sustained for 40+ years. Dr. Thenkabail has conducted pioneering research in the hyperspectral remote sensing of vegetation, global croplands mapping for water and food security, and crop water productivity. His work on hyperspectral remote sensing of agriculture and vegetation are widely cited. His papers on hyperspectral remote sensing are firsts of their kind and, collectively, they have: (1) determined optimal hyperspectral narrowbands (OHNBs) in the study of agricultural crops, (2) established hyperspectral vegetation indices (HVIs) to model and map crop biophysical and biochemical quantities, (3) created a framework and sample data for the global hyperspectral imaging spectral libraries of crops (GHISA), (4) developed methods and techniques of overcoming Hughes's phenomenon, (5) demonstrated the strengths of hyperspectral narrowband (HNB) data in advancing classification accuracies relative to multispectral broadband (MBB) data, (6) shown the advances one can make in modeling crop biophysical and biochemical quantities using HNB and HVI data relative to MBB data, and (7) created a body of work in understanding, processing, and utilizing HNB and HVI data in agricultural cropland studies. This body of work has become a widely referred reference worldwide. In studies of global croplands for food and water security, he has led the release of the world's first 30-m Landsat satellite-derived global cropland extent product at 30m (GCEP30; https://www.usgs.gov/apps/croplands/app/map); (Thenkabail et al., 2021; https://lpdaac.usgs.gov/news/release-of-gfsad-30-meter-cropland-extent-products/) and Landsat-derived global rainfed and irrigated area product at 30m (LGRIP30; https://lpdaac.usgs.gov/products/lgrip30v001/) (Teluguntla and Thenkabail et al., 2023). Earlier led producing the world's first global irrigated area map (https://lpdaac.usgs.gov/products/lgrip30v001/; https://lpdaac.usgs.gov/products/gfsad1kcdv001/) using multi-sensor satellite data that led to crop. The global cropland datasets using satellite remote sensing demonstrates a "paradigm shift" in global cropland mapping using remote sensing through big data analytics, machine learning, and petabyte-scale cloud computing on the Google Earth Engine (GEE). The LGRIP30 and GCEP30 products are released through NASA's LP DAAC and published in USGS professional paper 1868 (Thenkabail et al., 2021). He has been principal investigator of many projects over the years, including the NASA-funded global food security support analysis data in the 30-m (GFSAD) project (www.usgs.gov/wgsc/gfsad30).

His career scientific achievements can be gauged by making the list of the world's top 1% of scientists per the Stanford study ranking the world's scientists from across 22 scientific fields and 176 subfields based on deep analysis evaluating the Elsevier SCOPUS data of about 10 million scientists from 1996 to 2023 (Ioannidis, 2023; Ioannidis et al., 2020). Dr. Thenkabail was recognized as Fellow of the American Society of Photogrammetry and Remote Sensing (ASPRS) in 2023. He has published over 150 peer-reviewed scientific papers and edited 15 books. His scientific papers have won several awards over the years, demonstrating world-class research of the highest quality. These include: the 2023 Talbert Abrams Grand Award, the highest scientific paper award of the ASPRS (with Itiya Aneece), the 2015 ASPRS ERDAS award for best scientific paper in remote sensing (with Michael Marshall), the 2008 John I. Davidson ASPRS President's Award for Practical papers (with Pardha Teluguntla), and the 1994 Autometric Award for outstanding paper in remote sensing (with Dr. Andy Ward).

Dr. Thenkabail's contributions to series of leading edited books places him as a world leader in remote sensing science. There are three seminal book sets with a total of 13 volumes that he edited, and which have demonstrated his major contributions as an internationally acclaimed remote sensing scientist. These are: (1) *Remote Sensing Handbook* (Second Edition, six-volume book set, 2024) with 91 chapters and nearly 3000 pages and for which he is the sole editor, (2) *Remote Sensing Handbook* (First Edition, three-volume book set, 2015) with 82 chapters and 2304 pages and for which he is the sole editor, and (3) *Hyperspectral Remote Sensing of Vegetation* (four-volume book set, 2018) with 50 chapters and 1632 pages that he edited as the chief editor (co-editors: Prof. John Lyon and Prof. Alfredo Huete).

Dr. Thenkabail is at the center of rendering scientific service to the world's remote sensing community over long periods of service. This includes serving as Editor-in-Chief (2011–present) of *Remote Sensing Open Access Journal*; Associate Editor (2017–present) of *Photogrammetric Engineering and Remote Sensing (PE&RS)*, Editorial Advisory Board (2016–present) of the International Society of Photogrammetry and Remote Sensing (ISPRS), and Editorial Board Member (2007–2017) of *Remote Sensing of Environment* (2007–2016).

The USGS and NASA selected him as one of three international members on the Landsat Science Team (2006–2011). He is an Advisory Board member of the online library collection to support the United Nations' Sustainable Development Goals (UN SDGs), and currently a scientist for the NASA and ISRO (Indian Space Research Organization) Professional Engineer and Scientist Exchange Program (PESEP) program for 2022–2024. He was the Chair of the International Society of Photogrammetry and Remote Sensing (ISPRS) Working Group (WG) VIII/7 (land cover and its dynamics) from 2013 to 2016; played a vital role for USGS as Global Coordinator, Agricultural Societal Beneficial Area (SBA), Committee for Earth Observation (CEOS) (2010–2013), during which he co-wrote the global food security case study for the CEOS Earth Observation Handbook (EOS), Special Edition for the UN Conference on Sustainable Development, presented in Rio de Janeiro, Brazil; and was the Co-lead (2007–2011) of IEEE's "Water for the World" initiative, a nonprofit effort funded by IEEE that worked in coordination with the Group on Earth Observations (GEO) in its GEO Water and GEO Agriculture initiatives.

Dr. Thenkabail worked as a postdoctoral researcher and research faculty at the Center for Earth Observation (YCEO), Yale University (1997–2003) and led remote sensing programs in three international organizations, including:

- International Water Management Institute (IWMI), 2003–2008
- International Center for Integrated Mountain Development (ICIMOD), 1995–1997
- International Institute of Tropical Agriculture (IITA), 1992–1995

He began his scientific career as a Scientist (1986–1988) working for the National Remote Sensing Agency (NRSA) (now renamed the National Remote Sensing Center, or NRSC), the Indian Space Research Organization (ISRO), and the Department of State, Government of India.

Dr. Thenkabail's work experience spans over 25 countries, including East Asia (China), Southeast Asia (Cambodia, Indonesia, Myanmar, Thailand, and Vietnam), the Middle East (Israel, Syria), North America (United States, Canada), South America (Brazil), Central Asia (Uzbekistan), South Asia (Bangladesh, India, Nepal, and Sri Lanka), West Africa (Republic of Benin, Burkina Faso, Cameroon, Central African Republic, Cote d'Ivoire, Gambia, Ghana, Mali, Nigeria, Senegal, and Togo), and Southern Africa (Mozambique, South Africa). Dr. Thenkabail is regularly invited as keynote speaker or invited speaker at major international conferences and at other important national and international forums every year.

Dr. Thenkabail obtained his PhD in Agricultural Engineering from the Ohio State University, USA, in 1992 and has a master's degree in hydraulics and water resources engineering, as well as a bachelor's degree in civil engineering (both from India). He has 168 publications, including 15 books; 175+ peer-reviewed journal articles, book chapters, and professional papers/monographs; and 15+ significant major global and regional data releases.

REFERENCES

SCIENTIFIC PAPERS

https://scholar.google.com/citations?user=9IO5Y7YAAAAJ&hl=en

USGS Professional Paper, Data & Product Gateways, Interactive Viewers

Ioannidis, J.P.A. (2023). October 2023 data-update for "Updated science-wide author databases of standardized citation indicators". *Elsevier Data Repository, 6*. https://doi.org/10.17632/btchxktzyw.6.

Ioannidis, J.P.A., Boyack, K.W., and Baas, J. (2020). Updated science-wide author databases of standardized citation indicators. *PLoS Biol 18*(10), p. e3000918. https://doi.org/10.1371/journal.pbio.3000918.

Teluguntla, P., Thenkabail, P., Oliphant, A., Gumma, M., Aneece, I., Foley, D., and McCormick, R. (2023). Landsat-Derived Global Rainfed and Irrigated-Cropland Product @ 30-m (LGRIP30) of the World (GFSADLGRIP30WORLD). *The Land Processes Distributed Active Archive Center (LP DAAC) of NASA and USGS*. Pp.103. https://lpdaac.usgs.gov/news/release-of-lgrip30-data-product/ (download data, documents).

Thenkabail, P.S., Teluguntla, P.G., Xiong, J., Oliphant, A., Congalton, R.G., Ozdogan, M., Gumma, M.K., Tilton, J.C., Giri, C., Milesi, C., Phalke, A., Massey, R., Yadav, K., Sankey, T., Zhong, Y., Aneece, I., and Foley, D. (2021). Global Cropland-Extent Product at 30-m Resolution (GCEP30) Derived from Landsat Satellite Time-Series Data for the Year 2015 Using Multiple Machine-Learning Algorithms on Google Earth Engine Cloud: U.S. *Geological Survey Professional Paper 1868*, 63 p., https://doi.org/10.3133/pp1868 (research paper). https://lpdaac.usgs.gov/news/release-of-gfsad-30-meter-cropland-extent-products/ (download data, documents). https://www.usgs.gov/apps/croplands/app/map (view data interactively).

Books

Remote Sensing Handbook (Second Edition, Six Volumes, 2024)

Thenkabail, Prasad. 2024. Remote Sensing Handbook (Second Edition, Six Volume Book-set), *Volume I: Sensors, Data Normalization, Harmonization, Cloud Computing, and Accuracies*. Taylor and Francis Inc.\CRC Press, Boca Raton, London, New York. 978-1-032-89095-1—CAT# T132478. Print ISBN: 9781032890951. eBook ISBN: 9781003541141. Pp. 581.

Thenkabail, Prasad. 2024. Remote Sensing Handbook (Second Edition, Six Volume Book-set), *Volume II: Image Processing, Change Detection, GIS, and Spatial Data Analysis*. Taylor and Francis Inc.\CRC Press, Boca Raton, London, New York. 978-1-032-89097-5—CAT# T133208. Print ISBN: 9781032890975. eBook ISBN: 9781003541158. Pp. 464.

Thenkabail, Prasad. 2024. Remote Sensing Handbook (Second Edition, Six Volume Book-set), *Volume III: Agriculture, Food Security, Rangelands, Vegetation, Phenology, and Soils*. Taylor and Francis Inc.\CRC Press, Boca Raton, London, New York. 978-1-032-89101-9—CAT# T133213. Print ISBN: 9781032891019; eBook ISBN: 9781003541165. Pp. 788.

Thenkabail, Prasad. 2024. Remote Sensing Handbook (Second Edition, Six Volume Book-set), *Volume IV: Forests, Biodiversity, Ecology, LULC, and Carbon*. Taylor and Francis Inc.\CRC Press, Boca Raton, London, New York. 978-1-032-89103-3—CAT# T133215. Print ISBN: 9781032891033. eBook ISBN: 9781003541172. Pp. 501.

Thenkabail, Prasad. 2024. Remote Sensing Handbook (Second Edition, Six Volume Book-set), *Volume V: Water, Hydrology, Floods, Snow and Ice, Wetlands, and Water Productivity*. Taylor and Francis Inc.\CRC Press, Boca Raton, London, New York. 978-1-032-89145-3—CAT# T133261. Print ISBN: 9781032891453. eBook ISBN: 9781003541400. Pp. 516.

Thenkabail, Prasad. Remote Sensing Handbook (Second Edition, Six Volume Book-set), *Volume VI: Droughts, Disasters, Pollution, and Urban Mapping*. Taylor and Francis Inc.\CRC Press, Boca Raton, London, New York. 978-1-032-89148-4—CAT# T133267. Print ISBN: 9781032891484; eBook ISBN: 9781003541417. Pp. 467.

Hyperspectral Remote Sensing of Vegetation (First Edition, Four Volumes, 2018)

Thenkabail, P.S., Lyon, G.J., and Huete, A. (Editors) (2018). Book Title: *Hyperspectral Remote Sensing of Vegetation* (Second Edition, four-volume set).

Volume I Title: *Fundamentals, Sensor Systems, Spectral Libraries, and Data Mining for Vegetation*. Publisher: CRC Press- Taylor and Francis group, Boca Raton, London, New York. Pp.449, Hardback ID: 9781138058545; eBook ID: 9781315164151.

Volume II Title: *Hyperspectral Indices and Image Classifications for Agriculture and Vegetation*. Publisher: CRC Press- Taylor and Francis group, Boca Raton, London, New York. Pp.296. Hardback ID: 9781138066038; eBook ID: 9781315159331.

Volume III Title: *Biophysical and Biochemical Characterization and Plant Species Studies*. Publisher: CRC Press- Taylor and Francis group, Boca Raton, London, New York. Pp.348. Hardback: 9781138364714; eBook ID: 9780429431180.

Volume IV Title: *Advanced Applications in Remote Sensing of Agricultural Crops and Natural Vegetation*. Publisher: CRC Press- Taylor and Francis group, Boca Raton, London, New York. Pp.386. Hardback: 9781138364769; eBook ID: 9780429431166.

Remote Sensing Handbook (First Edition, Three Volumes, 2015)

Thenkabail, P.S., (Editor-in-Chief), 2015. "Remote Sensing Handbook"

Volume I: *Remotely Sensed Data Characterization, Classification, and Accuracies*. Taylor and Francis Inc.\CRC Press, Boca Raton, London, New York. ISBN 9781482217865—CAT# K22125. Print ISBN: 978-1-4822-1786-5; eBook ISBN: 978-1-4822-1787-2. Pp. 678.

Volume II: *Land Resources Monitoring, Modeling, and Mapping with Remote Sensing*. Taylor and Francis Inc.\CRC Press, Boca Raton, London, New York. ISBN 9781482217957—CAT# K22130. Pp. 849.

Volume III: *Remote Sensing of Water Resources, Disasters, and Urban Studies*. Taylor and Francis Inc.\CRC Press, Boca Raton, London, New York. ISBN 9781482217919—CAT# K22128. Pp. 673.

Hyperspectral Remote Sensing of Vegetation (First Edition, Single Volume, 2013)

Thenkabail, P.S., Lyon, G.J., and Huete, A. (Editors) (2012). Book entitled: *Hyperspectral Remote Sensing of Vegetation*. CRC Press\Taylor and Francis group, Boca Raton, London, New York. Pp.781 (80+ pages in color). http://www.crcpress.com/product/isbn/9781439845370

Remote Sensing of Global Croplands for Food Security (First Edition, Single Volume, 2009)

Thenkabail, P., Lyon, G.J., Turral, H., and Biradar, C.M. (Editors) (2009). Book entitled: *Remote Sensing of Global Croplands for Food Security*. CRC Press\Taylor and Francis Group, Boca Raton, London, New York. Pp.556 (48 pages in color). Published in June, 2009.

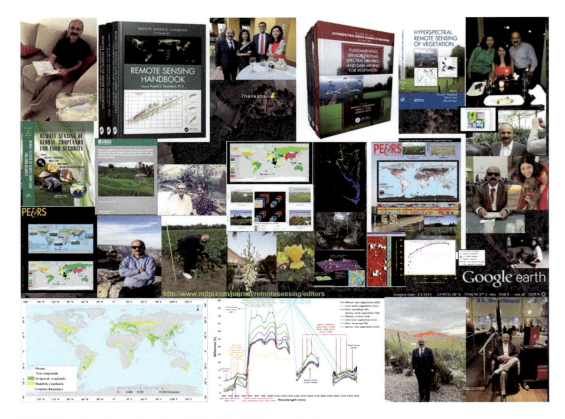

FIGURE: Snap shots of work and life of the Editor-in-Chief.

Contributors

Martha A. Anderson
Hydrology and Remote Sensing Laboratory (HRSL)
US Department of Agriculture, Agricultural Research Service (USDA ARS)
Beltsville, Maryland, USA

C. Atzberger
Institute for Surveying, Remote Sensing and Land Information
University of Natural Resources and Life Sciences (BOKU)
Vienna, Austria

Anushree Badola
Institute of Agriculture, Natural Resources and Extension
University of Alaska Fairbanks
Fairbanks, Alaska, USA

Hasi Bagan
School of Environmental and Geographical Sciences
Shanghai Normal University
Shanghai, China

Gabriel del Barrio
Estacion Experimental de Zonas Aridas (CSIC)
Almeria, Spain

Brittany Card
Program Coordinator, Signal Program on Human Security and Technology
Harvard Humanitarian Initiative
Cambridge, Massachusetts, USA

Christos Chalkias
Applied Geography and GIS, Department of Geography
Harokopio University of Athens
Athens, Greece

ChaominChen
School of Environmental and Geographical Sciences
Shanghai Normal University
Shanghai, China

Wade T. Crow
Hydrology and Remote Sensing Laboratory (HRSL)
US Department of Agriculture, Agricultural Research Service (USDA ARS)
Beltsville, Maryland, USA

Erwann Fillol
Action Contre la Faim—International
ACF West Africa Regional Office
Dakar, Senegal

Brian Fuchs
National Drought Mitigation Center (NDMC)
School of Natural Resources,
University of Nebraska-Lincoln
Lincoln, Nebraska, USA

Petra Füreder
Department of Geoinformatics—Z_GIS
Paris-Lodron University Salzburg
Salzburg, Austria

Wei Guo
IMSG NCWCP
College Park, Maryland
USA

Chris R. Hain
NASA Marshall Space Flight Center
Earth Science Branch
Madison County, Alabama, USA

F. Ham
Action Contre la Faim—International
ACF West Africa Regional Office
Dakar, Senegal

Joachim Hill
Estacion Experimental de Zonas Aridas (CSIC)
Almeria, Spain

Jungho Im
School of Urban and Environmental Engineering
Ulsan National Institute of Science and Technology (UNIST)
Ulsan, Republic of Korea

Norman Kerle
Faculty of Geo-Information Science and Earth Observation (ITC)
University of Twente
Enschede, Netherlands

Felix Kogan
National Oceanic and Atmospheric Administration
National Environmental Satellite Data and Information Services
Center for Satellite Applications and Research
College Park, Maryland, USA

Olaf Kranz
Research Field Representative Aeronautics, Space and Transport
Helmholtz-Association
Berlin, Germany

Claudia Kuenzer
Earth Observation Center
German Aerospace Center (DLR)
Oberpfaffenhofen Wessling, Germany

Stefan Lang
Division Head and Research Coordinator
Department of Geoinformatics—Z_GIS
Paris-Lodron University Salzburg
Salzburg, Austria

Kristofer Lasko
Department of Geographical Sciences
University of Maryland, College Park
College Park, Maryland, USA

Husi Letu
State Key Laboratory of Science and Remote Sensing
Aerospace Information Research Institute
Chinese Academy of Sciences
Beijing, China

Noam Levin
Department of Geography The Hebrew University Jerusalem, Israel

Di Liu
University of Oklahoma
Norman, Oklahoma, USA

M. Meroni
Institute for Environment and Sustainability
Joint Research Centre (JRC)
European Commission
Ispra, VA, Italy

Daisuke Murakami
Department of Statistical Data Science
Institute of Statistical Mathematics
Tachikawa City, Tokyo Prefecture, Japan

Santosh K. Panda
Institute of Agriculture, Natural Resources and Extension
University of Alaska Fairbanks
Fairbanks, Alaska, USA

Andreas Papp
Médecins Sans Frontières (MSF) Austria
Vienna, Austria

Seonyoung Park
Department of Applied Artificial Intelligence
Seoul National University of Science and Technology (SeoulTech)
Seoul, Republic of Korea

Anupma Prakash
Geophysical Institute
University of Alaska Fairbanks
Fairbanks, Alaska, USA

F. Rembold
Institute for Environment and Sustainability
Joint Research Centre (JRC)
European Commission
Ispra, VA, Italy

Jinyoung Rhee
Climate Services and Research Division
APEC Climate Center
Busan, Republic of Korea

Shadrock Roberts
University of Georgia
Center for Geospatial Research
Athens, Georgia, USA

Matt Rodell
Hydrologic Science Laboratory
NASA Goddard Space Flight Center (GSFC)

Contributors

Achim Röder
Department of Environmental Remote Sensing and Geoinformatics
Faculty Regional and Environmental Sciences
Trier University
Trier, Germany

O. Rojas
Food and Agriculture Organization of the United Nations (FAO)
Natural Resources Management and Environment Department
Rome, Italy

Hajime Seya
Graduate School of Engineering
Kobe University
Kobe City, Hyogo Prefecture, Japan

Ruth Sonnenschein
Institute for Earth Observation
European Academy of Bozen/Bolzano (EURAC)
Bolzano, Italy

Marion Stellmes
Remote Sensing and Geoinformatics
Department of Earth Sciences
Freie Universität Berlin
Berlin, Germany

Mark S. Svoboda
National Drought Mitigation Center (NDMC)
School of Natural Resources
University of Nebraska-Lincoln
Lincoln, Nebraska, USA

Tsegaye Tadesse
National Drought Mitigation Center (NDMC)
School of Natural Resources
University of Nebraska-Lincoln
Lincoln, Nebraska, USA

Prasad S. Thenkabail
US Geological Survey (USGS)
Flagstaff, Arizona, USA

Thomas Udelhoven
Department of Environmental Remote Sensing and Geoinformatics
Faculty Regional and Environmental Sciences
Trier University
Trier, Germany

Krishna Prasad Vadrevu
Department of Geographical Sciences
University of Maryland College Park
College Park, Maryland, USA

Christine F. Waigl
International Arctic Research Center
University of Alaska Fairbanks
Fairbanks, Alaska, USA

Brian D. Wardlow
Center for Advanced Land Management Technologies (CALMIT)
School of Natural Resources
University of Nebraska-Lincoln
Lincoln, Nebraska, USA
And
National Drought Mitigation Center (NDMC)
School of Natural Resources
University of Nebraska-Lincoln
Lincoln, Nebraska, USA

Robert Wright
Hawai'i Institute of Geophysics and Planetology
University of Hawai'i at Mānoa
Honolulu, Hawai'i, USA

Yoshiki Yamagata
Graduate School of System Design and Management
Keio University
Kanagawa, Japan

Takahiro Yoshida
Center for Spatial Information Science
The University of Tokyo
Kashiwa City, Chiba Prefecture, Japan

Qingling Zhang
School of Aeronautics and Astronautics
Sun Yat-sen University
Guangzhou, China

Acknowledgments

The *Remote Sensing Handbook* (Second Edition, Vols. I–VI) brought together a galaxy of highly accomplished, renowned remote sensing scientists, professionals, and legends from around the world. I chose the lead authors after careful review of their accomplishments and sustained publication record over the years. The chapters in the second edition were written/revised over a period of two years. All chapters were edited and revised.

Gathering such a galaxy of authors was the biggest challenge. These are all extremely busy people, and committing to a book project that requires substantial workload is never easy. However, almost all of those whom I requested agreed to write a chapter specific to their area of specialization, and only a few I had to convince to make time. The quality of the chapters should convince readers of why these authors are such highly rated professionals and why they are so successful and accomplished in their field of expertise. They not only wrote very high-quality chapters, but delivered them on time, addressed any editorial comments in a timely manner without complaints, and were extremely humble and helpful. Their commitment for quality science is what makes them special. I am truly honored to have worked with such great professionals.

I would like to mention the names of everyone who contributed and made the *Remote Sensing Handbook* (Second Edition, Vols. I–VI) possible. In the end, we had 91 chapters, a little over 3000 pages, and a little over 400 authors. My gratitude goes to each one of them. These are the well-known "who's who" in remote sensing science around the world. A list of all the authors is provided next. The names of the authors are organized chronologically for each volume and its chapters. Each lead author of the chapter is in bold. The names of the 400+ authors who contributed to the six volumes are as follows:

Volume I: Sensors, Data Normalization, Harmonization, Cloud Computing, and Accuracies: 18 chapters written by 53 authors (Editor-in-chief: Prasad S. Thenkabail):

Drs. Sudhanshu S. Panda, Mahesh Rao, Prasad S. Thenkabail, Debasmita Misra, and James P. Fitzerald; **Mohinder S. Grewal**; **Kegen Yu**, Chris Rizos, and Andrew Dempster; **D. Myszor**, O. Antemijczuk, M. Grygierek, M. Wierzchanowski, and K.A. Cyran; **Natascha Oppelt** and Arnab Muhuri; **Philippe M. Teillet**; **Philippe M. Teillet** and Gyanesh Chander; **Rudiger Gens** and Jordi Cristóbal Rosselló; **Aolin Jia** and Dongdong Wang; **Tomoaki Miura**, Kenta Obata, Hiroki Yoshioka, and Alfredo Huete; **Michael D. Steven**, Timothy J. Malthus, and Frédéric Baret; **Fabio Dell'Acqua** and Silvio Dell'Acqua; **Ramanathan Sugumaran**, James W. Hegeman, Vivek B. Sardeshmukh, Marc P. Armstrong; **Lizhe Wang**, Jining Yan, Yan Ma, Xiaohui Huang, Jiabao Li, Sheng Wang, Haixu He, Ao Long, and Xiaohan Zhang; **John E. Bailey** and Josh Williams; **Russell G. Congalton**; **P.J. Blount**; **Prasad S. Thenkabail**.

Volume II: Image Processing, Change Detection, GIS, and Spatial Data Analysis: 17 chapters written 64 authors (Editor-in-chief: Prasad S. Thenkabail):

Sunil Narumalani and Paul Merani; **Mutlu Ozdogan**; **Soe W. Myint**, Victor Mesev, Dale Quattrochi, and Elizabeth A. Wentz; **Jun Li**, Paolo Gamba, and Antonio Plaza; **Qian Du**, Chiranjibi Shah, Hongjun Su, and Wei Li; **Claudia Kuenzer**, Philipp Reiners, Jianzhong Zhang, Stefan Dech; **Mohammad D. Hossain** and Dongmei Chen; **Thomas Blaschke**, Maggi Kelly, Helena Merschdorf; **Stefan Lang** and Dirk Tiede; **James C. Tilton**, Selim Aksoy, and Yuliya Tarabalka; **Shih-Hong Chio**, Tzu-Yi Chuang, Pai-Hui Hsu, Jen-Jer Jaw, Shih-Yuan Lin, Yu-Ching Lin, Tee-Ann Teo, Fuan Tsai, Yi-Hsing Tseng, Cheng-Kai Wang, Chi-Kuei Wang, Miao Wang, and Ming-Der Yang; **Guiying Li**, Mingxing Zhou, Ming Zhang, and Dengsheng Lu; **Jason A. Tullis**, David P. Lanter, Aryabrata Basu, Jackson D. Cothren, Xuan Shi, W. Fredrick Limp, Rachel F. Linck, Sean G. Young, Jason Davis, and Tareefa S. Alsumaiti; **Gaurav Sinha**, Barry J. Kronenfeld, Jeffrey C. Brunskill;

May Yuan; **Stefan Lang**, Stefan Kienberger, Michael Hagenlocher, and Lena Pernkopf; **Prasad S. Thenkabail**.

Volume III: Agriculture, Food Security, Rangelands, Vegetation, Phenology, and Soils: 17 chapters written by 110 authors (Editor-in-chief: Prasad S. Thenkabail):

Alfredo Huete, Guillermo Ponce-Campos, Yongguang Zhang, Natalia Restrepo-Coupe, and Xuanlong Ma; **Juan Quiros-Vargas**, Bastian Siegmann, Juliane Bendig, Laura Verena Junker-Frohn, Christoph Jedmowski, David Herrera, and Uwe Rascher; **Frédéric Baret**; **Lea Hallik**, Egidijus Šarauskis, Ruchita Ingle, Indrė Bručienė, Vilma Naujokienė, and Kristina Lekavičienė; **Clement Atzberger** and Markus Immitzer; **Agnès Bégué**, Damien Arvor, Camille Lelong, Elodie Vintrou, and Margareth Simoes; **Pardhasaradhi Teluguntla**, Prasad S. Thenkabail, Jun Xiong, Murali Krishna Gumma, Chandra Giri, Cristina Milesi, Mutlu Ozdogan, Russell G. Congalton, James Tilton, Temuulen Tsagaan Sankey, Richard Massey, Aparna Phalke, and Kamini Yadav; **Yuxin Miao**, David J. Mulla, and Yanbo Huang; **Baojuan Zheng**, James B. Campbell, Guy Serbin, Craig S.T. Daughtry, Heather McNairn, and Anna Pacheco; **Prasad S. Thenkabail**, Itiya Aneece, Pardhasaradhi Teluguntla, Richa Upadhyay, Asfa Siddiqui, Justin George Kalambukattu, Suresh Kumar, Murali Krishna Gumma, and Venkateswarlu Dheeravath; **Matthew C. Reeves**, Robert Washington-Allen, Jay Angerer, Raymond Hunt, Wasantha Kulawardhana, Lalit Kumar, Tatiana Loboda, Thomas Loveland, Graciela Metternicht, Douglas Ramsey, Joanne V. Hall, Trenton Benedict, Pedro Millikan, Angus Retallack, Arjan J.H. Meddens, William K. Smith, and Wen Zhang; **E. Raymond Hunt Jr.**, Cuizhen Wang, D. Terrance Booth, Samuel E. Cox, Lalit Kumar, and Matthew C. Reeves; **Lalit Kumar**, Priyakant Sinha, Jesslyn F. Brown, R. Douglas Ramsey, Matthew Rigge, Carson A. Stam, Alexander J. Hernandez, E. Raymond Hunt, Jr., and Matt Reeves; **Molly E. Brown**, Kirsten de Beurs, and Kathryn Grace; **José A.M. Demattê**, Cristine L.S. Morgan, Sabine Chabrillat, Rodnei Rizzo, Marston H.D. Franceschini, Fabrício da S. Terra, Gustavo M. Vasques, Johanna Wetterlind, Henrique Bellinaso, and Letícia G. Vogel; **E. Ben-Dor**, J.A.M. Demattê; **Prasad S. Thenkabail**.

Volume IV: Forests, Biodiversity, Ecology, LULC, and Carbon: 12 chapters written by 71 authors (Editor-in-chief: Prasad S. Thenkabail):

E.H. Helmer, Nicholas R. Goodwin, Valéry Gond, Carlos M. Souza Jr., and
Gregory P. Asner; **Juha Hyyppä**, Xiaowei Yu, Mika Karjalainen, Xinlian Liang, Anttoni Jaakkola, Mike Wulder, Markus Hollaus, Joanne C. White, Mikko Vastaranta, Jiri Pyörälä, Tuomas Yrttimaa, Ninni Saarinen, Josef Taher, Juho-Pekka Virtanen, Leena Matikainen, Yunsheng Wang, Eetu Puttonen, Mariana Campos, Matti Hyyppä, Kirsi Karila, Harri Kaartinen, Matti Vaaja, Ville Kankare, Antero Kukko, Markus Holopainen, Hannu Hyyppä, Masato Katoh, and Eric Hyyppä; **Gregory P. Asner**, Susan L. Ustin, Philip A. Townsend, and Roberta E. Martin; **Sylvie Durrieu**, Cédric Véga, Marc Bouvier, Frédéric Gosselin, Jean-Pierre Renaud, and Laurent Saint-André; **Thomas W. Gillespie**, Morgan Rogers, Chelsea Robinson, and Duccio Rocchini; **Stefan Lang**, Christina Corbane, Palma Blonda, Kyle Pipkins, Michael Förster; **Conghe Song**, Jing Ming Chen, Taehee Hwang, Alemu Gonsamo, Holly Croft, Quanfa Zhang, Matthew Dannenberg, Yulong Zhang, Christopher Hakkenberg, and Juxiang Li; **John Rogan** and Nathan Mietkiewicz; **Zhixin Qi**, Anthony Gar-On Yeh, Xia Li, and Qianwen Lv; **R.A. Houghton**; **Wenge Ni-Meister**; **Prasad S. Thenkabail**.

Volume V: Water Resources: Hydrology, Floods, Snow and Ice, Wetlands, and Water Productivity: 13 chapters written by 60 authors (Editor-in-chief: Prasad S. Thenkabail):

James B. Campbell and Lynn M. Resler; **Sadiq I. Khan**, Ni-Bin Chang, Yang Hong, Xianwu Xue, and Yu Zhang; **Santhosh Kumar Seelan**; **Allan S. Arnesen**, Frederico T. Genofre, Marcelo P. Curtarelli, and Matheus Z. Francisco; **Sandro Martinis**, Claudia Kuenzer, and André Twele; **Le Wang**, Jing Miao, and Ying Lu; **Chandra Giri**; **D.R. Mishra**, X. Yan, S. Ghosh, C. Hladik,

Acknowledgments

J.L. O'Connell, H.J. Cho; **Murali Krishna Gumma**, Prasad S. Thenkabail, Pranay Panjala, Pardhasaradhi Teluguntla, Birhanu Zemadim Birhanu, and Mangi Lal Jat; **Trent W. Biggs**, Pamela Nagler, Anderson Ruhoff, Triantafyllia Petsini, Michael Marshall, George P. Petropoulos, Camila Abe, and Edward P. Glenn; **Antônio Teixeira**, Janice Leivas, Celina Takemura, Edson Patto, Edlene Garçon, Inajá Sousa, André Quintão, Prasad Thenkabail, and Ana Azevedo; **Hongjie Xie**, Tiangang Liang, Xianwei Wang, Guoqing Zhang, Xiaodong Huang, and Xiongxin Xiao; **Prasad S. Thenkabail**.

Volume VI: Droughts, Disasters, Pollution, and Urban Mapping: 14 chapters written by 53 authors (Editor-in-chief: Prasad S. Thenkabail):

Felix Kogan and Wei Guo; **F. Rembold**, M. Meroni, O. Rojas, C. Atzberger, F. Ham, and Erwann Fillol; **Brian D. Wardlow**, Martha A. Anderson, Tsegaye Tadesse, Mark S. Svoboda, Brian Fuchs, Chris R. Hain, Wade T. Crow, and Matt Rodell; **Jinyoung Rhee**, Jungho Im, and Seonyoung Park; **Marion Stellmes**, Ruth Sonnenschein, Achim Röder, Thomas Udelhoven, Gabriel del Barrio, and Joachim Hill; **Norman Kerle**; **Stefan Lang**, Petra Füreder, Olaf Kranz, Brittany Card, Shadrock Roberts, and Andreas Papp; **Robert Wright**; **Krishna Prasad Vadrevu** and Kristofer Lasko; **Anupma Prakash**, Claudia Kuenzer, Santosh K. Panda, Anushree Badola, and Christine F. Waigl; **Hasi Bagana**, Chaomin Chena, and Yoshiki Yamagata; **Yoshiki Yamagata**, Daisuke Murakami, Hajime Seya, and Takahiro Yoshida; **Qingling Zhang**, Noam Levin, Christos Chalkias, Husi Letu, and Di Liu; **Prasad S. Thenkabail**.

The authors not only delivered excellent chapters, but they also provided valuable insights and input for me in many ways throughout the book project.

I was delighted when Dr. Compton J. Tucker, Senior Earth Scientist, Earth Sciences Division, Science and Exploration Directorate, NASA Goddard Space Flight Center (GSFC) agreed to write foreword for the book. For anyone practicing remote sensing, Dr. Tucker needs no introduction. He has been a "godfather" of remote sensing and has inspired a generation of remote sensing scientists. I have been a student of his without ever really being one. I mean, I have not been his student in the classroom, but I have followed his legendary work throughout my career. I remember reading his highly cited paper (now with citations nearing 7700!):

- Tucker, C.J. (1979) 'Red and Photographic Infrared Linear Combinations for Monitoring Vegetation', *Remote Sensing of Environment,* 8(2),127–150.

I first read this paper in 1986 when I had just joined the National Remote Sensing Agency (NRSA; now NRSC) of the Indian Space Research Organization (ISRO). Dr. Tucker's pioneering works have been a guiding light for me ever since. After getting his PhD from Colorado State University in 1975, Dr. Tucker joined NASA GSFC as a postdoctoral fellow and became a full-time NASA employee in 1977. Since then, he has conducted several path-finding studies. He has used NOAA AVHRR, MODIS, SPOT Vegetation, and Landsat satellite data for studying deforestation, habitat fragmentation, desert boundary determination, ecologically coupled diseases, terrestrial primary production, glacier extent, and how climate affects global vegetation. He has authored or coauthored more than 280 journal articles that have been cited more than 93,000 times, is an adjunct professor at the University of Maryland, is a consulting scholar at the University of Pennsylvania's Museum of Archaeology and Anthropology, and has appeared in more than 20 radio and TV programs. He is a Fellow of the American Geophysical Union and has been awarded several medals and honors, including NASA's Exceptional Scientific Achievement Medal, the Pecora Award from the US Geological Survey, the National Air and Space Museum Trophy, the Henry Shaw Medal from the Missouri Botanical Garden, the Galathea Medal from the Royal Danish Geographical Society, and the Vega Medal from the Swedish Society of Anthropology and Geography. He was the NASA representative to the US Global Change Research Program from 2006 to 2009. He was instrumental in releasing the AVHRR 33-year (1982–2014) Global Inventory Monitoring and Modeling

Studies (GIMMS) data. I strongly recommend that everyone reads his excellent foreword before reading the book. In the foreword, Dr. Tucker demonstrates the importance of data from Earth Observation (EO) sensors from orbiting satellites to maintaining a reliable and consistent climate record. Dr. Tucker further highlights the importance of continued measurements of these variables of our planet in the new millennium through new, improved, and innovative EO sensors from sun synchronous and/or geostationary satellites.

I want to acknowledge with thanks for the encouragement and support received by my US Geological Survey (USGS) colleagues. I would like to mention the late Mr. Edwin Pfeifer, Dr. Susan Benjamin (my director at the Western Geographic Science Center), Dr. Dennis Dye, Mr. Larry Gaffney, Mr. David F. Penisten, Ms. Emily A. Yamamoto, Mr. Dario D. Garcia, Mr. Miguel Velasco, Dr. Chandra Giri, Dr. Terrance Slonecker, Dr. Jonathan Smith, Timothy Newman, and Zhouting Wu. Of course, my dear colleagues at USGS Dr. Pardhasaradhi Teluguntla, Dr. Itiya Aneece, Mr. Adam Oliphant, and Mr. Daniel Foley have helped me in numerous ways. I am ever grateful for their support and significant contributions to my growth and this body of work. Throughout my career, there have been many postdoctoral-level scientists who have worked with me closely and contributed to my scientific growth in different ways. They include Dr. Murali Krishna Gumma, Head of Remote Sensing at the International Crops Research Institute for the Semi-Arid Tropics; Dr. Jun Xiong, Geo ML ≠ ML with GeoData, Climate Corp.; Dr. Michael Marshall, Associate Professor, University of Twente, Netherlands; Dr. Isabella Mariotto, Former USGS postdoctoral researcher; Dr. Chandrashekar Biradar, Country Director, India for World Agroforestry, and numerous others. I am thankful for their contributions. I know I am missing many names: too numerous to mention them all, but my gratitude for them is the same as the names I have mentioned here.

There is a very special person I am very thankful for: the late Dr. Thomas Loveland. I first met Dr. Loveland at USGS, Sioux Falls for an interview to work for him as a scientist in the late 1990s when I was still at Yale University. But even though I was selected, I was not able to join him as I was not a citizen of the United States at that time and working for USGS required that. He has been my mentor and pillar of strength over two decades, particularly during my Landsat Science Team days (2006–2011) and later once I joined USGS in 2008. I have watched him conduct Landsat Science Team meetings with great professionalism, insights and creativity. I remember him telling my PhD advisor on me being hired at USGS: "we don't make mistakes!" During my USGS days, he was someone I could ask for guidance and seek advice from, and he would always be there to respond with kindness and understanding. Above all, he would share his helpful insights. It is sad that we lost him too early. I pray for his soul. Thank you, Tom, for your kindness and generosity.

Over the years, there are numerous people who have come into my professional life who have helped me grow. It is a tribute to their guidance, insights, and blessings that I am here today. In this regard I need to mention few names out of gratitude: (1) Prof. G. Ranganna, my master's thesis advisor in India at the National Institute of Technology (NIT), Surathkal, Karnataka, India. Prof. Ranganna is 92 years old (2024), and I met him few months back; to this day he is my guiding light on how to conduct oneself with fairness and dignity in professional and personal conduct. Prof. Ranganna's trait of selflessly caring for his students throughout his life is something that influenced me to follow. (2) Prof. E.J. James, former Director of the Center for Water Resources Development and Management (CWRDM), Calicut, Kerala, India. Prof. James was my master's thesis advisor in India, whose dynamic personality in professional and personal matters had an influence on me. Dr. James's always went out of his way to help his students in spite of his busy schedule. (3) The late Dr. Andrew Ward, my PhD advisor at The Ohio State University, Columbus, Ohio. He funded my PhD studies in the United States through grants. Through him I learned how to write scientific papers and how to become a thorough professional. He was a tough task maker, your worst critic (to help you grow), but also a perfectionist who helped you grow as a peerless professional and, above all, a very kind human being at the core. He would write you long memos on flaws in your research, but then help you out of it by making you work double the time! To make you work harder, he will tell you "You won't get my sympathy." Then when you accomplish your goal, he will tell

Acknowledgments

you "you have paid back for your scholarship many times over!!" (4) Dr. John G. Lyon, also my PhD advisor at the Ohio State University. He was a peerless motivator who encouraged you to believe in yourself. (5) Dr. Thiruvengadachari, Scientist at the National Remote Sensing Agency (NRSA), which is now the National Remote Sensing Center (NRSC), India. He was my first boss at the Indian Space Research Organization (ISRO), and through him I learned the initial steps in remote sensing science. I was just 25 years old then and had joined NRSA after my Master of Engineering (hydraulics and water resources) and Bachelor of Engineering (civil engineering) degrees. The first day in office Dr. Thiruvengadachari asked me how much remote sensing I knew. I told him "zero" and instantly thought he would ask me to leave the room. But his response was "very good!" and he gave me a manual on remote sensing from the Laboratory for Applications of Remote Sensing (LARS), Purdue University to study. Those were the days where there was no formal training in remote sensing in universities. So, my remote sensing lessons began by working practically on projects, and one of our first projects was "drought monitoring for India using NOAA AVHRR data." This was an intense period of learning the fundamentals of remote sensing science for me by practicing on a daily basis. Data came in 9 mm tapes and was read on massive computing systems, image processing was done mostly during night shifts by booking time on a centralized computer, fieldwork was conducted using false color composite (FCC) outputs and topographic maps (there was no global positioning systems or GPS), geographic information system (GIS) was in its infancy, a lot of calculations were done using calculators, and we had just started working in IBM 286 computers with floppy disks. So, when I decided to resign my NRSA job and go to the United States to do my PhD, Dr. Thiruvengadachari told me "Prasad, I am losing my right hand, but you can't miss opportunity." Those initial wonderful days of learning from Dr. Thiruvengadachari will remain etched in my memory. I am also thankful to my very good old friend Shri C. J. Jagadeesha, who was my colleague at NRSA/NRSC, ISRO. He was a friend who encouraged me to grow as a remote sensing scientist through our endless rambling discussions over tea in Iranian restaurants outside NRSA those days and elsewhere.

I am ever grateful to my former professors at the Ohio State University: the late Prof. Carolyn Merry, Dr. Duane Marble, and Dr. Michael Demers. They have taught and/or encouraged, inspired, or given me opportunities at the right time. The opportunity to work for six years at the Yale Center for Earth Observation (YCEO) was incredibly important. I am thankful to Prof. Ronald G. Smith, Director of YCEO, for the opportunity, guidance, and kindness. At YCEO I learned and advanced myself as a remote sensing scientist. The opportunities I got working for the International Institute of Tropical Agriculture (IITA) based in Nigeria and the International Water Management Institute (IWMI) based in Sri Lanka where I worked on remote sensing science pertaining to a number of applications such as agriculture, water, wetlands, food security, sustainability, climate, natural resources management, environmental issues, droughts, and biodiversity were extremely important in my growth as a remote sensing scientist—especially from the point of view of understanding the real issues on the ground in real-life situations. Finding solutions and applying one's theoretical understanding to practical problems and seeing them work has its own nirvana.

As is clear from the previous paragraphs, it is of great importance to have guiding pillars of light at crucial stages of your education. That is where you become what you become in the end, where you grow and make your own contributions. I am so blessed to have had these wonderful guiding lights come into my professional life at the right time of my career (which also influenced me positively in my personal life). From that firm foundation, I could build on what I learned and through the confidence of knowledge and accomplishments pursue my passion for science and do several significant pioneering studies throughout my career.

I mention all of the previous people out of gratitude for my ability today to edit such a monumental *Remote Sensing Handbook* (Second Edition, Vols. I–VI).

I am very thankful to Ms. Irma Shagla-Britton, Manager and Leader for Remote Sensing and GIS books at Taylor & Francis/CRC Press. Without her consistent encouragement to take on this responsibility of editing the *Remote Sensing Handbook*, especially in trusting me to accomplish this momentous work over so many other renowned experts, I would never have gotten to work on this in the first place. Thank you, Irma. Sometimes you need to ask several times before one can say yes to something!

I am very grateful to my wife (Sharmila Prasad), my daughter (Spandana Thenkabail), and my son-in-law (Tejas Mayekar) for their usual unconditional understanding, love, and support. My wife and daughter have always been pillars of my life, now joined by my equally loving son-in-law. I learned the values of hard work and dedication from my revered parents. This work wouldn't come through without their life of sacrifices to educate their children and their silent blessings. My father's vision in putting emphasis on education and sending me to the best of places to study despite our family's very modest income and my mother's endless hard work are my guiding light and inspiration. Of course, there are many, many others to be thankful for, but there are too many to mention here. Finally, it must be noted that a work of this magnitude, editing the monumental *Remote Sensing Handbook* (Second Edition, Vols. I–VI) continuing from the three-volume first edition, requires blessings of the almighty. I firmly believe nothing happens without the powers of the universe blessing you and providing needed energy, strength, health, and intelligence. To that infinite power my humble submission of everlasting gratefulness.

It has been my deep honor and great privilege to have edited the *Remote Sensing Handbook* (Second Edition, Vols. I–VI) after having edited the three-volume first edition that was published in 2014. Now after 10 years, we will have a six-volume second edition in 2024. A huge thanks to all the authors, the publisher, my family, friends, and everyone who made this huge task possible.

Dr. Prasad S. Thenkabail, PhD
Editor-in-Chief
Remote Sensing Handbook (Second Edition, Vols. I–VI)

Volume 1: Sensors, Data Normalization, Harmonization, Cloud Computing, and Accuracies
Volume 2: Image Processing, Change Detection, GIS, and Spatial Data Analysis
Volume 3: Agriculture, Food Security, Rangelands, Vegetation, Phenology, and Soils
Volume 4: Forests, Biodiversity, Ecology, LULC, and Carbon
Volume 5: Water Resources: Hydrology, Floods, Snow and Ice, Wetlands, and Water Productivity
Volume 6: Droughts, Disasters, Pollution, and Urban Mapping

Part I

Droughts and Drylands

1 Agricultural Drought Monitoring Using Space-Derived Biophysical Products
A Global Perspective

Felix Kogan and Wei Guo

ACRONYMS AND DEFINITIONS

BT	Brightness temperature
GAC	Global area coverage
GVH	Global vegetation health
GVI	Global Vegetation Index
LOM	Law of minimum
LOT	Law of tolerance
NIR	Near infrared
NDVI	Normalized Difference Vegetation Index
PCC	Principle of carrying capacity
VH	Vegetation health
VIS	Visible

1.1 INTRODUCTION

Drought is a part of Earth's climate, occurring every year without warning or recognizing borders or political and economic differences. Drought has wide-ranging impacts on water resources, ecosystems, energy, agriculture, forestry, human health, recreation, transportation, food supply and demands, and other resources and activities. Drought affects the largest number of people on the Earth and is a very costly disaster. In the United States, a country of high technology, drought is considered a "14 billion-dollar" annual event in terms of incurred losses (Kogan, 2022; NCDC, 2013). In the developing world, drought leads to food shortages, famine, population displacement, and death. Since drought is a very complex, and the least understood, phenomenon, drought prediction, what would be an effective way to fight drought consequences, is a very challenging task. Therefore, early detection and monitoring is currently an important way to deal with drought in assessing its impact and developing mitigation measures.

Weather data have been used traditionally as a tool for drought monitoring. However, weather station network is sparse, especially in climate, ecosystem, and population marginal areas. In the last 30 years, satellite technology has successfully filled this gap due to comprehensive coverage of high-quality data, its availability, and access as well as the ability to use the same data for multiple applications. Moreover, satellite indices and products provide cumulative approximation of weather impacts on the environment and economies, including estimation of drought-related losses in agriculture. Since the launch of the first operational weather satellites in the late 1960s, space-based remote sensing has proven to be a perfect method for operational drought management as a separate tool and were complemented weather data. In the past 20 years operational remote sensing has improved considerably since the introduction of the vegetation health (VH) method (Kogan, 2022; Kulkarni et al., 2020; Aswathi et al., 2018; Zhang et al., 2017; Kogan and Guo, 2014; Kogan et al., 2013; Kogan et al., 2012; Kogan, 1997, 2002). This chapter discusses the principal of the VH method, products, and how they are applied for monitoring global droughts and their impacts on agricultural losses.

1.2 DROUGHT: UNUSUAL WEATHER DISASTER

Drought is quite different than other weather and geophysical disasters because it starts unnoticeably; develops cumulatively and slowly; produces cumulative impacts, especially on agriculture; and by the time damages are visible, it is too late to mitigate the consequences (Kogan, 2022; Qu et al., 2019; Bento et al., 2018; Gidey et al., 2018; Zhang et al., 2017; Kogan, 1997). Besides, drought is a very complex and least understood phenomenon because in addition to be triggered by weather, other factors such as soils, ecosystems, climate, and human activities also contribute to drought initiation, development, and impacts. Although it is known that drought is a period with dry and hot weather, even a few months with such weather does not necessarily indicate drought, if the preceding period was wet. Owing to the creeping nature of drought, its effect often takes weeks, months, seasons, and even years to appear. Since drought is a multi-dimensional phenomenon by its appearance, properties, origination, and impacts, weather-based parameters and indices are insufficient to characterize special and temporal drought features, especially such characteristics as drought start, which is important for timely initiation of mitigating measures. Another specific drought feature is its cumulative development with the following cumulative impacts, which are not immediately observable (Kumar et al., 2024; Zeng et al., 2023; Zhang et al., 2017; Wilhite, 2000). By the time damages are visible, it is too late to mitigate the consequences. The latest is very important in regions and countries with limited and/or variable water supplies and whose economy is highly depended on agriculture.

1.3 HOW TO MEASURE DROUGHT

1.3.1 Traditional Approach

Traditionally, drought was measured with weather, soil moisture data, and their combination in the form of indices. The most widely used weather and soil parameters are precipitation, temperature, air humidity, and soil moisture. From weather indices, the most widely used are the Palmer Drought Severity Index (Palmer, 1965), the Standardized Precipitation Index (McKee et al., 1993), the US Drought Monitor Index (Svoboda et al., 2002), and a few others specific to a country environment and economic resources. Weather parameters and indices have many useful features in their applications (Bhushan et al., 2024; Kogan, 2022; Shahzaman et al., 2021; West et al., 2019; Zhang et al., 2017). However, the most important shortcomings are that they are calculated from a limited weather station network, especially in marginal areas and developing countries.

1.3.2 Remote Sensing Approach

Satellite indices were introduced in the late 1970s with the launch of operational weather satellites and started to be widely used from the 1980s with the accumulation of knowledge and satellite data. The indices were developed from the Earth surface reflectance measured by the operational environmental satellites' sensors. The most popular for land monitoring was the Normalized Difference Vegetation Index (NDVI) introduced by Deering (1978). The NDVI calculated as a ratio of visible (VIS) and near infrared (NIR) sun's reflectance (NDVI= (NIR-VIS)/(NIR+VIS)) is a dimensionless measure, used as a proxy to characterize land surface greenness and vigor (Cracknel, 1997). However, NDVI alone was not sufficient to monitor vegetation condition, drought, and variation in agricultural production. Therefore, in the 1990s, the new vegetation health (VH) method and indices were introduced. The new numerical method combined NDVI with thermal emission converted to brightness temperature (BT) of the land surface (Kogan, 1990). This method was built on the three basic environmental laws: law of minimum (LOM), law of tolerance (LOT), and the principle of carrying capacity (PCC), which provide a theoretical basis for the calculation of biophysical climatology (expressed as the lowest and the highest ecosystem level that the environmental resources can support every week inside 4 km² pixel). The satellite reflectance were pre- and post-launch calibrated, NDVI and BT were calculated, high-frequency noise was completely removed from NDVI and BT time series, ecosystems were stratified, and medium-to-low frequency vegetation fluctuations associated with weather variations were singled out (Kogan, 1997). Finally, three indices characterizing moisture (VCI), thermal (TCI), and vegetation health (VHI) conditions were derived by the following equations (Kogan, 2022):

$$VCI = 100*(NDVI - NDVI_{min})/(NDVI_{max} - NDVI_{min}) \quad (1.1)$$

$$TCI = 100*(BT_{max} - BT)/(BT_{max} - BT_{min}) \quad (1.2)$$

$$VTI = a*VCI + (1 - a)*TCI \quad (1.3)$$

where NDVI, $NDVI_{max}$, and $NDVI_{min}$ are no noise weekly NDVI, 32-year absolute maximum and minimum, respectively; BT, BT_{max}, and BT_{min} are similar data characterizing brightness temperature; *a* is a coefficient that quantifies a share of VCI and TCI contribution to the total vegetation health.

The three indices were scaled from 0 indicating extreme vegetation stress to 100 indicating optimal condition. Comparison of VH data with crops showed that 5–10% reduction in crop production signals about the beginning of drought, which corresponds to a reduction in indices (VCI, TCI, and VHI) values below 40. The new method was first applied to the NOAA Global Vegetation Index (GVI) data set (16 km²), issued routinely since 1985 (Kidwell, 1995). Currently, the method, called global vegetation health (GVH), is applied to the NOAA's global area coverage (GAC) data set produced from 1981 through the present (34 years) issued weekly for each 4 km² land surface between 75° N and 55° S. The data and products are delivered every week to the following NOAA web address: http://www.star.nesdis.noaa.gov/smcd/emb/vci/VH/index.php.

1.4 VEGETATION HEALTH INTERPRETATION

The VH method stems from the properties of green vegetation to reflect and emit incoming sunlight, which is converted to greenness (NDVI) and thermal (BT) indices (Wei et al., 2024; Hosen et al., 2023; Javed et al., 2021; Zhang et al., 2021; West et al., 2019). In drought-free years, vegetation is greener and cooler than in climate-normal and below-normal years stimulating NDVI increase and BT decrease compare to climatology (expressed by the denominators in (1.1) and (1.2), respectively). This is resulted in VCI, TCI, and VHI increase above 60. Drought normally depresses vegetation greenness and increases canopy temperature, resulting in NDVI decrease, BT increase, and decreases in VCI, TCI, and VHI to below 40 (Kogan, 2022; Kogan, 1997; Kogan et al., 2012).

Figure 1.1 demonstrates a color-coded map of VHI on October 9, 2012. Stressful, fair, and favorable conditions are represented by red, green/yellow, and blue colors, respectively. Large-area VHI-related intensive vegetation stress serves normally as an indicator of crop yield and pasture biomass reduction, intensive fire activities, reduction of water level in reservoirs and rivers (see USA, Kazakhstan, Ukraine, Brazil, and Argentina on the map). Oppositely, optimal vegetation health triggered by cool and wet weather indicates favorable conditions for above-average crop production, water availability, and no fire activity but is an indication for the development of mosquito-borne diseases because they require moist and cool conditions (see the blue area in sub-Sahara and southern Africa, northwestern India, and eastern China.

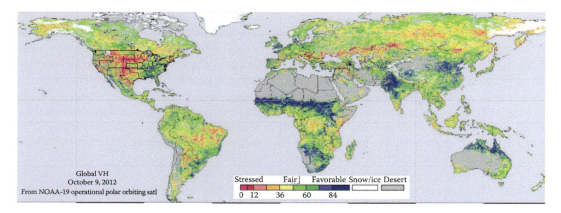

FIGURE 1.1 Color-coded global vegetation health index (VHI) map for October 9, 2012. An intensive October 2012 vegetation stress in the USA, southern Europe, east Siberia, and a part of the Amazon indicated potential for fire activity. In the USA, this stress started in April and continued through the entire summer, leading to considerable agricultural losses.

1.5 UNUSUAL DROUGHTS OF THE 21ST CENTURY

The 21st century began with a series of widespread, long, and intensive droughts around the world, continuing the tendencies of the previous two to three decades. Up to 16% of global land cover was affected by the severe droughts of 2000–2013 (Kogan, 2022; Zhang et al., 2021; West et al., 2019; Kogan et al., 2013). They reduced grain production in the important agricultural regions, leading to a negative balance between global food supplies and demands (PotashCorpo, 2012). The globe experienced the most dreadful droughts during the most recent three years (2010–2012). The USA was affected twice, in 2011 and 2012 (Figure 1.2). The worst drought of 2012 hit primary agricultural areas of the Great Plains, sharply reducing corn, soybean, hay, and pasture production, which resulted in price increases for food and farmland (U.S. Drought, 2012). This drought also affected grain production in other parts of the world, such as Kazakhstan and southern Russia, and resulted in 20% grain losses in Russia and almost 50% in Kazakhstan (UNDP, 2012). In the Southern Hemisphere, the 2012 drought affected agricultural regions of Brazil, northern Argentina, and southern Australia.

The 2011 US drought was much stronger than in 2012, but it only covered Texas, Oklahoma, New Mexico, and parts of the neighboring states (Figure 1.2). In the hardest-hit states, moderate-to-exceptional drought covered up to 100% of their territories. Another extremely strong drought developed in 2010, covering huge agricultural areas of Russia, Ukraine, and Kazakhstan (Figure 1.2). Russian grain production dropped to 75 million metric tons (versus 97 in 2009, FAO, 2012), forcing the Russian government to impose a grain embargo, which triggered a sharp increase in global wheat prices. This drought affected also the 2011 Russian harvest, since winter wheat was planted in dry soil. In addition to agriculture and water resources, the 2010 drought has also triggered hundreds of fires and heavy smoke, deteriorating human health and increasing the death rate in Russia.

Compared to 2010–2012, the 2013/14 agricultural year was favorable for global grain production, which was forecasted to exceed 8% of the 2012 level due to recovery of corn in the USA and wheat in the former Soviet Union (FAO, 2013). Meanwhile, as seen in Figure 1.3, the area affected by stronger than severe drought in 2013/14 is still large, especially in the Southern Hemisphere. The VHI estimated that 16% of the globe was affected by drought (stronger than severe) in 2013/14, which is typical for the 21st century (Kogan and Guo, 2014), except for 2012, when the drought area reached 21%. It is important to emphasize that the 2013/14 global drought impact on agriculture

Agricultural Drought Monitoring

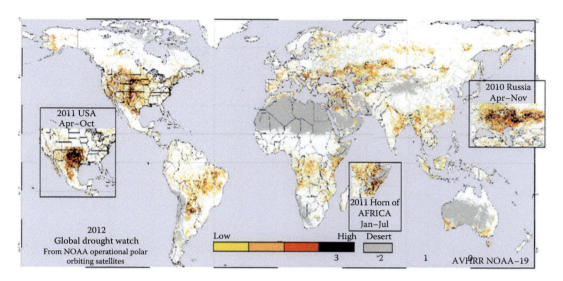

FIGURE 1.2 Drought area and intensity (derived from VHI < 40) in the major agricultural areas of the world during 2010–2012.

was much smaller than in the previous years since large grain-producing countries (China, USA, European Union, India, Former Soviet Union) were not affected (severe drought in the USA [Figure 1.3, July] covered only western states, causing intensive fire activity). Global grain production might have been even larger than projected for 2013/14 if severe drought had not affected grain-producing areas of Argentina, Australia, and eastern Brazil (Figure 1.3, February).

1.6 DROUGHT PRODUCTS

Vegetation health–based drought products include vegetation health, moisture and thermal stress, drought start/end, intensity, duration, magnitude, area, season, origination, and impacts (Kogan, 2022; Qu et al., 2019; Sur et al., 2019). Drought start and end is identified when VCI, TCI, and VHI decline below 40 thresholds, which are experimentally determined from *in situ* observations (crop yield reduction). Drought intensity (commonly referred to as the severity) was graded by a percent of yield reduction (dY) below the technological trend. Mild drought is identified if dY is 2–9% below trend, yield reduction of 10–14% indicates moderate drought, 15–24% is extreme, and below 24% is exceptional). Drought duration is measured in the number of days with drought in different VH intensities. Magnitude accounts for the combination of a drought's intensity and duration (example: the number of days with severe drought). Drought area can be measured as a percentage of an administrative region, with droughts of different intensity: this is a very important measure for developing mitigation strategies and evaluation of the budget (e.g., imposing some limitations for water use). Another criterion might be a combination of drought area, intensity, and duration. Drought timing is a very important component for determining if urgent measures are needed to mitigate drought impacts or if it is enough time for drought recovery with minimal losses to incur. Drought origination specifies if drought is moisture-based, thermal-based, or both, and the latter is the worst combination. Finally, drought impacts should be specified based on the type and extent of affected economy (e.g., crop yield reduction, water depletion). Each drought is unique, but common features of the most severe droughts that have far-reaching human and environmental impacts include time of drought start, intensity of it progress, long duration, large moisture deficits, severe thermal stress, large areal extent, and severe impact.

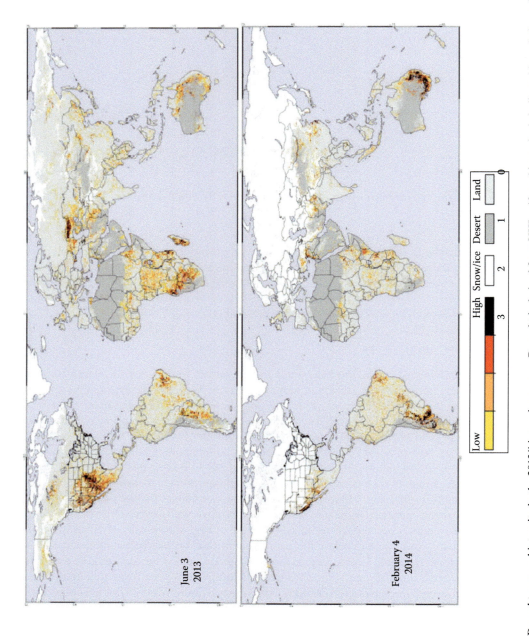

FIGURE 1.3 Drought area and intensity in the 2013/14 growing season. Drought is derived from VHI < 40, and intensity is indicated by a darker color. For example, the Central Valley in California, USA; the Pampas in Argentina; and southeast Australia are facing severe drought relative to the long-term climatology.

Agricultural Drought Monitoring

The following VH-based drought products are discussed next (Hosen et al., 2023; Kogan, 2022; Zhang et al., 2021; Badamassi et al., 2020; Aswathi et al., 2018; Bento et al., 2018): moisture stress from VCI, thermal stress from TCI, drought area, drought intensity, drought duration, fire risk, yield, and biomass reduction (Figures 1.4 and 1.5).

Figure 1.4 demonstrates moisture and thermal stress in the principal agricultural areas of southern and eastern Australia (Queensland, New South Wales, and Victoria). A very dramatic event occurred in 2006–2007, when severe drought reached the apogee, slashing water supplies, crops, and rangeland production severely. This was one of the worst droughts for agriculture due to the combined impacts of the extreme moisture and thermal stress. Following FAO (2012), Australian wheat yield dropped 46% in 2006 and 37% in 2007 (below the 1960–2010 yield's trend level). An example of localized VHI and TCI stress by different intensities and percent of the affected area is shown in Figure 1.5A and B. Drought intensity is demonstrated in Figure 1.5C during a five-year (2007–2011) period in Kenya (Horn of Africa) when the country was affected. Following moisture and thermal stress estimation, moderate-to-severe intensity droughts occurred in 2009 and 2011.

Thousands of acres of vegetative land are burned every year worldwide, leaving huge scars on the land, polluting the atmosphere, and affecting human health. An early assessment of fire risk can help to mitigate fire consequences. Drought is the principle factor creating fire risk conditions. In drought years, the burnt area increases twofold. VH fire risk monitoring is based on assessments of severity and duration of vegetation stress (Kogan, 2002). Figure 1.5D shows maximum fire risk area during the 2007 and 2010 seasons in Russia and Ukraine. These two droughts caused huge fires in both years. Based on VH fire risk product, Russia was the most affected in 2010 and Ukraine in 2007, although partial affects were observed in northern Ukraine as well. Although the fire risk area in the north was not large, it covered the region of the 1986 nuclear accident in Chernobyl. Soil and vegetation in that area still have radioactive remnants that were thrown into the air by wildfires.

Figure 1.5E shows the duration of droughts of different intensities in Kenya during the two most drought-affected years, 2009 and 2011. During March–July, the minor and major growing seasons

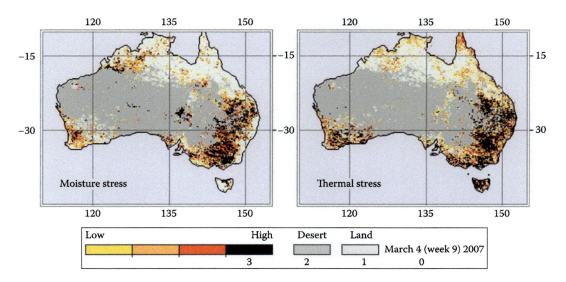

FIGURE 1.4 Moisture (VCI < 40) and thermal (TCI < 40) stress in agricultural areas of Australia, March 4, 2007. Numerical values of drought intensity are indicated by a darker color. Southern and eastern Australia are affected by both extreme moisture and thermal stress, which is the worst combination of drought severity and consequently expected considerable losses of crop production.

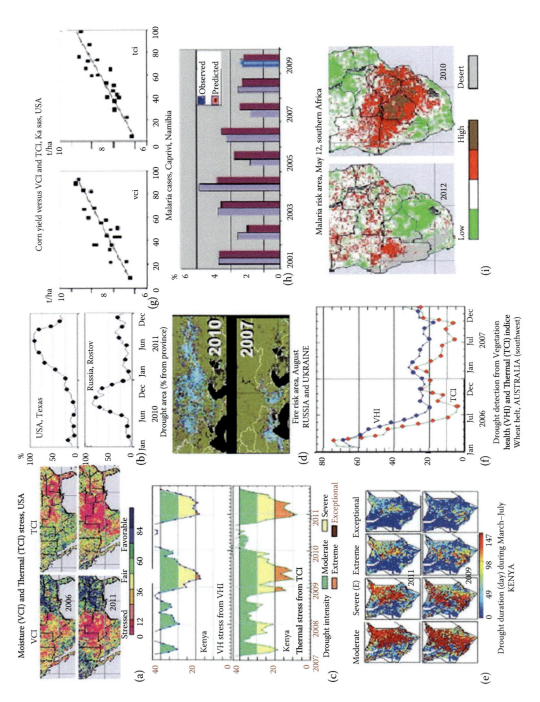

FIGURE 1.5 Drought products from VH available at NOAA: http://www.star.nesdis.noaa.gov/smcd/emb/vci/VH/index.php.

Agricultural Drought Monitoring 11

were affected by moderate and severe drought. Some differences between the years are observed for extreme and exceptional drought intensities, which affected southern Kenya in 2009 and central in 2011. Early drought detection is the most important part of drought management and developing mitigation strategies. Since there is no reliable drought prediction method, drought can be detected prior to its appearance from vegetation health dynamics. As seen in Figure 1.5F, the first signs of the approaching 2006–2007 drought in southern Australia (for both VHI and TCI) appeared in March–April 2006, two months prior to the time series crossed down the 40s threshold (indicating drought start) and vegetation was still in good health (VHI > 50).

Over the years of VH application for drought monitoring and verification, it has been revealed that VH methodology can be successfully applied for monitoring and predicting crop yield losses and risk of malaria (Kogan et al., 2012; Rahman et al., 2011). Testing VH indices in 34 countries of North America, South America, Africa, Europe, and Asia revealed that yield of such crops as wheat (both winter and spring), corn, sorghum, rice, and soybeans has strong correlation with the VH indices during the critical period of crops' growth, development, and reproduction. Figure 1.5G demonstrates some of these results. As seen, mean Kansas (USA) corn yield correlates strongly with VCI and TCI during the critical period of corn development in July. During drought years (the indices < 40) up to 40% of corn production could be lost (depending on drought severity) compared to normal and wet years. For example, in 2011, Texas drought slashed corn yield by 28% compared to the average for the last ten years and 36% relative to the highest yield received in the 2010 season (Taxes, 2012).

Malaria risk area and intensity is estimated from VCI and TCI values, which correlates with the number of malaria cases in some regions of India, Bangladesh, Colombia, and Namibia (Kogan, 2002; Rahman et al., 2011; Nizamuddin et al., 2013). An example from the Caprivi province of Namibia indicating high correlation between the observed and independently VH-simulated number of malaria cases is shown in Figure 1.5H. Finally, Figure 1.5I shows two images of VH-estimated malaria risk with a smaller area in 2012 (drought year) versus larger area in 2010 (wet year). This information is renewed at http://www.star.nesdis.noaa.gov/smcd/emb/vci/VH/index.php every week.

1.7 DROUGHTS IN A WARMER WORLD

The 2012 IPCC report stated that the average Earth surface temperature in the past 100 years increased 0.85° (IPCC, 2012). The warming trend has continued until 1998, after which the world temperature leveled off and for the past 15 years was remaining at that level (although in the Northern Hemisphere the warming trend is continuing). Based on IPCC (2012), it is anticipated that the risk of droughts in a warmer world will increase and they expand their area and intensify. Therefore, it is quite possible that such new tendencies have already started. We used 34-year VHI to estimate global and hemispheric drought dynamics (percent affected area) by the three intensities (Bhushan et al., 2024; Kogan, 2022; Shahzaman et al., 2021; Qu et al., 2019; Gidey et al., 2018): severe-to-exceptional (SE), extreme-to-exceptional (EE), and exceptional (E), following classification in Svoboda et al., 2002. Figure 1.6 presents these results for the entire world, Northern and Southern Hemispheres. Statistical assessment of drought dynamics presented in this figure have not supported upward trends in either of these regions for all three intensities. Meanwhile, visually, there is a very small reduction in the global and Northern Hemisphere percent drought area in SE and EE intensities during the 21st century and a slight increase in the Southern Hemisphere's percent drought area in all three intensities since 2005. These changes are so negligible they can be either incidental or related to local features of the regions. No upward trend conclusion is in line with the most recent revised analysis of the US Palmer Drought Severity Index over the past 60 years and drought trend analysis in the USA, Ukraine, Horn of Africa, and southern Australia (Sheffield et al., 2013; Kogan et al., 2013; Kogan and Guo, 2014). Figure 1.6 indicates also that droughts of SE, EE, and E intensities covered 25–35%, 15–20%, and up to 10%, respectively of these regions.

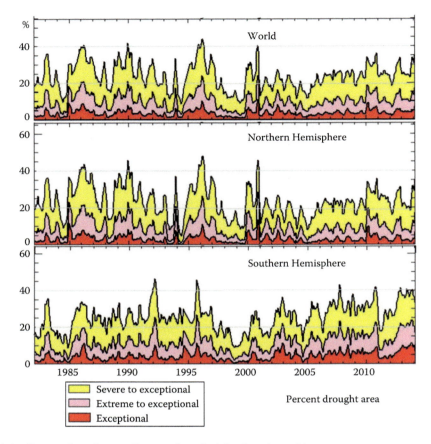

FIGURE 1.6 Percent drought area (from each region) for three intensities.

1.8 CONCLUSION

Drought is a part of Earth's climate, occurring every year without warning or recognizing borders or political and economic differences. Drought affects the largest number of people on the Earth and is a very costly disaster affecting water resources, ecosystems, agriculture, forestry, energy, human health, recreation, transportation, food supply and demands, and other resources and activities. Unlike other natural disasters, drought starts unnoticeably, develops cumulatively and slowly, produces cumulative impacts, and by the time damages are visible, it is too late to mitigate the consequences. Drought is characterized by the days of its start/end, intensity, duration, magnitude, area, season, origination, and impacts.

One of the main challenges in drought detection and monitoring is the sparse weather station network. Satellite data avoids this problem. Since the introduction of the vegetation health method, early drought detection and assessment of drought intensity, duration, origination, and impacts have become the widely used by the global community (Bhushan et al., 2024; Kumar et al., 2024; Wei et al., 2024; Hosen et al., 2023; Mullapudi et al., 2023; Zeng et al., 2023; Kogan, 2022; Javed et al., 2021; Khan and Gilani, 2021; Kloos et al., 2021; Lee et al., 2021; Shahzaman et al., 2021; Zhang et al., 2021; Badamassi et al., 2020; Ghazaryan et al., 2020; Kulkarni et al., 2020; Qu et al., 2019; Sur et al., 2019; West et al., 2019; Aswathi et al., 2018; Bento et al., 2018; Gidey et al., 2018; Zhang et al., 2017). VH method is accurate, simple to understand and interpret, available in real time, estimates drought every week for each 4 km² land surface, it was validated comprehensively against *in situ* data (climate, weather, agriculture, health, etc.) in 34 countries, including all major agricultural producers. A few drought products have been developed based on vegetation health methodology. The VH-based products include moisture index, thermal index, drought area, drought intensity, drought duration, soil wetness, fire risk, and others. The VH method and data set are used for the prediction of climate and land cover trends, crop yield, risk of malaria, climate forcing (ENSO) impacts on vegetation productivities, and others. The products and data are available on the NOAA website (http://www.star.nesdis.noaa.gov/smcd/emb/vci/VH/index.php), which is accessed by 3,000–4,500 users every month. The web provides assessments for the entire globe, continents, 194 countries, and around 4,000 first-order countries' administrative divisions. Finally, it is important to emphasize that drought detection and monitoring method will be considerably improved with the new generation of satellite technology that started in 2011. The new Visible Infrared Imager Radiometer Suite (VIIRS) on the NPOESS (National Polar-orbiting Operational Satellite System) Preparatory Platform (NPP) will provide observations for each 375 m of the globe, with a much sharper view at the edge of the scan and four times more spectral bands (compared to its predecessor).

REFERENCES

Aswathi, P.V., Nikam, B.R., Chouksey, A., and Aggarwal, S.P. 2018. Assessment and Monitoring of Agricultural Droughts in Maharashtra Using Meteorological and Remote Sensing Based Indices. *ISPRS Annals of the Photogrammetry, Remote Sensing and Spatial Information Sciences*, IV-5, 253–264. https://doi.org/10.5194/isprs-annals-IV-5-253-2018.

Badamassi, M.B.M., El-Aboudi, A., and Gbetkom, P.G. 2020. A New Index to Better Detect and Monitor Agricultural Drought in Niger Using Multisensor Remote Sensing Data. *The Professional Geographer*, 72(3), 421–432. https://doi.org/10.1080/00330124.2020.1730197.

Bento, V.A., Gouveia, C.M., DaCamara, C.C., and Trigo, I.F. 2018. A Climatological Assessment of Drought Impact on Vegetation Health Index. *Agricultural and Forest Meteorology*, 259(2018), 286–295. ISSN 0168-1923. https://doi.org/10.1016/j.agrformet.2018.05.014. https://www.sciencedirect.com/science/article/pii/S0168192318301667.

Bhushan, B., Dhurandher, A., and Sharma, A. 2024. Meteorological and Agricultural Drought Monitoring Using Geospatial Techniques. In: Shit, P.K., et al., eds. *Geospatial Practices in Natural Resources Management. Environmental Science and Engineering*. Springer, Cham. https://doi.org/10.1007/978-3-031-38004-4_13.

Cracknel, A.P. 1997. *The Advanced Very High Resolution Radiometer*. Taylor & Francis, Cambridge, UK, 350 pp.

Deering, D.W. 1978. *Rangeland Reflectance Characteristics Measured by Aircraft and Spacecraft Sensors*. Ph.D. Dissertation. Texas A&M University, College Station, TX, 388 p.

FAO. 2012. http://faostat.fao.org/?PageID=567#ancor

FAO. 2013. http://en.mercopress.com/2013/10/07/world-2013-14-cereal-production-forecasted-to-surpass-2012-level-by-8-says-fao. Accessed March 10.

Ghazaryan, G., König, S., Rezaei, E.E., Siebert, S., and Dubovyk, O. 2020. Analysis of Drought Impact on Croplands from Global to Regional Scale: A Remote Sensing Approach. *Remote Sensing*, 12(24), 4030. https://doi.org/10.3390/rs12244030.

Gidey, E., Dikinya, O., Sebego, R., et al. 2018. Analysis of the Long-Term Agricultural Drought Onset, Cessation, Duration, Frequency, Severity and Spatial Extent Using Vegetation Health Index (VHI) in Raya and Its Environs, Northern Ethiopia. *Environmental Systems Research*, 7, 13. https://doi.org/10.1186/s40068-018-0115-z.

Hosen, M., Alam, M., Chakraborty, T., et al. 2023. Monitoring Spatiotemporal and Seasonal Variation of Agricultural Drought in Bangladesh Using MODIS-Derived Vegetation Health Index. *Journal of Earth System Science*, 132, 188. https://doi.org/10.1007/s12040-023-02200-3.

IPCC. 2012. Summary for Policymakers. *Twelfth Session of Working Group 1*. WGI AR5. http://www.climatechange2013.org/images/uploads/WGIAR5-SPM_Approved27Sep2013.pdf. Accessed October 5, 36 p.

Javed, T., Li, Y., Rashid, S., Li, F., Hu, Q., Feng, H., Chen, X., Ahmad, S., Liu, F., and Pulatov, B. 2021. Performance and Relationship of Four Different Agricultural Drought Indices for Drought Monitoring in China's Mainland Using Remote Sensing Data. *Science of the Total Environment*, 759, 143530. ISSN 0048-9697. https://doi.org/10.1016/j.scitotenv.2020.143530. https://www.sciencedirect.com/science/article/pii/S0048969720370613.

Khan, R., and Gilani, H. 2021. Global Drought Monitoring with Big Geospatial Datasets Using Google Earth Engine. *Environmental Science and Pollution Research*, 28, 17244–17264. https://doi.org/10.1007/s11356-020-12023-0.

Kidwell, K.B. (Ed.). 1995. *NOAA Polar Orbiting Data User's Guide*. U.S. Department of Commerce Technical Report, Washington, DC, 92 p.

Kloos, S., Yuan, Y., Castelli, M., and Menzel, A. 2021. Agricultural Drought Detection with MODIS Based Vegetation Health Indices in Southeast Germany. *Remote Sensing*, 13(19), 3907. https://doi.org/10.3390/rs13193907.

Kogan, F. 1990. Remote Sensing of Weather Impacts on Vegetation in Nonhomogeneous Areas. *International Journal of Remote Sensing*, 11, 1405–1419.

Kogan, F. 1997. Global Drought Watch from Space. *Bulletin American Meteorological Society*, 78, 621–636.

Kogan, F. 2002. World Droughts from AVHRR-based Vegetation Health Indices. *Eos, Transactions American Geophysical Union*, 83(48), 557–564.

Kogan, F. 2022. New Remote Sensing Vegetation Health Technology. In *Remote Sensing Land Surface Changes*. Springer, Cham. https://doi.org/10.1007/978-3-030-96810-6_5.

Kogan, F., Adamenko, T., and Guo, W. 2013, Global and Regional Drought Dynamics in the Climate Warming Era. *Remote Sensing Letters*, 4, 364–372.

Kogan, F., and Guo, W. 2014. Early Twenty-First-Century Droughts During the Warmest Climate. *Geomatics, Natural Hazards and Risk*. http://doi.org/10.1080/19475705.2013.878399.

Kogan, F., Salazar, L., and Roytman, L. 2012. Forecasting Crop Production Using Satellite Based Vegetation Health Indices in Kansas, United States. *International Journal of Remote Sensing*, 3, 2798–2814. http://doi.org/10.1080/01431161.2011.621464.

Kulkarni, S.S., Wardlow, B.D., Bayissa, Y.A., Tadesse, T., Svoboda, M.D., and Gedam, S.S. 2020. Developing a Remote Sensing-Based Combined Drought Indicator Approach for Agricultural Drought Monitoring over Marathwada, India. *Remote Sensing*, 12(13), 2091. https://doi.org/10.3390/rs12132091.

Kumar, V., Sharma, K.V., Pham, Q.B., et al. 2024. Advancements in Drought Using Remote Sensing: Assessing Progress, Overcoming Challenges, and Exploring Future Opportunities. *Theoretical and Applied Climatology*. https://doi.org/10.1007/s00704-024-04914-w.

Lee, S.-J., Kim, N., and Lee, Y. 2021. Development of Integrated Crop Drought Index by Combining Rainfall, Land Surface Temperature, Evapotranspiration, Soil Moisture, and Vegetation Index for Agricultural Drought Monitoring. *Remote Sensing*, 13(9), 1778. https://doi.org/10.3390/rs13091778.

McKee, T.B., Doesken, N.J., and Kleist, J. 1993. The Relationship of Drought Frequency and Duration to Time Scale. *Preprints Eighth Conference on Applied Climatology*. American Meteorological Society, Boston, MA.

Mullapudi, A., Vibhute, A.D., Mali, S., et al. 2023. A Review of Agricultural Drought Assessment with Remote Sensing Data: Methods, Issues, Challenges and Opportunities. *Applied Geomatics*, 15, 1–13. https://doi.org/10.1007/s12518-022-00484-6.

NCDC (National Climatic Data Center). 2013. *Billion Dollar U.S. Weather Disasters*. http://www.ncdc.noaa.gov/oa/reports/billionz.html. Accessed July 25.

Nizamuddin, M., Kogan, F., Dihman, R., Guo, W., and Roytman, L. 2013. Modeling and Forecasting Malaria in Tripura, India Using NOAA/AVHRR-Based Vegetation Health Indices. *International Journal of Remote Sensing Applications*, 3(3), 108–116.

Palmer, W.C. 1965. *Meteorological Drought*. U.S. Weather Bureau, Res. Pap. No. 45, 58 p.

PotashCorpo. 2012. *Agriculture: Crop Overview*. http://www.potashcorp.com/industry_overview/2011/agriculture/16. Accessed November 22, 2012.

Qu, C., Hao, X., and Qu, J.J. 2019. Monitoring Extreme Agricultural Drought Over the Horn of Africa (HOA) Using Remote Sensing Measurements. *Remote Sensing*, 11(8), 902. https://doi.org/10.3390/rs11080902.

Rahman, A., Kogan, F., Roytman, L., Goldberg, M., and Guo, W. 2011. Modeling and Prediction of Malaria Vector Distribution in Bangladesh from Remote Sensing Data. *International Journal of Remote Sensing*, 32(5), 1233–1251.

Shahzaman, M., Zhu, W., Bilal, M., Habtemicheal, B.A., Mustafa, F., Arshad, M., Ullah, I., Ishfaq, S., and Iqbal, R. 2021. Remote Sensing Indices for Spatial Monitoring of Agricultural Drought in South Asian Countries. *Remote Sensing*, 13(11), 2059. https://doi.org/10.3390/rs13112059.

Sheffield, J., Wood, E.F., and Roderick, M.L. 2013. Little Change in Global Drought Over the Past 60 Years. *Nature*, 491, 435–438.

Sur, C., Park, S.Y., Kim, T.W., et al. 2019. Remote Sensing-based Agricultural Drought Monitoring Using Hydrometeorological Variables. *KSCE Journal of Civil Engineering*, 23, 5244–5256. https://doi.org/10.1007/s12205-019-2242-0.

Svoboda, M., LeComte, D., Hayes, M., Heim, R., Gleason, K., Angel, J., Rippey, B., Tinker, R., Palecki, M., Stooksbury, D., Miskus, D., and Stephens, S. 2002. The Drought Monitor. *Bulletin American Meteorological Society*, 83(8), 1181–1190.

Taxes. 2012. http://www.nass.usda.gov/Statistics_by_State/Texas/Charts_&_Maps/zcorn_y.htm. Accessed November 26.

UNDP. 2012. *Drought in Russia and Kazakhstan*. http://europeandcis.undp.org/aboutus/show/. Accessed September 10, 2012.

U.S. Drought. 2012. *New York Times, Science*. http://topics.nytimes.com/top/news/science/topics/drought/index.html. Accessed December 10, 2012.

Wei, W., Wang, J., Ma, L., Wang, X., Xie, B., Zhou, J., and Zhang, H. 2024. Global Drought-Wetness Conditions Monitoring Based on Multi-Source Remote Sensing Data. *Land*, 13(1), 95. https://doi.org/10.3390/land13010095.

West, H., Quinn, N., and Horswell, M. 2019. Remote Sensing for Drought Monitoring & Impact Assessment: Progress, Past Challenges and Future Opportunities. *Remote Sensing of Environment*, 232, 111291. ISSN 0034-4257. https://doi.org/10.1016/j.rse.2019.111291. https://www.sciencedirect.com/science/article/pii/S0034425719303104.

Wilhite, D.A. 2000. Drought as a Natural Disaster. In: Wilhite, D., ed. *Drought*. Routledge Hazards and Disasters Series, Routledge, Tailor and Francis Group, London and New York, pp. 3–19.

Zeng, J., Zhou, T., Qu, Y. et al. 2023. An Improved Global Vegetation Health Index Dataset in Detecting Vegetation Drought. *Scientific Data*, 10, 338. https://doi.org/10.1038/s41597-023-02255-3.

Zhang, X., Chen, N., Li, J., Chen, Z., and Niyogi, D. 2017. Multi-Sensor Integrated Framework and Index for Agricultural Drought Monitoring. *Remote Sensing of Environment*, 188, 141–163. ISSN 0034-4257. https://doi.org/10.1016/j.rse.2016.10.045. https://www.sciencedirect.com/science/article/pii/S0034425716304242.

Zhang, Z., Xu, W., Shi, Z., and Qin, Q. 2021. Establishment of a Comprehensive Drought Monitoring Index Based on Multisource Remote Sensing Data and Agricultural Drought Monitoring. *IEEE Journal of Selected Topics in Applied Earth Observations and Remote Sensing*, 14, 2113–2126. http://doi.org/10.1109/JSTARS.2021.3052194.

2 Agricultural Drought Monitoring Using Space-Derived Vegetation and Biophysical Products
A Global Perspective

F. Rembold, M. Meroni, O. Rojas, C. Atzberger, F. Ham, and Erwann Fillol

ACRONYMS AND DEFINITIONS

ACF	Action Contre la Faim
ASI	Agricultural Stress Index
ASIS	Agriculture Stress Index System
CART	Classification and regression tree
CDI	Combined Drought Index
CMI	Crop Moisture Index
DMP	Dry matter productivity
FAO	Food and Agriculture Organization
FAPAR	Fraction of absorbed photosynthetically active radiation
FEWSNET	Famine Early Warning System
GIEWS	Global Information and Early Warning System
JRC	Joint Research Centre
LAI	Leaf area index
MARS	Monitoring of Agricultural Resources
MSG	Meteosat Second Generation
NDVI	Normalized Difference Vegetation Index
NRT	Near real-time
PDI	Precipitation Drought Index
PDSI	Palmer Drought Severity Index
PRI	Photochemical Reflectance Index
RDI	Reclamation Drought Index
RS	Remote sensing
SPI	Standardized Precipitation Index
SWB	Small water bodies
SWSI	Surface Water Supply Index
TDI	Temperature Drought Index
USAID	United States Agency for International Development
VCI	Vegetation Condition Index
VD	Vegetation Drought Index

Agricultural Drought Monitoring

2.1 INTRODUCTION

For a long time, agricultural monitoring systems have used space remote sensing instruments to provide timely and synoptic information about drought. A variety of approaches are currently being used, and most of them are based on the analysis of remote sensing data in the optical domain (Bhushan et al., 2024; Chinembiri et al., 2024; Lee et al., 2023; Guo et al., 2023; Mulla, 2021; Oroud and Balling, 2021; Hoek et al., 2020; Meroni et al., 2019; Wu et al., 2019; Zhong et al., 2012, 2019; Kerdiles et al., 2017; Ahmadalipour et al., 2017; Du et al., 2017; Meroni et al., 2017; Rembold et al., 2017). They permit the mapping of vegetation vigor, as well as hydrological variables such as rainfall and evapotranspiration when using imagery in the thermal domain. Sensors operating in the microwave domain provide additional and valuable information regarding soil moisture. In this chapter, after providing background information about drought monitoring indices and systems in general, we focus on the current use of satellite-derived biophysical indicators of vegetation status from remote sensing (RS) in the optical domain.

2.1.1 What Is Drought?

Depending on the nature of drought and its impact, one can define meteorological, hydrological, or agricultural droughts, which all have different socioeconomic impacts. A meteorological drought is an extreme climate event over land characterized by below-normal precipitation over a period of time. This event may lead to what is generally defined as agricultural drought, a period with declining soil moisture and consequent crop failure (Hoek et al., 2020; Du et al., 2017; Rembold et al., 2017; Mishra and Singh, 2010). In this review we focus on agricultural drought, a phenomenon that is characterized by a severe reduction of the ratio between actual and potential evapotranspiration of crops (the so-called crop coefficient). Besides prevailing weather conditions, the crop coefficient is affected by other agriculture-specific characteristics such as stage of growth and the soil's physical and biological properties, among others.

Drought is part of the climate variability in arid regions, but it is different from aridity itself, which is a permanent climate characteristic, mainly defined by low average precipitation. Drought is a complex phenomenon that originates from anomalous rainfall deficiency. It results in low runoff, groundwater, and soil moisture and finally in the shortage of available water for plants, animals, and humans. However, drought does not depend only on precipitation but also on other factors, such as air temperature, humidity, wind speed, and soil properties. All these factors can substantially contribute to exacerbate drought severity.

2.1.2 How Does Drought Affect Agricultural Production and Food Security?

Drought, with its negative effects on agricultural production, is one of the main causes of food insecurity worldwide (Chinembiri et al., 2024; Hoek et al., 2020). Extreme droughts like those that hit the Sahel region in the 1970s and 80s, the Ethiopian drought in 1984, and the recent Horn of Africa drought in 2010/2011 have received extensive media attention because they directly caused hunger and the deaths of hundreds of thousands of people (Checchi and Courtland Robinson, 2013). With the recent trend of persistently high food prices and a continuously increasing demand for agricultural production to satisfy the food needs and dietary preferences of an increasing world population, drought is one of the climate-related factors with the highest potential of negative impact on food availability and societal development. Droughts aggravate the competition and conflicts for natural resources in those areas where water is already a limiting factor for agriculture, pastoralism, and human health. Climate change may further deteriorate this picture by increasing drought frequency and extent in many regions of the world due to the projected increased aridity in the next decades (IPCC AR5, 2013).

For drought to negatively affect agricultural production in a region, one has to consider both its spatial and temporal dimensions (Bhushan et al., 2024; Guo et al., 2023; Rojas et al., 2011). Drought is usually a slow onset problem that negatively impacts crop production and ultimately food security only if it persists for a period long enough to seriously reduce plant growth and health and if the area concerned is large enough to substantially reduce food production in a region. For estimating the drought impact on food security it is also important to take into account the level of vulnerability and coping capabilities of the exposed population as shown once again by the recent famine in Somalia (Maxwell and Fitzpatrick, 2012). This example shows clearly that for a crisis to evolve from a prolonged agricultural drought into a famine, many other factors are at play, such as high international food prices, limited access to the drought-affected area, civil conflicts, and political difficulties in organizing humanitarian interventions.

Crop failures and pasture biomass production losses are the primary direct impact of drought on the agricultural sector productivity. Drought-induced production losses cause negative supply shocks, but the amount of incurred economic impacts and distribution of losses depends on the market structure and interaction between the supply and demand of agricultural products (Ding et al., 2011). These adverse shocks affect households in a variety of ways, but typically the key consequences are on assets (UN, 2009). First, households' incomes are affected, as returns to assets (e.g., land, livestock, and human capital) tend to collapse, which may lead to or exacerbate poverty. Assets themselves may be lost directly due to the adverse shocks (e.g., loss of cash, live animals, and impacts on health or social networks) or may be used or sold in attempts to buffer income fluctuations, affecting the ability to generate income in the future.

Droughts are by their nature covariate phenomena, with many people affected at the same time, making their consequences even harder. For instance, rural populations affected by drought may be forced to opt for limiting or even disrupting copying strategies, including selling their production assets (e.g., livestock land and tools). Short-term impacts can last a few weeks or months. If response action (e.g., food relief or cash transfer) is taken to the household or community level right after the disaster, the consumption drops or income losses can be softened. On the contrary, if households have few assets to protect themselves during hardships and public protective measures come too late, the negative transitory effects on their members can deteriorate into more permanent disadvantages, for instance, migration or nutrition shortfalls in children that in turn could affect their development later in life.

Drought as a climate-related disaster is hard to prevent and main efforts toward reducing drought impacts traditionally focus on mitigation and on strengthening the resilience of drought-exposed livelihoods by efficient drought management. One way of mitigating drought impacts is by improving early warning and monitoring systems fed by objective and reliable drought indices (vegetation and precipitation) and near-real-time weather information (Boyd et al., 2013; Wilhite et al., 2007; Sheffield and Wood, 2011).

Obviously, even if the impact of a drought can be timely assessed, having an operational early warning system in place is only a first step toward ensuring rapid and efficient response (Hoek et al., 2020; Ahmadalipour et al., 2017; Du et al., 2017; Hillbrunner and Moloney, 2012).

For the implementation of programs that aim to increase food security, the identification of drought-prone areas and the estimation of the probability of drought occurrence are also fundamental. For example, knowing the probability of drought occurrence is of basic importance for risk management programs (e.g., crop and livestock insurances) as well as for planning efficient food-aid delivery. Furthermore, drought information is very important for a better interpretation of potential effects of climate change in Africa (Rojas et al., 2011) and elsewhere.

2.1.3 What Can Remote Sensing Do?

Most drought monitoring methods have focused mainly on rainfall and rainfall anomalies, and many indices have been developed over time. A significant number of indices for agricultural drought monitoring is based on a water balance approach, where several climatic and physical variables are observed over a certain period in order to estimate the soil moisture deficit. This family

Agricultural Drought Monitoring

of indices includes the Palmer Drought Severity Index (PDSI), The Crop Moisture Index (CMI), the Surface Water Supply Index (SWSI), and the Reclamation Drought Index (RDI) (Palmer, 1965; Shafer and Dezman, 1982).

Other indices are of more statistical nature and look at time series of precipitation, including or not temperature data. This is, for example, the case of the Standardized Precipitation Index (SPI) (Wu et al., 2013; WMO, 2012). To the same group belongs the more simple precipitation decile index (Gibbs and Maher, 1967) and the Percent of Normal Precipitation method. The latter index simply indicates the relative difference of current rainfall as compared to a long-term average. A good overview of agro-meteorological drought indices is made available by M.G. Hayes (http://www.civil.utah.edu/~cv5450/swsi/indices.htm).

At regional to continental scale, drought monitoring and drought risk assessment based on any of the aforementioned indices is often hampered by the scarcity of reliable rainfall data. In particular, the coverage of operational weather stations in many drought-prone countries shows large spatial gaps and individual stations often provide discontinuous data. Additionally, the spatial representativeness of in situ measurements is often very restricted and a continuous spatial description of precipitation is difficult to be achieved because of the known limitation in spatial interpolation of rainfall data from meteorological stations. Due to these reasons, rainfall measurements are commonly replaced by estimates generated by atmospheric circulation models or derived from meteorological satellite observations. Commonly used rainfall estimate datasets for drought monitoring are reported in Table 2.1.

Another approach to drought monitoring is to evaluate "vegetation health status" by using optical remote sensing (Meroni et al., 2019; Wu et al., 2019; Rembold et al., 2017). The large spatial coverage and high temporal revisit frequency of low spatial resolution satellite instruments such as MODIS or SPOT-VEGETATION makes them particularly useful for near-real-time information collection at the regional and global scales (Rembold et al., 2013). Thanks to their large swath width, low-resolution systems have currently a much better synoptic view and temporal revisit frequency compared to high spatial resolution sensors (e.g., Landsat instruments). The individual scenes span a width of up to 3,000 km, so that the entire Earth surface is scanned every day and the specific costs per ground area unit are very low.

A pragmatic and widespread approach to extract the relevant information from the various spectral bands of such satellite sensors relies on the computation of vegetation indices (VIs) (Mulla, 2021; Zhong et al., 2019; Rembold et al., 2017). Among the different VIs, the Normalized Difference Vegetation Index (NDVI, Rouse et al., 1974), based on the red and near-infrared reflectances, has become the most popular indicator for studying vegetation health and crop production. Research in

TABLE 2.1

Rainfall Estimate Datasets Used for Drought Monitoring in Africa.

Dataset	Reference	Website
ECMWF, European Centre for Medium-Range Weather Forecast	ECMWF, 2013	http://www.ecmwf.int/products/forecasts/d/charts
RFE, rainfall estimates of Climate Prediction Centre of the National Oceanic and Atmospheric Administration	NOAA, 2001	http://www.cpc.ncep.noaa.gov/products/fews/AFR_CLIM/afr_clim.html
TAMSAT, Tropical Applications of Meteorology using SATellite of Reading University	Grimes, 2003	http://www.met.reading.ac.uk/tamsat/about/
FAO-RFE rainfall estimates of Food and Agriculture Organization	Alessandrini and Evangelisti, 2011	http://geonetwork3.fao.org/climpag/FAO-RFE.php

vegetation monitoring has shown that NDVI is non-linearly related to the leaf area index (LAI) and linearly related to the fraction of absorbed photosynthetically active radiation (FAPAR), and hence the vegetation's photosynthetic activity. Vegetation indexes are subject to intrinsic limitations (e.g., saturation of the signal) and contaminations from different sources (e.g., illumination and observation geometry, 3D structure of the vegetated medium, and background reflectance) (Gobron et al., 1997). Alternative approaches make use of canopy radiative transfer models to derive key vegetation variables such as LAI and FAPAR from canopy reflectances (e.g., Myneni et al., 2002; Gobron et al., 2005; Baret et al., 2013). The advantage of these methods is that they provide access to inherent vegetation properties largely decontaminated of external factors (Pinty et al., 2009). In particular, FAPAR acts as an integrated indicator of the status and health of vegetation and plays a major role in driving gross primary productivity (Prince and Goward, 1995). These biophysical variables are also independent regarding the exact spectral band location and width of the satellite instrument used, thus potentially offering the possibility of building long-term datasets composed by multiple generations of instruments.

Time series of up to 30 years are nowadays available for a number of low-resolution satellite sensors, which makes time series analysis of biophysical indicators a common tool for drought monitoring (Table 2.2). Repeated acquisition of the same area is performed with hourly to daily frequency using coarse- to moderate-resolution instruments. Daily images can be used for specific purposes, but most of the monitoring systems described in this chapter make use of temporal synthesis of daily images (so-called composites), typically aggregating ten daily acquisitions into a ten-day composite image (also called dekadal composite). The maximum-value composite procedure (MVC, Holben, 1986) that retains the highest NDVI value within the compositing period is often used for the purpose, although several compositing techniques have been proposed (for an overview see Chuvieco et al., 2005). Compositing also reduces considerably the amount of noise that is present in the time series of NDVI or any other RS-derived biophysical indicator (due to different sources, including atmospheric perturbation, undetected clouds and anisotropy of the surface). In addition, temporal smoothing techniques are usually employed to further reduce the residual noise in the data (e.g., Chen et al., 2004; Atzberger and Eilers, 2011).

Table 2.2 is not taking into consideration low spatial resolution geostationary satellites that belong primarily to the meteorological domain like Meteosat and MSG (Meteosat Second Generation). Nevertheless, optical and thermal measurements offered by these satellites can be used for drought monitoring too (Fensholt et al., 2011). In addition, we acknowledge the use of radar RS for soil moisture estimation. For an introduction on the use of radar RS in drought monitoring, see Nghiem et al. (2012).

2.2 OPERATIONAL METHODS AND TECHNIQUES

In this section we describe four approaches/systems operationally used for drought monitoring with satellite-derived vegetation indices and biophysical products (Chinembiri et al., 2024; Guo et al., 2023; Mulla, 2021; Hoek et al., 2020; Meroni et al., 2019; Kerdiles et al., 2017):

1. NDVI-based vegetation anomaly indicators for drought and crop monitoring bulletins
2. FAO Agriculture Stress Index System (ASIS) approach targeting agriculture
3. Action Contre la Faim (ACF) approach targeting pastures
4. Joint Research Centre (JRC) approach for the early detection of biomass production deficit hot spots

We thus focus on agricultural drought monitoring systems (including pastoral systems) working in near real-time (NRT) at the global to regional scales. Several other approaches and systems exist and have been extensively described in recent books such as *Famine Early Warning Systems and Remote*

TABLE 2.2
Properties of the Most Common Optical Low- and Medium-Resolution Operational and Planned Sensors Relevant for Vegetation Monitoring. (The Following Abbreviations Are Used for Different Intervals of the Electromagnetic Spectrum Used for Optical Drought Monitoring: VIS = Visible (350–750 nm), NIR = Near Infrared (750–1400 nm), SWIR = Shortwave Infrared (1400–3000 nm). NB: The ENVISAT Mission Stopped Officially in May 2012. SPOT-VEGETATION Mission Stopped in May 2014 with PROBA_V, Ensuring the Continuity of Product Generation

Sensor	Platform	Spectral Range	Number of Bands	Resolution	Swath Width	Repeat Coverage	Launch
AVHHR	NOAA POES 6–19	VIS, NIR	5	1,100 m	2,400 km	12 h	1978
AVHRR	METOP	VIS, NIR, SWIR	5	1,100 m	2,400 km	12 h	2007
SEAWIFS	Orbview-2	VIS, NIR	8	1,100 m 4,500 m	1,500 km 2,800 km	1 day	1997
VEGETATION	SPOT 4, 5	VIS, NIR, SWIR	4	1,100 m	2,200 km	1 day	1998
MODIS	EOS AM1/PM1	VIS, NIR, SWIR	36	250–1,000 m	2,330 km	<2 days	1999
MERIS	ENVISAT	VIS, NIR	15	300 m (1,200 m)	1,150 km	<3 days	2000
PROBA-V	PROBA-V	VIS, NIR, SWIR	4	300 m (1,000 m)	2,250 km	1 day	2013
SENTINEL 3	SENTINEL	VIS, NIR	21	300 m	1,270 km	<2 days	Foreseen 2014

Sensing Data by Molly Brown in 2008 or *Remote Sensing of Drought: Innovative Monitoring Approaches* by Brian D. Wardlow et al. in 2012, as well as in other chapters of the present book. The first book focuses mainly on the use of remote sensing for food security early warning. It includes also a number of drought and hydrological stress monitoring applications. The second book explores a broad range of applications for monitoring and estimating vegetation health, soil moisture, precipitation, and evapotranspiration.

At the end of Section 2.2.4, Table 2.3 summarizes the main characteristics of the four drought-monitoring approaches described in this chapter and mentions main strengths and limitations for each of them.

2.2.1 NDVI-Based Vegetation Anomaly Indicators Used Mainly for Drought or Crop Monitoring Bulletins

Vegetation anomaly methods are mostly based on the qualitative (or semi-quantitative) interpretation of remote sensing–derived indicators (often NDVI). They generally compare the actual crop status to previous seasons or to what can be assumed to be the average or "normal" situation. Detected anomalies are then used to draw conclusions on possible vegetation "health" or yield limitations.

Simple, but timely and accurate, vegetation monitoring systems working both at the national and regional scale are particularly necessary in arid and semiarid countries, where temporal and geographic rainfall variability leads to high inter-annual fluctuations in primary production and to a large risk of famines (Hutchinson, 1991). These environmental situations, along with the wide extent of the areas to monitor and the generally poor availability of efficient ground data collection systems, represent a scenario where qualitative monitoring can quickly produce valid information for releasing early warnings about possible stress of crops and pastures. Such systems are typically used in many food insecure countries by FAO (Food and Agriculture Organization of the United Nations), FEWSNET (Famine Early Warning System) of USAID (United States Agency for International Development), and the MARS (Monitoring of Agricultural Resources) project of the Joint Research Centre of the European Commission. However, qualitative monitoring is not necessarily linked to an early warning context in arid areas but can also be very useful to get a quick overview of vegetation stress for large areas in different climatic zones of the world. An example is given in Figure 2.1, which depicts NDVI anomalies during the 2012 Northern Hemisphere crop growing season. The NDVI anomaly is computed as the difference between the mean NDVI value for the month of August 2012 and the mean NDVI value computed for the same month over the historical NDVI archive (in this case from 1998 to 2011 for a total of 14 years) (Bhushan et al., 2024; Hoek et al., 2020; Du et al., 2017). A clear stress for summer crops is visible in central parts of the United States and in southern parts of Russia due to poor rainfall distribution (data not shown), whereas favorable conditions are observed in large parts of China. In the Southern Hemisphere, negative vegetation anomalies are visible in northeastern Brazil and southern Africa, whereas favorable conditions can be observed in the southern part of Brazil.

In addition to analyzing anomaly images for qualitative vegetation growth monitoring, useful information can be derived from temporal profiles of remotely derived vegetation indices. These temporal profiles are extracted for the administrative area of interest by averaging all pixel values within the area or, to better focus on agricultural land, by considering only those pixels where crops are dominant. The profiles give a complete picture of the vegetation development during the seasonal cycle and can be compared with other (for example, previous) crop seasons and the long-term average vegetation profile. If available, information regarding the major phenological development stages can be considered during the evaluation as stress effects on crop growth differ for different phenological stages.

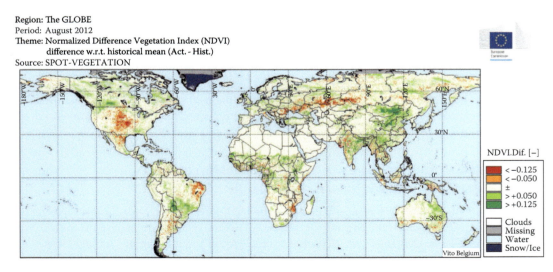

FIGURE 2.1 Global map of NDVI anomalies during the 2012 boreal hemisphere summer crop growing season (August). Negative anomalies are visible mainly in the central United States, central Asia, and northern Brazil, while a positive situation is evident in eastern China and southern Brazil. Data are from SPOT-VEGETATION. Anomalies are expressed as deviation in absolute NDVI units from the 1999–2011 average of NDVI for the month of August.

Signatures from coarse-resolution pixels usually represent mixtures of different land use classes (so-called mixed pixel). Several approaches have been elaborated for extracting land use specific signatures from the low-resolution pixel. A simple and common one is the crop-specific NDVI (CNDVI) method (Genovese et al., 2001). When computing the average NDVI value of an area of interest (e.g., a region, a department), the CNDVI approach adds proportional weights to the NDVI values based on the fraction of area covered by crop within each low-resolution pixel. More sophisticated methods are based on so-called un-mixing models, which consider the NDVI of a given low-resolution pixel as a linear mixture of so-called end-member spectral signatures (Busetto et al., 2008). Both CNDVI and un-mixing approaches require a land cover map at a spatial resolution higher than that of the coarse resolution time series. This land cover map is used to compute the relative presence of the classes of interest (e.g., crops). Relative advantages and limitations of CNDVI and un-mixing approaches have been recently studied in Atzberger et al. (2014).

For comparing recent VI images to "normal" conditions and its historical distribution, a variety of statistical indices have been proposed beyond the simple difference described so far. Following classical statistical theory, one approach is to calculate a so-called standard score (ZVI) by subtracting, for a given dekad (ten-day compositing period) d, the historical mean ($NDVI_{AVG}$) from the observed value and divide it by its standard deviation ($NDVI_{SD}$, also obtained from historical data):

$$ZVI(d) = \frac{NDVI(d) - NDVI_{AVG}(d)}{NDVI_{SD}(d)} \tag{2.1}$$

ZVI thus indicates how many standard deviations an observed NDVI value is below (or above) the historical average. To avoid the assumption of normal distribution of Equation (1), a non-parametric version of the index can be computed using, for instance, the median and the interquartile distance of the observed distribution.

Another example of a drought index is the Vegetation Condition Index (VCI) of Kogan (1995). The VCI locates the current VI value in the historical range of all preceding images acquired at the same time of the year, as shown by Equation (2).

$$VCI(d) = 100 \frac{NDVI(d) - NDVI_{MIN}(d)}{NDVI_{MAX}(d) - NDVI_{MIN}(d)} \tag{2.2}$$

where d refers to the dekad of the year at which the VCI is computed, subscripts *MIN* and *MAX* refer to the minimum and maximum value observed for dekad d in the historical archive, respectively. VCI should therefore be interpreted as a percentage expressing the vegetation status of a given pixel in relation to its historic range of variability represented by the minimum (worst conditions) and maximum (best conditions) NDVI over the years. VCI is, by definition, extremely sensitive to outliers in the series that would affect the maximum and minimum value of NDVI, which makes the use of temporal smoothing a basic requirement of the method (Bhushan et al., 2024; Meroni et al., 2019; Du et al., 2017; Rembold et al., 2017).

VCI can be combined with an analogous index built with remotely sensed surface temperature (i.e., the Temperature Condition Index, TCI) into the VHI (Vegetation Health Index) as proposed by Kogan (2001). Spatial and temporal aggregations of the VHI have been successfully used as an agricultural drought indicator by Rojas et al. (2011).

In a similar way, Balint et al. (2011) proposed the Combined Drought Index (CDI) that, besides NDVI and temperature, takes into account precipitations. In the CDI framework, drought (or vegetation stress) is conceived as a combination of magnitude and time persistence of the following factors: rainfall deficit, temperature excess, and soil moisture deficit. Because of limited availability of soil moisture observations at 1 km resolution, Balint et al. (2011) proposed to approximate the soil moisture component by NDVI deficits and deficit persistence. The three individual drought indices, i.e.,

PDI (Precipitation Drought Index), TDI (Temperature Drought Index), and VDI (Vegetation Drought Index—as a substitute for the Soil Moisture Drought Index), are computed by taking into account the magnitude of the anomaly and its duration and then combined as a weighted sum to yield the CDI index.

2.2.2 Agricultural Stress Index System Approach Targeting Agriculture

FAO is developing the Agriculture Stress Index System (ASIS) to detect agricultural areas with a high likelihood of water stress (drought) at the global level. Based on RS data, ASIS will support the vegetation monitoring activities of the FAO-Global Information and Early Warning System (GIEWS). The idea behind ASIS is to setup an agricultural drought monitoring system where the RS data processing part is highly automated and the end user can concentrate the analysis on the results of the system. ASIS provides maps of "drought hot spots" updated every ten days to the GIEWS officers and external users. To ensure that the system will not produce false alerts due to external factors such as atmospheric perturbations affecting low-resolution optical images, the officers then verify the "hot spots" with auxiliary information, for example, by contacting the Ministry of Agriculture of the affected country or by monitoring prices of the commodities. ASIS uses the VHI index as derived from the NDVI and surface brightness temperature (see Section 2.1) dekadal product from the METOP-AVHRR sensor at 1 km resolution. The first step consists in computing the temporal average of the VHI assessing the intensity and duration of the dry period(s) occurred during the crop cycle at pixel level. The second step consists in the calculation of the Agricultural Stress Index (ASI) as a percentage of agricultural area affected by drought (pixels with VHI < 35%—a value identified as critical in previous studies) to assess the spatial extent of the drought. Finally, each administrative area is classified into one of several drought severity classes according to the percentage of area affected.

By definition, VHI can potentially detect drought conditions at any time of the year (Guo et al., 2023; Mulla, 2021; Oroud and Balling, 2021; Hoek et al., 2020; Meroni et al., 2019). For agriculture, however, the analysis is restricted to the period between the start and end of the crop season. ASIS assesses the severity (intensity, duration, and spatial extent) of agricultural drought and provides summary statistics at the selected administrative level, allowing the comparison with the official agricultural statistics, where available. The full methodology is described in Rojas et al. (2011).

For the operational implementation the Flemish Institute for Technological Research (VITO) inter-calibrated the METOP-AVHRR data (available only since 2007) with the NOAA-AVHRR time series (since 1984) to produce a consistent long-term historical archive. The ASIS database thus allows the analysis of 30 years of potential agricultural hot spots, starting from 1984 when the Sahel was severely affected by drought. As an example of global hot spot maps produced, Figure 2.2 shows the ASI map for the year 1989, when a large fraction of the global agricultural land suffered from water scarcity (see also Figure 2.3).

2.2.3 Action Contre la Faim Approach Targeting Pastures

In Sahelian pastoral and agro-pastoral areas, livestock systems are fully dependent on the spatial repartition of pastoral resources, namely, pasture and water. Their scarcity and high spatio-temporal variability can lead to drought with consequent livestock movements and locally tuned adaptation strategies. In this semi-arid region, pastoral systems rely on rangelands and crop residues and are mainly nomadic or transhumant in the northern Sahel and more or less seasonally mobile in the southern part (rainfed and irrigated mixed crop–livestock systems). The unconditional need for mobility of livestock and families results from the scarce natural forage resources, while feed is unavailable or unaffordable owing to geographical isolation, high transportation costs, and low purchasing power (Ickowicz et al., 2012).

To address these specificities, ACF's approach consists of using a GIS and RS-based system designed to monitor pastoral resources on the one hand and livestock spatial adaptation strategies on the other hand. The full system—currently under development—ultimately aims at producing

Agricultural Drought Monitoring

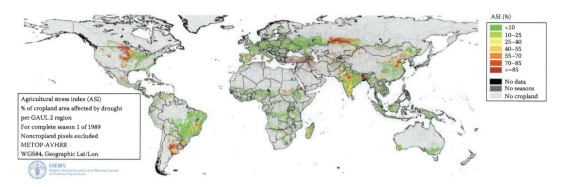

FIGURE 2.2 Agriculture Stress Index map for the first crop season of 1989. The ASI is an indicator developed by FAO that highlights anomalous vegetation growth and potential drought in arable land during a given cropping season. ASI assesses the temporal intensity and duration of dry periods and calculates the percentage of arable land affected by drought as pixels with a VHI value below 35% (reference threshold taken from literature). In 1989 large portions of agricultural land worldwide were affected by drought, with more than 85% of agricultural areas affected by drought in the United States, in Argentina, and in central Asia. Figure 2.3 provides an overview of the percentage of the agricultural land affected by drought in the time period 1984–2012, as derived from the ASIS method.

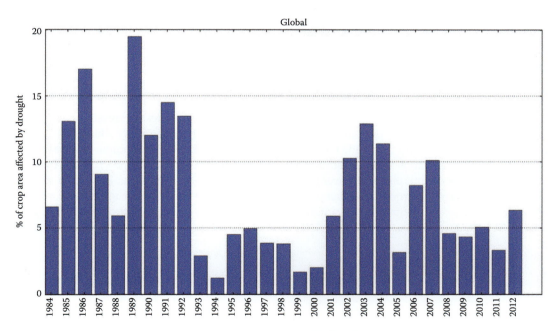

FIGURE 2.3 Percentage of agricultural land affected by drought according to ASIS by year (1984–2012). The total crop area used as a reference amounts to 23,440,622 km² and was derived from the crop mask shown in Figure 2.3 and obtained using the crop zones developed by FAO in the 1990s, and the Global Land Cover (GLC2000) (Bartholomé & Belward, 2005).

early alert and surveillance indicators in order to prevent and mitigate the impacts of drought events, targeting most affected areas and communities, anticipating tenses and possible conflicts for the access to the resources (Ham et al., 2011). The two main pastoral resources (pastures and surface water) are monitored through RS indicators.

2.2.3.1 Pasture Monitoring

This module is handled using the software package BioGenerator (Fillol, 2011a, 2011b), used to calculate dry matter biomass produced during the growing season by cumulating the DMP (Dry Matter Productivity) dekadal product at 1 km resolution produced by VITO from SPOT-VEGETATION data using a modified light-use efficiency approach (Eerens et al., 2004). The cumulative value obtained at the end of the growth period is used to produce maps of biomass production (Figure 2.4) and biomass production anomalies (expressed as % of average value) over the Sahel region (Figure 2.5). The output data of BioGenerator have been validated through several field studies, integrating field observations from Mali and Niger.

As compared to classical vegetation indices, the use of DMP product directly provide amounts of production in physical units (kg/ha). The DMP calculation approach is derived from the Monteith (1972) model and uses the Fraction of Photosynthetically Active Radiation (fAPAR), solar radiation, and temperature information retrieved from ECMWF (European Centre for Medium-Range Weather Forecasts) global climate model. The DMP model does not directly take into consideration water available to the plants and is therefore not extremely drought sensitive; also, it only provides total biomass production estimates and does not give any information on the actual pasture usability nor accessibility. Nevertheless, experience shows that the data are suitable and consistent for early warning purposes. GIS operations are then used to aggregate biomass production per geographical units (administrative, agro-ecological) in order to produce zonal balances. The output data of this module have already been used to assess drought impact at national and regional levels during known drought events in the Sahel region (2004–2005, 2009–2010, and 2011–2012) and are currently used operationally to analyze food security in pastoral areas (FAO-WFP, 2013). As an example, Figure 2.5 shows the biomass anomaly for 2013 compared to the 1999–2013 average, which appears to be an average year except in certain localized areas scattered along the pastoral zones bordering the desert to the south.

2.2.3.2 Surface Water Monitoring

The methodology is based on the use of the SWB (small water bodies) dekadal product available at a resolution of 1 km and providing information on the presence of surface water (Haas et al., 2009). SWB is produced by VITO from SPOT-VEGETATION data since 1999. This product has been validated using high-resolution satellite data and showed an overall accuracy of 95.4% considering only commission errors. Omission errors were significantly higher (about 30%) mainly due to the low spatial resolution of the product that hampers the detection of water bodies occupying only a minor fraction of the pixel. Integrating this information, the software package HydroGenerator (Fillol,

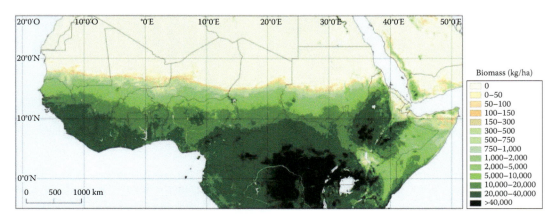

FIGURE 2.4 Average biomass annual productivity over the sub-Sahara area (1999–2013).

Agricultural Drought Monitoring

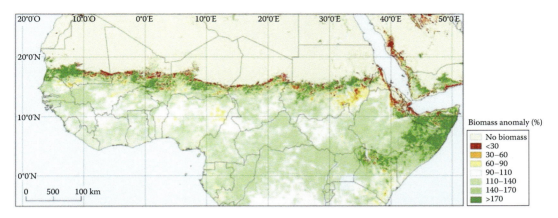

FIGURE 2.5 Biomass annual productivity anomaly for 2013 in comparison with the average (1999–2013).

2011a, 2011b) compiles annual statistical information on surface water. For each pixel and year, an index of water accessibility is calculated, using an integration of detections of all ponds within a 30 km circular buffer. The integration is weighted by the distance to these ponds. This monitoring of surface water should ideally be coupled with a consistent inventory of functioning underground water points (wells and boreholes). Having these two complementary pieces of information, it would be possible to make a realistic assessment of water availability and infer a differential accessibility of pastures. Nevertheless, the underground water inventory is not yet completed across the Sahel countries, and therefore the current product does not strictly reflect the actual situation of water availability. The current version of the product also does not account for security issues, social habits, livestock movements, or even financial constraints herders might face to access the water points.

As an example, Figure 2.6 shows the position and occurrence of surface water expressed in percentage of the normal dekadal occurrence calculated over the 1999–2013 period, and the position of boreholes and artificial water points in North Mali (Source: Direction Nationale Hydraulique Mali). Using this information it is possible to compute the accessible biomass over the area for each year as well as the inaccessible biomass defined as the difference between the total and the accessible biomasses (data not shown). This quantification of inaccessible or "lost" biomass could also support the expression of needed boreholes or water points in a way to optimize the biomass availability.

2.2.3.3 Integration with Livestock Movements

To estimate the evolution of the pastoral situation and assess the breeders' spatial strategy up to the next rainy season, it is necessary to get consistent knowledge of livestock movements. By combining pastoral resources monitoring data with grazing areas and projected movements, it would be possible to draw differential scenarios in order to target the most vulnerable areas and identify the plausible zones of tension and/or conflict for the access to the resources. This module is currently the least developed even if it is a crucial one. Research activities for the implementation of the module are currently focused on delineating "grazing shed" (i.e., the space used by a given herders group during a full pastoral cycle) and sub-zones used by animals during different periods; estimating the pastoral importance of each zone and calculating their respective load capacity; identifying and mapping transhumance routes in favorable and critical periods; and identifying changing movement patterns (difference in precocity or direction) as early warning indicators.

All these indicators will be part of the pastoral vulnerability model under construction. While the first module of the model is quite deeply used and diffused (pasture monitoring), the water and

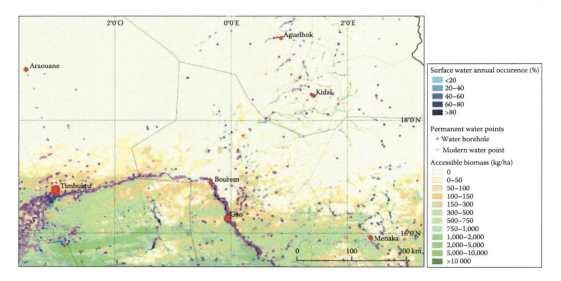

FIGURE 2.6 Surface water occurrence and position of permanent water point in North Mali and areas with accessible biomass for 2013.

the movement modules still need further research and require both technical developments and field data collection.

2.2.4 JRC Approach for the Early Detection of Biomass Production Deficit Hot Spots

Timely information on vegetation growth at the regional scale are needed in arid and semi-arid regions where rainfall variability leads to high inter-annual fluctuations in crop and pasture productivity and to high risk of food crisis in the presence of severe drought events (FAO, 2011). Monitoring systems should provide information on impending risks to national and international food security stakeholders as early as possible during the growing season. A regular update throughout the season is necessary to allow effective decision-making.

As mentioned in Section 2.1, current remote sensing monitoring methods are based on the computation of the anomaly of a vegetation status indicator (typically NDVI) with respect to a reference value (e.g., the historical mean). The interpretation of such anomalies is not straightforward, as the comparison is made at a fixed time of the year regardless of the actual plant development stage. Neglecting the actual development stage of the crop leads to a non-standardized spatial information. This is because anomalies at different locations may refer to different stages of development. For example, in one country the growing cycle could be nearly completed (high reliability of the information provided), while in another country it has just started (low reliability).

In addition, the traditional approach captures only a single time snapshot of vegetation status but misses an overall view of the entire seasonal development. To overcome such potential problems, a probabilistic approach has been recently proposed to estimate the probability of experiencing an end-of-season critical biomass production deficit during the ongoing growing season on the basis of the statistical analysis of long-term time series (from 1998 to today) of moderate-resolution SPOT-VEGETATION FAPAR observations (Meroni et al., 2014). The cumulative value of FAPAR during the growing season (CFAPAR) is used as a proxy of vegetation gross primary production (e.g., Bhushan et al., 2024; Hoek et al., 2020; Ahmadalipour et al., 2017; Du et al., 2017; Dardel et al., 2014; Fensholt et al., 2006; Jung et al., 2008) and of crop yield (e.g., Funk and Budde, 2009; Lobell et al., 2003; Meroni et al., 2013).

Agricultural Drought Monitoring

The method is applicable at the regional to continental scale and can be updated regularly during the season to provide a synoptic view of the hot spots of likely production deficit. The specific objective of the procedure is to deliver to the food security analyst, as early as possible during the season, only the relevant information (e.g., masking out areas without active vegetation at the time of analysis), expressed through a reliable and easily interpretable measure of impending risk.

Within-season forecasts of the final biomass production, expressed in terms of probability of experiencing a critical deficit, are based on a statistical approach taking into account the similarity between the current CFAPAR profile and past profiles observed in the time series and the uncertainty of past predictions of seasonal outcome (derived using jack-knifing technique). Processing is pixel-based and proceeds in five main steps: (1) retrieval of key phenological parameters (start and end of growing season); (2) computation of historical CFAPAR values and definition of a critical deficit based on the historical distribution; (3) definition of a metric estimating the likelihood of deficit occurrence; (4) assessment of its uncertainty in detecting the deficit occurrence; and (5) estimation of a formal deficit probability. Details about the processing algorithm can be found in Meroni et al. (2014).

The procedure is applied in near real-time whenever a new satellite observation becomes available to compute the probability of ending with a critical deficit. A critical production deficit is defined as the occurrence of a seasonal CFAPAR value below a certain threshold, conveniently set in relative terms as the *i-th* percentile of the distribution of the observed CFAPAR seasonal values. The method is thus aimed at detecting extremes that occur less than or equal to *i*% of the time. In addition, one can interpret the percentile in terms of return period (inverse of the frequency of occurrence) of an event of that magnitude or lower (e.g., the return period of a final CFAPAR falling in the first quartile is four years). In this way, the threshold value in CFAPAR units changes pixel by pixel, as it is the ranking with respect to the historical distribution that defines it.

As an example, Figure 2.7 shows the probability of deficit for the Sahel region as derived from the satellite observations available as of the first of September 2009, halfway through the season, a timing that can be considered appropriate for early warning analysis.

During 2009, poor rains were reported for a large fraction of the region shown in Figure 2.7. This is well captured by the estimated probability of deficit. For example, Figure 2.7 shows a widespread presence of hot spots of high deficit probability ranging from Mali, Niger, north Nigeria, Chad, and South Sudan. At the end of the 2009 season, significant reductions in grain harvest and pasture biomass production were reported in the Sahel, particularly in eastern Mali, Niger, Chad, northern Nigeria, and Burkina Faso (Fewsnet, 2009, 2010).

The method has been designed to timely provide an easily interpretable information: the probability of experiencing a critical deficit, a pragmatic indicator that can be easily understood despite the relative complexity involved in estimating it. Differently from standard anomaly products, the procedure maps the probability only where a growing season is actually ongoing, so that the analyst can focus on the relevant information only. Finally, the delivered information at a given time is already "weighted" for its reliability, as assessed using past estimations performed at that specific time.

In NRT operations, in the case where a problem may be emerging, the method would allow identifying its geographic dimensions and identifying where and how quickly it is developing. The appearance of deficit hot spots may also guide the decision on where to concentrate in-depth field assessments. In addition, starting from the probability map, country- or district-level key summary statistics can be extracted in tabular form, as for example, the fraction of the cropland or pasture area exposed to different levels of deficit risk (e.g., the fraction with a risk greater than the business-as-usual scenario, or greater than a selected threshold). This would allow assigning, for example, a risk score to each administrative unit taking into account the magnitude of the deficit probability and, at the same time, its spatial extent.

2.3 OTHER RECENT METHODOLOGICAL APPROACHES

In this section we highlight a non-exhaustive number of other and complementary recent advances made in drought monitoring using RS-based biophysical parameters (Bhushan et al., 2024;

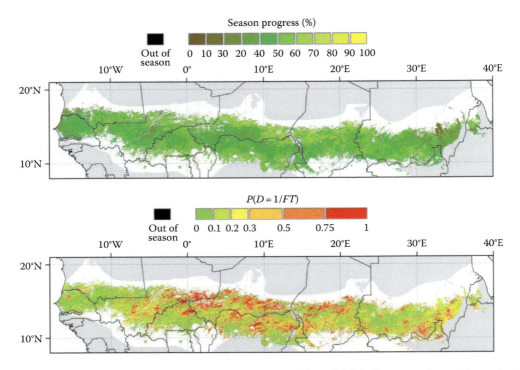

FIGURE 2.7 Progress of the season (upper panel) and probability of deficit (lower panel) as of September 1, 2009. Deficit threshold set to the first quartile of seasonal CFAPAR distribution. The analysis is limited to the herbaceous and cropland land covers (other classes in white) and the five main eco-regions of the Sahel (other regions in gray).

Chinembiri et al., 2024; Lee et al., 2023; Guo et al., 2023; Mulla, 2021; Oroud and Balling, 2021; Hoek et al., 2020; Meroni et al., 2019; Wu et al., 2019; Zhong et al., 2019; Kerdiles et al., 2017; Ahmadalipour et al., 2017; Du et al., 2017; Meroni et al., 2017; Rembold et al., 2017).

A first approach (Bhushan et al., 2024; Chinembiri et al., 2024; Lee et al., 2023; Guo et al., 2023; Mulla, 2021; Oroud and Balling, 2021; Hoek et al., 2020; Meroni et al., 2019; Wu et al., 2019; Zhong et al., 2019; Kerdiles et al., 2017; Ahmadalipour et al., 2017; Du et al., 2017; Meroni et al., 2017; Rembold et al., 2017; Tadesse et al., 2005; Brown et al., 2008, Wu et al., 2013) makes use of data-mining techniques to establish a relationship between a variety of inputs (including RS-derived variables) and observed historical droughts. The novelty of this approach is that the link between observed variables of different natures (RS, meteorological, and ancillary) and the occurrence of a drought is established using a multilevel decision tree (CART, Classification And Regression Tree). This allows the classification and stratification of the input variables before applying a linear regression scheme defining the proposed drought index. As a result, the index is defined by a set of rules that control the application of a specific regression involving one or more input variables. The regression tree nicely accommodates the fact that one input variable, for instance, the surface temperature, may contribute to explain the occurrence of a drought only when other conditions apply, for instance, if the precipitation was below a certain threshold. When such conditions do not apply, other variables may be picked up by the model. Rules and regression parameters are established in an iterative and automatic process that is finally cross-validated to ensure that the highly parameterized final model can provide the required predictive power. A drawback of the method may stand in the fact that it is not easy to interpret and gain physical insight from the generated rules.

A second approach developed by Cammalleri et al. (2023) and Sepulcre-Canto et al. (2012) at the European Drought Observatory (http://edo.jrc.ec.europa.eu) differs from the ones described so

TABLE 2.3
Summary of the Main Aspects of the Four Drought Monitoring Approaches Described in This Chapter

	Input Data	Methods	Strengths	Limitations
NDVI-based vegetation anomaly indicators for drought and crop monitoring	NDVI derived from medium- and low-resolution data. Usually weekly or ten-day composites.	Temporal comparisons and identification of anomalies for ongoing vegetative season.	Relatively simple and straightforward to be produced.	NDVI is only a proxy of vegetation vigor and health. Phenology not accounted for.
FAO Agriculture Stress Index System approach targeting agriculture	VHI derived from low-resolution NDVI and temperature data, ten-day composites.	Computation of cumulative value over time (crop season) and space (administrative areas) and use of thresholds to classify a drought event.	Provides information directly in terms of percentage of agricultural area affected.	Depends on the accuracy of the crop mask.
Action Contre la Faim approach targeting pastures	DMP based on Monteith approach. SWB (small water bodies) product.	Computation of seasonal cumulative values.	Simple calculation and easy to understand. Results expressed in biomass physical units.	DMP is not water limited, reducing sensitivity to drought.
Joint Research Centre approach for the early detection of biomass production deficit hot spots	FAPAR	Computation of seasonal cumulative values and computation of probability of final seasonal deficit	Provides a forecast for the end of the growing season adjusted for the uncertainty of the estimation.	More complex than the others.

far, as it is aimed to a qualitative classification of drought severity based on different data sources (including RS-based FAPAR) and a kind of "convergence of evidences" approach. The authors proposed a Combined Drought Indicator (CDI) built on an idealized cause-effect relationship for agricultural drought: a shortage of precipitation (formalized by the three-month Standardized Precipitation Index, SPI-3) leads to a soil moisture deficit (estimated through the LISFLOOD hydrological precipitation-runoff model) that in turn results in a reduction of vegetation productivity (monitored with FAPAR anomalies). By setting thresholds indicating extreme values for the three indicators and after having empirically defined a time lag between the three events, they combine the three time-lagged indicators (anomaly of FAPAR at dekad n, anomaly of soil moisture at n-1, and SPI-3 compute over the interval (n-9, n)) and classify such triplets into four CDI categories: "watch" when there is a precipitation deficit (SPI-3 is below the threshold), "warning" when there is also a soil moisture deficit (both SPI-3 and anomaly of soil moisture are below their thresholds), "alert 1" when there is vegetation stress following a precipitation deficit (both SPI-3 and anomaly of FAPAR are below their thresholds), and "alert 2" when vegetation stress follows a precipitation and soil moisture deficit (all indicators below their thresholds). This approach was applied in Europe to assess its reliability in spotting recent historical droughts, and it is currently being used operationally by the European Drought Observatory.

Another inspiring example of possible approach to drought detection is the work of Zscheischler et al. (2013). In this study they explore how to detect and quantify extreme events as spatiotemporal phenomena spanning over both the spatial and temporal dimensions. Working with a 30-year

FAPAR time series constructed with different satellite sensors they first define an extreme event in terms of FAPAR anomalies and then compute the contiguous volume occupied by the extreme event in the 3D space made up by latitude, longitude, and time. In this way, the severity of a drought is determined by its spatial magnitude and its persistency over time.

2.4 DISCUSSION AND FUTURE DEVELOPMENTS

Concerning coverage and quality in remote sensing data, the continuous trend in increasing spatial and temporal resolution of new satellite sensors (e.g., the recently launched PROBA-V instrument and the forthcoming Sentinel 2 and 3 ESA satellites) is expected to allow improvements of the remote sensing methods described. However, the availability of a long-term time series will remain a relevant issue, as it was shown to be a pre-condition in many of the drought monitoring methods described. In fact, in the absence of reliable ground measurements of crop yield and rangeland production in most of the food insecure regions of the world, most of monitoring systems are based on the use of long time series for comparison with previous years or with the average situation. Such approaches will not be able to profit immediately from the upcoming availability of higher spatial resolution sensors. This means that even when the next generation of Earth observing satellites with higher spatial ground sampling distance will be launched (e.g., Sentinel-2), a number of years will pass until the benefits of the increased spatial resolution will deploy their full impact on improving the quality of drought monitoring products. This lap time could only be reduced by un-mixing the signal recorded sensors of different resolution at administrative level to a common "endmember" signature. Increased research efforts on sensor inter-calibration and on methods for exploiting the current long-term archive for supporting the analysis of higher spatial resolution instruments are needed to simplify access to long time series of remotely sensed data from different sensors.

Ideally, one would have a suitable land process model at hand, describing the main processes with sufficient detail, and assimilate the remotely sensed observations to constrain the simulation outcomes. The climate community is actively working on such models (mainly for the purpose of improving weather forecasts and climate predictions), but nowadays such dynamic models are still too data demanding (e.g., relating to soil and vegetation properties) for being useful in operational programs.

Another critical issue in the use of RS time series that deserves further consideration refers to the trade-off between timeliness and noise reduction. Typically, noise reduction is obtained using temporal smoothing of the time series. Effective smoothing is in turn achieved when temporal observations before and after the data point being smoothed are available. As a result one has to choose between the exploitation of less reliable but more recent observation and the more reliable but not updated observations. An option to overcome this impasse may be represented by the use of more or less simple vegetation growth model and data assimilation techniques (for a review see Dorigo et al., 2007) for ingesting RS data. In such a way, the increasing uncertainty of the most recent observations may be taken into account.

Finally, it is worth mentioning that the tremendous advances made in vegetation stress detection using remote sensing techniques, developed mainly for precision farming applications using high spectral and spatial resolution data, has not been translated yet into operational and regional drought monitoring. Employed techniques span from the use of narrow-band vegetation indexes to detected leaf pigment concentration (e.g., Rossini et al., 2013; Haboudane et al., 2002) and leaf water content (Colombo et al., 2011 for a recent review), to the use of advanced optical indexes such as the Photochemical Reflectance Index (PRI), chlorophyll fluorescence, and thermal imagery (e.g., Meroni et al., 2009; Suárez et al., 2010; Zarco-Tejada et al., 2012) to characterize leaf physiology and detect vegetation stress in its early stages. There are many reasons for such a gap: technical constraints related to the unavailability of the required spectral and spatial resolution observations at the regional to global scale, relative novelty of such approaches as compared to more traditional ones based on well-known vegetation indices such as NDVI, complex effects of the 3D canopy structure and illumination and view angles on the advanced indices, difficulties in applying empirical methods tuned at the landscape scale to the regional scale, lack of calibration/validation data in drought-prone countries (e.g., Africa), among others. However, successful examples of application in carbon fluxes estimation over different biomes (see recent reviews and a discussions see Garbulsky et al., 2011; Penuelas et al., 2011) offer promising prospect for enhanced capability in regional to global drought monitoring, for instance, through the use of the PRI index available globally and with the required temporal frequency from MODIS instruments Terra and Aqua.

2.5 CONCLUSIONS

The operational agricultural drought monitoring systems using space observations and the recent methodological developments show that satellite data play a key role in drought surveillance and early warning at the global level. The examples of monitoring systems described in this chapter also show that the information derived from RS is directly used for drought management approaches that aim at designing response actions improving food and water availability in dry areas (Bhushan et al., 2024; Chinembiri et al., 2024; Lee et al., 2023; Guo et al., 2023; Mulla, 2021; Oroud and Balling, 2021; Hoek et al., 2020; Meroni et al., 2019; Wu et al., 2019; Zhong et al., 2019; Kerdiles et al., 2017; Ahmadalipour et al., 2017; Du et al., 2017; Meroni et al., 2017; Rembold et al., 2017). At the same time, monitoring and early warning systems are also becoming important components of more comprehensive (drought) risk management systems. These include crop and livestock insurance schemes that can be based on remotely sensed indices, and national risk management systems using RS-based drought information for early design of contingency plans for response actions.

It is also important to remember that RS-based methods observe primarily the physical signals of drought, while decision-makers and the general public are usually more interested in drought impact on local communities, which depend on many other factors, such as the vulnerability of the population, the economic and political situation at the national level and many others. Therefore it is important to combine RS information with other sources of data, including socioeconomic aspects addressing coping and adaptation strategies, to strengthen community's resilience while securing the production systems that are most exposed to uncertainty.

REFERENCES

Ahmadalipour, A., Moradkhani, H., Yan, H., & Zarekarizi, M. 2017. Remote sensing of drought: Vegetation, soil moisture, and data assimilation. In: Lakshmi, V. (ed) *Remote Sensing of Hydrological Extremes*. Springer Remote Sensing/Photogrammetry. Springer, Cham. https://doi.org/10.1007/978-3-319-43744-6_7

Alessandrini, S., & Evangelisti, M. 2011. FAO RFE the FAO African rainfall estimate. *Workshop "Rainfall Estimates for Crop Monitoring and Food Security"*, 22–23 October 2008, Ispra, VA, Italy. http://geonetwork3.fao.org/climpag/FAO-RFE-ISPRA2008.pdf.

Atzberger, C., & Eilers, P.H.C. 2011. Evaluating the effectiveness of smoothing algorithms in the absence of ground reference measurements. *International Journal of Remote Sensing*, 32, 13, 3689–3709.

Atzberger, C., Formaggio, A.R., Shimabukuro, Y.E., Udelhoven, T., Mattiuzzi, M., Sanchez, G.A., & Arai, E. 2014. Obtaining crop-specific time profiles of NDVI: The use of unmixing approaches for serving the continuity between SPOT-VGT and PROBA-V time series. *International Journal of Remote Sensing*, 35, 7, 2615–2638.

Balint, Z., Mutua, F.M., & Muchiri, P. 2011. *Drought Monitoring with the Combined Drought Index*. FAO-SWALIM, Nairobi, Kenya, 32 pp.

Baret, F., Weiss, M., Lacaze, R., Camacho, F., Makhmara, H., Pacholcyzk, P., & Smets, B. 2013. GEOV1: LAI and FAPAR essential climate variables and FCOVER global time series capitalizing over existing products. Part 1: Principles of development and production. *Remote Sensing of Environment*, 137, 299–309.

Bartholomé, E., & Belward, A.S. 2005. GLC2000: A new approach to global land cover mapping from earth observation data. *International Journal of Remote Sensing*, 26, 9, 1959–1977.

Bhushan, B., Dhurandher, A., & Sharma, A. 2024. Meteorological and agricultural drought monitoring using geospatial techniques. In: Shit, P.K., et al. (eds) *Geospatial Practices in Natural Resources Management. Environmental Science and Engineering*. Springer, Cham. https://doi.org/10.1007/978-3-031-38004-4_13

Boyd, E., Cornforth, R.J., Lamb, P.J., Tarhule, A., Lélé, M.I., & Brouder, A. 2013. Building resilience to face recurring environmental crisis in African Sahel. *Nature Climate Change*, 3, 631–637.

Brown, J., Wardlow, B., Tadesse, T., Hayes, M., & Reed, B. 2008. The Vegetation Drought Response Index (VegDRI): A new integrated approach for monitoring drought stress in vegetation. *GIScience Remote Sensing*, 45, 1, 16–46.

Busetto, L., Meroni, M., & Colombo, R. 2008. Combining medium and coarse spatial resolution satellite data to improve the estimation of sub-pixel NDVI time series. *Remote Sensing of Environment*, 112, 118–131.

Cammalleri, C., Acosta Navarro, J.C., Bavera, D., Diaz Mercado, D., Di Ciollo, C., Maetens, W., Masante, D., Spinoni, J., & Toreti, A. 2023. An event-oriented database of meteorological droughts in Europe based on spatio-temporal clustering. *Scientific Reports*, 13.

Checchi, F., & Courtland Robinson, W. 2013. *Mortality Among Populations of Southern and Central Somalia Affected by Severe Food Insecurity and Famine During 2010–2012*. FAO/FSNAU and Fewsnet, Rome and Washington, DC. www.fsnau.org/downloads/Somalia_Mortality_Estimates_Final_Report_8May2013_upload.pdf

Chen, J., Jonsson, P., Tamura, M., Gu, Z., Matsushita, B., & Eklundh, L. 2004. A simple method for reconstructing a high-quality NDVI time-series data set based on the Savitzky–Golay filter. *Remote Sensing of Environment*, 91, 332–344.

Chinembiri, T.S., Mutanga, O., & Dube, T. 2024. A multi-source data approach to carbon stock prediction using Bayesian hierarchical geostatistical models in plantation forest ecosystems. *GIScience & Remote Sensing*, 61, 1. http://doi.org/10.1080/15481603.2024.2303868

Chuvieco, E., Ventura, G., Pilar Martin, P., & Gomez, I. 2005. Assessment of multitemporal compositing techniques of MODIS and AVHRR images for burned land mapping. *Remote Sensing of Environment*, 94, 450–462.

Colombo, R., Meroni, M., Busetto, L., Rossini, M., & Panigada, C. 2011. Optical remote sensing of vegetation water content. In: Thenkabail, P.S., Lyon, J.G., & Huete, A. (eds) *Hyperspectral Remote Sensing of Vegetation*. CRC Press, Taylor and Francis Group, Boca Raton, FL, USA, pp. 227–244.

Dardel, C., Kergoat, L., Hiernaux, P., Mougine, E., Grippa, M., & Tucker, C.J. 2014. Re-greening of Sahel: 30 year of remote sensing data and field observations (Mali, Niger). *Remote Sensing of Environment*, 140, 350–364.

Ding, Y., Hayes, M., & Widhalm, M. 2011. Measuring economic impact of drought: A review and discussion. *Disaster Prevention and Management*, 20, 4, 434–446.

Dorigo, W.A., Zurita-Milla, R., de Wit, A.J.W., Brazile, J., Singh, R., & Schaepman, M.E. 2007. A review on reflective remote sensing and data assimilation techniques for enhanced agroecosystem modelling. *International Journal of Applied Earth Observation and Geoinformation*, 9, 2, 165–193.

Du, L., Song, N., Liu, K., Hou, J., Hu, Y., Zhu, Y., Wang, X., Wang, L., & Guo, Y. 2017. Comparison of two simulation methods of the temperature vegetation dryness index (TVDI) for drought monitoring in semi-arid regions of China. *Remote Sensing*, 9, 2, 177. https://doi.org/10.3390/rs9020177

ECMWF. 2013. User guide to ECMWF forecast products. *Technical Report*, 121 pp. http://old.ecmwf.int/products/forecasts/guide/user_guide.pdf

Eerens, H., Piccard, I., Royer, A., & Orlandi, S. 2004. *Methodology of the MARS Crop Yield Forecasting System. Vol. 3: Remote Sensing Information, Data Processing and Analysis*. Joint Research Centre European Commission, Ispra, Italy, 76 pp.

FAO, WFP. 2013. *Sécurité alimentaire et implications humanitaires en Afrique de l'ouest et au Sahel, N°51—Novembre/Décembre 2013*, 7 pp. http://reliefweb.int/sites/reliefweb.int/files/resources/Note%20conjointe%20FAO%20PAM%20%20N%2051%20Novembre%202013.pdf

FAO, Word Bank, UNSC. 2011. Global strategy to improve agricultural and rural statistics. *Report No.56719-GLB*. World Bank, Washington, DC, 40 pp.

Fensholt, R., Anyamba, A., Huber Gharib, S., Proud, S.R., Tucker, C., Small, J., Pak, E., Rasmussen, M.O., Sandholt, I., & Shisanya, C. 2011. Analysing the advantages of high temporal resolution geostationary MSG SEVIRI data compared to Polar operational environmental satellite data for land surface monitoring in Africa. *International Journal of Applied Earth Observation and Geoinformation*, 13, 5, 721–729.

Fensholt, R., Sandholt, I., Rasmussen, M.S., Stisen, S., & Diouf, A. 2006. Evaluation of satellite based primary production modelling in the semi-arid Sahel. *Remote Sensing of Environment*, 105, 173–188.

Fewsnet. 2009, November. Sahel and West Africa, food security update. *Technical Report*, 4 pp. http://www.fews.net/sites/default/files/documents/reports/West_Africa_FSU_2009_11_en.pdf

Fewsnet. 2010, February. Sahel and West Africa, food security update. *Technical Report*, 7 pp. http://www.fews.net/sites/default/files/documents/reports/West_FSU_2010_02_en.pdf

Fillol, E. 2011a. BioGenerator (v2.1), Guide de l'utilisateur. *Action Contre la Faim International*, 12 pp. http://www.accioncontraelhambre.org/publicaciones_biblioteca.php?sec=4#4, under Section "Manuales y Guias".

Fillol, E. 2011b. HydroGenerator (v1.1), Guide de l'utilisateur. *Action Contre la Faim International*, 15 pp. http://www.accioncontraelhambre.org/publicaciones_biblioteca.php?sec=4#4, under Section "Manuales y Guias".

Funk, C.C., & Budde, M.E. 2009. Phenologically-tuned MODIS NDVI-based production anomaly for Zimbabwe. *Remote Sensing of Environment*, 113, 115–125.

Garbulsky, M.F., Penuelas, J., Gamon, J., Inoue, Y., & Filella, I. 2011. The photochemical reflectance index (PRI) and the remote sensing of leaf, canopy, and ecosystem radiation use efficiencies. *Remote Sensing of Environment*, 115, 281–297.

Genovese, G., Vignolles, C., Nègre, T., & Passera, G. 2001. A methodology for a combined use of normalised difference vegetation index and CORINE land cover data for crop yield monitoring and forecasting. A case study on Spain. *Agronomie*, 21, 91–111.

Gibbs, W.J., & Maher, J.V. 1967. Rainfall deciles as drought indicators. *Bureau of Meteorology Bulletin, No. 48*. Commonwealth of Australia, Melbourne.

Gobron, N., Pinty, B., Taberner, M., Mélin, F., Verstraete, M., & Widlowski, J.-L. 2005. Monitoring the photosynthetic activity of vegetation from remote sensing data. *Advances in Space Research*, 38, 2196–2202.

Gobron, N., Pinty, B., & Verstraete, M.M. 1997. Theoretical limits to the estimation of the leaf area index on the basis of visible and nearinfrared remote sensing data. *IEEE Transaction on Geoscience and Remote Sensing*, 35, 1438–1445.

Grace, J., Nichol, C., Disney, M., Lewis, P., Quiafe, T., & Bowyer, P. 2007. Can we measure terrestrial photosynthesis from space, using spectral reflectance and fluorescence? *Global Change Biology*, 13, 1484–1497.

Grimes, D.I.F. 2003, January. Satellite-based rainfall monitoring for food security in Africa. In: Rijks, D., Rembold, F., Negre, T, Gommes, R., & Cherlet, M. (eds) *Crop and Rangeland Monitoring in Eastern Africa—for Early Warning and Food Security, Proceedings of the International Workshop on Crop and Rangeland Monitoring in East Africa*, European Commission, Nairobi.

Guo, Y., Han, L., Zhang, D., Sun, G., Fan, J., & Ren, X. 2023. The factors affecting the quality of the temperature vegetation dryness index (TVDI) and the spatial–temporal variations in drought from 2011 to 2020 in regions affected by climate change. *Sustainability*, 15, 14, 11350. https://doi.org/10.3390/su151411350

Haas, E.M., Bartholome, E., & Combal, B. 2009. Time series analysis of optical remote sensing data for the mapping of temporary surface water bodies in sub-Saharan western Africa. *Journal of Hydrology*, 370, 1–4, 52–63.

Haboudane, D., Miller, J.R., Tremblay, N., Zarco-Tejada, P.J., & Dextraze, L. 2002. Integrated narrow-band vegetation indices for prediction of crop chlorophyll content for application to precision agriculture. *Remote Sensing of Environment*, 81, 416–426.

Ham, F., Métais, T., Hoorelbeke, P., Fillol, E., Gómez, A., & Crahay, P. 2011. One horn of the cow, an innovative GIS-based surveillance and early warning system in pastoral areas of Sahel. In: *Risk Returns*, UNISDR, Geneva, Switzerland, pp. 127–131.

Hillbrunner, C., & Moloney, G. 2012. When early warning is not enough—Lessons learnt from the 2011 Somalia Famine. *Global Food Security*, 1, 1, 20–28.

Hoek, M.V., Zhou, J., Jia, L., Lu, J., & Zheng, C. 2020. A prototype web-based analysis platform for drought monitoring and early warning. *International Journal of Digital Earth*, 13, 7, 817–831. http://doi.org/10.1080/17538947.2019.1585978

Holben, B. 1986. Characteristics of maximum-value composite images from temporal AVHRR data. *International Journal of Remote Sensing*, 7, 11, 1417–1434.

Hutchinson, C.F.1991. Uses of satellite data for famine early warning in sub-Saharan Africa. *International Journal of Remote Sensing*, 12, 1405–1421.

Ickowicz, A., Ancey, V., Corniaux, C., Duteurtre, G., Poccard-Chappuis, R., Touré, I., Vall, E., & Wane, A. 2012. Crop–livestock production systems in the Sahel—increasing resilience for adaptation to climate change and preserving food security. In: *Proceedings of the Joint FAO/OECD Workshop Building Resilience for Adaptation to Climate Change in the Agriculture Sector*, Rome, 23–25 April, pp. 261–294. http://www.fao.org/docrep/017/i3084e/i3084e.pdf.

Intergovernmental Panel on Climate Change. 2013. *Climate Change 2013:The Physical Science Basis. Contribution of Working Group I to the Fifth Assessment Report of the Intergovernmental Panel on Climate Change*, Stocker, T.F., Qin, D., Plattner, G.-K., Tignor, M., Allen, S.K., Boschung, J., Nauels, A., Xia, Y., Bex, V., & Midgley, P.M. (eds). Cambridge University Press, Cambridge and New York, 1535 pp.

Jung, M., Verstraete, M., Gobron, N., Reichstein, M., Papale, D., Bondeau, A., Robustelli, M., & Pinty, B. 2008. Diagnostic assessment of European gross primary production. *Global Change Biology*, 14, 10, 2349–2364.

Kerdiles, H., et al. 2017. ASAP—Anomaly hot spots of agricultural production, a new global early warning system for food insecure countries. In *2017 6th International Conference on Agro-Geoinformatics*. Fairfax, VA, pp. 1–5. http://doi.org/10.1109/Agro-Geoinformatics.2017.8047072.

Kogan, F.N. 1995. Droughts of the late 1980s in the United States as derived from NOAA polar orbiting satellite data. *Bulletin of the American Meteorological Society*, 76, 655–668.

Kogan, F.N. 2001. Operational space technology for global vegetation assessment. *Bulletin of the American Meteorological Society*, 89, 1949–1964.

Lee, S.-J., Choi, C., Kim, J., Choi, M., Cho, J., & Lee, Y. 2023. Estimation of high-resolution soil moisture in Canadian croplands using deep neural network with Sentinel-1 and Sentinel-2 images. *Remote Sensing*, 15, 16, 4063. https://doi.org/10.3390/rs15164063

Lobell, D.B., Asner, G.P., Ortiz-Monasterio, J.I., & Benning, T.L. 2003. Remote sensing of regional crop production in the Yaqui Valley, Mexico: Estimates and uncertainties. *Agriculture, Ecosystems and Environment*, 94, 205–220.

Maxwell, D., & Fitzpatrick, M. 2012. The 2011 Somalia famine: Context, causes and complications. *Global Food Security*, 1, 1, 5–12.

Meroni, M., Fasbender, D., Kayitakire, F., Pini, G., Rembold, U., & Verstraete, M.M. 2014. Early detection of biomass production deficit hot-spots in semi-arid environment using FAPAR time series and a probabilistic approach. *Remote Sensing of Environment*, 142, 57–68.

Meroni, M., Fasbender, D., Rembold, F., Atzberger, C., & Klisch, A. 2019. Near real-time vegetation anomaly detection with MODIS NDVI: Timeliness vs. accuracy and effect of anomaly computation options. *Remote Sensing of Environment*, 221, 508–521. ISSN 0034–4257. https://doi.org/10.1016/j.rse.2018.11.041. https://www.sciencedirect.com/science/article/pii/S0034425718305509

Meroni, M., Rembold, F., Fasbender, D., & Vrieling, A. 2017. Evaluation of the Standardized Precipitation Index as an early predictor of seasonal vegetation production anomalies in the Sahel. *Remote Sensing Letters*, 8, 4, 301–310. http://doi.org/10.1080/2150704X.2016.1264020

Meroni, M., Rossini, M., Guanter, L., Alonso, L., Rascher, U., Colombo, R., & Moreno, J. 2009. Remote sensing of solar-induced chlorophyll fluorescence: Review of methods and applications. *Remote Sensing of Environment*, 113, 2037–2051.

Meroni, M., Verstraete, M., Marinho, M., Sghaier, N., & Leo, O. 2013. Remote sensing based yield estimation in a stochastic framework—Case study of Tunisia. *Remote Sensing*, 5, 2, 539–557.

Mishra, A.K., & Singh, V.P. 2010. A review of drought concepts. *Journal of Hydrology*, 391, 202–216.

Monteith, J.L. 1972. Solar radiation and productivity in tropical ecosystems. *Journal of Applied Ecology*, 9, 3, 747–766. https://www.jstor.org/stable/2401901. Accessed: 20-08-2024 20:34 UTC.

Mulla, D.J. 2021. Satellite remote sensing for precision agriculture. In: Kerry, R., & Escolà, A. (eds) *Sensing Approaches for Precision Agriculture. Progress in Precision Agriculture*. Springer, Cham. https://doi.org/10.1007/978-3-030-78431-7_2

Myneni, R.B., Hoffman, S., Knyazikhin, Y., Privette, J.L., Glassy, J., Tian, Y., Wang, Y., Song, X., Zhang, Y., Smith, G.R., Lotsch, A., Friedl, M., Morisette, J.T., Votava, P., Nemani, R.R., & Running, S.W. 2002. Global products of vegetation leaf area and fraction absorbed PAR from year one of MODIS data. *Remote Sensing of Environment*, 83, 214–231.

Nghiem, S.V., Wardlow, B.D., Allurer, D., Svoboda, M.D., LeComte, D., Rosencrans, M., Chan, S.K., & Neumann, G. 2012. Microwave remote sensing of soil moisture. In: Wralow, B.D., Anderson, M.C., & Verdin, J.P. (eds) *Remote Sensing of Drought: Innovative Monitoring Approaches*. CRC Press, Boca Raton, FL, USA, pp. 179–226.

NOAA. 2001. *The NOAA Climate Prediction Center African Rainfall Estimation Algorithm Version 2.0. Technical Report*, 4 pp. http://www.cpc.ncep.noaa.gov/products/fews/RFE2.0_tech.pdf

Oroud, I.M., & Balling, R.C. 2021. The utility of combining optical and thermal images in monitoring agricultural drought in semiarid mediterranean environments. *Journal of Arid Environments*, 189, 104499. ISSN 0140–1963. https://doi.org/10.1016/j.jaridenv.2021.104499. https://www.sciencedirect.com/science/article/pii/S0140196321000653

Palmer, W.C. 1965. *Meteorological Drought. Research Paper No. 45*. U.S. Department of Commerce Weather Bureau, Washington, DC.

Penuelas, J., Garbulsky, M., & Filella, I. 2011. Photochemical reflectance index (PRI) and remote sensing of plant CO2 uptake. *New Phytologist*, 191, 596–599.

Pinty, B., Lavergne, T., Widlowsky, J.L., Gobron, N., & Verstraete, M.M. 2009. On the need to observe vegetation canopies in the near-infrared to estimate visible light absorption. *Remote Sensing of Environment*, 113, 10–23.

Prince, S.D., & Goward, S.N. 1995. Global primary production: A remote sensing approach. *Journal of Biogeography*, 22, 815–835.

Rembold, F., Atzberger, C., Savin, I., & Rojas, O. 2013. Using low resolution satellite imagery for yield prediction and yield anomaly detection. *Remote Sensing*, 5, 4, 1704–1733.

Rembold, F., et al. 2017. ASAP—Anomaly hot Spots of Agricultural Production, a new early warning decision support system developed by the Joint Research Centre. In: *2017 9th International Workshop on the Analysis of Multitemporal Remote Sensing Images(MultiTemp)*. Brugge, Belgium, pp. 1–5. http://doi.org/10.1109/Multi-Temp.2017.8035205.

Rojas, O., Vrieling, A., & Rembold, F. 2011. Assessing drought probability for agricultural areas in Africa with coarse resolution remote sensing imagery. *Remote Sensing of Environment*, 115, 343–352.

Rossini, M., Fava, F., Cogliati, S., Meroni, M., Marchesi, A., Panigada, C., Giardino, C., Busetto, L., Migliavacca, M., Amaducci, S., & Colombo, R. 2013. Assessing canopy PRI from airborne imagery to map water stress in maize. *ISPRS Journal of Photogrammetry and Remote Sensing*, 86, 168–177.

Rouse, J.W., Haas, R.H., Schell, J.A., Deering, D.W., & Harlan, J.C. 1974. Monitoring the vernal advancement of retrogradation of natural vegetation. *Final Report, Type III*, NASA/GSFC, Greenbelt, MD, 371 pp.

Sepulcre, G., Horion, S., Singleton, A., Carrão, H., & Vogt, J.V. 2012. Development of a combined drought indicator to detect agricultural drought in Europe. *Natural Hazards and Earth System Sciences*, 12, 3519–3531.

Shafer, B.A., & Dezman, L.E. 1982. Development of a Surface Water Supply Index (SWSI) to assess the severity of drought conditions in snowpack runoff areas. In: *Proceedings of the Western Snow Conference*, pp. 164–175. http://www.westernsnowconference.org/sites/westernsnowconference.org/PDFs/1982Shafer.pdf

Sheffield, J., & Wood, E.F. 2011. *Drought: Past Problems and Future Scenarios*. Earthscan, London, p. 192.

Suárez, L., Zarco-Tejada, P.J., Gonzalez-Dugo, V., Berni, J.A.J., Sagardoy, R., Morales, F., & Fereres, E. 2010. Detecting water stress effects on fruit quality in orchards with time-series PRI airborne imagery. *Remote Sensing of Environment*, 114, 286–298.

Tadesse, T., Brown, J.F., & Hayes, M.J. 2005. A new approach for predicting drought-related vegetation stress: Integrating satellite, climate, and biophysical data over the U.S. central plains. *ISPRS Journal of Photogrammetry and Remote Sensing*, 59, 4, 244–253.

United Nations. 2009. *Global Assessment Report on Disaster Risk Reduction*. United Nations, Geneva, Switzerland. http://www.preventionweb.net/english/hyogo/gar/report/index.php?id=9413

Wilhite, D.A., Svoboda, M.D., & Hayes, M.J. 2007. Understanding the complex impacts of drought: A key to enhancing drought mitigation and preparedness. *Water Resources Management*, 21, 763–774.

WMO. 2012. Standardized precipitation index user guide. *Technical Document No. 1090*. Geneva, Switzerland. http://www.wamis.org/agm/pubs/SPI/WMO_1090_EN.pdf

Wu, J., Zhou, L., Liu, M., Leng, S., & Diao, C. 2013. Establishing and assessing the Integrated Surface Drought Index (ISDI) for agricultural drought monitoring in mid-eastern China. *International Journal of Applied Earth Observation and Geoinformation*, 23, 397–410.

Wu, Z., Lei, S., Bian, Z., et al. 2019. Study of the desertification index based on the albedo-MSAVI feature space for semi-arid steppe region. *Environmental Earth Sciences*, 78, 232. https://doi.org/10.1007/s12665-019-8111-9

Zarco-Tejada, P.J., Gonzalez-Dugo, V., & Berni, J.A.J. 2012. Fluorescence, temperature and narrow-band indices acquired from a UAV platform for water stress detection using a micro-hyperspectral imager and a thermal camera. *Remote Sensing of Environment*, 117, 322–337.

Zhong, L., Ostrenga, D., Teng, W., & Kempler, S. 2012. Tropical rainfall measuring mission (TRMM) precipitation data and services for research and applications. *Bulletin of the American Meteorological Society*, 93, 1317–1325.

Zhong, S., Di, L., Sun, Z., Xu, Z., & Guo, L. 2019. Investigating the long-term spatial and temporal characteristics of vegetative drought in the Contiguous United States. *IEEE Journal of Selected Topics in Applied Earth Observations and Remote Sensing*, 12, 3, 836–848. http://doi.org/10.1109/JSTARS.2019.2896159

Zscheischler, J., Mahecha, M.D., Harmeling, S., & Reichstein, M. 2013. Detection and attribution of large spatiotemporal extreme events in Earth observation data. *Ecological Informatics*, 15, 66–73.

3 Remote Sensing of Drought
Satellite-Based Monitoring Tools for the United States

*Brian D. Wardlow, Martha A. Anderson,
Tsegaye Tadesse, Mark S. Svoboda, Brian Fuchs,
Chris R. Hain, Wade T. Crow, and Matt Rodell*

ACRONYMS AND DEFINITIONS

ABL	Atmospheric boundary layer
ACIS	Applied climate information system
ALEXI	Atmosphere-land exchange inverse
AVHRR	Advanced very high resolution radiometer
BAI	Basal area increment
BLM	Bureau of Land Management
CMI	Crop Moisture Index
DOY	Day of year
EDDI	Evaporative Demand Drought Index
EROS	Earth resources observation science
ESA	European Space Agency
ESI	Evaporative Stress Index
FIA	Forest inventory and analysis
ForDRI	Forest Drought Index
GET-D	Geostationary evapotranspiration and drought
GLDAS	Global land data assimilation system
GRACE	Gravity recovery and climate experiment
GWS	Groundwater storage
HPRCC	High Plains Regional Climate Center
IDW	Inverse distance weight
LST	Land-surface temperature
LULC	Land use/land cover
MIR	Middle infrared
MODIS	Moderate resolution imaging spectroradiometer
NADM	North American Drought Monitor
NDDI	Normalized Difference Drought Index
NDMC	National Drought Mitigation Center
NDVI	Normalized Difference Vegetation Index
NDWI	Normalized Difference Water Index
NIR	Near-infrared
NLCD	National land cover dataset
NOAA	National Oceanic and Atmospheric Administration
NWS	National Weather Service

OS	Out of season
PASG	Percent annual seasonal greenness
PCA	Principal component analysis
PET	Potential evapotranspiration
RSAC	Remote Sensing Applications Center
RZSM	Root zone soil moisture
SAR	Synthetic aperture radar
SMAP	Soil moisture active passive
SMOS	Soil moisture and ocean salinity
SOSA	Start of season anomaly
SPEI	Standardized Precipitation Evapotranspiration Index
SPI	Standardized Precipitation Index
SVI	Standardized Vegetation Index
SWSI	Surface Water Supply Index
PSDI	Palmer Drought Severity Index
TCI	Temperature Condition Index
TRMM	Tropical rainfall measuring mission
TRW	Tree-ring widths
TSEB	Two-source energy balance
TWS	Terrestrial water storage
USDM	US drought monitor
USGS	US geological survey
VCI	Vegetation Condition Index
VegDRI	Vegetation Drought Response Index
VHI	Vegetation Health Index
VIIRS	Visible Infrared Imaging Radiometer Suite
VPD	Vapor pressure deficit

3.1 INTRODUCTION

Drought is a naturally recurring climatic feature that negatively impacts many sectors, including agriculture, water resources, energy, ecosystem services, and the economy. This climatic extreme is a common natural hazard in most regions of the world, often covering large geographic extents resulting in water scarcity, food insecurity, and major economic losses. In the United States, drought has been a frequent occurrence often impacting large areas of the country. Since the late 1890s, an average of ~20% of the country has been affected by drought at any given time, including key drought events that include the 1930s Dust Bowl and pronounced dry periods in the 1950s, late 1980s, and 2000s (Figure 3.1; NCDC, 2023). As a result, drought ranks as one of the costliest natural hazards in the United States, accounting for $7.7 billion in losses per year since 1980 (NOAA, 2023). Since 2000, 16-billion-dollar drought events have occurred in the United States (NOAA, 2023). The frequency and magnitude of drought is expected to increase under changing climatic conditions, as projected higher temperatures are expected to increase more than project precipitation increases leading to higher evapotranspiration and drier conditions (Collins et al., 2013). This will further exacerbate increasing pressures on finite water resources to support competing sectoral demands (e.g., agricultural, municipal, industrial, and ecological) and have significant environmental, social, and economic impacts (e.g., monetary losses, human health, and reduction in ecosystem services). Monitoring and early warning is a critical component of effective and proactive preparedness plans and actions by decision-makers to mitigate impacts and manage resources during drought events, and satellite remote sensing is increasingly being relied upon to provide timely information about drought conditions at regional, national, and global scales.

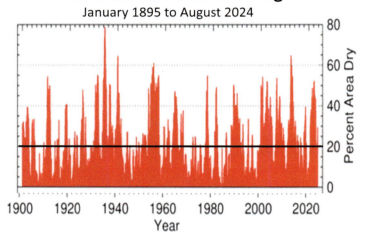

FIGURE 3.1 Percent area of the United States in moderate to severe drought as defined by the Palmer Drought Severity Index (PSDI) for 1895 to 2023 (modified from NCDC, 2023), with the horizontal line demarcating the 20.5% historical average area of the United States under this severity range of drought conditions.

3.1.1 Drought Definitions and Monitoring

Drought monitoring is complex and challenging because it lacks a single universal definition, which makes the identification and assessment of key drought characteristics such as duration, intensity (or severity), and geographic extent difficult (Mishra and Singh, 2010). In response, three physically based operational drought definitions have been established to characterize different types of drought (meteorological, agricultural, and hydrologic) (Wilhite and Glantz, 1985). The primary factor distinguishing these categories is the temporal length of dryness needed to initiate and recover from a drought, with the time period increasing in descending order of the three types of drought listed. As a result, a defined period of dryness may result in the occurrence of one type of drought (e.g., meteorological) but not others (e.g., hydrologic), whereas in the instance of longer, more severe periods, several types of drought may be occurring concurrently. Other defining factors of these three types of drought are the specific impacts and appropriate measures to monitor each drought type. For example, agricultural drought is commonly manifested through soil moisture depletion and observed plant stress on crops, whereas hydrologic drought may be best observed through declining stream flows and reservoir and groundwater levels.

3.1.2 Traditional, Station-Based Drought Monitoring Approaches

Traditional drought monitoring has relied upon *in situ*–based meteorological (i.e., precipitation and temperature) and hydrological (e.g., lake/reservoir level, stream flow, soil moisture, and groundwater elevation) measurements that represent discrete, point-based information for a specific geographic location. Typically, an extended historical record of these measurements are used to develop "anomaly" measures that represent deviations (i.e., post or negative and their magnitude) from some historical, average condition for drought detection. The *in situ*–based measurements are commonly spatially interpolated (e.g., kriging to estimate spatial patterns) or areally aggregated to

Remote Sensing of Drought   41

an administrative unit (e.g., county or crop reporting district) for decision-making purposes. Figure 3.2a shows an example of the modified Palmer Drought Severity Index (PDSI; Palmer, 1968) areally aggregated from weather station locations to the National Oceanic and Atmospheric Administration (NOAA) multi-county climate divisions, and Figure 3.2b shows a gridded, one-month Standardized

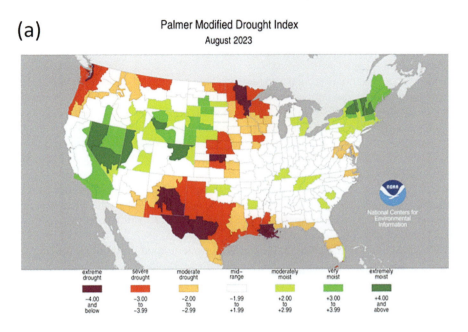

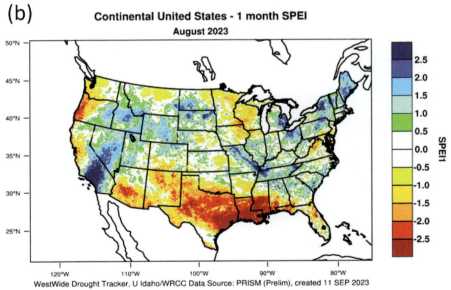

FIGURE 3.2 Examples from August 2023 of the modified Palmer Drought Severity Index (PDSI) map (NOAA, 2023) of station-based index values aggregated to the multi-county NOAA climate division (a) and a spatially interpolated, one-month Standardized Precipitation Evapotranspiration Index (SPEI) map (NOAA, 2023) from station-based precipitation observations (b).

Precipitation Evapotranspiration Index (SPEI; Vincente-Serrano et al., 2010) map that was spatially interpolated from station-based index values across the CONUS for August 2023.

The PDSI has been used since the 1960s to characterize agricultural drought and is calculated from precipitation, temperature, and soil available water content data using a supply-and-demand concept of a water balance equation. This index considers past conditions on the scale of nine to 12 months, and the intensity of drought for a specific time period and location is determined through estimates of evapotranspiration, soil moisture recharge, runoff, and moisture loss from the soil surface layer through the analysis of preceding precipitation and temperature observations in the water balance model. The primary limitation of the PDSI is the inherent lag in identifying the start of a drought because of the longer time scale of prior precipitation and temperature conditions incorporated into the model.

The SPEI is both a precipitation- and temperature-based index, which is calculated from a long-term record of precipitation and temperature observations (typically station-based, *in situ* measures) by fitting the data to a normal distribution and allowing the index to be normalized and calculated over various time scales (e.g., one, three, or six months) for a given location. The strengths of the SPEI are its flexibility to monitor both short- and long-term drought conditions and its ability to account for precipitation and potential evapotranspiration conditions in drought characterization. As a result, the SPEI is increasingly being used to monitor both agricultural and hydrologic drought conditions. The SPEI is an extension of another commonly calculated drought indicator from station data, the Standardized Precipitation Index (SPI, McKee et al., 1995), which can be calculated at various times scales using a similar approach to SPEI but only accounts for precipitation and does not include a temperature component.

Collectively, these indices have been widely used for operational drought monitoring with these types of spatially interpolated and aggregated map products. Products based on *in situ* observations are highly accurate for a specific location and local areas, but often do not adequately characterize spatial variations in drought conditions across large areas, particularly in regions with sparse observation networks such as parts of the western United States (e.g., Great Basin). They also do not reflect localized, spatial variations in drought conditions that may occur between in situ measurement locations. In addition to the PSDI, SPEI, and SPI; several other indices such as the Crop Moisture Index (CMI; Palmer, 1968), Keetch-Byram Drought Index (Keetch and Byram, 1968), and Surface Water Supply Index (SWSI, Shafer and Dezman, 1982) have been developed from station-based observations to monitor specific types of drought.

3.1.3 US Drought Monitor

The development of the US Drought Monitor (USDM) in 1999 marked a major advance in drought monitoring capabilities for the United States. The USDM represents a composite index tool that incorporates several commonly used drought indices such as PDSI, SPI, SPEI, and SWSI, along with other *in situ* measurements (e.g., stream flow and soil moisture) and remote sensing and modeled products, as well as guidance from a group of experts (e.g., climatologists, water and natural resource managers, and agriculturalists) to depict both short- and long-term drought conditions (defined by the "S" and "L" letter designations on the map, respectively) across the CONUS (Svoboda et al., 2002). Figure 3.3 presents the USDM map for August 22, 2023, when much of the central and south-central United States was experiencing moderate to exceptional drought conditions.

Until the early 2000s, the use of satellite remote sensing in the development of the USDM had been very limited, with the Vegetation Health Index (VHI; Kogan, 1995) being the only remote sensing–based input. The VHI incorporates both the normalized difference vegetation index (NDVI)–based Vegetation Condition Index (VCI) and thermal-based Temperature Condition Index (TCI) to produce an indicator suitable for agricultural drought monitoring. The VHI concept built upon well-established value of satellite-based NDVI data from the NOAA Advanced Very High Resolution Radiometer (AVHRR) for vegetation condition assessments, which was first

Remote Sensing of Drought 43

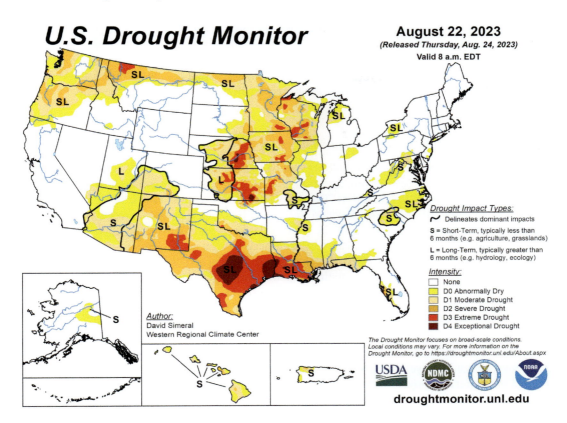

FIGURE 3.3 US Drought Monitor (USDM) map drought conditions over the CONUS for August 22, 2023 (*Note:* White areas represent areas that are not experiencing drought conditions on that date).

demonstrated for agricultural drought applications in sub-Saharan Africa in the 1980s by Tucker et al. (1986). The VHI also incorporated a thermal component through the TCI to represent changes in land surface temperature conditions related to plant stress-induced changes in ET rates and subsequent sensible heat fluxes. Although valuable, the VHI only characterizes agricultural drought conditions and has been shown to have limited utility in energy-limited environments, such as high elevation or high latitude locations (Karnieli et al., 2010).

The demand for more timely and spatially detailed drought-related data sets beyond the traditional NDVI and VHI measures that characterize different environmental conditions related to drought by the USDM and other drought monitoring activities in the United States (e.g., state disaster management planning and federal relief programs) and internationally (e.g., Famine Early Warning System [FEWS]) has led the development of several operational remote sensing–based tools since the early 2000s. The USDM represents a prime example of this need as several major decision-making activities (e.g., US Department of Agriculture [USDA] Farm Service Agency [FSA]) use the USDM as trigger for disaster assistance at a county-level scale. As a result, the USDM and other drought monitoring tools for the United States in general are increasingly being looked upon to provide drought information at a county to sub-county spatial scale where many administrative decisions are made. The majority of traditional drought monitoring tools, which are primarily based on *in situ* observational data, lack the necessary spatial resolution and scale to meet this demand. This is illustrated in the USDM map in Figure 3.3, where large groupings of counties within many states were assigned to a single drought category when there were likely local-scale variations in the severity of those conditions that could not be captured with the spatial resolution of most data inputs that were available for the USDM. Satellite-based remote sensing has increasingly been looked upon to provide more spatially detailed, drought-related information at local scales not only for the USDM but also for other drought-monitoring activities in the United States and internationally. Remote sensing provides higher spatial resolution observations across large areas that can fill in data gaps and complement traditional *in situ* measurements and existing drought assessment tools.

3.1.4 Traditional Remote Sensing of Drought

Historically, the use of remote sensing for operational drought monitoring has involved the application of historical NDVI-based products from the NOAA AVHRR dating back to the 1980s, as presented earlier with the VCI and VHI. This NDVI data record continues to be extended by observations of the more recent NASA Moderate Resolution Imaging Spectroradiometer (MODIS) and Visible Infrared Imaging Radiometer Suite (VIIRS) instruments. The lineage of AVHRR NDVI-based measures for this application can be traced back more than 30 years to the early work of Tucker et al. (1986), Hutchinson (1991), Kogan (1990), Burgan et al. (1996), and Unganai and Kogan (1998). The NDVI is a simple mathematical transformation developed in the early 1970s (Rouse et al., 1974) that incorporates spectral data from both the visible red and near-infrared (NIR) regions, which are sensitive to changes in plant chlorophyll content and inter-cellular spaces of the spongy mesophyll layers of the plants' leaves, respectively. NDVI could be readily calculated from early satellite-based Earth observing sensors like AVHRR, which had the requisite visible red and NIR bands for its calculation. Considerable research has shown that NDVI has a strong relationship with several biophysical vegetation characteristics (e.g., green leaf area and biomass) (Asrar et al., 1989; Baret and Guyot, 1991) and temporal changes in index values are highly correlated with inter-annual climate variations (Peters et al., 1991; Yang et al., 1998; McVicar and Bierwith, 2001; Ji and Peters, 2003). The value of using the AVHRR NDVI (and similar vegetation indices [VIs]) data for drought monitoring is the ability to calculate anomaly measures representing a departure from longer-term, historical average conditions from the extended AVHRR NDVI data time series dating back to the 1980s. As a result, AVHRR NDVI-based tools have been routinely used in several drought-related monitoring efforts in the United States (highlighted earlier), as well as internationally as part of systems like FEWS.

Building on the NDVI, researchers have developed other spectral-based indices over the past several decades that extend the initial NDVI concept with new types of mathematical transformations and/or data inputs from other spectral regions. The VCI discussed earlier in one example and was a precursor to the VHI now routinely used for global drought assessments. The VCI is based on the assumption that for a given location, the historical maximum and minimum NDVI values represent the upper and lower bounds of possible vegetation conditions, with abnormally low NDVI values on given data being indicative of drought-stressed vegetation (Kogan and Sullivan, 1993). A companion thermal-based index called the TCI was developed by Kogan (1995) based on data from the AVHRR thermal bands using the historical maximum (unfavorable vegetation conditions) and minimum (favorable conditions) to establish similar upper and lower bounds for vegetation conditions with drought being manifested as unusually high thermal values near the maximum. Kogan (1995) unified the VCI and TCI into a single index through the creation of the VHI. Other efforts such as the Standardized Vegetation Index (SVI; Peters et al., 2002), which describes the probability of NDVI variation from a normal (or average) NDVI value for a weekly time period of the extended AVHRR NDVI time series, with much lower than normal NDVI values indicative of drought. Others further refined the integration of thermal and NDVI data into indices such as the temperature/NDVI ratio (McVicar and Bierwirth, 2001) and two-dimensional geometric expressions (Karnieli and Dall'Olmo, 2003). As time-series, middle infrared (MIR) spectral data became available from NASA MODIS, other efforts have incorporated data from the MIR region, which is sensitive to plant water content, with other MODIS spectral bands to develop new indices representative of changes in moisture status of vegetation (and thus agricultural drought conditions). The Normalized Difference Water Index (NDWI), which integrated MIR and NIR data in a similar fashion to NDVI, was tested for drought monitoring by Gu et al. (2007). The NDWI was extended further by Gu et al. (2008) into the Normalized Difference Drought Index (NDDI), which incorporates NDVI and NDWI in order to leverage the relative strengths of both indices. Wang et al. (2007) also built upon the original NDWI concept by developing a three-band index called the Normalized Multi-band Drought Index (NMDI) that uses the NDVI in combination with the MIR bands available from MODIS, which are designed to be sensitive to soil and plant water content, respectively.

Collectively, these more traditional remote sensing efforts, utilizing primarily remotely sensed observations in the visible, NIR, MIR, and thermal infrared (TIR) regions, have attempted to develop improved and more effective drought assessment tools over the years. Some of these remote sensing–based tools (such as the NDVI and VHI) have been adopted for operational use in key monitoring systems highlighted earlier, while others were research efforts to test new approaches that have yet to be operationalized. This work has provided a valuable foundation for advancing the use of satellite-based tools for drought monitoring but has primarily focused on agricultural drought and the vegetation component of the landscape.

3.1.5 NEW ERA OF REMOTE SENSING OBSERVATIONS

Since the early 2000s, the launch of several new satellite-based earth observing systems coupled with advances in environmental models and algorithms as well as computing capabilities, has resulted in the rapid emergence of many new tools that monitor different aspects of the hydrologic cycle that influence drought conditions. Many of these sensors now have more than 15 years of observational data providing an extended historical baseline of data for characterizing drought conditions. Thermal infrared (TIR) satellite image data has provided the basis for developing evapotranspiration (ET)–related drought products, while satellite-based microwave and gravity-field observations from sensors such as the NASA Soil Moisture Active Passive (SMAP) and Gravity Recovery And Climate Experiment (GRACE) have provided new insights into sub-surface soil moisture and ground conditions, respectively (Mladenova et al., 2019; Rodell, 2012). Other recent efforts have developed combined drought indicators that integrate remote sensing data with other types of environmental datasets to monitor drought-specific stress conditions on vegetation (Brown et al., 2008; Tadesse et al., 2020).

This chapter will highlight several satellite-based remote sensing tools that have been developed to support and enhance operational drought monitoring in the United States. The focus is on tools that characterize terrestrial components of the hydrologic cycle related to drought, including vegetation health, ET, soil moisture, and groundwater. For each tool, the objectives and methods will be summarized, informational drought products highlighted, and other current/future work efforts related to them discussed. Ongoing research efforts and recent satellite missions that hold considerable potential to further advance drought monitoring will also be highlighted.

3.2 NEW DROUGHT MONITORING TOOLS

3.2.1 VEGETATION DROUGHT RESPONSE INDEX (VEGDRI)

The Vegetation Drought Response Index (VegDRI) is a hybrid VI designed to detect drought-related vegetation stress through integration of satellite, climate, and biophysical data (Figure 3.4; Brown et al., 2008). For the satellite input, VegDRI uses an NDVI-based measure to detect vegetation stress. Although NDVI has proven useful for this application, analysis of only NDVI information for drought monitoring is problematic because many types of environmental events (e.g., fire, flooding, pest infestation, plant disease, and land use/land cover change) can produce a decline in NDVI values that mimics drought stress. As a result, VegDRI incorporates climatic information to distinguish drought-impacted areas experiencing abnormally dry conditions from other areas impacted by these non-drought events. A biophysical component is added to VegDRI to include other environmental factors that can influence climate-vegetation interactions such as land use/land cover (LULC) type, soil characteristics, elevation, and rainfed versus irrigated systems. An empirical-based modeling approach is used to analyze the historical behavior of these data inputs over a 20-year period (1989–2008) to produce the VegDRI models. VegDRI (http://vegdri.unl.edu) has been operationally produced over the CONUS at a nominal 1-km spatial resolution since 2009 (with a complete history dating back to 1989) in a collaborative effort between the US Geological Survey (USGS) Earth Resources Observation Science (EROS) Center and the National Drought Mitigation Center (NDMC) at the University of Nebraska-Lincoln.

3.2.1.1 Input Data
3.2.1.1.1 Satellite Variables

Time-series AVHRR NDVI data were used as the data source to derive three satellite-based variables for VegDRI. The first NDVI-derived variable is Percent Annual Seasonal Greenness (PASG), which relates how vegetation conditions for a specific period during the growing season compare to the historical average conditions for the same period in the 20-year AVHRR NDVI record. A moving-window averaging technique (Reed et al., 1994) is used to determine the pixel-level start and end of season (SOS and EOS) day of year (DOY), and the NDVI date is summarized within these beginning and end dates to calculate the PASG, which is updated on a bi-weekly time step in concert with the temporal compositing period of the AVHRR NDVI data set (Eidenshink, 2006).

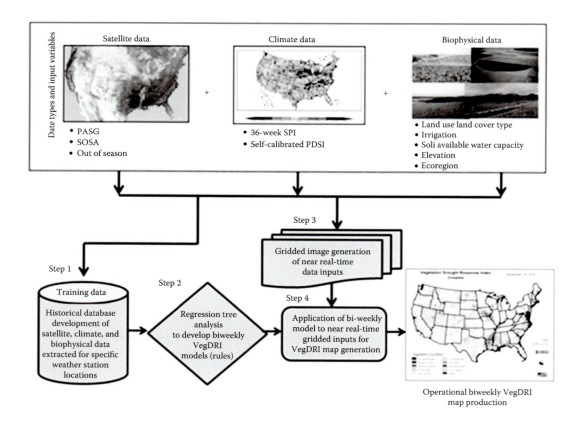

FIGURE 3.4 Overview of the VegDRI methodology (adapted from Wardlow et al. [2012]).

A second NDVI-based measure, the start of season anomaly (SOSA), is included in VegDRI to represent the departure in the pixel-level SOS for a specific year from the historical SOS date. SOSA was included to discriminate between areas that have a normal SOS with a low PASG because of inter-annual climate variations (e.g., late freeze from cold temperatures or drought) versus areas that have experienced non-climatic factors (e.g., LULC change such as a crop rotation), resulting in a comparably low PASG. A third NDVI-derived metric, out of season (OS), is included to represent the non-growing season period of the year when the vegetation is dormant and agricultural drought monitoring is not necessary. OS is calculated for each pixel using the SOS and EOS DOYs and is defined as the period for the EOS (e.g., December 1) to the SOS (e.g., March 1). The OS measure is used exclude VegDRI calculations during this time to avoid anomalous agricultural drought from being detected during periods when there is no actively growing vegetation and the NDVI signal is driven by non-vegetation factors (e.g., soils).

3.2.1.1.2 Climate Variables

VegDRI includes two commonly used climate indices for drought monitoring: the self-calibrated PDSI and the SPI. Historical data for both indices were taken from 2,417 weather station locations across the CONUS from the Applied Climate Information System (ACIS; http://rcc-acis.unl.edu/) with sufficiently long and relatively complete data records (i.e., 30-year length and less than 20% missing observations).

The self-calibrated PDSI (SC-PDSI) is a modified version of the original PDSI calculation developed by Palmer (1965), which was based on a simple supply-and-demand water balance model that considers precipitation, temperature, and the available water holding capacity of the soil at a given location. The SC-PDSI developed by Wells et al. (2004) calibrates the constants and duration factors in the original PDSI calculation to the local soil characteristics of each station. The PDSI is a well-established index for agricultural drought assessment in the United States, and a modified version of its drought severity classification scheme was adopted for VegDRI. The VegDRI classification scheme includes four drought classes (extreme, severe, moderate, and pre-drought), one normal class, and three classes of better-than-normal conditions (unusually, very, and extremely moist) as defined by the PDSI. An out-of-season class derived from the OS metric calculated from the AVHRR NDVI data is also included to mask areas where there is no actively growing vegetation and agricultural drought monitoring is not required at that time.

The SPI is designed to quantify precipitation anomalies over varying time periods (e.g., one-, three-, or six-month periods) based on fitting a long-term precipitation record for a station location over a specific time interval to a probability function, which is then transformed into a gamma distribution to standardize the mean SPI for all locations and time periods (McKee et al., 1985). A 36-week (or nine-month) SPI was selected to represent seasonal precipitation conditions in the VegDRI models.

3.2.1.1.3 Biophysical Variables

The dominant LULC type for each 1-km pixel across the CONUS is considered in VegDRI to account for the varying interactions of climate and vegetation response by different cover types. The LULC information is derived from the national-level 30-m National Land Cover Dataset (NLCD; Homer et al., 2004), which was aggregated to 1 km using a majority zonal function calculation. Further LULC information related to the irrigation status of each pixel is included through the incorporation of a national-level irrigated/rainfed agricultural lands data set called the MODIS Irrigated Agriculture Dataset for the United States (MIrAD-US; Pervez and Brown, 2010). The irrigation variable was included to represent the different effects that drought can have on crops and other land cover under irrigated and rainfed conditions. The soil available water holding capacity (SAWC) variable was derived from the State Soil Geographic (STATSGO) database (USDA, 1994) and incorporated into VegDRI to represent the potential of the soil to hold moisture that is available to plants, which is an indicator the susceptibility of vegetation to drought stress. An ecoregion variable derived from the Omernik Level II ecoregion data (Omernik, 1987) was added to the VegDRI model to represent a geographic framework across the CONUS that accounts for the considerable environmental variability encountered across the nation by grouping areas into regional ecosystems where vegetation has adapted to the local abiotic and biotic conditions and resources. Elevation is the final biophysical variable included in VegDRI to account for the elevational influence on vegetation and the different sensitivity to drought a vegetation type may experience across a vertical gradient. It should be noted that all biophysical variables were "static" across the historical record and do not reflect changes such as LULC conversion or crop rotations because CONUS-scale data sets with sufficient temporal resolution are not available.

3.2.1.2 Model Development and Implementation

The VegDRI methodology is presented in Figure 3.4. The initial step is the development of a training database of historical data for each input variable, which were extracted from 2,417 weather station locations from across the CONUS. Climate data were in a tabular format and could be incorporated directly into the database. For the gridded satellite and biophysical data sets, a zonal calculation within a 3-by-3 pixel window (i.e., 9-km^2 area) centered on each station location was used to populate the database for each of these variables. The zonal average was used for continuous variables (e.g., SAWC) and the zonal majority for thematic variables (e.g., LULC type). The final station locations used in model development were sited in non-urban areas and away from large water bodies to minimize the potential influence of urban areas (e.g., built structures) and water within the 1-km pixel in the NDVI data, which both could result in NDVI signals that are not representative of rainfed vegetation. The NLCD data set was used as the LULC reference and the majority LULC type within the 3-by-3 pixel window surrounding each station was analyzed to screen urban or water locations. All extracted historical data were incorporated into a SQL Server database to prepare the input data files for model development. The data were organized into 52 weekly periods spanning the calendar year, and for each period, the complete set of historical data for all stations was extracted for analysis. For the satellite- and climate-related variables, 14 years of observations (2004–2013; empirical-based VegDRI models were based on MODIS NDVI data with a plan to update the models in the near future with MODIS and VIIRS-based NDVI data to expand the historical record of data used for model development) for each specific bi-weekly period were included in this training data set along with the biophysical variables, which remain static across the historical record.

A commercial regression tree analysis algorithm called Cubist (Quinlan, 1993) was used to analyze the historical data for each specific bi-weekly period and generate a rule-based, piecewise linear regression VegDRI model for that period. The self-calibrated PDSI was the dependent variable for these empirical models, providing a well-established classification scheme of agricultural drought severity recognized within the drought community. The model development stage yielded 26 period-specific VegDRI models. Each Cubist-derived model consists of an unordered set of rules with each rule having the syntax "if x conditions are satisfied then use the associated linear multiple regression equation" to calculate the VegDRI value. The number of rule sets can vary by period and often more than one rule set may apply, resulting in multiple linear regression equations being applied and the average of those products used as the final VegDRI value.

For the mapping phase, period-specific VegDRI models are applied to the corresponding gridded data inputs using MapCubist software developed by the USGS EROS. An inverse distance weighted interpolation method is used to convert the point-based SPI data into a 1-km grid over the CONUS to match the other gridded satellite and biophysical data sets. During model implementation to the gridded data using MapCubist, the values of all input variables associated with a pixel location are considered to determine the specific rule set(s) and associated linear regression equation(s) to apply to calculate a VegDRI value. This process is repeated for all pixels across the CONUS, producing a 1-km resolution VegDRI map.

These VegDRI models were moved into operational production over the CONUS in 2009, producing national-level VegDRI maps of agricultural drought conditions in near real-time (i.e., less than 24 hours from the last satellite and climate data observation) that are updated on a weekly time step consistent with the expedited MODIS (eMODIS) (Jenkerson et al., 2010) and the more recent EROS Visible Infrared Imaging Radiometer Suite (eVIIRS) NDVI data production schedule. The operational VegDRI production system resides at USGS EROS, where the appropriate VegDRI model is applied to up-to-date eVIIRS NDVI-derived data inputs generated at USGS EROS, current SPI data acquired from the NOAA ACIS via the High Plains Regional Climate Center (HPRCC), and static biophysical information. Note that current VegDRI map production uses the VIIRS-based NDVI input because it was found to be comparable to MODIS-based NDVI data (Benedict et al., 2021), on which the empirical VegDRI models are based, and will be the long-term operational NDVI data input with the retirement of the MODIS sensors. A variety of value-added products (e.g., maps, tables, animations, and summaries) are staged for the public on the VegDRI webpage (http://vegdri.unl.edu/), as well as within a dynamic USGS VegDRI Viewer (http://vegdri.cr.usgs.gov/viewer/viewer.htm) that enables visualization of VegDRI maps in combination with other spatial data (e.g., USDM, geopolitical boundaries, and precipitation anomalies).

3.2.1.3 Examples of VegDRI

An example of a VegDRI map over the CONUS is presented in Figure 3.5 for August 11, 2011, when many parts of the southern United States were experiencing severe to extreme drought conditions. The most extreme drought conditions were in the south-central United States, centered on western Texas where many locations had the driest (or one of the driest) years in the instrumental data record for precipitation, which often spanned more than 100 years. The inset VegDRI map of the south-central US states shows that extreme drought was being detected by this mid-July period and conditions continued to worsen and expand over a larger geographic area in subsequent VegDRI maps as the growing season progressed, which was consistent with record crop and forage losses across this area. The severity and intensification of drought conditions over this area were further exacerbated by a prolonged period of days with air temperature exceeding 100 degrees Fahrenheit (F), with many locations within the core area of extreme drought characterized by VegDRI experiencing 70+ days of triple-digit temperatures (NCDC, 2014).

A comparison of the general drought patterns depicted in VegDRI and the USDM (Figure 3.5) reveals that similar patterns and severity levels of drought were detected by both tools over the CONUS. The severe to extreme droughts that were widespread across much of New Mexico, Oklahoma, and Texas are represented, as well as the extension of these drought conditions along the southern edge of the Gulf Coast region into the Atlantic Coast region. However, the higher spatial resolution of VegDRI compared to the USDM is quite apparent as subtle within-state variations in drought conditions are captured in VegDRI over areas where the USDM has assigned the same area to a single drought severity designation. A closer view of the VegDRI output for the southern Great Plains in Figure 3.5 illustrates the subtle spatial variations in drought conditions detected at a localized county to sub-country scale.

3.2.1.4 Applications of VegDRI

Information from VegDRI is increasingly being used for a range of drought monitoring applications and activities. The authors of the USDM routinely consult VegDRI in combination with other data sources to develop the USDM map. VegDRI is used to provide more local-scale information on spatial variations in agricultural drought conditions to provide more accurate spatial depictions of drought within the USDM. The Bureau of Land Management (BLM) has also adopted VegDRI as a response trigger and field visit priority guide within the Bureau's drought management plans (DOI, 2013). The National Weather Service (NWS) field offices are incorporating VegDRI information into drought impact statements. Several states such as Arizona, Montana, New Mexico,

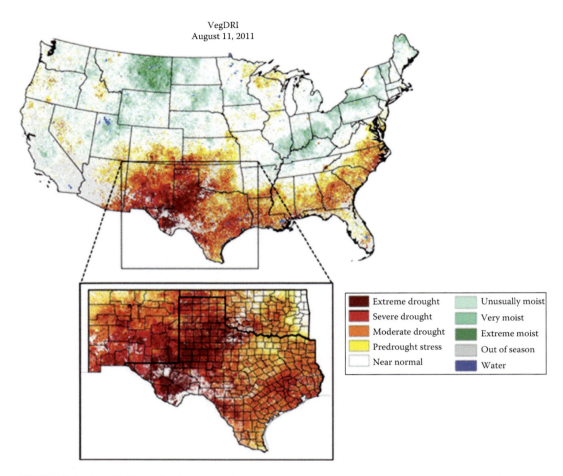

FIGURE 3.5 VegDRI map for August 11, 2011, over the CONUS (top) and the southern Great Plains region comprising New Mexico, Oklahoma, and Texas.

and Kansas have added VegDRI as a tool in their respective statewide drought monitoring systems and/or assessment reports. Currently, several countries (including Canada, China, Czech Republic, India, and Mexico) are attempting to develop modified versions of the VegDRI methodology for drought monitoring systems that are customized to their available data resources and specific information needs. The VegDRI approach provides a flexible framework within which various types of remote sensing, climatic, and biophysical data can be integrated to develop an agricultural drought-based indicator. For example, Canada is working to develop a VegDRI approach identical to the US version so the data sets produced in both countries can be spatially merged seamlessly to the multinational North American Drought Monitor (NADM; http://www.ncdc.noaa.gov/temp-and-precip/drought/nadm/). In comparison, a country such as India may have a different set of climate, satellite, and/or biophysical data inputs that can be integrated in a similar fashion to develop a customized agricultural drought monitoring tool for that specific area.

3.2.2 The Forest Drought Index (ForDRI)

Monitoring drought impacts in forest ecosystems is a complex process because forest ecosystems are composed of different species with heterogeneous structural compositions. Even though forest drought status is a key control on the carbon cycle, very few indices exist to monitor and predict forest drought stress. Most native trees in the United States have potential lifespans of at least several hundred years. Because they are long lived, trees have evolved to withstand even infrequent (e.g., 1-percentile) drought events. Adaptations include deep roots and woody water-conducting vessels resistant to drought-induced cavitation. In many cases these adaptations allow trees to continue to grow under conditions that would be detrimental to crops or pasture. As drought develops, however, trees increasingly limit water loss. Initially, trees employ short-term regulation of water loss via stomatal closure, and then longer-term changes such as increasing allocation of growth to roots. As drought intensity increases, trees may respond with the more extreme measure of leaf shedding. Exceptional drought can lead to cavitation in the tree water column and rapid death. The goal of a forest stress indicator is to identify when trees are experiencing sufficient stress to start limiting water loss because such actions in turn limit tree growth and increase susceptibility to a variety of other biotic and abiotic factors.

The Forest Drought Index (ForDRI) is a new monitoring tool developed by the NDMC to identify forest drought stress for the CONUS (Tadesse et al., 2020). The ForDRI tool is designed to improve drought monitoring efforts, specifically addressing the gap for better monitoring of the drought impacts on forests that may not be captured with current drought monitoring tools.

3.2.2.1 Input Data

ForDRI integrates 12 types of hydrometeorological variables and the satellite-based vegetation index data, including climate, evaporative demand, groundwater, and soil moisture, into a single hybrid index to estimate tree stress. Figure 3.6 shows the conceptual method and steps to develop

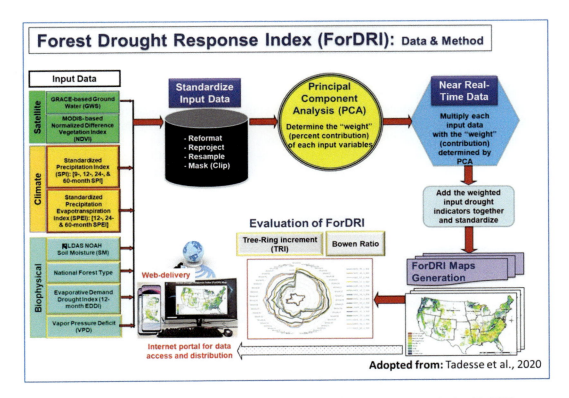

FIGURE 3.6 Conceptual method and steps to develop the Forest Drought Response Index (ForDRI).

the ForDRI. The concept of a ForDRI is to combine important atmospheric and hydrological variables together (i.e., using a multi-indicator or "convergence of evidence" approach) to produce a better drought index for forests than any of each individual drought indices used alone. The ForDRI model uses principal component analysis (PCA) to determine the contribution of each input variable based on its covariance in the historical records from 2003 to 2020. The input variables are described in additional detail next.

3.2.2.1.1 MODIS-Based NDVI

The NDVI information at 250-m spatial resolution is based on NASA MODIS data. The MODIS-based seven-day data from 2003 to 2020 were acquired from USGS (USGS, 2023) and resampled to a 1-km grid, and each dataset was standardized (Z-score) to be consistent with the other input variables. The MODIS data was available until October 2022. Thus, the operational ForDRI product continued using the comparable eVIIRS NDVI. For the ForDRI model, the Z-score of the NDVI is calculated using the formula weekly-observed value minus weekly-mean value divided by the standard deviation.

3.2.2.1.2 SPI

The SPI was calculated to quantify the precipitation anomaly for three specified time scales (previous 9, 12, 24, and 60 months) based on the long-term precipitation record over that specific time interval (Ruffault et al., 2018; Edwards and McKee, 1997). Since the SPI values are calculated by fitting the long-term record of precipitation over a specific time step to a probability distribution to standardize the values, we have used these three SPI values to represent different time scales of the rainfall conditions that would affect forest health. The four SPIs are selected to represent the long-term precipitation impact (from one year to five years) on tree stress. The rainfall data used to generate the time series of SPI were obtained from Applied Climate Information System (ACIS) meteorological stations data across the study region. We used the available daily long-term record of each station to generate SPI at 9-, 12-, 24-, and 60-month aggregate periods and interpolated using the inverse distance weight (IDW) method to produce 1-km resolution SPI maps.

3.2.2.1.3 SPEI

Unlike the SPI, which depends only on rainfall, the SPEI is designed to consider both precipitation and temperature. The time series of the SPEI were generated based on daily rainfall and temperature data acquired from ACIS meteorological stations data. The SPEI were generated at 24- and 60-month aggregate periods and interpolated (using the IDW method) to 12.5-km spatial resolution. With the temperature input, potential evapotranspiration (PET) is calculated and a historical time series of the simple water balance (precipitation—PET) is used in determining drought. Thus, the SPEI captures the main impact of increased temperatures on water demand (Vicente-Serrano et al., 2010). Two specified time periods of SPEI historical records (i.e., previous 24 and 60 months) that represent the temperature impact on water demand (rainfall) were used in building the ForDRI model to monitor forest drought response.

3.2.2.1.4 Evaporative Demand Drought Index (EDDI)

The EDDI indicates the anomalous condition of the atmospheric evaporative demand (also known as "the thirst of the atmosphere") for a given location and across a period of interest (Hobbins et al., 2016; McEvoy et al., 2016). The EDDI is expressed as (Eo) anomalies. The Eo is calculated using the Penman–Monteith FAO56 reference evapotranspiration formulation driven by temperature, humidity, wind speed, and incoming solar radiation from the North American Land Data Assimilation System datasets (NLDAS-2). EDDI is multi-scalar (i.e., captures drying dynamics that themselves operate at different timescales). We combined 12-month aggregated EDDI values with the other variables to monitor evaporative demand during forest drought.

3.2.2.1.5 Groundwater Storage (GWS)

GWS anomalies are calculated from Gravity Recovery and Climate Experiment (GRACE) observations (Bhanja et al., 2020; Li et al., 2019). Data from the Global Land Data Assimilation System (GLDAS), including terrestrial water storage (TWS), root zone soil moisture (RZSM) at 1-m depth, and snow water equivalence (SWE), were used to convert GRACE observations into a series of GWS anomalies (i.e., GWS = TWS—RZSM—SWE). NASA provided the data (2003–2017) at 12.5-km resolution for the United States. The groundwater product at 1-m depth represents deeper soil conditions that can be accessed by longer rooted tree species. The global GRACE data (2003–2020) is also available online by NASA GSFC Hydrological Sciences Laboratory at the NASA GESDISC data archive (NASA, 2023).

3.2.2.1.6 Noah Soil Moisture

The Noah soil moisture (SM) dataset used in this study is produced using a land surface model that forms a component of the GLDAS (Nearing et al., 2016; Xia et al., 2019; Kumar et al., 2014). The Noah soil moisture represents shallow soil depth conditions that can be accessed by short rooted species. Compared to other NLDAS-2 soil moisture products (e.g., VIC), Noah soil moisture shows the best performance in simulating shallow depth soil moisture (Cai et al., 2014). The Noah model uses a four-layered soil description with a 10-cm thick top layer and considers the fractions of sand and clay. Soil moisture dynamics of the top layer are governed by infiltration, surface and sub-surface runoff, gradient diffusion, gravity, and evapotranspiration (Liu et al., 2011). The model was forced by combination of NOAA/GLDAS atmospheric analysis fields, spatially and temporally disaggregated NOAA Climate Prediction Center Merged Analysis of Precipitation (CMAP) fields, and observation-based downward shortwave and longwave radiation fields derived using a method of the Air Force Weather Agency's Agricultural Meteorological system (NASA, 2023). The historical data (available since 2000) has a 25-km resolution (resampled to 1 km for combining with other model inputs). This dataset is also available as NOAA's NLDAS Drought Monitor Soil Moisture (NOAA, 2023).

3.2.2.1.7 Vapor Pressure Deficit

The vapor pressure deficit (VPD) represents the amount of water vapor deficit between the actual water vapor pressure in the air and vapor pressure when the air is saturated at a given temperature (Yuan et al., 2019). The VPD is one of the critical variables that controls control photosynthesis and water use efficiency of plants. The photosynthetic rates in leaves and canopies is inversely proportional to the atmospheric VPD (Fletcher et al., 2007). Thus, it is important for forest ecosystem structure and function (Li et al., 2017). Average daily VPD data using the PRISM model at 4-km resolution were retrieved from the PRISM Climate Group, Oregon State University (Daly et al., 2008; Daly et al., 2015; PRISM Climate Group, 2023).

3.2.2.1.8 National Forest Groups and Types

The US national forest types and forest groups geospatial dataset (1-km spatial resolution) used in this study was created by the USFS Forest Inventory and Analysis (FIA) program and the Remote Sensing Applications Center (RSAC) to show the extent, distribution, and forest type composition of the nation's forests. The dataset was created by modeling forest type from FIA plot data as a function of more than 100 geospatially continuous predictor layers. This process results in a view of forest type distribution in greater detail than is possible with the FIA plot data alone. The ForDRI model is calculated for forest areas based on this national forest type dataset acquired from the USDA USFS (USDA Forest Service, 2023).

3.2.2.2 Example and Evaluation of the ForDRI

The evaluation of the ForDRI models were conducted using three methods: (1) comparing the ForDRI with the USDM, (2) correlating with the normalized Bowen ratio, and (3) correlating with the tree ring data. Figure 3.7 shows the forest cover/type map that was produced by the US Forest Service. The ForDRI models are developed for only the forest cover across the CONUS. The red circles and green triangles on the map show the tree-ring and AmeriFlux sites, respectively, that were used for evaluation of ForDRI. We have used 22 Ameriflux sites that include 13 sites in the western and nine sites in the eastern United States for evaluation of the ForDRI.

3.2.2.2.1 Evaluation of ForDRI Using the USDM

Figure 3.8 shows an example of the qualitative comparison of the ForDRI with the USDM. The drought intensity estimates of ForDRI broadly agree with those for the same weekly period produced by the USDM. Note that ForDRI masks out non-forested lands (e.g., agricultural, rangelands, water, and urban) that are a focus of the USDM.

The qualitative comparison of the ForDRI and USDM maps showed that long-term severe and extreme droughts are highly correlated with the forest drought stress. For example, the 2012 ForDRI (Figure 3.8) shows severe to extreme drought over eastern Alabama and Georgia forests, which is also reflected on the USDM corresponding map. In contrast, the 2013 ForDRI map (Figure 3.3) shows moderate to severe drought (or forest stress) over the western half of the United States while the eastern-half US forests were not impacted by drought. Generally, it was observed that long-term, high-intensity droughts have significant impact on forests that are also detected by the ForDRI.

3.2.2.2.2 Bowen Ratio Data to Compare with ForDRI at the AmeriFlux Sites

Plant water stress is typically characterized by the water potential (ξ), which represents the tension in the water column and reflects the balance of free energy between atmospheric demand and soil water supply, modulated by leaf stomatal and hydraulic resistances (Philip, 1966). Plant water

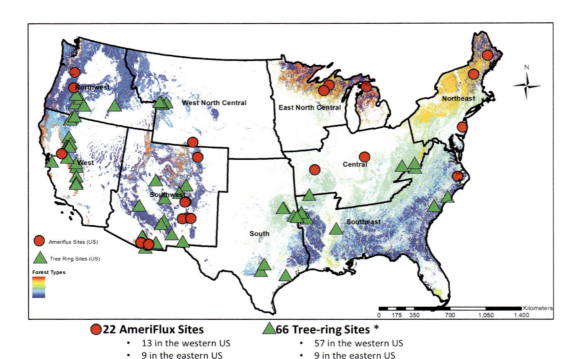

FIGURE 3.7 The forest cover map of the United States. The red circles and green triangles on the map show the tree-ring and AmeriFlux sites, respectively.

Remote Sensing of Drought

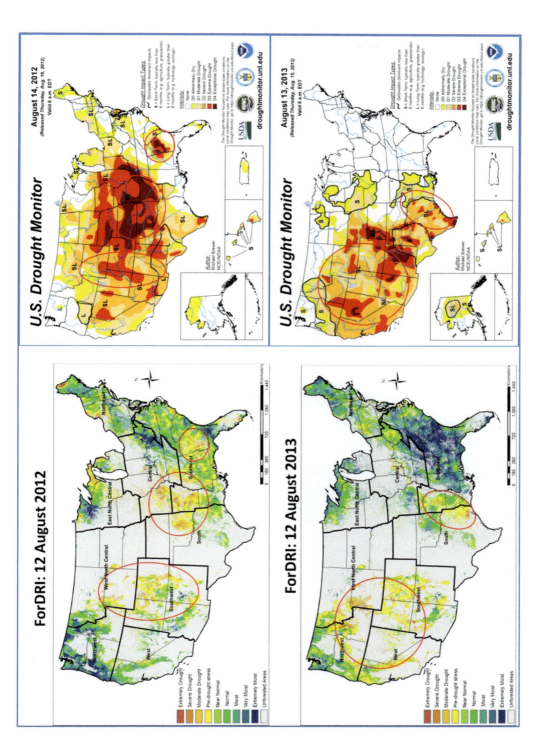

FIGURE 3.8 Comparison of the ForDRI with the USDM for August 12, 2012 and 2013.

potentials can be measured via pressure chamber (Scholander et al., 1965) or in situ hygrometer (Baughn and Tanner, 1976), but long-term observations across a range of sites are not available.

Forest water stress was assessed by using sensible (H) and latent heat (λE, evaporation) flux data measured at AmeriFlux network sites to calculate an integrated Bowen ratio (β_i):

$$\beta_i = \frac{\sum H}{\sum E} \tag{3.1}$$

Measured 30-minute H and λE fluxes (no gap-filled values) were summed over seven days, when both were > 50 W m^{-2}. The seven-day integration period was chosen to match the weekly timestep of ForDRI. The Bowen ratio in this context thus represents the weekly partitioning of the site net radiation. When a tree canopy is fully developed and water is passing through foliage on its way to the atmosphere, λE is generally greater than H, and $\beta < 1$. When water stress occurs, evaporation from a canopy is limited by stomatal closure and, potentially, reduced foliage area. These limits result in more of the incoming energy being converted to sensible heat, causing the Bowen ratio to increase.

Sensible (H) and latent (λE) heat data from 22 forested AmeriFlux eddy covariance sites in the United States were used to calculate the weekly Bowen ratio (β_t). These represented all forested sites in the United States with 12 or more years of H and λE data. Because there are seasonal as well as site-to-site variations in β, we normalized weekly, log-transformed integrated Bowen ratios $\log_{10}\beta_i$ by their standard deviations (σ) from the weekly mean over the full record ($\overline{\log_{10}\beta_i}$, where a negative value indicates a higher than average β_t and more drought-stressed conditions). This normalization (also referred to as a Z-score) occurs for each week of the growing season and helps highlight unusual behavior in the weekly β_t values consistently across sites.

$$Z-score(\beta_i) = \frac{\overline{\log_{10}\beta_i} - \log_{10}\beta_i}{\sigma} \tag{3.2}$$

This normalization also means that in a long enough record there is a direct, probabilistic interpretation of values based on characteristics of the normal distribution. Figure 3.9 shows box-plots of the Pearson's correlation between ForDRI and nine eastern US AmeriFlux sites in the growing season. The results showed that five out of nine AmeriFlux sites Bowen ratio values showed good correlation, with p-value less than 0.1. Figure 3.10 shows the box-plots of the Pearson's correlation between ForDRI and 13 western U.S. AmeriFlux sites Bowen ratio in the growing season. The result showed 9 out of 13 AmeriFlux sites across the western US indicated relatively strong correlation with ForDRI (with p-value < 0.1).

3.2.2.2.3 Tree Ring Data for ForDRI Evaluation

Because tree growth is sensitive to diverse local climates, tree-ring chronologies (widths) can provide valuable information for forestry management. Variations in the yearly climate (e.g., precipitation and temperature) time series are recorded by the sequence of tree rings in the forest. Thus, comparing the climate records with the tree-ring widths for the same period can provide valuable information for model evaluation. Efforts are being made to evaluate the ForDRI model using several ground observations, including tree rings. The evaluation includes comparing the ForDRI values with the tree-ring widths (TRW) and basal area increment (BAI) for the same period. The "Heat Map Chart" in Figure 3.11 shows where and when there is a significant person's correlation for the eastern and western US tree-ring sites. Our analysis showed that the western US sites are more correlated with the ForDRI values (i.e., the green and yellow colors). Generally, based on the qualitative and quantitative analyses of ForDRI models/maps, the results showed that the ForDRI had an excellent performance in predicting the likelihood of extreme forest drought events.

Remote Sensing of Drought

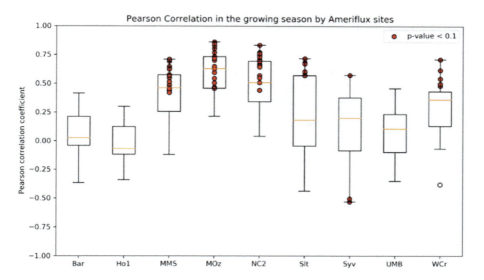

FIGURE 3.9 Pearson's correlation of nine AmeriFlux sites with the range (error bars) for the eastern United States.

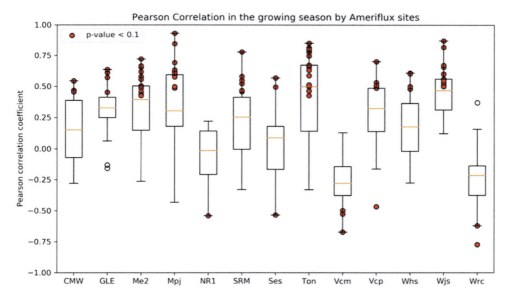

FIGURE 3.10 Pearson's correlation of 13 AmeriFlux sites with the range (error bars) for the western United States.

3.2.3 Evaporative Stress Index (ESI)

Of the various components of the hydrologic budget, ET is perhaps most directly tied to vegetation health. Canopy transpiration rates are regulated by root uptake and stomatal conductance and are strongly correlated with plant photosynthetic functioning, while the soil evaporation component of ET is directly related to the surface soil moisture status. The Evaporative Stress Index (ESI)

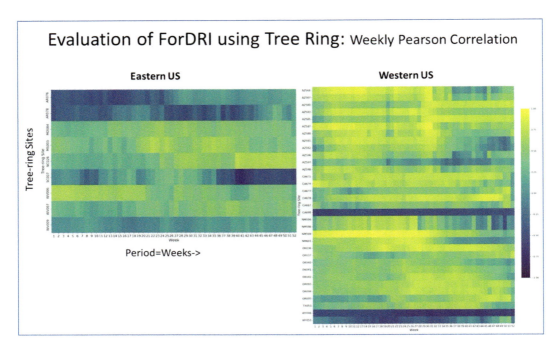

FIGURE 3.11 Heat map chart of person's correlation for the eastern and western US tree-ring sites.

reflects standardized anomalies in a ratio of actual-to-reference ET, with actual ET retrieved using the remote sensing–based Atmosphere-Land Exchange Inverse (ALEXI) model of surface energy balance (Anderson et al., 2007a, 2007b). Negative anomalies in ESI signify lower-than-normal consumptive water use, typically indicating depletion of root-zone soil moisture reserves (i.e., agricultural drought) and/or poor vegetation condition.

The primary diagnostic inputs to ALEXI are maps of land-surface temperature (LST) retrieved from satellite imagery collected in the TIR wavebands. Using principles of energy balance, ALEXI estimates the land evaporative flux required to keep the surface at the temperature observed from the satellite platform. One advantage of this diagnostic approach to estimating ET is that it does not require a priori information about precipitation or other ancillary sources of moisture—a necessity in prognostic land-surface models governed by water balance. In the energy balance approach, LST conveys proxy information about the surface moisture status, with moisture deficiencies reflected in elevated canopy and soil temperatures. Therefore, the LST-based ESI provides an independent assessment of drought conditions in comparison with standard precipitation-based indices and is arguably more directly related to actual stress in the vegetative canopy.

3.2.3.1 Model Development and Implementation
3.2.3.1.1 Two-Source Energy Balance (TSEB) Model

The ALEXI energy balance model is built on the Two-Source Energy Balance (TSEB) land-surface model of Norman et al. (1995), with subsequent revisions by Kustas and Norman (1999, 2000) and others. The TSEB treats the land surface as imaged by the remote TIR sensor as a composite of soil and canopy components, each with a characteristic temperature (T_S and T_C, respectively) that is constrained by the bulk radiometric surface temperature observation, $T_{RAD}(\theta)$, and the local fraction of green vegetation cover, $f(\theta)$, both apparent at the view zenith angle θ:

Remote Sensing of Drought

$$T_{RAD}(\theta) \approx \left[f(\theta) T_C^4 + \left[1 - f(\theta)\right] T_S^4 \right]^{1/4}. \tag{3.3}$$

These component temperatures in turn constrain fluxes of sensible heat from the soil and canopy components, H_S and H_C, computed using a series of resistance network regulating temperature gradients between the surface and above-canopy airspace, as described schematically in Figure 3.12.

Resistances factors diagrammed in Figure 3.12 are estimated based on local wind, stability, and surface roughness conditions following Norman et al. (1995). The net radiation flux above the canopy (RN) is likewise partitioned between energy divergence within the canopy (RN_C) and energy available at the soil surface (RN_S) based on a two-stream model of radiative transfer within the canopy (Anderson et al., 2000), with a diurnally varying fraction of RN_S conducted into the soil (G) following Santanello and Friedl (2003).

To solve the system of equations describing the energy balance associated with the two flux sources, an initial estimate of the canopy component of latent heat, λE_C, describing transpiration expected for a well-watered canopy is defined using either a modified Priestley–Taylor approximation or an analytical canopy conductance model (Anderson et al., 2008). The soil evaporation rate, λE_S, is then computed as an overall residual to the system energy balance:

$$\lambda E_S = (RN_S + RN_C) - (H_S + H_C) - G - \lambda E_C \tag{3.4}$$

If the canopy transpiration is less than potential (e.g., because of soil moisture restrictions and stress-induced stomatal closure), λE_C will be overestimated and Eq. 2 will lead to negative soil evaporation

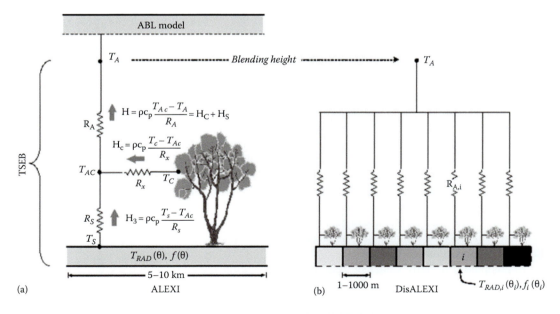

FIGURE 3.12 Schematic diagram representing the ALEXI (a) and DisALEXI (b) modeling schemes, highlighting fluxes of sensible heat (H) from the soil and canopy (subscripts s and c) along gradients in temperature (T), and regulated by transport resistances R_A (aerodynamic), R_x (bulk leaf boundary layer), and R_S (soil surface boundary layer). DisALEXI uses the air temperature predicted by ALEXI near the blending height (T_A) to disaggregate 10-km ALEXI fluxes, given vegetation cover ($f(q)$) and directional surface radiometric temperature ($T_{RAD}(q)$) information derived from high-resolution remote-sensing imagery at look angle q.

(i.e., condensation), which is not expected midday when polar orbiting thermal sensors typically overpass. In this case, λE_C is throttled back until reasonable λE_S is obtained. Evapotranspiration (E) is merely the total latent heat flux, $\lambda E = \lambda E_C + \lambda E_S$ (W m^{-2}), expressed in units of mass flux (kg s^{-1} m^{-2} or mm s^{-1}), where λ is the latent heat of vaporization (J kg^{-1}).

3.2.3.1.2 Regional Application of the TSEB Model (ALEXI and DisALEXI)

Although the TSEB has been demonstrated to work well in local applications using in situ measurements of T_{RAD} and air temperature (T_A in Figure 3.12), direct regional implementation is confounded primarily by (1) errors in LST retrieval due to atmospheric and emissivity corrections and (2) errors in gridded boundary conditions in near-surface air temperature, T_A. These lead to errors in H and therefore in λE (ET) by residual. The ALEXI model (Anderson et al., 1997, 2007a) was designed to provide a more robust framework for applying the TSEB regionally. This is accomplished by (1) applying TSEB in time differential mode, rendering it sensitive primarily to time changes in LST that can be retrieved much more accurately than absolute (instantaneous) LST; and (2) driving the time-differential TSEB with a modeled air temperature field governed by a simple slab model of morning atmospheric boundary layer (ABL) development. In (2), the air temperature boundary conditions are consistent with both the surface temperatures and modeled sensible heat flux, which serves to heat the air within the ABL over the morning hours.

ALEXI therefore requires at least two measurements of LST acquired during the morning, typically obtained from thermal imagers or sounders on geostationary satellite platforms with typical spatial resolution on the order of 3–10 km. For higher-resolution applications, requiring field or subfield scale sampling, the DisALEXI algorithm for spatially disaggregating ALEXI fluxes using TIR data from polar orbiting sensors or airborne systems was developed (Norman et al., 2003). DisALEXI constitutes a single time-of-day application of TSEB in gridded mode but using the ALEXI-modeled air temperature field T_A as a first guess (Figure 3.12). These air temperatures are iteratively modified until the DisALEXI sensible heat flux field aggregates to the ALEXI-derived field (Anderson et al., 2012). Again, no measurements of near-surface air temperature are required. Combined, ALEXI/DisALEXI provide multi-scale ET modeling capabilities from field to continental to global scales.

3.2.3.1.3 ESI Algorithm

The ESI represents standardized anomalies in a normalized clear-sky ET ratio, $f_{RET} = ETa/ET_{ref}$, where ETa is the actual ET retrieved at midday using ALEXI or DisALEXI and ET_{ref} is a reference ET scaling flux used to minimize impacts of non-moisture-related drivers on ET (e.g., seasonal variations in radiation load) (Anderson et al., 2007b, 2011, 2013). Here we use the FAO-96 Penman–Monteith reference ET for grass, as described by Allen et al. (1998). Anderson et al. (2013) compared several different scaling fluxes over the CONUS and found the FAO PM equation provided best agreement with drought classifications in the USDM and with soil moisture–based drought indices. It is hypothesized that use of clear-sky ET retrievals (as opposed to all-sky estimates) results in better separation of soil moisture–induced controls on ET from drivers related to variable radiation load such as cloud cover.

To compute ESI, daily values of clear-sky f_{RET} are composited over moving windows of 2-, 4-, 8-, and 12-week lengths, advancing in seven-day increments. Normal conditions (mean and standard deviation) are computed over the period of record (currently 2000–present) for each seven-day period and compositing interval. Finally, ESI is computed as a z-index, indicating the number of standard deviations (σ) the current composited f_{RET} value deviates from the normal moisture conditions for that period of time, with typical values +/- 3σ. Studies by Otkin et al. (2013a, 2013b) suggest that changes in ESI may have value as an early indicator of rapid drought development—capturing early thermal signals of increasing canopy stress during so-called flash drought events (Anderson et al., 2013).

Maps of ESI over the CONUS are distributed operationally by NOAA-NESDIS through the Geostationary Evapotranspiration and Drought (GET-D) product system at https://www.star.nesdis. noaa.gov/smcd/emb/droughtMon/products_drought Mon.php. The US product currently is at 2-km resolution using TIR data from the GOES Advanced Baseline Imager. Global maps at 5-km resolution are produced by NASA's Short-Term Prediction Research and Transition (SPoRT) Center and distributed by the NASA SERVIR project (https://servirglobal.net/Global/Evaporative-Stress-Index), created using MODIS/VIIRS/AVHRR day-night temperatures, as described next.

3.2.3.2 Input Data

As noted, ALEXI uses time-differential measurements of surface temperature change during the morning hours (between sunrise and local noon), typically obtained from geostationary platforms (ALEXI_GEO). This poses a challenge for international or global applications, requiring access to and integration of time-series thermal information from multiple satellites operated by different countries (e.g., the Geostationary Operational Environmental Satellites [GOES; covering North and South America], and the Meteosat satellites [Europe and Africa]). Additionally, GEO satellites lose effectiveness for ET retrieval at latitudes beyond approximately +/- 60° where the view angle becomes significantly oblique. To circumvent these problems, a version of ALEXI (ALEXI_POLAR) based on day-night LST observations from polar orbiting systems, such as MODIS on Terra or Aqua, was developed (Hain and Anderson, 2017). This version conveniently provides global land coverage from a single satellite system, including the near-polar regions, at workable view angles. Global ALEXI transitioning to VIIRS moving forward, and AVHRR in retrospective extend the historical ESI data record.

Using DisALEXI, ALEXI_GEO fluxes can be spatially disaggregated to 500 m using MODIS or VIIRS LST products (approximately daily depending on cloud cover), and to 30-m resolution using sharpened TIR imagery from Landsat (approximately bi-weekly to monthly). Data fusion algorithms have been developed to combine GEO/MODIS-VIISR/Landsat ET retrievals to approximate daily coverage at 30-m resolution—the spatiotemporal requirements for many field-scale water management applications and for early detection of developing crop stress (e.g., Cammalleri et al., 2013, 2014; Anderson et al., 2018; Knipper et al., 2019).

In addition to LST, gridded information on fraction cover ($f(\theta)$ in Eq. (2) is obtained from shortwave vegetation indices and/or standard leaf area index products (e.g., from MODIS or Meteosat). Meteorological inputs of windspeed (required for resistances) and atmospheric temperature profile (required for the ABL growth component of ALEXI) are obtained from specialized mesoscale analyses or standard regional or global reanalysis data sets (e.g., the Climate System Forecast Reanalysis). Land cover class in conjunction with cover fraction are used to specify canopy characteristics such as height and clumping factor, needed for roughness and canopy radiation transfer computations.

3.2.3.3 Examples of ESI

Figure 3.13 shows a three-month ESI composite ending August 31, 2023, generated at 4-km resolution using LST from the GOES-East and -West Imager systems over the CONUS. This can be compared to the USDM and VegDRI maps for mid-August shown in Figures 3.3 and 3.5. ESI shows elevated soil moisture and vegetation stress across much of the central and south-central United States, consistent with representations of drought within the USDM and VegDRI. This drought was a result not only of lower-than-normal precipitation, but also high winds and air temperature driving rapid depletion of remaining soil moisture reserves throughout the warm season. The ALEXI energy balance model used to construct the ESI considers each of these ancillary factors, and it captured the expansion of the drought-affected area with good time fidelity (Otkin et al., 2013a). Other areas of stress are shown in the ESI over the Pacific Northwest and sections of the Mid-Atlantic.

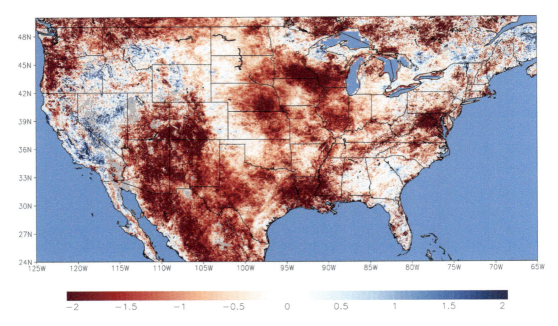

FIGURE 3.13 Three-month ESI composite over the continental United States ending August 30, 2023, generated using ALEXI-GEO applied to LST retrievals from the GOES East and West imager instruments at 4-km spatial resolution.

A comparable map of global ESI for this same period, created using ALEXI_MODIS, is shown in Figure 3.14. From a global perspective, we see that the 2023 drought in the south-central United States extended well into Mexico. Additionally, drought conditions in the northern tier of the United States extend well into Canada, where a drier and warmer period have led to widespread vegetation stress and a significant increase in wildfire occurrence. Finally, significant drought throughout the Iberian Peninsula and northern Africa is also shown in the global ESI product.

3.2.3.4 Applications of ESI

Given the strong physical relationship between the transpiration component of ET and vegetation health, ESI development to date has focused primarily on applications in the agricultural sector—in detecting crop stress, scheduling irrigation, and estimating yield. As a diagnostic indicator, ESI is also used as a comparative metric to drought indicators from prognostic land-surface modeling systems.

Global ESI products from NASA SPoRT/SERVIR are being ingested into the Geoglam Crop Monitor (https:/cropmonitor.org), the USDA Foreign Agricultural Service's Global Agricultural & Disaster Assessment System (GADAS) and NASA's DISASTERS Dashboard (https://disasters-nasa.hub.arcgis.com/). GET-D 2-km products over CONUS are used by NOAA for verification and improvement of numerical weather prediction models. In addition, 30-m ESI data sets generated using fusion of GEO/MODIS-VIIRS/Landsat and Landsat-like (e.g., ECOSTRESS) imagery are being used to forensically dissect coarse-scale drought signals apparent in maps such as those in Figures 3.13 and 3.14 (Xue et al., 2022). At this scale, stress signals can be segregated by crop type, which improves yield predictability, particularly when combined with field-scale remotely derived phenology information that can help to identify when different crops are in stages that are most susceptible to moisture stress (Yang et al., 2021). ESI maps at 30-m resolution have also been developed over forested systems as a predictor of drought-induced tree mortality (Yang et al., 2021), and are

Remote Sensing of Drought

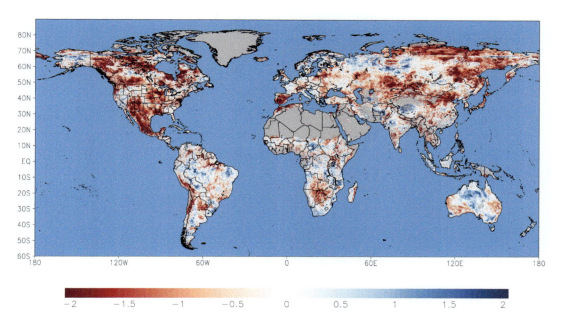

FIGURE 3.14 Three-month global ESI composite ending August 30, 2023, generated using ALEXI-MODIS applied to MODIS day/night LST and at 5-km spatial resolution.

currently being developed over grazing lands to investigate response to regenerative management practices. The goal is a multi-scale integrated drought assessment tool, with full global and regional coverage and the ability to map at high spatial resolution in drought-impacted areas.

The primary limitation on the TIR-based ET estimates currently used in the ESI products is temporal sampling—imposed by the inability to retrieve LST through clouds using thermal-band techniques. This can result in long periods of time where f_{RET} cannot be updated, particularly within persistently cloudy regions such as the Amazon and Equatorial Africa. This limitation may be addressed by combining with TIR-derived moisture estimates with microwave (MW) SM moisture retrievals within a data assimilation framework (as discussed in Section 3.2.4.3), or with ALEXI ET estimates generated using MW-based LST retrievals (Holmes et al., 2018).

3.2.4 Microwave-Based Surface Soil Moisture Retrievals

Soil moisture is a key parameter in drought monitoring because it strongly influences water availability to plants (reflecting agricultural drought conditions) and is also tied to longer-term hydrologic drought. The retrieval of soil moisture levels and the detection of negative anomalies representative of drought is important for drought monitoring and early warning. In situ observations of soil moisture conditions are challenging and existing networks are limited, particularly internationally. Satellite microwave remote sensing is being looked upon to help complement in situ observations and fill in the informational gaps on spatial variations in soil moisture conditions.

Because of the strong contrast between the dielectric properties of water versus dry soil, land surface emissivity in the 1–10 GHz microwave frequency range is strongly dependent on near-surface soil water content (Shutko, 1986). This sensitivity forms the physical basis of remote sensing algorithms that retrieve surface soil moisture content from satellite-based microwave measurements. In contrast to visible and thermal infrared-based sources of drought information discussed earlier, microwave-based observations provide a relatively direct measurement of soil moisture. In

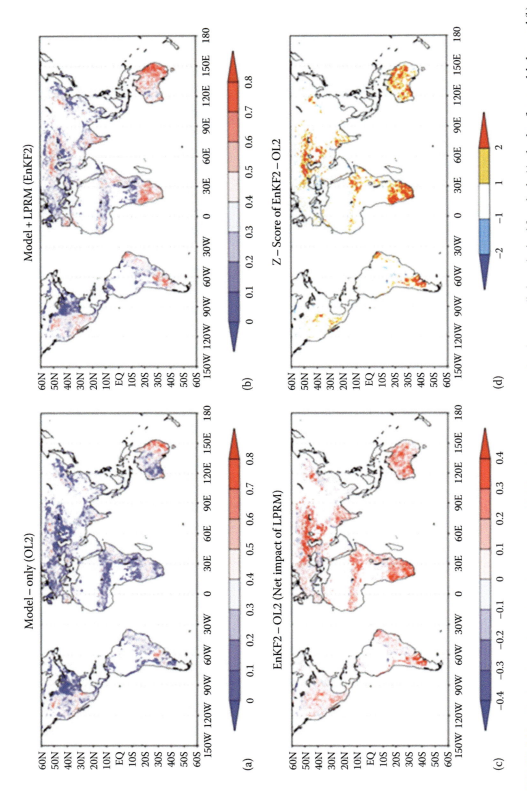

FIGURE 3.15 The lag +1 month correlation between future NDVI and current root-zone soil moisture predictions obtained from both: (a) a land surface model alone and (b) the land surface model enhanced by the assimilation of AMSR-E LPRM passive microwave surface soil moisture retrievals. Part (c) plots the difference between parts (b) and (a) (i.e., the increase in correlation associated with the assimilation of the retrievals). Part (d) relates the Z-score of the differences in part (c). Taken from Bolten and Crow (2012).

addition, satellite-based microwave retrievals can be acquired at dense temporal frequencies (on the order of three to five retrievals per week) because of the development of wide-swath satellite measurement techniques and the lack of interaction between 1–10 GHz microwave radiation and clouds. However, these advantages must be weighed against the relatively poor spatial resolution of satellite-based surface soil moisture retrievals (typically > 30 km), the shallow vertical support of microwave observations within the soil column (estimated to be on the order to 1–5 cm), and the tendency for decreased retrieval sensitivity under moderate and dense biomass conditions.

3.2.4.1 Input Data

Surface soil moisture retrieval algorithms have been developed for both active and passive microwave measurements of the land surface. Active measurements are based on measuring the fraction of transmitted energy scattered back toward a satellite-based radar emitter and a co-located antennae. The spatial resolution of observed backscatter is frequently enhanced via synthetic aperture radar (SAR) processing whereby the motion of the satellite, or airborne, platform is used to increase the effective size of the receiving antennae. In contrast, passive microwave techniques are based on the measurement of the natural gray-body emission of the land surface in the microwave spectral region.

Input requirements vary slightly between active- and passive-based surface soil moisture retrieval techniques. Active retrieval techniques are typically based on simplified forward backscatter modeling, and they attempt to model the scattering of an incident microwave energy pulse at the land surface as a function of land surface roughness, vegetation structure, and soil emissivity (i.e., soil wetness). The successful inversion of these models to retrieve absolute surface soil moisture values typically requires relatively detailed ancillary information concerning the statistical structure of surface roughness and the spatial distribution of water in the vegetation canopy (Wagner et al., 2007). However, the temporal contrast between the seasonal variation of soil roughness and vegetation properties and the relatively shorter time scales of soil moisture variability have enabled the development of effective active microwave-based change detection approaches with modest input requirements (Naeimi et al., 2009).

Relative to active measurements, passive microwave observations are generally considered to be less sensitive to variations in surface roughness and vegetation structure. For example, soil moisture retrieval from passive brightness temperature observations is commonly performed using a so-called zero-order tau-omega model for surface microwave emission that reduces the parameterization of soil/vegetation to the specification of bulk vegetation water content, a single-scattering albedo, and a root mean square indicator of surface roughness (Jackson et al., 1982). However, unlike active measurements, the retrieval of surface emissivity from passive microwave observations also requires an accurate estimate for the physical temperature of the land surface. Such temperatures are commonly obtained from either higher-frequency 37 GHz microwave observations or numerical weather prediction model output (Holmes et al., 2009, 2012). Finally, after land surface emissivity is estimated, it is typically converted into a volumetric soil moisture value based on knowledge of soil texture and the application of a soil moisture mixing model relating soil emissivity to volumetric soil water content.

It is also worth noting that some soil moisture retrieval algorithms can utilize multi-polarization and/or multi-incident angle observations to simultaneously estimate both soil moisture and vegetation canopy optical depth (Owe et al., 2008)—which, in turn, can be related to vegetation water content. While less frequently utilized than soil moisture retrievals, microwave-based vegetation optical depth estimates can also be applied to the detection of agricultural drought (Liu et al., 2013; Moesinger et al., 2022; Frappart et al., 2022).

3.2.4.2 Algorithm Development and Implementation

Early research on microwave remote sensing revealed the optimality of L-band (1–2 GHz) brightness temperature observations for soil moisture retrieval (Jackson, 1993). Higher-frequency 2–20

GHz microwave observations can also be used, but resulting soil moisture retrievals are subject to reduced sensitivity under moderate to dense vegetation cover (Jackson, 1993). Examples of higher-frequency passive microwave radiometers applied to soil moisture retrievals include the 19 GHz Special Sensor Microwave Imager (SSM/I) (Wen et al., 2005), 6.9 GHz Special Sensor Microwave Imager Sounder (SMMR) (Owe et al., 2008), and 10.7 GHz TRMM (Tropical Rainfall Measuring Mission) Microwave Imager (TMI) (Gao et al., 2006). However, it was not until the 2002 launch of AMSR-E observations that such observations were used to generate a true operational soil moisture product (Njoku et al., 2003). One disappointing development was the discovery of widespread radio frequency interference (RFI) in the AMSR-E 6.9 GHz (C-band) channel over the United States and Japan (Li et al., 2004). This discovery necessitated the more widespread use of the 10.7 (X-band) AMSR-E channel for soil moisture retrievals.

The first L-band satellite sensor designed specifically for soil moisture retrieval was the European Space Agency (ESA) Soil Moisture and Ocean Salinity (SMOS) mission launched in late 2009 (Kerr and Levine, 2008). The SMOS instrument mission concept is based on a passive microwave 2D interferometric polarimetric radiometer capable of making multi-angular measurements of L-band brightness temperature. It has been successfully generating a continuous L-band surface soil moisture product since early 2010; however, it has been challenged by the presence of unexpectedly high L-band RFI over areas of Europe and East Asia (Oliva et al., 2012). In addition, the multi-angular capability of the SMOS instrument has proven to be particularly useful for the extraction of vegetation optical depth products for vegetation monitoring applications (Wigneron et al., 2020).

The SMOS mission was followed by the launch of the L-band NASA Soil Moisture Active/Passive (SMAP) mission in early 2015. SMAP mission designers had the significant benefit of access to L-band RFI results from the earlier SMOS mission and were able to implement instrument and processing changes to provide more robust RFI mitigation (Mohammed et al., 2016). While originally designed based on the concept of simultaneous L-band radar and radiometer measurements, the SMAP radar failed in July 2015. However, the SMAP radiometer has performed exceptionally well and, like SMOS, supported the production of a long-term, high-quality soil moisture retrieval dataset (Colliander et al., 2021) at data latencies (i.e., roughly two days for Level 2 and 3 data products) that are suitable for operational drought monitoring.

For active microwave systems, the development of global data products has primarily been limited to the use of scatterometer (as opposed to SAR) observing systems. The first global-scale active microwave products were based on the use of retrospective 5.3 GHz European Remote Sensing (ERS)-1/2 scatterometer observations (Wagner et al., 1999). This system was later upgraded into a real-time operational system applied to 5.2 GHz METerological Operational (METOP) Advanced Scatterometer (ASCAT) retrievals (Naeimi et al., 2009). Despite theoretical concerns about the sensitivity of active remote sensing retrievals over moderate-to-dense vegetation areas, validation results suggest that scatterometer-based retrievals are surprisingly accurate over moderately vegetated areas (Crow and Zhan, 2007).

To date, the operational application of high-resolution, SAR-based soil moisture products has been slowed by their relatively sparse temporal coverage. However, due to the recent (and planned) deployment of SAR sensors with wider swath widths, we are now moving into an era of enhanced routine availability of high-resolution satellite SAR measurements. For example, plans for the L-band NASA NISAR mission, currently slated for launch in 2024, call for the routine generation of a global 200-m soil moisture product with a repeat time of approximately one week (Rajat Bindlish, personal communication).

3.2.4.3 Applications of Microwave-Based Surface Soil Moisture Retrieval Products

The shallow vertical penetration depth of microwave surface soil moisture retrievals is generally assumed to represent a significant barrier to their direct integration into agricultural drought monitoring activities (Albergel et al., 2008). For example, the assumed vertical support of these measurements (less than 5 cm) samples only a small fraction of the 30- to 150-cm total column rooting depth generally assigned to vegetation. As a result, the application of microwave-based soil moisture retrievals to agricultural drought commonly involves the use of land data assimilation systems. Such systems are designed to optimally integrate continuous (but error-prone) dynamic models with (presumably) more accurate, but less temporally/spatially complete, observations that can be related to the states of the model. Ideally, the relative weighting of both sources of information is based on the uncertainty of the continuous forecast model versus noncontinuous assimilated observations. By calculating error covariance information between multiple model states and background error states (typically via a Monte Carlo ensemble) and applying Kalman filtering concepts, such systems can also be used to update (unobserved) profile soil moisture states beyond the near surface using only surface soil moisture retrievals (Figure 3.15).

Multiple studies have demonstrated the ability of such systems to improve the accuracy and precision of modeled surface and root-zone soil moisture via the assimilation of surface soil moisture retrieval products—or the brightness temperature precursors of such products (Mladenova et al., 2019; Dong et al., 2019; Reichle et al., 2007, 2021). Soil moisture products generated by such systems are now operationally available for drought monitoring applications. Notably, the SMAP Level 4 Surface and Root-Zone Soil Moisture product (SMAP_L4; Reichle et al., 2022) currently produces a continuous, global, 9-km analysis surface and root-zone soil moisture product with less than three-day latency via the assimilation of SMAP brightness temperature observations into the NASA Catchment Land Surface model. The SMAP_L4 system has recently been upgraded to include near real-time precipitation data generated by the NASA Global Precipitation Mission (Reichle et al., 2023) and therefore now reflects the integrated use of best-available satellite precipitation and soil moisture retrieval products for agricultural drought monitoring. The SMAP L4 product is currently being routinely ingested into USDA agricultural drought monitoring systems (Zhang et al., 2022).

Data assimilation systems are also easily adapted to integrate multiple types of satellite-based soil moisture products. For example, the added value of simultaneously assimilating both active and passive-based soil moisture retrievals has been demonstrated (Draper et al., 2012). Data assimilation has also proven to be an effective technique for merging microwave-based soil moisture retrieval with TIR-based drought monitoring products (discussed earlier in Section 3.2.3). In general, the two methods have been shown to be highly complementary (Li et al., 2011). TIR methods (such as ESI) provide relatively high spatial resolution on the order of the resolution of the particular TIR sensor (~100 m to 10 km) but reduced temporal sampling due to constraints on TIR-based retrievals being limited to clear-sky conditions (Hain et al., 2011). In contrast, microwave methods provide soil moisture retrievals at relatively low spatial resolution (25–60 km) and high temporal resolution (retrievals possible through non-precipitating cloud cover every 2 to 3 days). The two techniques also differ in their sensitivity to vegetation cover; TIR methods provide the opportunity to attain a soil moisture signal over a wider range of cover regimes, while MW methods can suffer significant degradation in accuracy as the density of vegetation cover increases.

Hain et al. (2012) examined the efficacy of assimilating TIR-based ESI products and passive microwave-based soil moisture retrievals using an ensemble Kalman filter (EnKF) into the Noah land-surface model (LSM) to improve modeled soil moisture predictions. The methodology employed a data denial framework to quantify the ability of the two different soil moisture retrieval data sets to correct for errors in precipitation forcing (using a satellite-based precipitation product) in an open-loop simulation as compared to a control simulation forced with a higher-quality, gauge-based precipitation data set. As expected, the added value of TIR assimilation over microwave alone is most significant in areas of moderate to dense vegetation cover, where microwave retrievals have limited sensitivity to soil moisture at any depth.

The joint assimilation of TIR and MW soil moisture is of particular value in the case of fast-developing "flash drought" scenarios when signals of vegetation stress are apparent in the ESI signal (through elevated canopy temperatures) before a degradation in vegetation health occurs. Additionally, efforts are being made to assess the impact of SM information in a coupled land-atmosphere prediction system. Improvements in the representation initial soil moisture states within such a system have the potential to lead to improvements in temperature and precipitation forecasts produced by numerical weather prediction models (Munoz-Sabater et al., 2019). Joint SAR/TIR data assimilation systems are also now being applied for the operational monitoring of high-resolution (30-m) root-zone soil moisture for high-value irrigated fruit and nut crops (Chen et al., 2022). These systems currently rely on C-band SAR observations; however, they should benefit greatly from the increased availability of routine L-band SAR in the post-2025 period.

The ability of land data assimilation systems to improve the characterization of soil moisture anomalies appears to carry over to crop yield predictions. For example, Ines et al. (2013) documented improvements in corn yield predictions associated with the assimilation of microwave-based soil moisture retrievals into a crop system model. Finally, for applications in which a water balance approach is not utilized, land data assimilation concepts can also be applied to correct satellite-based rainfall accumulations using a time series of microwave surface soil moisture retrievals (Pellarin et al., 2008; Crow et al., 2011; Brocca et al., 2013).

In addition, a surprising result of the SMOS and SMAP L-band microwave era has been the degree to which the precise temporal measurement of even superficial soil moisture variations can accurately track the magnitude and progress of vegetation water stress. This capacity is based on the ability of remotely sensed soil moisture variations to track the nonlinear progression of soil moisture availability between water- and energy-limited flux regions (Sehgal et al., 2021). Such transitions are critical for detecting the rapid onset of soil moisture associated with flash drought. Importantly, it appears that such transitions can be adequately represented via the temporal monitoring of only surface soil moisture conditions (Dong et al., 2022). Therefore, there is growing evidence that the superficial nature of microwave-based soil moisture retrievals represents a less-significant barrier to agricultural drought applications than previously assumed (Qiu et al., 2014; Feldman et al., 2023). Reflecting this realization, surface (0–5 cm) SMAP SM soil moisture products are, by themselves, being utilized for agricultural drought monitoring. A notable example of this is the Crop CASMA product based on the downscaling of SMAP Level 3 soil moisture retrievals to 1-km (Liu et al., 2022). These downscaled soil moisture products are currently being used operationally by the USDA National Agricultural Statistics Service (NASS) (Zhang et al., 2022).

3.2.5 SATELLITE GRAVIMETRY-BASED SOIL MOISTURE AND GROUNDWATER

Although surface and root-zone soil moisture play an important role in the characterization of agricultural drought, a complete description of hydrologic drought requires more vertically integrated information. The NASA/German GRACE satellite mission (2002–2017) and its successor, GRACE-FO (2018-), uniquely provide global maps of anomalies (deviations from the long-term mean) of terrestrial water storage (TWS; the sum of groundwater, soil moisture, surface waters, snow, and ice). Thus, after accounting for natural seasonal variations in TWS, GRACE and GRACE-FO (collectively GRACE/FO) are well suited for drought assessment. This was recognized early in the GRACE mission, as several studies demonstrated the direct use of GRACE data to quantify drought (e.g., Andersen et al., 2005; Yirdaw et al., 2008; Chen et al., 2009; LeBlanc et al., 2009). In particular, Thomas et al. (2014) described how GRACE TWS data could be used to quantify the volume of water required by a region to return to normal water storage conditions and combined that storage deficit with event duration to assess drought intensity.

However, the spatial and temporal resolutions of GRACE/FO are low relative to other measurement types, and the data latency (typically two to four months) has made GRACE/FO data difficult to apply directly for operational drought monitoring. A solution is to integrate GRACE/FO TWS data with other hydrological observables that have higher resolutions and lower latency. A numerical LSM, which can be described as a sophisticated version of the land component of a weather or climate prediction model, is a useful vehicle for this sort of data integration. An added benefit is that data assimilation within an LSM allows vertical disaggregation of the GRACE TWS data into the TWS component contributions. As described next, such an approach has been in continuous operation since 2011 for drought monitoring in the United States. In 2019, it was expanded to a global scale.

3.2.5.1 Input Data
3.2.5.1.1 GRACE/FO Terrestrial Water Storage

GRACE, a joint mission of NASA and the German Space Agency, launched in 2002 (Tapley et al., 2004). Its successor, GRACE-FO, launched in 2018 and began delivering observations about 11 months after GRACE was decommissioned (Landerer et al., 2020). With an architecture nearly identical to that of GRACE, GRACE-FO comprises two satellites following each other around the Earth in a near-polar orbit, roughly 200 km apart, at an initial altitude of about 500 km. Unlike most Earth observing satellites, which measure light emitted or reflected from Earth's surface, GRACE-FO continuously and precisely measure the distance (and its rate of change) between the two satellites using a K-band microwave ranging system and (new for GRACE-FO) a laser ranging system. Heterogeneities in Earth's gravity field perturb the satellite orbits in a predictable manner, such that the inter-satellite range measurements can be used to construct a new numerical model of Earth's gravity field nominally once per month (Tapley et al., 2004). Global representations of Earth's gravity field are produced as sets of spherical harmonic coefficients. These can be manipulated using Gaussian averaging functions (Wahr et al., 1998) or the mass concentration technique (Rowlands et al., 2005) in order to isolate temporal mass anomalies over regions of interest. The primary sources of temporal changes in Earth's gravity field are mass variations in the oceans, atmosphere, and terrestrial water. By accounting for the first two using oceanic and atmospheric analysis models, it is thus possible to quantify monthly TWS anomalies (Wahr et al., 1998; Rodell and Famiglietti, 1999).

Challenges to using GRACE/FO terrestrial water storage anomaly data include their coarse spatial and temporal resolutions, the vertically integrated nature of the measurements, and data processing issues that include accounting for "leakage" of gravity signals from adjacent regions. There is a tradeoff between spatial resolution and accuracy, such that 100,000 km^2 is the approximate minimum area that can be resolved with a reasonable degree of certainty at mid-latitudes (Save et al., 2016; Wiese et al., 2016; Loomis et al., 2019). While GRACE/FO-based TWS fields are available from multiple sources, the following examples make use of gridded GRACE/FO data products from the University of Texas Center for Space Research (http://www2.csr.utexas.edu/grace/RL06_mascons.html; Save et al., 2016).

3.2.5.1.2 Meteorological Data

Land surface models are not prognostic and therefore rely on meteorological inputs to drive them forward in time. The required forcing variables are precipitation, downward shortwave and longwave radiation, near surface air temperature, specific humidity, windspeed, and surface pressure. In the example presented next, these forcing fields were obtained from two sources. One source is the Princeton Global Meteorological Forcing Dataset (Sheffield et al., 2006). The Princeton dataset spans 1948 to 2014 with three-hourly temporal and 1° spatial resolution. It was produced by combining observation-based data sets with a reanalysis product. The second source is the North American Land Data Assimilation System Phase 2 (NLDAS-2; Xia et al., 2012). The NLDAS-2 forcing data set covers central North America from 1979 to present and incorporates precipitation gauge observations, bias-corrected shortwave radiation, and surface meteorology reanalyses. The precipitation data have been temporally disaggregated using Stage II Doppler radar data and provided on a 0.125° resolution grid.

3.2.5.2 Algorithm Development and Implementation

Zaitchik et al. (2008) developed and tested an algorithm for assimilating GRACE TWS data into the Catchment LSM (CLSM; Koster et al., 2000). CLSM is well suited for GRACE/FO data assimilation because it simulates groundwater storage, which is requisite for properly representing TWS as a variable that can be updated using the GRACE/FO observations. The algorithm is a form of Ensemble Kalman Smoother (Evensen and van Leeuwen, 2000) that merges gridded monthly GRACE/FO-derived TWS anomalies into CLSM using information on the uncertainty in both the observation and the model (Kumar et al., 2016). Whereas Kalman filter data assimilation integrates observations as they become available and updates only the most recent model states, smoothers use information from a series of observations to update model states over a window of time. Smoothers are therefore well suited for GRACE/FO observations, which are non-instantaneous. Prior to assimilation, the GRACE/FO TWS anomalies are converted to absolute TWS values by adding the corresponding regional GRACE/FO-period mean TWS from the open loop (no assimilation) CLSM simulation.

Data assimilation enhances the value of GRACE/FO observations in three ways. First, the spatial resolution improves from river basin scale to the scale of counties or small catchments. Most practical applications benefit from data at the finest possible scales. Second, the resulting time series have sub-daily temporal resolution, as opposed to monthly, and the latest fields can be generated within a day of real time. Standard GRACE/FO hydrology products normally become available after a 2–4 month lag, which is unacceptable for operational applications. Third, GRACE/FO data assimilation output includes gridded maps of groundwater, soil moisture, and snow variations, which in most cases are easier to interpret than aggregate TWS. Hence, GRACE/FO assimilation output combines the veracity of an observation with the fine spatial and temporal resolutions of a model. Uncertainty in GRACE/FO data assimilation-based groundwater and soil moisture change estimates is highly variable and depends on a number of factors. Data assimilation seeks an optimal estimate based on two or more independent estimates (one from the model and the other from the observation) and knowledge of the errors in each. Hence the error in the result should be smaller than that in either of the input values. At the river basin scale (roughly 500,000 km^2 or larger), uncertainty in the monthly GRACE/FO terrestrial water storage anomalies is on the order of 1–2 cm equivalent height of water (Wahr et al., 2006). Uncertainty in the model estimates, which is determined based on ensemble spread, is similar in magnitude. At the fine spatial and temporal scales of the model output the errors are likely to be larger than those at the basin scale. Further, there is uncertainty associated with disaggregating terrestrial water storage into its components, although error in a given component is expected to be smaller than the error in the total. Determining actual errors in assimilated groundwater or soil moisture output in a specific region and at a certain scale requires comparison with independent data. For example, Zaitchik et al. (2008) showed that modeled groundwater agreed better with independent in situ measurements after the assimilation of GRACE TWS observations (Figure 3.16), with root mean square errors of 1.9–4.0 cm equivalent height of water in the Mississippi River basin and its four major sub-basins, compared with 2.4–6.3 cm for the open loop simulation. Using data from more than 4,000 wells in 11 countries, Li et al. (2019) determined that GRACE/FO data assimilation reduced errors in regional groundwater storage change estimates by 36% and improved correlation by 22% on average compared with open loop model output.

3.2.5.3 Application to Drought Monitoring

The ability to observe changes in deep soil moisture and groundwater from space is what makes satellite gravimetry unique as a tool for hydrology and drought monitoring. Unlike surface moisture conditions, which fluctuate rapidly with the weather, deep soil moisture and groundwater integrate meteorological conditions over weeks and months, making them natural gauges of drought severity (Rodell, 2012). Downscaling and extension to near real time via data assimilation, as described in the previous sub-section, make possible the application of GRACE/FO data for operational drought monitoring.

Premised on this potential, Houborg et al. (2012) developed surface and root zone soil moisture and groundwater drought/wetness indicators based on GRACE data assimilation results. They first

Remote Sensing of Drought

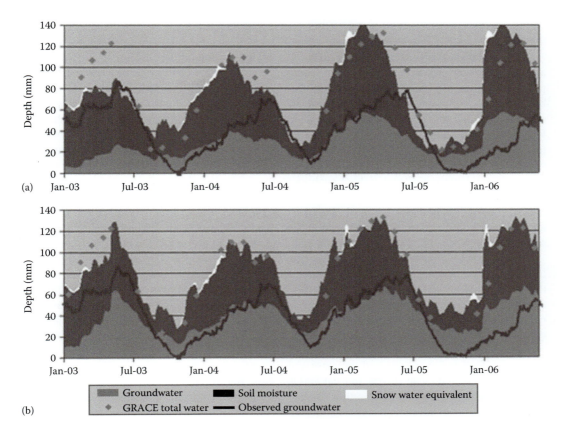

FIGURE 3.16 Groundwater, soil moisture, and snow water equivalent averaged over the Mississippi river basin from (A) open loop CLSM and (B) GRACE-DAS. Also shown are daily observation-based groundwater and monthly GRACE-derived TWS anomalies. GRACE and modeled TWS were adjusted to a common mean, as were observed and modeled groundwater. The correlation coefficient between simulated and observed groundwater improved from 0.59 (open loop) to 0.69 (GRACE-DAS). Unlike GRACE alone, the assimilated product is three-hourly and vertically distributed (From Zaitchik et al., 2008).

evaluated improvements in the output fields of soil moisture and groundwater storage that resulted from assimilation of GRACE TWS data using in situ groundwater observations from the USGS and soil moisture observations from the Soil Climate Analysis Network (SCAN). Houborg et al. (2012) then introduced a process for deriving drought/wetness indicators, which has since been refined and expanded to the global scale (Li et al., 2019). Both the US and global indicators are now being produced weekly at NASA Goddard Space Flight Center (GSFC) and distributed by the NDMC (https://nasagrace.unl.edu/), with accompanying 30-, 60-, and 90-day drought/wetness forecasts for the United States (Getirana et al., 2020). They are also consulted by the authors of the USDM and NADM.

The soil moisture and groundwater indicators are generated as follows. First, an open loop simulation of CLSM is executed for the period 1948–2014 using as input a meteorological forcing data set developed at Princeton University (Sheffield et al., 2006). The climatology of the open loop simulation is the basis for quantification of wetness/dryness conditions as a probability of occurrence during that 67-year period. The monthly GRACE/FO TWS anomalies are converted to absolute TWS by adding the temporal mean TWS from the open loop LSM output. Thus, the mean field of the assimilated TWS output is assured to be nearly identical to that of the open loop (i.e., GRACE/FO data assimilation does not introduce a bias).

Data assimilation begins in a separate branch of the CLSM simulation in 2002, when the GRACE data become available. At the same time in that branch simulation, NLDAS-2 meteorological forcing data replaced the Princeton forcing data. NLDAS-2 data are higher resolution, higher quality, and extend to near real time. However, while bias due to GRACE/FO data assimilation is avoided as previously described biases between the two forcing sources are significant enough that they must be accounted for and removed. Furthermore, the GRACE/FO data (and therefore the assimilation results), although unbiased, could still have a larger or smaller range of variability than the open loop LSM results at any given location. That is an important consideration because drought monitoring concerns the extremes. Therefore, statistical adjustments are applied based on the approach of Li et al. (2019). First, the NLDAS-2 meteorological data are scaled so that the temporal mean and standard deviation of each field at each location equals those of the Princeton data during the overlapping period (2003–2014). The output variables are similarly corrected for biases and differences in amplitude between the assimilation-mode NLDAS-2-forced model output and the open-loop Princeton-forced model output, using daily scale factors based on seven-day moving window averaged means and standard deviations from the same period of overlap (Li et al., 2019).

Finally, weekly drought/wetness indicator fields for surface (top several centimeters) soil moisture, root zone soil moisture, and groundwater are computed based on probability of occurrence at the same location (grid pixel) during the same time of year (i.e., eliminating seasonal variations) in the 1948–2014 record, using CDFs as before. For ease of comparison with the USDM, wetness (dryness) conditions are characterized from D0 (abnormally dry) to D4 (exceptional), corresponding to cumulative probability percentiles of 20–30%, 10–20%, 5–10%, 2–5%, and 0–2%.

Figure 3.14 compares May 2022 GRACE-FO data assimilation-based drought/wetness indicators for surface and root zone soil moisture and groundwater with the original GRACE-FO TWS anomalies (seasonal cycle intact) and the USDM. Note that the GRACE-FO-based indicators are used for reference by the USDM authors, hence they are not completely independent in this comparison. The level of agreement and disagreement typifies that displayed by the various GRACE-FO products and the USDM, though the patterns of correlation are variable. In this example, all five maps generally show major drought afflicting the southwestern United States. Wetness (lack of drought in the case of USDM) in western Washington State and the lower Mississippi River basin is also consistent among the five maps. Across most of the rest of the country, the GRACE-FO TWS anomaly map suggests that conditions were wetter than shown in the four indicator maps, which highlights the importance of accounting for the seasonal cycle of TWS. For example, if the long-term average TWS anomaly for May in a given location is 15 cm, then a TWS anomaly of 10 cm in May of 2022 would have been drier than normal, possibly equating to one of the drought categories. Differences between the GRACE-FO TWS anomaly map and the three data assimilation-based wetness indicator maps also reflect (1) their temporal mismatch (a monthly mean vs. a daily mean), (2) their differing spatial resolutions, (3) the vertically integrated nature of TWS vs. the three distinct layers represented by the indicator maps, and (4) the influence of meteorological inputs on the data assimilation output. Similarly, differences in spatial resolution, drought definition, and input information are responsible for many of the apparent discrepancies between the USDM and the wetness indicator maps. Discrepancies also may result from errors in the GRACE-FO observations, limitations of the LSM, or the fact that the USDM is conservatively adjusted each week by its authors and thus typically lags other indicators. For example, on May 17, 2022, the USDM displayed abnormally dry (D0) conditions in southern Tennessee and northern Alabama, while the three GRACE-FO-based indicator maps showed wetter than normal conditions (Figure 3.14). By May 31, the USDM authors had removed that designation (not shown).

The three GRACE-FO-based drought/wetness indicator maps are not identical because water in deeper layers responds more slowly to atmospheric conditions. In Figure 3.17 surface and root zone soils were relatively wet in northern North Dakota on May 16, 2022, while shallow groundwater had not yet completely recovered from drought. In Wisconsin, on the other hand, the surface soil was very dry, the root zone was somewhat dry, while shallow groundwater conditions were near normal.

Remote Sensing of Drought 73

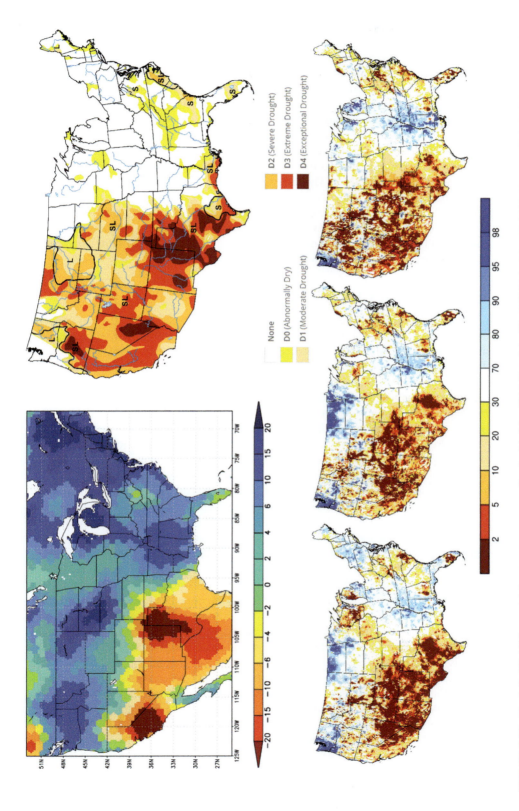

FIGURE 3.17 Top left: GRACE terrestrial water storage anomalies (cm) in May 2022. Top right: The USDM for May 17, 2022. Bottom: GRACE-FO data assimilation-based drought indicators for surface soil moisture (left), root zone soil moisture (middle), and groundwater (right) on May 16, 2022.

This suggests that a meteorological drought was beginning to take hold but had not yet caused hydrological drought and depressed groundwater levels. In general, the USDM tends to be more closely correlated with the root zone soil moisture drought/wetness indicator than with the other two. That does not mean that the root zone indicator is the best of the three; rather it reflects the type of information that is incorporated into the USDM and the needs of the end users that it serves. Part of the value of the GRACE-FO-based products is that they provide a type of drought information that previously did not exist (maps of groundwater percentiles), and they help the USDM authors to distinguish short-term (S on the map) from long-term (L) drought.

GRACE-FO launched in 2018 and has already surpassed its five-year mission objective. As the altitudes of the two satellites decline, increased atmospheric drag will cause larger errors in the retrieved mass anomaly fields, exacerbated by a malfunctioning accelerometer (intended to measure the drag) on one of the two satellites. The GRACE Continuity (GRACE-C) mission, formerly known as Mass Change, is currently under development (Wiese et al., 2022). Its design will be similar to that of GRACE-FO, with a launch date targeted for 2028. If a data gap occurs between the two missions, as happened between GRACE and GRACE-FO, the drought/wetness indicators described earlier can continue to be produced in the interim by running CLSM with the same atmospheric forcing inputs but without the benefit of TWS data assimilation.

3.3 CONCLUDING THOUGHTS

Since the early 2000s, a number of new satellite-based Earth observations have been collected establishing historical data records of 20+ years that can be used to monitor many components of the hydrologic cycle related to drought. In parallel, the methods and tools to analyze and retrieve estimates of key hydrologic parameters such as ET, soil moisture, and groundwater have matured and continue to improve. As a result, many innovative remote sensing–based tools are now available for operational drought monitoring in the United States. This chapter has reviewed several efforts to enhance drought monitoring capabilities for the United States using different types of satellite-based observations and modeling approaches that collectively provide insights into various aspects of the terrestrial hydrologic cycle relevant to the onset and intensification of drought conditions. These tools have also provided the basis to develop similar drought monitoring tools in other parts of the world or have had their geographic scope expanded to global coverage.

Other remote sensing efforts not summarized in this chapter provide additional opportunities to analyze other environmental parameters related to drought, such as vapor pressure deficit, new vegetation stress indicators related to solar-induced fluorescence (SIF), surface water conditions, snow cover and precipitation. Atmospheric vapor pressure deficit (VPD) is a relatively new remote sensing indicator calculated from satellite observations from NASA's Atmospheric Infrared Sounder (AIRS) for the early detection of drought onset and rapid severity intensification during flash drought events. VPD incorporates elements of both temperature and relative humidity and represents an indicator of difference between the amount of moisture in the atmosphere and the amount of moisture it can hold when fully saturated (Behrangi et al., 2016). As a result, VPD is a primary controlling factor of ET demand and has been found to be a valuable indictor for drought onset detection and rapid intensification (Behrangi et al., 2016; Farahmand et al., 2023). SIF is another potential early drought onset indicator that has been calculated from satellite observations form several sensors, including the European Space Agency's (ESA) Global Monitoring Ozone Experiment 2 (GOME-2; Joiner et al., 2013) and TROPOspheric Monitoring Instrument (TROPOMI; Kohler, 2018) and NASA's Greenhouse gases Observing Satellite (GOSAT; Frankenberg et al., 2011) and Orbiting Carbon Observatory-2 and -3 (OCO-2 and -3; Frankenberg et al., 2014; Taylor et al., 2020). The SIF emission rates change as part of a plant's response to stressors such as drought, which is a protective mechanism to dissipate excessive energy that cannot be used because of downregulated photosynthesis rates under stress conditions and has shown promise for early-stage drought detection (Jiao et al., 2019; Yoshida et al., 2015).

The first dedicated satellite mission for SIF, the FLuorescence EXplorer (FLEX) mission by ESA, is scheduled to be launched in 2025 (Buman et al., 2022) and holds considerable promise for providing SIF-related drought monitoring information (Damm et al., 2022). New perspectives on surface water conditions are provided by the NASA Surface Water and Ocean Topography (SWOT) instrument, which was launched in 2022 and is designed to estimate and monitor the water surface elevation of the ocean, rivers, and other inland water bodies (e.g., lakes and reservoirs). The SWOT mission uses an interferometric altimeter that includes a Ka-band synthetic aperture radar interferometer within two antennas to estimate water surface elevation and associated change over time (Durand et al., 2010). This new capability to monitor water elevation changes of rivers and other surface water bodies will generate critical new information for hydrologic drought monitoring and other water management activities by providing insights into locations and rates of surface water depletion. For precipitation, both satellite-based thermal and passive microwave observations have been used to provide rainfall estimates globally, which can be used to better spatially define areas of negative rainfall anomalies that are indicative of drought conditions. Prime examples include extended historical records of rainfall estimates from these types of satellite observations using established rainfall-retrieval algorithms such as the TRMM 3B42 (Huffman et al., 2007), the Climate Prediction Center (CPC) MORPHing (CMORPH; Joyce et al., 2004), and the Precipitation Estimation from Remotely Sensed Information using Artificial Neural Networks (PERSIANN; Sorooshian et al., 2000). Other innovative work has been done to improve the spatial resolution and precision of gridded climate data sets related to temperature and precipitation using ground-based meteorological station measurements in concert with remotely sensed rainfall estimates to produce precipitation grids and satellite land surface temperature data to generate air temperature grids, as demonstrated by Funk et al. (2012). For snow cover, the use of both optical and microwave satellite remote sensing have been used to characterize snow cover characteristics such as snow cover area and depth, as well as snow water equivalent (e.g., Kongoli et al., 2012). Snow cover information is an often overlooked but important parameter in drought assessments, as snow cover provides a critical water source for soil moisture and ground and surface water recharge. A lack of snow cover and water produced from snow melt and run-off can cause antecedent soil moisture and/or surface and groundwater conditions that make an area more susceptible for both agricultural and hydrologic drought if dry conditions persist into the growing season or sustain the multi-year drought from the previous year.

Long-term continuity of key satellite observations required to extend the long-term data record of key drought monitoring variables such as the NDVI, thermal-based ET, and TWS is critical. Given that drought monitoring primarily relies upon the detection of anomalous conditions relative to a longer-term historical baseline condition, extended multi-decadal records of consistent satellite observations and data products are needed. This often requires satellite observations from multiple sensors to extend the historical record and with sensors on new satellite missions. This can present a challenge given observational data characteristics may vary among sensors (e.g., spectral coverage of bands and spatial resolution) requiring inter-sensor observations and/or data products be converted into a consistent format to maintain data continuity across the historical record. A notable example is the development of a 35+ year historical record of NDVI data (1989–present) over the CONUS that has represented a core remote sensing data set for operational agricultural drought monitoring for several decades. This historical NDVI record comprises NDVI data derived from three different sensors that include AVHRR, MODIS, and VIIRS. The development of this long, inter-sensor NDVI data set required careful consideration of the varying spectral resolutions of the input bands among the three sensors, as well as sensor degradation within a specific sensor's data record, which can introduce anomalous trends/variations in the NDVI record that mimic a drought signal. Several efforts have been made to inter-calibrate the spectral data from the series of NOAA AVHRR sensors (Tucker et al., 2005; Eidenshink, 2006) and merge NDVI observations from AVHRR with MODIS (van Leeuwen et al., 2006) and VIIRS (Benedict et al., 2021) to create an extended, multi-sensor NDVI data record.

Similar multi-sensor retrospective analyses have been applied to other drought variables. For example, the ESA Climate Change Initiative has recently released a 30-year (1979–2009) surface soil moisture retrospective data product based on the merger of active and passive microwave observations obtained from a variety of satellite sensors (Wagner et al., 2012). Analogous efforts to extend the observational record of other remote sensing products such as ESI and TWS have also been possible with the launch of new spaceborne sensors collecting the required data inputs (e.g., thermal infrared and gravity, respectively) for these tools. Similar inter-sensor calibrations and/or model modifications will be necessary in the future as new sensor and data sets become available to ensure that a temporally consistent historical data record is maintained for operational drought monitoring.

Remote sensing of drought has rapidly evolved over the past two decades for operational drought monitoring in the United States. Many of the tools and methods presented in this chapter have expanded to global coverage (Rodell and Reager, 2023) or developed for other countries (Anderson et al., 2016; Tadesse et al., 2017). The complexity and multiple dimensions of this complex natural hazard require a set of tools rather than a single tool to effectively map and monitor drought conditions. The remote sensing products highlighted in this chapter illustrate a tool kit that exists to monitor many key components of the hydrologic cycle related to drought across a vertical continuum ranging from sub-surface groundwater and soil moisture to general plant health and surface-boundary layer conditions (i.e., ET). The role and contribution of satellite-based remote sensing tools continues to develop through improvements in existing tools, as well as the current efforts to build new tools presented earlier in this section. Collectively, continued work in this area by the remote sensing community is important to address the need for more local-scale drought information across multiple time scales (e.g., weeks, months, and seasons) to support a wide range of decision-making activities across multiple sectors (i.e., agriculture, water resources, energy, insurance, and public health).

REFERENCES

Albergel, C., C. Rüdiger, T. Pellarin, J.C. Calvet, N. Fritz, F. Froissard, D. Suquia, A. Petitpa, B. Piguet, and E. Martin. 2008. From near-surface to root-zone soil moisture using an exponential filter: An assessment of the method based on in situ observations and model simulations. *Hydrologic Earth System Science* 12:1323–1337.

Allen, R.G., L.S. Pereira, D. Raes, and M. Smith. 1998. *Crop Evapotranspiration: Guidelines for Computing Crop Water Requirements*. United Nations FAO, Irrigation and Drainage Paper 56, Rome, Italy.

Andersen, O.B., S.I. Seneviratne, J. Hinderer, and P. Viterbo. 2005. GRACE-derived terrestrial water storage depletion associated with the 2003 European heat wave. *Geophysical Research Letters* 32:1–4.

Anderson, M.C., F. Gao, K. Knipper, C. Hain, W. Dulaney, D.D. Baldocchi, E. Eichelmann, K.S. Hemes, Y. Yang, J. Medellín-Azuara, and W.P. Kustas. 2018. Field-scale assessment of land and water use change over the California Delta using remote sensing. *Remote Sensing* 10:889.

Anderson, M.C., C.R. Hain, J.A. Otkin, X. Zhan, K.C. Mo, M. Svoboda, B. Wardlow, and A. Pimstein. 2013. An intercomparison of drought indicators based on thermal remote sensing and NLDAS-2 simulations with U.S. Drought Monitor classifications. *Journal of Hydrometeorology* 14:1035–1056.

Anderson, M.C., C.R. Hain, B. Wardlow, J.R. Mecikalski, and W.P. Kustas. 2011. Evaluation of drought indices based on thermal remote sensing of evapotranspiration over the continental U.S. *Journal of Climate* 24:2025–2044.

Anderson, M.C., W.P. Kustas, J.G. Alfieri, C.R. Hain, J.H. Prueger, S.R. Evett, P.D. Colaizzi, T.A. Howell, and J.L. Chavez. 2012. Mapping daily evapotranspiration at Landsat spatial scales during the BEAREX'08 field campaign. *Advances in Water Resources* 50:162–177.

Anderson, M.C., J.M. Norman, G.R. Diak, W.P. Kustas, and J.R. Mecikalski. 1997. A two-source time-integrated model for estimating surface fluxes using thermal infrared remote sensing. *Remote Sensing of Environment* 60:195–216.

Anderson, M.C., J.M. Norman, W.P. Kustas, R. Houborg, P.J. Starks, and N. Agam. 2008. A thermal-based remote sensing technique for routine mapping of land-surface carbon, water and energy fluxes from field to regional scales. *Remote Sensing of Environment* 112:4227–4241.

Anderson, M.C., J.M. Norman, J.R. Mecikalski, J.A. Otkin, and W.P. Kustas. 2007a. A climatological study of evapotranspiration and moisture stress across the continental U.S. based on thermal remote sensing: I. Model formulation. *Journal of Geophysical Research* 112:D10117. http://doi.org/10110.11029/1200 6JD007506.

Anderson, M.C., J.M. Norman, J.R. Mecikalski, J.A. Otkin, and W.P. Kustas. 2007b. A climatological study of evapotranspiration and moisture stress across the continental U.S. based on thermal remote sensing: II. Surface moisture climatology. *Journal of Geophysical Research* 112:D11112. http://doi.org/11110.110 29/12006JD007507.

Anderson, M.C., J.M. Norman, T.P. Meyers, and G.R. Diak. 2000. An analytical model for estimating canopy transpiration and carbon assimilation fluxes based on canopy light-use efficiency. *Agricultural and Forest Meteorology* 101:265–289.

Anderson, M.C., C.A. Zolin, P.C. Sentelhas, C.R. Hain, K. Semmens, M. Tugrul Tilmaz, F. Gao, J.A. Otkin, and R. Terault. 2016. The Evaporative Stress Index as an indicator of agricultural drought in Brazil: An assessment based on crop yield analysis. *Remote Sensing of Environment* 174:82–99.

Asner, G.P., A.R. Townsend, and B.H. Braswell. 2000. Satellite observation of El Niño effects on Amazon forest phenology and productivity. *Geophysical Research Letters* 27:981–984.

Asrar, G., R.B. Myneni, and E.T. Kanemasu. 1989. Estimation of plant canopy attributes from spectral reflectance measurements. In G. Asrar (ed.), *Theory and Applications of Optical Remote Sensing* (pp. 252–296). New York: Wiley Publishers.

Baret, F., and G. Guyot. 1991. Potentials and limits to vegetation indices for LAI and APAR assessments. *Remote Sensing of Environment* 35:161–173.

Baughn, J.W., and C.B. Tanner. 1976. Leaf water potential: Comparison of pressure chamber and in situ hygrometer on five herbaceous species 1. *Crop Science* 16(2):181–184.

Behrangi, A., E.J. Fetzer, and S.L. Granger. 2016. Early detection of drought onset using near surface temperature and humidity observed from space. *International Journal of Remote Sensing* 37(16):3911–3923.

Benedict, T.D., J.F. Brown, S.P. Boyte, D.M. Howard, B.A. Fuchs, B.D. Wardlow, T. Tadesse, and K.A. Evenson. 2021. Exploring VIIRS continuity with MODIS in an expedited capability for monitoring drought-related vegetation conditions. *Remote Sensing* 13(6):2010.

Bhanja, S.N., A. Mukherjee, and M. Rodell. 2020. Groundwater storage change detection from in situ and GRACE-based estimates in major river basins across India. *Hydrological Sciences Journal* 65(4):650–659.

Biancamaria, S., D.P. Lettenmaier, and T.M. Pavelsky. 2016. The SWOT mission and its capabilities for land hydrology. In A. Cazenave, N. Champollion, J. Benveniste, and J. Chen (eds.), *Remote Sensing and Water Resources. Space Sciences Series of ISSI*, 55. Springer.

Bolten, J.D., and W.T. Crow. 2012. Improved prediction of quasi-global vegetation conditions using remotely-sensed surface soil moisture. *Geophysical Research Letters* 39:L19406. https://doi.org/10.1029/2012GL053470.

Brocca, L., F. Melone, T. Moramarco, and W. Wagner. 2013. A new method for rainfall estimation through soil moisture observations. *Geophysical Research Letters* 40(5):853–858. http://doi.org/10.1002/grl.50173.

Brown, J.F., B.D. Wardlow, T. Tadesse, M.J. Hayes, and B.C. Reed. 2008. The Vegetation Drought Response Index (VegDRI): A new integrated approach for monitoring drought stress in vegetation. *GIScience and Remote Sensing* 45(1):16–46.

Buman, B., A. Hueni, R. Colombo, S. Cogliati, M. Celesti, T. Julitta, A. Burkart, B. Siegmann, U. Rascher, M. Drusch, and A. Damm. 2022. Towards consistent assessments of in situ radiometric measurements for the validation of fluorescence satellite missions. *Remote Sensing of Environment* 274:112984.

Burgan, R.E., R.A. Hartford, and J.C. Eidenshink. 1996. *Using NDVI to Assess Departure from Average Greenness and Its Relation to Fire Business*. Gen. Tech. Rep. INT-GTR-333, U.S. Department of Agriculture, Forest Service, Intermountain Research Station, Ogden, Utah, 8 pp.

Cai, X., Z.L. Yang, Y. Xia, M. Huang, H. Wei, L.R. Leung, and M.B. Ek. 2014. Assessment of simulated water balance from Noah, Noah-MP, CLM, and VIC over CONUS using the NLDAS test bed. *Journal of Geophysical Research: Atmospheres* 119(24):13–751.

Cammalleri, C., M.C. Anderson, F. Gao, C.R. Hain, and W.P. Kustas. 2013. A data fusion approach for mapping daily evapotranspiration at field scale. *Water Resources Research* 49:1–15. http://doi.org/10.1002/wrcr.20349.

Cammalleri, C., M.C. Anderson, F. Gao, C.R. Hain, and W.P. Kustas. 2014. Mapping daily evapotranspiration at field scales over rainfed and irrigated agricultural areas using remote sensing data fusion. *Agricultural and Forest Meteorology* 186:1–11.

Chen, F., F. Lei, K. Knipper, F. Gao, L. McKee, M. del Mar Alsina, J. Alfieri, M. Anderson, N. Bambach, S.J. Castro, A.J. McElrone, K. Alstad, N. Dokoozlian, F. Greifender, W. Kustas, C. Notarnicola, N. Agam, J.H. Prueger, L.E. Hipps, and W.T. Crow. 2022. Application of the vineyard data assimilation (VIDA) system to vineyard root-zone soil moisture monitoring in the California Central Valley. *Irrigation Science*. http://doi.org/10.1007/s00271-022-00789-9.

Chen, J.L., C.R. Wilson, B.D. Tapley, Z.L. Yang, and G.Y. Niu. 2009. 2005 drought event in the Amazon River basin as measured by GRACE and estimated by climate models. *Journal of Geophysical Research* 114:B05404. http://doi.org/10.1029/2008JB006056.

Colliander, A., R.G. Reichle, W.T. Crow, M.H. Cosh, F. Chen, S. Chan, N.N. Das, R. Bindish, J. Chaubell, S. Kim, W. Liu, P.E. O'Neill, R.S. Dunbar, L.B. Dang, J.S. Kimball, T.J. Jackson, H.K. Al-Jassar, J. Asanuma, B. Bhattacharya, A.A. Berg, D.D. Bosch, L. Bourgeau-Chavez, T. Caldwell, J.-C. Calvet, C. Holifield Collins, K.H. Jensen, S. Livingston, J. Martinez-Fernandez, H. McNairn, M. Moghaddam, C. Montzka, C. Notarnicola, T. Pellarin, I. Greimeister-Pfeil, J. Pulliainen, J.G. Ramos Hernandez, M. Seyfried, P.J. Starks, Z. Su, Y. Zeng, M. Vreugdenhil, J.P. Walker, D. Entekhabi, and S.H. Yueh. 2021. Validation of soil moisture data products from the NASA SMAP mission. *IEEE Journal of Selected Topics in Applied Earth Observations and Remote Sensing* 15:364–392.

Collins, M., R. Knutti, J. Arblaster, J.-L. Dufresne, T. Fichefet, P. Friedlingstein, X. Gao, W.J. Gutowski, et al. (2013). Chapter 12—Long-term climate change: Projections, commitments and irreversibility. In: IPCC (ed.), *Climate Change 2013: The Physical Science Basis. IPCC Working Group I Contribution to AR5*. Cambridge: Cambridge University Press.

Crow, W.T., M.J. van Den Berg, G.F. Huffman, and T. Pellarin. 2011. Correcting rainfall using satellite-based surface soil moisture retrievals: The Soil Moisture Analysis Rainfall Tool (SMART). *Water Resources Research* 47:W08521. http://doi.org/10.1029/2011WR010576.

Crow, W.T., and X. Zhan. 2007. Continental-scale evaluation of remotely-sensed soil moisture products. *IEEE Geoscience and Remote Sensing Letters* 4(3):451–455.

Daly, C., M. Halbleib, J.I. Smith, J.W.P. Gibson, M.K. Doggett, G.H. Taylor, J. Curtisand, and P.A. Pasteris. 2008. Physiographically-sensitive mapping of temperature and precipitation across the conterminous United States. *International Journal of Climatology* 28:2031–2064.

Daly, C., J.I. Smith, and K.V. Olson. 2015. Mapping atmospheric moisture climatologies across the conterminous United States. *PloS ONE* 10(10):e0141140. http://doi.org/10.1371/journal.pone.0141140.

Damm, A., S. Cogliati, R. Colombo, L. Fritschem, A. Genangeli, L. Genesio, J. Hanus, A. Peressotti, P. Rademske, U. Rascher, and D. Schuettemeyer. 2022. Response times of remote sensing measured sun-induced chlorophyll fluorescence, surface temperature and vegetation indices to evolving soil water limitation in a crop canopy. *Remote Sensing of Environment* 273:112957.

Department of the Interior (DOI). 2013. Final environmental assessment—Carson city district drought management. *DOI-BLM-NV_C000–2013–0001-EA Report*, 236 pp.

Dong, J., R. Akbar, D.J. Short-Gianotti, A.F. Feldman, W.T. Cro, and D. Entekhabi. 2022. Can surface soil moisture information identify evapotranspiration regime transitions? *Geophysical Research Letters* 49:e2021GL097697.

Dong, J., W.T. Crow, R.H. Reichle, Q. Liu, F. Lei, and M. Cosh. 2019. A global assessment of added value in the SMAP Level 4 soil moisture product relative to its baseline land surface model. *Geophysical Research Letters* 46:6604–6613.

Draper, C.S., R.H. Reichle, G.J.M. De Lannoy, and Q. Liu. 2012. Assimilation of passive and active microwave soil moisture retrievals. *Geophysical Research Letters* 39:L04401.

Durand, M., L.L. Fu, D.P. Lettenmaier, D. Alsdorf, E. Rodriguez, and D. Esteban-Fernandez. 2010. The Surface Water and Ocean Topography mission: Observing terrestrial surface water and oceanic submesoscale eddies. *Proceedings of IEEE* 98(5):766–779.

Edwards, D.C., and T.B. McKee. 1997. Characteristics of 20th century drought in the United States at multiple time scales. *Climatology Report Number 97–2*, Department of Atmospheric Science, Colorado State University, Fort Collins, CO.

Eidenshink, J.C. 2006. A 16-year time series of 1 km AVHRR satellite data of the conterminous United States and Alaska. *Photogrammetric Engineering and Remote Sensing* 72:1027–1035.

Evensen, G., and P.J. van Leeuwen. 2000. An Ensemble Kalman smoother for nonlinear dynamics. *Monthly Weather Review* 128:1852–1867.

Farahmand, A., S. Ray, H. Thrastarson, S. Licata, S. Granger, and B. Fuchs. 2023. A workshop on using NASA AIRS data to monitor drought for the US Drought Monitor. *Bulletin of the American Meteorological Society* 104(1):E22–E30.

Feldman, A.F., D.J. Short Ginotti, J. Dong, R. Akbar, W.T. Crow, K.A. McCooll, A.G. Konigs, J.B. Nippert, S.J. Tumber-Davila, N.M. Holbrook, F.E. Rockwell, R.L. Scott, R.H. Reichle, A. Chatterjee, J. Joiner, B. Poulter, and D. Entekhabi. 2023. Remotely sensed soil moisture can capture dynamics relevant to plant water uptake. *Water Resources Research* 59:e2022WR033814.

Fletcher, A.L., T.R. Sinclair, and L.H. Allen Jr. 2007. Transpiration responses to vapor pressure deficit in well watered 'slow-wilting' and commercial soybean. *Environmental and Experimental Botany* 61(2):145–151.

Frankenberg, C., J.B. Fisher, J. Worden, G. Badgley, S.S. Saatchi, J.E. Lee, G.C. Toon, A. Butz, M. Jung, A. Kuze, and Y. Yokota. 2011. New global observations of the terrestrial carbon cycle from GOSAT: Patterns of plant fluorescence with gross primary productivity. *Geophysical Research Letters* 38(17).

Frankenberg, C., C. O'Dell, J. Berry, L. Guanter, J. Joiner, P. Köhler, R. Pollock, and T.E. Taylor. 2014. Prospects for chlorophyll fluorescence remote sensing from the Orbiting Carbon Observatory-2. *Remote Sensing of Environment* 147:1–12.

Frappart, F., J.-P. Wigneron, X. Li, X. Liu, A. Al-Yaari, L. Fan, M. Wang, C. Moisy, E. Le Masson, Z. Aoulad Lafkih, C. Valle, B. Ygorra, and N. Baghdadi. 2022. Global monitoring of the vegetation dynamics from the vegetation optical depth (VOD): A review. *Remote Sensing* 12(18):2915.

Funk, C., J. Michaelsen, and M.T. Marshall. 2012. Mapping recent decadal climate variations in precipitation and temperature across eastern Africa. In B.D. Wardlow, M.A. Anderson, and J. Verdin (eds.), *Remote Sensing of Drought: Innovative Monitoring Approaches* (pp. 331–358). Boca Raton, FL: CRC Press/Taylor and Francis.

Gao, H., E.F. Wood, T.J. Jackson, M. Drusch, and R. Bindlish. 2006. Using TRMM/TMI to retrieve surface soil moisture over the Southern United States from 1998 to 2002. *Journal of Hydrometeorology* 7:23–38.

Getirana, A., M. Rodell, S. Kumar, H.K. Beaudoing, K. Arsenault, B. Zaitchik, H. Save, and S. Bettadpur. 2020. GRACE improves seasonal groundwater forecast initialization over the United States. *Journal of Hydrometeorology* 21(1):59–71. http://doi.org/10.1175/JHM-D-19-0096.1.

Gu, Y., J.F. Brown, J.P. Verdin, and B. Wardlow. 2007. A five-year analysis of MODIS NDVI and NDWI for grassland drought assessment over the central Great Plains of the United States. *Geophysical Research Letters* 34. http://doi.org/10.1029/2006GL029127.

Gu, Y., E. Hunt, B. Wardlow, J.B. Basara, J.F. Brown, and J.P. Verdin. 2008. Evaluation of MODIS NDVI and NDWI for vegetation drought monitoring using Oklahoma Mesonet soil moisture data. *Geophysical Research Letters* 35. http://doi.org/10.1029/2008GL035772.

Hain, C.R., and M.C. Anderson. 2017. Estimating morning changes in land surface temperature from MODIS day/night observations: Applications for surface energy balance modeling. *Geophysical Research Letters* 44:9723–9733.

Hain, C.R., W.T. Crow, M.C. Anderson, and J.R. Mecikalski. 2012. Developing a dual assimilation approach for thermal infrared and passive microwave soil moisture retrievals. *Water Resources Research* 48:W11517. http://doi.org/11510.11029/12011WR011268.

Hain, C.R., W.T. Crow, J.R. Mecikalski, M.C. Anderson, and T. Holmes. 2011. An intercomparison of available soil moisture estimates from thermal-infrared and passive microwave remote sensing and land-surface modeling. *Journal of Geophysical Research* 116:D15107. http://doi.org/15110.11029/12011JD015633.

Hobbins, M.T., A. Wood, D.J. McEvoy, J.J. Huntington, C. Morton, M. Anderson, and C. Hain. 2016. The evaporative demand drought index. Part I: Linking drought evolution to variations in evaporative demand. *Journal of Hydrometeorology* 17(6):1745–1761.

Holmes, T., R. de Jeu, M. Owe, and A. Dolman. 2009. Land surface temperature from Ka band passive microwave observations. *Journal of Geophysical Research of the Atmosphere* 114:D04113. http://doi.org/10.1029/2008JD010257.

Holmes, T., C. Hain, W. Crow, M.C. Anderson, and W.P. Kustas. 2018. Microwave implementation of two-source energy balance approach for estimating evapotranspiration. *Hydrology and Earth System Science* 22:1351–1369.

Holmes, T., J. Jackson, R. Reichle, and J. Basara. 2012. An assessment of surface soil temperature products from numerical weather prediction models using ground-based measurements. *Water Resources Research* 48:W02531. http://doi.org/10.1029/2011WR010538.

Homer, C., C. Huang, L. Yang, B. Wylie, and M. Coan. 2004. Development of a 2001 national land cover database for the United States. *Photogrammetric Engineering and Remote Sensing* 70:829–840.

Houborg, R., M. Rodell, B. Li, R. Reichle, and B. Zaitchik. 2012. Drought indicators based on model assimilated GRACE terrestrial water storage observations. *Water Resources Research* 48:W07525. http://doi.org/10.1029/2011WR011291.

Huffman, G., R. Alder, D. Bolvin, G. Gu, E. Nelkin, K. Bowman, Y. Hong, E.F. Stocker, and D.B. Wolff. 2007. The TRMM multi-scale precipitation analysis: Quasi-global, multiyear combined-sensor precipitation estimates at fine scale. *Journal of Hydrometeorology* 8:38–55.

Hutchinson, C.F. 1991. Use of satellite data for famine early warning in sub-Saharan Africa. *International Journal of Remote Sensing* 12:1405–1421.

Ines, A.V.M., N.N. Das, J.W. Hansen, and E.G. Njoku. 2013. Assimilation of remotely sensed soil moisture and vegetation with a crop simulation model for maize yield prediction. *Remote Sensing of Environment* 138:149–164.

Jackson, T.J. 1993. Measuring surface soil moisture using passive microwave remote sensing. *Hydrological Processes* 7(2):139–152.

Jackson, T.J., T.J. Schmugge, and J.R. Wang. 1982. Passive microwave remote sensing of soil moisture under vegetation canopies. *Water Resources Research* 18:1137–1142.

Jenkerson, C., T. Maiersperger, and G. Schmidt. 2010. *eMODIS—A User-Friendly Data Source*. U.S. Geological Survey Open-File Report 2010–1055. http://pubs.usgs.gov/of/2010/1055/ (accessed September 29, 2023).

Ji, L., and A.J. Peters. 2003. Assessing vegetation response to drought in the northern great plains using vegetation and drought indices. *Remote Sensing of Environment* 87:85–98.

Jiao, W., G. Chang, and L. Wang. 2019. The sensitivity of satellite solar-induced chlorophyll fluorescence to meteorological drought. *Earth's Future* 7(5):558–573.

Joiner, J., G.R. Lindstrot, M. Voight, A.P. Casikov, E.M. Middleton, J.F. Hummerich, Y. Yoshida, and C. Frankenberg. 2013. Global monitoring of terrestrial chlorophyll fluorescence from moderate spectral resolution near-infrared satellite measurements: Methodology, simulations, and application to GOME-2. *Atmospheric Measurement Techniques* 6(2):2803–2823.

Joyce, R., J. Janowiak, P. Arkin, and P. Xie. 2004. CMORPH: A method that produces global precipitation estimates from passive microwave and infrared data at high spatial and temporal resolution. *Journal of Hydrometeorology* 5:487–503.

Karnieli, A., N. Agam, R.T. Pinker, M. Anderson, M.L. Imhoff, G.G. Gutman, N. Panov, and A. Goldberg. 2010. Use of NDVI and land surface temperature for drought assessment: Merits and limitations. *Journal of Climate* 23(3):618–633.

Karnieli, A., and G. Dall'Olmo. 2003. Remote-sensing monitoring of desertification, phenology, and drought. *Management of Environmental Quality: An International Journal* 41(1):22–38.

Keetch, J.J., and G.M. Byram. 1968. *A Drought Index for Forest Fire Control*. Research Paper SE-38. Asheville, NC: U.S. Department of Agriculture, Forest Service, Southeastern Forest Experiment Station, 32 pp. (Revised 1988).

Kerr, Y.H., and D. Levine. 2008. Forward to the special issue on the Soil Moisture and Ocean Salinity (SMOS) mission. *IEEE Transaction in Geoscience and Remote Sensing* 46(3):583–585.

Knipper, K.R., W.P. Kustas, M.C. Anderson, J.G. Alfieri, J.H. Prueger, C.R. Hain, F. Gao, Y. Yang, L.G. McKee, H. Nieto, L.E. Hipps, M.M. Alsina, and L. Sanchez. 2019. Evapotranspiration estimates derived using thermal-based satellite remote sensing and data fusion for irrigation management in California vineyards. *Irrigation Science* 37:431–449.

Kogan, E.N. 1990. Remote sensing of weather impacts on vegetation. *International Journal of Remote Sensing* 11:1405–1419.

Kogan, F. 1995. Application of vegetation index and brightness temperature for drought detection. *Advances in Space Research* 15:91–100.

Kogan, F., and J. Sullivan. 1993. Development of global drought-watch system using NOAA/AVHRR data. *Advances in Space Research* 13:219–222.

Kohler, P. 2018. Global retrievals of solar-induced chlorophyll fluorescence with TROPOMI: First results and intersensory comparison to OCO-2. *Geophysical Research Letters* 45(19):10:456–462.

Kongoli, C., P. Romanov, and R. Ferraro. 2012. Snow cover monitoring from remote-sensing satellites: Possibilities for drought assessment. In B.D. Wardlow, M.A. Anderson, and J. Verdin (eds.), *Remote Sensing of Drought: Innovative Monitoring Approaches* (pp. 359–386). Boca Raton, FL: CRC Press/Taylor and Francis.

Koster, R.D., M.J. Suarez, A. Duchame, M. Stieglitz, and P. Kumar. 2000. A catchement-based approach to modeling land surface processes in a general circulatin model: 1. Model structure. *Journal of Geophysical Research: Atmospheres* 105(D20):24809–24822.

Kumar, S.V., C.D. Peters-Lidard, D. Mocko, R. Reichle, Y. Liu, K.R. Arsenault, Y. Xia, M. Ek, G. Riggs, B. Livneh, and M. Cosh. 2014. Assimilation of remotely sensed soil moisture and snow depth retrievals for drought estimation. *Journal of Hydrometeorology* 15(6):2446–2469.

Kumar, S.V., B.F. Zaitchik, C.D. Peters-Lidard, M. Rodell, R. Reichle, B. Li, M. Jasinski, D. Mocko, A. Getirana, G. De Lannoy, and M.H. Cosh. 2016. Assimilation of gridded GRACE terrestrial water storage estimates in the North American land data assimilation system. *Journal of Hydrometeorology* 17(7):1951–1972.

Kustas, W.P., and J.M. Norman. 1999. Evaluation of soil and vegetation heat flux predictions using a simple two-source model with radiometric temperatures for partial canopy cover. *Agricultural and Forest Meteorology* 94:13–29.

Kustas, W.P., and J.M. Norman. 2000. A two-source energy balance approach using directional radiometric temperature observations for sparse canopy covered surfaces. *Agronomy Journal* 92:847–854.

Landerer, F.W., F.M. Flechtner, H. Save, F.H. Webb, T. Bandikova, W.I. Bertiger, . . . & D.N. Yuan. 2020. Extending the global mass change data record: GRACE follow-on instrument and science data performance. *Geophysical Research Letters* 47(12):e2020GL088306.

Leblanc, M.J., P. Tregoning, G. Ramillien, S.O. Tweed, and A. Fakes. 2009. Basin-scale, integrated observations of the early 21st century multiyear drought in southeast Australia. *Water Resources Research* 45:W04408. http://doi.org/10.1029/2008WR007333.

Li, B., M. Rodell, S. Kumar, H.K. Beaudoing, A. Getirana, B.F. Zaitchik, L.G. de Goncalves, C. Cossetin, S. Bhanja, A. Mukherjee, and S. Tian. 2019. Global GRACE data assimilation for groundwater and drought monitoring: Advances and challenges. *Water Resources Research* 55(9):7564–7586. http://doi.org/10.1029/2018WR024618.

Li, L., E.G. Njoku, E. Im, P.S. Chang, and K.S. Germain. 2004. A preliminary survey of radio-frequency interference over the US in Aqua AMSR-E data. *IEEE Transactions in Geoscience and Remote Sensing* 42:380–390.

Li, P., N. Omani, L. Chaubey, and X. Wei. 2017. Evaluation of drought implications on ecosystem services: Freshwater provisioning and food provisioning in the Upper Mississippi River Basin. *International Journal of Environmental Research and Public Health* 14(5):496. http://doi.org/10.3390/ijerph14050496.

Li, Q., R.H. Reichle, R. Bindish, M.H. Cosh, W.T. Crow, R. de Jeu, G. De Lannoy, G.J. Huffman, and T.J. Jackson. 2011. The contributions of precipitation and soil moisture observations to the skill of soil moisture estimates in a land data assimilation system. *Journal of Hydrometeorology* 12(5):750–765.

Liu, P.W., R. Bindlish, P.E. Oneill, B. Fang, V. Lakshmi, Z. Yang, M. Cosh, T. Bongiovanni, C.H. Collins, P. Starks, J. Prueger, D.D. Bosch, M. Seyfried, and M. Williams. 2022. Thermal hydraulic disaggregation of SMAP soil moisture over the continental United States. *IEEE Journal of Selected Topics in Applied Earth Observation Remote Sensing*. http://doi.org/10.1109/JSTARS.2022.3165644.

Liu, Y.Y., R.M. Parinussa, W.A. Dorigo, R.A. De Jeu, W. Wagner, A.L.J.M. Van Dijk, M.F. McCabe, and J.P. Evans. 2011. Developing an improved soil moisture dataset by blending passive and active microwave satellite-based retrievals. *Hydrology and Earth System Sciences* 15(2):425–436.

Liu, Y.Y., A.I.J.M. van Dijk, M.F. McCabe, J.P. Evans, and R.A.M. de Jeu. 2013. Drivers of global vegetation biomass trends (1988–2008) and attribution to environmental and human drivers. *Global Ecology and Biogeography*. http://doi.org/10.1111/geb.12024.

Loomis, B.D., S.B. Luthcke, and T.J. Sabaka. 2019. Regularization and error characterization of GRACE mascons. *Journal of Geodesy* 93:1381–1398.

McEvoy, D.J., J.L. Huntington, M.T. Hobbins, A. Wood, C. Morton, M. Anderson, and C. Hain. 2016. The evaporative demand drought index. Part II: CONUS-wide assessment against common drought indicators. *Journal of Hydrometeorology* 17(6):1763–1779.

McKee, T.B., N.J. Doesken, and J. Kleist. 1985. Drought monitoring with multiple time scales. *Ninth Conference on Applied Climatology*, Dallas, TX, January 15–20.

McVicar, T.R., and P.B. Bierwirth. 2001. Rapidly assessing the 1997 drought in Papua New Guinea using composite AVHRR imagery. *International Journal of Remote Sensing* 22:2109–2128.

Mishra, A.K., and V.P. Singh. 2010. A review of drought concepts. *Journal of Hydrology* 391(1):202–216.

Mladenova, I.E., J.D. Bolten, W.T. Crow, N. Sazib, M.H. Cosh, C.J. Tucker, and C. Reynolds. 2019. Evaluating the operational application of SMAP for global agricultural drought monitoring. *IEEE Journal of Selected Topics in Applied Earth Observations and Remote Sensing* 12(9):3387–3397.

Moesinger, L., R.-M. Zotta, R. van der Schalie, T. Scanlon, R. de Jeuand, and W. Dorigo. 2022. Monitoring vegetation condition using microwave remote sensing: The standardized vegetation optical depth index (SVODI). *Biogeosciences* 19:5107–5123.

Mohammed, P.N., M. Aksoy, J.R. Piepmeier, J.T. Johnson, and A. Bringer. 2016. SMAP L-band microwave radiometer: RFI mitigation prelaunch analysis and first year on-orbit observations. *IEEE Transactions on Geoscience and Remote Sensing* 54(10):6035–6047.

Muñoz-Sabater, J., H. Lawrence, C. Albergel, P. Rosnay, L. Isaksen, S. Mecklenburg, Y. Kerr, and M. Drusch. 2019. Assimilation of SMOS brightness temperatures in the ECMWF integrated forecasting system. *Quarterly Journal of the Royal Meteorological Society* 145:2524–2548.

Naeimi, V., K. Scipal, Z. Bartalis, S. Hasenauer, and W. Wagner. 2009. An improved soil moisture retrieval algorithm for ERS and METOP scatterometer observations. *IEEE Transactions on Geoscience and Remote Sensing* 47(7):1999–2013.

NASA GSFC. 2023. *Hydrological Sciences Laboratory—NASA Gesdisc Data Archive*. https://hydro1.gesdisc.eosdis.nasa.gov/data/GLDAS/GLDAS_CLSM025_DA1_D.2.2/ (accessed September 29, 2023).

National Climatic Data Center (NCDC). 2014. *Billion Dollar U.S. Weather Disasters*. http://www.ncdc.noaa.gov/oa/reports/billionz.html (accessed February 7, 2014).

National Oceanic and Atmospheric Administration (NOAA). 2023. *Climate Monitoring: Temperature, Precipitation, and Drought*. http://www.ncdc.noaa.gov/climate-monitoring/ (accessed September 20, 2023).

Nearing, G.S., D.M. Mocko, C.D. Peters-Lidard, S.V. Kumar, and T. Xia. 2016. Benchmarking NLDAS-2 soil moisture and evapotranspiration to separate uncertainty contributions. *Journal of Hydrometeorology* 17(3):745–759.

Njoku, E.G., T.J. Jackson, V. Lakshmi, T.K. Chan, and S.V. Nghiem. 2003. Soil moisture retrieval from AMSR-E. *IEEE Transactions on Geoscience and Remote Sensing* 41(2):215–229.

Norman, J.M., M.C. Anderson, W.P. Kustas, A.N. French, J.R. Mecikalski, R.D. Torn, G.R. Diak, T.J. Schmugge, and B.C.W. Tanner. 2003. Remote sensing of surface energy fluxes at 10^1-m pixel resolutions. *Water Resources Research* 39. http://doi.org/10.1029/2002WR001775.

Norman, J.M., W.P. Kustas, and K.S. Humes. 1995. A two-source approach for estimating soil and vegetation energy fluxes from observations of directional radiometric surface temperature. *Agricultural and Forest Meteorology* 77:263–293.

Oliva, R., E. Daganzo, Y.H. Kerr, S. Mecklenburg, S. Nieto, P. Richaume, and C. Gruhier. 2012. SMOS radio frequency interference scenario: Status and actions taken to improve the RFI environment in the 1400–1427-MHz passive band. *IEEE Transactions on Geoscience and Remote Sensing* 50(5):1427–1439.

Omernik, J.M. 1987. Ecoregions of the conterminous United States. *Annals of the Association of American Geographers* 77(1):118–125.

Otkin, J.A., M.C. Anderson, C.R. Hain, L.E. Mladenova, J.B. Basara, and M. Svoboda. 2013a. Examining rapid onset drought development using the thermal infrared based Evaporative Stress Index. *Journal of Hydrometeorology* 14:1057–1074.

Otkin, J.A., M.C. Anderson, C.R. Hain, and M. Svoboda. 2013b. Examining the relationship between drought development and rapid changes in the Evaporative Stress Index. *Journal of Hydrometeorology*. http://doi.org/10.1175/JHM-D-13-0110.1.

Owe, M., R. de Jeu, and T. Holmes. 2008. Multisensor historical climatology of satellite-derived global land surface moisture. *Journal of Geophysical Research* 113:F01002. http://doi.org/10.1029/2007JF000769.

Palmer, W.C. 1965. *Meteorological Drought*. Research Paper Number 4, Office of Climatology, U.S. Weather Bureau, Washington, DC, 58 pp.

Palmer, W.C. 1968. Keeping track of crop moisture conditions, nationwide: The new crop moisture index. *Weatherwise* 21(4):156–161.

Pellarin, T., A. Ali, F. Chopin, I. Jobard, and J.-C. Bergs. 2008. Using spaceborne surface soil moisture to constrain satellite precipitation estimates over West Africa. *Geophysical Research Lett*ers 35:L02813. http://doi.org/10.1029/2007GL032243.

Pervez, M.S., and J.F. Brown. 2010. Mapping irrigated lands at 250-m scale by merging MODIS data and national agricultural statistics. *Remote Sensing* 2:2388–2412.

Peters, A.J., D.C. Rundquist, and D.A. Wilhite. 1991. Satellite detection of the geographic core of the 1988 Nebraska drought. *Agricultural and Forest Meteorology* 57:35–47.

Peters, A.J., E.A. Walter-Shea, L. Ji, A. Vina, M. Hayes, and M.D. Svoboda. 2002. Drought monitoring with NDVI-based Standardized Vegetation Index. *Photogrammetric Engineering and Remote Sensing* 68(1):71–75.

Philip, J.R. 1966. Plant water relations: Some physical aspects. *Annual Review of Plant Physiology* 17(1):245–268.

PRISM Climate Group. Oregon State University. http://prism.oregonstate.edu (accessed September 29, 2023).

Qiu, J., W.T. Crow, G.S. Nearing, X. Mo, and S. Liu. 2014. The impact of vertical measurement depth on the information content of soil moisture times series data. *Geophysical Research Letters* 41(14):4997–5004.

Quinlan, J.R. 1993. *C4.5 Programs for Machine Learning*. San Mateo, CA: Morgan Kaufmann Publishers.

Reed, B.C., J.F. Brown, D. VanderZee, T.R. Loveland, J.W. Merchant, and D.O. Ohlen. 1994. Measuring phenological variability from satellite imagery. *Journal of Vegetation Science* 5:703–714.

Reichle, R.H., G. De Lannoy, R.D. Koster, W.T. Crow, J.S. Kimball, Q. Liu, and M. Bechtold. 2022. *SMAP L4 Global 3-Hourly 9 km EASE-Grid Surface and Root Zone Soil Moisture Geophysical Data, Version 7 [Data Set]*. Boulder, CO: NASA National Snow and Ice Data Center Distributed Active Archive Center. https://doi.org/10.5067/EVKPQZ4AFC4D (accessed September 20, 2023).

Reichle, R.H., R.D. Koster, P. Liu, S.P.P. Mahanama, E.G. Njoku, and M. Owe. 2007. Comparison and assimilation of global soil moisture retrievals from the Advanced Microwave Scanning Radiometer for the Earth Observing System (AMSR-E) and the Scanning Multichannel Microwave Radiometer (SMMR). *Journal of Geophysical Research* 112:D09108.

Reichle, R.H., Q. Liu, J.V. Ardizzone, W.T. Crow, G.J.M. De Lannoy, J. Dong, J.S. Kimball, and R.D. Koster. 2021. The contributions of gauge-based precipitation and SMAP brightness temperature observations to the skill of the SMAP Level-4 Soil Moisture Product. *Journal of Hydrometeorology* 22(2):405–424.

Reichle, R.H., Q. Liu, J.V. Ardizzone, W.T. Crow, G.J.M. De Lannoy, J.S. Kimball, and R.D. Koster. 2023. IMERG precipitation improves the SMAP level-4 soil moisture product. *Journal of Hydrometeorology*. http://doi.org/10.1175/JHM-D-23-0063.1.

Rodell, M. 2012. Satellite gravimetry applied to drought monitoring. In B. Wardlow, M. Anderson, and J. Verdin (eds.), *Remote Sensing of Drought: Innovative Monitoring Approaches* (pp. 261–280). Boca Raton, FL: CRC Press/Taylor and Francis.

Rodell, M., and J.S. Famiglietti. 1999. Detectability of variations in continental water storage from satellite observations of the time dependent gravity field. *Water Resources Research* 35:2705–2723.

Rodell, M., and J.T. Reager. 2023. Water cycle science enabled by the GRACE and GRACE-FO satellite missions. *Nature Water* 1(1):47–59. http://doi.org/10.1038/s44221-022-00005-0.

Rouse, J.W. Jr., R.H. Haas, J.A. Schell, D.W. Deering, and J.C. Harlan. 1974. Monitoring the vernal advancement and retrogradation (green wave effect) of natural vegetation. *NASA/GSFC Type III Final Report*, Greenbelt, MD.

Rowlands, D.D., S.B. Luthcke, S.M. Klosko, F.G. Lemoine, D.S. Chinn, J.J. McCarthy, and O.B. Anderson. 2005. Resolving mass flux at high spatial and temporal resolution using GRACE intersatellite measurements. *Geophysical Research Letters*, 32(4). http://doi.org/10.1029/2004GL021908.

Ruffault, J., N. Martin-StPaul, F. Pimont, and J.J. Dupuy. 2018. How well do meteorological drought indices predict live fuel moisture content (LFMC)? An assessment for wildfire research and operations in Mediterranean ecosystems. *Agricultural and Forest Meteorology* 262:391–401.

Santanello, J.A., and M.A. Friedl. 2003. Diurnal variation in soil heat flux and net radiation. *Journal of Applied Meteorology* 42:851–862.

Save, H., S. Bettadpur, and B.D. Tapley. 2016. High-resolution CSR GRACE RL05 mascons. *Journal of Geophysical Research: Solid Earth* 121:7547–7569.

Scholander, P.F., F.D. Bradstreet, E.A. Hemmingsen, and H.T. Hammel. 1965. Sap pressure in vascular plants: Negative hydrostatic pressure can be measured in plants. *Science* 148(3668):339–346.

Sehgal, V., N. Gaur, and B.P. Mohanty. 2021. Global flash drought monitoring using surface soil moisture. *Water Resources Research* 57:e2021WR029901.

Shafer, B.A., and L.E. Dezman. 1982. Development of a surface water supply index (SWSI) to assess the severity of drought conditions in snowpack runoff areas. *Preprints, Water Snow Conference*, Reno, NV, pp. 164–175.

Sheffield, J., G. Goteti, and E.F. Wood. 2006. Development of a 50-yr high-resolution global dataset of meteorological forcings for land surface modeling. *Journal of Climate* 19(13):3088–3111.

Shutko, A.M. 1986. *Microwave Radiometry of Water and Terrain Surfaces*. Moscow, Russia: Nauka Publisher.

Sorooshian, S., K.L. Hsu, X. guo, H.V. Gupta, B. Imam, and D. Braithwaite. 2000. Evaluation of PERSIANN system satellite-based estimates of tropical rainfall. *Bulletin of the American Meteorological Society* 81(9):2035–2046.

Svoboda, M., D. LeComte, M. Hayes, R. Heim, K. Gleason, J. Angel, B. Rippey, R. Tinker, M. Palecki, D. Stooksbury, D. Miskus, and S. Stephens. 2002. The drought monitor. *Bulletin of the American Meteorological Society* 83(8):1181–1190.

Tadesse, T., C. Champagne, B.D. Wardlow, T.A. Hadwen, J.F. Brown, G.B. Getachew, Y.A. Bayissa, and A.M. Davidson. 2017. Building the vegetation drought response index for Canada (VegDRI-Canada) to monitor agricultural drought: First results. *GIScience & Remote Sensing* 54(2):230–257.

Tadesse, T., D.Y. Hollinger, Y. Bayissa, M. Svoboda, B. Fuchs, B. Zhang, G. Demisse, B. Wardlow, G. Bohrer, K. Clark, L. Gu, A. Noormets, K. Novick, and A. Richardson. 2020. Forest Drought Response Index (ForDRI): A new combined model to monitor forest drought in the eastern United States. *Remote Sensing* 12(21):3605. https://doi.org/10.3390/rs12213605.

Tapley, B.D., S. Bettadpur, J.C. Ries, P.F. Thompson, and M.M. Watkins. 2004. GRACE measurements of mass variability in the Earth system. *Science* 305:503–505.

Taylor, T.E., A. Eldering, A. Merrelli, M. Kiel, P. Somkuti, C. Cheng, R. Rosenberg, B. Fisher, D. Crisp, R. Basilio, and M. Bennett. 2020. OCO-3 early mission operations and initial (vEarly) XCO2 and SIF retrievals. *Remote Sensing of Environment* 251:112032.

Thomas, A.C., J.T. Reager, J.S. Famiglietti, and M. Rodell. 2014. A GRACE-based water storage deficit approach for hydrological drought characterization. *Geophysical Research Letters* 41(5):1537–1545.

Tucker, C.J., C.O. Justice, and S.D. Prince. 1986. Monitoring the grasslands of the Sahel 1984–1985. *International Journal of Remote Sensing* 7:1571–1581.

Tucker, C.J., J.E. Pinzon, M.E. Brown, D.A. Slayback, E.W. Pak, R. Mahoney, E.F. Vermote, and N. El Saleous. 2005. An extended AVHRR 8-km NDVI dataset compatible with MODIS and SPOT Vegetation NDVI data. *International Journal of Remote Sensing* 26(20):4485–4498.

Unganai, L.S., and F.N. Kogan. 1998. Drought monitoring and corn yield estimation in southern Africa from AVHRR data. *Remote Sensing of Environment* 63:219–232.

USDA (United States Department of Agriculture). 1994. *State Soil Geographic (STATSGO) Data Base: Data Use Information*. USDA Miscellaneous Publication 1492:1–113.

USDA Forest Service. 2023. *National Forest Type Dataset*. https://data.fs.usda.gov/geodata/rastergateway/forest_type/ (accessed September 29, 2023).

USGS. 2023. *EROS Moderate Resolution Imaging Spectroradiometer (eMODIS) Digital Object Identifier*. http://doi.org/10.5066/F7H41PNT. https://www.usgs.gov/centers/eros/science/usgs-eros-archive-vegetation-monitoring-eros-moderate-resolution-imaging?qt-science_center_objects=0#qt-science_center_objects (accessed September 29, 2023).

Van Leeuwen, W.J.D., B.J. Orr, S.E. Marsh, and S.M. Hermann. 2006. Multi-sensor NDVI data continuity: Uncertainties and implications for vegetation monitoring applications. *Remote Sensing of Environment* 100(1):67–81.

Verdin, J., C. Funk, R. Klaver, and D. Robert. 1999. Exploring the correlation between Southern Africa NDVI and Pacific sea surface temperatures: Results for the 1998 maize growing season. *International Journal of Remote Sensing* 20:2117–2124.

Vicente-Serrano, S.M., S. Beguería, and J.I. López-Moreno. 2010. A Multi-scalar drought index sensitive to global warming: The standardized precipitation evapotranspiration index—SPEI. *Journal of Climate* 23:1696–1718.

Vicente-Serrano, S.M., C. Beguerìaand, and J.L. López-Moreno. 2010. A multiscalar drought index sensitive to global warming: The standardized precipitation evapotranspiration index. *Journal of Climate* 23(7):1696–1718.

Wagner, W., G. Blöschl, P. Pampaloni, J.-C. Calvet, B. Bizzarri, J.-P. Wigneron, and Y. Kerr. 2007. Operational readiness of microwave remote sensing of soil moisture for hydrologic applications. *Nordic Hydrology* 38(1):1–20.

Wagner, W., W. Dorigo, R. de Jeu, D. Fernandez, J. Benveniste, E. Haas, and M. Ertl. 2012. Fusion of active and passive microwave observations to create an Essential Climate Variable data record on soil moisture. *ISPRS Annuals of the Photeogrammetry, Remote Sensing and Spatial Information Sciences*, Volume I–7, XXII ISPRS Congress, Melbourne, Australia, 25 August–1 September, pp. 315–321.

Wagner, W., G. Lemoine, and H. Rott. 1999. A method for estimating soil moisture from ERS scatterometer and soil data. *Remote Sensing of Environment* 70(2):191–207.

Wahr, J., M. Molenaar, and F. Bryan. 1998. Time-variability of the Earth's gravity field: Hydrological and oceanic effects and their possible detection using GRACE. *Journal of Geophysical Research* 103(30):205–230.

Wang, L., J.J. Qu, and X. Hao. 2007. Forest fire detection using the normalized multi-band drought index (NMDI) with satellite measurements. *Agricultural and Forest Meteorology* 148:1767–1776.

Wahr, J., S. Swenson, and I. Velicogna. 2006. Accuracy of GRACE mass estimates. *Geophysical Research Letters* 33:L06401. http://doi.org/10.1029/2005GL025305.

Wells, N., S. Goddard, and M.J. Hayes. 2004. A self-calibrating Palmer Drought Severity Index. *Journal of Climate* 17(12):2335–2351.

Wen, J., T.J. Jackson, R. Bindlish, A.Y. Hsu, and Z.B. Su. 2005. Retrieval of soil moisture and vegetation water content using SSM/I data over a corn and soybean region. *Journal of Hydrometeorology* 6:854–863.

Wiese, D.N., B. Bienstock, C. Blackwood, J. Chrone, B.D. Loomis, J. Sauber, M. Rodell, R. Baize, D. Bearden, K. Case, and S. Horner. 2022. The mass change designated observable study: Overview and results. *Earth and Space Science* 9(8):e2022EA002311.

Wiese, D.N., F.W. Landerer, and M.M. Watkins. 2016. Quantifying and reducing leakage errors in the JPL RL05M GRACE mascon solution. *Water Resources Research* 52:7490–7502.

Wigneron, J.-P., X. Li, F. Frappart, L. Fan, A. Al-Yaari, G. De Lannoy, X. Liu, M. Wang, E. Le Masson, and C. Moisy. 2020. SMOS-IC data record of soil moisture and L-VOD: Historical development, applications and perspectives. *Remote Sensing of Environment* 254:112238.

Wilhite, D.A., and M.H. Glantz. 1985. Understanding the drought phenomenon: The role of definitions. *Water International* 10:111–120.

Xia, Y., Z. Hao, C. Shi, Y. Li, J. Meng, T. Xu, X. Wu, and B. Zhang. 2019. Regional and global land data assimilation systems: Innovations, challenges, and prospects. *Journal of Meteorological Research* 33(2):159–189.

Xia, Y., K. Mitchell, M. Ek, J. Sheffield, B. Cosgrove, E. Wood, L. Lifeng, C. Alonge, W. Helin, J. Meng, B. Livneh, D. Lettenmaier, V. Koren, D. Qingyun, K. Mo, F. Yun, and D. Mocko. 2012. Continental-scale water and energy flux analysis and validation for the North American Land Data Assimilation System project phase 2 (NLDAS-2): 1. Intercomparison and application of model products. *Journal of Geophysical Research* 117:D03109. http://doi.org/10.1029/2011JD016048.

Xue, J., M.C. Anderson, F. Gao, C. Hain, K.R. Knipper, Y. Yang, W.P. Kustas, N.E. Bambach, A.J. McElrone, S.J. Castro, J.G. Alfieri, J.H. Prueger, L.G. McKee, L.E. Hipps, and M. Alsina. 2022. Improving the spatiotemporal resolution of remotely sensed ET information for water management through Landsat, Sentinel-2, ECOSTRESS and VIIRS data fusion. *Irrigation Science* 40:609–634.

Yang, L., B. Wylie, L.L. Tieszen, and B.C. Reed. 1998. An analysis of relationships among climatic forcing and time-integrated NDVI of grasslands over the U.S. northern and central Great Plains. *Remote Sensing of Environment* 65:25–37.

Yang, Y., M.C. Anderson, F. Gao, D.M. Johnson, Y. Yang, L. Sun, W. Dulaney, C.R. Hain, J.A. Otkin, J. Prueger, T.P. Meyers, C.J. Bernacchi, and C.E. Moore. 2021. Phenological corrections to a field-scale, ET-based crop stress indicator: An application to yield forecasting across the U.S. Corn Belt. *Remote Sensing of Environment* 257:112337.

Yirdaw, S.Z., K.R. Snelgrove, and C.O. Agboma. 2008. GRACE satellite observations of terrestrial moisture changes for drought characterization in the Canadian Prairie. *Journal of Hydrology* 56:84–92.

Yoshida, Y., J. Joiner, C. Tucker, J. Berry, J.E. Lee, G. Walker, R. Reichle, R. Koster, A. Lyapustin, and Y. Wang. 2015. The 2010 Russian drought impact on satellite measurements of solar-induced chlorophyll fluorescence: Insights from modeling and comparisons with parameters derived from satellite reflectances. *Remote Sensing of Environment* 166:163–177.

Yuan, W., Y. Zheng, S. Piao, P. Ciais, D. Lombardozzi, Y. Wang, Y. Ryu, G. Chen, W. Dong, Z. Hu, and A.K. Jain. 2019. Increased atmospheric vapor pressure deficit reduces global vegetation growth. *Science Advances* 5(8):eaax1396.

Zaitchik, B.F., M. Rodell, and R.H. Reichle. 2008. Assimilation of GRACE terrestrial water storage data into a land surface model: Results for the Mississippi River Basin. *Journal of Hydrometeorology* 9(3):535–548.

Zhang, C., Z. Yang, H. Zhao, Z. Sun, L. Di, R. Bindlish, P.-W. Liu, A. Colliander, R. Mueller, W.T. Crow, R.H. Reichle, J. Bolten, and S.H. Yueh. 2022. Crop-CASMA: A web geoprocessing and map service-based architecture and implementation for serving soil moisture and crop vegetation condition data over U.S. Cropland. *International Journal of Applied Earth Observation and Geoinformation* 112:102902.

4 Regional Drought Monitoring Based on Multi-Sensor Remote Sensing

Jinyoung Rhee, Jungho Im, and Seonyoung Park

ACRONYMS AND DEFINITIONS

AET	Actual evapotranspiration
AMSR-E	Advanced microwave scanning radiometer for earth observing system
AOI	Arctic Oscillation Index
AVHRR	Advanced very high resolution radiometer
AWC	Available water capacity
CART	Classification and regression-tree
CMI	Crop Moisture Index
ConvLSTM	Convolutional long short-term memory
CPC	Climate Prediction Center
DSI	Drought Severity Index
ESI	Evaporative Stress Index
EVI	Enhanced Vegetation Index
GRACE	Gravity recovery and climate experiment
HSMDI	High resolution Soil Moisture Drought Index
IPAD	International production assessment division
KBDI	Keetch–Bryam Drought Index
LAI	Leaf Area Index
LST	Land surface temperature
MIDI	Microwave Integrated Drought Index
MJO	Madden-Julian oscillation
MODIS	Moderate resolution imaging spectroradiometer
NASS	National Agricultural Statistics Service
NCDC	National Climatic Data Center
NDDI	Normalized Difference Drought Index
NDII	Normalized Difference Infrared Index
NDMC	National Drought Mitigation Center
NDVI	Normalized Difference Vegetation Index
NDWI	Normalized Difference Water Index
NLCD	National land cover database
NMDI	Normalized Multiband Drought Index
NRCS	Natural Resources Conservation Service
OBDI	Objective Blend Drought Index
PASG	Percent of average seasonal greenness
PDSI	Palmer Drought Severity Index
PET	Potential evapotranspiration

DOI: 10.1201/9781003541417-5

SDCI	Scaled Drought Condition Index
SEBS	Surface energy balance system
SM	Soil moisture
SMOS	Soil moisture and ocean salinity
SOSA	Start of season anomaly
SPEI	Standardized Precipitation and Evapotranspiration Index
SPI	Standardized Precipitation Index
SWDI	Soil Wetness Deficit Index
SWI	Soil Wetness Index
TRMM	Tropical rainfall measuring mission
TVDI	Temperature-Vegetation Dryness Index
USDA	United Stated Department of Agriculture
USDM	United States Drought Monitor
USGS	US geological survey
VegDRI	Vegetation Drought Response Index
VHI	Vegetation Health Index
VSDI	Visible and Shortwave infrared Drought Index
VTCI	Vegetation Temperature Condition Index
WWAI	Wetland Water Area Index

4.1 INTRODUCTION

There are numerous definitions of drought since it has been recognized as a very costly natural disaster. The recent special report of the IPCC (2012) defines drought as "a period of abnormally dry weather long enough to cause a serious hydrological imbalance." Drought can also be defined for three types: a meteorological drought refers to "a period with an abnormal precipitation deficit," soil moisture drought or agricultural drought is "a deficit of soil moisture," and hydrological drought refers to "negative anomalies in streamflow, lake, and/or groundwater levels."

In order to monitor drought conditions, hydro-meteorological variables need to be estimated for each type of drought. Precipitation is an appropriate variable for monitoring meteorological drought, and hydrological variables such as streamflow and reservoir levels can be used to detect hydrological drought. Soil moisture drought or agricultural drought can be measured using variables of soil moisture condition and crop yield (Mishra and Singh, 2011). Variables such as precipitation and evapotranspiration can be considered the drivers of drought, while variables such as streamflow and reservoir levels are the response variables impacted by drought. Both types of variables may be used for efficient drought monitoring.

The type of drought is determined as a function of the spatial location of the drought, the time scale of it, as well as the stakeholders involved. Different types of drought are related, as one type of drought may develop into another, and the variables for quantifying drought or affected by drought are interconnected through land-atmosphere interactions. There exist couplings and feedbacks between variables, which may appear differently according to the regions of interests. For example, the coupling between soil moisture and evapotranspiration is controlled by soil moisture in dry regions, while it appears otherwise in wet regions (Seneviratne et al., 2010). The couplings and feedbacks of soil moisture-temperature and soil moisture-precipitation are more complex and intertwined between variables (Seneviratne et al., 2010; Figure 4.1).

Remote sensing techniques have been used to estimate variables related to each type of drought (Jensen, 2000). In some cases, however, the estimation of each variable may not fully represent the actual target when it is derived from algorithms using many other variables. The use of multiple variables in combined forms can improve the quantification of drought since the variables are related through the couplings and feedbacks. However, multicollinearity among the variables should

Regional Drought Monitoring

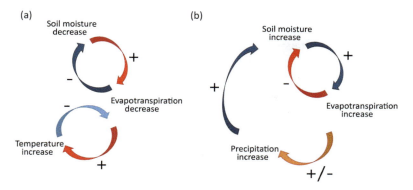

FIGURE 4.1 Processes contributing to (a) soil moisture-temperature, and (b) soil moisture-precipitation coupling and feedback loops (adapted from Seneviratne et al., 2010).

be kept in mind when a linear combination of variables is employed. Drought types and variables required to monitor each type of drought are listed in Table 4.1; socioeconomic drought is not included because it should be explained by many other factors than biophysical variables. It needs to be noted that the variables in Table 4.1 include both the drivers and impacts of drought, there exist overlaps of variables between drought types, and the list is not exhaustive. Generally, drought indices or combinations of the variables are useful for drought monitoring compared to single variables. Most variables can be estimated using remote sensing, which are well covered by Wardlow et al. (2012).

There are examples of using a variety of remote sensing–derived variables for drought monitoring, including the United States Drought Monitor (USDM; http://drougt.unl.edu/dm), Objective Blend Drought Index (OBDI) of the Climate Prediction Center (CPC), and Vegetation Drought Response Index (VegDRI; Brown et al., 2008).

The USDM is a drought monitoring system that provides maps with the spatial extent and severity of drought on a weekly basis for the conterminous United States. Drought information are derived mainly from six key physical indicators, which include observation-based indicators of the Palmer Drought Severity Index (PDSI; Palmer, 1965), Percent of Normal Precipitation (Willeke et al.,

TABLE 4.1
Drought Types and Variables Required to Monitor Each Type of Drought

Drought Type	Climatic/Hydrological Variables Required to Monitor Drought
Meteorological drought	Precipitation
	Evapotranspiration
Soil moisture drought	Precipitation
or agricultural drought	Evapotranspiration
	Soil moisture
	Vegetation activity
	Crop yields
Hydrological drought	Precipitation
	River discharge (streamflow)
	Reservoir storage
	Groundwater level

1994), Standardized Precipitation Index (SPI; McKee et al., 1993), US Geological Survey (USGS) Daily Streamflow Percentiles, model-based CPC Soil Moisture Model Percentiles (Huang et al., 1996), and satellite-based Vegetation Health Index (VHI; Kogan, 1995a, 1995b). Ancillary data are also used in the USDM, including, but not limited to, drought indices of the Palmer Crop Moisture Index (CMI; Palmer, 1968), the Keetch–Bryam Drought Index (KBDI; Keetch and Byram, 1968), as well as reservoir and lake levels, groundwater levels, and soil moisture field observations. For the western United States, the Natural Resources Conservation Service's (NRCS) Snowpack Telemetry (SNOTEL) observations to measure snow water equivalent for mountain sites are also additionally used (Svoboda et al., 2002).

The USDM integrates subject judgments of the "*authors*" who are drought experts in the National Drought Mitigation Center (NDMC), United Stated Department of Agriculture (USDA), CPC, and National Climatic Data Center (NCDC). Since the indicators in use and their weights are determined by the authors, the OBDI is also not entirely objective (Svoboda et al., 2002). The produced drought information from OBDI, however, is replicable because they are based on the fixed weights of determined indicators. The type of indicators and their weights are selected considering the time scale of drought of interest.

The VegDRI measures drought-induced vegetation stress by combining climate-based drought indices of SPI, self-calibrated PDSI (Wells et al., 2004) and USDM, and remote sensing–based vegetation index metrics, which are Normalized Difference Vegetation Index (NDVI), percent of average seasonal greenness (PASG), and start of season anomaly (SOSA). It also utilizes biophysical information, which includes land cover/land use, soil Available Water Capacity (AWC), irrigated agriculture, and ecoregions. The VegDRI adopted the supervised classification and regression-tree (CART) analysis to determine vegetation stress.

The examples listed earlier are based on various available data sources without limiting the indicators to remotely sensed ones. The combination of multiple variables can be applied for drought monitoring solely based on remote sensing. The combination of multiple indicators enables the production of customized drought indices that can be applied to specific regions of interests by determining the types and weights of the indicators.

This chapter investigates the regional drought monitoring methodology that utilizes multi-sensor remote sensing data. Recent trends in studies based on remote sensing–based drought monitoring are presented in the following section, and the development of the Scaled Drought Condition Index (SDCI) and its advances are introduced. Case studies of the Korean Peninsula and the United States are provided by applying the advanced SDCI for regional drought monitoring.

4.2 RECENT TRENDS OF REMOTE SENSING–BASED DROUGHT MONITORING

Since a multitude of factors are related with drought conditions, scientists have tried to incorporate such factors in remote sensing–based drought monitoring. Recent studies since 2010 show that the trends in remote sensing–based drought monitoring include (1) development of customized monitoring approaches to different types of drought, (2) development of new drought indices especially focusing on the incorporation of soil moisture and evapotranspiration, (3) data assimilation of remotely sensed products in process-based models for drought monitoring and prediction, and (4) the combination of machine learning and remotely sensed products for drought monitoring and prediction.

4.2.1 Customized Drought Monitoring Approaches

Since drought is a slow-onset event, it is hard to identify the exact start and end dates as well as types of drought. In particular, as drought has somewhat different characteristics by type, scientists have proposed customized monitoring approaches that are appropriately applicable to each type of drought. Because many drought indices are based on precipitation such as SPI, studies on monitoring meteorological drought typically investigated the relationship between remote sensing–derived products and *in situ* drought indices. For example, Caccamo et al. (2011) assessed the Moderate Resolution Imaging Spectroradiometer (MODIS)–based drought indices through comparison with SPI. They found that Normalized Difference Infrared Index (NDII) using band 6 ([band 2—band6]/[band2 + band6]) outperformed the other remote sensing-derived indices showing strong correlation with SPI, which implied that NDII can be operationally used for monitoring meteorological drought. Scaini et al. (2015) used Soil Moisture and Ocean Salinity (SMOS) soil moisture product to determine drought conditions by examining the relationships between soil moisture anomalies and SPI as well as Standardized Precipitation and Evapotranspiration Index (SPEI). They found that the short-term remotely sensed anomalies had a high response to precipitation events, and thus the optimal time scale for the SMOS values was one month.

Since drought directly affects crop yields, efforts have been made to monitor agricultural drought using remote sensing–derived products, including vegetation indices and soil moisture. Shaheen and Baig (2011) assessed drought conditions in arid areas using SPI, NDVI, crop yield anomaly, rainfall anomaly, and evapotranspiration focusing on meteorological and agricultural droughts. Remote sensing–based vegetation indices such as NDVI and VHI have been successfully used for agricultural drought monitoring (Anderson et al., 2010; Rojas et al., 2011; Gao et al., 2014). Soil moisture is one of the key indicators for monitoring agricultural drought as it is closely related with vegetation growth. Bolten et al. (2010) incorporated the Advanced Microwave Scanning Radiometer for Earth Observing System (AMSR-E) soil moisture into the International Production Assessment Division (IPAD) Water Balance Model to operationally monitor agricultural drought. Crop yield data has been commonly used to assess agricultural drought (Rhee et al., 2010; Leng and Hall, 2019). However, correlation between crop yield and drought severity may not be always high, especially for irrigated agricultural lands (Park et al., 2016).

Vegetation phenology has been examined with drought conditions because drought during growing seasons can significantly affect crop yields. Ivits et al. (2014) investigated spatiotemporal patterns of drought conditions focusing on vegetation phenology using SPEI and NDVI based on major European bioclimatic zones. Garcia et al. (2014) evaluated the Temperature-Vegetation Dryness Index (TVDI) to estimate water deficits using MODIS data and found that the best conditions for TVDI performance agreed with the growing season that typically showed higher soil water content and lower vapor pressure deficit.

Hydrologic variables such as streamflow, soil moisture, evapotranspiration, and water level have been used as reference indicators to identify hydrologic drought. For example, Choi et al. (2013) compared remote sensing–based drought indices with standard indices using streamflow and soil moisture measurements. They found that Evaporative Stress Index (ESI) successfully captured severe drought conditions to be a promising drought index for characterizing streamflow and soil moisture anomalies. Since ESI calculation uses a thermal remote sensing energy balance framework, it is closely related with evapotranspiration deficits, which can be used as a diagnostic fast-response indicator. A combination of ESI and soil moisture can be used as a valuable early warning tool for rapidly evolving flash drought conditions (Anderson et al., 2013). Seitz et al. (2014) examined water mass from Gravity Recovery and Climate Experiment (GRACE), water stage from Envisat, and water extent from Phased Array type L-band Synthetic Aperture Radar (PALSAR) to identify their spatial and temporal variability. They found that water extent and water level measurements in heavy drought conditions could be used to identify water volume changes that are closely related with hydrological drought.

4.2.2 Development of New Remote Sensing–Based Drought Indices

Many studies have recently proposed new drought indices that use multi-sensor satellite products related with potential drought indicators such as precipitation, temperature, vegetation healthiness, soil moisture, and evapotranspiration. Table 4.2 summarizes recently introduced drought indices from multi-sensor satellite data. For example, Mu et al. (2013) proposed a new Drought Severity Index (DSI) that uses MODIS evapotranspiration, potential evapotranspiration, and NDVI products. Huang et al. (2011) developed a Wetland Water Area Index (WWAI) from PDSI, NDVI, and Normalized Difference Water Index (NDWI) to predict water surface area, and they were able to identify intra- and inter-annual water change between 1910 and 2009. They also proposed a water allocation model to simulate spatial distribution of water bodies, which can be used to simulate major changes in wetland water surface for ecosystem service. Wang et al. (2013) utilized passive microwave AMSR-E data to develop drought indices in the Huaihe River basin focusing on soil moisture.

As drought has different characteristics by type, drought indices have been developed for a specific type of drought. Rhee et al. (2010) proposed the SDCI that uses MODIS Land Surface Temperature (LST), NDVI, and Tropical Rainfall Measuring Mission (TRMM) precipitation through linear combination and found that the index could be useful for agricultural drought monitoring in both arid and humid regions. Zhang and Jia (2013) proposed a Microwave Integrated Drought Index (MIDI) to monitor short-term drought (i.e., meteorological drought) over semi-arid regions using TRMM-derived precipitation, AMSR-E-derived soil moisture and land surface temperature. Keshavarz et al. (2014) proposed Soil Wetness Deficit Index (SWDI) based on Soil Wetness Index (SWI) from MODIS satellite data to examine agricultural drought conditions. Zhang et al. (2013) proposed a Visible and Shortwave infrared Drought Index (VSDI) for monitoring moisture in soil and vegetation. VSDI is expected to be efficient for agricultural drought monitoring over different land cover types during the plant-growing season.

4.2.3 Data Assimilation for Drought Monitoring and Prediction

Process-based physical models have been used to quantify various drought-related factors such as soil moisture and evapotranspiration. Products generated from such models can be combined with remote sensing–derived drought indices to better document drought conditions. For example, Anderson et al. (2011) investigated drought conditions using ESI calculated based on evapotranspiration and potential evapotranspiration produced from the ALEXI model. They found that ESI performed similarly to short-term precipitation-based indices such as SPI at higher spatial resolution without using any precipitation data. Anderson et al. (2012) used METRIC, ALEXI, and SEBAL models with Landsat 7 ETM+, MODIS, and GOES satellite data to explore the utility of moderate-resolution thermal satellite imagery in water resources management. They found that the fusion of the multi-sensor data at different scales could be effective for drought monitoring.

Zhong et al. (2014) investigated Soil Water Deficit (SWD) from the Surface Energy Balance System (SEBS) model using Advanced Very High Resolution Radiometer (AVHRR) and MODIS data. They validated the SWD index using AMSR-E soil moisture data and found that soil moisture may have diurnal variations. Sakamoto et al. (2014) examined daily AMSR-E soil moisture data to derive soil moisture anomalies based on four spatial aggregation approaches. They found the spatial aggregation approaches could provide useful information on soil moisture anomalies for a relatively short period of remote sensing data available.

Remote sensing–derived products can be combined with forecasting models to predict future drought conditions. Han et al. (2010) used an AutoRegressive Integrated Moving Average (ARIMA) model with Vegetation Temperature Condition Index (VTCI) data produced between 1999 and 2006 to examine the feasibility of drought forecasting based on the proposed approach. Remote sensing–based approaches are yet limited in drought forecasting. Remote sensing data have been used to provide input values as initial conditions for drought forecasting models through data assimilation, as well as to produce ancillary data such as land cover. As remote sensing–derived drought information documents past and present drought conditions, it can be used to regularly update modeling results to increase the reliability of drought forecasting models.

TABLE 4.2
Summary of the Recently Proposed Satellite-Based Drought Indices

Drought Index	Description	Strength	Weakness	Reference
Normalized Difference Drought Index (NDDI)	(NDVI−NDWI)/(NDVI + NDWI)	• Contains strength of both NDVI and NDWI. • Works well for identifying agricultural drought.	• Less reliable for short-term drought monitoring	Gu et al. (2007)
Normalized Multiband Drought Index (NMDI)	(NIR−(1640nm−2130nm)/(NIR−(1640nm+2130nm)	• Works well for identifying agricultural drought in areas with a high vegetation rate	• Less reliable for short-term drought monitoring	Wang and Qu (2007)
Normalized Difference Water Index (NDWI)	(NIR−SWIR)/(NIR+SWIR)	• Responds to drought faster than NDVI		Gao (1996)
Scaled Drought Condition Index (SDCI)	0.25* scaled LST + 0.5* scaled TRMM + 0.25* scaledNDVI	• Works for both meteorological and agricultural drought • Works for both arid and humid regions	• Needs refinement to optimize weights of variables	Rhee et al. (2010)
Microwave Integrated Drought Index (MIDI)	a* PCI+b* SMCI + (1−a−b)* TCI (PCI = scaled TRMM SMCI = scaled Soil Moisture)	• Appropriate for short-term drought monitoring • Provides high temporal resolution products based on microwave data	• Less reliable for long-term drought monitoring	Zhang and Jia (2013)
Visible and shortwave infrared drought index (VSDI)	1−[(SWIR−Blue)+(Red−Blue)]	• Appropriate for use as a real-time drought indicator • Can be applied to various land covers	• Considers the very limited number of drought factors	Zhang et al. (2013)
Evaporative Stress Index (ESI)	Based on the ratio of ET and PET, and anomalies	• Monitors drought without using antecedent precipitation and subsurface soil characteristics	• Can be affected by local factors such as groundwater	Anderson et al. (2010)
Vegetation Drought Response Index (VegDRI)	Calculate index values using climatic, satellite, and biophysical components in rule-based models	• Considers various drought factors • Provides high spatial resolution products	• Does not consider some important variables, such as land surface temperature	Brown et al. (2008)
Vegetation Outlook (VegOut)	Combine satellite, climatic, oceanic data as well as biophysical data to monitor drought biweekly	• Able to predict drought using historical data	• Focus only on agricultural drought based on vegetation greenness	Tadesse et al. (2010)

4.2.4 Machine Learning for Drought Monitoring and Prediction

Machine learning can be used to downscale coarse remote sensing data for local or regional drought monitoring (Im et al., 2016; Park et al., 2017). Park et al. (2017) adopted the method of Im et al. (2016) to downscale AMSR-E soil moisture data into 1km soil moisture; they used MODIS LST, ET, vegetation indices, albedo, and TRMM precipitation data as input and generated 1km soil moisture through the random forest machine learning. The High resolution Soil Moisture Drought Index (HSMDI) was developed from the normalized 1km soil moisture, and it showed good performance for depicting meteorological and agricultural drought in terms of three-month SPI and crop yield in the Korean peninsula.

Rule-based machine learning models have also been directly used for drought monitoring and forecasting. Rhee et al. (2020) successfully used rule-based models for detecting hydrological droughts in ungauged areas from remotely sensed hydro-meteorological variables of precipitation, actual evapotranspiration, NDVI, LST, and soil moisture. Rhee and Im (2017) used various rule-based models for drought forecasting from remotely sensed data of LST, NDVI, NDWI, large-scale climate indices of Multivariate El Nino-Southern Oscillation Index (MEI) and Arctic Oscillation Index (AOI), as well as long-range climate forecast data for precipitation and potential evapotranspiration. The use of machine learning models in the study helped overcoming the limitations of a simple bias-correction method for drought forecasting in ungauged areas. Short-term drought prediction models based on random forest were developed for predicting flash droughts over East Asia using remote sensing data and Madden-Julian Oscillation (MJO) index (Park et al., 2018). Short-term forecasting of drought indices based on the convolutional long short-term memory (ConvLSTM) and random forest approaches showed promising results when using temporal patterns and upcoming weather conditions from numerical model outputs (Park et al., 2020).

4.2.5 Issues of Multi-Sensor Data Combination

The integration of multiple variables from different sensors for drought monitoring can be viewed as a special case of data fusion. Data fusion combines data from multiple sensors to produce improved information, and is known to have diverse challenges, including conflicting data, data modality, data correlation, and data association (Khaleghi et al., 2013). The combination of multi-sensor-derived variables does not share all the challenges of data fusion since it is based on the "use of best available data" for the detection of drought of the areas of interests, and uses multiple variables to extract useful drought information rather than dealing with single variables that may be conflicting or inconsistent. However, the combination of multiple variables still has some issues, such as remote sensing data imperfection and outliers, and different spatial resolutions of data.

Remote sensing data have impreciseness (Khaleghi et al., 2013) compared to observation data since remote sensing data rely heavily on estimation algorithms. Outliers also exist due to many factors interfering with the acquisition of high-quality data such as atmospheric conditions. These issues become sources of uncertainty. Use of artificial intelligence can help relieving such problems (Mishra and Desai, 2006). Such techniques can recognize the relationships between variables without considering the physics explicitly, and performs well even with outliers (Mishra and Desai, 2006). The issue regarding different spatial resolutions is commonly ignored based on the 'use of best available data" for the detection of drought of the areas of interest, as previously mentioned. Downscaling approaches can relieve the spatial resolution issue, though.

… Regional Drought Monitoring

4.3 DEVELOPMENT OF THE SDCI

4.3.1 Drought in Humid Regions

Droughts in arid regions have drawn much attention historically compared to humid regions due to the arid climatological characteristic. Since drought is a natural phenomenon and relative in nature, drought also occurs in humid regions. Many existing studies using remote sensing–based drought indices were mostly performed for arid/semi-arid regions (e.g., Ji and Peters, 2003; Wan et al., 2004), and only a limited number of studies considered humid/sub-humid regions (e.g., Kogan, 1995a, 1995b).

There is historical documentation of drought in humid/sub-humid regions; in the southeastern United States, it is known that drought occurred almost every decade since the 1920s (Knutson and Hayes, 2002; Weaver, 2005). In 1986, South Carolina, USA, experienced a drought with the return period of over 100 years (Cook et al., 1988; Karl and Young, 1987), and the economic loss by the four year drought of 1998–2002 in the Carolinas was enormous (Carbone et al., 2007).

4.3.2 Introduction of SDCI, Solely Based on Remote Sensing Data

The SDCI has been proposed for use in humid regions as well as arid regions proving the performance of multi-sensor data fusion mainly for agricultural drought monitoring (Rhee et al., 2010). It is composed of three components based on remote sensing data: a temperature component using LST, a vegetation component using one of the vegetation-related indices, and a precipitation component using remotely sensed rainfall data. The Terra MODIS LST data and the TRMM monthly rainfall data were used for remotely sensed LST and precipitation data, respectively. Compared were various vegetation indices, including the Terra MODIS NDVI, and relatively new indices, including the NDWI, the Normalized Difference Drought Index (NDDI), and the Normalized Multiband Drought Index (NMDI) calculated based on the Terra MODIS surface reflectance data. The land cover from the National Land Cover Database (NLCD) 2001 product was used as ancillary data and only locations with grassland or cropland areas were used for the analyses. All analysis was performed for the period 2000–2009.

Additive linear combinations of three variables as scaled from zero to one, each from one of the three components, were tested with three sets of weights against three- and six-month SPI as well as PDSI values obtained from observation data. Among the tested sets, the combination of scaled LST, scaled TRMM, and scaled NDVI where the weight for the precipitation component is twice of the others was selected as an optimum index that performed great in both arid and humid regions.

4.3.3 Validation of the SDCI

The developed SDCI was validated in two ways—the drought conditions derived by the SDCI were compared to USDM maps for well-known drought events (Figures 4.2 and 4.3), and the gridded values were merged for each county to be compared to crop yield.

The states of Arizona and New Mexico, USA (95 °W–122 °W longitude/31 °N–37 °N latitude) were selected as an arid region, while the states of North Carolina and South Carolina, USA (the 71°W–87.5 °W longitude/32 °N–37 °N latitude) were chosen as a humid region. The year-to-year changes of drought conditions of the region for the period 2000–2009 were examined and compared to USDM maps.

Drought conditions based on SDCI for all land cover types were compared to USDM in Figures 4.2 and 4.3, despite that SDCI is optimized only for grassland and cropland land cover types. It is

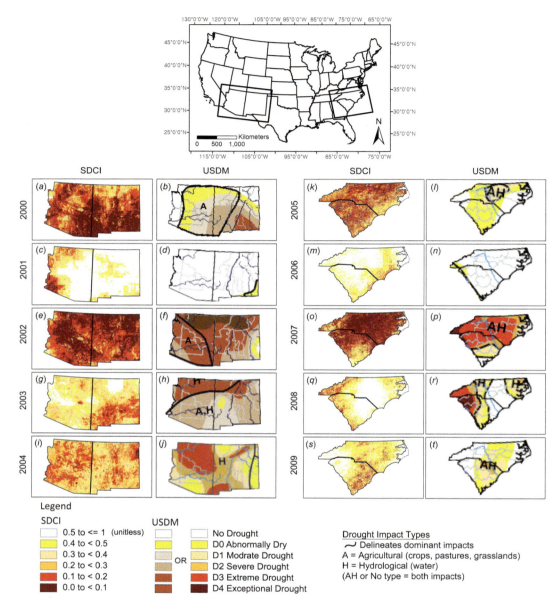

FIGURE 4.2 Year-to-year comparisons of the SDCI and USDM (a)–(j) in the arid region (AZ and NM, USA) during May, and (k)–(t) in the humid region (NC and SC, USA) during September for 2000–2009 (adapted from Rhee et al., 2010).

not appropriate to compare SDCI and USDM for this reason, and also because the USDM includes subject judgments of drought authors and utilizes numerous inputs.

The areas marked with "A (agricultural drought)" and "AH (agricultural and hydrological drought)" in USDM maps are examined more closely for comparisons. The months of May and September for arid and humid regions, respectively, were selected since those are when SDCI showed the highest correlations with observational drought indices. The year-to-year changes of spatial distribution and severity of drought generally agreed between SDCI and USDM (Figures 4.2 and 4.3).

Regional Drought Monitoring

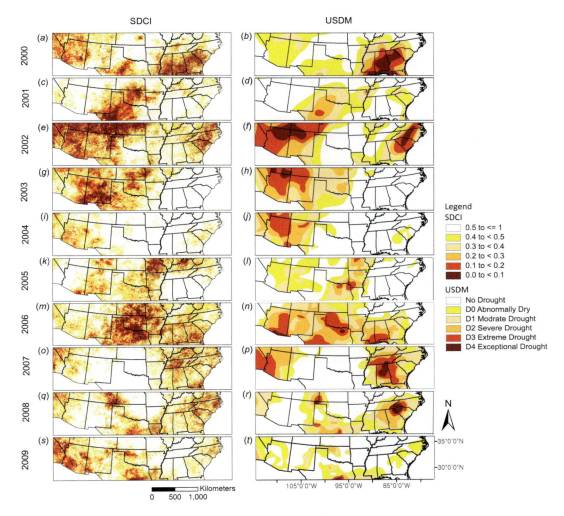

FIGURE 4.3 Year-to-year comparisons of the SDCI and USDM in the southern part of the USA during July for 2000–2009.

The SDCI was also tested against crop yield data. The main cultivated crops for the areas are cotton wheat, corn, and sorghum for arid region, and soybean, corn, and cotton for humid region based on 2007 census of agriculture (http://www.agcensus.usda.gov). Crop yield data were obtained from the National Agricultural Statistics Service (NASS; http://www.nass.usda.gov) of USDA, and Curry County of New Mexico; Lee County of South Carolina; and Edgecombe, Greene, Lenoir, Pitt, Robeson, Sampson, Wayne, and Wilson Counties of North Carolina were chosen with larger than 33% total area of croplands.

Regression analyses were performed between monthly SDCI and yearly crop yield data. The remote sensing index values with 2001 NLCD cultivated crops land cover type for each county were averaged and correlated with the crop yield data for the county. In the arid region, only May SDCI showed statistically significant correlation with yearly cotton yield. In the humid region, June and July SDCI showed high correlations with corn yield for some counties, August SDCI with cotton and soybean yield, and September SDCI with soybean yield. The months showing

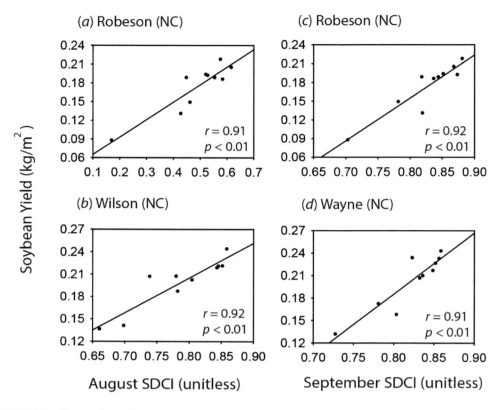

FIGURE 4.4 Scatterplots of the SDCI versus soybean yield data in NC and SC, USA for 2000–2009. The SDCI values with 2001 NLCD cultivated crops land cover type for each county were averaged and correlated with the crop yield data for the county (adapted from Rhee et al., 2010).

high correlations with yearly crop yield are the active growing periods during the phonological cycles of the crops (Jensen, 2000; Savitsky, 1986). Scatterplots of some examples are shown in Figure 4.4.

4.3.4 Development of the Advanced SDCI

Although SDCI with linear combinations of three components can be successfully used to monitor agricultural drought, it has some limitations: the weights are arbitrarily determined, and it includes only limited types of variables. These issues can be solved by integrating more variables related to drought and combining artificial intelligence, such as neural networks or machine learning techniques.

As previously mentioned, soil moisture is the most important variable for agricultural drought monitoring. Since there exist couplings and feedback between soil moisture and precipitation, temperature, and evapotranspiration, the inclusion of the three components may explain other variables that are not considered. In order to more fully represent the responses of the variables during drought, more variables are added: actual evapotranspiration (AET) and potential evapotranspiration (PET), enhanced vegetation index (EVI), and leaf area index (LAI) estimated from MODIS and soil moisture (SM) from AMSR-E sensor. The SDCI was improved by providing multiple blends of multi-sensor indices for different types of drought.

The blends of multi-sensor indices can be used in two ways—linear combinations of variables can be used as SDCI but with weights determined using a machine learning technique, or nonlinear

TABLE 4.3
Strengths and Limitations of SDCI Compared to USDM

	SDCI	USDM
Data source	Utilizes only remote sensing data	Combines observation-based indicators, satellite-based indicators, model-based data, ancillary data, and expert judgments
Strengths	Can be applied to any region, especially to areas without observation data	• Maximizes the use of available data/information • Currently provides optimized information for the conterminous US • Provides information for various types of drought
Limitations	• Currently optimized for agricultural drought monitoring	• Requires long-term *in situ* data to derive observation-based indicators • Needs subjective inputs from experts when applied to additional regions other than the conterminous US

combinations of variables can be used as trained using the machine learning technique though a semi-"blackbox." Strengths and limitations of SDCI compared to USDM are shown in Table 4.3. Two multi-sensor blending examples are described in the following two sections.

4.4 CASE STUDY: LINEARLY COMBINED SDCI IN THE KOREAN PENINSULA

The advanced SDCI with a linear combination was applied to the Korean peninsula, including North and South Korea (Figure 4.5), for agricultural drought monitoring. The study area has four distinct seasons over the course of the year and contains complex topography with a variety of land cover types. Remote sensing data were obtained for the period 2003 to 2011, and observation data from 39 out of 106 weather stations with grassland or cropland land cover types were used for training to obtain weights (Figure 4.5).

Weights for variables (LST, EVI, LAI, PET, SM, and precipitation [PRCP]) were determined using a random forest machine learning approach. Random forest uses an ensemble approach to predict a target variable through combining predictions returned by multiple CART using the Gini index for selecting an attribute at a node (Han and Kamber, 2011; Li et al., 2014). Random forest introduces two levels of randomness into CART to improve the weaknesses of CART, such as overfitting to samples and dependency to training data configuration: (1) a random subset of training samples and (2) a random subset of candidate variables at each node. The randomness allows to generate numerous de-correlated trees, which can deliver robustness to noise, outliers, and overfitting. Random forest provides relative variable importance using out-of-bag data (i.e., accuracy change when a variable is out-of-bag). The training was performed for SPI and SPEI with 1–12 months of time scales (Figure 4.6).

The advanced SDCI and a couple of remote sensing–based indices, including NMDI and NDWI, were tested using the do (province)-level yield data of Highland Radish for two weather station locations of Jangsu, Jeollabuk-do and Daegwallyeong, Kangwon-do, located above the 400m level. (Figure 4.7). The June SDCI showed statistically significant high correlations with SPEI12 (Figure 4.7a, r = 0.71, p = 0.031) and SPI12 (Figure 4.7b, r = 0.70, p = 0.035) in Jangsu, Jeollabuk-do, as well as July NMDI (Figure 4.7d, r = 0.68, p = 0.042). The May NDWI (Figure 4.7c, r = 0.82, p = 0.007) showed good correlations to yield data in Daegwallyeong, Kangwon-do. The Kendall's tau values were also examined since the number of samples is small; only the May NDWI and July NMDI showed statistical significance with the significance level of 0.05 (tau = 0.67, p = 0.012, and tau = 0.56, p = 0.037, respectively).

The scatterplots in Figure 4.7 are for the comparisons between the do (province)-level yearly Highland Radish yield and the SDCI values for weather station locations, not averaged value for the highland areas of the corresponding province. It explains somewhat smaller correlation coefficient values compared to Figure 4.4. The comparisons between the averaged grid values versus do (province)-level yield data are being performed in the following study. Since the case study did not use the full range of variables, a further study is required to explore the effect of the use of each variable.

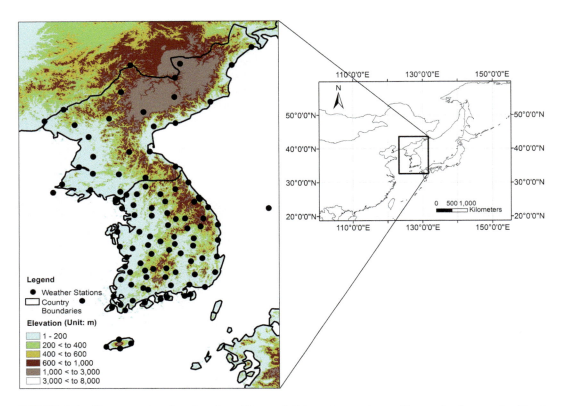

FIGURE 4.5 Korean peninsula—data from 39 out of 106 weather stations with grassland or cropland land cover types were used for training to derive weights.

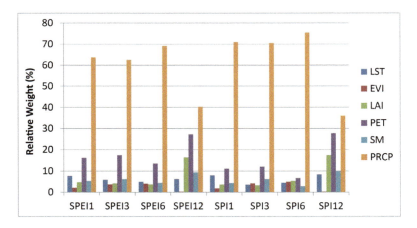

FIGURE 4.6 Weights for each variable were obtained to fit the widely used drought indices; the SPI and SPEI with 1–12 months of time scales.

Regional Drought Monitoring

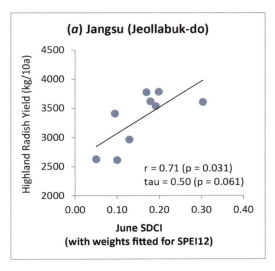
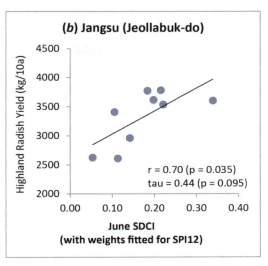
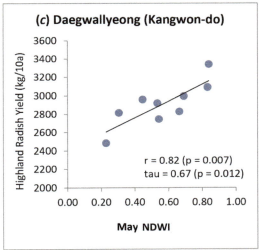
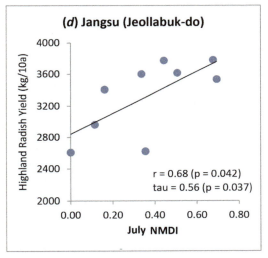

FIGURE 4.7 Scatterplots with correlation coefficient r as well as Kendall's tau values between the advanced SDCI, NDWI, and NMDI at two weather stations locations versus do (province)-level Highland Radish yield.

4.5 CASE STUDY: NONLINEARLY COMBINED SDCI IN THE USA

The drought indicators that are used to produce SDCI can be nonlinearly combined to increase the predictability of drought. Machine learning approaches were evaluated to estimate SPI using SDCI drought indicators, including MODIS LST, NDVI, NDWI, NDMI, NDDI, evapotranspiration, and TRMM-derived precipitation over both arid and humid regions in the United States, the same study regions in Rhee et al. (2010) (Park et al., 2016). The arid region includes Arizona and New Mexico, and the humid region includes North and South Carolina. MODIS and TRMM products (i.e., a total of 15 variables) from 2000 and 2012 were scaled from 0 to 1 to temporally normalize the data (Rhee et al., 2010) (Table 4.4). SPI was calculated for 54 weather stations (28 stations in the arid region and 26 in the humid region) at the accumulated 1-, 3-, 6-, 9-, and 12-month time scales (i.e., SPI1, SPI3, SPI6, SPI9, and SPI12).

TABLE 4.4
Variables Used for Nonlinear Combination of SDCI

Satellite Sensor	Variable	Description
MODIS	Land Surface Temperature (LST)	
	Evapotranspiration (ET)	
	Normalized Difference Vegetation Index (NDVI)	
	NDVI500	Using 500m data
	Normalized Difference Water Index (NDWI5)	Using band 5
	Normalized Difference Water Index (NDWI6)	Using band 6
	Normalized Difference Water Index (NDWI7)	Using band 7
	Normalized Difference Drought Index (NDDI5)	Using band 5
	Normalized Difference Drought Index (NDDI6)	Using band 6
	Normalized Difference Drought Index (NDDI7)	Using band 7
	Normalized Multiband Drought Index (NMDI)	
TRMM	TRMM1	For 1 month
	TRMM3	For 3 months
	TRMM6	For 6 months
	TRMM9	For 9 months
	TRMM12	For 12 months

Three machine learning techniques were adopted in estimating SPI at each time scale, including boosted regression trees, random forest, and Cubist. Boosted regression trees are similar with random forest in that an ensemble approach based on CART to predict a target variable is used. Boosted regression trees produce a series of weighted predictions from individual trees. Weights assigned to each sample are updated at every iteration based on how the sample was explained in the previous iteration. That way, misinterpreted samples will get more attention in the next iteration, which can increase overall accuracy, but risks overfitting the model to such samples (Han and Kamber, 2011). Boosted regression trees and random forest were implemented in R (Robert Gentleman and Ross Ihaka, version 2.7.2) and its contributed packages (R Core Development Team, 2008) ("gbm" and "randomForest" packages, respectively) with default settings, except that 1000 trees were used instead of the default tree numbers.

Cubist uses a modified regression tree system developed by Quinlan (1993) to build rule-based predictive models. Cubist is a commercial product, and has been proved useful in various remote sensing–based regression tasks (Im et al., 2009; Im et al., 2012; Li et al., 2014). Each rule generated from Cubist is associated with a multivariate regression model to estimate a target variable. When multiple rules are applied to a sample, the output is averaged from the associated regression models. Cubist was applied using the Cubist software package (Rulequest Research, 2012).

Results show that random forest among the three machine learning approaches outperformed the other two methods regardless of the target variable and region used (Figure 4.8). Prediction of SPI in the arid region was slightly better than the humid region for all three approaches used. All three machine learning approaches provide relative variable importance: boosted regression trees and random forest document change in accuracy when a variable is out-of-bag. Cubist documents how many times a variable is used in rules and multivariate regression models.

Relative variable importance by machine learning approach when all variables were used against SPI1, SPI6, and SPI12 for the arid region is summarized in Table 4.5. For the arid region,

Regional Drought Monitoring

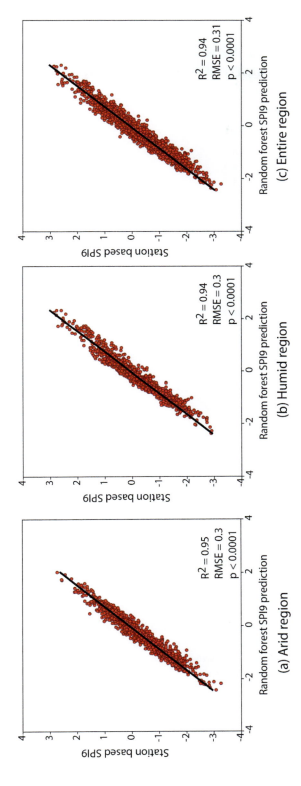

FIGURE 4.8 Scatterplots between random forest–based and station-based SPI9 values for (a) the arid region, (b) the humid region, and (c) the entire region.

TABLE 4.5
Relative Variable Importance in Percentage by Random Forest (RF) and Boosted Regression Trees (BRT) against SPI1, SPI6, and SPI12 for the Arid Region. The Top Five Important Variables Are in Bold for Each Case

	SPI1		SPI6		SPI12	
Variable	RF	BRT	RF	BRT	RF	BRT
LST	**6.4**	**8.5**	5.4	4.8	4.0	2.5
ET	**8.7**	**6.3**	6.0	4.0	6.4	2.8
NDVI	5.8	4.7	**7.8**	5.5	7.8	**5.9**
NDVI 500	4.0	1.3	6.9	1.0	7.5	2.5
NDWI5	2.0	1.2	3.4	1.8	4.3	1.9
NDWI6	5.0	2.7	**9.6**	**7.8**	**9.0**	3.1
NDWI7	5.2	1.4	**9.3**	**10.5**	**9.5**	**8.6**
NDDI5	2.9	1.7	4.6	1.0	5.9	1.3
NDDI6	4.5	3.2	**9.1**	**7.3**	**10.8**	**7.9**
NDDI7	2.6	1.2	5.2	2.6	7.1	3.6
NMDI	1.2	0.6	4.2	0.6	5.2	1.3
TRMM1	**21.3**	**38.0**	3.5	1.2	1.8	0.9
TRMM3	**9.6**	**12.0**	5.3	2.7	5.8	2.5
TRMM6	**10.3**	**11.6**	**17.8**	**36.6**	**14.2**	**13.3**
TRMM9	4.8	1.9	7.2	**7.5**	7.4	1.8
TRMM12	5.5	3.7	6.0	5.0	**20.0**	**40.2**

precipitation was the most important variable when random forest and boosted regression trees were used. Vegetation indices were also very useful for estimating SPI12 when Cubist was used (not shown). For both arid and humid regions, precipitation was dominantly important for short-term drought (i.e., SPI1, SPI3), while vegetation indices became important for long-term drought (i.e., SPI9, SPI12). It should be considered that some variables have high multi-collinearity, and thus further analysis is required to determine weights for drought indicators to identify each type of drought.

4.6 CONCLUSIONS

In this chapter, we examined recent trends in remote sensing–based drought monitoring and investigated the regional drought monitoring methodology based on the use of multi-sensor remote sensing data. The development of the SDCI, which integrates three components of temperature, precipitation, and vegetation-related stress, was introduced along with its further improvements by integrating more pivotal variables, including soil moisture and evapotranspiration, and utilizing machine learning techniques. Case studies of the Korean peninsula and the United States were presented.

In the case study for the Korean peninsula, the advanced SDCI with a linear combination of variables showed fairly good correlations with the yield of Highland Radish, which is generally cultivated without irrigation, proving the usability for monitoring and assessment of agricultural drought. The June SDCI showed statistically significant high correlations with Highland Radish yield data with weights fitted for SPEI12 ($r = 0.71$, $p = 0.031$) and SPI12 ($r = 0.70$, $p = 0.035$) in Jangsu, Jeollabuk-do. In the case study for the arid and humid regions in the United States, machine learning approaches were successfully applied to estimate SPI at different time scales and to examine the variable importance for documenting drought. Such variable importance can be used to determine weights for key variables to monitor drought conditions by drought type.

The development and advancement of SDCI follows the trend of recent studies of providing customized monitoring approaches to different types of drought, and developing new indices focusing on the consideration of more variables, including soil moisture and evapotranspiration. Currently, the integration of remote sensing data–based SDCI with process-based Land Surface Models is under development, differentiating the use of the SDCI from existing ones emphasizing its use for regions with limited observation data.

Since the coupling and feedback between variables are intertwined, it is not appropriate to investigate only limited variables to monitor each type of drought. Multiple indices need to be examined together for better drought monitoring and assessment. The use of drought indicators solely from globally available satellite data in obtaining SDCI enables the proposed index to be used in areas with limited regional in situ data. Drought now-cast and forecast information is especially important for those areas, since it helps stakeholders make informed decisions on the allocation of water resources and the determination of planting dates. Great cost may be avoided by the introduction of multi-sensor remote sensing drought indicators, which are based on globally available data and can be easily customized.

4.7 ACKNOWLEDGMENTS

This research was supported by the National Space Lab Program through the National Research Foundation of Korea (NRF) funded by the Ministry of Science, ICT, & Future Planning (Grant: NRF-2013M1A3A3A02042391).

REFERENCES

Anderson, L., Y. Malhi, L. Aragão, R. Ladle, E. Arai, N. Barbier, and O. Phillips. 2010. Remote sensing detection of droughts in Amazonian forest canopies. *New Phytol* 187:733–750.

Anderson, M., R. Allen, A. Morse, and W. Kustas. 2012. Use of Landsat thermal imagery in monitoring evapotranspiration and managing water resources. *Rem Sens Environ* 122:50–65.

Anderson, M., C. Hain, J. Otkin, X. Zhan, K. Mo, M. Svoboda, et al. 2013. An intercomparison of drought indicators based on thermal remote sensing and NLDAS-2 simulations with US drought monitor classifications. *J Hydrometeo* 14:1035–1056.

Anderson, M., C. Hain, B. Wardlow, A. Pimstein, J. Mecikalski, and W. Kustas. 2011. Evaluation of drought indices based on thermal remote sensing of evapotranspiration over the continental United States. *J Climate* 24:2024–2044.

Bolten, J., W. Crow, X. Zhan, T. Jackson, and C. Reynolds. 2010. Evaluating the utility of remotely sensed soil moisture retrievals for operational agricultural drought monitoring. *IEEE J Sel Top Appl Earth Obs Rem Sens* 3:57–66.

Brown, J. F., B. D. Wardlow, T. Tadesse, M. J. Hayes, and B. C. Reed. 2008. The vegetation drought response index (VegDRI): A new integrated approach for monitoring drought stress in vegetation. *GIScience Rem Sens* 45:16–46.

Caccamo, G., L. Chisholm, R. Bradstock, and M. Puotinen. 2011. Assessing the sensitivity of MODIS to monitor drought in high biomass ecosystems. *Rem Sens Environ* 115:2626–2639.

Carbone, G., J. Rhee, H. Mizzell, and R. Boyles. 2007. A regional–scale drought monitoring tool for the Carolinas. *Bull Am Meterol Soc* 89:20–28.

Choi, M., J. Jacobs, M. Anderson, and D. Bosch. 2013. Evaluation of drought indices via remotely sensed data with hydrological variables. *J Hydro* 476:265–273.

Cook, E., M. Kablack, and G. Jacoby. 1988. The 1986 drought in the southeastern United States: How rare an event was it? *J Geophy Res* 93:14257–14260.

Gao, B. 1996. NDWI—A normalized difference water index for remote sensing of vegetation liquid water from space. *Rem Sens Environ* 58: 257–266.

Gao, Z., Q. Wang, X. Cao, and W. Gao. 2014. The responses of vegetation water content (EWT) and assessment of drought monitoring along a coastal region using remote sensing. *GIScience Rem Sens* 51:1–16.

Garcia, M., N. Fernández, L. Villagarcía, F. Domingo, J. Puigdefábregas, and I. Sandholt. 2014. Accuracy of the temperature–vegetation dryness index using MODIS under water-limited vs. energy-limited evapotranspiration conditions. *Rem Sens Environ* 149:100–117.

Gu, Y., J. F. Brown, J. P. Verdin, and B. Wardlow. 2007. A five-year analysis of MODISNDVI and NDWI for grassland drought assessment over the central Great Plains of the United States. *Geophys Res Lett* 34:L06407.

Han, J., and M. Kamber. 2011. *Data Mining: Concepts and Techniques*. Burlington, MA: Elsevier.

Han, P., P. Wang, S. Zhang, and D. Zhu. 2010. Drought forecasting based on the remote sensing data using ARIMA models. *Math Comp Model* 51:1398–1403.

Huang, J., H. Van den Dool, and K. P. Georgakakos. 1996. Analysis of model-calibrated soil moisture over the United States (1931–93) and application to long-range temperature forecasts. *J Clim* 9:1350–1362.

Huang, S., D. Dahal, C. Young, G. Chander, and S. Liu. 2011. Integration of Palmer Drought Severity Index and remote sensing data to simulate wetland water surface from 1910 to 2009 in Cottonwood Lake area, North Dakota. *Rem Sens Environ* 115:3377–3389.

Im, J., J. R. Jensen, M. Coleman, and E. Nelson. 2009. Hyperspectral remote sensing analysis of short rotation woody crops grown with controlled nutrient and irrigation treatments. *Geocarto Int* 24:293–312.

Im, J., Z. Lu, J. Rhee, and L. J. Quackenbush. 2012. Impervious surface quantification using a synthesis of artificial immune networks and decision/regression trees from multi-sensor data. *Rem Sens Environ* 117:102–113.

Im, J., S. Park, J. Rhee, J. Baik, and M. Choi. 2016. Downscaling of AMSR-E soil moisture with MODIS products using machine learning approaches. *Environ Earth Sci* 75(15):1120.

IPCC. 2012. *Managing the Risks of Extreme Events and Disasters to Advance Climate Change Adaptation*. A Special Report of Working Groups I and II of the Intergovernmental Panel on Climate Change [Field, C. B., V. Balrros, T. F. Stocker, D. Qin, D. J. Dokken, K. L. Ebi, M. D. Mastrandrea, K. J. Mach, G.-K. Plattner, S. K. Allen, M. Tignor, and P. M. Midgley (eds.)]. Cambridge University Press, Cambridge and New York, 582 pp.

Ivits, E., S. Horion, R. Fensholt, and M. Cherlet. 2014. Drought footprint on European ecosystems between 1999 and 2010 assessed by remotely sensed vegetation phenology and productivity. *Glob Chan Bio* 20:581–593.

Jensen, J. R. 2000. *Remote Sensing of the Environment: An Earth Resource Perspective* (Chapter 10). Upper Saddle River, NJ: Pearson Prentice Hall.

Ji, L., and A. Peters. 2003. Assessing vegetation response to drought in the northern Great Plains using vegetation and drought indices. *Rem Sens Environ* 87:85–98.

Karl, T., and P. Young. 1987. The 1986 Southeast drought in historical perspective. *Bull Am Meterol Soc* 68:773–778.

Keetch, J., and G. Byram. 1968. A drought index for forest fire control. *Forest Service Research Paper SE-38*. U. S. Department of Agriculture – Forest Service Southeastern Forest Experiment Station Asheville, North Carolina, USA, 32 pp.

Keshavarz, M., M. Vazifedoust, and A. Alizadeh. 2014. Drought monitoring using a Soil Wetness Deficit Index (SWDI) derived from MODIS satellite data. *Agric Water Manag* 132:37–45.

Khaleghi, B., A. Khamis, F. O. Karray, and S. N. Razavi. 2013. Multisensor data fusion: A review of the state-of-the-art. *Inform Fusion* 14(1):28–44.

Knutson, C., and M. Hayes. 2002. South Carolina drought mitigation and response assessment: 1998–2000 drought. *Quick Response Research Report #136, National Hazards Research and Applications Information Center*. University of Colorado, Boulder, CO.

Kogan, F. N. 1995a. Drought of the late 1980s in the United States as derived from NOAA polar–orbiting satellite data. *Bull Am Meterol Soc* 76:655–668.

Kogan, F. N. 1995b. Application of vegetation index and brightness temperature for drought detection. *Adv Space Res* 15:91–100.

Leng, G., J. Hall. 2019. Crop yield sensitivity of global major agricultural countries to droughts and the projected changes in the future. *Sci Total Environ* 654: 811–821.

Li, M., J. Im, L. Quackenbush, and L. Tao. 2014. Forest biomass and carbon stock quantification using airborne LiDAR data: A case study over Huntington Wildlife Forest in the Adirondack Park. *IEEE J Sel Top Appl Earth Obs Rem Sens* 7(7):3143–3156.

McKee, T. B., N. J. Doesken, and J. Kleist. 1993. The relationship of drought frequency and duration to time scales. Preprints, Eighth Conference on Applied Climatology, Dallas, TX. *Am Meterol Soc*:233–236.

Mishra, A. K., and V. R. Desai. 2006. Drought forecasting using feed-forward recursive neural network. *Ecol Modell* 198:127–138.

Mishra, A. K., and V. Singh. 2011. Drought modeling—A review. *J Hydrol* 403:157–175.

Mu, Q., M. Zhao, J. Kimball, N. McDowell, and S. Running. 2013. A remotely sensed global terrestrial drought severity index. *Bull Am Meterol Soc* 94:83–98.

Palmer, W. C. 1965. Meteorological drought. *Research Paper No. 45*. Office of Climatology, U.S. Weather Bureau, U.S. Department of Commerce, Washington, DC, USA, 58 pp.

Palmer, W. C. 1968. Keeping track of crop moisture conditions, nationwide: The new crop moisture index. *Weatherwise* 21:156–161.

Park, S., J. Im, D. Han, and J. Rhee. 2020. Short-term forecasting of drought using numerical model output and temporal patterns of satellite-based drought indices. *Remote Sens* 12:3499.

Park, S., J. Im, E. Jang, and J. Rhee. 2016. Drought assessment and monitoring through blending of multi-sensor indices using machine learning approaches for different climate regions. *Agric For Meteoral* 216:157–169.

Park, S., J. Im, S. Park, and J. Rhee. 2017. Drought monitoring using high resolution soil moisture through multi-sensor satellite data fusion over the Korean peninsula. *Agric For Meteoral* 237:257–269.

Park, S., E. Seo, D. Kang, J. Im, and M.-I. Lee. 2018. Prediction of drought on pentad scale using remote sensing data and MJO index through random forest over East Asia. *Remote Sens* 10:1811.

Quinlan, J. 1993. *C4.5: Programs for Machine Learning*. San Mateo, CA: Morgan Kaufman.

R Core Development Team. 2008. *R: A Language and Environment for Statistical Computing*. http://www.R-project.org/.

Rhee, J., and J. Im. 2017. Meteorological drought forecasting for ungauged areas based on machine learning: Using long-range climate forecast and remote sensing data. *Agric For Meteoral* 237:105–122.

Rhee, J., J. Im, and G. J. Carbone. 2010. Monitoring agricultural drought for arid and humid regions using multi-sensor remote sensing data. *Remote Sens Environ* 114:2875–2887.

Rhee, J., K. Park, S. Lee, S. Jang, and S. Yoon. 2020. Detecting hydrological droughts in ungauged areas from remotely sensed hydro-meteorological variables using rule-based models. *Nat Hazards* 103:2961–2988.

Rojas, O., A. Vrieling, and F. Rembold. 2011. Assessing drought probability for agricultural areas in Africa with coarse resolution remote sensing imagery. *Remote Sens Environ* 115:343–352.

Rulequest Research. 2012. *Data Mining with Cubist*. http://www.rulequest.com/cubist-info.html

Sakamoto, T., A. Gitelson, and T. Arkebauer. 2014. Near real-time prediction of US corn yields based on time-series MODIS data. *Remote Sens Environ* 147:219–231.

Savitsky, B. G. 1986. *Agricultural Remote Sensing in South Carolina: A Study of Crop Identification Capabilities Utilizing Landsat Multispectral Scanner Data*, unpublished master thesis. Columbia: University of South Carolina Geography Department, 78pp.

Scaini, A., N. Sánchez, S. Vicente-Serrano, and J. Martínez-Fernández. 2015. SMOS-derived soil moisture anomalies and drought indices: A comparative analysis using in situ measurements. *Hydro Proc* 29(3):373–383.

Seitz, F., K. Hedman, F. Meyer, and H. Lee. 2014. Multi-sensor space observation of heavy flood and drought conditions in the Amazon region. In *Earth on the Edge: Science for a Sustainable Planet* (pp. 311–317). Berlin and Heidelberg: Springer.

Seneviratne, S. I., T. Corti, E. L. Davin, M. Hirschi, E. B. Jaeger, I. Lehner, B. Orlowskly, and A. J. Teuling. 2010. Investigating soil moisture-climate interactions in a changing climate: A review. *Earth-Sci Rev* 99:125–161.

Shaheen, A., and M. Baig. 2011. Drought severity assessment in arid area of Thal Doab using remote sensing and GIS. *Int J Water Resour Arid Environ* 1:92–101.

Svoboda, M., D. LeComte, M. Hayes, R. Heim, K. Gleason, J. Angel, B. Rippey, R. Tinker, M. Lalecki, D. Stooksbury, D. Miskus, and S. Stephens. 2002. The drought monitor. *Bull Am Meterol Soc* 83:1181–1190.

Tadesse, T., B.D. Wardlow, M.J. Hayes, and M.D. Svoboda. 2010. The vegetation outlook (VegOut): A new method for predicting vegetation seasonal greenness. *GISci Remote Sens* 47(1):25–52.

Wan, Z., P. Wang, and X. Li. 2004. Using MODIS land surface temperature and normalized difference vegetation index products for monitoring drought in the southern Great Plains, USA. *Int J Rem Sens* 25:61–72.

Wang, L., and J. J. Qu. 2007. NMDI: A normalized multi-band drought index for monitoring soil and vegetation moisture with satellite remote sensing. *Geophys Res Lett* 34: L20405.

Wang, R., J. Xu, D. Wang, X. Xie, and P. Wang. 2013. Construction of drought indices from passive microwave remote sensing AMSR-E data in Huaihe River Basin. *Appl Mech Mat* 397:2503–2506.

Wardlow, B. D., M. C. Anderson, and J. P. Verdin. Eds. 2012. *Remote Sensing of Drought: Innovative Monitoring Approaches*. Boca Raton, FL: CRC Press, p. 422.

Weaver, J. C. 2005. The drought of 1998–2002 in North Carolina—precipitation and hydrologic conditions. *Scientific Investigations Report, 2005–5053*. U.S. Geological Survey, U.S. Department of the Interior, Reston, VA, 98 pp.

Wells, N., S. Goddard, and M. J. Hayes. 2004. A self-calibrating palmer drought severity index. *J Clim* 17:2235–2351.

Willeke, G., J. R. M. Hosking, J. R. Wallis, and N. B. Guttman. 1994. The National Drought Atlas. *Institute for Water Resources Rep.94-NDS-4*. U. S. Army Corps of Engineers, CD-ROM.

Zhang, A., and G. Jia. 2013. Monitoring meteorological drought in semiarid regions using multi-sensor microwave remote sensing data. *Rem Sens Environ* 134:12–23.

Zhang, N., Y. Hong, Q. Qin, and L. Liu. 2013. VSDI: A visible and shortwave infrared drought index for monitoring soil and vegetation moisture based on optical remote sensing. *Int J Rem Sens* 34:4585–4609.

Zhong, L., Y. Ma, Y. Fu, X. Pan, W. Hu, Z. Su, et al. 2014. Assessment of soil water deficit for the middle reaches of Yarlung-Zangbo River from optical and passive microwave images. *Rem Sens Environ* 142:1–8.

5 Land Degradation Assessment and Monitoring of Drylands

Marion Stellmes[1], Ruth Sonnenschein[2], Achim Röder[3], Thomas Udelhoven[3], Gabriel del Barrio[4], and Joachim Hill[3]*

ACRONYMS AND DEFINITIONS

AVHRR	Advanced very high resolution radiometer
BRDF	Bidirectional reflectance distribution function
CBD	Convention on biological diversity
CCD	Convention to combat desertification
CHIME	Copernicus hyperspectral imaging mission for the environment
CNES	Centre National d'Études Spatiales
CNNs	Convolutional neural networks
EC-JRC	European Commission Joint Research Center
ETM	Enhanced Thematic Mapper
EVI	Enhanced Vegetation Index
faPAR	Fraction of absorbed photosynthetic active radiation
FORCE	Framework for operational radiometric correction for environmental monitoring
GPCC	Global Precipitation Climatology Centre
LEDAPS	Landsat ecosystem disturbance adaptive processing system
LSTM	Land surface temperature monitoring
MERIS	Medium resolution imaging spectrometer
MODIS	Moderate RESOLUTION IMAGING SPECTRORADIOMETER
NDVI	Normalized Difference Vegetation Index
NPP	net primary productivity
OLI/TIRS	Operational Land Imager and Thermal Infrared Sensors
PET	Potential evapotranspiration
SAVI	Soil Adjusted Vegetation Index
SDG	Sustainable development goal
SeaWIFS	Sea-viewing wide field-of-view sensor
SMA	Spectral mixture analysis
SPOT	Satellite Pour l'Observation de la Terre
STARFM	Spatial and temporal adaptive reflectance fusion model
UAS	Unmanned aerial systems
UNCCD	United Nations convention to combat desertification
UNCOD	United Nations conference on desertification

[1] Remote Sensing and Geoinformatics, Department of Earth Sciences, Freie Universität Berlin, 12249 Berlin, Germany
[2] Institute for Earth Observation, European Academy of Bozen/Bolzano (EURAC), 39100 Bolzano, Italy
[3] Dep. Environmental Remote Sensing and Geoinformatics, Fac. Regional and Environmental Sciences, Trier University, 54286 Trier, Germany
[4] Estacion Experimental de Zonas Aridas (CSIC), 04120 Almeria, Spain
* Corresponding author: marion.stellmes@fu-berlin.de

UNEP	United Nations Environment Programme
USGS	United States Geological Survey
VOD	Vegetation optical depth

5.1 INTRODUCTION

5.1.1 DRYLANDS

Drylands cover about 40% of the Earth's land surface, comprising hyper-arid to dry sub-humid climate zones, which are defined by low mean annual precipitation amounts compared to potential evaporation, i.e., the ratio of precipitation (P) to potential evapotranspiration (PET) is less than 0.65 (Safriel et al., 2005; Thomas and Middleton, 1994; see figure 5.1). They include a large number of ecosystems that belong to the four broad biomes—forests, Mediterranean, grasslands, and deserts (Safriel et al., 2005)—and are home to about three billion humans (Mirzabaev et al., 2019). Many of these residents directly depend on dryland ecosystem services (Table 5.1), including the provision of food, forage, water, and other resources (Millenium Ecosystem Assessment, 2005a; Cherlet et al., 2018). Drylands also provide ecosystem services of global significance, such as climate regulation by sequestering and storing vast amounts of carbon (Lal, 2004) and provide, for instance, about 40% of the global net primary productivity (NPP) (Wang et al., 2022).

Drylands are characterized by high variability in both rainfall amounts and intensities and the occurrence of cyclic and prolonged periods of drought. Most frequently, soils contain low nutritious reserves and have low contents of organic matter and nitrogen (Skujins, 1991). In addition, surface runoff events, soil-moisture storage, and groundwater recharge in drylands are generally more variable and less reliable than in more humid regions (Koofhafkan and Stewart, 2008).

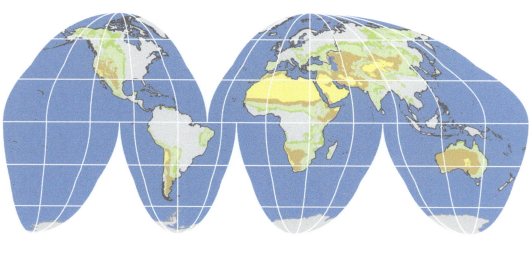

FIGURE 5.1 The spatial extent of drylands based on the aridity index (AI equals ratio of rainfall [P] and potential evapotranspiration [PET] for the period 1970–2000). Hyper-arid: P/PET < 0.05; arid: 0.05 ≤ P/PET < 0.20; semi-arid: 0.20 ≤ P/PET < 0.50; dry sub-humid: 0.50 ≤ P/PET < 0.65. Projection Goode Homolosine (Sources: based on Zomer et al. (2022) and ESRI Data & Maps).

TABLE 5.1
Key Dryland Ecosystem Services (after Millenium Ecosystem Assessment, 2005a)

Supporting Services
Services That Maintain the Conditions for Life on Earth
- Soil development (conservation, formation)
- Primary production
- Nutrient cycling
- Biodiversity

Provisioning Services	Regulating Services	Cultural Services
Goods produced or provided by ecosystems	Benefits obtained from regulation of ecosystem processes	Nonmaterial benefits obtained from ecosystems
• Provision derived from biological productivity: food, fiber, forage, fuelwood, and biochemicals • Fresh water	• Water purification and regulation • Pollination and seed dispersal • Climate regulation (local through vegetation cover and global through carbon sequestration)	• Recreation and tourism • Cultural identity and diversity • Cultural landscapes and heritage values • Indigenous knowledge systems • Spiritual, aesthetic, and inspirational services

Water availability and the tolerance to periods of water scarcity are key factors in dryland productivity (Stafford Smith et al., 2009). In response to water scarceness and prolonged drought periods, fauna and flora of dryland ecosystems have adapted to these conditions following manifold morphological, physical, and chemical strategies, such as the development of drought-avoiding (i.e., ephemeral annual grasses) or drought-enduring (i.e., xerophytes) plant species and plant adaptations such as xeromorphological leaf structures. Fire is a further important element in the functioning and maintenance of dryland ecosystems (Bond and Keeley, 2005). The review of Grünzweig et al. (2022) focuses on strategies on how terrestrial ecosystems respond to climate change on a global scale. They identify and discuss 12 "dryland mechanisms" that affect processes of ecosystem functioning in this context, including vegetation growth, water dynamics, energy balance, cycles of carbon and nutrients, plant productivity, and the decomposition of organic matter.

Climate change is expected to increase the aridity of existing drylands and is projected to expand by 11% under RCP4.5 and 23% under RCP8.5 (RCP: Representative Concentration Pathways), compared to the 1961–1990 baseline by the end of the 21st century (Huang et al., 2016). But quantification of effects of climate change on the productivity and extent of drylands is still an ongoing debate. Lian et al. (2021) and Wang et al. (2022) have compiled and summarized the findings from current studies on this topic. Berdugo et al. (2020) highlight the largely unknown aspect of whether aridification leads to gradual or abrupt and systemic or specific ecosystem changes. Yet, there is also evidence of dryland greening and increased productivity due to higher atmospheric CO_2 levels (Lu et al., 2016). While warming is projected to increase PET more than P, leading to an expansion of drylands, recent studies challenge this view, arguing that traditional aridity indices may overestimate aridity and dryland expansion by not accounting for vegetation responses to higher CO_2. Climate warming might not significantly change the spatial extent of drylands, but increasing water constraints could still impact vegetation productivity in these areas. In turn, this can put a strain on land use sustainability. Besides the common definition of drylands Sörensen (2007) introduced a broader definition where dry and sub-humid tropical forests in accordance with paragraph 13 of the Convention on Biological Diversity (CBD) were included.

5.1.2 LAND USE IN DRYLAND AREAS

Over millennia humans developed diverse strategies for sustainable use of drylands, adapting to varying levels of aridity. These strategies encompass shifting agriculture systems, annual croplands, home gardens, and mixed agriculture–livestock systems, including nomadic pastoral and transhumant practices (Koofhafkan and Stewart, 2008). The vast majority of drylands that support vegetation are used as rangelands (69%), which sustain about 50% of the world's total livestock population, whereas 25% of the dryland areas are used as croplands (Reid et al., 2004). However, land use varies largely among dryland climates. The proportion of rangeland increases with aridity, from 34% in sub-humid regions to 97% in hyper-arid areas (Millenium Ecosystem Assessment, 2005b), whereas arable cultivation is restricted to semi-arid and dry sub-humid regions (Koofhafkan and Stewart, 2008). Also, the use of fire as a land use management tool has a history of millennia in drylands and includes pastoral burning to improve rangeland conditions (Naveh, 1975), but also slash and burn agriculture, honey collection, charcoal production, and opening landscapes to facilitate hunting as practiced in African savannas (Mbow et al., 2000). Even though dryland ecosystems are adapted to fires, changing fire regimes may cause land degradation and loss of biodiversity as they impact species composition and vegetation structure and severely affect nutrient cycling (e.g. Trapnell, 1959; Anderson et al., 2003).

Countries with drylands differ in their socioeconomic development. Differences range from agrarian via industrialized to service-oriented societies, whereby at least 90% of the dryland population lives in developing countries (Safriel et al., 2005). The development stage defines to a large extent the land use systems and the corresponding process framework of land use/land cover changes (DeFries et al., 2004). Even though land use changes are affecting almost all terrestrial ecosystems, drylands are considered as those most vulnerable to degradation processes. Thus, water scarcity, overuse of resources, and climate change are a much greater threat for dryland ecosystems than for non-dryland systems (Millenium Ecosystem Assessment, 2005a).

5.1.3 Land Degradation and Desertification

Degradation of terrestrial dryland ecosystems, caused by desertification, is recognized as one of the major threats to the global environment directly impacting human well-being (Millenium Ecosystem Assessment, 2005a) and threatening to reverse the gains in human development in many parts of the world (UNU, 2006). The terms "land degradation" and "desertification" received worldwide attention following the prolonged Sahel drought during the 1970s and 1980s, which caused a humanitarian catastrophe. As a result of the United Nations Conference on Desertification (UNCOD) in 1977 a "Plan of action to combat desertification" was approved. Limited progress in reducing the problem of desertification since then led the Rio Conference in 1992 to call on the United Nations General Assembly to prepare a Convention to Combat Desertification (CCD) through intergovernmental negotiation. Thus, in 1994, the United Nations Convention to Combat Desertification (UNCCD) was adopted and brought into force in 1996, having received notification of the 50th ratification of the Convention, which by now has 193 signatory parties. The definition of both terms was subject to highly controversial debates (Hermann and Hutchinson, 2005).

A nowadays widely accepted definition of land degradation and desertification is provided by the UNCCD (1994), who defined land degradation as

> the reduction or loss, in arid, semi-arid and dry sub-humid areas, of the biological or economic productivity and complexity of rainfed cropland, irrigated cropland, or range, pasture, forest and woodlands resulting from land uses or from a process or combination of processes, including processes arising from human activities and habitation patterns.

In turn, desertification is defined as "land degradation in arid, semi-arid and dry sub-humid areas, resulting from various factors, including climatic variations and human activities" (UNCCD, 1994).

These definitions aim to cover, at large, the broad range of complex processes that cause a sustained decrease of ecosystem services throughout all terrestrial ecosystems in drylands. Nevertheless, they also leave room for interpretation and uncertainties concerning the terminology (Vogt et al., 2011). However, even admitting inherent imprecision of those definitions, they make clear that: (1)

desertification causes land degradation, not the other way around and (2) desertification is always caused by human activities, hence climate may become an aggravating factor, not a cause by itself.

Recognizing the critical issue of land degradation, the United Nations General Assembly endorsed the sustainable development goals (SDGs) in September 2015 (Wunder et al., 2018). SDG 15 "Life on land" included the aim of Land Degradation Neutrality (LDN) (SDG 15.3). LDN is an integral concept, which focuses on protecting, restoring, and promoting sustainable land use and thereby preventing land degradation.

5.1.4 Scientific Perception of Land Degradation

In the past decades, the scientific communities' understanding has undergone a shift concerning the key factors that are required to allow for adequate assessment and monitoring of land degradation. The assessment of land degradation changed from a mere biophysical perception to a more holistic approach where human-induced or climate-driven underlying forces as well as spatial and temporal scale issues have been recognized as factors that should be considered to understand and identify land degradation processes (Vogt et al., 2011).

The understanding of land degradation processes, including their causes and consequences on ecosystem functioning as well as the identification of affected areas and regions at risk, are a prerequisite to develop strategies to mitigate and avoid land degradation. Accordingly, over the past decades many national and international research initiatives reviewed the status of land degradation sciences and identified gaps and developed strategies to assess and monitor land degradation and desertification.

This chapter provides an overview of important Earth observation data and studies to assess land degradation in drylands. Section 5.2 presents general considerations regarding the assessment and monitoring of land degradation, including suitable indicators as well as sensor systems. The following sections review the state of the art on the assessment of land condition (Section 5.3), the monitoring of land use/land cover changes to assess land degradation processes (Section 5.4), and the identification of human-induced drivers of land degradation using integrated concepts (Section 5.5), whereas Section 5.6 describes limits and uncertainties regarding dryland observation. This chapter concludes with a summary of land degradation assessment and monitoring by remote sensing techniques (Section 5.7).

5.2 REMOTE SENSING OF DRYLAND DEGRADATION PROCESSES

Various scientific disciplines contribute valuable information that enhances the understanding of land degradation and desertification at different temporal and spatial scales. These include studies ranging from the plot scale to global assessments as well as the collection of biophysical or socioeconomic data and the implementation of models to predict land use changes in future decades.

Earth observation is a tool that essentially contributes to the assessment and monitoring of ecosystems from a local to a global scale. Hence, information extracted from remote sensing data can be employed to: (1) assess the extent and condition of ecosystems and (2) monitor changes of ecosystem's conditions and services over long time periods and large areas (Foley et al., 2005; Turner II et al., 2007). The use of Earth observation data fundamentally contributes to the understanding of dynamics and responses of vegetation to climate and human interactions (DeFries, 2008).

Monitoring drylands requires observation data that are able to observe long-term trends and short-term disturbances across large areas. For this reason, remote sensing data are important components of monitoring strategies, as they provide objective, repetitive, and synoptic observations across large areas (Graetz, 1996; Hill et al., 2004). Three major components are particularly important to provide (1) a comprehensive observation of dryland areas, (2) ensure their relevance for policy and management, and (3) help preventing unsustainable use of ecosystems goods and services:

1. Assessment of actual land condition, i.e., the capacity of an ecosystem to provide goods and services (compare 3),

2. Monitoring of land cover changes and assessment of their implications for land condition, separating natural processes, i.e., climate variability and fire, from human-induced land use/land cover–related processes (compare 4), and
3. Integrated concepts that link remotely sensed results to the human dimension in order to identify drivers of land degradation (compare 5).

Neither the condition of ecosystems nor the processes affecting them can directly be measured by Earth observation data. Rather, suitable indicators have to be identified (Verstraete, 1994) that (1) can be related to the status and processes and (2) can be derived in a standardized and replicable way.

5.2.1 Suitable Remote Sensing Indicators for Dryland Observation

A range of approaches and models has been developed allowing researchers to derive a variety of biophysical parameters appropriate for the observation of drylands (Hill, 2008; Lacaze, 1996). Depending on the spatial and spectral characteristics of the remote sensing data these qualitative and quantitative measures include vegetation indices related to greenness, vegetation cover, pigment and water content, soil organic matter of the topsoil, landscape metrics, etc. (e.g., Blaschke and Hay, 2001; Hill et al., 2004).

The biological productivity of ecosystems is one of the key factors that describe the functioning of an ecosystem, and it is also explicitly stated in the definition of desertification and land degradation of the UNCCD (del Barrio et al., 2010). Parameters related to productivity such as greenness, vegetation cover, and biomass can therefore serve as proxies to assess and monitor land degradation. These parameters are especially suitable for Earth observation methods due to the distinct spectral signature of vegetation. A commonly used vegetation index calculated from the red and near-infrared spectral information is the Normalized Difference Vegetation Index (NDVI) (Rouse et al., 1974; Tucker, 1979). NDVI is a proxy for greenness and linearly related to the fraction of absorbed photosynthetic active radiation (faPAR) (Myneni and Williams, 1994; Fensholt et al., 2004), which in itself is an important factor of assessing the net primary productivity (NPP). However, the NDVI has well-known weaknesses due to its sensitivity to soil background, especially when vegetation cover is low (Price, 1993; Elmore et al., 2000). Advanced vegetation indices overcome these problems, like the Enhanced Vegetation Index (EVI) (Huete et al., 2002) and the Soil Adjusted Vegetation Index (SAVI) (Huete, 1988). More advanced methods to derive parameters that are related to vegetation are the Tasseled Cap Transformation (Kauth and Thomas, 1976) and Spectral Mixture Analysis (SMA) (Adams et al., 1986; Smith et al., 1990). The latter directly provides vegetation cover if correctly parameterized and is often used for Landsat-based land degradation assessment in drylands (Sonnenschein et al., 2011). Especially in dryland regions used for grazing, mapping of woody cover and the separation of tree, shrub, and grass layers has frequently been identified as a key requirement to support rangeland management (Abdi et al., 2022). Nevertheless, for long-term temporal analysis it seems to be of more decisive importance to employ a robust and consistent measure (Udelhoven and Hill, 2009; Sonnenschein et al., 2011).

Even though land degradation indicators related to soil have proven to provide important information on land degradation (Wang et al., 2023), vegetation cover hampers the remotely sensed assessment of soil properties. Thus, soil properties can only be reliably assessed at low vegetation cover (Jarmer et al., 2009). Furthermore, many of the proposed indicators, e.g., grain size distribution, mineral content, and soil organic carbon, presence of biological soil crusts, or specific vegetation-related parameters (e.g., leaf water content) require data that is currently only experimental or not available at sufficient spatial resolution, e.g., (hyper)thermal, hyperspectral, fluorescence, or LiDAR. To date, these data were mostly acquired using airborne systems, making them costly and only available for small areas. As a result, only few studies exist that use hyperspectral imagery for land degradation assessment (e.g., Shrestha et al., 2005; De Jong and Epema, 2011).

Land Degradation Assessment and Monitoring of Drylands

Smith et al. (2019) provide an excellent overview of the state of the art of remote sensing of dryland ecosystem structure and function as well as potential future uses.

5.2.2 Earth Observation Platforms Used in Dryland Observation

High variability of precipitation amounts infers also a high variability of vegetation cover and its vitality. Moreover, disturbances like fires create abrupt changes of vegetation cover. Dryland observation requires us to consider these variations by using long-term observation to separate gradual long-term trends from short-term variations as well as to discriminate inherent trends reflecting human-induced degradation (Evans and Geerken, 2004). Among the numerous space-borne sensors, only a few satellites are fulfilling the two criteria of collecting data that (1) cover a long time period (e.g., more than 20 years) and (2) provide a systematic global coverage. These systems can be distinguished into two major groups: the first provides a medium spatial resolution but has a limited temporal resolution, the second provides coarse scale resolution, but has high temporal resolution. Figure 5.2 gives information of important sensors systems applied in dryland monitoring and derived archives, which are presented in more detail in the following sections.

5.2.2.1 Medium Spatial Resolution Sensors

The Landsat program consists of a series of multi-spectral optical sensors that record the reflected radiance in the visible to middle infrared domain (complemented by band(s) in the thermal domain) which allows the derivation of several surrogates related to vegetation properties (Fang et al., 2005). Landsat Thematic Mapper (TM), Enhanced Thematic Mapper (ETM+), and Operational Land Imager and Thermal Infrared Sensors (OLI/TIRS), respectively are providing data of the Earth's

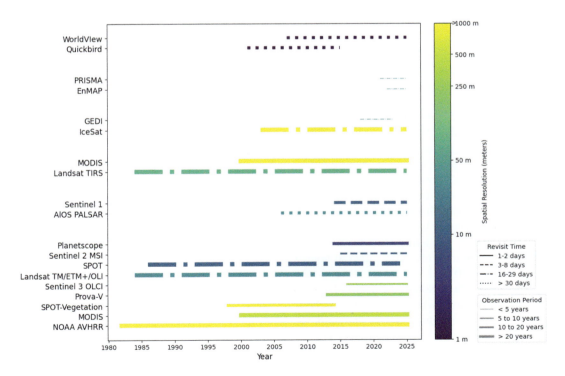

FIGURE 5.2 Overview of common Earth observation sensors and their specifications used in dryland monitoring.

surface with a spatial resolution of 30 m x 30 m since 1982 (Goward and Masek, 2001). The temporal revisit rate of the sensor is 16 days and could theoretically provide a time series of Earth observations with similar density compared to those composite products provided by coarse-scale sensors, but also in many dryland areas cloud cover impedes the acquisition of utilizable images. Thus, often only a few images of sufficient quality can be acquired per season.

The SPOT (Satellite Pour l'Observation de la Terre) satellites operated by Centre National d'Études Spatiales (CNES) provide multi-spectral data since 1986 with a spatial resolution of 6 m x 6 m up to 20 m x 20 m, a revisit rate of 26 days and a secured data continuity until 2024. In contrast to Landsat sensors the possibility of off-nadir-acquisition allows prioritization and increases data acquisition in specific areas but at the expense of other regions.

Long-term monitoring requires accurate geometric and radiometric correction of the data to reduce noise that originates from observational conditions, including observation geometry, atmospheric conditions and sensor degradation.

The opening of the Landsat archives distributed by the United States Geological Survey (USGS) has not only enabled new opportunities to assess land cover changes based on the full range of available data from the archive but has also fostered pre-processing and data analyses in an automated way together with better computational capacities and new methodologies. This includes the provision of geometrically corrected Landsat data by USGS, cloud detection via Fmask (Zhu and Woodcock, 2012; Qiu et al., 2019), and automated radiometric correction schemes like the Landsat Ecosystem Disturbance Adaptive Processing System (LEDAPS) (Masek et al., 2006), the Australian Bidirectional Reflectance Distribution Function (BRDF) correction scheme (Flood et al., 2013), or the Framework for Operational Radiometric Correction for Environmental monitoring (FORCE; Frantz, 2019). Today a number of operational processing frameworks exist that provide analysis-ready time series of data at different levels of pre-processing (Doxani et al., 2018, 2023). The introduction of cloud computing platforms, notably the Google Earth Engine since its inception in 2010, has marked a paradigm shift in the automation of image pre-processing tasks (Gorelick et al., 2017). This advancement has not only streamlined the methodological framework for handling large-scale datasets but has also democratized the accessibility of advanced data analysis techniques. Consequently, this technological evolution has extended the capability of conducting intricate big data analyses to a broader spectrum of users, encompassing both professional researchers and the general public.

5.2.2.2 Coarse Spatial Scale Satellite Sensors

Regional and global dryland studies often utilize coarse-scale, high-temporal resolution imagery, with the NOAA Advanced Very High Resolution Radiometer (AVHRR) sensor series playing a crucial role due to its extensive historical data and despite the challenges in setting up a homogenous time series. Pre-processing of these archives involves correcting orbital drift effects, and inter-calibrating spectral channels and creating multi-day composites across different AVHRR sensors. Moreover, the limited spectral properties of AVHRR sensors restrict the biophysical parameter derivation but typically rely on NDVI. The most recent version of the database is the Global Inventory Modeling and Mapping Studies-3rd Generation V1.2 NDVI3g+ (GIMMS-3G+), which offers bi-monthly NDVI data at approximately 8 km x 8 km resolution for 1981 to 2022 (Pinzon et al., 2023).

The Moderate Resolution Imaging Spectroradiometer (MODIS) provides a better spatial and spectral resolution, which allows researchers to derive more enhanced biophysical surrogates. NDVI and EVI are provided as standard vegetation parameter products. Moreover, the sensor properties facilitate the provision of a consistent high-quality data archive, including the possibility to derive Bidirectional Reflectance Distribution Function (BRDF) corrected data (Strahler et al., 1999). Other sensors delivering time series suitable for land degradation assessment are, e.g., Satellite Pour l'Observation de la Terre (SPOT) Vegetation, Sea-Viewing Wide Field-of-View Sensor (SeaWIFS), Medium Resolution Imaging Spectrometer (MERIS), and Sentinel-3 OLCI.

However, in comparison to the NOAA AVHRR data sets these archives are still confined to rather short observation periods.

5.2.2.3 Recent Developments for Obtaining Earth Observation Time Series

Although both coarse and medium sensor types provide data that allow for adequate dryland observation, there is a trade-off between the geometric and spectral levels of detail, areas covered, temporal resolution, and temporal coverage that needs to be considered. Several studies aimed at combining different data archives to overcome the different spectral responses, differing observation characteristics, including observation geometry and diverging spatial and temporal resolutions of the sensor systems (Ceccherini et al., 2013). There are basically two major approaches, (1) the fusion of data, i.e., improve the spatial, spectral, and/or temporal resolution by integrating the data of two sensors, and (2) harmonization of data, i.e., combining data from two or more sensors to create a harmonized data set.

One technique is the fusion of Landsat and MODIS images with the Spatial and Temporal Adaptive Reflectance Fusion Model (STARFM) (Gao et al., 2006), aiming at providing time series with a temporal resolution of MODIS but the spatial resolution of Landsat. The approach was applied successfully to dryland areas (Schmidt et al., 2012; Walker et al., 2012) and offers the possibility to monitor land degradation processes in more detail. One drawback of this procedure is that the fusion can only be performed after the launch of MODIS Terra in the year 2000.

The launch of ESA's Sentinel-2 satellites in 2015 and 2017 has significantly enhanced global data acquisition capabilities, complemented Landsat's Operational Land Imager (OLI), and improved the repetition rate for capturing global data, thus increasing the chances of obtaining cloud-free observations. The HLS (Harmonized Landsat and Sentinel-2) project by NASA (Claverie et al., 2018) and the Sen2Like initiative from ESA (Saunier et al., 2019) are collaborative efforts aiming at creating a unified, high-quality record of surface reflectance, leveraging data from Landsat-8/9 and Sentinel-2A/B satellites through the OLI and MSI instruments. This integration provides and ensures a comprehensive and consistent data set for global environmental monitoring and analysis.

Various hyperspectral, spaceborne missions were recently launched, e.g., EnMAP (Environmental Mapping and Analysis Program) under the lead of the German Aerospace Center (DLR), or PRISMA (PRecursore IperSpettrale della Missione Applicativa) of the Agenzia Spaziale Italiana (ASI). First demonstration studies have been published, e.g., Barnes et al. (2017) evaluate the sensitivity of narrowband indices to track biophysical parameters, such as predawn leaf water potential or leaf gas exchange, Leitao et al. (2015) used airborne hyperspectral and Hyperion data to demonstrate the potential of EnMAP data of characterizing ecological gradients in dryland systems. It is to be expected that the utilization of hyperspectral imagery in the context of land degradation assessment will increase in the near future, also with the advent of the satellite mission Copernicus Hyperspectral Imaging Mission for the Environment (CHIME) in 2028 (Rast et al., 2021). Further, the establishment of spaceborne missions acquiring hitherto experimental-only data will open up new domains for analysis, such as energy fluxes, biological soil crusts, or plant traits (Smith et al., 2019).

With an increasing availability of SAR data, also other multi-sensor approaches are driven forward and improve the mapping capabilities in dryland systems. Combining optical with SAR data, Higginbottom et al. (2018) have mapped fractional woody cover in the Limpopo region of South Africa by fusing Landsat and ALOS-PALSAR L-band data. In a more experimental setup, Wessels et al. (2019, 2023) trained ALOS-PALSAR with woody cover estimates derived from airborne LiDAR acquired across northern Namibia to successfully provide wall-to-wall woody cover maps for different time steps. But even with optical data alone, scaling up was shown to work well for different vegetation types provided adequate ground calibration and validation data and suitable image data at each scale level are available (Gessner et al., 2013).

5.2.2.4 Additional Remote Sensing Data for Dryland Assessment

In addition to the already mentioned sensor systems, a range of studies have used remote sensing data to obtain information on dryland ecosystems. Guirado et al. (2020) presented a novel approach using convolutional neural networks (CNNs) for estimating tree cover in global drylands from satellite and aerial images. The study addresses the challenges of accurately estimating tree cover in drylands and compares the performance of CNN-based models with traditional methods, demonstrating the potential of deep learning techniques for more precise and automated tree cover estimation.

In recent years, unmanned aerial systems (UAS) have increasingly found their way in many dryland-related studies. With increasing loading capacities, enhanced sensors (multi- and hyperspectral, thermal, LiDAR) and ease of use they have been used to directly map target variables. Amputu et al. (2023) mapped plant traits related to rangeland condition using multi-spectral information and a canopy height model derived from the point cloud, while Sankey et al. (2017) explored the combination of UAS-derived hyperspectral and LiDAR data to map dryland vegetation at the plant species level. Barnetson et al. (2020) were able to even predict highly specific rangeland-related indicators, such as grass biomass, crude protein content, and acid detergent fiber using machine learning models with hyperspectral UAS data and structure from motion-derived height information. In the second half of the 2020 decade ESA will expand the existing Sentinel missions with six high-priority candidate missions. Next to CHIME the Land Surface Temperature Monitoring (LSTM) will be of particular relevance in the context of land degradation assessment.

5.2.3 Time Series Analysis Techniques

The creation of a medium-resolution time series is challenging because images should originate from comparable phenological stages or represent appropriately land surface phenology. Because of limited data availability many of the early studies investigating time trajectories of vegetation based on Landsat time series are confined to only one observation per season and were based on linear trend analyses (e.g., Hostert et al., 2003; Röder et al., 2008a). With pre-processing automation, time series approaches were developed that allow for the detection of gradual or abrupt changes, or both simultaneously. Several methodologies and tools were published, e.g., Landsat-based Detection of Trends in Disturbance and Recovery—LandTrendr (Kennedy et al., 2010), the Vegetation Change Tracker—VCT (Huang et al., 2010), Breaks For Additive Seasonal and Trend—BFAST (Verbesselt et al., 2010), Continuous Monitoring of Forest Disturbance Algorithm—CMFDA (Zhu et al., 2012), and Continuous Change Detection and Classification—CCDC (Zhu and Woodcock, 2014). Based on CCDC, Zhu et al. (2020) developed the COntinuous monitoring of Land Disturbance (COLD) algorithm to provide a Landsat-tailored, large-scale, continuous, and accurate detection of land disturbance. Ye et al. (2023) introduced OB-COLD (Object-Based COLD) that enhances the pixel-based approach of COLD to an object-based analysis technique.

Many of these approaches were implemented and tested in boreal and temperate forest ecosystems (e.g., Griffiths et al., 2011; Schroeder et al., 2011). In such ecosystems the vegetation signal is high and yearly variations are small compared to dryland areas. Moreover, vegetation communities in drylands are often very complex and the spatial arrangement of the landscape very heterogeneous. These factors plus the occurrence of fires hamper the detection of subtle modifications of vegetation cover due to land degradation processes. Therefore, enhanced time-series analyses tools gained more and more importance as they allow for monitoring not only the overall increase or decrease of greenness, but also more complex change patterns including its character, i.e., gradual and abrupt changes (De Jong et al., 2012). This represents reality better, as trends are rarely uniform during a long observation period, e.g., due to droughts, fire events, and macro weather situations.

Dense temporal resolution time series can additionally be used to portray the land surface phenology of vegetation and its changes by deriving phenological metrics using for instance Timesat (Jönsson and Eklundh, 2002), Timestats (Udelhoven, 2010), or based on polar coordinate transformation (Brooks et al., 2020). With the higher temporal resolution afforded by joint use of the Landsat and Sentinel-2 archives, it was even possible to predict phenological parameters at higher resolution from image fusion (Frantz et al., 2016b) or infer them directly from medium-resolution sensors (Frantz et al., 2022).

5.3 ASSESSING LAND CONDITION

Land degradation may be defined as a long-term loss of an ecosystem's capacity to provide goods and services. Rather than being a landscape type or an isolated feature, degradation is a (often terminal) level within a full spectrum of states of ecological maturity, referred to as land condition. Ecological maturity in turn refers to ecological succession and can be defined in terms of energy flow by using biomass, NPP, and turnover ratio (NPP / biomass) to deal with ecosystem maturity, exploitation, and degradation (del Barrio et al., 2021). Therefore, a major component of a comprehensive dryland observation is the assessment of land condition which can be linked to the ecosystem status. Even though land degradation is recognized as a severe threat, only few global land degradation assessments have been carried out up to this point (Millenium Ecosystem Assessment, 2005a; Vogt et al., 2011).

The first global assessment of land quality was provided in the framework of the GLASOD project (Global Assessment of Human-Induced Soil Degradation, 1987–1990) where human-induced soil degradation (extent, type, and grade) was mapped at a scale of 1:10 million based on expert judgment (Oldeman et al., 1990). Another global assessment was provided by Dregne and Chou (1992) who also integrated information on vegetation status based on secondary sources. Whereas the map provided by GLASOD indicated that 20% of soils in drylands were degraded, Dregne and Chou estimated that 70% of dryland areas were affected either by degradation of soil or vegetation. A more recent study (Lepers, 2003) prepared for the Millennium Ecosystem Assessment covered over 60% of all dryland areas. Several data sources, including remote sensing data, were integrated in the analyses and indicated that 10% of the observed area was affected by land degradation. One of the major points of criticisms related to the subjectivity of the studies, which impede operational use or comparability (Millennium Ecosystem Assessment, 2005a). Different concepts were developed and implemented to assess land conditions, which will be described in the following section.

5.3.1 Assessment of Land Condition Related to the Biological Productivity of Ecosystems

In recent years, the assessment of land condition has been primarily related to the biological productivity of ecosystems. The concept is based on the fact that land degradation, which might be caused by a wide variety of climate- and human-induced processes, results in a decline of the potential of the soil to sustain plant productivity (del Barrio et al., 2010). Using the example of rangelands, Figure 5.3 illustrates the dependence of biological productivity on grazing pressure, rainfall, and soil properties. In this respect, soil properties like water holding capacity and nutrient supply are essential factors that directly affect primary productivity. Ongoing overgrazing drives feedback loops between vegetation and soil, resulting in a degradation of these soil properties and triggers a sustained decrease of the soil's capacity to sustain primary productivity. Consequently, the ecosystem's capacity to utilize local resources (such as soil nutrients and water availability) in relation to its potential capacity may be defined as land condition. This in turn allows drawing conclusions on the degradation status of observed areas (Boer and Puigdefabregas, 2005). Hence, biological productivity is considered a suitable surrogate to assess land condition and surrogates derived from remote sensing are predestined to support this assessment.

At the local scale Boer and Puigdefabrégas (2005) conceptualized and implemented a spatial modeling framework to assess land condition based on climate data as well as on NDVI data derived

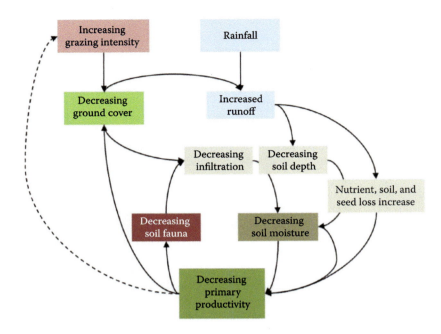

FIGURE 5.3 Aspects of landscape function using the example of grazing. Changes of ground cover, which are at short time scales mainly driven by rainfall variability and grazing pressure, can affect soil properties negatively. If thresholds are crossed the cycle moves toward a new state that is characterized by degraded soil properties and a long-term loss in productivity. Even though negative feedback exists to grazing intensity, management interventions often weaken this mechanism by maintaining constant stock numbers (modified from Stafford Smith et al., 2009).

from the Landsat sensor, which served as a proxy for primary productivity. This approach assumes that in arid and semi-arid areas water availability is the major limiting factor of productivity and, furthermore, that the water balance, which depends on rainfall, soil properties (evaporation), vegetation (interception and transpiration), and discharge, reflects land condition. Based on this theoretical concept they proposed a long-term ratio of mean actual evapotranspiration and precipitation to assess land condition.

Prince (2004) and Prince et al. (2009) introduced the Local Net Primary Productivity Scaling (LNS) method where the actual NPP is compared to the potential NPP of the corresponding Land Capability Class (LCC). The LCCs are homogenous areas that are determined by climate, soils, land cover and land use, and are independent of actual NPP. The magnitude of the difference provides a measure of land degradation and at the same time the loss of carbon sequestration. The actual NPP is derived for each pixel from multi-temporal Earth observation data. The potential NPP, i.e., the NPP that could be expected without human land use, equals the maximum NPP found in the corresponding LCC and enables the implementation of this approach for large physical heterogeneous areas which is exemplarily shown in Figure 5.4 for a test area in Australia (Jackson, and Prince, 2016). Similar methodologies were developed by Bastin et al. (2012) and Reeves and Baggett (2014) to identify rangeland conditions in Queensland, Australia, and the southern and northern Great Plains, USA, respectively.

5.3.2 Assessment of Land Condition Including Climate and Its Variability

Wessels et al. (2007) used a residual trend analysis (RESTREND) to identify potentially degraded areas by decoupling the NDVI signal from rainfall variability based on NOAA AVHRR data. This

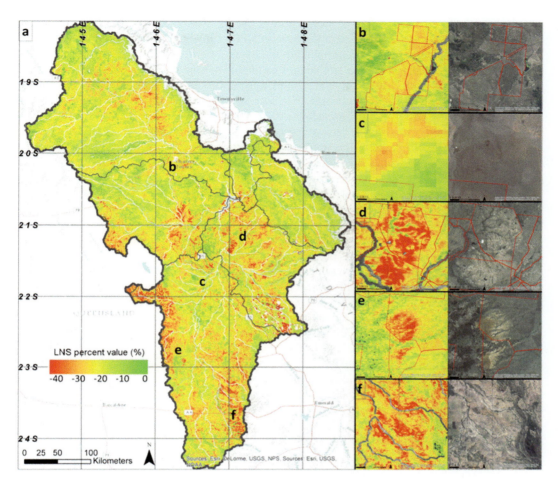

FIGURE 5.4 Local net production scaling (LNS) in the Burdekin Dry Tropics (a) and enlargements of the areas indicated in (a): (b) high and low LNS values on either side of a station boundary, (c) variation within a single station showing gradients from low to high, (d) low LNS in eroded drainage area, (e) hillslope erosion resulting in bare surface with little to no vegetation cover, and (f) area of tree removal with visible erosion and reduced cover. Black lines are the boundaries of river basins and red lines are station boundaries (from Jackson and Prince, 2016).

methodology identifies areas where a reduction in productivity per unit rainfall has occurred by comparing modeled accumulated NDVI values based on rainfall data to the observed NDVI. While the method proved capable of identifying potentially degraded areas in South Africa, Wessels et al. (2007) stressed that the cause of the negative trend cannot be explained solely by this approach, but needs detailed investigation. Li et al. (2012) transferred the RESTREND methodology to a rangeland area in Inner Mongolia, China. Their results showed that until the year 2000 heavy overgrazing deteriorated rangelands in this area, but grasslands recovered afterwards due to the implementation of new land use polices. The authors concluded that the methodology is useful to identify human-induced changes in drylands, but also underlined that the results need careful interpretation. Burrell at al. (2019) introduced TSS-RESTREND, which includes temperature and improves the detection of dryland degradation.

Coupling the RESTREND methodology with the BFAST time series segmentation, Burrell et al. (2019) showed improved performance over using RESTREND alone in an analysis across the whole of Australia based on the GIMMS3g NDVI dataset.

Other approaches make use of the concept of Rain Use Efficiency (RUE), which was introduced by Le Houérou (1984). RUE is defined as the ratio of NPP to precipitation over a given time period and may be interpreted as being "proportional to the fraction of precipitation released to the atmosphere" (del Barrio et al., 2010). Several studies explored RUE in dryland areas based on remote sensing (e.g., Prince et al., 1998; Bai et al., 2008) causing debates between scientists due to supposed weaknesses in the rationale (Hein and de Ridder, 2006; Prince et al., 2007; Wessels, 2009). In the framework of the LADA (Land Degradation Assessment in Drylands) project Bai et al. (2008) proposed a methodology to assess and monitor land conditions by deriving RUE based on the global NOAA AVHRR GIMMS dataset. The implemented methodology and the results were criticized (Wessels, 2009) because rainfall is not a limiting factor in more humid areas and, moreover, RUE values are dependent on precipitation amounts and thus impede the direct comparison of RUE values from regions of diverging aridity level. Fensholt et al. (2013) proposed to only use NPP proxies that are positively linearly correlated to precipitation and to only consider the rainy-season variation of NDVI for those areas where the correlation between RUE and annual precipitation is close to zero.

RUE has two further shortcomings, in addition to its dependency on aridity: first, it is usually computed for annual periods, which is proportional to stable biomass but does not account for rapid and ephemeral vegetation growth such as that of annuals; and second, it requires reference values to allow for a proper interpretation. Del Barrio et al. (2010) presented the 2dRUE method to deal with the aforementioned shortcomings. First, RUE is implemented for each pixel at two temporal scales, a long-term one based on averaging annual RUE from a ten-year time series, and a short-term one based on a single value computed by dividing the maximum NDVI of the time series by the precipitation of the six antecedent months. This yields proxies to biomass and productivity, respectively. Both RUE implementations are then detrended for aridity by finding statistical maximum and minimum values of all pixels for each aridity level. These are then used as respective references to scale the rest of the RUE values relative to them, thus enabling their comparison across very large study areas. The combination of relative proxies to biomass and productivity is finally made according to ecological succession theory to yield a legend of states comprehending the full range of land condition (del Barrio et al., 2016; see also Figure 5.5).

5.4 MONITORING OF LAND USE/LAND COVER CHANGES TO ASSESS LAND DEGRADATION PROCESSES

Land cover is defined by the attributes of the land surface, including all aspects such as flora, soil, rock, water, and anthropogenic surfaces, whereas land use has been defined as the purpose for which humans employ land cover (Lambin et al., 2006). Changes in land use are often accompanied by alterations in land cover that always imply changes in ecosystem functions, such as primary productivity, soil quality, water balance, and climatic regulation (e.g., Foley et al., 2005; Turner II et al., 2007). Land use maintains concrete interactions with land condition, i.e., the poorer the condition, the less land use change options are available, which creates a positive feedback that, in extreme cases, may hamper escape from land degradation (del Barrio et al., 2021). This is why land use changes are explicitly rated with respect to land degradation at evaluating LDN in SDG indicator 15.3.1 (Cowie et al., 2018).

The monitoring of landscape dynamics forms therefore an essential component for dryland observation as it provides information about the nature and extent of the changes and allows for the evaluation of the consequences for ecosystem functions.

Land cover changes can be distinguished in two major groups: (1) conversion and (2) modification (Lambin et al., 2006). Land use conversion commonly involves the replacement of one land use/land cover class by another (e.g., shrublands with arable land), whereas modification is usually related to gradual changes within one thematic class (e.g., shrub encroachment within natural ecosystems). The assessment of both conversion and modification is important to provide a

Land Degradation Assessment and Monitoring of Drylands

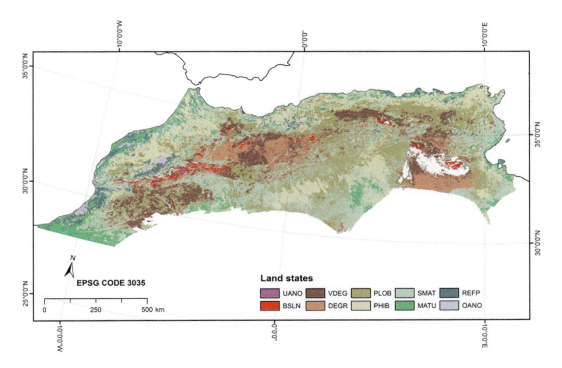

FIGURE 5.5 Land condition states in the NW Maghreb drylands (1998–2008): Legend abbreviations: UANO: underperforming anomaly; BSLN: baseline performance; VDEG: very degraded; DEGR: degraded; PLOB: productive with low biomass; SMAT: submature; MATU: mature; REFP; reference performance; OANO: over-performing anomaly. Lambert Azimuthal Equal Area projection, ETRS89 Datum, EPSG Code 3035 (from del Barrio et al., 2016).

comprehensive picture of land use/land cover changes. The assessment of land use/land cover conversion is often based on land use change detection performed at defined years of interest. Several strategies and methods were developed to optimize the results of change detection analyses. A detailed overview of change detection techniques and their application, potentials, and limits is given in Hecheltjen et al. (2014).

The assessment of modifications is a crucial element in dryland areas, because land cover changes related to land degradation are often associated with a modification of the landscape (Lambin et al., 2006). These include, for instance, vegetation cover loss due to overgrazing or primary or secondary succession on abandoned fields and rangelands. The detection and monitoring of a modification is often more challenging as changes of biophysical properties must be observed and distinguished from inter-annual variability. This is especially important for dryland areas where primary productivity is dependent on the highly variable climatic conditions in terms of rainfall (Turner II et al., 2007). Time series analysis of remote sensing archives is a suitable methodology to assess gradual changes of land cover (Udelhoven, 2010), providing means to delineate inter-annual variability from long-term trends. This requires consistent long-term data of biophysical parameters connected to surface properties, such as those provided by the broad remote sensing data sources described in the previous chapter.

The high geometric detail of Landsat data often matches the scale of land management decisions (Cohen and Goward, 2004; Lambin et al., 2006), whereas coarse-scale data such as those provided by MODIS are more suitable to cover large areas while providing a much higher temporal repetition rate. These data archives therefore permit the detection of changing parameters connected to vegetation cover as well as the deduction of changes in phenology (e.g., Andres et al., 1994; Brunsell and Gillies, 2003, Stellmes et al., 2013).

5.4.1 LOCAL-SCALE STUDIES TO DETECT LAND DEGRADATION-RELATED MODIFICATIONS

At a local scale, many early studies focused on monitoring both long-term and abrupt modifications using annual Landsat time series, as images had to be paid for per-image and therefore series had been limited to one per year at best. In many of these studies the impact of grazing pressure on vegetation cover has been analyzed in different parts of the world, e.g., in Bolivia (Washington-Allen et al., 2008), Greece (Hostert et al., 2003; Röder et al., 2008a; Sonnenschein et al., 2011), and Nepal (Paudel and Andersen, 2010). These studies used either vegetation indices like the NDVI or enhanced parameters such as proportional vegetation cover derived from SMA. Degradation processes were identified in all study areas, and often additional information layers were used to explain these findings. Washington-Allen et al. (2008) assessed the effect of an El Niño-Southern Oscillation (ENSO)–induced drought on a rangeland system in Bolivia employing Landsat time series. This study showed that the decrease in vegetation cover of the rangelands resulted in an increased risk of soil erosion. In northern Greece (Röder et al., 2008a) patterns of over- and under-grazing were identified following changed rangeland management practices from transhumance to sedentary pastoralism.

Similar patterns were observed on the island of Crete, Greece (Hostert et al., 2003). Frantz et al. (2022) have revisited this study of Hostert et al. with full archive access in the island of Crete. They showed that time series analysis may yield largely different trend patterns when calculated across the full stack of images available today and using novel time series segmentation techniques (Figure 5.6).

Another important dimension in land degradation science is the understanding of impacts of land use/land cover changes on ecosystems. Hill et al. (2014) used the ecosystem services concept (Millennium Ecosystem Assessment, 2005a) to estimate changes in ecosystems services introduced by land use/land cover changes detected using a Landsat TM/ETM+ time series in Inner Mongolia, China, between 1987 and 2007. Other studies focused on abrupt changes caused by fires, including studies on mapping fire patterns (Diaz-Delgado and Pons, 2001; Bastarrika et al., 2011) or post-fire recovery (e.g., Viedma et al., 1997; Röder et al., 2008b). Assessments of the relationship of gradual and abrupt vegetation changes in the Mediterranean are rather seldom (Sonnenschein et al., 2011). In grazed rangelands on Cyprus, von Keyserlingk et al. (2021) evaluated the resilience of vegetation to drought. Using all available Landsat images, they applied the BFAST time-series decomposition algorithm to identify breakpoints during drought periods and used them as an inverted proxy for vegetation resilience, while regression slopes for the post-drought segments were interpreted as proxies for recovery rate.

While many studies focused on local areas, i.e., covering one Landsat scene, operational systems for the monitoring of rangeland areas have been set up in Australia during the last decades (Wallace et al., 2006, 2004). Several regional projects use parameters derived from Landsat time series to monitor land cover changes and land condition, which are integrated in the Australian Collaborative Rangeland Information System (ACRIS) on a nationwide level. The software tool VegMachine was developed where satellite imagery and expert knowledge are combined to assess the health status of grazing grounds and to support pastoral producers as well as management decisions (CSIRO, 2009). Furthermore, Rangeland and Pasture Productivity (RaPP) Map is one activity within the Group on Earth Observations Global Agricultural Monitoring initiative (GEOGLAM). It is a freely accessible online platform provided by Australia's national science agency CSIRO (Commonwealth Scientific and Industrial Research Organisation) dedicated to the systematic monitoring and evaluation of global rangeland and pastureland conditions (see https://map.geo-rapp.org/). This tool is instrumental in facilitating sustainable land management practices and optimizing livestock production systems.

Land Degradation Assessment and Monitoring of Drylands   125

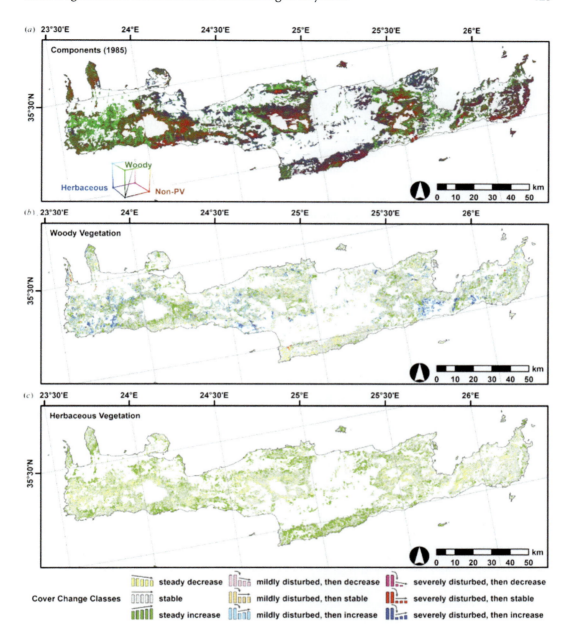

FIGURE 5.6 Long-term change of vegetation components over the study period 1984–2006. (a) RGB representation of the major vegetation components for 1985; stretched from 0–100%. (b,c) Syndrome-based labeling of land-use change classes for woody and herbaceous vegetation trajectories, respectively. White areas are non-natural areas or above 1500 m a.s.l. (from Frantz et al., 2022).

5.4.2 Regional- to Global-Scale Studies to Detect Land Degradation-Related Modifications

Most studies covering large dryland areas (including continental and global studies) are based on NOAA AVHRR archives or similar sensor systems such as MODIS. Many of these studies focused on ordinary least-square regression or non-parametric trend tests such as the Mann–Kendall test on NDVI values, which were often seasonally aggregated and served as a proxy for NPP (e.g., Eklundh and Olsson, 2003; Anyamba and Tucker, 2005) or other parameters related to greenness (e.g., Lambin and Ehrlich, 1997; Cook and Pau, 2013). Also changes in phenological metrics were analyzed to monitor dryland areas (e.g., Heumann et al., 2007; Stellmes et al., 2013; Hilker et al., 2014) and change detection techniques were applied to also describe non-linear trends (e.g., Jamali et al., 2014). Lewińska et al. (2020) derived annual cumulative endmember fractions instead of single date vegetation cover fractions and combined them with LandTrendr to distinguish short-term vegetation loss and decadal degradation of grasslands in the Caucasus. In the past years, the access to large data archives have spurred large-area applications at higher resolutions and partially overcame the traditional divide between medium resolution/local scale and coarse resolution/regional to continental scale. Synergistic use of these frameworks with multiple sensors has enabled mapping approaches at Landsat—or even Sentinel-2—level resolutions. For instance, Venter et al. (2020) used GEE to calculate EVI-based vegetation trends across South Africa at 30m resolution, showing both greening and browning patterns that were interpreted in relation to the different biomes. In a multi-scale and multi-sensor approach, Brandt et al. (2018) used Quickbird and WorldView high-resolution images to train boosted decision tree regression models with Proba-V and ALOS-PALSAR data. They were able to show that despite widespread reduction of tree cover in West African woodlands, in farmland areas villagers seem to promote and conserve trees.

A major concern of monitoring dryland areas is the distinction between land cover changes driven by climatic fluctuations and those caused by human intervention. Various techniques were employed for this purpose, such as the before-mentioned RUE (Geerken and Ilaiwi, 2004; Fensholt et al., 2013), 2dRUE (del Barrio et al., 2016), and RESTREND (Wessels et al., 2007). Also, there are methods aiming at detecting variations in the vegetation cover irrespective of the cause, such as linear regression analysis (Helldén and Tottrup, 2008), distributed lag models (Udelhoven et al., 2009), multiple stepwise regression (del Barrio et al., 2010; Zeng et al., 2013), and dynamic factor analysis (Campo-Bescós et al., 2013), teleconnections of macro weather situations (Williams and Hanan, 2011) and global sea surface temperature (Huber and Fensholt, 2011). Recently, Zhou et al. (2021) suggested a Transfer Function Analysis (TFA) method in the frequency domain in the context of studying the relationship between vegetation and rainfall across multiple temporal scales. Key components are the coherence to confirm the existence of a response of vegetation to rainfall, the gain to measure the strength of this relationship, and the phase to measure the time-lag between changes in vegetation and rainfall.

Many of the large-scale studies focused on Africa, especially on sub-Saharan Africa, including the Sahel region. Droughts in the first decades of the 20th century as well as in the 1960s to 1980s caused disastrous famines in the Sahel zone and had a strong impact on vegetation cover. Yet, resilience in these systems often led to recovery under more profitable climatic conditions, while the term "desertification" involves a permanent and irreversible reduction in vegetation productivity. In the 1990s remote sensing studies started to support the analysis based on time series analyses. Recent studies dealing with greening trends in the Sahel found vegetation recovery in most parts of the Sahel (e.g., Eklundh and Olsson, 2003; Herman et al., 2005). Heumann et al. (2007) showed that both annual and perennial vegetation recovery processes drive the observed greening and Dardel et al. (2014) demonstrated that soil type and soil depth are important factors for recovery. Jamali et al. (2014) implemented an automated approach to account for non-linear changes. Results showed a dominance of positive linear trends distributed in an east-west band across the Sahel, whereas regions of non-linear change occur only in limited areas, mostly on the peripheries of larger regions of linear change (see Figure 5.7).

These studies all implied that vegetation recovered after the severe droughts in the 1970s and 1980s and that land degradation not related to water availability/droughts is not a widespread phenomenon but is confined to smaller areas (Fensholt et al., 2013). Also, in other parts of the world a large proportion of dryland areas showed "greening-up" trends (e.g., Helldén and Tottrup, 2008; Hill et al., 2008; De Jong et al., 2012; Fensholt et al., 2012; Stellmes et al., 2013). The global study

Land Degradation Assessment and Monitoring of Drylands

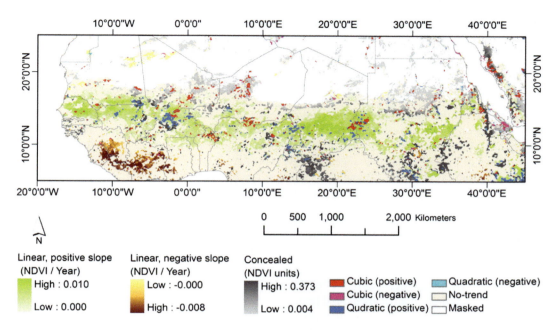

FIGURE 5.7 Results of a polynomial fitting-based approach to account also for non-linear trends (Jamali et al., 2014). Trend slope for the linear trends, range of annual variations of NDVI for the concealed trends and trend sign for the cubic and quadratic trends obtained by using the annual GIMMS–NDVI data series for the Sahel (1982–2006) in the trend classification scheme. Concealed trends are indicating that no net change in vegetation productivity has occurred, but the curve exhibits at least one minimum or maximum. Areas with a mean yearly NDVI < 0.1 were masked out (from Jamali et al., 2014).

of Cook and Pau (2013) focused on rangeland productivity between 1982 and 2008 and indicated that almost 25% of the rangelands were affected by significant trends. These trends were found to be mostly with increasing productivity, whereas decreasing productivity related to land degradation was found in rather isolated spots, mainly in China, Mongolia, and Australia. However, whether increased productivity means land recovery by itself, without describing the corresponding ecosystem, is yet to be determined, as hypothesized by Verón et al. (2006) and found in degradation-triggered landscape changes in the Maghreb (Hirche et al., 2011) and New Mexico (Browning et al., 2014). Whereas in many other regions rainfall was the dominant factor influencing NDVI, in Mongolia 80% of the decline in greenness was attributed to an increase in livestock by Hilker et al. (2014), while Yin et al. (2018) evaluated the impact of China's re-vegetation program and found a decrease in forest loss rates and cropland retirement rates based on trajectory-based land use and land cover change analysis.

In a review article, Wang et al. (2022) highlight that satellite remote sensing has revealed an overall trend of greening in global drylands over the past 30 years in regions like the Sahel, the Tibetan Plateau, and the western United States. In contrast, extensive regions, such as the southwestern United States, southern Argentina, Kazakhstan, Mongolia, Afghanistan, and parts of Australia have experienced a reduction in vegetation cover.

Generally, a comprehensive analysis of land degradation needs to include, also at regional to global scale, the fire regime, and possible inter-linkages to land use and land cover, e.g., by analyzing recovery after fire events (Katagis et al., 2014). The two MODIS fire products, Active Fire and Burned Area, allow monitoring of important variables of fire regimes (Justice et al., 2006; Loboda et al., 2012), such as fire frequency, fire seasonality, and fire intensity and allow for identifying drivers (Archibald et al., 2009) and model potential changes (Batllori et al., 2013). Besides pixel-based analysis fire segmentation algorithms were developed that allow researchers to analyze fire-related parameters such as the perimeter, ignition points, etc. (Frantz et al., 2016b; Balch et al., 2020; Mahood et al., 2022).

5.5 INTEGRATED CONCEPTS TO ASSESS LAND DEGRADATION

The previous sections have illustrated that time series analysis allows researchers to discriminate human-induced land cover changes and changes caused by inter-annual climatic variability. Beyond this, a crucial element of land degradation assessment is the identification of underlying and proximate causes of human-induced changes (e.g., Reynolds et al., 2007). Only in this manner the coupled human-natural character of land cover changes can be understood and an identification of the mechanisms that drive land degradation is possible. This knowledge provides the foundation to support the development of sustainable land management strategies.

A comprehensive framework designed to capture the complexity of land degradation and desertification was provided by Reynolds et al. (2007). They introduced the term "Drylands Development Paradigm" (DDP), which "represents a convergence of insights and key advances drawn from a diverse array of research on desertification, vulnerability, poverty alleviation, and community development" (Reynolds et al., 2007). The DDP aims at identifying and synthesizing those dynamics central to research, management, and policy communities (Reynolds et al., 2007). The essence of this paradigm, which consists of five principles, builds on the assumption that desertification cannot be measured by solitary variables but that it has to consider biophysical and socioeconomic data at the same time (Vogt et al., 2011). A limited number of "slow" variables (e.g., soil fertility) are usually sufficient to explain the human-natural system dynamics. These slow variables possess thresholds and if these thresholds are exceeded the system moves to a new state. "Fast" variables, for instance, climatic variability, often mask the slow variables and thus aggravate the assessment of the slow variables, which is a prerequisite to understanding the ecosystem behavior. Moreover, it is important to consider that human-natural systems are "hierarchical, nested, and networked across multiple scales" (Reynolds et al., 2007). Accordingly, both the human component, e.g., stakeholders at different levels, and the biophysical component, e.g., slow variables at one scale can be affected by the change of slow variables operating at another scale (Reynolds et al., 2007). Stringer et al. (2017) updated the DPP by empirically deriving eight characteristics which were distilled into three integrative principles. These principles aim to transform dryland science and development, extending and advancing the DDP to make it actionable and relevant in the current global dryland context.

Prior to the DDP, Geist and Lambin (2004) examined the main mechanisms that trigger land degradation processes and conclude that these processes, which often manifest in land use/land cover changes, are governed by proximate causes (immediate human and biophysical actions), which in turn are depending on underlying drivers (fundamental social and biophysical processes). Figure 5.8 illustrates the dependencies of land use/land cover changes from proximate causes and underlying drivers. Furthermore, alterations of ecosystem services caused by land use/land cover changes can again alter underlying drivers, proximate causes, and even external constraints, hence resulting in a feedback loop. Policy plays an important role in avoiding positive feedback mechanisms, which can accelerate unsustainable land use (Reid et al., 2006).

5.5.1 Integrated Studies at Local Scale

Several local studies linked the biophysical dimension of land use/land cover changes to the human dimension in various dryland areas such as Spain (Álvarez-Martinez et al., 2014; Améztegui et al., 2010; Serra et al., 2008), Brazil (De Marzo et al., 2022), Greece (Lorent et al., 2008), Kenya (Were et al., 2014), China (Li et al., 2012), Mongolia (Hilker et al., 2014), and Uzbekistan (Dubovyk et al., 2013). Regression-based models are the most widely used approach to identify the major drivers

Land Degradation Assessment and Monitoring of Drylands

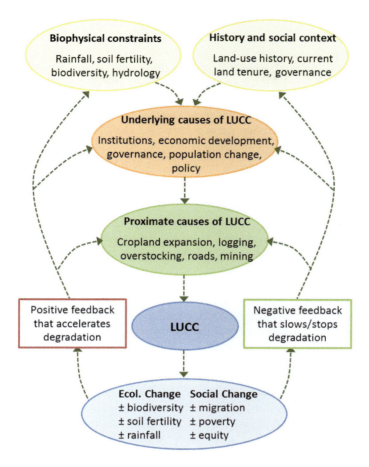

FIGURE 5.8 Conceptual model illustrating the feedback loop of Land Use/Land Cover Changes (LUCC), its consequences, and the underlying and proximate causes (modified from Reid et al., 2006).

of change (Were et al., 2014) and mostly rely on land cover changes derived from land use/land cover classifications at several time steps. However, time series of remotely sensed data were only rarely used (Lorent et al., 2008; Dubovyk et al., 2013). The drivers of land use/land cover change depend very much on the contextual framework of the study area, including physical and socioeconomic characteristics. Therefore, it is essential to first set up a hypothesis that identifies major underlying drivers of land use/land cover change. For instance, in Spain and Greece the Common Agricultural Policy (CAP) subsidies of the European Union (EU) were identified as one of the important drivers. These largely influenced agricultural developments like intensification and land abandonment, where abandonment of marginal areas involved forest expansion and bush encroachment (Améztegui et al., 2010; Lorent et al., 2008; Serra et al., 2008). In the grasslands of Inner Mongolia/China many factors explained observed grassland degradation between 1990 and 2000 and the reduced degradation rate between 2000 and 2005, which were altitude, slope, annual rainfall, distance to highway, soil organic matter, sheep unit density, and fencing policy. Fencing policy was negatively correlated, suggesting that fencing of sensitive areas can reduce land degradation. The analysis of cropland degradation in the Khorezm region, Usbekiztan, based on MODIS time series (Dubovyk et al., 2013) revealed that one-third of the area was characterized by a decline of greenness between 2000 and 2010. Groundwater table, land use intensity, low soil quality, slope and salinity of the ground water were identified as the main drivers of degradation. These examples show that the combination of remote sensing supported land use/land cover change assessment and underlying proximate causes may reveal the most important drivers of land degradation. However, such analyses are often hampered by the fact that for each study area (1) all potential and relevant drivers must be identified and (2) spatially explicit information of each driver or a proxy has to be available with a sufficient spatial resolution. In this context, one challenge is also to link observed dynamics to meaningful process descriptions, which may be achieved by place-based analysis using ancillary data, higher-resolution images and local knowledge, and which often still requires considerable expert-based interpretation of apparent patterns (e.g., Munawar et al., 2022).

5.5.2 INTEGRATED STUDIES AT REGIONAL TO GLOBAL SCALE

Another approach capable of supporting land degradation assessment is the syndrome approach which has been developed in the context of global change research (Cassel-Gintz and Petschel-Held, 2000; Petschel-Held et al., 1999). It aims at a place-based, integrated assessment by describing global change by archetypical, dynamic, co-evolutionary patterns of human-nature interactions instead of regional or sectoral analyses. In this framework, syndromes (as a "combination of symptoms") describe bundles of interactive processes ("symptoms"), which appear repeatedly and in many places in typical combinations and patterns. Sixteen global change syndromes were suggested and distinguished into utilization, development, and sink syndromes. Downing and Lüdeke (2002) applied the approach to land degradation. Based on the set of global change syndromes they identified the syndromes that are of relevance in dryland areas and linked vulnerability concepts to degradation processes. The syndrome concept is considered a suitable interpretation framework that allows for an integrated assessment of land degradation (Sommer et al., 2011; Verstraete et al., 2011) and is, in fact, at the basis of the Convergence of Evidence concept applied by the latest World Atlas of Desertification (Cherlet et al., 2018). This concept was transferred to Earth observation–based studies and implemented for Spain based on NOAA AVHRR data between 1989 and 2004 (Hill et al., 2008; Stellmes et al., 2013), thus enabling researchers to monitor changes in land cover after the accession of Spain to the European Union (see Figure 5.9). In these studies, the focus was not on the identification of land cover changes, but also on the link of these findings to underlying causes enabling the designation of syndromes of land use change. The main findings of the two studies comprise three major land cover change processes caused by human interaction: shrub and woody vegetation encroachment in the wake of land abandonment of marginal areas; intensification of non-irrigated and irrigated, intensively used fertile regions; and urbanization trends along the coastline caused by migration and the increase of mass tourism.

At a global scale LADA has implemented a Global Land Degradation Information system (GLADIS), which provides information on land degradation with a spatial resolution of 8 km x 8 km.

The interpretation of ecosystem changes in GLADIS includes RUE, NPP, and climatic variables and is based on an integrated land use system map. This map entails information about the main proximate causes of LUCCs such as livestock pressure and irrigation. The major constraints of this approach concern the derivation of the RUE and the NPP from the GIMMS NOAA AVHRR dataset (compare Section 3) (Wessels, 2009) and the coarse spatial resolution that hampers the detection of land cover changes (Vogt et al., 2011). Nevertheless, Vogt et al. (2011) emphasized that this assessment is a first step toward an integrated assessment.

Another spatially explicit assessment concept that was not specifically designed in the context of land degradation, but was adapted and implemented, is the Human Appropriation of Net Primary Production (HANPP, Erb et al., 2009; Haberl et al., 2007). HANPP represents the aggregated impact of land use on biomass available each year in ecosystems as a measure of the human domination of the biosphere. Global maps of the parameter were prepared based on vegetation modeling, agricultural and forestry statistics and geographical information systems data on land use, land cover, and soil degradation (Erb et al., 2009; Haberl et al., 2007). In a global study Zika and Erb

Land Degradation Assessment and Monitoring of Drylands

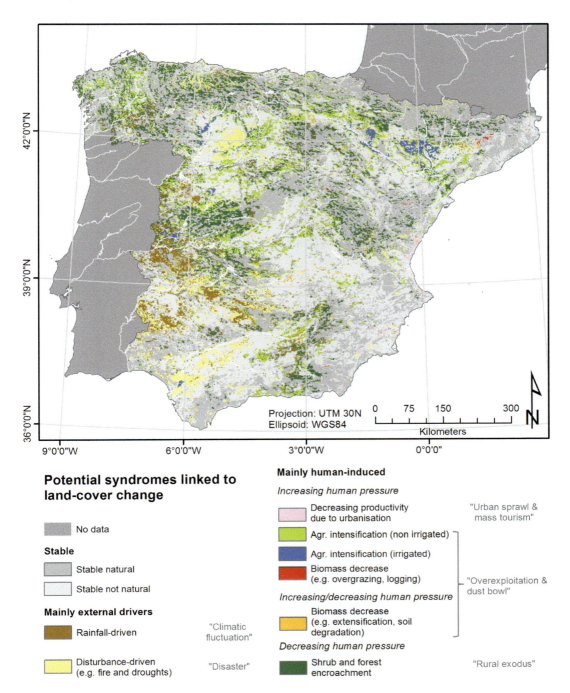

FIGURE 5.9 Syndromes and main drivers of the identified land cover changes in Spain derived from MEDOKADS NDVI data, 1989–2004 (from Stellmes et al., 2013).

(2009) estimated the annual loss of NPP due to land degradation at 4–10% of the potential NPP of drylands, ranging up to 55% in some degraded agricultural areas.

Orr et al. (2017) provide an extensive guideline for LDN assessments. Three indicators serve as a set of global metrics, i.e., land cover change, land productivity (assessed as NPP) and carbon stocks (assessed as soil organic carbon), which were adopted by the UNCCD for reporting and as a means to understanding the status of degradation. Von Maltitz et al. (2019) provide insights in their experiences from setting up LDN monitoring in Southern Africa. Akinyemi et al. (2020) have assessed LDN for Botswana using remote sensing-based indicators. Similar studies were carried out by Reith et al. (2021) in the Kiteto and Kongwa districts of Tanzania and by Pan et al. (2023) for the Three-Rivers Headwater Region of China. The study of Fan et al. (2020) focuses on discerning the patterns of desertification changes in the China-Mongolia-Russia Economic Corridor (CMREC) from 2000 to 2015 by using the classification and regression tree (CART) method, which they applied to a variety of remote sensing data, such as Landsat data and MODIS land surface temperature available in the Google Earth Engine (GEE).

5.6 UNCERTAINTIES AND LIMITS

Manifold methods were developed for assessing and monitoring land degradation, ranging from detailed local to broad global studies. When considering remote sensing of drylands, it is essential to acknowledge and understand the uncertainties associated with the data and models used. Traditionally, errors in measurements are categorized as either "random" or "systematic." Random errors are those that would be expected to average out to zero when a large number of measurements are taken under the same conditions. In contrast, systematic errors are those that imply a consistent deviation in a particular direction (Povey and Grainger, 2015).

Also, Fagan (2020) emphasized the importance of understanding the limitations and uncertainties of remote sensing products. The heterogeneity of vegetation composition along with sparse and dry vegetation poses significant challenges to remote sensing analyses (Ganem et al., 2022). One aspect concerns the different sensors with different observation periods, spatial resolutions, the use of several proxies and methodologies used in the studies introduced in the previous sections. This caution is crucial in acknowledging the potential biases and uncertainties that may arise from the practical use of remote sensing data in, for instance, dryland land use management. Some of these aspects shall be discussed in more detail in the following sections.

5.6.1 UNCERTAINTIES REGARDING THE DEFINITION AND DELINEATION OF LAND DEGRADATION

Monitoring of drylands is often based on analyzing indicators related to the productivity of vegetation. Therefore, the loss of productivity is considered to be linked to degradation processes. However, it should be stressed that the decrease of primary productivity does not necessarily imply land degradation processes. This was illustrated by an example in Syria where unsustainable irrigation agriculture was transformed to near-natural rangelands in Syria (Udelhoven and Hill, 2009). In turn, a positive trend of productivity is not always an indicator for improving land condition, a greening-up of, for instance, rangelands, does not necessarily imply an improvement of pastures (Miehe et al., 2010). In marginal areas of the European Mediterranean, greening-up has been shown to be caused by bush encroachment due to land abandonment and the consequences for both local ecosystems (Stellmes et al., 2013) and remotely tele-coupled ones (Martínez-Valderrama et al., 2021) are heavily discussed. Thus, on the one hand soils can be stabilized and soil erosion can be reduced (Thomas and Middleton, 1994), more carbon can be sequestered (Padilla et al., 2010), but on the other hand run-off and groundwater recharge is reduced (Beguería et al., 2003), biodiversity is altered (Forman and Collinge, 1996), and the fire regime changes (Duguy et al., 2007). Thus, including additional information sources, for instance, on land use, is required to allow a meaningful interpretation of time-series results (Vogt et al., 2011). The same is also true in case of fires, which strongly affect the time-series signal, e.g., induce short-term decreases in productivity and subsequent increase in productivity due to vegetation recovery.

5.6.2 Uncertainties Regarding Remote Sensing Data

5.6.2.1 Remote Sensing Archives and Their Analysis

Uncertainties in remote sensing observations pose a set of methodological and practical challenges for both the analysis of long-term trends and the comparison between different data archives. Creating consistent remote sensing time series is challenging and the prerequisite for a meaningful trend analysis. Using combined data from different sensors affording high temporal resolution, such as AVHRR, MODIS, and SPOT-VGT, in principle allow for the construction of time series in surface reflectance and related changes back to the early 1980s. However, this is hampered by several sources of uncertainties in the comparability between different sensor products (Yin et al., 2012). Comparison of the absolute NDVI values from different archives as well as the derived trends showed strong differences; a good correspondence of derived NDVI trends was found for some regions but also high discrepancies that often depend on the environmental setting and vegetation cover (Beck et al., 2011; Fensholt and Proud, 2012; Hall et al., 2006; Yin et al., 2012; Bai et al., 2019). Data as provided by NOAA AVHRR and MODIS are therefore enhanced continuously, and new data archives are published in irregular time intervals.

As an example, Figure 5.10 shows trends derived from NOAA NDVI3g+ (Pinzon et al., 2023) and MODIS MOD13A1 (Didan, 2021) NDVI data covering the same observation period (2002–2021) of the eastern Sahel. Even though the general picture is quite similar for the mean annual NDVI trends, a more detailed analysis reveals a disagreement between both datasets that also addresses the temporal trends for a phenological parameter (i.e., the amplitudes of the annual NDVI cycle). Possible explanations for these incoherencies include different data pre-processing schemes for different sensors. The effects of sensor degradation on the captured signal are different and AVHRR data need to be additionally corrected for orbital drift effects that introduce systematic changes in the bidirectional characteristics of surfaces. With the new NOAA NDVI3g+ archive many of the problems of older versions seem to be eliminated. Another factor of disagreement is different spectral mixture effects in heterogeneous regions that arise from the different spatial resolutions of the GIMMS and MODIS data products.

The comparability of many studies is additionally hampered by the fact that the used methods and techniques, vegetation proxies and thresholds to exploit the time series are very diverse, since the implemented methods are often adapted to specific objectives and certain study areas. This is often necessary as drylands are very diverse concerning the degradation processes and the environmental settings including climate, soil, geology, fauna, and flora. The study of Taylor et al. (2021) focused on understanding the opportunities and limitations of monitoring land surface phenology in drylands. One of the main conclusions was that fractional vegetation cover and the seasonal amplitude of the vegetation index cycle are the most important factors to successfully monitor phenological features.

5.6.2.2 Observation Period

As outlined before, rainfall variability is a key driver of variability of vegetation productivity within drylands. In consequence the observation period will substantially influence the derived trends depending on the assembly of drier and wetter periods. Figure 5.11 illustrates the difference of trends for different observation periods derived from the NOAA NDVI3g+ archive for two non-overlapping 20-year NDVI time series and the overall time series of 40 years.

This underlines that dryland monitoring should always consider rainfall variability, e.g., implemented in the RESTREND method (Wessels et al., 2007). Hereby, similar to remote sensing archives, the homogeneity and reliability of the precipitation time series is of utmost importance. Even though some authors generated interpolated precipitation fields for their studies themselves (Wessels et al., 2007; del Barrio et al., 2010), diverse global and regional gridded precipitation data are available, e.g., Global Precipitation Climatology Centre (GPCC) (Meyer-Christoffer et al., 2011), African Rainfall Climatology (ARC2) (Novella and Thiaw, 2013), Climate Hazards Group InfraRed Precipitation with Station data (CHIRPS) (Funk et al., 2015) and Climatic Research Unit Timeseries (CRU TS) (Harris et al., 2020). The choice of an appropriate dataset should be based on plausibility checks (e.g., Anyamba et al., 2014). Tozer et al. (2012) demonstrated for three-monthly gridded Australian rainfall datasets that interpolated data are rather restricted as a useful proxy for observed point data, although these grids are "based" on observed data. Gridded datasets often significantly vary from gauged rainfall datasets, and they do not capture gauged extreme events. Apart from observation errors these uncertainties are mainly introduced by the spatial interpolation algorithms, which always introduce some artificiality. Furthermore, it is difficult to verify the "ground truth" of the gridded data in areas or epochs with sparse observation gauges. Tozer et al. (2012) recommend always acknowledging these uncertainties in using gridded rainfall data and to try to quantify and account for it in any study, if possible.

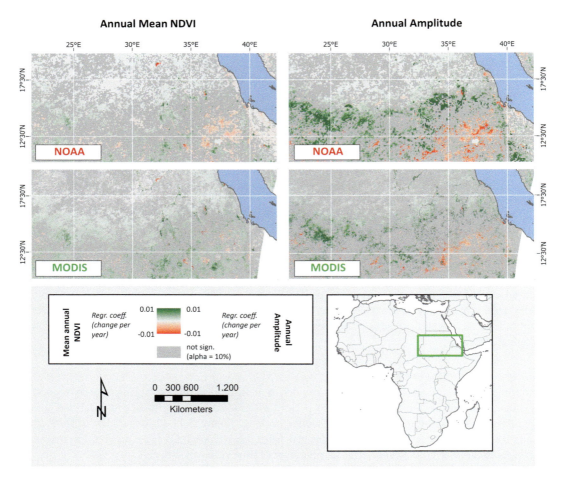

FIGURE 5.10 Trends derived from linear regression analysis for the annual total sum of NDVI based on the NOAA NDVI3g+ archive (upper panels) and MODIS MOD13A1 NDVI time series (lowest panel) for the Eastern Sahel from 2002 to 2021. Time series analysis performed with Timestats (Udelhoven, 2010).

Land Degradation Assessment and Monitoring of Drylands 135

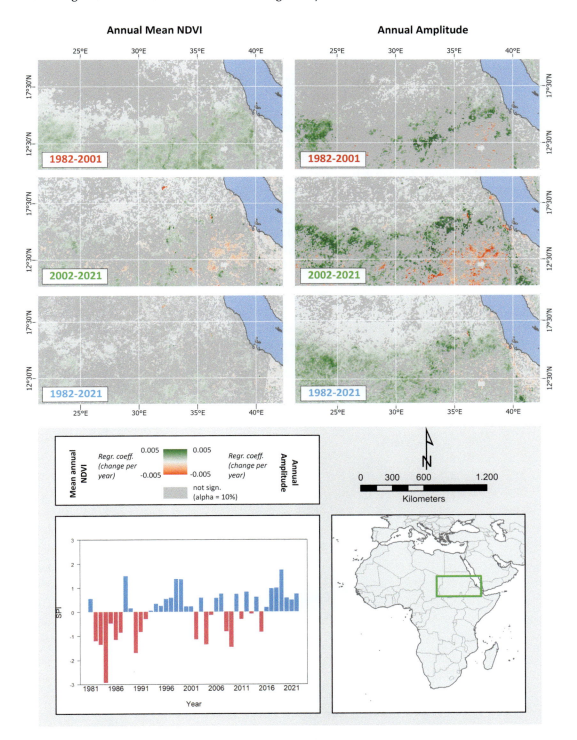

FIGURE 5.11 Change of the NDVI derived for three different observation periods based on the NOAA NDVI3g+ archive and Standardized Precipitation Index from 1982 to 2021 (based on CHIRPS). Time series analysis performed with Timestats (Udelhoven, 2010).

5.6.2.3 Spatial Scale

One drawback of regional to global studies is the coarse pixel resolution, which often impedes the monitoring of fine-detail land degradation processes (e.g., Stellmes et al., 2010; Fensholt et al., 2013) as illustrated in Figure 5.12. Coarse pixels implicitly assume that there is no topographic redistribution of water. Therefore, the underlying paradigm of the analysis method must be consistent with this. For example, RUE is appropriate for coarse pixels because all the considered water inputs and outputs are vertical, and vegetation must be aggregated at a consistent scale. If RUE were applied at very detailed pixels, vegetation differences due to topographic variations would be evident, and horizontal water redistribution (e.g., runoff) should be accounted for. In other words, the process and the spatial resolution must match.

At coarse resolutions species composition cannot be identified and vegetation structure is not resolved, and often the focus is put on green vegetation cover, even though dry vegetation is an important component in drylands. Some approaches try to solve some of these gaps, e.g., decomposition of time series to assess woody and herbaceous components (Lu et al., 2001) or using a clumping index to estimate woody cover from MODIS data (Hill et al., 2011; Zhou et al., 2019). Other methods make use of alternative sensor systems such as passive microwave radar to derive Vegetation Optical Depth (VOD), which is sensitive to both photosynthetic active and non-active biomass (Andela et al., 2013) or combine the analysis of optical and radar imagery (Bucini et al., 2010).

Methods like STARFM or the improved availability of Landsat-like medium-resolution data (e.g., Sentinel-2 mission) will only partially solve the problem, since dryland observation requires long-term archives. However, these sensors and methods will improve the situation over the long term. The same is true for operational satellite-based hyperspectral data that will allow for the development and application of enhanced indicators for dryland observation.

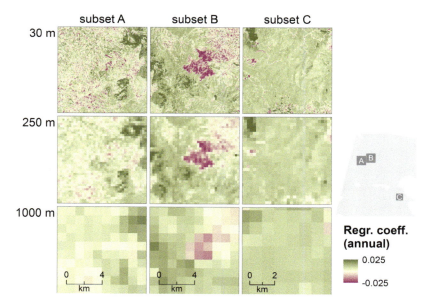

FIGURE 5.12 Effects of spatial degradation of Landsat TM/ETM+ time series (1990–2000) from a geometric resolution of 30 m x 30 m to 1000 m x 1000 m on the derived regression coefficient of a linear trend analysis. The three presented subsets represent different types and scales of land cover change (modified from Stellmes et al., 2010).

5.7 SUMMARY AND CONCLUSIONS

The definition and perception of land degradation and desertification have undergone a substantial transformation within the past decades. While in the beginning the biophysical assessment of degradation processes, which often focused on soil degradation, was the primary objective of many research initiatives, the necessity to investigate the mechanisms of human-environmental systems as a prerequisite to create a comprehensive understanding of desertification has been recognized recently. This is considered essential to understand the impacts of land degradation on the provision of ecosystem goods and services and thus, its impact on human well-being (Millennium Ecosystem Assessment, 2005a). Great efforts were put into developing methodologies to enhance the understanding of coupled human-environmental systems and the influence of natural climatic variations.

One of the major challenges that remains is to link these observations with socioeconomic data, thus connecting biophysical and socioeconomic information to yield combined information of land change processes and their underlying causes. This proves especially crucial for large-scale assessments of land degradation from national to global scales. This intricacy even increases if degradation is not only defined as a loss of productivity of ecosystems but as the decline of important ecosystem services as suggested by the Millennium Ecosystem Assessment (2005a). Even though this definition further increases the complexity of dryland assessment, it might be more compliant with the needs of policymakers and land management to develop and establish sustainable land use practices.

This complexity might also explain that until today no comprehensive picture of dryland condition is available, even though manifold methods were developed for assessing and monitoring land degradation, ranging from detailed local to broad global studies. Moreover, dryland studies differ in implemented techniques, indicators, observation periods, thresholds, and significance levels, as well as the spatial and temporal resolution and the spectral characteristics of the sensor, hampering a comparison of the studies to form a picture of global dryland condition.

Therefore, it is of utmost interest to promote international cooperation in order to harmonize dryland studies, such as the initiative to compile the World Atlas of Desertification (Cherlet et al., 2018) under the lead of the United Nations Environment Programme (UNEP) and the European Commission Joint Research Center (EC-JRC). It provides comprehensive, detailed visual documentation and analysis of global desertification trends, offering valuable insights also on the drivers and other background information. Nevertheless, the latest World Atlas of Desertification (Cherlet et al., 2018) paradoxically acknowledges that no general maps of desertification are included. There are several reasons for this: desertification syndromes overlapping in the same area; telecouplings by which causes and effects may be simultaneous apparent but separated in space, often involving long distances; and degradation process hierarchy not always having a direct spatial expression. Ultimately, the problem is that desertification is polythetic (i.e., classifying a syndrome as desertification involves a list of symptoms, a relevant number of which must be observed, but none of which being determinant). This is the reason why most mapping approaches only show partial aspects of land degradation (e.g., indicators based on NPP), and the paradigm of Convergence of Evidence is recommended in parallel (Gianoli et al., 2023). Land degradation mapping therefore provides a main spatial platform to address desertification. Summarizing, one singular methodology will be insufficient to comprehensively analyze drylands; rather, depending on the respective physical and socioeconomic framework, complementary approaches must be applied, as introduced in the preceding sections.

The land degradation and desertification topic is part of the more broadly perceived debate on global change. Climate change and its environmental and economic consequences therefore are major environmental issues of global interest. Human activities have transformed a major part of the Earth's terrestrial ecosystems to meet rapidly growing demands for food, fresh water, timber, fiber, and fuel. Land use practices have not only affected global and regional climate due to the emission of relevant greenhouse gases, but also by altering energy fluxes and water balances (Foley et al., 2005). Additionally, even seemingly "unaffected" areas are also influenced and altered indirectly through pollutants and climate change (DeFries et al., 2004; Foley et al., 2005).

Whereas in the past, conservation of ecosystems was given priority to maintain ecosystem services, in the face of global change it cannot be assumed that the future behavior of ecosystem responses to changes will be the same as in the past (Chapin III et al., 2010). Instead, the challenge of future land use management will include the assessment of trade-offs between acute human needs and the long-term capacity of ecosystems to provide goods and services (DeFries et al., 2004; Foley et al., 2005).

It is essential to consider that ecosystem responses to land use changes vary in time and space and, moreover, analysis should encompass larger areas with sufficient spatial resolution to ensure that on- and off-site ecosystem responses are detected. Sustainable management of ecosystems requires information concerning the actual conditions and, furthermore, alterations of ecosystems in relation to reference states. Such information allows for a thorough analysis of ecosystem functionality and enables rating trade-offs between ecosystem services which policy decisions (where necessary considering climate change scenarios) could impose by inducing land use changes (DeFries et al., 2004). The understanding of the impact of land use/land cover changes is even more urgent in the context of climate change, and the prospect land use will be further intensified to satisfy humanities' growing demand for resources (Foley et al., 2011). Especially when considering the expected rise to 10 billion people by the end of the 21st century (Lee, 2011), pressure on dryland ecosystems could further increase, making the development of integrated, multi-component dryland observation and management even more important.

REFERENCES

Abdi, A.M., Brandt, M., Abel, C., and Fensholt, R. 2022. Satellite remote sensing of savannas: Current status and emerging opportunities. *Journal of Remote Sensing*. http://doi.org/10.34133/2022/9835284

Adams, J.B., Smith, M.O., and Johnson, P.E. 1986. Spectral mixture modeling: A new analysis of rock and soil types at the Viking Lander 1 site. *Journal of Geophysical Research*, 91(B8): 8098–8112.

Akinyemi, F.O., Ghazaryan, G., and Dubovyk, O. 2020. Assessing UN indicators of land degradation neutrality and proportion of degraded land for Botswana using remote sensing based national level metrics. *Land Degradation and Development*, 32: 158–172. https://doi.org/10.1002/ldr.3695

Álvarez-Martínez, J.M., Suárez-Seoane, S., Stoorvogel, J.J., and de Luis Calabuig, E. 2014. Influence of land use and climate on recent forest expansion: A case study in the Eurosiberian-Mediterranean limit of northwest Spain. *Journal of Ecology*. http://doi.org/10.1111/1365-2745.12257

Améztegui, A., Brotons, L., and Coll, L. 2010. Land-use changes as major drivers of mountain pine (Pinus uncinata Ram.) expansion in the Pyrenees. *Global Ecology and Biogeography*, 19: 632–664.

Amputu, V., Knox, N., Braun, A., Heshmati, S., Retzlaff, R., Röder, A., and Tielbörger, K. 2023. Unmanned aerial systems accurately map rangeland condition indicators in a dryland savannah. *Ecological Informatics*, 75: 102007.

Andela, N., Liu, Y.Y., van Dijk, A.I.J.M., de Jeu, R.A.M., and McVicar, T.R. 2013. Global changes in dryland vegetation dynamics (1988–2008) assessed by satellite remote sensing: Comparing a new passive microwave vegetation density record with reflective greenness data. *Biogeosciences*, 10: 6657–6676.

Anderson, A.N., Cook, G.D., and Williams, R.J. 2003. *Fire in Tropical Savannas*. New York: Springer.

Andres, L., Salas, W.A., and Skole, D. 1994. Fourier analysis of multi-temporal AVHRR data applied to land cover classification. *International Journal of Remote Sensing*, 15: 1115–1121.

Anyamba, A., Small, J.L., Tucker, C.J., and Pak, E.W. 2014. Thirty-two years of sahelian zone growing season non-stationary NDVI3g patterns and trends. *Remote Sensing*, 6: 3101–3122

Anyamba, A., and Tucker, C.J. 2005. Analysis of Sahelian vegetation dynamics using NOAA-AVHRR NDVI data from 1981—2003. *Journal of Arid Environments*, 63: 596–614.

Archibald, S., Roy, D.P., van Wilgen, B.W., and Scholes, R.J. 2009. What limits fire? An examination of drivers of burnt area in Southern Africa. *Global Change Biology*, 15(3): 613–630.

Bai, Y., Yang, Y., and Jiang, H. 2019. Intercomparison of AVHRR GIMMS3g, terra MODIS, and SPOT-VGT NDVI products over the Mongolian Plateau. *Remote Sensing*, 11: 2030. https://doi.org/10.3390/rs11172030

Bai, Z.G., Dent, D.L., Olsson, L., and Schaepman, M.E. 2008. Proxy global assessment of land degradation. *Soil Use and Management*, 24: 223–234.

Balch, J.K., St. Denis, L.A., Mahood, A.L., Mietkiewicz, N.P., Williams, T.M., McGlinchy, J., and Cook, M.C. 2020. FIRED (fire events delineation): An open, flexible algorithm and database of US fire events derived from the MODIS burned area product (2001–2019). *Remote Sensing*, 12: 3498. https://doi.org/10.3390/rs12213498

Barnes, M.L., Breshears, D.D., Law, D.J., van Leeuwen, W.J.D., Monson, R.K., Fojtik, A.C., Barron-Gafford, G.A., and Moore, D.J.P. 2017. Beyond greenness: Detecting temporal changes in photosynthetic capacity with hyperspectral reflectance data. *PLoS ONE*, 12: e0189539. https://doi.org/10.1371/journal.pone.0189539

Barnetson, J., Phinn, S., and Scarth, P. 2020. Estimating plant pasture biomass and quality from UAV imaging across Queensland's rangelands. *AgriEngineering*, 2: 523–543.

Bastarrika, A., Chuvieco, E., and Martin, M.P. 2011. Mapping burned areas from Landsat TM/ETM+ data with a two-phase algorithm: Balancing omission and commission errors. *Remote Sensing of Environment*, 115: 1003–1012.

Bastin, G., Scarth, P., Chewings, V., Sparrow, A., Denham, R., Schmidt, M., O'Reagain, P., Shepherd, R., and Abbott, B. 2012. Separating grazing and rainfall effects at regional scale using remote sensing imagery: A dynamic reference-cover method. *Remote Sensing of Environment*, 121: 443–457. https://doi.org/10.1016/j.rse.2012.02.021

Batllori, E., Parisien, M.-A., Krawchuk, M.A., and Moritz, M.A. 2013. Fire shifts in Mediterranean ecosystems. *Global Ecology and Biogeography*, 22: 1118–1129. https://doi.org/10.1111/geb.12065

Beck, H.E., McVicar, T.R., van Dijk, I.J.M., Schellekens, J., de Jeu, R.A.M., and Bruijnzeel, L.A. 2011. Global evaluation of four AVHRR–NDVI data sets: Intercomparison and assessment against Landsat imagery. *Remote Sensing of Environment*, 115(10): 2547–2563.

Beguería, S., López-Moreno, J.I., Lorente, A., Seeger, M., and García-Ruiz, J.M. 2003. Assessing the effect of climate oscillations and land-use changes on streamflow in the Central Spanish Pyrenees. *Ambio*, 32: 283–286.

Berdugo, M., Delgado-Baquerizo, M., Soliveres, S., Hernández-Clemente, R., Zhao, Y., and Gaitán, J.J., Gross, N., Saiz, H., Maire, V., Lehmann, A., Rillig, M.C., Solé, R.V., and Maestre, F.T. 2020. Global ecosystem thresholds driven by ardity. *Science*, 367: 787–790. http://doi.org/10.1126/science.aay5958

Blaschke, T., and Hay, G.J. 2001. Object-oriented image analysis and scale-space: Theory and methods for modeling and Evaluating multiscale landscape structure. *International Archives of Photogrammetry and Remote Sensing*, 34: 22–29.

Boer, M.M., and Puigdefabregas, J. 2005. Assessment of dryland condition using spatial anomalies of vegetation index values. *International Journal of Remote Sensing*, 26: 4045–4065.

Bond, W.J., and Keeley, J.E. 2005. Fire as a global 'herbivore': The ecology and evolution of flammable ecosystems. *Trends in Ecology & Evolution*, 20: 387–394.

Brandt, M., Rasmussen, K., Hiernaux, P., Herrmann, S., Tucker, C.J., Tong, X., Tian, F., Mertz, O., Kergoat, L., Mbow, C., David, J., Melocik, K., Dendoncker, M., Vincke, C., and Fensholt, R. 2018. Reduction of tree cover in West African woodlands and promotion in semi-arid farmlands. *Nature Geoscience*, 11: 328–333.

Brooks, B.-G.J., Lee, D.C., Pomara, L.Y., and Hargrove, W.W. 2020. Monitoring broadscale vegetational diversity and change across North American landscapes using land surface phenology. *Forests*, 11: 606. https://doi.org/10.3390/f11060606

Browning, D.M., Franklin, J., Archer, S.R., Gillan, J.K., and Guertin, D.P. 2014. Spatial patterns of grassland–shrubland state transitions: A 74-year record on grazed and protected areas. *Ecological Applications*, 24: 1421–1433. https://doi.org/10.1890/13-2033.1

Brunsell, N.A., and Gillies, R.R. 2003. Determination of scaling characteristics of AVHRR data with wavelets: Application to SGP97. *International Journal of Remote Sensing*, 24: 2945–2957.

Bucini, G., Hanan, N.P., Boone, R.B., Smit, I.P.J., Saatchi, S., Lefsky, M.A., and Asner, G.P. 2010. Woody fractional cover in Kruger National Park, South Africa: Remote-sensing-based maps and ecological insights. In M.J. Hill and N.P. Hanan (eds.), *Ecosystem Function in Savannas: Measurement and Modeling at Landscape to Global Scales*. Boca Raton, FL: CRC Press, pp. 219–237.

Burrell, A.L., Evans, J.P., and Liu, Y. 2019. The addition of temperature to the TSS-RESTREND methodology significantly improves the detection of dryland degradation. *IEEE Journal of Selected Topics in Applied Earth Observations and Remote Sensing*, 12(7): 2342–2348. http://doi.org/10.1109/JSTARS.2019.2906466

Campo-Bescós, M.A., Muñoz-Carpena, R., Southworth, J., Zhu, L., Waylen, P.R., and Bunting, E. 2013. Combined spatial and temporal effects of environmental controls on long-term monthly NDVI in the Southern Africa Savanna. *Remote Sensing*, 5(12): 6513–6538.

Cassel-Gintz, M., and Petschel-Held, G. 2000. GIS-based assessment of the threat to world forests by patterns of non-sustainable civilisation nature interaction. *Journal of Environmental Management*, 59: 279–298.

Ceccherini, G., Gobron, N., Robustelli, M. 2013. Harmonization of fraction of absorbed photosynthetically active radiation (FAPAR) from sea-viewingwide field-of-view sensor (SeaWiFS) and medium resolution imaging spectrometer instrument (MERIS). *Remote Sensing*, 5(7): 3357–3376.

Chapin III, F.S., Carpenter, S.R., Kofinas, G.P., Folke, C., Abel, N., Clark, W.C., Olsson, P., Smith, D.M.S., Walker, B., Young, O.R., Berkes, F., Biggs, R., Grove, J.M., Naylor, R.L., Pinkerton, E., Steffen, W., and Swanson, F.J. 2010. Ecosystem stewardship: Sustainability strategies for a rapidly changing planet. *Trends in Ecology & Evolution*, 25: 241–249.

Cherlet, M., Hutchinson, C., Reynolds, J., Hill, J., Sommer, S., and Von Maltitz, G. (eds.). 2018. *World Atlas of Desertification*. Luxembourg: Publication Office of the European Union.

Claverie, M., Ju, J., Masek, J.G., Dungan, J.L., Vermote, E.F., Roger, J.-C., Skakun, S.V., and Justice, C. (2018). The Harmonized Landsat and Sentinel-2 surface reflectance data set. *Remote Sensing of Environment*, 219: 145–161. https://doi.org/10.1016/j.rse.2018.09.002

Cohen, W.B., and Goward, S.N. 2004. Landsat's role in ecological applications of remote sensing. *BioScience*, 54: 535–545.

Cook, B.I., and Pau, S. 2013. A global assessment of long-term greening and browning trends in pasture lands using the GIMMS LAI3g dataset. *Remote Sensing*, 5: 2492–2512.

Cowie, A.L., Orr, B.J., Castillo Sanchez, V.M., Chasek, P., Crossman, N.D., Erlewein, A., Louwagie, G., Maron, M., Metternicht, G.I., Minelli, S., Tengberg, A.E., Walter, S., and Welton, S. 2018. Land in balance: The scientific conceptual framework for Land Degradation Neutrality. *Environmental Scieince & Policy*, 79: 25–35. https://doi.org/10.1016/j.envsci.2017.10.011

CSIRO. 2009. *Property-Scale Solution for Environmental Monitoring and Land Management*. http://www.csiro.au/solutions/Vegmachine.html

Dardel, C., Kergoat, L., Hiernaux, P., Mougin, E., Grippa, M., and Tucker, C.J. 2014. Re-greening Sahel: 30 years of remote sensing data and field observations (Mali, Niger). *Remote Sensing of Environment*, 140: 350–364.

De Jong, R., Verbesselt, J., Schaepman, M.E., and de Bruin, S. 2012. Trend changes in global greening and browning: Contribution of short-term trends to longer-term change. *Global Change Biology*, 18: 642–655.

De Jong, S., and Epema, G.F. 2011. Imaging spectrometry for surveying and modelling land degradation. In F.D. van der Meer and S.M. De Jong (eds.), *Imaging Spectrometry*. Dordrecht: Springer.

De Marzo, T., Gasparri, N.I., Lambin, E.F., and Kuemmerle, T. 2022. Agents of forest disturbance in the argentine dry chaco. *Remote Sensing*, 14: 1758. https://doi.org/10.3390/rs14071758

DeFries, R.S. 2008. Terrestrial vegetation in the coupled human-earth system: Contributions of remote sensing. *Annual Review of Environment and Resources*, 33: 369–390.

DeFries, R.S., Foley, J.A., and Asner, G.P. 2004. Land-use choices: Balancing human needs and ecosystem function. *Frontiers in Ecology and the Environment*, 2: 249–257.

del Barrio, G., Puigdefabregas, J., Sanjuan, M.E., Stellmes, M., and Ruiz, A. 2010. Assessment and monitoring of land condition in the Iberian Peninsula, 1989–2000. *Remote Sensing of Environment*, 114: 1817–1832.

del Barrio, G., Sanjuan, M.E., Hirche, A., Yassin, M., Ruiz, A., Ouessar, M., Martinez Valderrama, J., Essifi, B., and Puigdefabregas, J. 2016. Land degradation states and trends in the northwestern Maghreb drylands, 1998–2008. *Remote Sensing*, 8: 603. https://doi.org/10.3390/rs8070603

del Barrio, G., Sanjuán, M.E., Martínez-Valderrama, J., Ruiz, A., and Puigdefábregas, J. 2021. Land degradation means a loss of management options. *Journal of Arid Environments*, 189: 104502. https://doi.org/10.1016/j.jaridenv.2021.104502

Diaz-Delgado, R., and Pons, X. 2001. Spatial patterns of forest fires in Catalonia (NE of Spain) along the period 1975–1995—Analysis of vegetation recovery after fire. *Forest Ecology and Management*, 147: 67–74.

Didan, K. 2021. *MODIS/Terra Vegetation Indices 16-Day L3 Global 500m SIN Grid V061*. NASA EOSDIS Land Processes DAAC. https://doi.org/10.5067/MODIS/MOD13A1.061

Downing, T.E., and Lüdeke, M. 2002. International desertification. Social geographies of vulnerability and adaptation. In J.F. Reynolds and D.M. Stafford-Smith (eds.), *Global Desertification. Do Humans Cause Deserts?* Berlin: Dahlem University Press, pp. 233–252.

Doxani, G., Vermote, E.F., Roger, J.-C., Gascon, F., Adriaensen, S., Frantz, D., Hagolle, O., Hollstein, A., Kirches, G., Li, F., et al. 2018. Atmospheric correction inter-comparison exercise. *Remote Sensing*, 10: 352. https://doi.org/10.3390/rs10020352.

Doxani, G., Vermote, E.F., Roger, J.-C., Skakun, S., Gascon, F., Collison, A., De Keukelaere, L., Desjardins, C., Frantz, D., Hagolle, O., Kim, M., Louis, J., Pacifici, F., Pflug, B., Poilvé, H., Ramon, D., Richter, R., and Yin, F. 2023. Atmospheric correction inter-comparison eXercise, ACIX-II land: An assessment of atmospheric correction processors for Landsat 8 and Sentinel-2 over land. *Remote Sensing of Environment*, 285: 113412. https://doi.org/10.1016/j.rse.2022.113412.

Dregne, H.E., and Chou, N.-T. 1992. Global desertification dimensions and costs. In H.E. Dregne (ed.), *Degradation and Restoration of Arid Lands*. Lubbock: Texas Tech University.

Dubovyk, O., Menz, G., Conrad, C., Kan, E., Machwitz, M., and Khamzina, A. 2013. Spatio-temporal analyses of cropland degradation in the irrigated lowlands of Uzbekistan using remote sensing and logistic regression modelling. *Environmental Monitoring and Assessment*, 185(6): 4775–4790.

Duguy, B., Alloza, J.A., Röder, A., Vallejo, R., and Pastor, F. 2007. Modeling the effects of landscape fuel treatments on fire growth and behaviour in a Mediterranean landscape (eastern Spain). *International Journal of Wildland Fire*, 16: 619–632.

Eklundh, L., and Olsson, L. 2003. Vegetation trends for the African Sahel in 1982–1999. *Geophysical Research Letters*, 30: 1430–1434.

Elmore, A.J., Mustard, J.F., Manning, S.J., and Lobell, D.B. 2000. Quantifying vegetation change in semiarid environments: Precision and accuracy of spectral mixture analysis and the Normalized Difference Vegetation Index. *Remote Sensing of Environment*, 73: 87–102.

Erb, K.-H., Krausmann, F., Gaube, V., Gingrich, S., Bondeau, A., Fischer-Kowalski, M., and Haberl, H. 2009. Analyzing the global human appropriation of net primary production—processes, trajectories, implications. An introduction. *Ecological Economics*, 69: 250–259.

Evans, J., and Geerken, R. 2004. Discrimination between climate and human-induced dryland degradation. *Journal of Arid Environments*, 57(4): 535–554. https://doi.org/10.1016/S0140-1963(03)00121-6

Fagan, M.E. 2020. A lesson unlearned? Underestimating tree cover in drylands biases global restoration maps. *Global Change Biology*, 26: 4679–4690. https://doi.org/10.1111/gcb.15187

Fan, Z., Li, S., and Fang, H. 2020. Explicitly identifying the desertification change in CMREC area based on multisource remote data. *Remote Sensing*, 12: 3170. https://doi.org/10.3390/rs12193170

Fang, H., Liang, S., McClaran, M.P., Van Leeuwen, W., Drake, S., Marsh, S.E., Thomson, A.M., Izaurralde, R.C., and Rosenberg, N.J. 2005. Biophysical characterization and management effects on semiarid rangeland observed from Landsat ETM+ data. *IEEE Transactions on Geoscience and Remote Sensing*, 43: 125–134.

Fensholt, R., Langanke, T., Rasmussen, K., Reenberg, A., Prince, S.D., Tucker, C., Scholes, R.J., Le, Q.B., Bondeau, A., Eastman, R., Epstein, H., Gaughan, A.E., Hellden, U., Mbow, C., Olsson, L., Paruelo, J., Schweitzer, C., Seaquist, J., and Wessels, K. 2012. Greenness in semi-arid areas across the globe 1981–2007—an Earth Observing Satellite based analysis of trends and drivers. *Remote Sensing of Environment*, 121: 144–158.

Fensholt, R., and Proud, S.R. 2012. Evaluation of earth observation based global long term vegetation trends—Comparing GIMMS and MODIS global NDVI time series. *Remote Sensing of Environment*, 119: 131–147.

Fensholt, R., Rasmussen, K., Kaspersen, P., Huber, S., Horion, S., and Swinnen, E. 2013. Assessing land degradation/recovery in the African Sahel from long-term earth observation based primary productivity and precipitation relationships. *Remote Sensing*, 5(2): 664–686.

Fensholt, R., Sandholt, I., and Rasmussen, M. 2004. Evaluation of MODIS LAI, fAPAR and the relation between fAPAR and NDVI in a semi-arid environment using in situ measurements. *Remote Sensing of Environment*, 91(3–4): 490–507.

Flood, N., Danaher, T., Gill, T., and Gillingham, S. 2013. An operational scheme for deriving standardised surface reflectance from Landsat TM/ETM+ and SPOT HRG imagery for Eastern Australia. *Remote Sensing*, 5: 83–109.

Foley, J.A., DeFries, R., Asner, G.P., Barford, C., Bonan, G., Carpenter, S.R., Chapin, F.S., Coe, M.T., Daily, G.C., Gibbs, H.K., Helkowski, J.H., Holloway, T., Howard, E.A., Kucharik, C.J., Monfreda, C., Patz, J.A., Prentice, I.C., Ramankutty, N., and Snyder, P.K. 2005. Global consequences of land use. *Science*, 309: 570–574.

Foley, J.A., Ramankutty, N., Brauman, K.A., Cassidy, E.S., Gerber, J.S., Johnston, M., Mueller, N.D., O'Connell, C., Ray, D.K., West, P.C., Balzer, C., Bennett, E.M., Carpenter, S.R., Hill, J., Monfreda, C., Polasky, S., Rockstrom, J., Sheehan, J., Siebert, S., Tilman, D., and Zaks, D.P.M. 2011. Solutions for a cultivated planet. *Nature*. http://doi.org/10.1038/nature10452.

Forman, R.T.T., and Collinge, S.K. 1996. The "spatial solution" to conserving biodiversity in landscapes and regions. In R.M. DeGraaf and R.I. Miller (eds.), *Conservation of Faunal Diversity in Forested Landscapes*. London: Chapman & Hall, pp. 537–568.

Frantz, D. 2019. FORCE—Landsat + Sentinel-2 analysis ready data and beyond. *Remote Sensing*, 11: 1124. https://doi.org/10.3390/rs11091124.

Frantz, D., Hostert, P., Rufin, P., Ernst, S., Röder, A., and van der Linden, S. 2022. Revisiting the past: Replicability of a historic long-term vegetation dynamics assessment in the era of big data analytics. *Remote Sensing*. https://doi.org/10.3390/rs14030597.

Frantz, D., Stellmes, M., Röder, A., and Hill, J. 2016a. Fire spread from MODIS burned area data: Obtaining fire dynamics information for every single fire. *Journal of the International Association of Wildland Fire*, 25(12): 1228–1237. https://doi.org/10.1071/WF16003

Frantz, D., Stellmes, M., Röder, A., Udelhoven, T., Mader, S., and Hill, J. 2016b. Improving the spatial resolution of land surface phenology by fusing medium- and coarse-resolution inputs. *IEEE Transactions on Geoscience and Remote Sensing*, 54(7): 4153–4164. https://doi.org/10.1109/TGRS.2016.2537929

Funk, C., Peterson, P., Landsfeld, M., Pedreros, D., Verdin, J., Shukla, S., Husak, G., Rowland, J., Harrison, L., Hoell, A., and Michaelsen, J. 2015. The climate hazards infrared precipitation with stations—a new environmental record for monitoring extremes. *Science Data*, 2: 150066. https://doi.org/10.1038/sdata.2015.66

Ganem, K., Xue, Y., de Almeida Rodrigues, A., Rocha, W., Terra de Oliveira, M., Carvalho, N., Turpo Cayo, E.Y., Rosa, M., Cerqueira Dutra, A., and Shimabukuro, Y. 2022. Mapping South America's drylands through remote sensing—A review of the methodological trends and current challenges. *Remote Sensing*, 14: 736. http://doi.org/10.3390/rs14030736

Gao, F., Masek, J., Schwaller, M., and Hall, F. 2006. On the blending of the Landsat and MODIS surface reflectance: Predicting daily Landsat surface reflectance. *IEEE Transactions on Geoscience and Remote Sensing*, 44(8): 2207–2218.

Geerken, R., and Ilaiwi, M. 2004. Assessment of rangeland degradation and development of a strategy for rehabilitation. *Remote Sensing of Environment*, 90: 490–504.

Geist, H.J., and Lambin, E.F. 2004. Dynamic causal patterns of desertification. *BioScience*, 54: 817–829.

Gessner, U., Machwitz, M., Conrad, C., and Dech, S. 2013. Estimating the fractional cover of growth forms and bare surface in savannas. A multi-resolution approach based on regression tree ensembles. *Remote Sensing of Environment*, 129: 90–102.

Gianoli, F., Weynants, M., and Cherlet, M. 2023. Land degradation in the European Union—Where does the evidence converge? *Land Degradation & Development*, 34: 2256–2275. https://doi.org/10.1002/ldr.4606

Gorelick, N., Hancher, M., Dixon, M., Ilyushchenko, S., Thau, D., and Moore, R. 2017. Google Earth Engine: Planetary-scale geospatial analysis for everyone. *Remote Sensing of Environment*, 202: 18–27. https://doi.org/10.1016/j.rse.2017.06.031.

Goward, S.N., and Masek, J.G. 2001. Landsat—30 years and counting. *Remote Sensing of Environment*, 78: 1–2.

Graetz, R.D. 1996. Empirical and practical approaches to land surface characterization and change detection. In J. Hill and D. Peter (eds.), *The Use of Remote Sensing for Land Degradation and Desertification Monitoring in the Mediterranean Basin—State of the Art and Future Research. Proceedings of a Workshop, Jointly Organized by JRC/IRSA and DGXII/D-2/D-4, Valencia, 13–15 June 1994, Valencia* (pp. 9–23). Luxembourg: Office for Official Publications of the European Communities.

Griffiths, P., Kuemmerle, T., Kennedy, E.R., Abrudan, I.V., Knorn, J., and Hostert, P. 2011. Using annual time-series of Landsat images to assess the effects of forest restitution in post-socialist Romania. *Remote Sensing of Environment*, 118: 199–214.

Grünzweig, J.M., De Boeck, H.J., Rey, A., Santos, M.J., Adam, O., Bahn, M., . . . and Yakir, D. (2022). Dryland mechanisms could widely control ecosystem functioning in a drier and warmer world. *Nature Ecology & Evolution*, 6(8): 1064–1076.

Guirado, E., Alcaraz-Segura, D., Cabello, J., Puertas-Ruíz, S., Herrera, F., and Tabik, D. 2020. Tree cover estimation in global drylands from space using deep learning. *Remote Sensing*, 12: 343. https://doi.org/10.3390/rs12030343

Haberl, H., Erb, K.H., Krausmann, F., Gaube, V., Bondeau, A., Plutzar, C., Gingrich, S., Lucht, W., and Fischer-Kowalski, M. 2007. Quantifying and mapping the human appropriation of net primary production in earth's terrestrial ecosystems. *PNAS*, 104: 12942–12947.

Hall, F., Masek, J.G., and Collatz, G.J. 2006. Evaluation of ISLSCP initiative II FASIR and GIMMS NDVI products and implications for carbon cycle science. *Journal of Geophysical Research*, 111: D22S08. http://doi.org/10.1029/2006JD007438.

Harris, I., Osborn, T.J., Jones, P., and Lister, D. 2020. Version 4 of the CRU TS monthly high-resolution gridded multivariate climate dataset. *Science Data*, 7: 109. https://doi.org/10.1038/s41597-020-0453-3

Hecheltjen, A., Thonfeld, F., and Menz, G. 2014. Recent advances in remote sensing change detection—a review. In I. Manakos and M. Braun (eds.), *Land Use and Land Cover Mapping in Europe. Practices and Trends*. Dordrecht, Heidelberg, New York, and London: Springer, pp. 145–178.

Hein, L., and de Ridder, N. 2006. Desertification in the Sahel: A reinterpretation. *Global Change Biology*, 12: 751–758.

Helldén, U., and Tottrup, C. 2008. Regional desertification: A global synthesis. *Global and Planetary Change*, 64(3–4): 169–176.

Herman, S.M., Anyamba, A., and Tucker, C.J. 2005. Recent trends in vegetation dynamics in the African Sahel and their relationship to climate. *Global Environmental Change*, 15: 394–404.

Hermann, S.M., and Hutchinson, C.F. 2005. The changing contexts of the desertification debate. *Journal of Arid Environments*, 63: 538–555.

Heumann, B.W., Seaquist, J.W., Eklundh, L., and Jönnson, P. 2007. AVHRR derived phenological change in the Sahel and Soudan, Africa, 1982–2005. *Remote Sensing of Environment*, 108: 385–392.

Higginbottom, T.P., Symeonakis, E., Meyer, H., and van der Linden, S. 2018. Mapping fractional woody cover in semi-arid savannahs using multi-seasonal composites from Landsat data. *ISPRS Journal of Photogrammetry and Remote Sensing*, 139: 88–102.

Hilker, T., Natsagdorj, E., Waring, R.H., Lyapustin, A., and Wang, Y. 2014. Satellite observed widespread decline in Mongolian grasslands largely due to overgrazing. *Global Change Biology*, 20: 418–428.

Hill, J. 2008. Remote sensing techniques for monitoring desertification. In *LUCINDA Booklet Series A*. Number 3. http://geografia.fcsh.unl.pt/lucinda/booklets/Booklet%20A3%2-0EN.pdf

Hill, J., Hostert, P., and Röder, A. 2004. Long-term observation of mediterranean ecosystems with satellite remote sensing. In S. Mazzoleni, G. di Pasquale, M. Mulligan, P. di Martino and F. Rego (eds.), *Recent Dynamics of the Mediterranean Vegetation and Landscape*. Chichester: John Wiley & Sons Ltd., pp. 33–43.

Hill, J., Stellmes, M., Udelhoven, T., Röder, A., and Sommer, S. 2008. Mediterranean desertification and land degradation Mapping related land use change syndromes based on satellite observations. *Global and Planetary Change*, 64: 146–157.

Hill, J., Stellmes, M., and Wang, C. 2014. Land transformation processes in NE China: Tracking trade-offs in ecosystem services across seceral decades with Landsat-TM/ETM+ time series. In I. Manakos and M. Braun (eds.), *Land Use and Land Cover Mapping in Europe—Practices and Trends*. Heidelberg: Springer, pp. 383–410.

Hill, M.J., Roman, M.O., Schaaf, C.B., Hutley, L., Brannstrom, C., Etter, A., and Hanan, N.P. 2011. Characterizing vegetation cover in global savannas with an annual foliage clumping index derived from the MODIS BRDF product. *Remote Sensing of Environment*, 115: 2008–2024.

Hirche, A., Salamani, M., Abdellaoui, A., Benhouhou, S., and Valderrama, J.M. 2011. Landscape changes of desertification in arid areas: The case of south-west Algeria. *Environmental Monitoring and Assessment*, 179: 403–420. https://doi.org/10.1007/s10661-010-1744-5

Hostert, P., Röder, A., Hill, J., Udelhoven, T., and Tsiourlis, G. 2003. Retrospective studies of grazing-induced land degradation: A case study in central Crete, Greece. *International Journal of Remote Sensing*, 24: 4019–4034.

Huang, C.Q., Coward, S.N., Masek, J.G., Thomas, N., Zhu, Z.L., and Vogelmann, J.E. 2010. An automated approach for reconstructing recent forest disturbance history using dense Landsat time series stacks. *Remote Sensing of Environment*, 114: 183–198.

Huang, J., Yu, H., Guan, X., Wang, G., and Guo, R. 2016. Accelerated dryland expansion under climate change. *Nature Climate Change*, 6: 166. https://doi.org/10.1038/nclimate2837

Huber, S., and Fensholt, R. 2011. Analysis of teleconnections between AVHRR-based sea surface temperature and vegetation productivity in the semi-arid Sahel. *Remote Sensing of Environment*, 115: 3276–3285.

Huete, A.R. 1988. A soil-adjusted vegetation index (SAVI). *Remote Sensing of Environment*, 25: 295–309.

Huete, A.R., Didan, K., Miura, T., Rodriguez, E.P., Gao, X., and Ferreira, L.G. 2002. Overview of the radiometric and biophysical performance of the MODIS vegetation indices. *Remote Sensing of Environment*, 83: 195–213.

Jackson, H., and Prince, S. D. 2016. Degradation of net primary production in a semiarid rangeland. *Biogeosciences*, 13, 4721–4734, https://doi.org/10.5194/bg-13-4721-2016.

Jamali, S., Seaquist, J., Eklundh, L., and Ardö, J. 2014. Automated mapping of vegetation trends with polynomials using NDVI imagery over the Sahel. *Remote Sensing of Environment*, 141: 79–89.

Jarmer, T., Lavée, H., Sarah, P., and Hill, J. 2009. Using reflectance spectroscopy and Landsat data to assess soil inorganic carbon in the Judean Desert (Israel). In A. Röder and J. Hill (eds.), *Recent Advances in Remote Sensing and Geoinformation Processing for Land Degradation Assessment*. London: CRC Press/Balkema (Taylor and Francis Group), pp. 227–241.

Jönsson, P., and L. Eklundh. 2002. Seasonality extraction by function fitting to time-series of satellite sensor data. *IEEE Transactions on Geoscience and Remote Sensing*, 40: 1824–1832.

Justice, C., Giglio, L., Boschetti, L., Roy, D., Csiszar, I., Morisette, J., and Kaufman, Y. 2006. *MODIS Fire Products—Version 2.3*. Algorithm Technical Background Document.

Katagis, T., Gitas, I.Z., Toukiloglou, P., Veraverbeke, S., and Goossens, R. 2014. Trend analysis of medium- and coarse-resolution time series image data for burned area mapping in a Mediterranean ecosystem. *International Journal of Wildland Fire*. http://doi.org/10.1071/WF12055

Kauth, R.J., and Thomas, G.S. 1976. *The Tasselled Cap—A Graphic Description of the Spectral-Temporal Development of Agricultural Crops as Seen by LANDSAT*. LARS Symposia. Paper 159. New York, NY: IEEE.

Kennedy, R.E., Yang, Z., and Cohen, W.B. 2010. Detecting trends in forest disturbance and recovery using yearly Landsat time series: 1. LandTrendr—Temporal segmentation algorithms. *Remote Sensing of Environment*, 114: 2897–2910.

Koofhafkan, P., and Stewart, B.A. 2008. *Water and Cereals in Drylands*. FAO Publications. EarthScan, London, Sterling VA. http://www.fao.org/docrep/012/i0372e/i0372e00.htm (accessed 10 March 2014).

Lacaze, B. 1996. Spectral characterisation of vegetation communities and practical approaches to vegetation cover changes monitoring. In J. Hill and D. Peter (eds.), *The Use of Remote Sensing for Land Degradation and Desertification Monitoring in the Mediterranean Basin—State of the Art and Future Research. Proceedings of a Workshop, Jointly Organized by JRC/IRSA and DGXII/D-2/D-4, Valencia, 13–15 June 1994* (pp. 149–166). Valencia: Office for Official Publications of the European Communities.

Lal, R. 2004. Carbon sequestration in dryland ecosystems. *Environmental Management*, 33: 528–544.

Lambin, E.F., and Ehrlich, D. 1997. Land-cover changes in sub-Saharan Africa (1982–1991): Application of a change index based on remotely sensed surface temperature and vegetation indices at a continental scale. *Remote Sensing of Environment*, 61: 181–200.

Lambin, E.F., Geist, H.J., and Rindfuss, R.R. 2006. Introduction: Local processes with global impacts. In E.F. Lambin and H. Geist (eds.), *Land-Use and Land-Cover Change—Local Processes and Global Impacts*. Berlin, Heidelberg, and New York: Springer.

Lee, R. 2011. The outlook for population growth. *Science*, 333: 569–573.

Le Houérou, H.N. 1984. Rain use efficiency: A unifying concept in arid-land ecology. *Journal of Arid Environments*, 7: 213–247.

Leitão, P., Schwieder, M., Suess, S., Okujeni, A., Galvão, L., Linden, S., and Hostert, P. 2015. Monitoring natural ecosystem and ecological gradients: Perspectives with EnMAP. *Remote Sensing*, 7: 13098–13119. https://doi.org/10.3390/rs71013098.

Lepers, E. (2003). Synthesis of the main areas of land-cover and land-use change. *Final Report*. In Millennium Ecosystem Assessment. http://www.geo.ucl.ac.be/LUCC/lucc.html

Lewińska, K.E., Hostert, P., Buchner, J., Bleyhl, B., and Radeloff, V.C. 2020. Short-term vegetation loss versus decadal degradation of grasslands in the Caucasus based on Cumulative Endmember Fractions. *Remote Sensing of Environment*, 248: 111969. https://doi.org/10.1016/j.rse.2020.111969.

Li, S., Verburg, P.H., Lv, S., Wu, J., and Li, X. 2012. Spatial analysis of the driving factors of grassland degradation under the conditions of climate change and intensive use in Inner Mongolia, China. *Regional Environmental Change*, 12: 461–474.

Lian, X., Piao, S., Chen, A, Huntingford, C., Fu, B., Li, L.Z.X., Huang, J., Sheffield, J., Berg, A.M., Keenan, T.F., McVicar, T.R., Wada, Y., Wang, X., Wang, T., and Yang, Y. 2021. Multifaceted characteristics of dryland aridity changes in a warming world. *Nature Reviews Earth & Environment*, 2: 232–250. https://doi.org/10.1038/s43017-021-00144-0

Loboda, T.V., Giglio, L., Boschetti, L., and Justice, C.O. 2012. Regional fire monitoring and characterization using global NASA MODIS fire products in drylands of Central Asia. *Frontiers of Earth Science*, 6(2): 196–205.

Lorent, H., Evangelou, C., Stellmes, M., Hill, J., Papanastasis, V., Tsiourlis, G., Roeder, A., and Lambin, E.F. 2008. Land degradation and economic conditions of agricultural households in a marginal region of northern Greece. *Global and Planetary Change*, 64: 198–209.

Lu, H., Raupach, M.R., and McVicar, T.R. 2001. Decomposition of vegetation cover into woody and herbaceous components using AVHRR NDVI time series. *CSIRO Land and Water Technical Report 35/01*. CSIRO Land and Water. Canberra, Australia. http://www.clw.csiro.au/publications/technical2001/tr35-01.pdf

Lu, X., Wang, L., and McCabe, M.F. 2016. Elevated CO2 as a driver of global dryland greening. *Scientific Reports*, 6: 20716. https://doi.org/10.1038/srep20716

Mahood, A.L., Lindrooth, E.J., Cook, M.C., and Balch, J.K. 2022. Country-level fire perimeter datasets (2001–2021). *Science Data*, 9: 458. https://doi.org/10.1038/s41597-022-01572-3

Martínez-Valderrama, J., Sanjuán, M.E., del Barrio, G., Guirado, E., Ruiz, A., and Maestre, F.T. 2021. Mediterranean landscape re-greening at the expense of South American agricultural expansion. *Land*, 10: 204. https://doi.org/10.3390/land10020204

Masek, J.G., Vermote, E.F., Saleous, N., Wolfe, R., Hall, F.G., Huemmrich, K.F., Gao, F., Kutler, J., and Lim, T.K. 2006. A Landsat surface reflectance data set for North America, 1990–2000. *Geoscience and Remote Sensing Letters*, 3: 68–72.

Mbow, C., Nielsen, T.T., and Rasmussen, K. 2000. Savanna fires in east-central Senegal: Distribution patterns, resource management and perceptions. *Human Ecology*, 28(4): 561–583.

Meyer-Christoffer, A., Becker, A., Finger, P., Rudolf, B., Schneider, U., and Ziese, M. 2011. *GPCC Climatology Version 2011 at 0.25°: Monthly Land-Surface Precipitation Climatology for Every Month and the Total Year from Rain-Gauges Built on GTS-based and Historic Data*. http://doi.org/10.5676/DWD_GPCC/CLIM_M_V2011_025.

Miehe, S., Kluge, J., von Wehrden, H., and Retzer, V. 2010. Long-term degradation of Sahelian rangeland detected by 27 years of field study in Senegal. *Journal of Applied Ecology*, 47: 692–700.

Millennium Ecosystem Assessment. 2005a. *Ecosystems and Human Well-Being: Desertification Synthesis*. Washington, DC: World Resources Institute.

Millennium Ecosystem Assessment. 2005b. *Ecosystems and Human Well-Being: Synthesis*. Washington, DC: Island Press.

Mirzabaev, A., Wu, J., Evans, J., García-Oliva, F., Hussein, I.A.G., Iqbal, M.H., Kimutai, J., Knowles, T., Meza, F., Nedjraoui, D., Tena, F., Türkeş, M., Vázquez, R.J., and Weltz, M. 2019. Desertification. In P.R. Shukla, J. Skea, E. Calvo Buendia, V. Masson-Delmotte, H.-O. Pörtner, D.C. Roberts, P. Zhai, R. Slade, S. Connors, R. van Diemen, M. Ferrat, E. Haughey, S. Luz, S. Neogi, M. Pathak, J. Petzold, J. Portugal Pereira, P. Vyas, E. Huntley, K. Kissick, M. Belkacemi, and J. Malley (eds.), *Climate Change and Land: An IPCC Special Report on Climate Change, Desertification, Land Degradation, Sustainable Land Management, Food Security, and Greenhouse Gas Fluxes in Terrestrial Ecosystems*. https://doi.org/10.1017/9781009157988.005

Munawar, S., Röder, A., Syampungani, S., and Udelhoven, T. 2022. Place-based analysis of satellite time series shows opposing land change patterns in the copperbelt region of zambia. *Forests*, 13: 1–21. https://doi.org/10.3390/f13010134.

Myneni, R.B., and Williams, D.L. 1994. On the relationship between FAPAR and NDVI. *Remote Sensing of Environment*, 49(3): 200–211.

Naveh, Z. 1975. The evolutionary significance of fire in the Mediterranean region. *Vegetati*, 29: 199–208.

Novella, N.S., and Thiaw, W.M. 2013. African rainfall climatology version 2 for famine early warning systems. *Journal of Applied Meteorology and Climatology*, 52: 588–606.

Oldeman, L.R., Hakkeling, R.T.A., and Sombroek, W.G. 1990. *World Map on Status of Human-Induced Soil Degradation (GLASOD)*. Wageningen, Netherlands: ISRIC.

Orr, B.J., Cowie, A.L., Castillo Sanchez, V.M., Chasek, P., Crossman, N.D., Erlewein, A., Louwagie, G., Maron, M., Metternicht, G.I., Minelli, S., Tengberg, A.E., Walter, S., and Welton, S. 2017. Scientific conceptual framework for land degradation neutrality. *A Report of the Science-Policy Interface*. United Nations Convention to Combat Desertification (UNCCD), Bonn.

Padilla, F.M., Vidal, B., Sánchez, J., and Pugnaire, F.I. 2010. Land-use changes and carbon sequestration through the twentieth century in a Mediterranean mountain ecosystem: Implications for land management. *Journal of Environmental Management*, 91: 2688–2695.

Pan, Y., Yin, Y., and Cao, W. 2023. Integrated assessments of land degradation in the three-rivers headwater region of China from 2000 to 2020. *Remote Sensing*, 15: 4521. https://doi.org/10.3390/rs15184521

Paudel, K.P., and Andersen, P. 2010. Assessing rangeland degradation using multi temporal satellite images and grazing pressure surface model in Upper Mustang, Trans Himalaya, Nepal. *Remote Sensing of Environment*, 114: 1845–1855.

Petschel-Held, G., Lüdeke, M.K.B., and Reusswig, F. 1999. Actors, structures and environment: A comparative and transdisciplinary view on regional case studies of global environmental change. In B. Lohnert and H. Geist (eds.), *Coping with Changing Environments: Social Dimensions of Endangered Ecosystems in the Developing World*. London: Ashgate, pp. 255–291.

Pinzon, J.E., Pak, E.W., Tucker, C.J., Bhatt, U.S., Frost, G.V., and Macander, M.J. 2023. *Global Vegetation Greenness (NDVI) from AVHRR GIMMS-3G+, 1981–2022*. ORNL DAAC, Oak Ridge, TN. https://doi.org/10.3334/ORNLDAAC/2187.

Povey, A.C., and Grainger, R.G. 2015. Known and unknown unknowns: Uncertainty estimation in satellite remote sensing. *Atmospheric Measurement Techniques*, 8(11): 4699–4718. https://doi.org/10.5194/amt-8-4699-2015

Price, J.C. 1993. Estimating of leaf area index from satellite data. *IEEE Transactions on Geoscience and Remote Sensing*, 31: 727–734.

Prince, S.D. 2004. Mapping desertification in southern Africa. In G. Gutman, A. Janetos, C.O. Justice, E.F. Moran, J.F. Mustard, R.R. Rindfuss, D. Skole, and B.L. Turner II (eds.), *Land Change Science: Observing, Monitoring, and Understanding Trajectories of Change on the Earth's Surface*. Dordrecht, NL: Kluwer, pp. 163–184.

Prince, S.D., Becker-Reshef, I., and Rishmawi, K. 2009. Detection and mapping of long-term land degradation using local net production scaling: Application to Zimbabwe. *Remote Sensing of Environment*, 113: 1046–1057.

Prince, S.D., De Colstoun, E.B., and Kravitz, L.L. 1998. Evidence from rain-use efficiencies does not indicate extensive Sahelian desertification. *Global Change Biology*, 4: 359–374.

Prince, S.D., Wessels, K.J., Tucker, C.J., and Nicholson, S.E. 2007. Desertification in the Sahel: A reinterpretation of a reinterpretation. *Global Change Biology*, 13: 1308–1313.

Qiu, S., Zhu, Z., and He, B. 2019. Fmask 4.0: Improved cloud and cloud shadow detection in Landsats 4–8 and Sentinel-2 imagery. *Remote Sensing of Environment*, 231: 111205. https://doi.org/10.1016/j.rse.2019.05.024

Rast, M., Nieke, J., Adams, J., Isola, C., and Gascon, F. 2021. Copernicus hyperspectral imaging mission for the environment (chime). In *2021 IEEE International Geoscience and Remote Sensing Symposium IGARSS*. Brussels, Belgium: IEEE, pp. 108–111.

Reeves, M.C., and Baggett, L.S. 2014. A remote sensing protocol for identifying rangelands with degraded productive capacity. *Ecological Indicators*, 43: 172–182.

Reid, R.S., Thornton, P.K., McCRabb, G.J., Kruska, R.L., Atieno, F., and Jones, P.G. 2004. Is it possible to mitigate greenhouse gas emissions in pastoral ecosystems of the tropics? *Development and Sustainability*, 6: 91–109.

Reid, R.S., Tomich, T.P., Xu, J., Geist, H., Mather, A., DeFries, R., Liu, J., Alves, D., Agbola, B., Lambin, E.F., Chabbra, A., Veldkamp, T., Kok, K., van Noordwijk, M., Thomas, D., Palm, C., and Verburg, P.H. 2006. Linking land-change science and policy: Current lessons and future integration. In E.F. Lambin and H. Geist (eds.), *Land-Use and Land-Cover Change—Local Processes and Global Impacts*. Berlin, Heidelberg, and New York: Springer.

Reith, J., Ghazaryan, G., Muthoni, F., and Dubovyk, O. 2021. Assessment of land degradation in Semiarid Tanzania—Using multiscale remote sensing datasets to support sustainable development goal 15.3. *Remote Sensing*, 13: 1754. https://doi.org/10.3390/rs13091754

Reynolds, J.F., Stafford-Smith, D.M., Lambin, E.F., Turner II, B.L., Mortimore, M., Batterbury, S.P.J., Downing, T.E., Dowlatabadi, H., Fernandéz, R.J., Herrick, J.E., Huber-Sannwald, E., Jiang, H., Leemans, R., Lynam, T., Maestre, F.T., Ayarza, M., and Walker, B. 2007. Global Desertification: Building a science for dryland development. *Science*, 316: 847–851.

Röder, A., Hill, J., Duguy, B., Alloza, J.A., and Vallejo, R. 2008b. Using long time series of Landsat data to monitor fire events and post-fire dynamics and identify driving factors. A case study in the Ayora region (eastern Spain). *Remote Sensing of Environment*, 112: 259–273.

Röder, A., Udelhoven, T., Hill, J., Del Barrio, G., and Tsiourlis, G.M. 2008a. Trend analysis of Landsat-TM and -ETM+ imagery to monitor grazing impact in a rangeland ecosystem in Northern Greece. *Remote Sensing of Environment*, 112: 2863–2875.

Rouse, J.W., Haas, R.H., Schell, J.A., and Deering, D.W. 1974. Monitoring vegetation systems in the Great Plains with ERTS. In *Third ERTS Symposium: NASA SP-351I*. Washington, DC: NASA, pp. 309–317.

Safriel, U., Adeel, Z., Niemeijer, D., Puigdefabregas, J., White, R., Lal, R., Winslow, M., Ziedler, J., Prince, S., Archer, E., King, C., Shapiro, B., Wessels, K., Nielsen, T., Portnov, B., Reshef, I., Thonell, J., Lachman, E., and McNab, D. 2005. Dryland systems. In R.M. Hassan, R.J. Scholes and N. Ash (eds.), *Millenium Ecosystem Assessment: Ecosystems and Human Well-Being: Current State and Trends: Findings of the Condition and Trends Working Group*. Washington, DC: Island Press, pp. 623–662.

Sankey, T.T., McVay, J., Swetnam, T.L., McClaran, M.P., Heilman, P., Nichols, M., Pettorelli, N., and Horning, N. 2017. UAV hyperspectral and lidar data and their fusion for arid and semi-arid land vegetation monitoring. *Remote Sensing in Ecology and Conservation*, 4: 20–33.

Saunier, S., Louis, J., Debaecker, V., Beaton, T., Cadau, E.G., Boccia, V., and Gascon, F. 2019. Sen2like, a tool to generate Sentinel-2 harmonised surface reflectance products—first results with Landsat-8. *IGARSS 2019–2019 IEEE International Geoscience and Remote Sensing Symposium*. Yokohama, Japan, pp. 5650–5653. http://doi.org/10.1109/IGARSS.2019.8899213.

Schmidt, M., Udelhoven, T., Gill, T., and Röder, A. 2012. Long term data fusion for a dense time series analysis with modis and landsat imagery in an australian savanna. *Journal of Applied Remote Sensing*, 6(1): 063512–1-063512–18.

Schroeder, T.A., Wulder, M.A., Healey, S.P., and Moisen, G.G. 2011. Mapping wildfire and clearcut harvest disturbances in boreal forests with Landsat time series data. *Remote Sensing of Environment*, 115: 1421–1433.

Serra, P., Pons, X., and Saurí, D. 2008. Land-cover and land-use change in a Mediterranean landscape: A spatial analysis of driving forces integrating biophysical and human factors. *Applied Geography*, 28: 189–209.

Shrestha, D.P., Margate, D.E., van der Meer, F., and Anh, H.V. 2005. Analaysis and classification of hyperspectral data for mapping land degradation: An application in southern Spain. *International Journal of Applied Earth Observation and Geoinformation*, 7: 85–96.

Skujins, J. 1991. *Semiarid Lands and Deserts: Soil Resource and Reclamation*. Boca Raton: CRC Press.

Smith, M.O., Ustin, S.L., Adams, J.B., and Gillespie, A.R. 1990. Vegetation in deserts: I. A regional measure of abundance from multispectral images, Remote Sens. *Environ*, 31: 1–26.

Smith, W.K., Dannenberg, M.P., Yan, D., Herrmann, S., Barnes, M.L., Barron-Gafford, G.A., Biederman, J.A., Ferrenberg, S., Fox, A.M., Hudson, A., Knowles, J.F., MacBean, N., Moore, D.J.P., Nagler, P.L., Reed, S.C., Rutherford, W.A., Scott, R.L., Wang, X., and Yang, J. 2019. Remote sensing of dryland ecosystem structure and function: Progress, challenges, and opportunities. *Remote Sensing of Environment*, 233: 1114012–31. https://doi.org/10.1016/j.rse.2019.111401

Sommer, S., Zucca, C., Grainger, A., Cherlet, M., Zougmore, R., Sokona, Y., Hill, J., Peruta, R.D., Roehrig, J., and Wang, G. 2011. Application of indicator systems for monitoring and assessment of desertification from national to global scales. *Land Degradation & Development*. http://doi.org/10.1002/ldr.1084

Sonnenschein, R., Kuemmerle, T., Udelhoven, T., Stellmes, M., and Hostert, P. 2011. Differences in Landsat-based trend analyses in drylands due to the choice of vegetation estimate. *Remote Sensing of Environment*, 115(6): 1408–1420.

Sörensen, L. 2007. *A Spatial Analysis Approach to the Global Delineation of Dryland Areas of Relevance to the CBD Programme of Work on Dry and Subhumid Lands*. Cambridge: UNEP-WCMC.

Stafford Smith, M., Abel, N., Walker, B., and Chapin III, F.S. 2009. Drylands: Coping with uncertainty, thresholds, and changes in state. In F.S. Chapin III, G.P. Kofinas and C. Folke (eds.), *Principles of Ecosystem Stewardship*. New York: Springer, pp. 171–195.

Stellmes, M., Röder, A., Udelhoven, T., and Hill, J. 2013. Mapping syndromes of land change in Spain with remote sensing time series, demographic and climatic data. *Land Use Policy*, 30: 685–702.

Stellmes, M., Udelhoven, T., Röder, A., Sonnenschein, R., and Hill, J. 2010. Dryland observation at local and regional scale—comparison of Landsat TM/ETM+ and NOAA AVHRR time series. *Remote Sensing of Environment*, 114(10): 2111–2125.

Strahler, A.H., Lucht, W., Schaaf, C.B., Tsang, T., Gao, F., Li, X., Muller, J.-P., Lewis, P., and Barnsley, M.J. 1999. *MODIS BRDF/Albedo Product: Algorithm Theoretical Basis Document Version5.0*. NASA. http://modis.gsfc.nasa.gov/data/atbd/atbd_mod09.pdf (accessed 5 March 2014).

Stringer, L.C., Reed, M.S., Fleskens, L., Thomas, R.J., Le, Q.B., and Lala-Pritchard, T. 2017. A new dryland development paradigm grounded in empirical analysis of dryland systems science. *Land Degradation & Development*, 28: 1952–1961. http://doi.org/10.1002/ldr.2716

Taylor, S.D., Browning, D.M., Baca, R.A., and Gao, F. 2021. Constraints and opportunities for detecting land surface phenology in drylands. *Journal of Remote Sensing*. http://doi.org/10.34133/2021/9859103

Thomas, D.S.G., and Middleton, N.J. 1994. *Desertification: Exploding the Myth*. Chichester: Wiley.

Tozer, C.R., Kiem, A.S., and Verdon-Kidd, D.C. 2012. On the uncertainties associated with using gridded rainfall data as a proxy for observed. *Hydrology and Earth System Sciences*, 16: 1481–1499. http://doi.org/10.5194/hess-16-1481-2012

Trapnell, C.G. 1959. Ecological results of woodland burning experiments in Northern Rhodesia. *Journal of Ecology*, 47: 129–168.

Tucker, C.J. 1979. Red and photographic infrared linear combinations for monitoring vegetation. *Remote Sensing of Environment*, 8: 127–150.

Turner II, B.L., Lambin, E.F., and Reenberg, A. 2007. The emergence of land change science for global environmental change and sustainability. *PNAS*, 104: 20666–20671.

Udelhoven, T. 2010. TimeStats: A software tool for the retrieval of temporal patterns from global satellite archives. *IEEE Journal of Selected Topics in Applied Earth Observations and Remote Sensing (J-STARS)*. http://doi.org/10.1109/JSTARS.2010.2051942

Udelhoven, T., and Hill, J. 2009. Change detection in Syria's rangelands using long-term AVHRR data (1982–2004). In A. Röder and J. Hill (eds.), *Recent Advances in Remote Sensing and Geoinformation Processing for Land Degradation Assessment*. London: Taylor and Francis.

Udelhoven, T., Stellmes, M., del Barrio, G., and Hill, J. 2009. Assessment of rainfall and NDVI anomalies in Spain (1989–1999) using distributed lag models. *International Journal of Remote Sensing*, 30: 1961–1976.

UNCCD. 1994. *United Nations Convention to Combat Desertification in Countries Experiencing Serious Drought and/or Desertification, Particularly in Africa*. http://www.unccd.int/en/about-the-convention/Pages/Text-overview.aspx

UNU. 2006. *International Year of Deserts and Desertification*. United Nations University—Institute for Water, Environment and Health. http://inweh.unu.edu/desertification06/ (accessed 15 March 2014).

Venter, Z.S., Scott, S.L., Desmet, P.G., and Hoffman, M.T. 2020. Application of Landsat-derived vegetation trends over South Africa: Potential for monitoring land degradation and restoration. *Ecological Indicators*, 113: 106206. https://doi.org/10.1016/j.ecolind.2020.106206

Verbesselt, J., Hyndman, R., Newnham, G., and Culvenor, D. 2010. Detecting trend and seasonal changes in satellite image time series. *Remote Sensing of Environment*, 114: 106–115.

Verón, S.R., Paruelo, J.M., and Oesterheld, M. 2006. Assessing desertification. *Journal of Arid Environments*, 66: 751–763. https://doi.org/10.1016/j.jaridenv.2006.01.021

Verstraete, M.M. 1994. The contribution of remote sensing to montior vegetation and to evaluate its dynamic aspects. In F. Veroustrate and R. Ceulemans (eds.), *Vegetation, Modeling and Climage Change Effects*. The Hague: SPB Academic Publishing, pp. 207–212.

Verstraete, M.M., Hutchinson, C.F., Grainger, A., Stafford-Smith, M., Scholes, R.J., Reynolds, J.F., Barbosa, P., Léon, A., and Mbow, C. 2011. Towards a global drylands observing system: Observational requirements and institutional solutions. *Land Degradation & Development*, 22: 198–213.

Viedma, O., Melia, J., Segarra, D., and GarciaHaro, J. (1997). Modeling rates of ecosystem recovery after fires by using Landsat TM data. *Remote Sensing of Environment*, 61: 383–398.

Vogt, J.V., Safriel, U., Von Maltitz, G., Sokona, Y., Zougmore, R., Bastin, G., and Hill, J. 2011. Monitoring and assessment of land degradation and desertification: Towards new conceptual and integrated approaches. *Land Degradation & Development*. http://doi.org/10.1002/ldr.1075

von Keyserlingk, J., de Hoop, M., Mayor, A.G., Dekker, S.C., Rietkerk, M., and Foerster, S. 2021. Resilience of vegetation to drought: Studying the effect of grazing in a Mediterranean rangeland using satellite time series. *Remote Sensing of Environment*, 255: 112270. https://doi.org/10.1016/j.rse.2020.112270

Von Maltitz, G.P., Gambiza, J., Kellner, K., Rambau, T., Lindeque, L., and Kgope, B. 2019. Experiences from the South African land degradation neutrality target setting process. *Environmental Science & Policy*, 101: 54–62. https://doi.org/10.1016/j.envsci.2019.07.003

Walker, J.J., de Beurs, K.M., Wynne, R.H., and Gao, F. 2012. Evaluation of Landsat and MODIS data fusion products for analysis of dryland forest phenology. *Remote Sensing of Environment*, 117: 381–393.

Wallace, J., Behn, G., and Furby, S. 2006. Vegetation condition assessment and monitoring from sequences of satellite imagery. *Ecological Managment and Restoration*, 7: 31–36.

Wallace, J., Caccetta, P.A., and Kiiveri, H.T. 2004. Recent developments in analysis of spatial and temporal data for landscape qualities and monitoring. *Australian Ecology*, 29: 100–107.

Wang, J., Zhen, J., Hu, W., Chen, S., Lizaga, I., Zeraatpisheh, M., and Yang, X. 2023. Remote sensing of soil degradation: Progress and perspective. *International Soil and Water Conservation Research*, 11(3): 429–454. https://doi.org/10.1016/j.iswcr.2023.03.002

Wang, L., Jiao, W., MacBean, N., Rulli, M.C., Manzoni, S., Vico, G., and D'Odorico, P. 2022. Dryland productivity under a changing climate. *Nature Climate Change*, 12(11): 981–994.

Washington-Allen, R.A., Ramsey, R.D., West, N.E., and Norton, B.E. 2008. Quantification of the ecological resilience of drylands using digital remote sensing. *Ecology and Society*, 13(1): 33.

Were, K., Dick, Ø.B., and Singh, B.R. 2014. Exploring the geophysical and socio-economic determinants of land cover changes in Eastern Mau forest reserve and Lake Nakuru drainage basin, Kenya. *Geojournal*. http://doi.org/10.1007/s10708-014-9525-2.

Wessels, K. 2009. Letter to the Editor: Comments on 'Proxy global assessment of land degradation' by Bai et al. (2008). *Soil Use and Management*, 25: 91–92.

Wessels, K.J., Li, X., Bouvet, A., Mathieu, R., Main, R., Naidoo, L., Erasmus, B., and Asner, G.P. 2023. Quantifying the sensitivity of L-Band SAR to a decade of vegetation structure changes in savannas. *Remote Sensing of Environment*, 284: 113369. https://doi.org/10.1016/j.rse.2022.113369.

Wessels, K.J., Mathieu, R., Knox, N., Main, R., Naidoo, L., and Steenkamp, K. 2019. Mapping and monitoring fractional woody vegetation cover in the arid savannas of Namibia using LiDAR training data, machine learning, and ALOS PALSAR data. *Remote Sensing*, 11: 2633. https://doi.org/10.3390/rs11222633.

Wessels, K.J., Prince, S.D., Malherbe, J., Small, J., Frost, P.E., and van Zyl, D. 2007. Can human-induced land degradation be distinguished from the effects of rainfall variability? A case study in South Africa. *Journal of Arid Environment*, 68(2): 271–297.

Williams, C.A., and Hanan, N.P. 2011. ENSO and IOD teleconnections for African ecosystems: Evidence of destructive interference between climate oscillations. *Biogeosciences*, 8: 27–40.

Wunder, S., Kaphengst, T., and Frelih-Larsen, A. 2018. Implementing land degradation neutrality (SDG 15.3) at national level: General approach, indicator selection and experiences from Germany. In H. Ginzky, E. Dooley, I. Heuser, E. Kasimbazi, T. Markus and T. Qin (eds.), *International Yearbook of Soil Law and Policy2017*, 1st ed. Berlin, Germany: Springer-Verlag.

Ye, S., Zhu, Z., and Cao, G. 2023. Object-based continuous monitoring of land disturbances from dense Landsat time series. *Remote Sensing of Environment*, 287: 113462. https://doi.org/10.1016/j.rse.2023.113462.

Yin, H., Pflugmacher, D., Li, A., Li, Z., and Hostert, P. 2018. Land use and land cover change in Inner Mongolia—understanding the effects of China's re-vegetation programs. *Remote Sensing of Environment*, 204: 918–930.

Yin, H., Udelhoven, T., Fensholt, R., Pflugmacher, D., and Hostert, P. 2012. How normalized difference vegetation index (NDVI) trendsfrom advanced very high resolution radiometer (AVHRR) and Système

Probatoire d'Observation de la Terre VEGETATION (SPOT VGT) time series differ in agricultural areas: An inner mongolian case study. *Remote Sensing*, 4(11): 3364–3389. http://doi.org/10.3390/rs4113364.

Zeng, F.-W., Collatz, G.J., Pinzon, J.E., and Ivanoff, A. 2013. Evaluating and quantifying the climate-driven interannual variability in global inventory modeling and mapping studies (GIMMS) normalized difference vegetation index (NDVI3g) at global scales. *Remote Sensing*, 5: 3918–3950.

Zhou, L., Jia, L., Menenti, M., van Hoek, M., Lu, J., Zheng, C., Wu, H., and Yuan, X. 2021. Characterizing vegetation response to rainfall at multiple temporal scales in the Sahel-Sudano-Guinean region using transfer function analysis. *Remote Sensing of Environment*, 252: 112108. https://doi.org/10.1016/j.rse.2020.112108.

Zhou, Q., Liu, S., and Hill, M.J. 2019. A novel method for separating woody and herbaceous time series. *Photogrammetric Engineering & Remote Sensing*, 85: 509–520. https://doi.org/10.14358/PERS.85.7.509

Zhu, Z., and Woodcock, C.E. 2012. Object-based cloud and cloud shadow detection in Landsat imagery. *Remote Sensing of Environment*, 118: 83–94.

Zhu, Z., and Woodcock, C.E. 2014. Continuous change detection and classification of land cover using all available Landsat data. *Remote Sensing of Environment*, 144: 152–171.

Zhu, Z., Woodcock, C.E., and Olofsson, P. 2012. Continuous monitoring of forest disturbance using all available Landsat imagery. *Remote Sensing of Environment*, 122: 75–91.

Zhu, Z., Zhang, J., Yang, Z., Aljaddani, A.H., Cohen, W.B., Qiu, S., and Zhou, C. 2020. Continuous monitoring of land disturbance based on Landsat time series. *Remote Sensing of Environment*, 238: 111116. https://doi.org/10.1016/j.rse.2019.03.009.

Zika, M., and Erb, K.-H. 2009. The global loss of net primary production resulting from human-induced soil degradation in drylands. *Ecological Economics*, 69: 310–318.

Zomer, R.J., Xu, J., and Trabucco, A. 2022. Version 3 of the global aridity index and potential evapotranspiration database. *Scientific Data*, 9: 409. https://doi.org/10.1038/s41597-022-01493-1.w

Part II

Disasters

6 Disasters

Risk Assessment, Management, and Post-disaster Studies Using Remote Sensing

Norman Kerle
University of Twente

ACRONYMS AND DEFINITIONS

AA	Anticipatory actions
AVHRR	Advanced very high resolution radiometer
CEMS	Copernicus emergency management service
CNN	Convolutional neural networks
DEM	Digital elevation model
DMC	Disaster monitoring constellation
DNSS	Defense navigation satellite system
DRM	Disaster risk management
EMS	European Macroseismic Scale
EO	Earth observation
FEWS	Famine early warning system
GEO-CAN	Global Earth Observation-Catastrophe Assessment Network
GIS	Geographic information system
GOES	Geostationary weather satellite
GPS	Global positioning system
HAPS	High altitude pseudo satellite
HOT	Humanitarian OpenStreetMap Team
LiDAR	Light detection and ranging
MACS	Modular aerial camera system
ML	Machine learning
MSS	Multispectral scanner
OBIA	Object-based image analysis
OMI	Ozone monitoring instrument
SAR	Synthetic aperture radar
SfM	Structure from motion
SMCE	Spatial multi criteria evaluation
TIROS	Television infrared observation satellite
TOMS	Total ozone mapping spectrometer
TRMM	Tropical rainfall measuring mission
UAV	Unmanned aerial vehicles
VGI	Volunteered geographic information
VHRR	Very high resolution radiometer
VIIRS	Visible infrared imaging radiometer suite

DOI: 10.1201/9781003541417-8

6.1 INTRODUCTION

The number and consequences of natural disasters have shown dramatic developments in the past decades. The number of disaster events rose from annually less than ten in the first decade of the 19th century, to about 30 per year by the early 1960s. From then onwards, in the 60-year timeframe that is the focus of this book, numbers first rose rapidly to more than 500 events by the year 2000, but in the following 12 years showed a less clear, and at times declining trend (Figure 6.1). The annual total economic damage figures are more erratic, on one hand also showing a strong increase since about 1970, but with increasing high inter-annual variation owing to individual, exceptionally costly events, such as Hurricane Katrina in 2005, or the 2011 Tohoku (Japan) earthquake and tsunami event that caused damage in excess of US$300 billion, the costliest event to date. Category 4 or 5 hurricanes now frequently exceed the US$100 billion mark (e.g., Harvey [2017, 125 bn], Ian [2022, 112 bn]). However, all damage statistics must be considered with care. For one, to be included in the CRED EM-DAT database (CRED 2023) that is the source of the preceding numbers, events either need to cause at least ten fatalities, affect at least 100 people, or lead to a state of emergency or a call for international assistance, an arbitrary definition. Furthermore, with reporting, especially of smaller, more local events getting increasingly sketchy the further back we look (or indeed still today in more remote places, or those suffering from weak governments or from conflicts), the aforementioned numbers suffer from variable reliability and completeness, as is true for all such event databases (Mazhin et al. 2021; Panwar and Sen 2020). It is nevertheless interesting to note that the time frame of this book appears to coincide with a new, more intense, epoch of natural disaster occurrence. The increase clearly reflects global population numbers that also started to rise rapidly around 1960 (Kerle and Alkema 2011), and a strong growth in wealth and thus the basis for economic damage. The trend also owes to growing transparency and interest in global affairs, better reporting, and clearly also remote sensing has played a role. While disasters have been a focus area as long as remote sensing has existed, in particular the advent of space-based remote sensing in the 1960s (military) and 1970s (civilian) has led to better technical means to detect and asses disaster events, including of events national rulers would have preferred not to be seen by the world

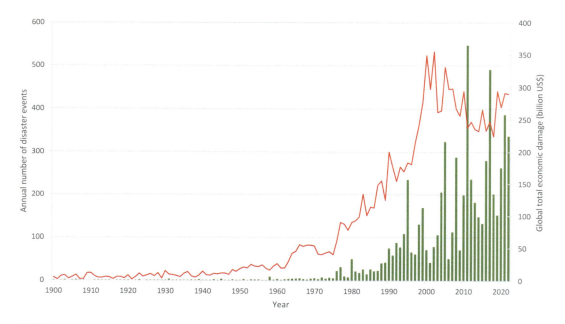

FIGURE 6.1 Number of annual natural disasters between 1900 and 2001 in bold (primary axis), and total annual economic disaster damage for the same period in hatched line (secondary axis; CRED 2023).

(e.g., earthquake or flooding disasters in China in the 1970s, or the Chernobyl nuclear power plant disaster in 1986). Hence, remote sensing has been playing a pivotal role in allowing increasingly informed disaster risk management, and in achieving a global inventory of disaster events.

There has been enormous progress in remote sensing in the last 60 years that has vastly benefited disaster risk management (DRM), the summary term used here to address risk and management of, and response to disasters. These developments are the focus of this chapter. On one hand this includes the many technical advances made—new and better sensor types and platforms, the move from analog to digital data capture and distribution, as well as better computing resources and software (Izumi et al. 2019; Sarker et al. 2020). But better data have also led to an equally fundamental shift in our understanding of what causes disasters, and has led to new ways to assess and quantify risk, to anticipate events with disaster potential and provide early warning, but also to respond rapidly to damaging events. In each of the previous categories a host of conceptual and methodological advances can be identified. They are flanked by significant organizational developments that include better international cooperation (leading to global monitoring and response systems), but also a growing involvement of civil society and laypeople (Noran 2014; Lassa 2018).

This chapter first introduces several key DRM terms and concepts before discussing selected domain developments, and the role remote sensing has played in those. The second part of the chapter details a range of relevant trends and developments, as well as remaining gaps and limitations where more work is needed. The chapter only addresses natural risks and disasters, excluding technological events or industrial accidents, as well as complex humanitarian emergencies linked to political or ethnic conflicts.

6.2 FROM HAZARDS TO DISASTER RISK—TERMS AND CONCEPTS

The very use of the term *disaster risk management*, and that it has largely replaced the more historic term *disaster management*, is evidence of profound developments in our understanding of the nature of disasters, and our ability to deal with them. Until well into the 20th century, disasters were widely seen as the result of a violent nature that somehow had to be tamed, increasingly with engineering measures (Smith 2004, a mindset that still frequently appears today). Events that did occur were responded to and managed. The middle of the century saw a fundamental shift in perspective, one marked by political ecology views (Kerle and Alkema 2011). Those not only considered nature as a hazard, but added social, economic, and political elements to the equation in an attempt to understand why nature and society intersected disastrously at ever shorter intervals. This led to the concept of vulnerability, meaning the capacity for loss or damage to people, systems, or material assets (Blaikie et al. 1994; White 1974). Hence, it became clear that, for a disaster to occur, a number of factors had to be in place: a hazardous event or process, and so-called elements at risk (EaR) that are located in the area exposed to the hazard, and that are vulnerable to the hazard in question, at a given magnitude. This can be summed up in a risk equation that typically takes the form of: Risk = Hazard x Value x Vulnerability, the latter two terms referring to all EaR present. This provides the basis for a quantitative assessment of risk in the form of expected losses per time period considered, typically per year. The hazard is thus treated as a process or phenomenon of a certain type, magnitude, and probability of occurrence, while the value and vulnerability are, respectively, the total economic value of a given EaR, and the fraction that is expected to be lost in case of the hazard event being considered. For example, if an event has a 10% probability of occurrence in a given year, and the capacity to damage an EaR by 10%, the expected loss, or risk, in that year is 1% of the total value of the EaR. Increasingly, vulnerability, defined as capacity for loss, is offset by coping or adaptive capacity (vulnerability ÷ capacity), which effectively reduces the risk (for more background on risk assessment see van Westen et al. 2011; Fuchs et al. 2019).

This conceptualization contains a number of aspects that are directly relevant from a remote sensing perspective: (1) all elements are spatial in nature, meaning that they have a geographic location and characteristics such as extent, proximity, or adjacency to other risk elements, and can thus be associated with attributes that are linked to a geographic place or area; (2) for each EaR a number of characteristics must be known (e.g., location, size, and occurrence interval of a hazard, or the location, type, and vulnerability of an EaR), and for the assessment of many of those remote sensing–based approaches have been developed; and (3) a careful match of the specific risk element with a suitable remote sensing data type or processing method is needed. At the same time, though, remote sensing is versatile and can address various needs. DRM is often conceptualized as a cycle that includes mitigation, preparedness, and the disaster event, followed by response, recovery, and rehabilitation, which then leads again to mitigation and preparedness (Figure 6.2). However, the cyclical conceptualization has been increasingly questioned in recent years. This is because the simplistic views suggests an inevitability of a disaster eventually being followed by the next one, while omitting the role of disaster risk reduction measures (Bosher, Chmutina and van Niekerk 2021). A depiction of DRM as a spiral or a helix that implies a continuous process and that can lead to a largely disaster-free situation, or at least one with a substantial reduction in the damaging consequences, is thus seen as preferable. The value of a DRM depiction that focuses on both the distinct aspects of risk and how they can be influenced but that also considers dynamic developments, such as urbanization, allows for more targeted risk reduction measures to be identified (Michellier et al. 2020). However, no clear consensus within the community exists yet. It is further important to note that remote sensing–based methods can often benefit different phases in the cycle. For example, methods designed to map EaR (buildings, infrastructure, etc.), can be adapted to map post-event structural damage.

Since the first edition of this book (2015) significant changes have taken place, both on the technical and the organizational side of DRM. Those include a drastically increased sophistication and availability of unmanned aerial vehicles (UAVs/drones), machine learning (ML), cloud-based platforms for big data processing, or publicly available spatial decision support systems for DRR, but also an increase in commercial and civilian actors using geodata in the disaster field, and not least growing attention on anticipatory actions (AA) that aim at reducing the consequences of an impending disaster event. Those developments are now also addressed in this chapter.

6.3 DOMAIN DEVELOPMENTS AND SELECTED TECHNICAL ADVANCES

6.3.1 Early Disaster Mapping with Remote Sensing

The unique perspective afforded by remote sensing quickly led to many application areas to be identified. While initially still too limited in terms of available and suitable instruments and their deployability to provide a useful tool to respond to disasters, this soon changed. Early DRM was

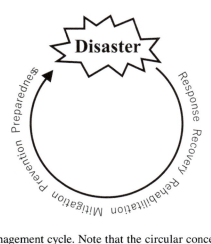

FIGURE 6.2 Disaster risk management cycle. Note that the circular concept has been criticized, as it does not reflect post-disaster risk reduction measures. Spiral- or helix-shaped alternative visualizations have been proposed, but a consensus within the DRM community does not yet exist.

Disasters

largely limited to providing information following disaster events. Not counting balloon-based visual assessment of damage caused by military action during the US Civil War in the 1860s, the first documented deployment of an airborne sensing instrument following a natural disaster occurred in 1906, when George R. Lawrence launched a 20 kg camera on a series of kites some 600 m above San Francisco to map the damage caused by a severe earthquake only days earlier (see also O'Rourke et al. 2006). Developments were subsequently spurred primarily by World War I, leading to a situation where aerial imagery would be routinely acquired, including for environmental damage assessment. As early as 1919 an image-based forestry mapping program was established in Canada, with the first airborne thermal infrared images being acquired in the same year. Methods and scope of operations continuously improved until the field of remote sensing changed fundamentally in 1960, when the first satellite-based image was successfully taken. This marks the beginning of the period reviewed in this chapter.

6.3.2 Dawn of the Satellite Era

Technological developments for military purposes, which continue to be a key driver in Earth observation (EO), and geoinformation science in general, led to the acquisition of the first satellite-based images in 1960. At the height of the Cold War, the Corona program, declassification of which finally began in 1995 and only finished in 2002, saw the launch of nearly 150 satellite missions aimed at image capture over areas relevant for the national security of the United States. For DRM purposes, Corona, as well as sister programs Argon and Lanyard, were primarily relevant insofar as they advanced satellite and sensor developments, since the image acquisition was aimed at areas of military relevance, and data were not made available outside the intelligence community. However, following declassification the data acquired by Corona over its 12-year mission duration have been used for a variety of studies related to DRM. Those range from geological mapping to population growth studies (Dashora, Lohani and Malik 2007), time series analysis of glacier retreat (Narama et al. 2010), volcanic hazard assessment (Karakhanian et al. 2003), or deforestation mapping for landslide hazard assessment (Kerle, de Vries and Oppenheimer 2003). The majority of the images were taken with black-and-white film (with a few being taken in color or infrared), with later missions delivering images with spatial resolutions of up to 1.8 m (Galiatsatos, Donoghue and Philip 2008). It took three decades for satellite images of comparable resolution to become available for civilian users. The data had two additional advantages. The film strips covered a very narrow (14 km) but also very long (188 km) corridor (Figure 6.3), explaining why often also areas that do not seem to be of much military relevance were imaged. Some of the Corona missions also carried stereo cameras, the data from which, following declassification, have been used to create digital elevation models (DEMs). Especially prior to the more widespread availability of detailed Light Detection and Ranging (LiDAR) data or images from modern high-resolution stereo satellites, the resulting DEMs with resolutions approximately ranging between 9 m (Schmidt, Goossens and Menz 2001) and 17 m (Galiatsatos et al. 2008), were a valuable data source, for example, to study glacier hazards (Lamsal, Sawagaki and Watanabe 2011).

The study of hydrometeorological hazards, such as tropical storms, benefits tremendously from Earth observation. First experiments with spaceborne meteorological observations and measurements also date back to 1960, the beginning of the Television Infrared Observation Satellite (TIROS) program. The data from these missions were unclassified, making TIROS the first civilian EO satellite, though providing images with a rather low spatial resolution of less than 1 km. This continued when the satellites started to be named NOAA in 1970, which, after initially operating the Very High Resolution Radiometer (VHRR), since 1978 have been flying the Advanced Very High Resolution Radiometer (AVHRR). The series continues to this day, providing images with a spatial resolution ranging from 1.1 km (at nadir) up to a maximum of 3.8 km for peripheral pixels (https://www.eumetsat.int/avhrr). The visible and infrared data are particularly useful for regional analysis and monitoring of global cloud cover, sea surface temperatures, as well as ice, snow, and vegetation cover characteristics, and hence of relevance for a variety of hazards. In particular the monitoring

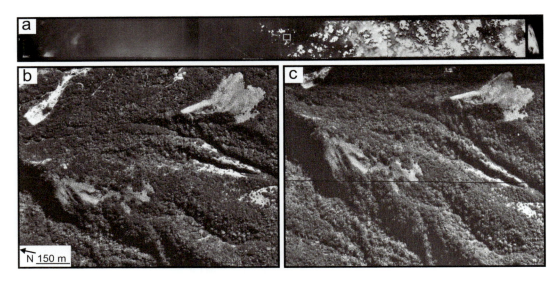

FIGURE 6.3 Example of a Corona satellite photograph from 1967 of parts of Casita volcano, Nicaragua, covering an area of 14 x 188 km, and part of a stereo pair (a). Close-up of the two stereo scenes to illustrate the amount of detail visible (b and c). Approximate location shown by box in (a).

of slow-onset events affecting large regions, such as droughts, can be effectively done with AVHRR data (Adedeji et al. 2020; Ansari Amoli, Aghighi and Lopez-Baeza 2022). Also impending hazard events can be detected: TIROS III was the first satellite to capture images of a developing hurricane (Esther in September 1961; Perlroth 1962), and such data have since then been of vital importance both to understand atmospheric hazards, and to provide early warning. Despite their modest spatial resolution the NOAA satellites became valuable due to their high temporal resolution, broad swath width (nearly 2400 km), as well as additional provision of infrared information (Hastings and Emery 1992). The daily coverage and moderate spatial detail started to allow the monitoring of hazardous events such as large floods (e.g., the Mississippi flood of 1973), tropical storms, as well as of sea ice dynamics (Space Science Board—National Research Council 1974). One of the NOAA satellites also provided first repeat coverage of a large volcanic eruption (St. Augustine, Alaska, in 1976).

Spatial resolution of the NOAA satellites was improved by an order of magnitude by the launch of ERTS-1 (Earth Resources Technology Satellites) in 1972, which marked the beginning of the Landsat series, the longest and most significant civilian EO program. The spatial resolution was initially limited to 100 m for national security reasons (Ramachandran, Justice and Abrams 2010). The principal instrument on ERTS-1 was the four-channel multispectral scanner (MSS) that had a nominal spatial resolution of 80 m (Mika 1997). The same Mississippi flooding observed by the NOAA-2 satellite, and that lasted approximately three months, was mapped in 16-day intervals, and with considerably more detail, by ERTS-1. The satellite operated for nearly six years, and the successful MSS sensor continued to be deployed on the follow-up missions, up to the launch of the currently operating Landsat-8 in 2013. While the daily coverage of the VHRR system allowed for monitoring of dynamic events, ERTS-1's better spatial resolution started to allow the more detailed study of spatial phenomena related to natural hazards. This included the mapping of volcanic areas and other major geological units (Space Science Board—National Research Council 1974), as well as geological evidence of seismic activity, such as fault lines (Gedney and VanWormer 1973). The lifetime of ERTS-1 also coincided with two major earthquakes, the 1972 magnitude 5.6 event that destroyed much of Managua, Nicaragua, and the magnitude 7.0 earthquake in August 1973 near Mexico City. The ERTS-1 satellite was rapidly programmed to obtain data of both events (Carter and Rinker 1976), marking the beginning of dedicated post-disaster surveillance with satellite technology.

The year the lifespan of ERTS-1 ended, 1978, saw another remote sensing milestone with the launch of SeaSat, the first synthetic aperture radar (SAR) instrument. Though operating for less than six months, it laid the technical and application foundations for subsequent radar missions. The high spatial resolution images (ca. 25 m) were primarily aimed at answering oceanographic questions. However, they were also used for terrestrial studies, including those related to DRM, such as changes in forest cover (Hoffer and Lee 1989), or to study volcanic features (Mackenzie and Ringrose 1986), and also its potential to address agricultural disasters was explored (Howard, Barrett and Heilkema 1978). The ability of satellite radar data to map inundated areas was also realized (National Research Council (U.S.) Committee on Remote Sensing Programs for Earth Resource Surveys 1977), which was to become the most commonly used application area for all later SAR instruments.

Several other technical developments of relevance to DRM occurred in the 1970s. In 1975 the first geostationary weather satellite (GOES-1) was launched, providing hemispheric meteorological data. The GOES satellites were later joined by comparable instruments from Japan, Russia, Europe, India, and China, to provide global coverage, and collectively proving data that are critical for the study of atmospheric hazards. In 1978 the polar orbiting Nimbus 7 satellite was launched that carried the Total Ozone Mapping Spectrometer (TOMS), which collected global ozone data until 1994, and allowed the detection of the ozone hole (Herman, McPeters and Larko 1993), a global health hazard. The 1970s also saw the groundwork being laid for the Defense Navigation Satellite System (DNSS), later renamed NAVSTAR and subsequently Global Positioning System (GPS). Though not a remote sensing system, it has nevertheless been playing a critical supporting role in the use of EO data in nearly all DRM application fields (Kerle 2013a).

The 1960s and 1970s were thus the pioneering decades in spaceborne remote sensing, and included many milestones in optical and radar EO, developments that fundamentally and lastingly defined the principles of monitoring and mapping of the Earth's land, oceans, and atmosphere. All subsequent spaceborne missions can be characterized by continuity and refinement of the early missions detailed previously. The growing amount of instruments in orbit, and more rapidly available data of a range of spatial, spectral, and temporal resolutions, led to hazard and disaster research, making increasing use of such data. Two additional developments advanced this process: the launch and operation of EO satellites by countries other than the United States, and later also by private companies, both developments with profound effects for EO and DRM. Already in 1979 India launched the polar orbiting Bhaskara-I experimental satellite, a first step that later morphed into one of the most extensive and successful EO programs in the world. In 1986 SPOT-1 (Satellite Pour l'Observation de la Terre) was launched by France, providing images with a ground-breaking 10 m (pan-chromatic) resolution. This was followed by Japan's first EO satellite (MOS-1) in 1987. The US Land Remote Sensing Policy Act of 1992 provided the basis for a development that has changed the remote sensing field in equal measure. For the first time private companies in the United States were allowed to build and operate civilian remote sensing satellites, leading to the launch of OrbView-1 in 1995, the world's first commercial EO satellite. This led to a race to provide ever more sophisticated imagery, but also undermined governmental EO efforts, and hence hurt data acquisition continuity. It has also been shifting the focus to targets that are commercially valuable, which carries the risk that fewer data are acquired of areas for which no commercial market exists.

6.3.3 Disaster Risk as a Multi-Faceted Spatial Phenomenon, and the Role of Operational Remote Sensing

As explained earlier, disaster risk is understood to be a function of a hazardous process or phenomenon that spatially intersects with vulnerable EaR. With increasing availability of remote sensing data of various types, as well as better geographic information system (GIS) tools and advanced spatial modeling, a fundamental shift has occurred from detection and mapping of hazards or disaster events, toward more comprehensive analysis of disaster risk. In the following sections the significance of remote sensing for the different aspects of DRM is discussed in more detail. Table 6.1 provides definitions for major elements of DRM, and aspects that determine the utility of remote sensing data.

TABLE 6.1
Definition of Key DRM Terms, and Aspects That Determine the Utility of Remote Sensing

DRM Parameter	Definition	Aspects That Determine the Use of EO Data
Hazard	Two distinct ways of defining a hazard exist. It can be seen (1) as a potentially damaging physical event, phenomenon, or human activity that may cause loss of life or injury, property damage, social and economic disruption, or environmental degradation; or (2) the probability of occurrence within a specified period of time and within a given area of a potentially damaging phenomenon. The types of natural processes that can cause damage are manifold and can be broadly grouped into hydro-meteorological and geophysical. Hazards are typically assessed using one of three approaches: heuristically (based on expert knowledge), statistically (based on evaluation of past events and causative factors), and deterministically (based on physically-based process modeling).	Hazards vary tremendously in terms of spatial extent, duration, frequency, presence of direct and indirect indicator, as well as the nature of the indicators (e.g., physical, chemical), with profound consequences for the utility of EO data in terms of their spatial, spectral, and temporal resolutions. The value of image data also strongly depends on the type of hazard assessment method. The main contributions of EO data are the identification of past hazard events (e.g., floods, landslides) and the characterization of relevant parameters (e.g., slope angle, land cover).
Disaster	A serious disruption of the functioning of a community or a society causing widespread human, material, economic, or environmental losses that exceed the ability of the affected community or society to cope using its own resources, and necessitating external assistance. Thresholds for what constitutes a disaster vary. For inclusion in the EM-DAT database one of the following are required: at least ten fatalities, at least 100 people affected, or declaration of a state of emergency or a call for international assistance. We can further distinguish slow- and fast-onset events, and while disasters are typically of limited duration, some event types (e.g., drought, desertification, or even climate change) that can also be considered to be disasters can last years or decades.	Disasters come in many forms and lead to highly variable consequences. Some are physical, and may thus be directly detectable in image data (see damage). Damage to systems and processes may be detectable through proxies (e.g., reduction of ships in a harbor). It is important to distinguish between an area being affected by a hazard event (e.g., through flood extent) and actual damage.
Risk	The probability of harmful consequences, or expected losses (deaths, injuries, property, livelihoods, economic activity disrupted, or environment damaged), resulting from interactions between (natural, human-induced, or man-made) hazards and vulnerable elements at risk.	Risk cannot be directly detected from remote sensing data, though the principal factors determining risk—hazard and vulnerability—can be assessed (see entries for Hazard and Vulnerability).

Disasters

Element at risk	Population, properties, economic activities, including public services, or any other defined values exposed to hazards in a given area, also referred to as "assets." The amount of elements at risk can be quantified either in numbers (of buildings, people, etc.), area, in monetary value (replacement costs, market costs, etc.), or perception (importance of elements at risk).	EO data are very well suited to assess and characterize elements at risk, in particular, physical ones. This includes using high spatial resolution imagery (optical images, radar, LiDAR) to identify and characterize (e.g., type, size, height) infrastructure elements (e.g., buildings, bridges), but also natural entities (wetlands, national parks, etc.). Systemic elements such as networks (road/infrastructure) can be identified based on the detection of their constituting components. Some not directly visible elements (e.g., economic activity) can be detected using physical proxies.
Vulnerability	Capacity for loss or damage to people, systems, or material assets, typically ranging from 0 (no loss) to 1 (complete loss). Since damage can be inflicted on physical elements at risk as well as systems and processes, different types of vulnerability exist. Typically considered in DRM are physical, social, environmental, and economic. However, also the vulnerability of other systems, such as political or institutional, can be considered. Vulnerability should always be assessed as a function of a given hazard type and magnitude.	Assessing how a physical entity or system will fare in a specific hazard situation is challenging, in particular considering a range of scenarios, and the utility of EO data is often limited. For different building types vulnerability curves are created based on the relationship of hazard intensities and observed damage. Image data can then be used, within limits, to identify buildings of a given type and vulnerability. For non-physical or process vulnerabilities physical proxies can be extracted from image data (e.g., physical state of neighborhoods, road networks, or model bottlenecks).
Resilience	"Capacity of a system, community or society potentially exposed to hazards to adapt, by resisting or changing in order to reach and maintain an acceptable level of functioning and structure" UNISDR (2004). At times resilience is seen as the inverse of vulnerability.	Direct measurement with EO data is not possible. However, by measuring recovery (both of physical assets and systems/functions) from image parameters, resilience of an affected area can to some extent be assessed.
Damage	Impairment or harm inflicted on an element at risk by a hazardous event or phenomenon, and can be either physical or functional. For different hazard types different damage scales have been created, such as the European Macroseismic Scale (EMS-98), that classes building damage from D1 (none or negligible damage) to D5 (total collapse). For system-related damage also a percentage of capacity or performance reduction can be used.	EO data of many different types have been used to detect and characterize disaster damage, in particular for physical assets. Focus has been on building damage mapping, in particular with very high-resolution optical satellite images, LiDAR, high-resolution SAR, and airborne oblique imagery. Detailed, per-building assessment, especially on intermediary damage, remains difficult. EO are very suitable for other damage types (e.g., forests, wetlands), while methods continue to be largely lacking for systemic damage assessment
Recovery/Reconstruction	Decisions and actions taken after a disaster with a view to restoring or improving the pre-disaster living conditions of the affected community, while encouraging and facilitating necessary adjustments to reduce disaster risk.	Image data are well suited to assess physical post-disaster reconstruction, also using semantic or ontological analysis methods to account for reconstruction that can occur in different ways. Assessment of system or process recovery (e.g., economic activity, ecosystem performance) with image data can be done via physical proxies, though fewer methods exist.

6.3.3.1 Hazard Assessment with EO Data

Apart from the multitude of man-made hazards that are not the focus of this chapter, many processes in nature can be harmful to people, the infrastructure they build, and the systems (e.g., environmental, social, economic, political, transport) they create. Those hazards can be characterized by their origin and extent, as well as their rate of occurrence and intensity/magnitude. A common approach to assess those parameters is via a frequency-magnitude relationship that is based on the realization that low-magnitude events occur more frequently than larger events. A frequency-magnitude analysis rests on two assumptions: that the history of events of a given type in a given area is well known (e.g., all flooding events to affect a given area, and their magnitude as expressed by duration and spatial extent), and that the processes that led to the hazardous events remain largely unchanged, and that thus the past can be seen as a guide to the future. This type of analysis is an indicator of the probability of occurrence of a hazard of a given type in a given area, and thus critical for risk assessment. A frequency analysis requires knowledge of the existence of the hazard in the first place, something remote sensing frequently provides. For example, Landsat data were used to identify previously unknown volcanoes in the Andes (Francis and De Silva 1989). Similarly, airborne laser scanning (LiDAR) data, with the ability to survey the ground beneath a forest, have been used to identify landslides through the (revegetated) scars they leave (Razak et al. 2013; Van Den Eeckhaut et al. 2012), thus allowing a more complete hazard assessment than would otherwise be possible. Unfortunately the record for LiDAR and other airborne data typically flown for dedicated missions also remains patchy for many parts of the worlds, and repeat datasets are rare.

A remote sensing archive of some 60 years now exists for many parts of the world, useful in particular to create inventories of larger events, such as flooding. Flood frequency assessment has been carried out with optical images (e.g., Huang et al. 2012) as well as radar data (Hoque et al. 2011). Similar time-series analyses have been done to understand drought frequencies (e.g., Heumann et al. 2007; Ghazaryan et al. 2020) and trends in desertification (Dardel et al. 2014; Rivera-Marin, Dash and Ogutu 2022), though especially for studies related to climate change the extent of our archive, and how reliably it allows the detection of frequencies and trends, has been called into question (Loew 2014). More recently, however, in particular long-term observations of the oceans have revealed mounting evidence of climate change (Pisano et al. 2020; Kulk et al. 2020).

Given its rich archive that has been expanding since the launch of the first satellite in the series in 1972, Landsat data have been employed in a range of hazard-related time-series studies, ranging from glacier-lake expansion (Nie, Liu and Liu 2013) to multi-decadal wild fire frequency analysis (Oliveira, Pereira and Carreiras 2012) or changes in wildfire hazard over time (Quintero et al. 2019). For events of lower spatial extent, and thus requiring data with higher spatial resolution, a shorter record exists, and assessments into the occurrence of past hazard events are limited to more recent times. Nevertheless, remote sensing data have been instrumental in the assessment of smaller events, such as landslides, including the dynamics of continuously developing mass movements (e.g., Martha et al. 2012). Also the temporal behavior of phenomena that constitute no direct hazard to people but to their livelihood, such as erosion, has been effectively analyzed with multi-temporal data (Shruthi et al. 2015). Table 6.2 provides a comprehensive overview of the utility of different types of remote sensing data to assess a range of hazards, as well as other elements of the DRM cycle.

6.3.3.2 Hazard Assessment—A Multi-Faceted Problem

Talking of hazards in a broad sense—e.g., *the* flood hazard, or *the* volcanic hazard—may be meaningful when describing general situations, or when communicating with decision-makers or potentially exposed people. From a risk assessment and planning perspective a more nuanced approach is needed, since those broad hazards, in a strict sense, do not exist. Flooding comes in many different forms (riverine, coastal, flash flooding, among others), each with a unique behavior, and each requiring a different approach for their assessment and monitoring. Likewise, rock falls and debris flows are both gravity-driven mass movements, but both their genesis and behavior of movement differ dramatically. *The* volcanic hazard also does not exist; instead we can distinguish about 20 different

Disasters 163

TABLE 6.2
Utility of Remote Sensing for Different DRM Aspects of Different Natural Hazard Types

Hazard type	Subtype	Hazard Assessment		Prevention/[1] Mitigation		Monitoring/ Early Warning		Syn-event Monitoring		Damage Assessment		Recovery/ Rehabilitation
Flood	Riverine	++	HRV, DEM, LiDAR	+	HRV, DEM, LiDAR	++	Met	++	HRV, MRV	++	(V)HRV	++	(V)HRV, LiDAR
	Flash flood	+	HRV, DEM	+	HRV, DEM	+	Met	–		+	VHRV	+	VHRV
	Coastal	++	MRV, HRV, DEM	+	MRV, DEM	+	Met	+	MRV	++	MRV, HRV	++	MRV, HRV
Tsunami		–		–		–		–		++	MRV, (V)HRV	++	MRV, (V)HRV
Storm[2]	Regional (cyclones)	++	Met	–		++	Met	++	Met	++	(V)HRV, Obl	++	(V)HRV
	Local (tornado)	+	Met	–		+	Met, DR	+	Met, DR	++	(V)HRV, Obl	++	(V)HRV
Earthquake		+	SAR	–		+[3]	SAR	–		++	IR	+	IR
Drought		++	IR, Met	+	IR, Met	++	IR, Met	++	IR	++	(V)HRV	+	MRV
Volcanic	Magmatic activity	+	HRV, SAR, TRS	–		++	TRS	++	TRS	++	MRV, HRV, IR	++	MRV, HRV, IR
	Lahar (mudflow)	+	HRV, DEM	+	HRV, DEM	+[4]	Met	–		+[5]	IR	+	IR
	Gas emission (local)	+	Hyp[6]	–		+	TRS, Hyp	+	Hyp	+[7]	Met	+	Met
	Gas/ash emission (stratospheric)	+	Hyp	–		+	TRS, Hyp	++	Hyp				
Landslide	Crater lake breach	++	VHRV, LiDAR	–		+	Met, HRV	–		++	(V)HRV	++	MRV
	Fast slope failures	++	HRV, DEM, IR	+	HRV, DEM	+	Met	–		+	HRV	+	HRV
	Slow-moving slides	++	SAR, HRV, SD, LiDAR	+[8]	HRV, IR	+	Met	+	SAR, SD	+	HRV	+	HRV
	Re-vegetated slides	+	LiDAR	–		–		–		–			
Wild fire		++	IR, DEM, Met	+	IR, Met	++	TRS, Met	++	TRS, Met, MRV	++	MRV, HRV, IR	++	MRV, HRV, IR
Desertification		+	LRV, IR, Met	+	LRV, IR	+	IR	–		+[9]	LRV, MRV	+	LRV, MRV
Erosion		+	VHRV, LiDAR	+	VHRV	–		–[10]		+	(V)HRV	+	(V)HRV

(Continued)

Key: - limited or no use
+ moderate utility
++ high utility
Remote sensing data
VHRV—very high resolution visible (spatial resolution better than 1 m)
Obl—oblique imagery, often from UAVs
HRV—high resolution visible (spatial resolution > 1–10 m)
SD—stereo imagery
MRV—moderate resolution visible (> 10–30 m)

TABLE 6.2 (Continued)
Utility of Remote Sensing for Different DRM Aspects of Different Natural Hazard Types

LiDAR—airborne laser scanning
LRV—low resolution visible (> 30 m)
Hyp—hyperspectral imagery
TRS—thermal imagery
Met—meteorological[11]
SAR—imaging radar, interferometric
SARIR—infrared (typically a NIR band in optical sensors)
DR—Doppler radar

Notes:

1—mitigation or prevention in a sense of remote sensing data allowing strategies or activities to be identified that reduce the hazard or the probability of an event occurring
2—storms come in a large variety of forms and area affected, ranging from small water spouts to winter storms and tropical cyclones. Here two main categories in terms of area affected are discussed: tornadoes (local) and cyclones (which include tropical and extra-tropical/mid-latitude storms) with more regional range
3—through SAR interferometry revealing surface deformation indication seismic stress build-up; actual remote sensing–based detection of earthquake precursors continues to be investigated
4—e.g., through monitoring of crater lake volumes, rainfall amounts, or remaining amount of mobilizable ash deposits
5—through detection of vegetation damage caused by acid rain; health damage cannot be detection in image data
6—e.g., mapping of SO_2 concentrations with the hyperspectral (ultraviolet-visible range) Ozone Monitoring Instrument (OMI) on NASA's EOS AURA mission, or similar approaches based on absorption spectroscopy
7—large amounts of volcanic gases injected into the stratosphere can cause global cooling, which can be considered a form of damage, and which can be measured indirectly with weather satellites. Recovery would then be a return to pre-event temperatures
8—e.g., through soil moisture data allowing a better understanding of slide dynamics and triggers, allowing intervention and stabilization measures to be defined
9—in the sense of desertified areas
10—erosion progresses both continuously and through individual, high intensity rainfall events. Remote sensing can only monitor changes in erosion prevalence at different times.
11—including from geostationary meteorological satellites and from specific weather-related mission such as the Tropical Rainfall Measuring Mission (TRMM).

hazardous processes that are associated with a volcano, some of which relate to magmatic or eruptive activity, while others are typical for dormant or extinct structures. Some of those specific or sub-hazards tend to affect a small area on the volcanic edifice itself (e.g., volcanic bombs, or fumaroles), while others, such as gas and ash injected into the stratosphere, have global consequences. Some, such as seismic activity related to ascending magma, may last hours or days, while degassing can continue over decades or longer. A clear understanding of those specific processes is thus always needed before deciding on a suitable remote sensing strategy. This particularly requires clarity on how a specific hazard indicator manifests itself. For example, surface changes such as lava flows or dying vegetation require optical or infrared image data, while warming of a volcanic edifice or crater lake is detectable in thermal bands. Physical changes, on the other hand, such as surface uplift or the smoothing of a surface due to pyroclastic flow or lahar deposits are detectable in radar data.

Remote sensing research has made immense progress at this more detailed level that is the basis for actual decision-making and planning. For example, while flood hazards are typically assessed using 1D or 2D flood models (Kerle and Alkema 2011), remote sensing provides critical data on land cover (to estimate evapotranspiration and surface roughness; van der Sande, de Jong and de Roo 2003; Straatsma and Baptist 2008), as well as on soil type (to allow estimation of water infiltration capacity), and topography (e.g., Kraus and Pfeifer 1998). In particular the combination of multispectral information with airborne LiDAR data has allowed both a detailed characterization of the vegetation in flood-prone areas (which has a strong effect on flow behavior), as well as the underlying topography (Straatsma 2008; Geerling et al. 2009). The same multi-parameter assessment is done for wild fire hazard assessment and monitoring, as was reviewed by Chuvieco (2003). Remote sensing data are being used to assess the amount of burnable vegetation matter, its dryness, and the topography of the area, but also meteorological information is used to monitor atmospheric parameters such as wind and temperature that directly affect wild fire potential. In addition, LiDAR data, whose use for detailed vegetation mapping was already mentioned, also allow a better wildfire hazard assessment (Newnham et al. 2012). Radar data have also been shown to allow a detailed characterization of biomass, canopy height, or forest types (Balzter 2001; Choi et al. 2023). As something of an exception for hazards, a comparably comprehensive assessment for seismic hazard is less straightforward. Again, for detailed planning the seismic movement is only the starting point: the actual hazard to people and infrastructure is determined by geology, soil type and thickness, and the topography. Remote sensing data have allowed the mapping and characterization of fault lines (Ramasamy 2006), regolith thickness (Shafique, van der Meijde and Rossiter 2011), as well as detailed topography that informs the amount of site-specific ground amplification (Shafique, van der Meijde and van der Werff 2012; Tronin 2010; van der Meijde et al. 2020). In particular interferometric SAR (InSAR) data have allowed the assessment of fault line dynamics and crustal deformation (Hooper et al. 2012). Specific forms of InSAR support the detection of ground deformations over time, such as caused by seismic movements (differential InSAR/DInSAR; Markogiannaki et al. 2020), or focus on the detection of movement of stable points such as building corners (permanent scatterer/PSInSAR; Kothyari et al. 2022). In a similar fashion the relevant parameters for other hazard types are being assessed with remote sensing (see, for example, Joyce et al. 2009; Kerle 2013b).

Both flooding and (some) volcanic hazards are good examples of a distance effect that can be well addressed with remote sensing data. In both cases, but also for other hazard types, large distances may lie between the hazard source and the area where damage may be caused. The melt or rain water may originate in mountains far from settlements, yet pose a downstream flood hazard, as do very mobile mass movements, such as volcanic lahars that can reach distances of 50–100 km from the point of origin. Similarly, a remote volcano has the potential to alter weather patterns elsewhere on the globe, at times leading to a global temperature drop of 0.5°C or more, effects that can last for years. Remote sensing constitutes a vital information source to detect and characterize those hazard sources, and to link them to EaR. To that effect the information derived from remote sensing is frequently combined with models or other forms of GIS analysis, including to model hazards responding to climate change (Tsatsaris et al. 2021). Table 6.2 summarizes the utility of remote sensing for different part of the DRM cycle for a range of natural hazards.

6.3.3.3 Mapping of Elements at Risk

Anything of value, be it monetary, historical, sentimental, cultural, or otherwise, can get adversely affected by hazardous events or processes. Such EaR can thus be highly diverse, and range from the obvious—buildings or infrastructure—to features less frequently considered in a risk analysis, such as places of cultural or natural heritage, national parks, sites of high biological diversity such as some tropical forests or coral reefs, or even beaches or other coastal assets, such as cliffs. In particular natural sites of value to the tourism industry are frequently taken for granted, with their significance, and susceptibility to damage, only being appreciated after a disaster strikes and income is lost (see also Liu 2014). All of the above are primarily physical features. However, in DRM we can also consider valuable non-physical systems, processes or functions that may get affected by hazardous events, be it cultural diversity, political systems, or economic processes. Not surprisingly, remote sensing has been predominantly used to map and characterize physical EaR, in particular buildings and infrastructure (Ferro, Brunner and Bruzzone 2013). Those include physical assets exposed to flooding (e.g., Müller 2013), or specific building types. Vertical optical satellite images have been useful for building footprint detection. To provide extra information, such as the height of a structure, shadow analysis has been used (Lee and Kim 2013). Alternatively, data from instruments with stereoscopic capabilities, such as Ikonos, Cartosat-1, SPOT, or GeoEye, can be processed photogrammetrically (Poli and Caravaggi 2013). In particular the growing availability of LiDAR data has been a great asset for building detection, since it not only allows effective discrimination of vegetation that plagues other image processing methods, but also because it very readily allows the ground surface to be determined, resulting in building height (Sithole and Vosselman 2004; Chen, Gao and Devereux 2017). Since LiDAR data tend to be very detailed, with modern systems collecting dozen of points per m^2, they provide useful information that allows characterization of building types (Alexander et al. 2009), especially when adding information from high-resolution optical images (Huang et al. 2017), or of entire urban landscapes, for example, to support informed flood risk assessment and management (Talebi et al. 2014; Muhadi et al. 2020).

Detecting specific building types and similar forms of investigation go in the direction of semantic analysis. For decades image processing was primarily focused on using spectral information, which gives insight into the material exposed on the ground. However, many hazard parameters, as well as EaR, must be defined—or indeed can only be fully recognized—from their context, i.e., spectral information needs to be combined with other forms of information, be it topographic, topological, or contextual. In particular the detection of many EaR that are not clearly defined physical units must be approached in such a fashion. For example, parameters related to biodiversity in forests have been identified with remote sensing data in such a manner (Torontow and King 2011; Reddy 2021). Ontological frameworks that encode feature knowledge have also been used as a basis for image-based detection of specific urban units such as slums, which can also be considered as EaR in specific hazard scenarios (Kohli et al. 2012).

The success and utility of remote sensing is always a function of the data and how they are processed. Detection and characterization of EaR in particular has been benefiting from a shift away from pixel-based processing toward object-based image analysis (OBIA). Fundamentally a two-stage technique that combines image segmentation with subsequent classification, it is far more than that, with dramatic consequences for DRM, including the detection and characterization of EaR. The strength of OBIA is that it allows an effective incorporation of process and features knowledge, different data types, and its flexibility in handling 2D–4D data (the latter being multi-temporal data cubes). This has become an asset in the detection of specific hazards, such as different landslide types (Martha et al. 2010; Figure 6.4) or forms of erosion (Shruthi, Kerle and Jetten 2011), but also to characterize EaR. For example, Kohli et al. (2013) used the ontological framework mentioned earlier to provide the knowledge basis to detect slum units using OBIA. Refugee dwellings, associated with another highly vulnerable group of people, have been identified in image data in a similar way (Gao et al. 2022). Such knowledge-driven approaches are also the key to detecting more complex EaR, such as more complex processes (level of economic activity, or the state of a transport

Disasters

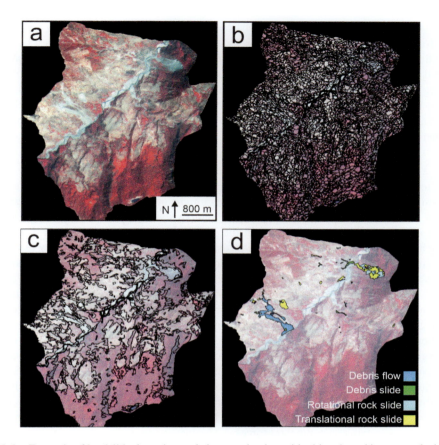

FIGURE 6.4 Example of landslide detection and characterization with object-based image analysis (OBIA), which combines image segmentation and feature or process knowledge–driven analysis of the resulting segments. Image (a) shows part of a Resourcesat-1 image (5.8 m multi-spectral resolution) of the Indian Himalayas (Okhimath) that experienced a variety of landslide types. The image was first finely segmented, with (b) and (c) showing the effect of different segmentation scales. Following the identification off all bare areas that may be landslides, a number of false positives (e.g., shadow, river sands, clear-cuts) were sequentially eliminated based on image spectral and textural characteristics, as well as information from a DEM. The resulting landslides objects were classified, again using knowledge of their different spatial and contextual characteristics, into different landslide types (d). For more details see Martha et al. (2010).

system), though research is only starting to address these questions. OBIA, for all its strengths, also has limitations. Segments can be formed in a number of ways, and > 100 features and characteristics are readily calculated for those units by software tools such as eCognition. However, which features (and in which combination) best identify a given feature or object class, what thresholds to use in the classification, etc., remain significant challenges. Here, computer vision and ML have strongly gained in influence, evident for example in the use of random forest approaches (e.g., Stumpf and Kerle 2011; Shruthi et al. 2014). In particular the ability of deep learning with convolutional neural networks (CNN) to identify and recognize features and patterns in potentially highly variable remote sensing data has massively accelerated the use of such models, including for the mapping and characterization of buildings (Han et al. 2022; Alidoost and Arefi 2018). The rapid rise of tech companies has also had unexpected consequences that are relevant here: for example, Microsoft and Facebook collaborated on creating a global database comprising about 1.3 billion building footprints extracted from Bing Imagery, which are being gradually augmented with building height estimates (https://github.com/microsoft/GlobalMLBuildingFootprints). However, semantic image analysis, for all its strengths, is no match for local knowledge, for example, when it comes to specific building use. Here collaborative image analysis methods, most prominently OpenStreetMap (OSM), fill an important gap. The efforts by more than 1 million mappers has led to a database comprising approximately 0.5 billion building footprints (Biljecki, Chow and Lee 2023), in addition to detailed local information. However, the case of the 2010 Haiti earthquake showed that detailed OSM data are often only generated in the aftermath of a disaster event (Ajmar, Boccardo and Tonolo 2011), when they would have already been useful in risk reduction and disaster avoidance.

6.3.3.4 Vulnerability Assessment

In simple terms, vulnerability is defined as the capacity for loss, i.e., how much damage a given EaR will sustain in a specific hazard scenario (type, magnitude, and duration). In practical terms vulnerability is a more subtle phenomenon, one where remote sensing can also help, but within limits. The concept of vulnerability has its roots in the social sciences of the 1970s and is a response to the purely hazard-oriented understanding of disaster risk at that time. Vulnerability assessment has been primarily applied to physical features—buildings and infrastructure—and how those may get affected by hazards, due to physical forces exerted by ground motion, water, wind, etc. Vulnerability ranges between 0 (no damage) and 1 (total loss). Within the scientific community definitions of vulnerability, and optimal ways to analyze and quantify it, continue to be discussed (see, for example, Birkmann 2007; Galderisi and Ferrara 2013). A number of characteristics make vulnerability assessment particularly challenging, which need to be considered when assessing the utility of EO data. (1) Vulnerability has different facets. The ones commonly considered today are physical, social, economic, and environmental (United Nations Office for Disaster Risk Reduction [UNISDR] 2004), yet an inclusion of other types, such as political, ecological, or institutional, is equally valid. (2) Vulnerability is dynamic, hence it changes over time. For example, as a building ages its vulnerability to certain hazards can increase, while a human being, as it grows from child to adult, may become less vulnerable to certain environmental forces. (3) Vulnerability is strongly scale-dependent, meaning that it can, and often must, be assessed on a scale that ranges from the individual (building, person, road, etc.) up to a city/country, community, or network level. (4) Vulnerability is a function of hazard type and magnitude. The same building that might withstand flooding well (low vulnerability) might suffer extensively during an earthquake (high vulnerability), and more so during a strong than a weaker tremor. Hence, in principle every EaR has as many vulnerabilities as the number of hazard types it is exposed to. Spatial Multi Criteria Evaluation (SMCE) is often used in vulnerability assessment, providing a way to include different vulnerability indicators. Those can be physical (e.g., building material), social (e.g., ethnic or age distribution), or related to capacity (e.g., distance to hospital, risk awareness levels). Those indicators are weighted, using a pairwise criteria comparison, arriving at a vulnerability index (for more details see van Westen et al. 2011).

To complicate matters further, the natural counter-weight to vulnerability is resilience, which must be known for a comprehensive analysis (Galderisi and Ferrara 2013). While the concept has also been applied to physical structures to describe their ability to bounce back after a shock, for communities it relates to the strengths and resources available that can collectively reduce the effect of a hazardous event. Hence, it relates to social, political, and organizational parameters, and the different conceptional origins of resilience and vulnerability have led some researchers to question their compatibility (Miller et al. 2010). There has also been a relevant conceptual development of resilience, one that reflects better the "build-back-better" concept that focuses on learning from previous disasters and allows for adaptation and transformation (Graveline and Germain 2022). For a more in-depth introduction to vulnerability, and its assessment with geospatial data, see van Westen et al. (2011), or Ghaffarian et al. (2018) for a detailed review of the role of remote sensing data.

The previous discussion explains why in geoinformation science, and in remote sensing in particular, the main focus to date has been on physical vulnerability. The key to its assessment lies in so-called vulnerability curves that express the relationship between hazard intensity and the resulting damage. (Note the focus on intensity; as stated in Section 6.3.3.2 the actual magnitude of an event, for example, an earthquake, is only the starting point. What matters in risk assessment is the actual location-specific hazard intensity.) Vulnerability curves are constructed for different types of structures (e.g., adobe brick buildings vs. reinforce concrete structures of different heights). We also distinguish between relative curves, which indicate what percentage of the value of a structure will be lost, and absolute curves that show the actual losses, since the value of the assets being considered is already incorporated in the analysis. Hence, from a remote sensing perspective physical vulnerability assessment is largely reduced to identifying EaR and determining their respective category, in terms of type, material, etc. For examples, see studies by Ehrlich et al. (2013) and Harb et al. (2015). The value of OBIA for EaR detection was already described. The concept is also useful to assess the physical vulnerability of the detected features (e.g., Wu et al. 2014; Englhardt et al. 2019). Remote sensing–based physical vulnerability data have also been analyzed for their accuracy and cost-effectiveness, with positive results (Torres et al. 2023).

Other types of vulnerability are no less important in comprehensive risk assessment, yet less progress has been made in terms of remote sensing–based methodologies. This is primarily due to the fact that a direct detection is typically not possible; instead, physical proxies must be employed (Ghaffarian et al. 2018). For example, social vulnerability can be defined as "people's differential incapacity to deal with hazards, based on the position of the groups and individuals within both the physical and social worlds" (Clark et al. 1998), which allows the definition of relevant physical indicators that can explain the social vulnerability of people living in a given place. The possibility of such an assessment based on satellite images, and also making use of OBIA techniques, was shown by Ebert et al. (2009). Similar approaches can also be used for environmental vulnerability assessment (see, for example, Petrosillo, Zaccarelli and Zurlini 2010; Poompavai and Ramalingam 2013; Kamran and Yamamoto 2023). To assess to what extent complex phenomena, such as economic or political systems, can be affected by hazardous processes, remote sensing data will be of less use; instead, more domain-specific modeling must be employed (such as macro-economic agent-based modeling to understand linkages in economic systems, and how adverse effects might spread within them; e.g., Kromker, Eierdanz and Stolberg 2008; Colon, Hallegatte and Rozenberg 2021; Schlögl et al. 2019).

6.3.3.5 Monitoring and Early Warning to Detect Potentially Hazardous Situations

Once hazards are understood in terms of their spatio-temporal characteristics, efforts can be made to mitigate them where possible (e.g., by giving rivers more flood plain space to spread into when needed, re-establishing mangrove forests for coastal protection, or by removing dry vegetation close to buildings to create a wildfire buffer). Another strategy is to monitor hazards. Due to the continuity provided by EO systems—as well as the availability of an archive for reference and benchmarking purposes—their data are exceptionally well suited to keep a permanent eye on many potentially hazardous situations, and provide advance warning in case of a looming threat. In particular for the detection of anomalies, and the automated routine processing of large amounts of data, machine learning has become increasingly important, for example, for the monitoring of volcanic activity (Romano et al. 2022) or to identify developing drought situations (Houmma et al. 2023). For every hazard type monitoring methodologies have been developed, and for nearly all of them remote sensing plays a major role. An exception is tsunamis, where buoy systems installed on the ocean floor are used. Also the image-based real-time monitoring of seismic activity remains limited. While interferometric SAR can be used to monitor surface deformation that can be related to seismic potential (Hooper et al. 2012), research is still ongoing to identify precursor signals of earthquakes that can be extracted from satellite data (Tronin 2010; Picozza, Conti and Sotgiu 2021). The focus here is less on image data, but instead on satellites that assess geomagnetic (Yao, Wang and Teng 2022) or ionospheric parameters (Budak and Gider 2023).

The principal challenge of hazard monitoring is to understand the natural variability of parameters linked to hazards, to be able to spot unusual developments. For example, keen process understanding is needed to pick up early drought indicators from within a potentially complex phenological cycle of a crop. Existing agriculture monitoring schemes based on remote sensing were recently reviewed by Weiss et al. (2020) and Khanal et al. (2020). Such data play a critical role in combating hunger in the severely drought-affected parts of the world, especially the region around the Horn of Africa. AVHRR data provide daily information, the use of which to link drought and crop yields was already shown many years ago by Unganai and Kogan (1998). Other systems that provide very frequent information of use for detailed vegetation state mapping are MODIS and SPOT Vegetation. Therefore, remote sensing is a core ingredient in operational drought monitoring systems such as FEWS (Famine Early Warning System Network; http://www.fews.net/), that has been in operation already since 1985. Remote sensing data are the key to successful monitoring of a variety of other atmospheric hazards, due to the long history of such instrumentation, the strength stemming from highly frequent geostationary observation (e.g., nearly 100 scenes per day of Africa and most of Europe from the current Meteosat satellite), coupled with better than daily and more detailed images from polar orbiters (e.g., AVHRR) and land-based instruments (Doppler radar), and rapid and wide availability of the data. As a consequence, in most parts of the world the developments of broad atmospheric hazards can now be monitored. Countries that also possess a detailed network of ground stations can go beyond that and also assess the developments of more local and short-lived phenomena such as tornados or flash floods. Indeed it was remote sensing—the study of Doppler radar data—that nearly 50 years ago allowed the signature fingerprint of developing tornadoes to be identified (Ray and Hane 1976), paving the way for modern early warning systems.

As explained earlier in this chapter, the suitability of remote sensing is tightly coupled with the specific characteristics of a given hazard. In broad terms, volcanoes are probably the most rewarding hazard setting for remote sensing to be applied to. This is because nearly all of the many subhazards previously mentioned emit some form of measurable signal prior to the commencement of a threatening process. The consequent potential for EO technology was quickly noticed, and concepts for satellite-based volcano surveillance appeared soon after the launch of the first civilian satellite, ERTS-1 (Endo et al. 1974; Cochran and Pyle 1978). The major detectable signs are (1) seismic activity that signals magma movement within the edifice (though this is primarily monitored with ground-based seismic instruments), (2) thermal signals on flanks or in crater lakes that indicate magma approaching the surface, (3) topographic changes due to magma movement or structural deformation that may lead to mass movements, and (4) gas emissions. As such, virtually all types of ground-, air-, and spaceborne EO instruments at our disposal today have been successfully deployed on volcanoes (for a more detailed review see, for example, Joyce et al. 2009; Tralli et al. 2005). In particular, satellites with hemispheric or global coverage are being used for operational volcano monitoring. For example, the MODVOLC system (http://modis.higp.hawaii.edu/) operated by the University of Hawaii used to provide global volcanic hot spot information based on MODIS imagery since 2003. Recently the monitoring was shifted to use data from the Visible Infrared Imaging Radiometer Suite (VIIRS; Campus et al. 2022). Other researchers have focused on the combined strength of different systems to optimize monitoring (Cochran and Pyle 1978; Murphy et al. 2013; Marchese et al. 2021). Satellite data are also a principal input to Volcanic Ash Advisories issued by the International Civil Aviation Organization to alert aircraft pilots to volcanic aviation hazards stemming from eruption clouds. Examples of recent remote sensing developments that have been benefiting the monitoring of volcanoes include PSInSAR for ground deformation detection, and the deployment of UAVs, for example, for in situ gas measurements (Karbach, Bobrowski and Hoffmann 2022) or for surface monitoring (Granados-Bolaños, Quesada-Román and Alvarado 2021).

A very recent development in early warning is to use remote sensing data not only to identify impending hazard events, but to provide actionable information to allow effective disaster preparedness, and to reduce losses due to the expected event, termed early (or anticipatory) action. The idea is to use remote sensing data to identify and monitor developing hazardous situations, most commonly related to hydrometeorological hazards, to couple those observations with models that not

Disasters 171

only predict developments such as storm tracks over the coming days, but also to model expected damages. Such impact-based forecasting allows early action to be taken, including timely evacuation, house-strengthening, or emergency cash provision (Nauman et al. 2021). The challenge is that not only near-real time image analysis results are needed, but those must be coupled with suitable forward-modeling and anticipatory damage estimation, which in turn requires readily available EaR and vulnerability data. In addition, capacity by relevant governmental agencies or the humanitarian sector must be available on the ground to provide the necessary rapid action (Sedhain et al. in press).

6.3.3.6 Post-disaster Response and Damage Assessment

Despite the many efforts to mitigate hazards, prepare for disasters, and to provide early warning, the reality shown in Figure 6.1 is that we need to deal with several hundred disasters worldwide every year. Due to site accessibility problems, potentially wide-spread damage, and health hazards that damage sites themselves pose for first responders, remote sensing has become an indispensable tool to provide first intelligence on the nature and consequences of a disaster. Given the very long history of image-based damage assessment that stretches back to Lawrence's efforts in 1906 (e.g., Figure 6.5a), and the many potentially useful tools and methods available, a very rich body of literature

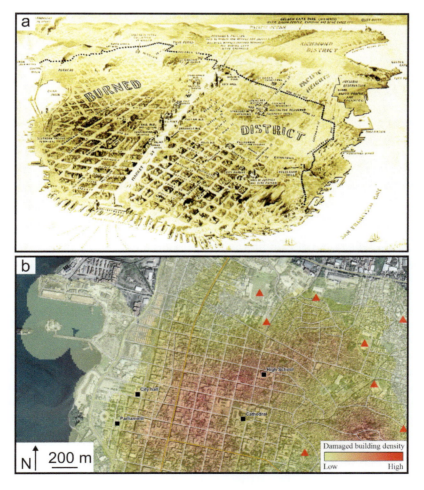

FIGURE 6.5 Damage map of San Francisco following the 1906 earthquake, based on kite-borne imagery by George R. Lawrence (a). Damage map of Port-au-Prince, Haiti, following the 2010 earthquake, prepared by SERTIT by visual interpretation of 50 cm resolution GeoEye imagery (b).

exists on this topic. Without a doubt decades of research have led to significant developments. At the same time a more sober assessment would lead to the following conclusion: remote sensing–based damage mapping has seen the least progress of all the DRM aspects reviewed in this chapter. The first edition of this book phrased it drastically: in some ways image-based damage mapping is done as it was in 1906 (Figure 6.5). To a large extent this description is still accurate, though technological advances, in particular concerning UAV and ML, but also experiments with citizen science/crowdsourcing, have changed the picture, as explained at the updated end of this section.

This negative statement on damage mapping requires qualification and explanation. Lawrence flew his kites at a time when no airplanes, let alone satellites, existed. In terms of technology, of course, there has been enormous progress, both on the sensor and platform side, and we have seen dozens of combinations of the two. Virtually any type of sensor in existence, be it optical, thermal or radar, mono or stereo, active or passive, has been fitted to satellites of different types, to airplanes large or small, or flying at low or stratospheric heights, or to balloons, kites or UAVs, in short any platform that can be brought into the air or space, and aimed at a disaster site, the typical operational altitude range for some of which is shown in Figure 6.6. The use of air- and space-based remote sensing for emergency response and damage mapping was extensively reviewed by Kerle et al. (2008, 2020) and Zhang and Kerle (2008; Matin and Pradhan 2022; Ge, Gokon and Meguro 2020), respectively, showing a rich palette of possibilities, as well as technical advances. This technology provides us with spectrally diverse images of high spatial resolution. Indeed, very significant progress has been made in terms of temporal and spatial resolution, and current delays to initial image acquisition after an event are measured in hours, instead of days or weeks. Nevertheless, those high-quality images continue to be primarily analyzed manually, as described in more detail later in this section.

Several issues require further explanation. First, we must be clear on what we mean by disaster response and damage mapping. Detecting disaster sites as anomalies, as a departure from the normal, tends to be quite straightforward, in particular when suitable pre-event reference data exist, and when the event leads to significant physical changes on the ground. However, caution must be used to separate evidence of an event from evidence of damage. For example, water extent provides a clear delineation of the reach of a flooding event and can be automatically detected in image data, yet it is not synonymous with damage. It makes sense here to distinguish between something being affected (by flooding, wind, etc.) and being damaged. From a damage assessment perspective it is also important to know if damage is permanent or not. Crops that suffer short-term flooding may survive, experience partial loss, or be completely destroyed (Rahman and Di 2020). Likewise, we can readily identify the extent of wild fires in remote sensing data. However, how specifically the vegetation fared, for example, if only brush was lost and trees largely survived, requires a more detailed analysis, or subsequent imagery to track recovery (Kurbanov et al. 2022).

When we talk about disaster damage mapping, almost invariably we talk about structural damage assessment, which is precisely where the main challenges lie. Ground-based structural damage mapping frequently makes use of damage categorizations such as the European Macroseismic Scale of 1998 (EMS98; Grünthal 1998), and image-based assessment tries to emulate that process. In terms of vagueness of the damage signal, the damage scale that ranges from D1 (no damage) to D5 (complete destruction) shows a Gaussian distribution: the extreme ends, D1 and D5, are comparatively easy to detect, provided the spatial resolution of the image is suitable with respect to the average size of the structural objects. Hence, for places that suffered blanket destruction, and where damage is quite well delimited from intact areas, such as caused by strong tornados (Figure 6.7), even automated damage mapping is possible with good accuracies (see, for example, Brown, Liang and Womble 2012; Womble, Wood and Mohammadi 2018).

However, for sites characterized by more intermediate damage states, automatic assessment methods quickly find their limits. This is true for virtually any image type that has been tried, be it airborne TV footage (Mitomi, Yamzaki and Matsuoka 2000), radar data (see Uprety, Yamazaki, and Dell'Acqua 2013 for results with coherence and intensity data, respectively, Arciniegas et al. 2007),

Disasters

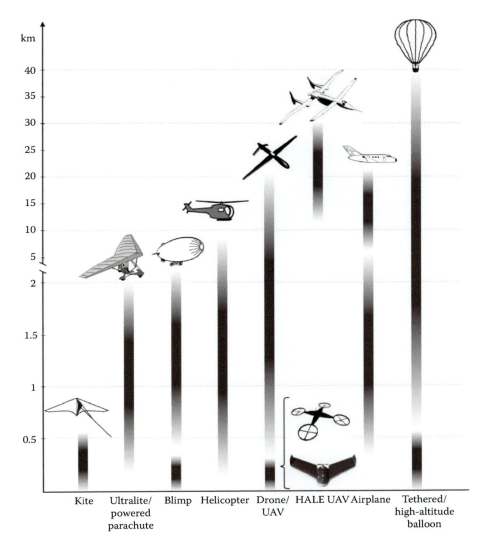

FIGURE 6.6 Overview of approximate operating altitude ranges of different airborne platforms (shaded bars). For a number of platform types, such as blimps or balloons, both low altitudes (typically tethered) and sophisticated polyurethane ones that reach stratospheric heights exist. Similarly, low-flying airplanes can carry out surveys at altitudes of approximately 1–5 km, while other planes, in particular military surveillance aircraft, can fly as high as about 21 km. Drones/UAVs have seen the most dynamic development in recent years. Those, too, include high-flying, mostly military, versions, and numerous low-flying types that can be classed into fixed-wing and multi-rotor devices. HALE and HAPS UAVs focus on both high altitude and long-term autonomous deployment. Modified from Kerle et al. (2008).

and various types of vertical optical images (e.g., Ehrlich et al. 2009). Image-based damage mapping has been a particular challenge for earthquake sites, and the state of the art was reviewed by Dell'Acqua and Gamba (2012), Kerle (2010), and more recently by Contreras et al. (2021). A widespread assumption in the remote sensing community has long been that better data, in particular with increasing spatial resolution, would allow more detailed and more accurate analysis, yet that has only been partly the case. Following the 2010 Haiti earthquake damage analysis was first based on GeoEye-1 images (0.41 m panchromatic resolution), and repeated later using airborne

FIGURE 6.7 Subset of a GeoEye image of Washington, Illinois, that suffered severe damage during a series of tornados on November 18, 2013. Damage patterns are nearly binary (no/little damage versus complete destruction), typical for tornados.

images with approximately 15 cm resolution. This difference led to the amount of damage being identified to increase by a factor of ten (see Gerke and Kerle 2011; van Aardt et al. 2011). Overall, however, even such high-quality vertical image data have turned out to be of limited utility for structural damage mapping, since they provide a very limited perspective (mostly only of the roof) of what is in reality a very complex 3D scene, or even 4D if changes, such as caused by seismic aftershocks, are included. Hence, analysis strongly relies on proxies such as blow-out debris, or changes in shadow or building position, the latter being of particular use when reference images exist (Kerle and Hoffman 2013). In recent years high-quality multi-perspective oblique imaging solutions have become available, such as the Pictometry© system, now part of EagleView (Monfort, Negulescu and Belvaux 2019). The camera system acquires images in five directions (nadir and the four cardinal directions at a 45° angle), in principle providing a very comprehensive view of the disaster scene (Figure 6.8). Similar cameras have also become available for multicopter UAVs with > 1 kg carrying capacity, such as the PSDK 102S by ShareUAV. The images are not only taken in five directions, but also acquired in stereo, allowing for photogrammetric processing. Pictometry data of Port-au-Prince following the 2010 earthquake were used by Gerke and Kerle (2011) to map damage, in an approach based on OBIA and ML. Damage indicators were classified with accuracies between 60% and 70%, values comparable with what studies using other remote sensing have found. However, the work has confirmed a problem: most research efforts on image-based structural damage mapping are experimental, and tend to work well on specific test data but have insufficiently proven ready transferability to other sites and data types (see also Booth et al. 2011). This explains why organizations charged with rapid post-disaster damage mapping, e.g., by mandate of the International Charter "Space and Major Disasters" (www.disasterscharter.org), or through the Copernicus Emergency Management Service (https://emergency.copernicus.eu) continue with manual damage mapping. While those efforts lead to maps of high cartographic quality, map styles, and data accuracy are quite variable (Kerle 2010), with the resulting map variety not only leading to interpretation challenges for the users, but with maps often also not adhering to cartographic standards and conventions (Divjak and Lapaine 2018; Thomas et al. 2022).

The ultimate reason for the limited progress in image-based damage mapping can be readily identified: images depict damage as a physical entity; however, damage is better thought of as a complex concept with different facets. While damage can justifiably be seen as a physical state,

Disasters 175

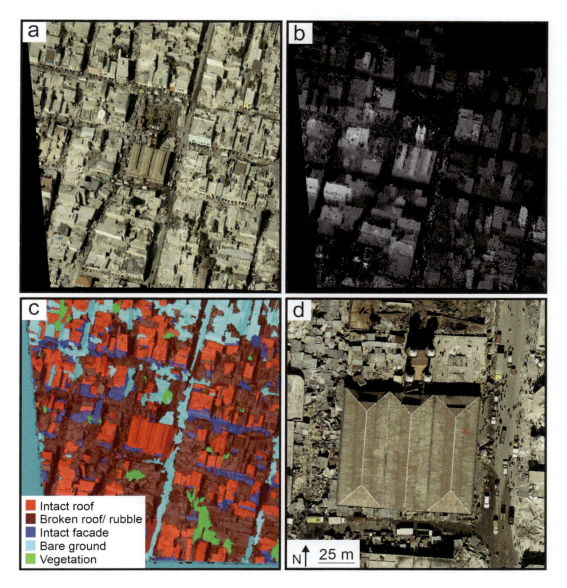

FIGURE 6.8 Oblique Pictometry image of part of Port-au-Prince, Haiti, following the 2010 Earthquake (a), derived depth information that shows geometric homogeneity (b), classified damage features based on segmentation-based supervised classification and machine learning (for details see Gerke and Kerle 2011; (c) and a vertical image of part of the area to show the limitations of only that perspective (d). Images (a) and (c) ©Pictometry International Corp., now part of EagleView.

where damage can be described in absolute or relative terms with respect to the entire structure or the pre-event state, this says little about internal damage and, more significantly, even less about its functional damage. A nuclear power station may appear to be structurally largely intact yet may leak radiation, while a hospital that suffered disruption to its electricity or water supply may be structurally sound, but severely incapacitated in functional terms. More work is needed to semantically link physical external damage indicators to functional capacity. In recent years relevant research that considers functionality in the context of damage and resilience has been emerging (Zhang, Zhang and Li 2021a; Wang et al. 2020).

6.3.3.7 Post-disaster Recovery

Damages caused by a disaster may be limited or vast, which is mirrored by the time and effort needed to rebuild. While for locally confined or overall limited damage the term *reconstruction* is appropriate, the recovery process that follows large events such as the 2010 Haiti or the 2011 Tohoku earthquake and tsunami disasters falls into an entirely different category, and can last years or even decades. The remote sensing community has been slow in addressing this phase, or where it has, work was largely limited to the detection or quantification of physical reconstruction (Guo et al. 2010), or perhaps extended to vegetation recovery (Li et al. 2016). The principal reason that a more functional and system-oriented focus of recovery assessment with remote sensing has only been more recently emerging is similar to the dearth of work on functional damage: it is very difficult to assess the functional state of individual sectors and changes therein. Direct observations, such as of the state of port or rail installations, need to be coupled with proxy indicators of recovery. Remote sensing can indeed provide those, and recent work has shown that a multi-temporal recovery assessment is possible (Ghaffarian et al., 2020; Li et al. 2023a). Since recovery patterns tend to be complex and difficult to predetermine, ML has been a particular asset in recovery assessment from imagery (Sheykhmousa et al. 2019; Ghaffarian and Emtehani 2021). However, while image data can reveal recovery patterns, they cannot explain them. Yet, to allow decision-makers to guide an effective recovery a detailed understanding of the effects of various push and pull factors or the choices of individuals and economic sectors is needed. Recent research has attempted to integrate such information with remote sensing image analysis, for example, through macro-economic agent-based modeling (Ghaffarian et al. 2021).

6.4 TRENDS AND DEVELOPMENTS

In the final section of this chapter a number of noteworthy developments with relevance to the use of remote sensing data in DRM will be assessed, covering both technical and organizational aspects. Several of the trends described as novel in the firsts edition of this book are by now well-established, while others have lost relevance and new ones have emerged.

6.4.1 NEW PLATFORMS

Throughout most of its history, remote sensing has been a high-tech and costly endeavor, limiting it to a few, mostly government and military, players. When commercial opportunities became apparent, the private sector joined in, though costs of infrastructure and operation remained high, both in air- and spaceborne remote sensing. A number of significant changes have taken place. As mentioned before, many countries have launched their own satellites, owing in part to the development of micro- and nanosatellites that combine good performance with relative affordability, which also paved the way for novel sensing solutions (Sandau, Briess and D'Errico 2010). A series of small satellites was launched by Surrey Satellite Technology Limited (SSTL) in the UK between 2002 and 2011, together forming a virtual Disaster Monitoring Constellation (DMC). Taking advantage of a very large swath width of more than 600 km and a spatial resolution that improved from 32 m in the first generation to 5 m in the later models, the idea was to enable countries such as Algeria, Nigeria, or Turkey to own a satellite and through that build up Earth observation capacity, while also providing for more rapid global response in case of a disaster. Indeed, one of the DMC satellites (NigeriaSat-1) was the first to provide high-resolution spaceborne images of the area affected by Hurricane Katrina in 2005 (Salami, Akinyede and de Gier 2010). While the DMC was discontinued after 2015, the trend for smaller satellites continued. Nanosatellites, which include the popular CubeSat category, are defined as platforms with a weight of less than 10 kg, and have gained capabilities to allow disaster monitoring and response (Santilli et al. 2018; Chadalavada and Dutta 2022). Small size also implies lower production and launch costs, which has allowed the creation of larger constellations. Planet Labs, for example, aims a daily global image coverage through hundreds of nanosatellites called doves that can provide ca. 3 m spatial resolution images (Frazier and

Hemingway 2021). This makes them particularly suitable for hazard or disaster situations where frequent observations of large areas are useful, such as to monitor volcanic activity (Aldeghi et al. 2019) or extensive wildfires (Michael et al. 2018).

While nanosatellites are a popular trend, also conventional Earth observation satellites continue to be built. The most significant new addition since the first edition of this book is Copernicus, the Earth observation program of the European Union established in 2014, and new satellites are currently scheduled up to 2028. Implemented through a comprehensive constellation of Sentinel satellites that carry both optical and radar instruments, the program addresses six different domains: atmosphere, marine, land, climate, emergency and security. Data are made available free of charge, and also a large number of other constellations, called contributing missions, feed into the Copernicus program (Jutz and Milagro-Pérez 2020). Copernicus follows in the environmental sensing tradition of ENVISAT or Landsat, with missions dedicated to long-term monitoring of landmasses, oceans, polar regions or the atmosphere, and does not compete with commercial providers of high spatial resolution imagery.

The other significant platform development is that of UAVs. On one hand this closes the circle back to Lawrence's kites, allowing nearly anyone do-it-yourself remote sensing. The other end of the UAV spectrum leads closer to the stratosphere, with the development of high elevation—long endurance (HALE) instruments. While some technical and legislation issues still need to be sorted out, the idea from a DRM perspective is that such instruments, as part of a network, can either provide continuous, high spatial resolution coverage of significant places (large urban areas, etc.), or can be relatively rapidly deployed to a disaster area to provide comprehensive and continuous coverage of the post-event situation. Early scientific excitement about the prospect of HALE UAVs resulted in many design proposals (e.g., Cestino 2006; Frulla and Cestino 2008). However, technical and regulatory obstacles have resulted in none of them becoming operationally available for the envisioned disaster response yet. In 2023 BAE systems successfully brought its solar-powered PHASA-35 platform to a flying height of 20 km, hence development of such systems, now commonly referred to as High Altitude Pseudo Satellite (HAPS), continues.

Also the utility of small UAVs that fly at low elevations (a maximum of 100–200 m above the ground) continue to be hampered by regulations, or often rather the lack of them. In many countries it is simply not allowed to operate them, though pressure from industry and the many emerging potential users has been forcing a chance to that. Successful operations in countries such as Ghana or Rwanda to expedite automated UAV-based blood supplies to remote hospitals (Ackerman and Koziol 2019; Umlauf and Burchardt 2022) also illustrate the commercial potential of the technology. In Europe the new regulation 2019/947 came into effect on January 1, 2024, and governs the steps toward safe incorporation of civil UAVs into European airspace.

In terms of platforms and sensing technology, UAVs have seen the most dramatic development since the first edition of this book. In 2013 DJI launched its first consumer drone, the Phantom, and has since been market leader for consumer drones of which several million are sold every year. Not only has this created wide awareness and acceptance of the technology, but has also brought robust and easily usable sensing technology to stakeholders such as emergency responders. However, most of those users reduce UAVs to eyes-in-the-sky, placed at a useful vantage point from which they can provide image or video data. However, very extensive research on how best to exploit the data has accompanied the proliferation of UAVs, though solutions for more advanced image processing have been slow in becoming operationally adopted by responders. In early work UAV imagery were processed with OBIA methods mentioned before, making use of detail-rich 3D point clouds and the oblique perspective to detect disaster damage (Fernandez Galarreta, Kerle and Gerke 2015) (Figures 6.9 and 6.10). Also, other DRM areas have started to see the benefit of UAV instruments, for example, for landslide monitoring and glacier change mapping (Immerzeel et al. 2014). The first edition of this book stated that "ongoing developments in miniaturization, autonomous navigation and swarm intelligence mean that small UAVs will soon be deployable *en masse* for disaster site surveillance, including the ability to enter structures to assess damage and search for victims or survivors." This prediction proved to be largely true, with technological advances even surpassing

FIGURE 6.9 UAV-based photo of a church in Miranello, Italy, destroyed in an 2012 earthquake (a), and detailed textured point cloud (b).

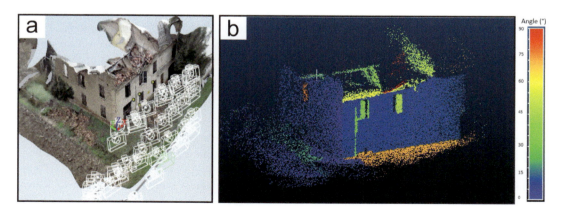

FIGURE 6.10 Textured 3D point cloud of a building damaged by an earthquake constructed from a multitude of UAV and ground-based images (Camera positions shown in white in (a)). The model provides detailed geometric information (b), such as angle information (vertical sections in blue), or rubble piles (orange).

expectations. They were substantially fueled by progress in photogrammetry that allows images from diverse angles and distances to be processed to create very detailed and geometrically robust 3D data of the scene. Taking advantage of the structure from motion (SfM) technique (Jiang, Jiang and Jiang 2020), software such as the market-leading Pix4D (www.pix4d.com) makes it possible to extract 3D data from UAV images without specialist knowledge. The second development of fundamental importance is progress in AI/ML, as explained in more detail in Section 6.4.3. As expected, UAVs are also beginning to enter structure, allowing indoor damage mapping (Steenbeek and Nex 2022) or the search for victims (Zhang et al. 2021b; Zhang et al. 2022). Another relevant recent development is that of hybrid UAVs that combine the ability for vertical take of an landing that characterizes most modern UAVs with flying characteristics of a fixed-wing aircraft. The design allows deployment from confined spaces, the ability to approach and circle an object, as well as to hover, but also to follow linear flight paths. The efficiency of fixed-wing flight allows such platforms to operate longer than multicopter UAVs, but at the expense of carrying capacity: while larger multicopters can carry several kg worth of instruments, even larger hybrids are limited to smaller sensors. Nevertheless, such hybrid UAVs have started to become available in disaster response. The German space agency DLR operates a Modular Aerial Camera System (MACS) on board a hybrid drone, which was deployed over one of the towns affected by the 2023 Turkey earthquake. With a flight duration of up to 90 minutes, and time image streaming and mosaicking, the UAV can provide a real-time orthorectified base map to support the emergency response (Hein et al. 2019).

Table 6.3 provides a detailed overview of recent and upcoming novel platforms useful for DRM.

6.4.2 New Sensors

A number of recent and upcoming sensors are of interest from a DRM perspective. While new and more advanced optical satellites continued to get launched with some regularity since Landsat 1, the situation for radar instruments was different. For one, their number has always been much lower, and after an initial phase in the early 1990s where several instruments were launched (e.g., ERS-1 [1991], JERS-1 [1992], ERS-2 [1995], RADARSAT-1 [1995]), follow-up missions were scarce. A noteworthy, but quite isolated new radar mission was ASAR on ENVISAT (launched in 2002, and operating for ten years). Spaceborne radar remote sensing saw a renaissance with the launch of PALSAR on the ALOS satellite (2006, working until 2011), and followed by missions such as COSMO-SkyMed (constellation with four radar satellites; 2007), TerraSar-X (2007), RISAT-2 (2009), and TanDEM-X (2010). In particular the high spatial resolution data from the TerraSAR-X and TanDEM-X couple that reaches 1 m in Spotlight mode have been shown to be useful for structural damage mapping (Ferro et al. 2013; Uprety et al. 2013). The data were also used to create a new global digital elevation model (WorldDEM) with a resolution of 12 m x 12 m, a substantial improvement on currently available ASTER-based GDEM and the SRTM dataset (Riegler, Hennig and Weber 2015). Yet another radar instrument is part of the Copernicus constellation: Sentinel-1 comprises two C-band SAR satellites (Torres et al. 2012) that were launched in 2014 and 2016, respectively.

Hyperspectral remote sensing, in particular spaceborne, has always been a niche technology, though data with up to 250 narrow spectral bands are useful in many domains, including disaster risk management. Experimental missions such as Hyperion (flying on EO-1 from 2000–2017), the CHRIS instrument on PROBA (launch in 2001 and still operating in early 2024) and the short-lived Indian Hyper Spectral Imager (2008) were later joined by the Chinese HJ-1A instrument, and demonstrated the utility of such data. Only in recent years have developments on more mature and operational hyperspectral systems started. The Japanese Hyperspectral Imager Suite (HISUI) flies on board the ALOS-3 satellite. Launched in 2023 it provides 30 m resolution data. After years of delays the Italian PRISMA mission was finally launched in 2019. Also the German-led EnMAP mission that provides 30 m resolution data was eventually launched in 2022. NASA's HyspIRI mission has also been many years in the planning, and the launch has still not happened.

An original new addition to the sensor-platform menu is real-time spaceborne video, offered by a SkySat, part of Planet Labs. The system can acquire high-definition video clips of maximum 90 seconds, with pixel resolutions better than 1 m. Such video streams have the potential to become a valuable tool in disaster response scenarios (Li et al. 2023b). Table 6.4 summarizes recent and upcoming new sensors of relevance for DRM.

TABLE 6.3
Selected Recent and Upcoming Novel Platforms of UAVs for DRM

Platform Type/Name		Maker/Operator	Description	Technical Parameters	Utility for DRM	Limitations	Notes	Example Literature
Airborne	Pictometry ©	Pictometry, Inc., EagleView	Piloted aircraft with five cameras; since 2002	One nadir camera, four oblique at 45° in each cardinal direction). Typical flying height: 1000 m, resulting in stereo imagery of approximately 15 cm. Images georeferenced, visible range only	Detailed vertical and oblique imagery of a structure from five perspectives, allowing structural elements at risk to be characterized, including in 3D	Expensive data; manned aircraft with limited survey flexibility; Limited spatial resolution/detail		Booth et al. (2011), Gerke and Kerle (2011), Monfort, Negulescu and Belvaux (2019)
	HALE/HAPS UAV	AeroVironment (Global Observer), BAE Systems	High-flying UAV; developments and tests since 2010	Approx. 160 kg in weight; flying at max. 20 km for up to seven days; image spatial resolution of ca. 30 cm); providing real-time image data downlink; solar-powered	Highly valuable for sustained syn- and post-disaster surveillance of extensive and/or dynamic events, such as wild fires or flooding	Expensive system, predominantly developed for military applications. Difficult to deploy quickly in disaster area	HALE UAVs have been in development since about 2000; both technical hurdles and legislation remain challenging	Van Achteren et al. (2013)
	UAVs	Various	Fixed-wing or multi-rotor platforms, ranging from ca. 3 g to several kg; hybrid versions for mapping of line features or larger areas	Dozens of types available, including experimental devices (e.g., Delfly or Crazyflie, at < 20 g, including a camera), inexpensive platforms with limited performance, and survey-grade devices with live-data downlink and carrying	Excellent for detailed multi-perspective building and infrastructure damage assessment; experiments ongoing with UAVs entering damaged structures. Ready	Legislation varies greatly among countries. Civilian UAVs often limited to a total weight of 5 kg and a maximum flying height of 200–300 m; battery capacity	Most dynamic remote sensing segment, with many solutions available; rapidly growing maturity. Processing of uncalibrated stereo imagery readily possible	Fernandez Galarreta et al. (2015), Hinkley and Zajkowski (2011, Kerle et al. (2020)

				capacity of several kg. Stereo data can be processed photogrammetrically. Increasingly onboard (edge) processing of images, including with ML is possible	repeat acquisitions for change detection	of multi-rotors limited, allowing only local surveys	
Spaceborne Micro-/Nano-constellations	DMC	Satellites made by Surrey Satellite Technology Ltd. (SSTL), owned by individual countries	Microsatellites owned by Algeria, China, Nigeria, Thailand, Turkey, UK, Vietnam	Different satellite generations have been produced, starting with AlSAT-1 (Algeria) in 2002 (32 m resolution), and similar instruments for Turkey and Nigeria. Beijing-1 (launched in 2005) added a 4 m pan band. Most recent satellite (NigeriaSAT-2; 2011) added a 2.5 m pan, and 5 m multispectral band. New constellation of three satellites ((DMC-3;1 m pan, 4 m MS) is being planned	DMC satellites are noted for their moderate spatial resolution, but very extensive swath width (up to 650 km), allowing most parts on Earth to be imaged daily, ideal for extensive disasters. High-resolution instruments offer lower coverage but more details.	High data costs (marketed by DMC International Imaging Ltd. (DMCii), now defunct)	DMC satellites provided the first satellite data of areas devastated by the Asian tsunami (2004) and Hurricane Katrina (2005). The constellation was discontinued after 2013	Sandau et al. (2010),
	Dove, SkySat	Planet Labs	Constellation of hundreds of nanosatellites [Doves]): SkySat provides high-resolution	The more than 200 Doves are nanosatellites that only weigh 4 kg each, but are planned to provide daily global	The combination of relatively high spatial resolution and nearly daily imaging makes the constellation	The spatial resolution of Doves is not sufficient to capture small-scale	Some 430 Dove and Superdove satellites are planned for daily global monitoring	Frazier and Hemingway (2021)

(*Continued*)

TABLE 6.3 (Continued)

Selected Recent and Upcoming Novel Platforms of uUe for DRM

Platform Type/Name	Maker/Operator	Description	Technical Parameters	Utility for DRM	Limitations	Notes	Example Literature
		images and space-based video sequences	image coverage at 3–5 m spatial resolution; SkySat provides 1 m resolution images, and up to 90-second video sequences	unique. Dynamically changing areas such as large disaster sites can be effectively monitored; video data capture rapidly changing processes or phenomena (though only for short periods)	hazard or damage features (e.g., soil erosion, detailed building damage)		
Copernicus	European Space Agency	Six Sentinel missions with two satellites each planned (Copernicus EO program)	Different mission focusing on land, ocean, and atmosphere, to be launched successively until ca. 2028. Includes SAR instruments (two Sentinel-1 satellites launched in 2014 and 2016), high-resolution optical instruments (S-2), altimeters for ocean studies (S-3/6), spectrometers and sounders for atmospheric studies (S-4/5)	Sentinel was designed for comprehensive environmental monitoring purposes, with double-satellites ensuring revisits of six days (or better). Many risk- and disaster-related problems can be addressed with Sentinel	Sentinel is not specifically dedicated to DRM	Copernicus is the new name for the Global Monitoring for Environment and Security (GMES) program that started in 1998. Sentinel data access is free of charge	Torres et al. (2012), Jutz and Milagro-Pérez (2020)

| TerraSar-X and TanDEM-X | German Aerospace Center (DLR) | Radar twin-satellite system for high-resolution DEM generation | Satellites flying as close as 250 m to each other, working to create WorldDEM, a global DEM with up to 12 m resolution; satellites provide SAR data or up to 0.25 m resolution (in Staring SpotLight (ST) mode) | High-resolution SAR imagery is useful for hazard assessment (e.g., elements at risk, floodplain characterization); Topographic information from WorldDEM is very useful for process modeling (flooding, landslides) | Expensive data (15 km^2 0.25 m resolution data for 7,000 Euro) | Ferro et al. (2013), Rossi and Gernhardt (2013), Riegler et al. (2015) |

TABLE 6.4
Overview of Selected Recent and Upcoming Sensors of Relevance for DRM. For Details on TerraSAR-X/TanDEM-X, and Sentinel, Also Relevant from a Sensor Perspective, See Table 6.3

Sensor Name	Operator, Launch	Sensor Type	Technical Specifications	Utility for DRM	Limitations	Example Literature
COSMO-SkyMed	Italian Space Agency; four missions, launched between 2007 and 2010; COSMO second generation being developed (two SAR satellites, launch from 2016)	SAR	X-band instruments (3.1 cm); constellation complete since 2011, resolution varying from 15 to 100 m, depending on mode; field of view ranging from 10 x 10 km to 200 x 200 km	Suitable for landslide and ground deformation studies, volume estimation (e.g., of landslides), damage mapping, volcanic hazard and post-eruption studies (e.g., flow emplacement)	Latitudinal coverage of only ±20°–60° (Mediterranean focus)	Bovenga et al. (2012), Covello et al. (2010)
RISAT-2	Indian Space Research Organization (ISRO); 2012	SAR	C-band (5.6 cm), spatial resolution up to 3 m, depending on mode, swath width ranging from 30 to 240 km; several follow-up missions planned to ensure data continuity	Data are useful for hazard assessment, element at risk mapping, flood mapping, estimation for near-shore bathymetry		Hegde et al. (2009), Mishra et al. (2014)
PRISMA	Italian Space Agency/Temporary Industrial Group; 2019	Hyper-spectral	Precursor mission; hyperspectral data with 20–30 m, pan data with 2.5–5 m resolution	Useful for hazard studies where detailed spectral information or surface materials is needed (e.g., volcanology, wild fire)	Only a precursor for future operational missions	Sacchetti et al. (2010)
EnMAP	German Space Agency; 2022	Hyper-spectral	VNIR and SWIR coverage; 30 m resolution, 30 km swath width; steerable for off-nadir acquisition, four-day temporal resolution	Same as for PRISMA; in addition more useful for post-event imaging due to higher temporal resolution		Stuffler et al. (2009)
HyspIRI	NASA; 2024 (expected)	Hyper-spectral	VNIR-SWIR sensor, as well as mid-TIR instrument, with temporal resolution of 19 and 5 days, respectively, and 60 m spatial resolution	Very suitable for studies of volcanic, wild fire, and drought hazards		Abrams et al. (2013)

6.4.3 BETTER DATA ANALYSIS METHODS

Many methodological advances have already been mentioned in previous sections, such as OBIA, photogrammetry with uncalibrated imagery using SfM, or ML, and a very long list of technical advances could be named here. Data analysis, however, has also developed in non-technical ways. Just like multi-core or distributed computing has accelerated data processing, also people have been being increasingly made use of to support data analysis. Crowdsourcing, or citizen sensing, has become a broad area of many subtypes where people, typically, but not exclusively, laypeople, actively or passively provide geographical information. From a DRM perspective such Volunteered Geographic Information (VGI; e.g., Goodchild 2007) offers many useful contributions, ranging from better base data (e.g., via Google Map Maker or OSM) to better post-disaster response. The latter has been benefiting from people located in affected areas providing information, including imagery, of the disaster site. Other efforts tried to pool remote sensing expertise by getting analysts to perform rapidly a shared image-based structural damage assessment. Following the 2010 Haiti earthquake more than 600 volunteers with remote sensing expertise participated in what became known as the Global Earth Observation-Catastrophe Assessment Network (GEO-CAN; see Ghosh et al. 2011). While this allowed an area of > 1000 km^2 to be mapped in detail within a few weeks, the effort raised questions such as how best to instruct volunteers for this type of work, how to get them to map consistently as a group, and how best to process and validate their contributions (Kerle and Hoffman 2013). Collaborative damage mapping is still being done, such as by the Humanitarian OpenStreetMap Team (HOT). However, a crowdsourcing effort by satellite operator DigitalGlobe called Tomnod that was started in 2010 ceased operations in 2019, and it has become increasingly clear that the success of collaborative image analysis strongly depends on the cognitive level of a given mapping objective: digitizing and annotating buildings as is done in OSM can lead to accurate results, while more demanding tasks such as damage mapping do not.

The phenomenal progress of ML since the publication of the first edition of this book was already mentioned earlier, and its potential for faster and more reliable signal processing and image analysis means that all aspects of DRM supported by geodata have seen the emergence of ML-based solutions. It supports hazard and risk assessment (Díez-Herrero and Garrote 2020; Liu et al. 2023), early warning of hydrometeorological hazard events (Chen, Zhang and Wang 2020), and more effective disaster response. The main strength of ML has been to identify anomalies by training models predominantly with examples of normal situations, eliminating the need to specifically encode target features such as specific damage indicators. For example, Vetrivel et al. (2018) trained a CNN to identify highly diverse seismic damage, overcoming the need for so-called handcrafted features such as texture layers that formed the basis for older methods. Also many of the earlier cited works on post-disaster recovery assessment or victim detection make use of ML, in particular CNN-based methods (e.g., Ghaffarian and Emtehani 2021; Steenbeek and Nex 2022; Han et al. 2022; Zhang et al. 2022). A comprehensive review of the potential of ML in DRM was recently provided by Linardos et al. (2022).

6.4.4 ORGANIZATION

Disasters create empathy, especially with larger events creating much visibility and an international response. While at a practical level much has been done to allow fast response of search and rescue teams, also higher organizational levels have seen strong improvements. The already mentioned Disaster Charter is the most prominent example, which since 2000 has been providing rapid post-disaster support based on satellite imagery. The protocols provide for data acquisition on a priority basis, and drawing on the space assets of virtually all space agencies, but also commercial entities such as Digital Globe and GeoEye. The data from the more than 800 Charter activations to date are primarily processed (and, as mentioned before, largely in a manual manner) by three main agencies, UNOSAT, DLR-ZKI, and SERTIT (for details see Kerle 2013b), though increasingly regional institutions have been taking over project management. A similar support effort is the Copernicus Emergency Management Service (CEMS), though in addition to emergency response also addressing early warning. Launched in 2012, CEMS also provides image-based damage maps to support emergency response agencies. Both Charter and CEMS are successful examples of a global effort to use collective resources to address disaster events. However, they also illustrate a situation where various organizations are now routinely engaging in image-based damage mapping, leading to duplication of efforts, and creating an overwhelming amount of damage map products that becomes counter-productive (see also Kerle 2011; Voigt et al. 2011). In addition, both services continue to base their mapping on visual image interpretation, despite substantial progress in automated, ML-based methods that are being increasingly used by commercial companies. For example, Microsoft's AI for Good lab provides freely available tools (https://github.com/microsoft) (Manzini et al. 2023), while Google SKAI (https://github.com/google-research/skai) provides similar automated tools (Kirkpatrick 2023).

In addition to the Charter other global efforts are working toward a better use of EO assets for DRM. The United Nations Platform for Space-based Information for Disaster Management and Emergency Response (UN-SPIDER) operates a Knowledge Portal that aims at providing background information, as well as guidance and best practice information on how best to use (spaceborne) remote sensing for disasters. In that context training sessions are organized, and national governments advised on how best to work toward remote sensing–based DRM.

The Sentinel Asia program also focuses on coordinating the use of existing space infrastructure, together with auxiliary spatial data, for DRM, rather than promoting the development of new instruments. It is coordinated by JAXA, the Japan Aerospace Exploration Agency.

6.5 GAPS AND LIMITATIONS

Since the early days of spaceborne remote sensing in the 1960s we have witnessed a multitude of exciting developments that continue to provide us with more and better data that benefit every aspect of DRM, and ever more rapidly. The previous sections are an attempt to show how technical advances, deeper insights in disaster risk as a complex phenomenon, but also better international cooperation, have led to a situation where remote sensing has morphed into a vital instrument in our efforts to understand and reduce risk, but also to warn of events and respond effectively where they do happen. In the process much has been achieved, and gaps have been reduced or closed. Nevertheless, a number of limitations remain that are briefly discussed in this section.

6.5.1 THE MILITARY SIDE

Military interests can safely be seen as the driving force behind all significant remote sensing developments, usually with military developments later leading to civil adaptations. Hence, the many attempts and efforts originating from the military/national security community—both successes and failures—over decades have led to today's arsenal of remote sensing instruments and tools that benefits DRM in countless ways.

At the same time, the military has always kept a skeptical eye on the civilian use of their assets, and frequently imposed limitations and restrictions. Several of those have already been mentioned in this chapter. This has included limiting the detail that remote sensing instruments were allowed to acquire (e.g., the 100 m maximum resolution of ERTS-1), or the artificial signal limitation of GPS accuracy until selective availability was abolished in 2000. Other datasets continue to suffer those restrictions. Until as recently as 2013 the US military imposed a limit to the maximum spatial resolution (0.5 m) that data by commercial companies could provide, and only recently those rules have started to get relaxed. Likewise, DEM data based on the Shuttle Radar Topography Mission flown in 2000 for many years were available for non-US territories only at a reduced resolution of 90 m (versus the 30 m available for the United States). Only since 2014 the higher-resolution data are available with near-global coverage (SRTM-GL1; O'Loughlin et al. 2016; Satge et al. 2016). Also other countries have shown a very restrictive data sharing policy, amongst the major EO nations notably India and China.

In sum, military/national security interests have been both a blessing and a burden for the civil remote sensing community and its efforts to use EO information.

6.5.2 Methodological Gaps in DRM

Disaster risk management, both in theoretical and practical terms, has seen tremendous advances. It has moved away from seeing disasters as a result of a violent nature, toward a realization that, while nature indeed contains destructive forces and processes, it is usually human beings that move into hazardous terrain, or who change the probability of events occurring, be it through poor land management or excessive greenhouse gas emissions. This means both nature and human society with its assets and activities must be jointly considered if the aim is to reduce costs and damage. Modern risk assessment is based on a complex understanding of the role of the various parameters that influence the outcome of an event, and EO data have helped to understand those processes and connections. In practical terms, those data are also a critical asset to help manage the delicate balance of an ever-growing society living in a closely coupled and complex manner with nature.

Our understanding and abilities have limitations, though. While risk assessment methods are mathematically flexible and extendable, they have been optimized for single-hazard situations. Dealing with multiple, or with cascading hazard situations, remains a challenge. For example, an area hit by a seismic event may suffer direct structural damage due to ground shaking. The same force may lead to the rupture of gas or water mains, potentially resulting in explosions/fires or flooding, respectively. At the same time the earthquake may lead to landslides, which can block waterways, subsequently resulting in dangerous breakout floods. The water left in the area quickly poses a health hazard due to waterborne diseases. Assessing the total risk in situations with this type of complexity remains a methodological challenge, though relevant work has been emerging, such as for complex seismic (Nishino 2023) or hydrological emergencies (Biondi, Scarcella and Versace 2023). Likewise, assessing the total risk, including to remote places, or accounting for secondary or tertiary consequences, remains difficult (Sangha et al. 2020). For example, the relatively minor 2010 eruption of the Eyjafjallajökull volcano in Iceland shut down production at several BMW plants in Bavaria (Jones and Mendoza Bolivar 2011), since the closed airspace prevented needed parts to be brought in from Japan. A complete risk assessment would have to account for damage from the lost production, as well as financial consequences airlines and stranded passengers suffered.

We have a good understanding of the role vulnerability plays in risk, and that we need to distinguish between different vulnerability types. However, risk, as expected losses, has traditionally been quantified in monetary terms. How, for example, social or political vulnerability can be reconciled with those traditional approaches still requires a consensus. Some researchers (e.g., Adger and Kelly 1999; Smit and Pilifosova 2003) have thus focused more on understanding functional relationships.

6.5.3 Lack of Standards and Suitable Legislation

Both points have already been mentioned in earlier sections. Lack of standards relates primarily to post-disaster damage mapping that continues to see a largely redundant creation of many maps (more than 2000 damage map products were created following the 2010 Haiti earthquake), using a number of different styles and nomenclatures (Kerle 2011). Also for damage maps created in a crowdsourcing manner, operational standards continue to be lacking. At this point it has not been conclusively shown that such a collaborative effort, even when recruiting volunteers only from a pool of people with remote sensing experience, can provide results that are sufficiently accurate. Also operating procedures for how to engage with volunteers, how to provide instructions for the specific job, including corrective feedback (Kerle and Hoffman 2013), do not yet exist. However, given the unlimited pool of volunteers that can report on an impending hazard event, or provide useful information directly from a disaster site, interest in the topic remains high (Feng, Huang and Sester 2022; Kankanamge et al. 2019).

6.6 SUMMARY

The purpose of this chapter was to assess the development of remote sensing over the last 60 years in terms of its value for disaster risk management, that is for the entire spectrum from hazard and risk assessment to post-disaster response and recovery. Disaster risk with all its components is a spatial phenomenon, hence, for all its social and organizational facets, can only be effectively addressed with spatial data. Therefore, the main source that feeds DRM models and analysis tools is remote sensing information, and without such data meaningful risk assessment, early warning/action or effective post-disaster response would simply not be possible. This chapter has highlighted that in DRM a vast number of methods and techniques have been developed that make use of virtually any remote sensing sensor or platform ever developed. At the same time the analysis also showed that we often have very suitable or interesting data, but not yet adequate ways to use them, for example, in damage mapping or recovery assessment. The chapter also made the point that DRM is an application field that can be seen as native to EO, in that its potential was one of the first uses of remote sensing data to be realized, and where essentially continuous developments to maximize the utility of EO data have been taking place.

The process of making EO data a critical pillar in DRM has been a very dynamic one. It reflects as much the race for military supremacy during the two world wars and the subsequent Cold War, as the more recent, more concerted international efforts aimed at finding global solutions, such as the Disaster Charter or UN efforts mentioned. Despite being strongly driven by military spending, remote sensing has benefited in large measure from the effort of individuals, be it pioneers who showed what can be done (such as Lawrence in 1906), or the people who transitioned remote sensing from a governmental activity into private businesses. In particular the latter has created a new race, a commercial one, which too has led to rapid advanced in technology and use of EO data, but that also carries the risk that governments leave important services to commercial interests.

The chapter finished reviewing a number of remaining gaps and limitations, and further work for the different communities has been identified. For a very long time disasters were seen, in a fatalistic fashion, as unavoidable acts of nature (or perhaps divine punishment), which later gave way to a more reasoned analysis of the role society plays. EO technology now provides us with all the data and analysis means we need to make informed decisions. This, however, means that this must be followed by an adequate socio-political decision-making process. Technology, in EO or otherwise, is not going to eliminate disasters. What is needed is a sound understanding—among all stakeholders—of the state of nature and societal processes, as provided in large part by EO instruments, but also how human activity relates with natural systems and processes. Hence, remote sensing is an effective means to reveal the relevant connections, but a sustained reduction of the number of annual disasters can only be achieved if the insights a risk assessment process provides are converted into effective risk reduction measures.

REFERENCES

Ackerman, E. & M. Koziol (2019) The blood is here: Zipline's medical delivery drones are changing the game in Rwanda. *IEEE Spectrum*, 56, 24–31.

Adedeji, O., A. Olusola, G. James, H. A. Shaba, I. R. Orimoloye, S. K. Singh & S. Adelabu (2020) Early warning systems development for agricultural drought assessment in Nigeria. *Environmental Monitoring and Assessment*, 192, 798.

Adger, W. N. & P. M. Kelly (1999) Social vulnerability to climate change and the architecture of entitlements. *Mitigation and Adaptation Strategies for Global Change*, 4, 253–266.

Ajmar, A., P. Boccardo & F. G. Tonolo (2011) Earthquake damage assessment based on remote sensing data. The Haiti case study. *Italian Journal of Remote Sensing-Rivista Italiana Di Telerilevamento*, 43, 123–128.

Aldeghi, A., S. Carn, R. Escobar-Wolf & G. Groppelli (2019) Volcano monitoring from space using high-cadence Planet CubeSat images applied to Fuego volcano, Guatemala. *Remote Sensing*, 11.

Alexander, C., S. Smith-Voysey, C. Jarvis & K. Tansey (2009) Integrating building footprints and LiDAR elevation data to classify roof structures and visualise buildings. *Computers Environment and Urban Systems*, 33, 285–292.

Alidoost, F. & H. Arefi (2018) A CNN-based approach for automatic building detection and recognition of roof types using a single aerial image. *Pfg-Journal of Photogrammetry Remote Sensing and Geoinformation Science*, 86, 235–248.

Ansari Amoli, A., H. Aghighi & E. Lopez-Baeza (2022) Drought risk evaluation in Iran by using geospatial technologies. *Remote Sensing*, 14, 3096.

Arciniegas, G., W. Bijker, N. Kerle & V. A. Tolpekin (2007) Coherence- and amplitude-based analysis of seismogenic damage in Bam, Iran, using Envisat ASAR data. *IEEE Transactions on Geoscience and Remote Sensing*, 45, 1571–1581.

Balzter, H. (2001) Forest mapping and monitoring with interferometric synthetic aperture radar (InSAR). *Progress in Physical Geography*, 25, 159–177.

Biljecki, F., Y. S. Chow & K. Lee (2023) Quality of crowdsourced geospatial building information: A global assessment of OpenStreetMap attributes. *Building and Environment*, 237, 110295.

Biondi, D., G. E. Scarcella & P. Versace (2023) CERCA (Cascading Effects in Risk Consequences Assessment): An operational tool for geo-hydrological scenario risk assessment and cascading effects evaluation. *Hydrology Research*, 54, 189–207.

Birkmann, J. (2007) Risk and vulnerability indicators at different scales: Applicability, usefulness and policy implications. *Environmental Hazards*, 7, 20–31.

Blaikie, P., T. Cannon, I. Davis & B. Wisner. 1994. *At Risk: Natural Hazards, People's Vulnerability, and Disasters*. London and New York: Routledge.

Booth, E., K. Saito, R. Spence, G. Madabhushi & R. T. Eguchi (2011) Validating assessments of seismic damage made from remote sensing. *Earthquake Spectra*, 27, S157–S177.

Bosher, L., K. Chmutina & D. van Niekerk (2021) Stop going around in circles: Towards a reconceptualisation of disaster risk management phases. *Disaster Prevention and Management*, 30, 525–537.

Bovenga, F., J. Wasowski, D. O. Nitti, R. Nutricato & M. T. Chiaradia (2012) Using COSMO/SkyMed X-band and ENVISAT C-band SAR interferometry for landslides analysis. *Remote Sensing of Environment*, 119, 272–285.

Brown, T. M., D. A. Liang & J. A. Womble (2012) Predicting ground-based damage states from windstorms using remote-sensing imagery. *Wind and Structures*, 15, 369–383.

Budak, C. & V. Gider (2023) LSTM based forecasting of the next day's values of ionospheric total electron content (TEC) as an earthquake precursor signal. *Earth Science Informatics*, 16, 2323–2337.

Campus, A., M. Laiolo, F. Massimetti & D. Coppola (2022) The transition from MODIS to VIIRS for global volcano thermal monitoring. *Sensors*, 22, 1713.

Carter, W. D. & J. N. Rinker. 1976. Structural features related to earthquakes in Managua, Nicaragua, and Cordoba, Mexico. In *ERTS-1, a New Window on Our Planet. U.S. Geological Survey Professional Paper*, eds. R. S. Williams Jr. & W. D. Carter, 123–128. Washington, DC: U.S. Geological Survey.

Cestino, E. (2006) Design of solar high altitude long endurance aircraft for multi payload & operations. *Aerospace Science and Technology*, 10, 541–550.

Chadalavada, P. & A. Dutta (2022) Regional CubeSat constellation design to monitor hurricanes. *IEEE Transactions on Geoscience and Remote Sensing*, 60.

Chen, R., W. M. Zhang & X. Wang (2020) Machine learning in tropical cyclone forecast modeling: A review. *Atmosphere*, 11.

Chen, Z. Y., B. B. Gao & B. Devereux (2017) State-of-the-art: DTM generation using airborne LIDAR data. *Sensors*, 17.

Choi, C., M. Pardini, J. Armston & K. P. Papathanassiou (2023) Forest biomass mapping using continuous InSAR and discrete waveform Lidar measurements: A TanDEM-X/GEDI test dtudy. *IEEE Journal of Selected Topics in Applied Earth Observations and Remote Sensing*, 16, 7675–7689.

Chuvieco, E. 2003. *Wildland Fire Danger Estimation and Mapping: The Role of Remote Sensing Data.* Singapore: World Scientific Publishing.

Clark, G. E., S. C. Moser, S. J. Ratick, K. Dow, W. B. Meyer, S. Emani, W. Jin, J. X. Kasperson, R. E. Kasperson & H. E. Schwarz (1998) Assessing the vulnerability of coastal communities to extreme storms: the case of Revere, MA., USA. *Mitigation and Adaptation Strategies for Global Change*, 3, 59–82.

Cochran, D. R. & R. L. Pyle (1978) Volcanology via satellite. *Monthly Weather Review*, 106, 1373–1375.

Colon, C., S. Hallegatte & J. Rozenberg (2021) Criticality analysis of a country's transport network via an agent-based supply chain model. *Nature Sustainability*, 4, 209–215.

Contreras, D., S. Wilkinson & P. James (2021) Earthquake reconnaissance data sources, a literature review. *Earth*, 2, 1006–1037.

Covello, F., F. Battazza, A. Coletta, E. Lopinto, C. Fiorentino, L. Pietranera, G. Valentini & S. Zoffoli (2010) COSMO-SkyMed an existing opportunity for observing the Earth. *Journal of Geodynamics*, 49, 171–180.

CRED (2023). *EM-DAT: The OFDA/CRED International Disaster Database.* https://www.emdat.be/

Dardel, C., L. Kergoat, P. Hiernaux, E. Mougin, M. Grippa & C. J. Tucker (2014) Re-greening Sahel: 30 years of remote sensing data and field observations (Mali, Niger). *Remote Sensing of Environment*, 140, 350–364.

Dashora, A., B. Lohani & J. N. Malik (2007) A repository of earth resource information—CORONA satellite programme. *Current Science*, 92, 926–932.

Dell'Acqua, F. & P. Gamba (2012) Remote sensing and earthquake damage assessment: Experiences, limits, and perspectives. *Proceedings of the IEEE*, 100, 2876–2890.

Díez-Herrero, A. & J. Garrote (2020) Flood risk analysis and assessment, applications and uncertainties: A bibliometric review. *Water*, 12.

Divjak, A. K. & M. Lapaine (2018) Crisis maps—observed shortcomings and recommendations for improvement. *ISPRS International Journal of Geo-Information*, 7.

Ebert, A., N. Kerle & A. Stein (2009) Urban social vulnerability assessment with physical proxies and spatial metrics derived from air- and spaceborne imagery and GIS data. *Natural Hazards*, 48, 275–294.

Ehrlich, D., H. D. Guo, K. Molch, J. W. Ma & M. Pesaresi (2009) Identifying damage caused by the 2008 Wenchuan earthquake from VHR remote sensing data. *International Journal of Digital Earth*, 2, 309–326.

Ehrlich, D., T. Kemper, X. Blaes & P. Soille (2013) Extracting building stock information from optical satellite imagery for mapping earthquake exposure and its vulnerability. *Natural Hazards*, 68, 79–95.

Endo, E. T., P. L. Ward, D. H. Harlow, R. V. Allen & J. P. Eaton (1974) A prototype global volcano surveillance system monitoring seismic activity and tilt. *Bulletin Volcanologique*, 38, 315–344.

Englhardt, J., H. de Moel, C. K. Huyck, M. C. de Ruiter, J. Aerts & P. J. Ward (2019) Enhancement of large-scale flood risk assessments using building-material-based vulnerability curves for an object-based approach in urban and rural areas. *Natural Hazards and Earth System Sciences*, 19, 1703–1722.

Feng, Y., X. Huang & M. Sester (2022) Extraction and analysis of natural disaster-related VGI from social media: Review, opportunities and challenges. *International Journal of Geographical Information Science*, 36, 1275–1316.

Fernandez Galarreta, J., N. Kerle & M. Gerke (2015) UAV-based urban structural damage assessment using object-based image analysis and semantic reasoning. *Natural Hazards and Earth System Sciences*, 15, 1087–1101.

Ferro, A., D. Brunner & L. Bruzzone (2013) Automatic detection and reconstruction of building radar footprints from single VHR SAR images. *IEEE Transactions on Geoscience and Remote Sensing*, 51, 935–952.

Francis, P. W. & S. L. De Silva (1989) Application of the Landsat Thematic Mapper to the identification of potentially active volcanos in the Central Andes. *Remote Sensing of Environment*, 28, 245–255.

Frazier, A. E. & B. L. Hemingway (2021) A technical review of Planet Smallsat data: Practical considerations for processing and using PlanetScope imagery. *Remote Sensing*, 13, 3930.

Frulla, G. & E. Cestino (2008) Design, manufacturing and testing of a HALE-UAV structural demonstrator. *Composite Structures*, 83, 143–153.

Fuchs, S., M. Keiler, R. Ortlepp, R. Schinke & M. Papathoma-Köhle (2019) Recent advances in vulnerability assessment for the built environment exposed to torrential hazards: Challenges and the way forward. *Journal of Hydrology*, 575, 587–595.

Galderisi, A. & F. F. Ferrara. 2013. Resilience. In *Encyclopedia of Natural Hazards*, ed. P. T. Bobrowsky, 849–850. Dordrecht: Springer.

Galiatsatos, N., D. N. M. Donoghue & G. Philip (2008) High resolution elevation data derived from stereoscopic CORONA imagery with minimal ground control: An approach using Ikonos and SRTM data. *Photogrammetric Engineering and Remote Sensing*, 74, 1093–1106.

Gao, Y. Y., S. Lang, D. Tiede, G. W. Gella & L. Wendt (2022) Comparing OBIA-generated labels and manually annotated labels for semantic segmentation in extracting refugee-dwelling footprints. *Applied Sciences*, 12, 17.

Ge, P. L., H. Gokon & K. Meguro (2020) A review on synthetic aperture radar-based building damage assessment in disasters. *Remote Sensing of Environment*, 240.

Gedney, L. & J. VanWormer. 1973. *ERTS-1, Earthquakes, and Tectonic Evolution in Alaska*. Sect. A, 745–756. Washington, DC.

Geerling, G. W., M. J. Vreeken-Buijs, P. Jesse, A. M. J. Ragas & A. J. M. Smits (2009) Mapping river floodplain ecotopes by segmentation of spectral (CASI) and structural (LiDAR) remote sensing data. *River Research and Applications*, 25, 795–813.

Gerke, M. & N. Kerle (2011) Automatic structural seismic damage assessment with airborne oblique Pictometry © imagery. *Photogrammetric Engineering and Remote Sensing*, 77, 885–898.

Ghaffarian, S. & S. Emtehani (2021) Monitoring urban deprived areas with remote sensing and machine learning in case of disaster recovery. *Climate*, 9.

Ghaffarian, S., N. Kerle & T. Filatova (2018) Remote sensing-based proxies for urban disaster risk management and resilience: A review. *Remote Sensing*, 10, 30.

Ghaffarian, S., A. Rezaie Farhadabad & N. Kerle (2020) Post-disaster recovery monitoring with Google Earth Engine. *Applied Sciences*, 10, 4574.

Ghaffarian, S., D. Roy, T. Filatova & N. Kerle (2021) Agent-based modelling of post-disaster recovery with remote sensing data. *International Journal of Disaster Risk Reduction*, 60, 102285.

Ghazaryan, G., O. Dubovyk, V. Graw, N. Kussul & J. Schellberg (2020) Local-scale agricultural drought monitoring with satellite-based multi-sensor time-series. *GIScience & Remote Sensing*, 57, 704–718.

Ghosh, S., C. K. Huyck, M. Greene, S. P. Gill, J. Bevington, W. Svekla, R. DesRoches & R. T. Eguchi (2011) Crowdsourcing for rapid damage assessment: The Global Earth Observation Catastrophe Assessment Network (GEO-CAN). *Earthquake Spectra*, 27, S179–S198.

Goodchild, M. F. (2007) Citizens as sensors: The world of volunteered geography. *GeoJournal*, 69, 211–221.

Granados-Bolaños, S., A. Quesada-Román & G. E. Alvarado (2021) Low-cost UAV applications in dynamic tropical volcanic landforms. *Journal of Volcanology and Geothermal Research*, 410, 107143.

Graveline, M. H. & D. Germain (2022) Disaster risk resilience: conceptual evolution, key issues, and opportunities. *International Journal of Disaster Risk Science*, 13, 330–341.

Grünthal, G. 1998. *European Macroseismic Scale 1998 (EMS-98). 99*. Luxembourg: Cahiers du Centre Européen de Géodynamique et de Séismologie, Centre Européen de Géodynamique et de Séismologie.

Guo, H., L. Liu, L. Lei, Y. Wu, L. Li, B. Zhang, Z. Zuo & Z. Li (2010) Dynamic analysis of the Wenchuan Earthquake disaster and reconstruction with 3-year remote sensing data. *International Journal of Digital Earth*, 3, 355–364.

Han, Q. Z., Q. Yin, X. Zheng & Z. Y. Chen (2022) Remote sensing image building detection method based on Mask R-CNN. *Complex & Intelligent Systems*, 8, 1847–1855.

Harb, M., D. De Vecchi & F. Dell'Acqua (2015) Physical vulnerability proxies from remote sensing: Reviewing, implementing and disseminating selected techniques. *IEEE Geoscience and Remote Sensing Magazine*, 3, 20–33.

Hastings, D. A. & W. J. Emery (1992) The advanced very high-resolution radiometer (AVHRR)—A brief reference guide. *Photogrammetric Engineering and Remote Sensing*, 58, 1183–1188.

Hegde, V. S., V. Jayaraman & S. K. Srivastava (2009) India's EO infrastructure for disaster reduction: Lessons and perspectives: *Acta Astronautica*, 65 (9–10), 1471–1478.

Hein, D., T. Kraft, J. Brauchle & R. Berger (2019) Integrated UAV-based real-time mapping for security applications. *ISPRS International Journal of Geo-Information*, 8.

Herman, J. R., R. McPeters & D. Larko (1993) Ozone depletion at northern and southern latitudes derived from January 1979 to December 1991 Total Ozone Mapping Spectrometer data. *Journal of Geophysical Research-Atmospheres*, 98, 12783–12793.

Heumann, B. W., J. W. Seaquist, L. Eklundh & P. Jonsson (2007) AVHRR derived phenological change in the Sahel and Soudan, Africa, 1982–2005. *Remote Sensing of Environment*, 108, 385–392.

Hinkley, E. A. & T. Zajkowski (2011) USDA forest service-NASA: Unmanned aerial systems demonstrations—pushing the leading edge in fire mapping. *Geocarto International*, 26, 103–111.

Hoffer, R. M. & K. S. Lee. 1989. Forest change classification using Seasat and SIR-B satellite SAR data. In *Geoscience and Remote Sensing Symposium (IGARSS '89)/12th Canadian Symposium on Remote Sensing*, 1372–1375, Vancouver, Canada. Piscataway NJ: Institute of Electrical and Electronics Engineers (IEEE).

Hooper, A., D. Bekaert, K. Spaans & M. Arikan (2012) Recent advances in SAR interferometry time series analysis for measuring crustal deformation. *Tectonophysics*, 514, 1–13.

Hoque, R., D. Nakayama, H. Matsuyama & J. Matsumoto (2011) Flood monitoring, mapping and assessing capabilities using RADARSAT remote sensing, GIS and ground data for Bangladesh. *Natural Hazards*, 57, 525–548.

Houmma, I. H., L. El Mansouri, S. Gadal, E. Faouzi, A. A. Toure, M. Garba, Y. Imani, M. El-Ayachi & R. Hadria (2023) Drought vulnerability of central Sahel agrosystems: A modelling-approach based on magnitudes of changes and machine learning techniques. *International Journal of Remote Sensing*, 44, 4262–4300.

Howard, J. A., E. C. Barrett & J. U. Heilkema (1978) The application of satellite remote sensing to monitoring of agricultural disasters. *Disasters*, 2, 231–240.

Huang, C., J. P. Wu, Y. Chen & J. Yu. 2012. Detecting floodplain inundation frequency using MODIS time-series imagery. In *2012 First International Conference on Agro-Geoinformatics (Agro-Geoinformatics)*, 349–354, Shanghai, PR China. Piscataway, NJ: Institute of Electrical and Electronics Engineers (IEEE).

Huang, Y., L. Zhuo, H. Tao, Q. Shi & K. Liu (2017) A novel building type classification scheme based on integrated LiDAR and high-resolution images. *Remote Sensing*, 9, 679.

Immerzeel, W. W., P. D. A. Kraaijenbrink, J. M. Shea, A. B. Shrestha, F. Pellicciotti, M. F. P. Bierkens & S. M. de Jong (2014) High-resolution monitoring of Himalayan glacier dynamics using unmanned aerial vehicles. *Remote Sensing of Environment*, 150, 93–103.

Izumi, T., R. Shaw, R. Djalante, M. Ishiwatari & T. Komino (2019) Disaster risk reduction and innovations. *Progress in Disaster Science*, 2, 8.

Jiang, S., C. Jiang & W. S. Jiang (2020) Efficient structure from motion for large-scale UAV images: A review and a comparison of SfM tools. *ISPRS Journal of Photogrammetry and Remote Sensing*, 167, 230–251.

Jones, S. & E. Mendoza Bolivar. 2011. *Natural Disasters and Business: The Impact of the Icelandic Volcano of April 2010 on European Logistics and Distribution – A Case Study of Malta. Working Papers* 2011/20. The Netherlands: Maastricht School of Management. https://ideas.repec.org/p/msm/wpaper/2011-20.html

Joyce, K. E., S. E. Belliss, S. V. Samsonov, S. J. McNeill & P. J. Glassey (2009) A review of the status of satellite remote sensing and image processing techniques for mapping natural hazards and disasters. *Progress in Physical Geography*, 33, 183–207.

Jutz, S. & M. P. Milagro-Pérez (2020) Copernicus: The European Earth observation programme. *Revista De Teledeteccion*, V-XI.

Kamran, M. & K. Yamamoto (2023) Evolution and use of remote sensing in ecological vulnerability assessment: A review. *Ecological Indicators*, 148.

Kankanamge, N., T. Yigitcanlar, A. Goonetilleke & M. Kamruzzaman (2019) Can volunteer crowdsourcing reduce disaster risk? A systematic review of the literature. *International Journal of Disaster Risk Reduction*, 35.

Karakhanian, A., R. Jrbashyan, V. Trifonov, H. Philip, S. Arakelian, A. Avagyan, H. Baghdassaryan, V. Davtian & Y. Ghoukassyan (2003) Volcanic hazards in the region of the Armenian Nuclear Power Plant. *Journal of Volcanology and Geothermal Research*, 126, 31–62.

Karbach, N., N. Bobrowski & T. Hoffmann (2022) Observing volcanoes with drones: Studies of volcanic plume chemistry with ultralight sensor systems. *Scientific Reports*, 12, 17890.

Kerle, N. (2010) Satellite-based damage mapping following the 2006 Indonesia earthquake—How accurate was it? *International Journal of Applied Earth Observation and Geoinformation*, 12, 466–476.

Kerle, N. (2011) Remote sensing based post—disaster damage mapping: Ready for a collaborative approach? *Earthzine: Fostering Earth Observation and Global Awareness*. IEEE Oceanic Engineering Society. https://earthzine.org/remote-sensing-based-post-disaster-damage-mapping-ready-for-a-collaborative-approach/

Kerle, N. (2013a) Global positioning systems (GPS) and natural hazards. In *Encyclopedia of Natural Hazards*, ed. P. T. Bobrowsky, 416–417. Dordrecht: Springer.

Kerle, N. (2013b) Remote sensing of natural hazards and disasters. In *Encyclopedia of Natural Hazards*, ed. P. T. Bobrowsky, 837–847. Dordrecht: Springer.

Kerle, N. & D. Alkema. 2011. Multiscale flood risk assessment in Urban Areas—A geoinformatics approach. In *Applied Urban Ecology,* eds. M. Richter & U. Weiland, 93–105. Hoboken, NJ: Wiley-Blackwell.

Kerle, N., B. V. de Vries & C. Oppenheimer (2003) New insight into the factors leading to the 1998 flank collapse and lahar disaster at Casita volcano, Nicaragua. *Bulletin of Volcanology*, 65, 331–345.

Kerle, N., S. Heuel & N. Pfeifer. 2008. Real-time data collection and information generation using airborne sensors. In *Geospatial Information Technology for Emergency Response*, eds. S. Zlatanova & J. Li, 43–74. London: Taylor & Francis.

Kerle, N. & R. R. Hoffman (2013) Collaborative damage mapping for emergency response: The role of Cognitive Systems Engineering. *Natural Hazards and Earth System Sciences (NHESS)*, 13, 97–113.

Kerle, N., F. Nex, M. Gerke, D. Duarte & A. Vetrivel (2020) UAV-based structural damage mapping: A review. *ISPRS International Journal of Geo-Information*, 9, 14.

Khanal, S., K. C. Kushal, J. P. Fulton, S. Shearer & E. Ozkan (2020) Remote sensing in agriculture—accomplishments, limitations, and opportunities. *Remote Sensing*, 12.

Kirkpatrick, K. (2023) Using algorithms to deliver disaster aid. *Communications of the ACM*, 66, 17–19.

Kohli, D., R. Sliuzas, N. Kerle & A. Stein (2012) An ontology of slums for image-based classification. *Computers Environment and Urban Systems*, 36, 154–163.

Kohli, D., P. Warwadekar, N. Kerle, R. Sliuzas & A. Stein (2013) Transferability of object-oriented image analysis methods for slum identification. *Remote Sensing*, 5, 4209–4228.

Kothyari, G. C., K. Malik, R. K. Dumka, S. P. Naik, R. Biswas, A. K. Taloor, K. Luirei, N. Joshi & R. S. Kandregula (2022) Identification of active deformation zone associated with the 28th April 2021 Assam earthquake (Mw 6.4) using the PSInSAR time series. *Journal of Applied Geophysics*, 206.

Kraus, K. & N. Pfeifer (1998) Determination of terrain models in wooded areas with airborne laser scanner data. *ISPRS Journal of Photogrammetry and Remote Sensing*, 53, 193–203.

Kromker, D., F. Eierdanz & A. Stolberg (2008) Who is susceptible and why? An agent-based approach to assessing vulnerability to drought. *Regional Environmental Change*, 8, 173–185.

Kulk, G., T. Platt, J. Dingle, T. Jackson, B. F. Jönsson, H. A. Bouman, M. Babin, R. J. W. Brewin, M. Doblin, M. Estrada, F. G. Figueiras, K. Furuya, N. González-Benítez, H. G. Gudfinnsson, K. Gudmundsson, B. Huang, T. Isada, Ž. Kovač, V. A. Lutz, E. Marañón, M. Raman, K. Richardson, P. D. Rozema, W. H. V. D. Poll, V. Segura, G. H. Tilstone, J. Uitz, V. V. Dongen-Vogels, T. Yoshikawa & S. Sathyendranath (2020) Primary production, an index of climate change in the ocean: Satellite-based estimates over two decades. *Remote Sensing*, 12, 826.

Kurbanov, E., O. Vorobev, S. Lezhnin, J. M. Sha, J. L. Wang, X. M. Li, J. Cole, D. Dergunov & Y. B. Wang (2022) Remote sensing of forest burnt area, burn severity, and post-fire recovery: A review. *Remote Sensing*, 14.

Lamsal, D., T. Sawagaki & T. Watanabe (2011) Digital terrain modelling using Corona and ALOS PRISM data to investigate the distal part of Imja Glacier, Khumbu Himal, Nepal. *Journal of Mountain Science*, 8, 390–402.

Lassa, J. A. 2018. Roles of non-government organizations in disaster risk reduction. In *Oxford Research Encyclopedia of Natural Hazard Science*, 19. Oxford: Oxford University Press.

Lee, T. & T. Kim (2013) Automatic building height extraction by volumetric shadow analysis of monoscopic imagery. *International Journal of Remote Sensing*, 34, 5834–5850.

Li, L., A. Chang-Richards, M. Boston, K. Elwood & C. Molina Hutt (2023a) Post-disaster functional recovery of the built environment: A systematic review and directions for future research. *International Journal of Disaster Risk Reduction*, 95, 103899.

Li, S., X. Sun, Y. Gu, Y. Lv, M. Zhao, Z. Zhou, W. Guo, Y. Sun, H. Wang & J. Yang (2023b) Recent advances in intelligent processing of satellite video: Challenges, methods, and applications. *IEEE Journal of Selected Topics in Applied Earth Observations and Remote Sensing*, 16, 6776–6798.

Li, X., L. Yu, Y. Xu, J. Yang & P. Gong (2016) Ten years after Hurricane Katrina: Monitoring recovery in New Orleans and the surrounding areas using remote sensing. *Science Bulletin*, 61, 1460–1470.

Linardos, V., M. Drakaki, P. Tzionas & Y. L. Karnavas (2022) Machine learning in disaster management: Recent developments in methods and applications. *Machine Learning and Knowledge Extraction*, 4, 446–473.

Liu, S. L., L. Q. Wang, W. A. Zhang, Y. W. He & S. Pijush (2023) A comprehensive review of machine learning-based methods in landslide susceptibility mapping. *Geological Journal*, 58, 2283–2301.

Liu, T. M. (2014) Analysis of the economic impact of meteorological disasters on tourism: The case of typhoon Morakot's impact on the Maolin National Scenic Area in Taiwan. *Tourism Economics*, 20, 143–156.

Loew, A. (2014) Terrestrial satellite records for climate studies: How long is long enough? A test case for the Sahel. *Theoretical and Applied Climatology*, 115, 427–440.

Mackenzie, J. S. & P. S. Ringrose (1986) Use of SeaSat SAR imagery for geological mapping in a volcanic terrain—Askja-caldera, Iceland. *International Journal of Remote Sensing*, 7, 181–194.

Manzini, T., R. R. Murphy, E. Heim, C. Robinson, G. Zarrella & R. Gupta (2023) Harnessing AI and robotics in humanitarian assistance and disaster response. *Science Robotics*, 8.

Marchese, F., C. Filizzola, T. Lacava, A. Falconieri, M. Faruolo, M. Genzano, G. Mazzeo, C. Pietrapertosa, N. Pergola, V. Tramutoli & M. Neri (2021) Mt. Etna paroxysms of February-April 2021 monitored and quantified through a multi-platform satellite observing system. *Remote Sensing*, 13.

Markogiannaki, O., A. Karavias, D. Bafi, D. Angelou & I. Parcharidis (2020) A geospatial intelligence application to support post-disaster inspections based on local exposure information and on co-seismic DInSAR results: the case of the Durres (Albania) earthquake on November 26, 2019. *Natural Hazards*, 103, 3085–3100.

Martha, T. R., N. Kerle, V. G. Jetten, C. J. van Westen & K. Vinod Kumar (2010) Characterising spectral, spatial and morphometric properties of landslides for semi—automatic detection using object—oriented methods. *Geomorphology*, 116, 24–36.

Martha, T. R., N. Kerle, C. J. van Westen, V. Jetten & K. V. Kumar (2012) Object-oriented analysis of multi-temporal panchromatic images for creation of historical landslide inventories. *ISPRS Journal of Photogrammetry and Remote Sensing*, 67, 105–119.

Matin, S. S. & B. Pradhan (2022) Challenges and limitations of earthquake-induced building damage mapping techniques using remote sensing images-A systematic review. *Geocarto International*, 37, 6186–6212.

Mazhin, S. A., M. Farrokhi, M. Noroozi, J. Roudini, S. A. Hosseini, M. E. Motlagh, P. Kolivand & H. Khankeh (2021) Worldwide disaster loss and damage databases: A systematic review. *Journal of Education and Health Promotion*, 10, 13.

Michael, Y., I. M. Lensky, S. Brenner, A. Tchetchik, N. Tessler & D. Helman (2018) Economic assessment of fire damage to urban forest in the wildland-urban interface using planet satellites constellation images. *Remote Sensing*, 10.

Michellier, C., P. Pigeon, A. Paillet, T. Trefon, O. Dewitte & F. Kervyn (2020) The challenging place of natural hazards in disaster risk reduction conceptual models: Insights from Central Africa and the European Alps. *International Journal of Disaster Risk Science*, 11, 316–332.

Mika, A. M. (1997) Three decades of Landsat instruments. *Photogrammetric Engineering and Remote Sensing*, 63, 839–852.

Miller, F., H. Osbahr, E. Boyd, F. Thomalla, S. Bharwani, G. Ziervogel, B. Walker, J. Birkmann, S. van der Leeuw, J. Rockström, J. Hinkel, T. Downing, C. Folke & D. Nelson (2010) Resilience and vulnerability: complementary or conflicting concepts? *Ecology and Society*, 15.

Mishra, M. K., D. Ganguly, P. Chauhan & Ajai (2014) Estimation of coastal Bathymetry using RISAT-1 C-band microwave SAR data. *IEEE Geoscience and Remote Sensing Letters*, 11 (3), 671–675.

Mitomi, H., F. Yamzaki & M. Matsuoka. 2000. Automated detection of building damage due to recent earthquakes using aerial television images. In *21st Asian Conference on Remote Sensing*, 401–406. Taipei, Taiwan: GIS Development.

Monfort, D., C. Negulescu & M. Belvaux (2019) Remote sensing vs. field survey data in a post-earthquake context: Potentialities and limits of damaged building assessment datasets. *Remote Sensing Applications-Society and Environment*, 14, 46–59.

Muhadi, N. A., A. F. Abdullah, S. K. Bejo, M. R. Mahadi & A. Mijic (2020) The use of LiDAR-derived DEM in flood applications: a review. *Remote Sensing*, 12.

Müller, A. (2013) Flood risks in a dynamic urban agglomeration: A conceptual and methodological assessment framework. *Natural Hazards*, 65, 1931–1950.

Murphy, S. W., R. Wright, C. Oppenheimer & C. R. Souza (2013) MODIS and ASTER synergy for characterizing thermal volcanic activity. *Remote Sensing of Environment*, 131, 195–205.

Narama, C., A. Kääb, M. Duishonakunov & K. Abdrakhmatov (2010) Spatial variability of recent glacier area changes in the Tien Shan Mountains, Central Asia, using Corona (~1970), Landsat (~2000), and ALOS (~2007) satellite data. *Global and Planetary Change*, 71, 42–54.

National Research Council (U.S.). Committee on Remote Sensing Programs for Earth Resource Surveys. 1977. *Microwave Remote Sensing from Space for Earth Resource Surveys*. Washington, DC: National Academy of Sciences.

Nauman, C., E. Anderson, E. C. de Perez, A. Kruczkiewicz, S. McClain, A. Markert, R. Griffin & P. Suarez (2021) Perspectives on flood forecast-based early action and opportunities for Earth observations. *Journal of Applied Remote Sensing*, 15.

Newnham, G. J., A. S. Siggins, R. M. Blanchi, D. S. Culvenor, J. E. Leonard & J. S. Mashford (2012) Exploiting three dimensional vegetation structure to map wildland extent. *Remote Sensing of Environment*, 123, 155–162.

Nie, Y., Q. Liu & S. Y. Liu (2013) Glacial lake expansion in the Central Himalayas by Landsat images, 1990–2010. *PLoS ONE*, 8, 8.

Nishino, T. (2023) Post-earthquake fire ignition model uncertainty in regional probabilistic shaking-fire cascading multi-hazard risk assessment: A study of earthquakes in Japan. *International Journal of Disaster Risk Reduction*, 98.

Noran, O. (2014) Collaborative disaster management: An interdisciplinary approach. *Computers in Industry*, 65, 1032–1040.

Oliveira, S. L. J., J. M. C. Pereira & J. M. B. Carreiras (2012) Fire frequency analysis in Portugal (1975–2005), using Landsat-based burnt area maps. *International Journal of Wildland Fire*, 21, 48–60.

O'Loughlin, F. E., R. C. D. Paiva, M. Durand, D. E. Alsdorf & P. D. Bates (2016) A multi-sensor approach towards a global vegetation corrected SRTM DEM product. *Remote Sensing of Environment*, 182, 49–59.

O'Rourke, T. D., A. L. Bonneau, J. W. Pease, P. Shi & Y. Wang (2006) Liquefaction and ground failures in San Francisco. *Earthquake Spectra*, 22, 91–112.

Panwar, V. & S. Sen (2020) Disaster damage records of EM-DAT and DesInventar: A systematic comparison. *Economics of Disasters and Climate Change*, 4, 295–317.

Perlroth, I. (1962) Relationship of central pressure of hurricane Esther (1961) and the sea surface temperature field. *Tellus*, 14, 403–408.

Petrosillo, I., N. Zaccarelli & G. Zurlini (2010) Multi-scale vulnerability of natural capital in a panarchy of social-ecological landscapes. *Ecological Complexity*, 7, 359–367.

Picozza, P., L. Conti & A. Sotgiu (2021) Looking for earthquake precursors from space: A critical review. *Frontiers in Earth Science*, 9.

Pisano, A., S. Marullo, V. Artale, F. Falcini, C. Yang, F. E. Leonelli, R. Santoleri & B. Buongiorno Nardelli (2020) New evidence of mediterranean climate change and variability from sea surface temperature observations. *Remote Sensing*, 12, 132.

Poli, D. & I. Caravaggi (2013) 3D modeling of large urban areas with stereo VHR satellite imagery: Lessons learned. *Natural Hazards*, 68, 53–78.

Poompavai, V. & M. Ramalingam (2013) Geospatial analysis for coastal risk assessment to cyclones. *Journal of the Indian Society of Remote Sensing*, 41, 157–176.

Quintero, N., O. Viedma, I. R. Urbieta & J. M. Moreno (2019) Assessing landscape fire hazard by multitemporal automatic classification of Landsat time series using the Google Earth Engine in West-Central Spain. *Forests*, 10, 518.

Rahman, M. S. & L. P. Di (2020) A systematic review on case studies of remote sensing-based flood crop loss assessment. *Agriculture*, 10.

Ramachandran, B., C. O. Justice & M. J. Abrams. 2010. *Land Remote Sensing and Global Environmental Change: NASA's Earth Observing System and the Science of ASTER and MODIS*. New York: Springer.

Ramasamy, S. M. (2006) Remote sensing and active tectonics of South India. *International Journal of Remote Sensing*, 27, 4397–4431.

Ray, P. S. & C. E. Hane (1976) Tornado-parent storm relationship deduced from a dual-Doppler radar analysis. *Geophysical Research Letters*, 3, 721–723.

Razak, K. A., M. Santangelo, C. J. Van Westen, M. W. Straatsma & S. M. de Jong (2013) Generating an optimal DTM from airborne laser scanning data for landslide mapping in a tropical forest environment. *Geomorphology*, 190, 112–125.

Reddy, C. S. (2021) Remote sensing of biodiversity: what to measure and monitor from space to species? *Biodiversity and Conservation*, 30, 2617–2631.

Riegler, G., S. D. Hennig & M. Weber. 2015. WorldDEM—a novel global foundation layer. In *Joint ISPRS Conference on Photogrammetric Image Analysis (PIA) and High Resolution Earth Imaging for Geospatial Information (HRIGI)*, 183–187. Munich, Germany: Technische University Munchen.

Rivera-Marin, D., J. Dash & B. Ogutu (2022) The use of remote sensing for desertification studies: A review. *Journal of Arid Environments*, 206, 104829.

Romano, P., B. Di Lieto, S. Scarpetta, I. Apicella, A. T. Linde & R. Scarpa (2022) Dynamic strain anomalies detection at Stromboli before 2019 vulcanian explosions using machine learning. *Frontiers in Earth Science*, 10.

Sacchetti, A., A. Cisbani, G. Babini & C. Galeazzi. 2010. *The Italian Precursor of an Operational Hyperspectral Imaging Mission*. Berlin: Springer-Verlag Berlin.

Salami, A. T., J. Akinyede & A. de Gier (2010) A preliminary assessment of NigeriaSat-1 for sustainable mangrove forest monitoring. *International Journal of Applied Earth Observation and Geoinformation*, 12, S18–S22.

Sandau, R., K. Briess & M. D'Errico (2010) Small satellites for global coverage: Potential and limits. *ISPRS Journal of Photogrammetry and Remote Sensing*, 65, 492–504.

Sangha, K. K., J. Russell-Smith, J. Evans & A. Edwards (2020) Methodological approaches and challenges to assess the environmental losses from natural disasters. *International Journal of Disaster Risk Reduction*, 49, 101619.

Santilli, G., C. Vendittozzi, C. Cappelletti, S. Battistini & P. Gessini (2018) CubeSat constellations for disaster management in remote areas. *Acta Astronautica*, 145, 11–17.

Sarker, M. N. I., Y. Peng, C. Yiran & R. C. Shouse (2020) Disaster resilience through big data: Way to environmental sustainability. *International Journal of Disaster Risk Reduction*, 51, 8.

Satge, F., M. Denezine, R. Pillco, F. Timouk, S. Pinel, J. Molina, J. Garnier, F. Seyler & M. P. Bonnet (2016) Absolute and relative height-pixel accuracy of SRTM-GL1 over the South American Andean Plateau. *ISPRS Journal of Photogrammetry and Remote Sensing*, 121, 157–166.

Schlögl, M., G. Richter, M. Avian, T. Thaler, G. Heiss, G. Lenz & S. Fuchs (2019) On the nexus between landslide susceptibility and transport infrastructure—an agent-based approach. *Natural Hazards and Earth System Sciences*, 19, 201–219.

Schmidt, M., R. Goossens & G. Menz. 2001. Processing techniques for CORONA satellite images in order to generate high-resolution digital elevation models. In *Observing Our Environment from Space: New Solutions for a New Millennium*, ed. G. Bégni, 191–196. Lisse, The Netherlands: Swets & Zeitlinger.

Sedhain, S., M. van den Homberg, A. Teklesadik, M. van Aalst & N. Kerle (in press) Explainable impact-based forecasting for tropical cyclones *Progress in Disaster Science*.

Shafique, M., M. van der Meijde & D. G. Rossiter (2011) Geophysical and remote sensing-based approach to model regolith thickness in a data-sparse environment. *Catena*, 87, 11–19.

Shafique, M., M. van der Meijde & H. M. A. van der Werff (2012) Evaluation of remote sensing-based seismic site characterization using earthquake damage data. *Terra Nova*, 24, 123–129.

Sheykhmousa, M., N. Kerle, M. Kuffer & S. Ghaffarian (2019) Post-disaster recovery assessment with machine learning-derived land cover and land use information. *Remote Sensing*, 11, 1174.

Shruthi, R. B. V., N. Kerle & V. Jetten (2011) Object-based gully feature extraction using high spatial resolution imagery. *Geomorphology*, 134, 260–268.

Shruthi, R. B. V., N. Kerle, V. Jetten, L. Abdellah & I. Machmach (2015) Quantifying temporal changes in gully erosion areas with object oriented analysis. *Catena*, 128, 262–277.

Shruthi, R. B. V., N. Kerle, V. Jetten & A. Stein (2014) Object-based gully system prediction from medium resolution imagery using Random Forests. *Geomorphology*, 216, 283–294.

Sithole, G. & M. G. Vosselman (2004) Experimental comparison of filter algorithms for bare-earth extraction from airborne laser scanning point clouds. *ISPRS Journal of Photogrammetry & Remote Sensing*, 59, 85–101.

Smit, B. & O. Pilifosova (2003) From adaptation to adaptive capacity and vulnerability reduction. In *Enhancing the Capacity of Developing Countries to Adapt to Climate Change*, eds. S. Huq, J. Smith & R. T. J. Klein, 9–25. London: Imperial College Press.

Smith, K. (2004) *Environmental Hazards*. London and New York: Routledge.

Space Science Board—National Research Council (1974). *United States Space Science Program: Report to COSPAR 86*. Washington, DC: National Academy of Sciences—National Research Council.

Steenbeek, A. & F. Nex (2022) CNN-based dense monocular visual SLAM for real-time UAV exploration in emergency conditions. *Drones*, 6, 79.

Straatsma, M. W. (2008) Quantitative mapping of hydrodynamic vegetation density of floodplain forests under leaf-off conditions using airborne laser scanning. *Photogrammetric Engineering and Remote Sensing*, 74, 987–998.

Straatsma, M. W. & M. J. Baptist (2008) Floodplain roughness parameterization using airborne laser scanning and spectral remote sensing. *Remote Sensing of Environment*, 112, 1062–1080.

Stuffler, T., K. Forster, S. Hofer, M. Leipold, B. Sang, H. Kaufmann, B. Penne, A. Mueller & C. Chlebek (2009) Hyperspectral imaging-An advanced instrument concept for the EnMAP mission (Environmental Mapping and Analysis Programme). *Acta Astronautica*, 65, 1107–1112.

Stumpf, A. & N. Kerle (2011) Object-oriented mapping of landslides using Random Forests. *Remote Sensing of Environment*, 115, 2564–2577.

Talebi, L., A. Kuczynski, A. J. Graettinger & R. Pitt (2014) Automated classification of urban areas for storm water management using aerial photography and LiDAR. *Journal of Hydrologic Engineering*, 19, 887–895.

Thomas, C., P. Matthieu, S. Arnaud, R. Nancy, L. Christian & L. Frédéric (2022) Visualising post-disaster damage on maps: A user study. *International Journal of Geographical Information Science*, 36, 1364–1393.

Torontow, V. & D. King (2011) Forest complexity modelling and mapping with remote sensing and topographic data: A comparison of three methods. *Canadian Journal of Remote Sensing*, 37, 387–402.

Torres, R., P. Snoeij, D. Geudtner, D. Bibby, M. Davidson, E. Attema, P. Potin, B. Rommen, N. Floury, M. Brown, I. N. Traver, P. Deghaye, B. Duesmann, B. Rosich, N. Miranda, C. Bruno, M. L'Abbate, R. Croci, A. Pietropaolo, M. Huchler & F. Rostan (2012) GMES Sentinel-1 mission. *Remote Sensing of Environment*, 120, 9–24.

Torres, Y., S. Martínez-Cuevas, S. Molina-Palacios, J. J. Arranz & A. Arredondo (2023) Using remote sensing for exposure and seismic vulnerability evaluation: Is it reliable? *Giscience & Remote Sensing*, 60.

Tralli, D. M., R. G. Blom, V. Zlotnicki, A. Donnellan & D. L. Evans (2005) Satellite remote sensing of earthquake, volcano, flood, landslide and coastal inundation hazards. *ISPRS Journal of Photogrammetry and Remote Sensing*, 59, 185–198.

Tronin, A. A. (2010) Satellite remote sensing in seismology. A review. *Remote Sensing*, 2, 124–150.

Tsatsaris, A., K. Kalogeropoulos, N. Stathopoulos, P. Louka, K. Tsanakas, D. E. Tsesmelis, V. Krassanakis, G. P. Petropoulos, V. Pappas & C. Chalkias (2021) Geoinformation technologies in support of environmental hazards monitoring under climate change: An extensive review. *ISPRS International Journal of Geo-Information*, 10.

Umlauf, R. & M. Burchardt (2022) Infrastructure-as-a-service: Empty skies, bad roads, and the rise of cargo drones. *Environment and Planning a-Economy and Space*, 54, 1489–1509.

Unganai, L. S. & F. N. Kogan (1998) Drought monitoring and corn yield estimation in Southern Africa from AVHRR data. *Remote Sensing of Environment*, 63, 219–232.

UN-ISDR (United Nations International Strategy for Disaster Reduction). 2004. *Living with Risk: A Global Review of Disaster Reduction Initiatives*. Geneva, Switzerland: UN/ISDR. https://press.un.org/en/2004/iha922.doc.htm

Uprety, P., F. Yamazaki & F. Dell'Acqua (2013) Damage detection using high-resolution SAR imagery in the 2009 L'Aquila, Italy, earthquake. *Earthquake Spectra*, 29, 1521–1535.

van Aardt, J. A. N., D. McKeown, J. Faulring, N. Raqueno, M. Casterline, C. Renschler, R. Eguchi, D. Messinger, R. Krzaczek, S. Cavillia, J. Antalovich, N. Philips, B. Bartlett, C. Salvaggio, E. Ontiveros & S. Gill (2011) Geospatial disaster response during the Haiti earthquake: A case study spanning airborne deployment, data collection, transfer, processing, and dissemination. *Photogrammetric Engineering and Remote Sensing*, 77, 943–952.

Van Achteren, T., B. Delauré, J. Everaerts, N. Lewyckyj & B. Michiels (2013) A lightweight and wide-swath UAV camera for high-resolution surveillance missions. *Proc. SPIE 8713, Airborne Intelligence, Surveillance, Reconnaissance (ISR) Systems and Applications X*, 871309. https://doi.org/10.1117/12.2025192

Van Den Eeckhaut, M., N. Kerle, J. Poesen & J. Hervas (2012) Object-oriented identification of forested landslides with derivatives of single pulse LiDAR data. *Geomorphology*, 173, 30–42.

van der Meijde, M., M. Ashrafuzzaman, N. Kerle, S. Khan & H. van der Werff (2020) The influence of surface topography on the weak ground shaking in Kathmandu Valley during the 2015 Gorkha Earthquake, Nepal. *Sensors*, 20.

van der Sande, C. J., S. M. de Jong & A. P. J. de Roo (2003) A segmentation and classification approach of IKONOS-2 imagery for land cover mapping to assist flood risk and flood damage assessment. *International Journal of Applied Earth Observation and Geoinformation*, 4, 217–229.

van Westen, C. J., D. Alkema, M. C. J. Damen, N. Kerle & N. C. Kingma. 2011. *Multi-Hazard Risk Assessment*. Enschede, The Netherlands: UNU-ITC DGIM.

Vetrivel, A., M. Gerke, N. Kerle, F. Nex & G. Vosselman (2018) Disaster damage detection through synergistic use of deep learning and 3D point cloud features derived from very high resolution oblique aerial images, and multiple-kernel-learning. *ISPRS Journal of Photogrammetry and Remote Sensing*, 140, 45–59.

Voigt, S., T. Schneiderhan, A. Twele, M. Gahler, E. Stein & H. Mehl (2011) Rapid damage assessment and situation mapping: Learning from the 2010 Haiti earthquake. *Photogrammetric Engineering and Remote Sensing*, 77, 923–931.

Wang, J., Y. Yu, Q. H. Gong & S. X. Yuan (2020) Debris flow disaster risk analysis and modeling via numerical simulation and land use assessment. *Arabian Journal of Geosciences*, 13.

Weiss, M., F. Jacob & G. Duveiller (2020) Remote sensing for agricultural applications: A meta-review. *Remote Sensing of Environment*, 236.

White, G. F. 1974. Natural hazards research: Concepts, methods, and policy implications. In *Natural Hazards: Local, National and Global*, ed. G. F. White, 3–16. New York: Oxford University Press.

Womble, J. A., R. L. Wood & M. E. Mohammadi (2018) Multi-scale remote sensing of tornado effects. *Frontiers in Built Environment*, 4.

Wu, H., Z. P. Cheng, W. Z. Shi, Z. L. Miao & C. C. Xu (2014) An object-based image analysis for building seismic vulnerability assessment using high-resolution remote sensing imagery. *Natural Hazards*, 71, 151–174.

Yao, X. Y., W. Q. Wang & Y. T. Teng (2022) Detection of geomagnetic signals as precursors to some earthquakes in China. *Applied Sciences*, 12.

Zhang, J., M. Y. Zhang & G. Li (2021a) Multi-stage composition of urban resilience and the influence of pre-disaster urban functionality on urban resilience. *Natural Hazards*, 107, 447–473.

Zhang, N., F. Nex, N. Kerle & G. Vosselman (2021b) Towards learning low-light indoor semantic segmentation with illumination-invariant features. *The International Archives of the Photogrammetry, Remote Sensing and Spatial Information Sciences*, XLIII-B2–2021, 427–432.

Zhang, N., F. Nex, G. Vosselman & N. Kerle (2022) Training a disaster victim detection network for UAV search and rescue using harmonious composite images. *Remote Sensing*, 14, 2977.

Zhang, Y. & N. Kerle. 2008. Satellite remote sensing for near-real time data collection. In *Geospatial Information Technology for Emergency Response*, eds. S. Zlatanova & J. Li, 75–102. London: Taylor & Francis.

7 Humanitarian Emergencies
Causes, Traits, and Impacts as Observed by Remote Sensing

Stefan Lang, Petra Füreder, Olaf Kranz, Brittany Card, Shadrock Roberts, and Andreas Papp

ACRONYMS AND DEFINITIONS

ATCOR	Atmospheric/topographic correction for satellite imagery
CEMS	Copernicus Emergency Management Service
CNL	Cognition network language
DAC	Development assistance committee
DL	Deep learning
DRC	Democratic Republic of the Congo
EO	Earth observation
ESA	European space agency
FDP	Forcibly displaced people
HCS	Hyperspherical color sharpening
HDP	Humanitarian development peace
HDX	Humanitarian Data Exchange
HOT	Humanitarian OpenStreetMap Team
ICCM	International network of crisis mappers
IDP	Internally displaced persons
IPIS	International peace information service
LRRD	Linking relief, rehabilitation and development
LSHTM	London School of Hygiene and Tropical Medicine
OBIA	Object-based image analysis
OCHA	United Nations Office for the Coordination of Humanitarian Affairs
MK	Mann-Kendall
NDVI	Normalized Difference Vegetation Index
NIR	Near infrared
SAF	Sudan armed forces
SESA	Copernicus Service on Support to EU External and Security Actions
SK	Seasonal Kendall
SPLA	Sudan People's Liberation Army
UAV	Unmanned aerial vehicles
UN	United Nations
UNMIS	United Nations Mission in Sudan
UNOOSA	United Nations Office for Outer Space Affairs
VHR	Very high resolution
VGI	Volunteered geographic information
V&TC	Volunteer and Technical Communities

DOI: 10.1201/9781003541417-9

7.1 INTRODUCTION

7.1.1 Humanitarian Disasters: A Particular Case?

Drawing a sharp conceptual line between natural and humanitarian disasters is difficult because of the mutual relationships amongst them that all too often lead to humanitarian crises (Lang et al., 2020; Lang et al., 2018). In fact, the notion of any "disaster" has a human component, not necessarily in a causal, but always is an affected sense. Without human reference or any impact to the anthropogenic sphere, a natural event like an earthquake, a landslide, or a river flood, would rather be considered a disturbance in the sense of an episodic event inherent to an (eco-)system's integrity. Moreover, due to climate change and other large-scale anthropogenic effects, natural disasters apparently increase both in terms of occurrence and severity (Munyaka et al., 2024; Jiang et al., 2017; IPCC, 2001), thus the term "natural" is even more deceptive. In order to find a (pragmatic) borderline to literature dealing with natural disasters in a stricter sense, we shall concentrate on disasters that are either caused or reinforced by crises or conflicts, whether they ultimately root in natural (e.g., a drought spanning over several years) or societal causes (any type of aggression, fights over resources, etc.). This corresponds, again with many transitions and uncertainties, to the field of humanitarian action. Unlike natural disaster response, which operates in fairly distinct phases in a rather distinct disaster management cycle (Huang et al., 2023; Avtar et al., 2021; Lang et al., 2018; Joyce et al., 2009), humanitarian action faces more gradual, at times protracted, response phases (see later in this chapter). In particular, humanitarian conflict situations often lack a distinct peak situation (as compared to catastrophic events such as floods, earthquakes, or wildfires); thus it is hard to pinpoint the exact point in time when a man-made conflict leads to a humanitarian disaster (Sadiq et al., 2022; Zheng et al., 2021; Jiang et al., 2017; Winkler et al., 2017). While conceptually disconnected here, natural disasters or resource scarcity / abundance may overlay and reinforce conflict situations or contribute to secondary risks through the outbreak of a disease or other calamities (as, for example, in the case of the Haiti earthquake in 2010).

7.1.2 Forced Migrations and Regional Conflicts

The most obvious, and often the most adverse, impact of humanitarian emergencies is the forced displacement of large numbers of people. Next to natural disasters and changing environmental conditions (land degradation, desertification, large-scale land investments, etc.), violent regional conflicts are among the main drivers that make people flee their homes (Louw et al., 2022; Lang et al., 2020). As indicated earlier, the causes are often multi-layered (Humanitarian Coalition, n.d.), with systemic and reinforcing cycles, and ultimately lead to high amounts of population displacements (see Figure 7.1).

Among the total crowd of forcibly displaced people (FDP), we differentiate between internally displaced persons (IDPs) and refugees, depending on whether persons were displaced within their country of origin or crossed an international border. Currently, there are about 117.3 million people forcibly displaced, which is a number that has doubled compared to ten years ago, among them 68.3 million IDPs due to conflicts and violence, 31.6 million refugees, 6.9 million asylum seekers, 6 million Palestine refugees, and 5.8 million people in need of international protection (UNHCR, 2023; IDMC, 2024). In addition, 7.7 million people were living in internal displacement in 2023 as a result of disasters (IDMC, 2024). Most of these people gather in camps or informal settlements. The prevalence of human-made crises in the shadow of armed conflicts and wars has led to an ever-increasing number of displaced people in the last decades. The total number of people living in internal displacement increased by 51% over the past five years. Nearly half of the world's internally displaced people are hosted by Sudan, with 9.1 million the highest number of IDPs ever reported, Syria, the Democratic Republic of the Congo (DRC), Colombia, and Yemen (IDMC, 2024).

Humanitarian Emergencies

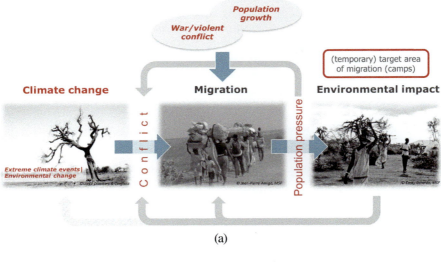

(a)

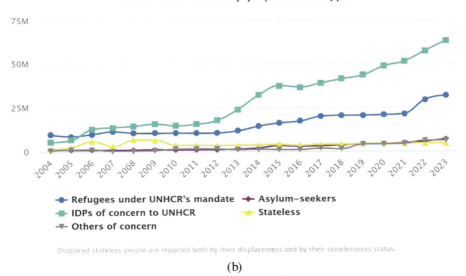

(b)

FIGURE 7.1 Above: Conflict-related migrations and environmental impact of large-scale spontaneous settlements influencing each other in a reinforcing feedback loop potentially reinforced by climate change. Below: numbers of internally displaced persons (IDPs) and refugees since the 1990s.

Source: UNHCR (2024)unhcr.org/refugee-statistics/download/?url=xB3e9E.

Forced migrations, the raising of large spontaneous settlements, most often in areas scarce in resources anyway, is a global problem with strong local implications. In turn, increasing pressure on scare resources is a common factor for intensification or even the outbreak of new local conflicts. The UNEP Expert Advisory Group on Environment, Conflict and Peacebuilding stated in 2009 that "there is significant potential for conflict over natural resources to intensify in the coming decades" (UNEP, 2009) (p. 5, see also Walker et al. (2012) for facts and figures on this issue). Conflicts over natural resources are one of the most frequent conflict items in sub-Saharan Africa

(HIIK, 2013), but they do not only occur in countries that are scarce in natural resources. Resource abundance might also lead to violent conflicts—the so-called resource curse (Ye, 2022; Zheng et al., 2021; Lang et al., 2020; Le Billon, 2001; Mildner et al., 2011), as, for example, related to the financing of weapons, supplies, and corruption through revenues from the exploitation of natural resources. Extraction and trade of minerals, timber, oil, and other resources are often controlled or even conducted by armed opposition groups or private security companies. As a result, the conflict situation intensifies, eventually leading to a complex emergency.

7.1.3 THE ROLE OF SATELLITE REMOTE SENSING IN HUMANITARIAN ACTION

Due to its observational power, ubiquitous usage, and availability, remote sensing in combination with advanced analysis techniques have become a decision-supporting tool for humanitarian professionals (Bjorgo, 2001). Remote sensing and related derived information products are complementary data sources to field-based surveys enriching the pool of spatially aware technologies (Sadiq et al., 2022; Zheng et al., 2021; Koshimura et al., 2020; Cowan, 2011; Verjee, 2011) for humanitarian relief support. This chapter focuses on the use of remotely sensed data from Earth observation (EO) satellites, i.e., orbital sensors in support to civilian applications in various societal benefit areas.

7.1.3.1 Objectivity, Sensitivity, and the Crowd

Above all, remote sensing has the advantage of an objective imaging device that captures data over large areas under equal conditions. This is much like a "neutral" observer's camera, just covering a much larger extent and taken from an orbital view. Thus, remote sensing entails a trend for a "democratizing tool" (Lang et al., 2020; Lang et al., 2018; Lang et al., 2013), designed to reveal the situation "as is," non-distorted, non-manipulated, and potentially accessible to everyone.

The potential for this accessibility of remote sensing data to act as a convening point among humanitarian actors was noted as commercial satellite imagery was becoming available in the early 2000s, with one significant concern: it was thought that the high cost and limited availability could mitigate its wide-spread use (Aimaiti et al., 2022; Kucharczyk and Hugenholtz, 2021; Quinn et al., 2018; Bjorgo, 2001). However, there is a large variety of different optical as well as radar sensor types available today that can provide information at all kinds of spatial, temporal, and spectral resolution applicable for many different humanitarian disaster scenarios. For slow-onset disasters such as droughts, floods, or diseases, frequent coverage of large areas is ensured through high- to medium-resolution satellites (e.g., MODIS, Landsat, Sentinel-2). Analyses of soil moisture in monitoring the emergence of droughts medium-resolution radar data (e.g., ScanSAR modes of TerraSAR-X, Radarsat, Cosmo-SkyMed). The same applies to floods (Sentinel-1), which frequently occur in the context of cyclones and other extreme seasonal rainfall patterns. The derived information is crucial for early warning, for better response to and mitigation (or even prevention) of humanitarian disasters.

Since very high resolution (VHR) imagery have become commercially available more than two decades ago, they are a key source humanitarian cartographic products as a powerful means of collaboration between humanitarian agencies and the broader public: especially through crowdsourcing—the distribution of a task to a large group of undefined people.[9] Advancements in web technology allow a wide range of nontraditional actors to engage during a humanitarian response. Indeed new forms of online participants—often called "volunteer and technical communities" (V&TC)—have emerged with the specific goal of supporting humanitarian response, see Capelo et al. (2012). Because many of the features found in VHR imagery are recognizable by non-specialists there have been several cases of humanitarian organizations experimenting with "volunteered geographic information" (VGI) derived by visual image interpretation.

This phenomenon became obvious in the creation of vector data, manually digitized from satellite imagery, in Port-au-Prince, Haiti, in the hours following the devastating earthquake of 2010. Using OpenStreetMap (OSM, an online mapping platform and database that allows users to trace satellite imagery to create open data), several hundred online volunteers created more than 1.4 million edits in the weeks following the quake, making the OSM of Haiti the primary source of cartographic data for humanitarian responders. Since then, a Humanitarian OpenStreetMap Team (HOT) has formed to act as a bridge between the humanitarian responders and the OSM community. Other online platforms that make VHR imagery available for interpretation, such as Tomnod, have experimented with the enumeration of IDP dwelling units in Somalia (SBTF, 2011).

This type of public engagement, whether solicited or not by humanitarian agencies, presents both opportunities and challenges for the humanitarian sector. The data quality of any type of VGI is a long-standing research focus (Elwood, 2008); ensuring that the fundamentals of imagery interpretation, such as color interpretation, minimum mapping unit or other functions of scale, are understood and consistently applied by the general public, is a challenge. Interestingly research concerning the spatial accuracy and precision of data found within OSM indicate no reason, *a priori*, to dismiss these types of data since their accuracy can be quite good compared to "authoritative" data sets (Hakley, 2010). However, a comparison of a crowdsourced damage assessment with field-based assessments in the Philippines found that, the OpenStreetMap community did reasonably well at identifying affected buildings, but were less accurate when reporting the type of damage sustained. Overall, when compared to field data, completely destroyed buildings were overestimated by 134%, while major damaged and partial / undamaged buildings were underestimated by 25% and 18%, respectively. The highest uncertainty revealed the class "destroyed building," where only 16% were actually destroyed, around 70% were actually majorly or partially damaged. Conversely, buildings tagged as "undamaged" actually had major damage or were destroyed 50% of the time. Overall, the proportion of buildings that were accurately tagged by OSM contributors was only 36%. The assessment concluded that the OSM data was not reliable enough to utilize for damage analysis and recovery planning, but still sees OSM as strong platform for damage assessments in the future, if modest technological investments and better coordination mechanisms are provided (ARC, 2014).

The potential surge of unsolicited, crowdsourced data that may be of a type or format not completely fitting to existing humanitarian data pipelines or workflows has created challenges for humanitarian responders and the simple fact that non-professionals are engaged in any aspect of humanitarian response may cause consternation[10] (HHI, 2011). However, some humanitarian institutions are paving the way for the public use of VHR imagery to support humanitarian operations. The Humanitarian Information Unit of the US Department of State, together with representatives from the National Geospatial Intelligence Agency, and the US Agency for International Development, met with the Humanitarian OpenStreetMap Team to establish the necessary policies and protocols for federally purchased VHR imagery to be served via OSM precisely for the engagement of the V&TC. This process, known as "Imagery to the Crowd" has been used to map refugee camps in Kenya and Ethiopia. The increasing availability of refugee camp data available in OSM together with emerging disaster risk reduction projects that rely heavily upon it (Soden et al., 2014) and provide guidance for its use, suggest that leveraging the crowd to interpret VHR imagery will become increasingly common.

While the use of remote sensing has great advantages for direct observations there are some issues from an ethical point of view. This particularly applies to the use of VHR sensors (Slonecker et al., 1998). Such imagery, provides a level of detail that can detect features relating to individual privacy (e.g., housing, cars). While individual persons cannot be identified on current satellite imagery, larger groups of people such as a refugee trails can be traced (Ehrlich et al., 2009). The sensitivity of remotely sensed image data increases with greater resolution towards the level of standard digital cameras, in the case of any kind of in situ, i.e., near-ground, remote sensing devices. The usage of unmanned aerial vehicles (UAVs, a.k.a. drones) is currently discussed from this perspective. There are ethical issues attached to the usage of UAVs, which often resemble small military drones and might be mistaken.[11] Still, as technology as such is neutral, they offer a wide range of options not limited to imaging but also in support to logistical tasks or even medical treatment, some in operational stage (see, for example, WFP[12]).

7.1.3.2 Indication Based on Time Series

Remote sensing technology provides critical information in each phase of humanitarian crisis response (see Section 7.1.4). Before discussing technical details, we broadly reflect on the general capacity of EO systems in response to information needs by humanitarian actors. Satellite imagery captures a "snapshot in time" capable to represent the dynamic of a given phenomenon in steps only (Avtar et al., 2021; Zheng et al., 2021; Lang et al., 2018). To an increasing degree, satellite sensors provide sequential imagery under comparable conditions, a crucial prerequisite to any kind of monitoring activity (Aschbacher, 2002). As in any other application domain, the time span between two captured scenes is critical to understanding changes or directions of non-static events accordingly. Thus, the frequency at which imagery is captured needs to be commensurate to the dynamics of a situation (see Section 7.3).

Just like other remote sensing applications, EO-based crisis response relies on suitable indicators, as usually the phenomenon under concern is not directly observable. Depending on the information need (see Section 7.1.4) and the respective type (spatial/spectral resolution, passive/active, etc.), such indicators can be derived in different scales and extent, as the following examples illustrate: (1) the prevalence of a specific dwelling type (e.g., a tent) derived within a refugee camp area from sub-meter WorldView-2 data may represent vivid temporary living conditions amount of people based on an average occupation rate. (2) A high NDVI (Normalized Difference Vegetation Index) derived on the district level from medium-resolution MODIS data may point to intensified agricultural activities, suggesting increasing human presence; or in contrast a decreasing vegetation index over time might indicate extraction of wood resources due to locally increasing population in the camps. (3) A specific pattern of linear structures in vegetation-scarce areas observed on Landsat imagery may indicate geological fault zones likely to store groundwater.

7.1.3.3 Remote Sensing vs. Field Mapping

In comparison to conventional terrestrial field mapping and observations on the ground, EO-based humanitarian response benefits from general assets of remotely sensed data (Lang et al., 2020; Braun et al., 2019; Lang et al., 2018; Quinn et al., 2018):

- **From a distance**. The core principle of remote sensing data, its obtainment from indirect contact with the object of concern, is critical to crisis-related applications. Often the area affected by the crisis is inaccessible or difficult to reach or from a security point of view too dangerous to enter (Aschbacher, 2002). This means that information derived from remote sensing is the only information available.
- **Area-wide coverage**. Depending on the granularity (spatial resolution), areas can be covered with variable extent, under the same imaging conditions and characteristics. The trade-off between resolution and extent is thereby a limiting factor that is also reflected in costs and timeliness of data provision.
- **Global availability**. Satellite data are globally available with a theoretical cover rate of some 95% of the inhabitable space of the globe and factually 100% of the permanent settlement area. Note that cloud cover (e.g., in tropical latitudes) is a limiting factor for data acquisition and analysis (see Section 7.3.2).
- **Retrospective view**. Time series not only enable constant monitoring in future time steps, but also ex-post assessments by past sequences. This is a key factor for estimating detected trend patterns in a more reliable matter (see Figure 7.4).

Humanitarian Emergencies

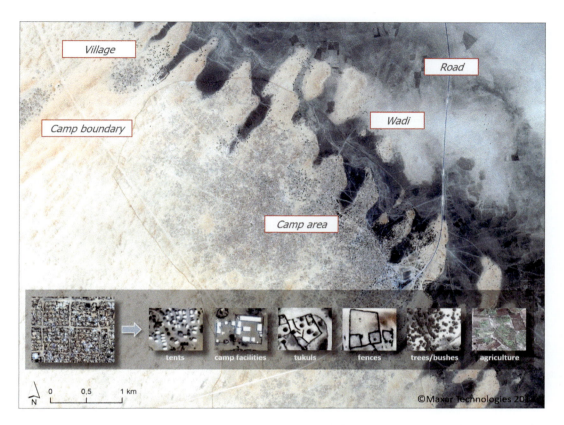

FIGURE 7.2 Detail and overview of a camp area provided by VHR data (here: Zam Zam IDP camp, Darfur, Sudan, WorldView-2). Depending on the type of camp, different dwelling structures can be identified besides camp infrastructure and surrounding land cover. The number of single dwellings, derived by visual interpretation or automated feature extraction, can be used for population estimations.

7.1.3.4 Space Policy and Regulations

In support to disaster management using information derived from remotely sensed data there are several international, regional, and national mechanisms in place. One of them is the International Charter on Space and Major Disasters. This is an agreement among space agencies around the globe to provide satellite-based data and information in support of relief operations during emergencies and crises caused by major disasters (Voigt et al., 2007). Such disasters can either be natural or man-made, while the latter only include oil spills and industrial accidents and not conflict or crisis situations. Several national space agencies as well as the European Space Agency (ESA) participate in this endeavor and provide critical information resources to mitigate the effects of disasters. The Charter can be activated by dedicated authorized users, such as disaster management authorities from countries of Charter members. Up to now, through this mechanism, imagery free of charge has been provided in more than 900 cases since the first activation in 2000, cf. www.disasterscharter.org.

Next to the provision of (raw) imagery there are organizations or institutionalized services to provide dedicated information products derived from satellite imagery: Since 2012 the European Copernicus[13] Emergency Management Service (CEMS) is operational. This service provides timely geospatial information generated from satellite remote sensing images and in situ or open data sources to actors involved in the management of natural disasters, man-made emergencies and humanitarian crises. Besides a rush mode providing information rapidly for immediate response for emergency management a non-rush mode is dedicated to post-disaster needs assessments and FDP camp monitoring. In addition to the EMS, the European Space program Copernicus has established a service providing geospatial intelligence on support to EU External and Securtiy Action (SESA). Information needs are—amongst others—related to conflict prevention and mitigation also including support to humanitarian crisis management.

REACH is a globally operating, joint initiative of international nongovernmental organizations to "to strengthen evidence based decision-making by humanitarian actors through efficient data collection, management and analysis in contexts of crisis, disaster and displacement" (https://www.impact-initiatives.org/what-we-do/reach/). One of the key players, UNITAR's operational satellite application program (UNOSAT), delivers satellite solutions and analysis to relief and development organizations within and outside the United Nations (UN). The aim is to provide reliable information to those who work at reducing the impact of crises and disasters and help nations plan for sustainable development (www.unitar.org).

A critical issue during disasters is the collection and provision of access to timely and reliable information. When it comes to this exchange of disaster-related information three networks need to be mentioned: ReliefWeb (www.reliefweb.int) and the Humanitarian Data Exchange (HDX)Platform (https://data.humdata.org) with more than 20,000 open datasets, both managed by UNOCHA and UN-SPIDER (www.un-spider.org). These platforms focus on providing fast and efficient access to space-based information to involved actors by compiling links, information and data on major disasters.

Along with the aforementioned institutional entities, new networks of humanitarian actors have formed. The International Network of Crisis Mappers (ICCM) was founded in 2009 and is composed of more than 9,600 members worldwide whose experience with remote sensing in the humanitarian sector ranges from professional to volunteer.[14] ICCM maintains a "Google Group" that has become an important forum for ad-hoc coordination of geospatial information and remotely sensed data. Its members often point one another to existing sources of remotely sensed information and the various V&TC that are part of the network have, in some cases, engaged directly with satellite imagery providers to acquire imagery. These providers often alert ICCM if they are releasing satellite imagery without charge, as was the case in Haiti when Maxar Technologies (formerly DigitalGlobe) made a significant number of VHR images publicly available at no cost. The last annual conference was held in 2016.

Questions often arise regarding the availability of satellite technology, whether anyone (including NGOs) can access satellite imagery? The general answer is yes—provided the technical and financial means are available for data handling and licensing costs. Neither the operation of satellites nor the acquisition of satellite data is bound to territorial sovereignty, as, for example, applicable to air-space control. The International Space Law and related principles pursued by the United Nations Office for Outer Space Affairs (UNOOSA) underlines that remote sensing is

> for the benefit and in the interests of all countries, irrespective of their degree of economic, social or scientific and technological development, and taking into particular consideration the needs of the developing countries. [It shall promote] the protection of the Earth's natural environment [and] the protection of mankind from natural disasters.

This was phrased in a time (in the 1980s), where VHR data did not play a role yet in civil applications. Today, with sub-meter resolutions available, there are ethical issues arising whose regulations are being discussed or yet to come (Slonecker et al., 1998).

In the last decade(s), volunteered geographic information (VGI; Goodchild, 2007) has become an invaluable data source and instrument to engage laypersons in support to geohumanitarian action in crises and disasters. The aforementioned Humanitarian OpenStreetMap Team (HOT) and the NGO-conjoint initiative Missing Maps stimulate crowd-mapping for humanitarian mapping. Volunteers from all over the world contribute to EO image interpretation from their home or any other place. To ensure the required high level of data quality, there are rigid editing and checking mechanisms in place.

7.1.4 Information Needs in Humanitarian Action

In general, humanitarian action relies on firsthand, reliable information on the development of the situation within a certain region of interest. These areas are often (too) remote or (too) insecure or the situation is too unstable to gather the required information timely in the field. Through satellite-based assessments information can be provided for supporting the strategic planning of humanitarian relief missions (Fakhri and Gkanatsios, 2021; Lang et al., 2020; Winkler et al., 2017; Bjorgo, 2001; Kranz et al., 2010; Tiede et al., 2013). This kind of information is generally (highly) sensitive, as it might be used for help and support, but likewise misused for any tactic movements or strategic decisions in the conflict itself.

With respect to the phases of humanitarian crisis response, remote sensing can provide the following critical information:

1. **Information in support of early warning or disaster preparedness** (see Section 7.2.1). Prior to the outbreak of a humanitarian disaster, remote sensing provides the necessary synoptic viewpoint that is strategically critical to better understand the severity, geographical focus, and characteristics of a developing emergency situation, thereby improving preparedness for a disaster (e.g., drought). Likewise, the dynamics and directions of conflict-related destabilizing trends can be identified[15] to support planning of appropriate countermeasures (e.g., outbreak of armed violence, mass displacement). Situational awareness also includes the maintenance and cohesion of geospatial data layers for ensuring an adequate preparedness level.
2. **Information about the current situation** (see Section 7.2.2) **in support of crisis monitoring and humanitarian action**. In the course of a conflict or during the peak of a disastrous event, remote sensing can be used to direct humanitarian response activities,[16] including indications (and verifications) of the affected population and settlements, destroyed infrastructure and other assets, or large-scale displacements and potential secondary crisis scenarios due to emergence of spontaneous settlements and pressure on local resources. Humanitarian actors are often being forced to leave—partially or fully, and often unclear for how long—their emergency response program in a conflict zone due to the unacceptable security situation for their own personnel, highlighting the importance of remote sensing services as the only reliable source of information about the developments on the ground.
3. **Information on the mid- and long-term effects of humanitarian disasters (see Section 7.2.3) in support of potential integration and rehabilitation**. Remote sensing data and image analysis help understand the effects of displacements and the impact on environmental conditions and resources, livelihoods and land-use practices, as well as resettlement and repatriation scenarios with (potentially) recursive consequences.

In the following, the example of EO-based refugee/IDP camp monitoring (Bjorgo, 2000; Giada et al., 2003; Kranz et al., 2010; Lang et al., 2010; Kemper and Heinzel, 2014) illustrates the information needs according to the underlying questions of humanitarian relief organization during the different phases of disaster management. In order to prepare for a humanitarian crisis situation in a certain region, indicators need to be investigated that are suitable for risk assessment and early warning. The monitoring of large population movements and respective agglomeration might be a sign for increasing potential of local conflict. The same is true for decreasing availability of natural resources in the vicinity of camps due to increased extraction of water and firewood. During humanitarian disasters information about the development of the situation in and around certain camps is required, focusing on an effective camp management or even mission planning. Generated information includes population dynamics, camp development and structure as well as, impact on the environment, including potential pressure on natural resources. Underlying questions are related to the treatment of the inhabitants with food, water, and medicine as well as shelter and where the supply is most needed. In addition, it is important to gather information about the security situation in the area. Related to this issue, it is important to get information about arising local stress factors that might lead to conflicts through certain military or other violent attacks on camps and settlements in the region. The phase of integration and rehabilitation requires information suitable for supporting progressive stabilization of the situation, such as conflict sensitive programming of the emergency response measures and the following rehabilitation phase in a conflict-prone region. Besides a comprehensive picture about the effects of displacement and related environmental impact, this includes also reliable information about the effects and possibilities of long-term integration of migrants. Local governments and relief organizations require information that allow for estimating or even modeling the sustainability of the entire region with respect to natural resources (water, firewood, and building timber), setup of infrastructure, traditional power structures, tribal identities, and development of socioeconomic parameters.

7.2 CRISIS-RELATED EARTH-OBSERVABLE INDICATORS

Satellite Earth observation is said to be "ubiquitous," which relates to the general capacity to cover any point on Earth under orbit, but unfortunately not at the same time. Remote sensing data are limited to snapshots in time, even though the time spans in between can be quite low, up to the range of a few days or wtih new satellite generation (e.g. Worldview Legion) even hours (Louw et al., 2022; Sadiq et al., 2022; Ye, 2022; Zheng et al., 2021; Lang et al., 2020). Here we have to distinguish between two cases: EO satellites of moderate (medium to high) spatial resolution cover that record permanently while orbiting the Earth. The USGS Landsat 8 program as well as the Copernicus Sentinel-2 program, for example, delivers data in a fixed, sequential tracking mode, covering any particular location on Earth in a regular interval, e.g., every 16 days for Landsat and five days for Sentinel (twin constellation). On the other hand, VHR sensors, like WorldView or Pléiades, capture and deliver data on demand. Depending on the technical setup (skewing sensor, sensor constellation, etc.), a higher frequency of coverage can be achieved as the orbital revisiting rate allows. The design-depending availability makes remote sensing a responsive, reactive tool that either requires searching (in the archives), waiting or tasking. In other words, whether or not remotely sensed imagery is available is a function of ideal capturing conditions ("chance") and a certain trigger ("action"). Again, this is very much like a journalist's camera capturing a certain event, while on a different scale. In reality, even if the sensor would be ready and in place, with an ideal temporal matching scenario, other constraints may hamper the availability of a "perfect scene" in the end: those are atmospheric conditions, natural conditions (e.g., seasonal vegetation cycles), shutter controls, and, last but not least, costs involved.

7.2.1 "Early Warning"

Any capability of remote sensing to act as an "early warning tool" is highly desirable when preventing conflict from escalating or a natural event from turning into a catastrophe (Huang et al., 2023; Avtar et al., 2021; Kucharczyk and Hugenholtz, 2021; Zheng et al., 2021; Lang et al., 2018). While there are many encouraging examples for this potential (c.f. the Famine Early Warning Systems Network FEWSNET[17]), remote sensing data have limited predictive power due to their indirect measurements, availability, and continuity, and also because of the unpredictability inherent to the systemic effects of such events, in particular of the human behavior. To consider the limited foresight potential, the term "prediction" will be replaced with "indication" in this context, as it is difficult to tell whether any observed parameter is an indication of something already going on or yet to come. Examples of the indicative power of remote sensing that can affect levels of sensitivity and preparedness include:

- Anomalies in time series of soil-moisture data as compared to normal seasonal variability as an indication for a forthcoming drought threatening food security (Wagner et al., 2003; Kuenzer et al., 2008; Rhee et al., 2010).
- Monitored rapid growth of a refugee/IDP camp may indicate the rise of additional security and safety issues within the camp, as well as the conflicts related to the diminishing surrounding resource supply (Hagenlocher et al., 2012) (see Section 7.5.2).

Humanitarian Emergencies

- A damage density map showing different magnitudes of building damages and their distribution caused by an earthquake, flood, or conflict-related destruction, may indicate areas most affected and in most urgent need to response (Pesaresi et al., 2007; Tiede et al., 2011).
- Detected areas where illegal or informal activities (logging, mining, cropping, etc.) are carried out in an increasing scale, can be an indicator for potentially upcoming regional instabilities and conflicts (Schöpfer and Kranz, 2010; Luethje et al., 2014).
- Decreasing vegetation cover near camps might indicate increasing pressure on natural wood resources, longer distances for the local population to walk for collecting firewood (and consequently growing insecurity especially for women) and potentially leading to regional conflicts.
- Rapid conversion of a stretch of land from unmanaged savannah or forest into a large agricultural complex may represent large-scale land investments that potentially imply the displacement of smallholders and pastorals, destruction of villages, higher pressure on local resources due to intensification of production and the like (FAO, 2013) (see Figure 7.3).

The initial statement on the limited predictive power of remote sensing may sound a bit discouraging. This is indeed the most challenging and most research-intensive aspect of this technology. Herein this application domain competes, as many others, with the legitimate power of operational weather forecast. The key ingredient of any indication that shall gain more robustness in terms of evidence (backward-looking) or prediction (forward-looking), is repeated analysis based on time series. Time series capture the dynamics of a phenomenon and allow filtering out anomalies from the regular case or a higher degree of change than usual. Notably, all the aforementioned examples contain some changing conditions, indicated by the word "rapid." A single snapshot in time needs either subsequent, recursive information, or reference information from a time slot in the past. In addition, as in many other technological fields, the complementarity of tools and available data collection devices may be the key to trigger efficient action "ahead of time."

7.2.2 Crisis Monitoring

Again, the exact stage when a humanitarian crisis begins is often difficult to determine, and the period prior to a humanitarian crisis, during and afterwards, is mostly rather a gradual transition than point in time clearly to be spotted out. Here we examine the immediate effects that accompany such disasters and how the crisis can be monitored and further escalation mitigated (Lang et al., 2020; Braun et al., 2019; Lang et al., 2018; Quinn et al., 2018).

 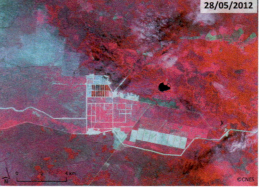

FIGURE 7.3 Conversion of a piece of land from forest, shrubland, and grassland to large-scale agricultural fields, including construction of roads and irrigation systems in Gambella, Ethiopia, visible on VHR data from SPOT-5 between 2005 and 2012.

- The provision of drinking water is the first and foremost key prerequisite when maintaining a camp for forcibly displaced people. Remote sensing can help to narrow down the range of (if not allocate) potential groundwater supply (Drury and Deller, 2002).
- Understanding the population dynamics within a camp including the general growth of the camp (extent), camp structure, densification of camp sections, partly de- or reconstruction, etc. Remote sensing can help estimate the overall number of people present in a camp as well as provide additional indications like population densities, camp management facilities, etc.
- Monitoring the impact of refugee or IDP camps on the surrounding environment, e.g., based on changes in wood resources or agricultural activities, helps assessing the increasing pressure on natural resources and the negative impact for the so-called host communities living in the wider camp area and might support decisions for mitigation and even prevention of emerging conflicts (transitions to mid-term impact).

7.2.3 Mid- to Long-Term Impact

Humanitarian crises, in particular complex or protracted crises, can have a long-term impact on the societal and environmental integrity of the affected area. Remote sensing can help assess this impact through the use of time series, including in a retrospective view (Lang et al., 2020; Lang et al., 2010; Hagenlocher et al., 2012) (see Figure 7.4).

- The prolonged existence of temporary camps[18] may transition to a semi-permanent settlement with critical effects on both the societal integrity of the hosting community(ies) and the carrying capacity of the environment (Martin, 2005). Remote sensing can help to analyze the spatial impact of such large-scale, long-lasting settlements in terms of infrastructure, food, and resource supply and environment.
- The impact of resource extraction, such as logging and exploitation of minerals on the environment is a critical issue towards establishing a sustainable management of natural resources and to mitigate negative impact on the local population and through this a contribution to mid- to long-term stabilization of a certain conflict region.

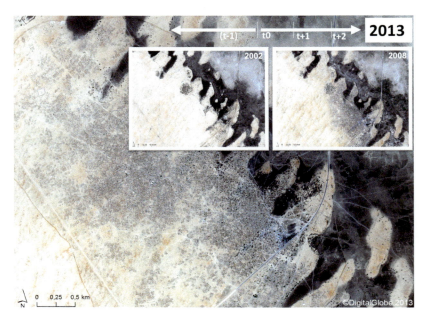

FIGURE 7.4 Retrospective view using VHR satellite time series over Zam Zam IDP camp, Darfur, Sudan.

- The mid- to long-term monitoring of natural resources in conflict regions provides important information for international transparency initiatives[19] to be set in place. This is also contributing to stabilization and peace-building in the affected region.

7.3 (SATELLITE) EARTH OBSERVATION CAPACITIES

This section provides a broad synopsis of satellite sensors which are currently used in the context of humanitarian action and human disaster response (see also the section on case studies, where the actual usage is demonstrated in various examples). As in all other application domains of remote sensing, there are evolving dynamics on at play (Avtar et al., 2021; Zheng et al., 2021; Lang et al., 2020; Winkler et al., 2017), including two major components. (1) Technical developments with new sensor technology, image retrieval and analysis, as well as mobile communication devices. (2) User uptake with changing attitudes in acceptance and debate of such technological assets over recent years. A match between these two, the technical capacity and the trust and willingness to use it, ensures technology may turn out to be a clear asset for the user demand.

7.3.1 Usage of EO Data—Chances and Challenges in the Humanitarian Development Peace (HDP) Nexus

While talking about assets and advantages of this technology, we acknowledge that there is still a gap between the capacity and the actual usage. But what are the main concerns that the humanitarian community of practice brings about the usage of remote sensing? It is hard to find those clearly reflected in scientific publications; they appear inherently during workshops and discussion rounds.

However, today's crises are increasingly long-lasting, recurring, complex, and interdependent, which makes a more integrated, efficient, and sustainable way of working necessary. The greatest challenges for NGOs is to simultaneously alleviate recurring humanitarian needs, achieve longer-term development goals, and maintain peace or support peaceful conflict transformation processes. Against this background, and in view of the growing gap between humanitarian needs and the resources provided, Ban Ki-Moon, the then Secretary-General of the United Nations, called for a paradigm shift and a new way of working to make the international system more efficient and effective at the first Humanitarian World Summit in Istanbul in 2016. As a result, the humanitarian-development-peace (HDP) nexus concept was created to improve the interlinking of humanitarian aid with long-term development cooperation. This approach has already started in the 1990s through the Linking Relief, Rehabilitation and Development (LRRD) concept. What is particularly new in terms of the triple nexus concept is the inclusion of the peace dimension. Humanitarians as well as development and peace actors are called upon to better coordinate their work to more effectively promote the transformation of crises and conflicts into sustainable peace. Greater coordination, more changes of perspective, and dedicated cooperation at all levels are paramount. The recommendations published by the Development Assistance Committee (DAC) of the OECD in 2019 are central to the implementation of the HDP nexus (see ANNEX No. 1: OECD, DAC Recommendation on the Humanitarian Development Peace Nexus, OECD/LEGAL/5019). EO data can offer a holistic approach to cross-phase analysis beyond the classic operational phases of disaster relief in order to make the added value of the HDP-Nexus approach possible. Satellite-generated data and in-situ developments, as well as top-down and bottom-up analysis are automatically processed over longer periods of time to enable a seamless transition between operational phases. Over time, they can create use cases that not only allow them to react in the event of a disaster but also take action in advance to avoid crises or minimize their negative consequences. We should aim for new digital platforms and EO data to be a bridge between Northern and Southern practitioners, between technology providers and NGO workers, between the formal policy talks process and civil society advocacy, between supportive international actors, and between divided people in conflict situations.

7.3.2 Different Tasks—Different Sensors

In general, most of the available sensors are useful in the humanitarian context, depending on the nature of the task. We generally divide sensors into two major types: optical (passive) and microwave (active) sensors. Both can be further grouped by spatial resolution into low (> 300m), medium (30 to < 300m), high (HR2: 10 to < 30m; HR1: 4 to < 10m), and very high (VHR2: 1 to 4m; VHR1: < 1m) spatial resolution sensors. Besides spatial resolution, sensors are characterized by other specifications in terms of spectral bands, repetition rate, etc. Data acquisition is—amongst the underlying purpose— a matter of these characteristics along with costs, acquisition time, and effort. For example, in the case of environmental impact assessment of camps, a medium- to high-resolution sensor may be sufficient. If the focus is on actual population and dynamics, the extraction of individual housing in temporary settlements may require VHR imagery. In the first example the temporal resolution might be the key factor in the sense that frequent availability over a longer time period is necessary to draw conclusions about the causes and trends of certain patterns. The cost factor counts in the case of long-time monitoring in large areas. When it comes to retrospective analyses the decision between different sensors within the same family (HR1, HR2, VHR1, VHR2) is mostly determined by available archive data.

7.3.2.1 Optical Sensors

Optical satellite sensors are most commonly used for the "direct mapping" of geographical features relevant to a (potential) humanitarian crisis scenario.

Direct mapping means that features represented on optical images are comprehensible to human vision and thus can be mapped. Examples include (see also Section 7.2) (Huang et al., 2023; Fakhri and Gkanatsios, 2021; Zheng et al., 2021; Koshimura et al., 2020; Lang et al., 2020):

- The recent conversion and current use of a piece of land as large complex of agricultural fields as compared to a previous state when this piece of land was still forest (e.g., in the course of large-scale land investments);

FIGURE 7.5 Selection of HR and VHR optical and SAR satellite systems in support of humanitarian action and disaster response.

Humanitarian Emergencies

- The prevalence of unusual color (i.e., spectral) characteristics of cropland during a vegetative drought;
- The emergence of a temporary settlement on previously agricultural land;
- The presence of different dwelling types and infrastructure inside a camp (e.g., tents, huts, camp management facilities, or pathways);
- An apparent shape of a quarry, indicating illegal mining activities or a clear-cut suggesting illegal logging; and
- The inconsistent shape of buildings and the presence of clutter around their footprints, assuming heavy damage or collapse.

All of these features appear intuitively to a skilled interpreter, who is able to abstract features from different scale of representation ("remote" sensing), different perspective ("top"-view) and (most often) different spectral characteristics. Optical imagery can thus be interpreted visually (some say "manually," as usually there is some kind of tool, a pencil or mouse that is operated manually) or (semi-)automatically by the use of dedicated algorithms. While computer algorithms prove to be stronger in discerning different spectral behavior, the human eye is usually superior in identifying specific structures or patterns in multiple scales across the image, but is highly subjective and unrepeatable (Jensen, 2005). Nevertheless, some experiences in interpretation are required, especially when it comes to false-color imagery. The trait of human perception is difficult to automate, in particular on a reasonable level of efficiency and reliability. Recently great efforts went into the automation of visual processes, e.g., using object-based image analysis (OBIA) (Blaschke, 2010) or deep learning (DL) techniques. While DL is based on a foundation model, which does not always cope with the high dynamic in humanitarian applications, the OBIA approach requires operator responsibility considering the complexity and high level of sensitivity of information extraction from VHR EO data (Lang, 2008) (see also Section 7.4).

Table 7.3 gives an overview of optical sensor characteristics, including spatial, spectral, radiometric, and temporal resolution as well as indicative costs, and some of the key features and application areas in the context of humanitarian crisis response.

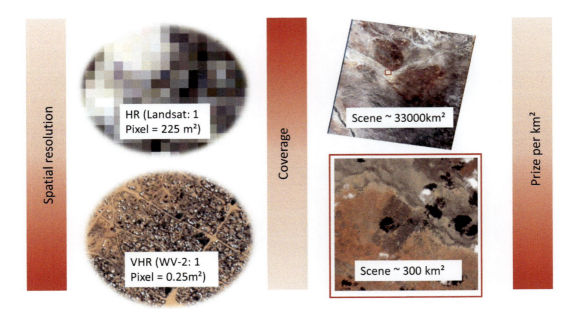

FIGURE 7.6 Acquiring satellite data is a trade-off between resolution (here: pan-sharpened), extent, and prize.

TABLE 7.1
Important Refugee Figures and Causes of Displacement during the Last Two Decades Due to Armed Conflict, Human Rights Violations, or Generalized Violence

Country of Origin	Duration	IDPs	IDPs (peak)	Refugees	Refugees (peak)
Syria	2011–ongoing	7.2 million (Jul 2023)	7.6 million (2014)	6.4 million (2023)	6.8 million (2021)
Afghanistan	1978–ongoing (start IDPs: 1993)	3.2 million (2023)	3.5 million (2021)	6.4 million (2023)	6.4 million (2023)
Colombia	1960–ongoing	7 million (May 2024)	8.3 million (2020)	116,000 (2023)	550,000 (2007)
Mozambique	1981–1995	–	4–4.5 million[1]	–	1.4 million (1992)
Iraq	1968–ongoing	1.1 million (Apr 2024)	4.4 million (2015)	329,000 (2023)	2.3 million (2007)
Democratic Republic of the Congo	1996–ongoing	7.3 million (Apr 2024)	7.3 million (Apr 2024)	978,000 (2023)	978,000 (2023)
Nigeria	1999–ongoing	3.6 million (Jul 2024)	3.3 million (20123)	410,000 (2023)	410,000 (2023)
Sudan	1984–ongoing	10.5 million (Jun 2024)	10.5 million (Jun 2024)	1.5 million (2023)	1.5 million (2023)
South Sudan[2]	2011–ongoing	2 million (Jan 2024)	2 million (2021)	2.3 million (2023)	2.4 million (2017)
Central African Republic	2005–ongoing	450,000 (Jun 2024)	940,000 (2013)	759,000 (2023)	759,000 (2023)
Rwanda	1994–1996[3]	–	650,000[4] (1998–1999)	250,000 (2023)	2.3 million (1994)
Somalia	1991–ongoing	3.9 million (May 2024)	3.8 million (2023)	840,000 (2023)	1.1 million (2015)
Burundi	1972–ongoing[5]	80,000 (Jul 2024)	800,000 (1996)	320,000 (2023)	870,000 (1993)
Palestine	Late 19th/early 20th century–ongoing	1.7 million (2023)	1.7 million (2023)	6 million (2023)	6 million
Ukraine	2022–ongoing	3.5 million (Apr 2024)	5.9 million (2022)	6 million (2023)	6 million (2023)
Myanmar	2021–ongoing	3.3 million (Aug 2024)	2.6 million (2023)	1.3 million (2023)	1.3 million (2023)
Ethiopia	Late 1970 to 1990 (mainly refugees); 2017–ongoing (mainly IDPs)	4.4 million (Feb 2024)	3.6 million (2021)	158,000 (2023)	2.6 million (1980)

Sources: UNHCR Refugee Data Finder (unhcr.org/refugee-statistics), UNHCR Operational Data Portal (https://data.unhcr.org), IDMC, 2014, UNHCR Population Statistics,[6] UNHCR Historical Refugee Data & Information Portals,[7] UNHCR Global Trends, 2013[8]

TABLE 7.2
Overview of Disaster and Crisis Initiatives (Selection)

Disaster Initiative by Agency	Type of Service or Data	Specificities
International Charter on Space and Major Disasters [24/7 operational service]	Satellite-based data and information during major natural or man-made disasters	• Satellite data free of charge for authorized users • Only for major natural disasters and limited to accidents with respect to man-made disasters • Not for conflict regions or humanitarian crisis situations
Copernicus Emergency Management Service [Operational service] rush mode: within hours or days non-rush mode: weeks/months	Geospatial information derived from satellite remote sensing images and in situ or open data sources in the course of natural disasters, man-made emergencies, and humanitarian crises *rush mode:* standardized products: Reference maps, delineation maps (providing an assessment of the event extent) and grading maps (providing an assessment of the damage grade and its spatial distribution *non-rush mode:* prevention, preparedness, disaster risk reduction, and recovery phases (reference maps, pre-disaster situation maps, and post-disaster situation maps)	• Free of charge for users • Not for conflict regions and limited with respect to humanitarian crisis situations (depending on future strategic direction) • Results publicly available at http://emergency.copernicus.eu/mapping/
Copernicus Service on support to EU External and Security Action [Operational service] rush mode: within hours or days non-rush mode: weeks/months	Geospatial information in support of EU External Action (e.g., conflict prevention and mitigation, support to humanitarian crisis management climate security, security of EU citizens or cultural heritage), based on pre-operational research projects	• provides three service types: Geospatial Analysis, Mapping for Situational Awareness and Support to Planning
REACH Initiative	Information products for aid actors during emergency, recovery, and development phase; combines fieldwork and satellite imagery analysis	• Explicitly incorporates humanitarian disasters, products are open to all aid actors
UN-SPIDER	Fast and efficient access to space-based information by compiling links and information on major disasters	• Collection of links and information provides an overview about different activities during certain disasters • No own mapping and analysis capabilities
ReliefWeb	Largest humanitarian information portal, compiling links, reports, maps, guidelines, assessments, infographics, etc. on global crises and disasters	• Public access • Overview about different activities during certain disasters; provision of several background and additional information, reports and publications about certain regions • No own mapping and analysis capabilities
International Network of Crisis Mappers (ICCM)	Community platform	• At the intersection of humanitarian crises, new technology, crowd-sourcing, and crisis mapping

(Continued)

TABLE 7.2 (Continued)
Overview of Disaster and Crisis Initiatives (Selection)

Disaster Initiative by Agency	Type of Service or Data	Specificities
Humanitarian Open Street Map (HOT)	Open community platform	• International team dedicated to humanitarian action and community development • Volunteers can work both remotely and on-site to collect relevant data
Missing Maps	Community platform	• Tries to preventively map places in the world where the most vulnerable people live • In contrast to a response to natural disasters, conflicts, or epidemics, regions are identified in which crises are expected and mapped

TABLE 7.3
Optical Sensors: Key Characteristics and Application-Relevant Features

Platform/Sensor	Resolution (Spatial/Spectral/Radiometric/Temporal)	Humanitarian Application Domain (Including Examples of Detectable Features)	Indicative Costs (Archive/Tasking) (EUR per km²)[20]	Minimum Order Size (km²)
MODIS	250m [ground sample distance]/36 bands/16 bit [quantization]/16 days [to revisit]	The NDVI product allows the analysis of time series for vegetation trend patterns	0	1 scene
Landsat-8	15m/11 bands/16 bit/16 days	Vegetation trend patterns	0	1 scene
DMCii	22m/3 bands/8 or 10 bit/daily	Logging activities, clearings, road infrastructure built-up	0.01–0.14 (minimum price for archive: 1,260)	25,600
RapidEye	6.5m/5 band/16 bit/daily	Logging activities, clearings, general infrastructure built-up Bare soil detection (as indication for mining sites)	1	500 for archive, 3,500 for tasking
SPOT-4	10m/5 bands/8 bit/2–3 days	Land cover change analysis Vegetation trend patterns	1–2.5 (depending on size)	400
DMCii	2.5m/4 bands/16bit/2 days	Road network, detecting larger buildings, detailed land cover information	1.9–2.5	1600
SPOT-5	2.5m/5 bands/8 bit/2–3 days		2–8 (depending on size)	400
FORMOSAT-2	2m/5 bands/8 bit/daily		3–4.5	576
SPOT-6/7	1.5m/5 bands/12 bit/daily if using both satellites		4–4.5	25 for archive, 100 for tasking
IKONOS (decommissioned 2015)	1m/5 bands/11 bit/3 days	Feature extraction Single dwellings/huts/tents	7–14	25 for archive, 100 for tasking

TABLE 7.3 (Continued)
Optical Sensors: Key Characteristics and Application-Relevant Features

Platform/Sensor	Resolution (Spatial/ Spectral/Radiometric/ Temporal)	Humanitarian Application Domain (Including Examples of Detectable Features)	Indicative Costs (Archive/Tasking) (EUR per km²)[20]	Minimum Order Size (km²)
QuickBird (decommissioned 2015)	0.6m/5 bands/11 bit/1–3.5 days	Single trees Fences, walls	12-18	25 for archive, 100 for tasking
GeoEye-1	0.5m/5 bands/11 bit/2–8 days	Groups of people Mining site detection		25 for archive, 100 for tasking
WorldView-1	0.5m/1 band (pan)/11 bit/1.7 days	Stereo capability: DEM, 3D structures	12.5–21.5	25 for archive, 100 for tasking
WorldView-2	0.5m/8 bands/11 bit/1.1 days		4 bands: 16–25 8 bands: 17–26	25 for archive, 100 for tasking
Pléiades 1/2	0.5m/5 bands/12bit/ daily		10–17	25 for archive, 100 for tasking
WorldView-3	0.31m/17 bands/11 bit/ daily		4 bands: 20/29 8 bands: 22/31	25 for archive, 100 for tasking
Skysat	0.5m/5 bands/11 bit/6-7 times per day, constellation of 21 satellites	Same as above + monitor population movements and infrastructure development	6	50
Dove satellite constellation (~180 satellites):	3m/4 bands, 8 bands (SuperDoves)/11 bit/ daily	land cover information; settlement detection	2	250
WorldView Legion (6 planned in total)	0.3/8 bands/12 bit/ up to 15 revisits per day			25 for archive, 100 for tasking
Pléidades Neo (2 satellites)	0.3m/6 bands/12 bits/2 times per day		18–26	25 for archive, 100 for tasking
SuperView	0.5m/4 bands/11 bit/ daily		12–21	25 for archive, 100 for tasking
SuperView-Neo (16 satellites)	0.3m/4 bands/11 bit/ daily		18	25 for archive, 100 for tasking
Beijing-3 (BJ3A, BJ3N)	0.5m (BJ3A), 0.3m (BJ3N)/4 bands/ 11 bit/5 days		14	25 for archive, 100 for tasking
Satellogic (up to 200 satellites)	1m-0.7m/4bands/8 or 16 bits/several daily revisits		4–8	50 for tasking
SkySat (21 satellites)	0.5 m/4 bands/11 bit/6-7 times per day on average		6	25 for archive, 100 for tasking

The dependence on atmospheric conditions, even more than the match of sensor specifics to extractable features in terms of spectral, spatial, and temporal resolutions, constrains the usability of optical sensors. Optical sensors record sunlight reflected by the Earth's surface, thus in a "passive" mode. Atmospheric conditions, in particular the presence of water vapor and other gas molecules, interact with the wavelengths, to which optical sensors are sensitive. These influences may hamper the quality of images as clouds may partly or fully obscure features on the ground. When critical features are not visible, an entire satellite scene may be rendered useless, as the required information simply cannot be obtained. A class of sensors not affected by atmospheric conditions, but with other constraints, are radar sensors.

7.3.2.2 Radar Sensors

As opposed to optical imagery, radar sensors are active devices, recording microwave radiation that they emit. Due to their specific wavelength range, microwaves (or radio waves) do not interact with water vapor or other atmospheric particles, and thereby also penetrate clouds (Lang et al., 2020; Lang et al., 2018; Quinn et al., 2018). The generated signal is independent of any external energy source, allowing radar image acquisition under bad weather conditions; over clouded, tropical forests or temperate climates; and at night. While this independence from weather and daylight sounds like an unbeatable pro for their usability, radar data are more complex to process and analyze, and much less intuitive to interpret. In operational application scenarios, radar data are mostly used to the detection and extraction of specific, well-defined features, such as a water mask in the aftermath of a flood or the accessibility of a road network after the peak of the flood has receded. Other applications utilize standardized measurements of biophysical parameters such as soil moisture to better understand and predict drought occurrence (Wagner et al., 2003). Further research looks into the transfer of radar technology to optical image analysis domains such as dwelling extraction in refugee camps (Bernhard, 2013), by combining the extractability of features using higher spatial resolutions with the ubiquity of light-independent radar technology.

In Table 7.4, some of the most prominent radar sensors are listed and characterized, again with some potential (or actual) application domains.

7.3.2.3 Nano-/Microsatellites

There is a substantial growing market of small satellites designed to reduce costs by minimizing mass. Depending on the mass, several categories are distinguished: small satellites (100–500 kg), microsatellites (10–100 kg), and nanosatellites (1–10 kg). Beyond that there are picosatellites (< 1 kg) and femtosatellites (10–100 g) in production. In 2013, for example, almost 100 micro- and nanosatellites were launched. Many of them were built in the CubeSat standard format, with a volume of exactly 1 l (10 cm cube) and a mass of no more than 1.33 kg. The spatial resolution is up to 1 m with revisiting times of up to several hours. Companies like Planet Labs, Spire (formerly Nanosatisfi), Surrey Satellite Technology, Dauria Aerospace, or Skybox Imaging have launched nano- and microsatellites, to name a few.

7.3.2.4 Unmanned Aerial Vehicles (UAVs)

Intuitive in usage ("toy-factor") and potentially unlimited in their usability, unmanned aerial vehicles (UAVs) were initially disputed in the humanitarian response domain. In a forward-looking manner, the Humanitarian UAV network aims to bridge humanitarian and UAV communities and to establish clear standards for humanitarian use. In addition to the sensitivity of the tool, the actual advantage over satellite data is the flexibility and controllability of the tool, and the lack of additional costs once a device is purchased. However, legal and factual constraints may limit the usage of UAVs due to accessibility or risk. Operators need to be physical on site: in the course of natural disasters this is more feasible as NGOs may obtain their immediate information needs with first responders who have their own UAVs. For (protracted) humanitarian disasters this may be more difficult, when restricted access and security issues do not allow acquisitions, as the hallmarking of UAVs as a weapon of war simply does not allow it, as opposed by the people on the ground, whether civilians or especially among armed groups. While the general strength of UAVs is their universality, the need for careful operation and handling, as well as the limited coverage makes it complementary, but not superior, to VHR data, from the group of micro/nanosatellites in particular.

7.4 IMAGE ANALYSIS TECHNIQUES

7.4.1 GENERAL WORKFLOW—EXAMPLE POPULATION MONITORING

Image analysis entails a consolidated workflow (Kucharczyk and Hugenholtz, 2021; Zheng et al., 2021; Lang et al., 2020; Braun et al., 2019; Lang et al., 2018; Lang et al., 2006). Figure 7.7 shows the principal workflow, as illustrated by the case of population monitoring. Situational awareness is a

TABLE 7.4
Radar Sensors and Their Specifics

	Band	Polarization	Spatial Resolution/Repeat Cycle	Application Domains
Sentinel-1	C-band	HH, VV, HV, VH	5*20m	Flood detection, oil spills, sea ice monitoring, ship detection, forest monitoring (forest cover, vertical structure, biomass), surface movement monitoring, daily revisits
1. Interferometric wide-swath mode			5m	
2. Wave mode			5m	
3. Strip map			20*40m	
4. Extra-wide swath				
TerraSAR-X:	X-band	HH, VV, HV, VH	18m/2.5 days	
1. ScanSAR mode			3m/2.5 days	
2. Strip map mode			1m/2.5 days	
3. Spotlight mode				
Radarsat-1	C-band	HH	8m	
1. Fine res.			30m	
2. Standard			50–100m	
3. ScanSAR				
Radarsat-2	C-band	HH, VV, HV, VH	3*1m	
1. Spotlight			3*3m	
2. Ultra fine			10*9m	
3. Fine res.			25*28m	
4. Standard			50*50m	
5. ScanSAR			25*28m	
6. Fine Quad-pol				
ALOS-PALSAR	L-band	HH, VV, HV, VH	10m	
1. Fine beam single			20m	
2. Fine beam dual			20m	
3. Direct downlink			100m	
4. ScanSAR wide beam			30m	
5. Polarimetric				
COSMO-SkyMed	X-band	HH, VV, HV, VH	30m	
1. ScanSAR mode			3–15m/	
2. Strip map mode			1m	
3. Spotlight mode				
Capella-2	X-Band	HH, VV	2m	
1. Strip map mode			0.5m	
2. Spotlight mode				
ICEEYE	X-band	HH or VV, HH/VV	1m	daily revisits
1. ScanSAR mode				
2. Strip map mode				
3. Spotlight mode				

crucial first step to understanding the information needs and the required products to be obtained by any kind of survey or user request form. Accordingly, imagery is acquired—from archive if less time critical or newly tasked when the actual situation requires it. In addition to image acquisition any available auxiliary data (such as local reports, data on camp infrastructure, etc.) needs to be collated and integrated. Image pre-processing like spatial referencing, orthographic, atmospheric correction, etc. are required to provide a proper data source for the subsequent image analysis procedures, including validation of the results in both technical (accuracy) and usability aspects. The final step of the workflow comprises the user-targeted information delivery by using appropriate cartographic means and geo-visualization techniques as well as suitable geomedia tools or platforms.

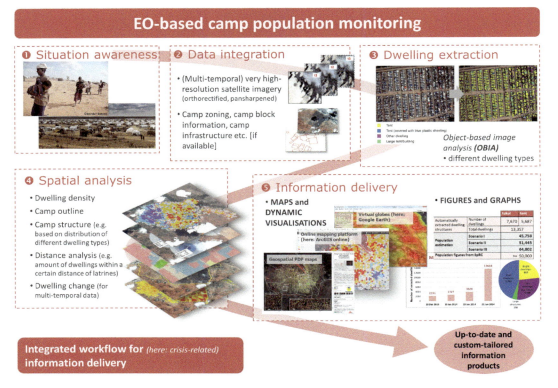

FIGURE 7.7 Workflow for EO-based information delivery in the context of camp population monitoring (Z_GIS, 2015).

7.4.2 Visual Image Interpretation

The majority of EO-based image analysis tasks in the context of disaster management so far has been conducted by visual image interpretation (Zheng et al., 2021; Lang et al., 2020). The focus is often on detailed analysis, requiring interpretation of specific features (e.g., buildings, roads, infrastructure facilities). Visual interpretation is still considered most reliable for such complex interpretations, as human vision usually outperforms algorithmic approaches in identifying complex features and structures, even at different scales, image contexts, image quality, or seasonal effects (Checchi et al., 2013). Visual perception can be trained by experience and thus adapt to various situations. There is an "in-built" capacity of human interpreters to delineate features in a given scale domain.

However, when it comes to repetitive, standardized tasks (e.g., the delineation of single dwellings in a large camp setting), visual inspection turns increasingly inefficient and prone to errors or rash generalizations. Here, automated classification and feature extraction techniques (see Section 7.4.3) start to become more efficient and superior in operational, analogous settings (Tiede et al., 2013). Ideally, human vision and automated techniques are combined in a hybrid approach (Füreder et al., 2014), each contributing its specific strengths in a complementary manner. Still, there are hardly any automated techniques that may interpret a complex scene better than humans do, and there are similar challenges faced by human and computer vision.

For example, when looking at a VHR satellite image of a refugee camp, the focus for the interpretation would—after a first assessment of the situation including accessibility, water sources, and land use/land cover—most likely be the camp itself. A human interpreter may first assess the extent of the settled areas and focus the subsequent interpretation on that area. This minimizes the detection of false positives in the outskirts of the camp (e.g., dry vegetation, tree shadows, or rocks with similar spectral and geometric characteristics like traditional huts). Depending on the type of camp and geographic area, there are different dwelling structures (e.g., tents covered with white or

Humanitarian Emergencies

blue plastic sheeting with rectangular or tunnel shape, traditional round huts, shelters made of local materials, dwellings with metal roof, makeshift shelters), fences to separate family compounds and larger camp facility areas. The human eye can differentiate easily between such different camp structures and dwelling types. It becomes trickier in complex situations, e.g., with connected dwellings, dense complex pattern of roofs or dwellings of little contrast to bare soil or when dwellings are covered by vegetation (Checchi et al., 2013).

Prior to the interpretation, a set of target features needs to be defined, including types of buildings and other geographic features of relevance in scene understanding. Scenarios do exist where routine-based automated techniques may add important information. Humanitarian disasters caused by droughts, famine, or slow-onset floods are examples where the recursive analysis of satellite time series provides important information about the dynamic of the situation in the region of interest. In the context of slow-onset floods it is important to receive information beyond flooded areas, including accessible roads and intact infrastructure to serve as safe locations, potential evacuation routes, or shelter areas. Furthermore, such analyses can be used for an effective early warning system of the indications on the onset of drought conditions.

7.4.3 Automated Feature Extraction and Image Classification

The use of automated feature extraction methods in an operational mode supports targeted analyses (e.g., extraction of flood mask, extraction of distinct dwelling structures) (Ye, 2022; Zheng et al.,

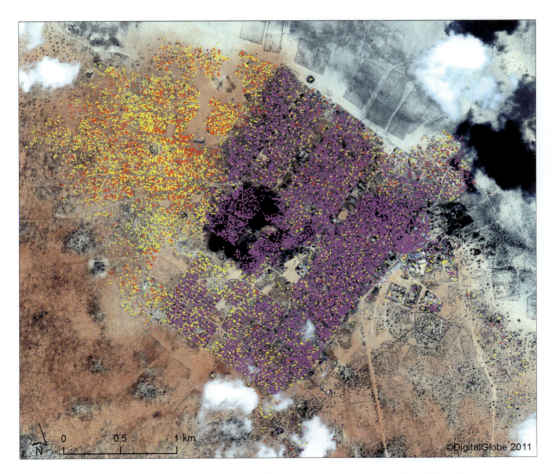

FIGURE 7.8 Automated distinction between dwelling structures based on a WorldView-2 image: existing dwellings with metal roof (violet) and new dwellings (yellow = white plastic sheeting, red = makeshift shelter).

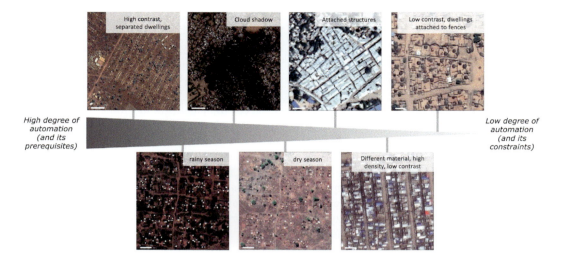

FIGURE 7.9 Automation level for dwelling extraction in relation to degree of complexity of image data (Füreder et al., 2014, adapted).

2021; Lang et al., 2018; Jiang et al., 2017; Winkler et al., 2017). Despite the success of deep learning methods, more complex feature extraction is still considered as "experimental."

Automated methods are advantageous in being scalable and transferable. This makes them suitable for monitoring analyses of large areas or large amounts of features, respectively. Various methods for automated feature extraction have been reported (Tiede et al., 2010; Kemper and Heinzel, 2014). Object-based image analysis (OBIA) techniques (Lang, 2008) are expert-based techniques, which—compared to manual interpretation—are flexible to adopt complex classification schemes and include a high degree of detailed class descriptors (e.g., shape and size of buildings, specific arrangements, the inclusion of biophysical parameters and vegetation indices, etc.). Increasingly, deep learning techniques are being used for routinized image analysis techniques. The current challenge is the transferability of foundational models trained on standard settings, to specific situations (Ghorbanzadeh et al., 2020; Gella et al., 2023). Specific transfer learning strategies and self-learning techniques gradually compensate the lack of samples (Gella et al., 2023).

In complex ground situations, where automated feature extraction methods reach their limits, it makes sense to combine automated and manual interpretations. This reduces analysis time compared to visual interpretation.

7.4.4 Spatial Analysis and Modeling

The results of image interpretation and classification can be enhanced by geospatial analysis techniques (Fakhri and Gkanatsios, 2021; Lang et al., 2020; Jiang et al., 2017). Depending on the information needed, aggregated information allows a better overview of a situation, whenever there is a high amount of features or classes present in an image. Such hot spot maps are based on density measurers calculated from extracted point data. For example, dwelling density resembles the distribution of population in a refugee camp; density of damaged buildings is a first indication of most affected areas; aggregating land-cover classification results to larger reporting units (administrative units, regular grids, hexagons, etc.) show predominant classes.

Combining the extracted information from EO data with other data allows further insights. For camp planning, for example, it is useful to have additional information on how many tents or other buildings are within a specific distance to boreholes, hospitals, latrines, etc. (see Figure 7.10).

For monitoring purposes, multi-temporal analyses provide changes between two or more time stamps. A post classification comparison considers directional changes, meaning that change dynamic is made explicit for each possible transition from one class to another. However, a large

Humanitarian Emergencies

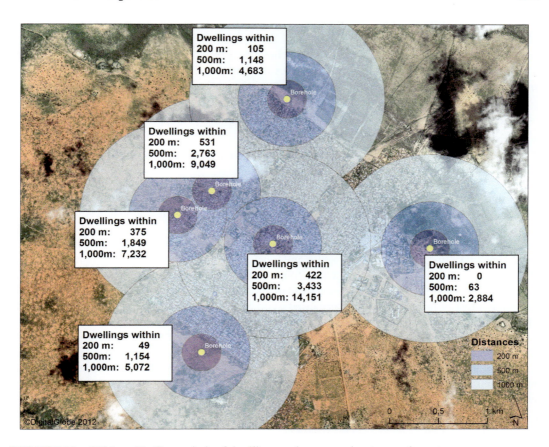

FIGURE 7.10 GIS-based buffer analysis of dwelling numbers around water supply posts.

number of classes (e.g., land cover) such multi-directional changes may be difficult to interpret. Readability can be significantly improved if relevant changes are aggregated to regular units.

7.4.5 Validation and Ground Reference Information

The technical part of validation contains an assessment of the classification accuracy. A classification error is defined as the discrepancy between the classification and reality (Kucharczyk and Hugenholtz, 2021; Lang et al., 2018; Quinn et al., 2018). The results of an accuracy assessment are only meaningful when accurate ground data exists (Zheng et al., 2021; Lang et al., 2020) In reality, ground reference data may also contain errors, often related to subjective interpretations. In most of the presented case studies, spatially explicit ground data were not available and, for security and other reasons, could not be collected by NGOs in the field. In a comparative study the London School of Hygiene and Tropical Medicine (LSHTM) and the University of Salzburg assessed the estimation of population sizes using two different image analysis methods in Am Timan city, Chad (unpublished). The comparison of automated and manual count was benchmarked at a standard population survey (quadrat method), carried out in January 2012. It turned out that the mean total structure count for the automated method was some 10% lower than the manual method, mainly because the automated method could not discern between close-by buildings identified as a single larger building. Manual post-processing led to very similar results, with a difference of some 2% at a total amount of some 12,000 dwelling structures. In terms of time required for the analysis, the automated method did not involve much less effort compared to the manual one. This applies to the initial stage at least, when setting up the methodology, while the transferability to similar tasks improves the performance of the automated method in a repetitive mode.

7.5 CASE STUDIES

In this chapter, we illustrate the concepts and principles mentioned earlier on selected case studies. These showcases cover each of the three phases discussed earlier (early warning, crisis monitoring, and mid- to long-term impact) and demonstrate the variety of data usage including free data sources. All case studies reflect "real cases," i.e., analyses that were conducted on concrete scenarios, including the information provided and data sets used.

7.5.1 Monitoring Population Dynamics in the Refugee Camp Dagahaley, Kenya

7.5.1.1 Background

The refugee camp Dagahaley is located in the Garissa district in Kenya, around 100 km from the border to Somalia. The camp was established by the UN Refugee Agency (UNHCR) in 1992 around the town Dadaab together with the camps Ifo and Hagadera. Initially planned to host up to 90,000 people, mainly refugees from the civil war in Somalia, it has been extended by two additional camps in 2011 (Ifo 2 East and West and Kambioos) and is now the world's largest refugee camp complex with about 350,000 people. Until 2006 the population of the Dagahaley camp was quite stable with about 30,000. New refugees arrived in 2006, 2008, 2010, and 2011 (UNHCR, 2014). The last large influx of refugees in 2011 has caused a combination of a severe drought and famine at the Horn of Africa and ongoing conflicts and violence in Somalia. New arrivals settled in the outskirts of Dagahaley, where in July 2011 around 25,000 people lived (MSF, 2011). The enormous influx of people to the Dagahaley refugee camp brought the camp registration to a halt and revealed the need for a more efficient camp monitoring. Moreover, emerging violence against Kenyan security forces but also humanitarian workers at the end of 2011 hampered the ability of aid agencies to provide relief assistance, protection, and essential services (UNHCR, 2012).

7.5.1.2 Data Used

For monitoring the development of the Dagahaley camp WorldView-2 images from July 2011 and December 2011 with eight multispectral bands (2 m spatial resolution) and a panchromatic band (0.5 m spatial resolution) were available. Due to cloud cover in the December image, the eastern part of the image was replaced by an additional WorldView-2 image from January 2012.

The images have been pan-sharpened using the HCS (Hyperspherical Color Sharpening) method, designed for sharpening WorldView-2 images. This method can handle any input bands and produces acceptable colour and spatial recovery (Padwick et al., 2010). The December and January images were co-registered on the July image to prepare for change analysis. Due to different conditions in image acquisition (off-nadir angle, target azimuths), a minor offset between the images remained.

7.5.1.3 Methods Applied

Dwellings and camp buildings were semi-automatically delineated using an OBIA-based algorithm for dwelling extraction. This rule-set, coded in CNL (Cognition Network Language) within eCognition (Trimble Geospatial Imaging), uses edge filtering to extract settlement areas in a first step. For dwelling extraction, image segments were classified by relative spectral differences and spatial characteristics (Lang et al., 2010). The underlying rule-set, which was developed in a different geographical area (Tiede et al., 2010), is transferable as a general model to analyze other time slices or other areas with similar camp settings. For the Dadaab region, the rule-set was adapted regarding dwelling types and the increased spectral information of WorldView-2.

For the analysis of the July image three dwelling types were distinguished: tents, huts, and buildings with a metal roof. The analysis was conducted in several steps, starting with a rapid assessment of the newly settled areas in the western outskirts of Dagahaley concentrated on tents, the prevailing dwelling type there. Makeshift huts were delineated manually due to their small structures and similar signature to the surrounding dry vegetation. First results were produced within one day after pre-processing the data. Subsequently, the algorithm was adjusted to the whole camp area, where also dwellings with metal roof, the predominant dwelling type within the camp, were extracted. The rule-set applied to the July image was then transferred to the December image. At this time, clear indications of newly settled areas nearly have disappeared, meaning that very few makeshift huts were still present and many dwellings with metal roof, which were abundant in the July image, have been covered with white plastic sheeting in the December/January images due to the rainy season, which made a differentiation to white tents unfeasible. Therefore only one class "dwelling" was extracted for the second timeslot. Finally, manual refinement eliminated most obvious classification errors. The rule-sets of both images were slightly adapted to camp areas covered by cloud shadows. Thereby we merged the classification of shaded and unshaded areas and the combination of the results into one rule-set to increase the degree of automation (see Figure 7.10).

7.5.1.4 Delivered Information

The key information is the amount of single dwellings distinguished between different dwelling types. For July 2011 about 23,400 dwellings were extracted: 13,950 dwellings with metal roofs, 6,650 tents, and 2,800 huts. The combined results of the December and January images revealed 21,950 dwellings. An additional information product based on single dwellings comprises dwelling density (dwellings/km^2), calculated using Kernel density methods (Lang et al., 2010) to provide a better overview of the spatial distribution of dwellings. A camp outline is generated from dwelling density, which is important for rapidly expanding camps, where the extent can hardly be observed on the ground. A change analysis of dwellings between July 2011 and December 2011/January 2012 aggregated to a hexagonal grid provides an overview of major changes within the camp. It showed a decrease of dwellings in the western outskirts of the camp and a slight increase of dwellings in the main part of the camp. Areas covered by clouds in either of the two images were excluded from the change analysis.

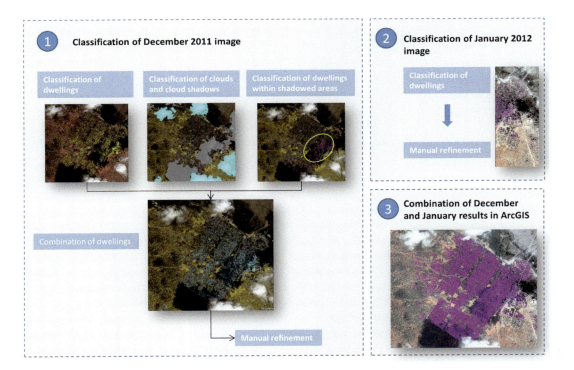

FIGURE 7.11 Workflow of dwelling extraction (here: example with cloud shadows and different images) (Füreder et al., 2012).

7.5.1.5 Impact

In extreme events, with enormous influx of refugees, reliable information on population is critical to complement population figures maintained by the local camp managing NGO or UNHCR. In the case of Dagahaley, limited access to the outskirts of the camp due to security reasons required information about new arrivals and their need for assistance. Monitoring changes of dwelling structures showed that population moved from the outskirts, partly relocated inwards or to the neighboring camp, Ifo 2.

Methodologically, the cloud-shadowed areas are challenging. Due to the high radiometric resolution of the WorldView-2 images, areas under cloud shadow were analyzed, but the lower accuracy led to an underestimation of dwellings as compared to illuminated regions.

7.5.2 Evolution and Impact of the IDP Camp Zam Zam, Sudan at Local and Regional Scale

7.5.2.1 Background

The IDP camp Zam Zam is located to the southwest of Wadi El Ko and Wadi Golo, approximately 15 km south of El Fasher, the capital city of Northern Darfur, between longitudes 25°17' and 25°19' east and latitudes 13°28' and 13°30' north. A semi-arid environment with low annual rainfall and high seasonal variability characterizes the region. The land surrounding Zam Zam is used for cultivating crops and vegetables. When the first IDPs came in 2003 they settled around existing villages; since then the population number rose from approximately 18,900 in 2004 to over 50,000 in 2008 with continuing trend. In 2013 UN agencies and NGOs working in the camp estimated a number of 164,000 IDPs living in Zam Zam (UNOCHA, 2013).

7.5.2.2 Data Used

For detailed analysis (Lang et al., 2010), two QuickBird scenes were ordered from DigitalGlobe image archive (June 18, 2002, and December 20, 2004). A third one was tasked and recorded on May 8, 2008. QuickBird, which is no longer operating, had a spatial resolution of up to 0.6 m, and multispectral images are collected at a resolution of 2.4 m. The three QuickBird scenes have three optical bands and a near infrared (NIR).

For the analysis of the entire region around Zam Zam MODIS Vegetation Indices (MOD13Q1) over a time period of more than one decade has been acquired starting on February 18, 2000, and ending on May 25, 2010. The MODIS image data sets are provided every 16 days as a gridded level 3 product at a spatial resolution of 250 m and thus the entire time series contains 237 data sets.

7.5.2.3 Methods Applied

The workflow contains two analysis steps: (1) analyzing camp extent and population dynamics in several time slices and (2) assessing likely impacts on the surrounding environmental conditions—at local and regional scale. The first part aims to perform an ex-post analysis, looking back in time in a retrospective view (Lang et al., 2010). The study was conducted in 2008, looking at the evolvement of the Zam Zam IDP camp over a period of up to six years. Since QuickBird imagery was available from 2001 onwards, we were able to cover the entire time span of camp evolution from 2002 until 2008. Later, the analysis of camp evolution was extended with two additional time slices from 2010 and 2013.

For an integrated assessment of the changes in the immediate surroundings of the camp, Hagenlocher et al. (2012) used a Weighted Natural Resource Depletion index. This index reflects the relative importance of single land use classes in the context of human security and ecosystem integrity, as judged by experts. The calculation was done using weighted overlay techniques. Extending the view to the entire region around Zam Zam trend patterns resulting from time series analyses of MODIS Vegetation Indices between 2000 and 2010 were generated. The trend analysis was based on the Seasonal Kendall test, deriving significance and slope values of detected trend patterns. This variation of Mann-Kendall was applied to incorporate the high seasonality of the climate in North Darfur (Kranz et al., 2010). Finally, based on the derived significance and curve slope values a trend classification has been set up that allows for a more detailed separation of positive and negative changes.

7.5.2.4 Delivered Information

The results of the medium-resolution analysis can be described as a detection of hot spots of significant positive and negative trends in vegetation cover over the period considered. Due to their impact on the surrounding environment, significant negative trend patterns are located in the vicinity of camps such as Zam Zam, Abu Shouk, and Al Salaam (Figure 7.12). This information on a large scale indicates the influence of human activities on the vegetation cover for large regions. This evokes increasing pressure on natural resources (e.g., firewood) and consequently is one indicator showing increasing potential for local conflict.

7.5.2.5 Impact

Population displacement often results in large-scale settlements with a potential impact on the local environment. The dynamics apply both to the structural changes of the camp area (extent and the population density), and the wider impacts on the surroundings. Environmental deterioration can result in violent conflicts over access to, or control of, scarce natural resources, or in renewed migration of the population (Hagenlocher et al., 2012). Remote sensing can capture actual population dynamics, and hint to environmental changes in the surrounding areas. While the analysis of VHR satellite imagery provides detailed information about camp structure and local changes of the environment, medium-resolution data has the capability to provide trend patterns of environmental impact on a regional scale.

7.5.3 INDICATIONS OF DESTRUCTIONS—THE CASE OF ABYEI, DISPUTED BORDER REGION BETWEEN SUDAN AND SOUTH SUDAN

7.5.3.1 Background

On May 21, 2011, Sudan Armed Forces (SAF) invaded Abyei Town, located in the Abyei Administrative Area. The Abyei Area is a contested region on the border of Sudan and South Sudan, then Republic of South Sudan. SAF's invasion of Abyei town followed multiple skirmishes between SAF, SAF-aligned forces, and the Sudan People's Liberation Army (SPLA). As fighting broke out between SAF-aligned and SPLA-aligned forces, civilians fled the area, and NGO personnel were evacuated to Southern Sudan. The United Nations Mission in Sudan (UNMIS) was also present in Abyei Town at the time of the attack, although their freedom of movement was restricted. During fighting, civilian dwellings, infrastructure, and humanitarian compounds were attacked. Figures estimate that between 30,000 and 80,000 civilians, primarily Dinka Ngok, were forcefully displaced from their homes (Al Achkar et al., 2013).

7.5.3.2 Data Used

Two data sets were used to visually confirm reported events that impacted the civilian and humanitarian communities in Abyei Town: VHR satellite imagery and open-source data from news reports and UN reports. This case study examines the Harvard Humanitarian Initiative's use of satellite imagery to confirm events reported by UNMIS Human Rights. An UNMIS report dated May 29, 2011, claims that the

> The Abyei market and all the shops have been looted and burned. The World Food Programme [WFP] and UNICEF warehouse, which were fully stocked before the SAF attack, as well as the MSF (Médecines Sans Frontières), AECOM and Humanitarian Affairs compounds have also been completely plundered.

(UNMIS, 2011)

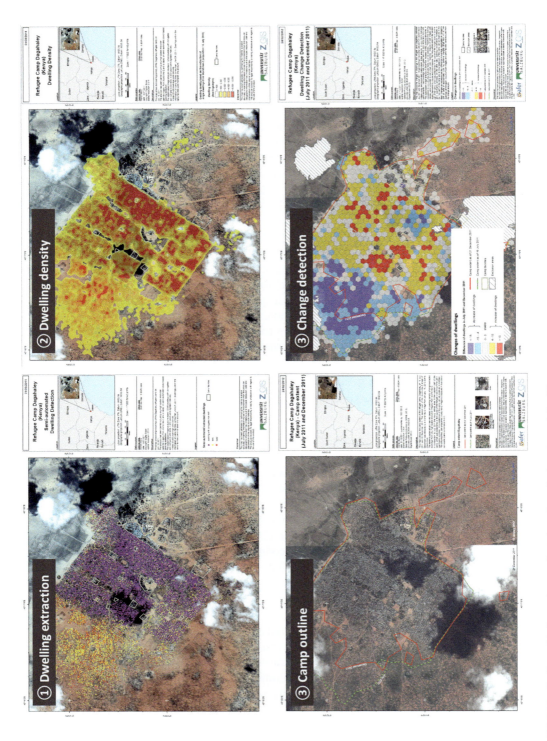

FIGURE 7.12 Delivered information products based on single extracted dwellings.

Humanitarian Emergencies 229

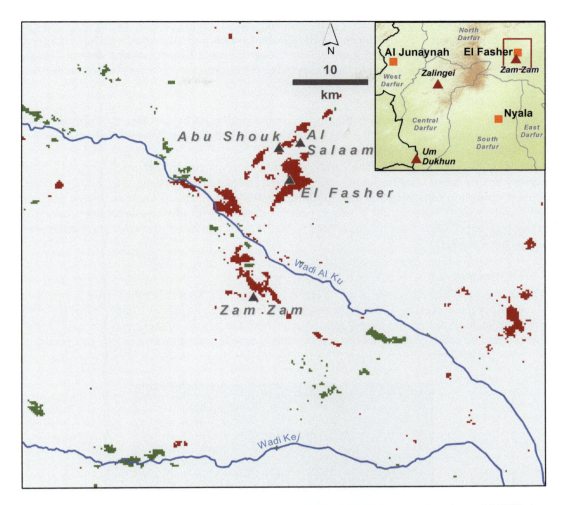

FIGURE 7.13 Negative (red) and positive (green) trend pattern of NDVI values based on a MODIS time series analysis with 237 data sets between 2000 and 2010 in the area of El Fasher/Zam Zam. Trend patterns have been derived using Seasonal Kendall test.

To confirm these reported events, analysis was conducted using two archive VHR images. The pre-event image, captured in Feb, 2011, is from WorldView-2 with a resolution of 50 cm (commercial use). The post-event image, from QuickBird with a resolution of 61 cm, was acquired five days after the events on May 26, 2011. Images were obtained from DigitalGlobe in geo-referenced and pan-sharpened mode.

7.5.3.3 Methods Applied

Two adjustments were made to the raw images in ERDAS Imagine before analysis was performed. Cubic convolution resampling was conducted to refine the pixels image to allow for better analysis. Min-max stretch rescaling was also applied to balance the color and contrast in the image. The areas of interest, the Abyei Market, WFP, and MSF compounds, were then located in the satellite image using open-source mapping sources, like OpenStreetMap and Wikimapia.

Manual analysis was conducted on both images. Damage to and the destruction of structures was identified by conducting change detection between the two images in geographically synced windows in ERDAS Imagine. Burned structures were identified due to their darker color, similar appearance to other burned structures, and the apparent presence of ash. Razed structures appeared to be damaged, with their original structure size undetectable due to apparent dismantling. Ground photographs of looting were used to help assess what the results of this act look like in imagery.

7.5.3.4 Delivered Information

Analysis of the post-event satellite imagery corroborates UNMIS reports that the WFP and MSF compounds were looted and that the Abyei Market was looted and razed. Additional indications of destructions are also apparent in the imagery.

At both the WFP and MSF compounds, scattered debris is visible both inside and outside the perimeter. This debris is indicative of looting. A cited UN Media report states that "800 metric tons of food, enough to feed 50,000 people for three months" and other supplies were looted from the WFP compound (UNRadio, 2011). Imagery analysis also reveals 12 razed structures at the MSF compound, ten of which appear to have been burned. Debris and razed structures are also apparent at the Abyei Market.

7.5.3.5 Impact

Visually documenting the results of the attacks on the Abyei Market and WFP and MSF compounds is critical for many reasons. First, these attacks are documented as part of a larger effort to understand the scale of destruction throughout Abyei Town. The results of these acts not only signify events that immediately impacted the civilian and humanitarian communities but also presented

(a)

FIGURE 7.14 WFP and MSF compounds after the attacks.

Humanitarian Emergencies

(b)

FIGURE 7.14 (Continued)

FIGURE 7.15 Abyei Market before (left) and after (right) the attacks.

long term challenges. Documenting the destruction of critical civilian and humanitarian infrastructure provides insights into impediments that may prohibit the local population from returning to their community. This point is exemplified and amplified by the report that the WFP compound, prior to the looting, held enough food to feed 50,000 for three months.

The proliferation of this technology has also raised questions concerning the possible role remote sensing may play in documenting violations of international humanitarian law. Current widespread standards for the use of remote sensing in this capacity do not exist. Current research reflects that if this particular use is to occur then imagery acquisition, analysis, and documentation must occur within a proper chain-of-custody due to potential perceived politicization of the analysis.

7.5.4 LOGGING AND MINING ACTIVITIES IN RELATION TO THE CONFLICT SITUATION IN THE DEMOCRATIC REPUBLIC OF THE CONGO (DRC)

7.5.4.1 Background

The Democratic Republic of the Congo (DRC) is one of the most resource-rich countries in the world but economically one of the poorest and on the edge of a failed state (UNDP, 2011; Haken et al., 2013). Natural resources locally and abundantly available do not only cover minerals but also wood—especially high-value tree species. These resources are known to fuel the conflict in the country either by directly investing revenues into weapons or by conflict amongst the resources itself (USAID, 2005; Le Billon, 2006; GlobalWitness, 2013). Furthermore, armed groups are taking the control over wood and mineral resources as well as their trade. The insecure and instable situation especially in the eastern provinces of the DRC (North and South Kivu) caused significant numbers of internally displaced persons (IDPs). With 2.6 million IDPs and another 500,000 refugees the displaced population was one of the largest in the world back then (IDMC, 2014).

Mining in the DRC is mainly carried out by civilians using artisanal and small-scale mining techniques (ASM), which in fact means working with shovel, pickaxes, and hammers or even using bare hands (Garrett, 2008). With respect to logging, one can distinguish three types: industrial logging, informal logging, and slash-and-burn activities. In general, the fragmentation of rainforests by logging results in well-analyzed secondary effects, such as increasing agricultural land use alongside the logging roads (UNEP, 2006), charcoal production, poaching, as well as mining activities (Potapov et al., 2012).

Detecting land cover changes related to mining and logging activities provides information about the situation on the ground in the insecure and often remote areas of interest.

7.5.4.2 Data Used

With respect to the monitoring of logging and slash-and-burn activities DMCii images of three points of time (January, May, and September 2010) with a spatial resolution of 22 m have been analyzed, covering an area of 50*50 km in the Orientale province in DRC. As a reference for the identification of the defined land cover classes VHR satellite scenes from IKONOS (March 2011) and GeoEye-1 (June 2009) have been used.

Data of the same VHR sensors have been applied for the monitoring of mining activities in North and South Kivu provinces of the DRC. Of particular interest was the most well-known cassiterite mining site of Bisie, located in North Kivu. For this area three scenes were available comprising a period of April and September 2010 (GeoEye-1) to March 2011 (IKONOS). For the aim of covering a large region to identify hot spots of mining activities RapidEye data with a spatial resolution of 6.5 m has been acquired. For pre-processing ortho-rectification and atmospheric correction have been applied to all images with Atmospheric/Topographic Correction for Satellite Imagery (ATCOR) (Richter, 1997). The GeoEye-1 and IKONOS data were pan-sharpened and additional thematic layers were used as ancillary data.

7.5.4.3 Methods Applied

The overall methodological procedure follows the multi-scale approach developed by Schöpfer and Kranz (2010). The initial step is the analysis of HR satellite data with the aim of identifying hot spots of mining/logging activities on large survey extends. In order to reach this goal a transferable feature extraction scheme is generated building upon OBIA concepts. In the case of the analysis of DMCii imagery pixel-based approaches have proven to be very efficient as well for the detection of logging and shifting cultivation. These activities as well as the extraction of minerals imply complete clearing of the sites, a fact that can be used for focusing the detection of corresponding activities on bare soil areas only.

With respect to mining the following assumptions as basis for further investigations can be made: all exploitation activities are related to small-scale artisanal mining. Thus, major interventions into the environment through large excavation, soil accumulation, or pollution are not expected. With RapidEye data a further distinction between bare soil and potential mining site is only possible through secondary indicators, such as distance to settlements/dwellings, roads, and water bodies/rivers. The identified hot spots of resource extraction activities are visually assessed to decide whether VHR data should be acquired in order to investigate the most salient areas in more detail. Subsequently, for the selected sites a multi-temporal change detection analysis is conducted. In the case of the mining site at Bisie a monitoring has been conducted on GeoEye/IKONOS data to evaluate the evolution of the detected mining area.

The outlined methodology has several advantages: (1) the coverage of large survey extents for the identification of hot spots of resource exploitation activities, (2) highly detailed analyses of the detected hotspot areas, and (3) the multi-temporal monitoring resulting in land cover changes for more qualitative insights about developments and trends in relation to the conflict situation in the region.

7.5.4.4 Delivered Information

The outcome of the entire approach is twofold, referring to the different scales the analyses are based on. Firstly, hot spots of mining and logging activities are detected on HR data, which allows not only the coverage of a large region but provides indications for efficiently focusing on the most important sites of interest. As such, areas are identified that are characterized by significant land cover changes potentially related to the conflict situation on the ground. The information resulting from the VHR data–based monitoring provides complementary information to reports and statistics about the extraction of conflict resources. These documents frequently give only a limited insight about ongoing activities in a certain area of interest that is often too remote or too insecure or even both to be investigated by field surveys. As an example, a brief summary of a monitoring conducted during a mining ban (between 2010 and 2011) for the mining site at Bisie in the North Kivu province, DRC, is given in the following.

The analysis is based on two VHR satellite images acquired three days prior to the announcement of the ban by the president and exactly at the date the ban has been lifted. As a result an expansion of the mining site indicates a continuation of exploitation activities during the mining ban, although this does not provide confirming evidence (see Figure 7.16).

Linking these results with statistical data and information gathered from reports, we reached a good level of reliability considering further indications supporting the hypothesis of continued mining during the ban (Zingg-Wimmer and Hilgert, 2011). This underlines the benefit of an integrated assessment of satellite-based information and socioeconomic data, especially in unstable regions where only a convergence of evidence may facilitate a comprehensive picture of the situation in the region. For future analysis, the assessment of the settlement structure at the mining site has been identified as an important indicator for both population dynamics and mining activities through the monitoring of mine workers.

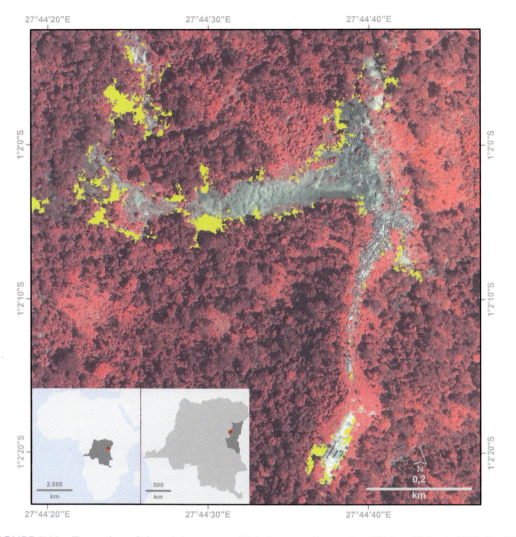

FIGURE 7.16 Expansion of the mining area at Bisie between September 2010 and March 2011 (Lüthje et al., 2014) (modified).

7.5.4.5 Impact

The expansion detected trough the analysis of satellite imagery has been called "the clearest sign of continued mining activities at Bisie" (Zingg Wimmer and Hilgert, 2011) by the International Peace Information Service (IPIS)—a research center supporting governmental, non-governmental, and intergovernmental development actors. In an additional study, IPIS compared official statistics about cassiterite trade in the region with the expansion of the mining area. The overall aim of the presented analysis was to raise the awareness of the real consequences of the mining ban that are: (1) mining continued, (2) production volumes went down considerably, and (3) the military increased their control over mineral resources in the region (which is contrary to the objectives of setting up the ban). Based on the study, general recommendations referred to strengthening the demilitarization of mines and trading routes, deployment of disciplined regiments for securing the territory, and strengthening the mining police. Actions derived from these insights may provide a stable basis for international transparency initiatives to continue to be set in place. At last, the business around

7.6 CONCLUSION

This chapter has shown what great potential remote sensing and Earth observation may provide for humanitarian emergency response but also which challenges are faced when using it (Munyaka et al., 2024; Huang et al., 2023; Aimaiti et al., 2022; Louw et al., 2022; Sadiq et al., 2022; Ye, 2022; Avtar et al., 2021; Fakhri and Gkanatsios, 2021; Kucharczyk and Hugenholtz, 2021; Zheng et al., 2021; Koshimura et al., 2020; Lang et al., 2020; Braun et al., 2019; Lang et al., 2018; Quinn et al., 2018; Jiang et al., 2017; Winkler et al., 2017). Restrictions are, to a lesser degree, of a technical nature; more frequently, they relate to institutional, operational, or political constraints. The authors tried to discuss in a reflective but optimistic and forward-looking way the areas where remotely sensed data, along with the appropriate analysis techniques, can enhance existing workflows. The convergence of both technological and operational aspects stimulates mutual trust among those who deliver and those who receive, such that Earth observation will support where help is needed most—ultimately saving lives.

NOTES

1. http://www.un.org/en/peacekeeping/missions/past/onumozFT.htm
2. South Sudan gained independence from Sudan in 2011.
3. War 1990–1994.
4. Norwegian Refugee Council (2005): Profile of Internal Displacement: Rwanda.
5. Conflict 1993–2005.
6. http://popstats.unhcr.org
7. http://data.unhcr.org
8. Numbers > 1 million rounded to the nearest 100,000, numbers < 1 million rounded to the nearest 10,000.
9. While not yet fully explored and applied by NGOs, it should be noted that the authors see large and unexplored potential in various ways of support from the "crowd." We see similar acceptance and proliferation patterns as with remote sensing data, even more rapid and wide-ranging.
10. One of the reasons why NGOs are sometimes "consternated" is that EO-based data may include sensitive (i.e., political, military) information, and NGOs might lose control on how it is used or abused—with adverse consequences, such as accusations of spying, not being impartial, etc.
11. Practically, NGOs have just started to look into the option of applying UAVs, and the problem is the hallmarking as a military tool, in fact a weapon to assassinate the enemy. At this stage UAVs are hardly appropriate for humanitarian activities within complex emergencies. Instead, other types of natural disasters, chemical and nuclear accidents, etc., may greatly benefit from this right now.
12. https://drones.wfp.org/updates/using-drones-deliver-critical-humanitarian-aid
13. Formerly GMES (Global Monitoring for Environment and Security): the first operational services were called "GIO" (GMES Initial Operations).
14. http://crisismapping.ning.com/
15. For example, a negative trend in vegetation cover as monitored by remote sensing may signalize increasing pressure on the local food production that may cause migration or conflicts about this increasingly scarcer resource. Even a hunger crisis might be predicted using indices derived from remote sensing data.
16. As the timing of aid delivery is always crucial, EO-based information can help identify the spots where humanitarian needs are biggest. It does not replace missing movement clearances from other actors.
17. www.fews.net
18. What sounds like a contradiction in terms, is owed to the fact that politically speaking there is no such thing as a semi-permanent camp or any such transition. While (especially IDP) camps may de-facto exist over time spans of more than ten years, such development is neither intended nor supported by NGOs, the UN, or the host communities. According to UNHCR the term "protracted refugee situation" is used for long(er)-term camps, which are defined as "one where more than 25,000 refugees have been in exile for more than five years." According to this definition around 6.3 million refugees were in a protracted situation by the end of 2013 (UNHCR, Global Trends, 2013)

19 https://eur-lex.europa.eu/eli/dir/2024/1760/oj
20 The information on costs is only a rough indication dated 2015. Actual costs depend on priority and pre-processing levels, whether from archive or tasked and other parameters. Details to be obtained from data provider catalogues, Handling or administrative fees are not considered. Prices refer to Bundle (panchromatic and multispectral bands), where applicable.

REFERENCES

Al Achkar, Z., et al. 2013. *Sudan: Anatomy of a Conflict*. Harvard Humanitarian Initiative: 31–41. https://hhi.harvard.edu/publications/sudan-anatomy-conflict

ARC. 2014. *OpenStreetMap Damage Assessment Review: Typhoon Haiyan (Yolanda) Interim Report*. A. R. C. A. a. REACH. https://americanredcross.github.io/OSM-Assessment/

Aschbacher, J. 2002. Monitoring environmental treaties using earth observation. *VERTIC Verification Yearbook*: 171–186. https://www.vertic.org/media/Archived_Publications/Yearbooks/2002/VY02_Aschbacher.pdf

Bernhard, E. 2013. *Refugee Camp Mapping in Jordan Using TerraSAR-X Data*. https://www.researchgate.net/publication/259898098_Refugee_camp_mapping_in_Jordan_using_TerraSAR-X_data

Bjorgo, E. 2000. Refugee camp mapping using very high spatial resolution satellite sensor images. *Geocarto International* 15(2): 79–88.

Bjorgo, E. 2001. Supporting humanitarian relief operations. In *Commercial Observation Satellites: At the Leading Edge of Global Transparency*. Eds. J. C. Baker, K. M. O'Connell and R. A. Williamson. Santa Monica, CA: ASPRS Rand: 403–427.

Blaschke, T. 2010. Object based image analysis for remote sensing. *ISPRS International Journal of Photogrammetry and Remote Sensing* 65(1): 2–16.

Capelo, L., N. Chang and A. Verity. 2012. *Guidance for Collaborating with Volunteer and Technical Communities*. D. H. Network. https://www.urban-response.org/system/files/content/resource/files/main/guidance-for-collaborating-with-volunteer-and-technical-communities.pdf

Checchi, F., B. T. Stewart, J. J. Palmer and C. Grundy. 2013. Validity and feasibility of a satellite imagery-based method for rapid estimation of displaced populations. *International Journal of Health Geographics* 12(4).

Cowan, N. M. 2011. A geospatial data management framework for humanitarian response. *8th International ISCRAM Conference Lisbon*. Portugal: 1–5. https://idl.iscram.org/files/cowan/2011/418_Cowan2011.pdf

Drury, S. A. and M. E. A. Deller. 2002. Remote sensing and locating new water sources. *The Use of Space Technology for Disaster Management for Africa*. Vienna: UNOOSA. https://www.unoosa.org/pdf/sap/2002/ethiopia/presentations/12speaker01_1.pdf

Ehrlich, D., S. Lang, G. Laneve, S. Mubareka, S. Schneiderbauer and D. Tiede. 2009. Can Earth observation help to improve information on population? Indirect population estimations from EO derived geospatial data. In *Remote Sensing from Space Supporting International Peace and Security*. Eds. B. Jasani, M. Pesaresi, S. Schneiderbauer and G. Zeug. Berlin: Springer: 211–237.

Elwood, S. 2008. Volunteered geographic information: Key questions, concepts and methods to guide emerging research and practice. *GeoJournal* 72: 133–135.

FAO. 2013. *Trends and Impacts of Foreign Investment in Developing Country Agriculture: Evidence from Case Studies*. F. a. A. O. o. t. U. N. (FAO). Rome.

Füreder, P., D. Hölbling, D. Tiede, S. Lang and P. Zeil. 2012. Monitoring refugee camp evolution and population dynamics in Dagahaley, Kenya, based on VHSR satellite data. *9th International Conference African Association of Remote Sensing of the Environment (AARSE)*. El Jadida, Morocco.

Füreder, P., D. Tiede, F. Lüthje and S. Lang. 2014. Object-based dwelling extraction in refugee/IDP camps—challenges in an operational mode. *South-Eastern European Journal of Earth Observation and Geomatic* 3(2s): 539–544.

Garrett, N. W. 2008. *Artisanal Cassiterite Mining and Trade in North Kivu—Implications for Poverty Reduction and Security*. Washington, DC: World Bank, Communities and Small-Scale mining Project [CASM].

Gella, G. W., L. Wendt, S. Lang, D. Tiede, B. Hofer, Y. Gao and A. Braun. 2022. Mapping of dwellings in IDP/refugee settlements from very high-resolution satellite imagery using a mask region-based convolutional neural network. *Remote Sensing* 14(3): 689.

Ghorbanzadeh, O., D. Tiede, L. Wendt, M. Sudmanns and S. Lang. 2020. Transferable instance segmentation of dwellings in a refugee camp—integrating CNN and OBIA. *European Journal of Remote Sensing*: 1–14.

Giada, S., T. de Groeve and D. Ehrlich. 2003. Information extraction from very high resolution satellite imagery over Lukole refugee camp, Tanzania. *International Journal of Remote Sensing* 24(22): 4251–4266.

GlobalWitness. 2013. Breaking the links between natural resources and conflict: The case for EU regulation. *A Civil Society Position Paper*. G. W. Publication. https://www.cidse.org/2013/09/16/breaking-the-links-between-natural-resources-and-conflict-the-case-for-eu-rloegulation/

Goodchild, M. F. 2007. Citizens as sensors: The world of volunteered geography. *GeoJournal* 69: 211–221.

Hagenlocher, M., S. Lang and D. Tiede. 2012. Integrated assessment of the environmental impact of an IDP camp in Sudan based on very high resolution multi-temporal satellite imagery. *Remote Sensing of Environment* 126: 27–38.

Haken, N., J. J. Messner, K. Hendry, P. Taft, K. Lawrence and F. Umaa. 2013. Failed state index IX 2013. *Technical Report. F. F. Peace*. https://fundforpeace.org/2013/06/24/failed-states-index-2013-the-book/

Hakley, M. 2010. How good is volunteered geographical information? A comparative study of OpenStreetMap and ordnance survey datasets. *Environment and Planning B: Planning and Design* 37(4): 682–703.

HHI. 2011. *Disaster Relief 2.0: The Future of Information Sharing in Humanitarian Emergencies*. Washington, DC and Berkshire, UK: UN Foundation and Vodafone Foundation Technology Partnership.

HIIK. 2013. *Conflict Barometer 2013. H. I. f. I. C. R. (HIIK)*. Heidelberg.

Humanitarian Coalition. n.d. *What Is a Humanitarian Emergency?* Retrieved 24 September 2024, from https://www.humanitariancoalition.ca/what-is-a-humanitarian-emergency

IDMC. 2014. *Global Overview 2014. People Internally Displaced by Conflict and Violence*. I. D. M. Centre. https://www.internal-displacement.org/publications/global-overview-2014-people-internally-displaced-by-conflict-and-violence/

IDMC. 2024. *GRID 2024: Global Report on Internal Displacement*. IDMC-GRID-2024-Global-Report-on-Internal-Displacement.pdf

IPCC. 2001. *Climate Change 2001: The Scientific Basis. Contribution of Working Group I to the Third Assessment Report of the Intergovernmental Panel on Climate Change*. Cambridge: Cambridge University Press.

Jensen, J. R. 2005. *Introduction to Digital Image Processing. A Remote Sensing Perspective*. Upper Saddle River, NJ: Prentice-Hall.

Joyce, K. E., S. E. Belliss, S. V. Samsonov, S. J. McNeill and P. J. Glassey. 2009. A review of the status of satellite remote sensing and image processing techniques for mapping natural hazards and disasters. *Progress in Physical Geography* 33(2): 183–207.

Kemper, T. and J. Heinzel. 2014. Mapping and monitoring of refugees and internally displaced people using EO data. In *Global Urban Monitoring and Assessment through Earth Observation*. CRC Press: 195–216. https://www.taylorfrancis.com/chapters/mono/10.1201/b17012-15/mapping-monitoring-refugees-internally-displaced-people-using-eo-data-qihao-weng

Kranz, O., et al. 2010. Monitoring refugee/IDP camps to support international relief action. In *Geoinformation for Disaster and Risk Management—Examples and Best Practices. Joint Board of Geospatial Information Societies (JB GIS)*. Eds. O. Altan, R. Backhaus, P. Piero Boccardo and S. Zlatanova. Vienna: United Nations Office for Outer Space Affairs (UNOOSA): 51–56.

Kuenzer, C., M. Bartalis, M. Schmidt, D. Zhaoa and W. Wagner. 2008. *The International Archives of the Photogrammetry, Remote Sensing and Spatial Information Sciences* XXXVII. https://www.unoosa.org/documents/pdf/psa/activities/2008/graz/presentations/07-02.pdf

Lang, S. 2008. Object-based image analysis for remote sensing applications: Modeling reality—dealing with complexity. In *Object-Based Image Analysis—Spatial Concepts for Knowledge-Driven Remote Sensing Applications*. Eds. T. Blaschke, S. Lang and G. J. Hay. Berlin: Springer: 3–28.

Lang, S., C. Corbane and L. Pernkopf. 2013. Earth observation for habitat and biodiversity monitoring. In *Ecosystem and Biodiversity Monitoring: Best Practice in Europe and Globally [section title]*. Eds. S. Lang and L. Pernkopf. Heidelberg: Wichmann: 478–486.

Lang, S., D. Tiede and F. Hofer. 2006. Modeling ephemeral settlements using VHSR image data and 3D visualization—the example of Goz Amer refugee camp in Chad. *PFG—Photogrammetrie, Fernerkundung, Geoinformatik* 4: 327–337.

Lang, S., D. Tiede, D. Hölbling, P. Füreder and P. Zeil. 2010. EO-based ex-post assessment of IDP camp evolution and population dynamics in Zam Zam, Darfur. *International Journal of Remote Sensing* 31(21): 5709–5731.

Le Billon, P. 2001. The political ecology of war: Natural resources and armed conflicts. *Political Geography* 20: 561–584.

Le Billon, P. 2006. *Fuelling War: Natural Resources and Armed Conflict*. Abingdon: Routledge.

Luethje, F., Kranz, O., & Schoepfer, E. (2014). Geographic object-based image analysis using optical satellite imagery and GIS data for the detection of mining sites in the Democratic Republic of the Congo. *Remote Sensing*, 6(7), 6636–6661.

Martin, A. 2005. Environmental conflict between refugee and host communities. *Journal of Peace Research* 42(3): 329–347.
Mildner, S. A., G. Lauster and W. Wodni. 2011. Scarcity and abundance revisited: A literature review on natural resources and conflict. *International Journal of Conflict and Violence* 5(1): 155–172.
MSF. 2011. *Kenya: Humanitarian Crisis on the Outskirts of Overcrowded Dadaab Camp*. https://www.doctorswithoutborders.org/latest/kenya-humanitarian-crisis-outskirts-overcrowded-dadaab-camp
Padwick, C., M. Deskevich, F. Pacifiä and S. Smallwood. 2010. WorldView-2 pan-sharpening. *Proc. American Society for Photogrammetry and Remote Sensing* 13.
Pesaresi, M., A. Gerhardinger and F. Haag. 2007. Rapid damage assessment of built-up structures using VHR satellite data in tsunami-affected areas. *International Journal of Remote Sensing* 28(13): 3013–3036.
Potapov, P. V., et al. 2012. Quantifying forest cover loss in Democratic Republic of the Congo, 2000–2010, with Landsat ETMC data. *Remote Sensing of Environment* 122: 106–116.
Rhee, J., I. Jungho and G. J. Carbone. 2010. Monitoring agricultural drought for arid and humid regions using multi-sensor remote sensing data. *Remote Sensing of Environment* 114(12): 2875–2887.
Richter, R. 1997. Correction of atmospheric and topographic effects for high spatial resolution satellite imagery. *International Journal of Remote Sensing* 18: 1099–1111.
Schöpfer, E. and O. Kranz. 2010. Monitoring natural resources in conflict using an object-based multiscale image analysis approach. *International Archives of Photogrammetry, Remote Sensing, and Spatial Information Sciences* XXXVIII-4(C7).
Slonecker, E. T., D. M. Shaw and T. M. Lillesand. 1998. Emerging legal and ethical issues in advanced remote sensing technology. *Photogrammetric Engineering and Remote Sensing* 64(6): 589–595.
Soden, R., N. Budhathoki and L. Palen. 2014. Resilience-building and the crisis informatics agenda: lessons learned from Open Cities Kathmandu. In *11th International ISCRAM Conference*. Eds. S. R. Hiltz, M. S. Pfaff, L. Plotnick and A. C. Robinson. Pennsylvania. https://cmci.colorado.edu/~palen/palen_papers/SodenBudhathokiPalen-ISCRAMKathmandu.pdf
Standby Task Force. 2011. *Crowdsourcing Satellite Imagery Analysis for UNHCR-Somalia: Latest Results*. Standby Task Force. https://standbytaskforce.org/2011/11/10/unhcr-somalia-latest-results/
Tiede, D., P. Füreder, S. Lang, D. Hölbling and P. Zeil. 2013. Automated analysis of satellite imagery to provide information products for humanitarian relief operations in refugee camps—from scientific development towards operational services. *PFG Photogrammetrie—Fernerkundung—Geoinformation*: 185–195.
Tiede, D., S. Lang, P. Füreder, D. Hölbling, C. Hoffmann and P. Zeil. 2011. Automated damage indication for rapid geospatial reporting. An operational object-based approach to damage density mapping following the 2010 Haiti earthquake. *Photogrammetric Engineering and Remote Sensing* 9: 933–942.
Tiede, D., S. Lang, D. Hölbling and P. Füreder. 2010. Transferability of OBIA rule sets for IDP camp analysis in Darfur. *International Archives of Photogrammetry, Remote Sensing, and Spatial Information Sciences* XXXVIII-4(C-7).
UNDP. 2011. *Human Development Report 2011*. Technical Report. U. N. D. P. (UNDP). https://hdr.undp.org/content/human-development-report-2011
UNEP. 2006. *Africa Environment Outlook 2. Our Environment, Our Wealth*. D. o. E. W. a. A. United Nations Environment Programme (UNEP). Nairobi, Kenya.
UNEP. 2009. *From Conflict to Peacebuilding*. The Role of Natural Resources and the Environment. U. N. E. P. (UNEP). https://www.iisd.org/publications/conflict-peacebuilding-role-natural-resources-and-environment
UNHCR. 2012. *Global Trends 2011. A Year of Crisis*. https://www.unhcr.org/statistics/country/4fd6f87f9/unhcr-global-trends-2011.html
UNHCR. 2013. *Global Trends 2013: War's Human Cost*. https://www.unhcr.org/media/unhcr-global-trends-2013
UNHCR. 2014. *Refugees in the Horn of Africa: Somali Displacement Crisis*. https://data.unhcr.org/en/documents/details/31994
UNHCR. 2023. *Global Trends: Forced Displacement in 2023*. Global Trends report 2023 | UNHCR
UNHCR. 2024. *Refugee Data Finder*. Retrieved 19 August 2024, from https://www.unhcr.org/refugee-statistics/download/?url=IAr67y
UN Mission in Sudan. 2011. *Update on the Attack and Occupation of Abyei by SAF. U. N. M. i. Sudan.*
UNOCHA. 2013. *Sudan: Zamzam IDP Camp Profile*. Retrieved 29 May 2014, from http://reliefweb.int/sites/reliefweb.int/files/resources/sud13_North%20Darfur_Zamzam%20IDP%20Camp%20Profile_a3_09may13.pdf.
UNRadio. 2011. *Humanitarian Supplies Looted in Abyei Town of Sudan*. U. N. Radio.

USAID. 2005. *Forest and Conflict. A Toolkit for Intervention*. O. o. C. M. a. M. U.S. Agency for International Development. Washington, DC.

Verjee, F. 2011. *GIS Tutorial for Humanitarian Assistance*. Redlands, CA: ESRI Press.

Voigt, S., T. Kemper, T. Riedlinger, R. Kiefl, K. Scholte and H. Mehl. 2007. Satellite image analysis for disaster and crisis-management support. *IEEE Transactions on Geoscience and Remote Sensing* 45(6): 1520–1528.

Wagner, W., K. Scipal, C. Pathe, D. Gerten, W. Lucht and B. Rudolf. 2003. Evaluation of the agreement between the first global remotely sensed soil moisture data with model and precipitation data. *Journal of Geophysical Research* 108: 4611.

Walker, P., J. Glasser and S. Kambli. 2012. *Climate Change as a Driver of Humanitarian Crises and Response*. https://fic.tufts.edu/publication-item/climate-change-as-a-driver-of-humanitarian-crises-and-response/

Zingg Wimmer, S. and F. B. Hilgert. 2011. A one-year snapshot of the DRCs principal cassiterite mine. *Technical Report I. P. I. S. (IPIS)*. https://ipisresearch.be/wp-content/uploads/2011/11/20111128__Bisie_FHilgert_SZingg.pdf

Part III

Volcanoes

8 Remote Sensing of Volcanoes

Robert Wright

ACRONYMS AND DEFINITIONS

ASTER	Advanced spaceborne thermal emission and reflection radiometer
ATSR	Along-track scanning radiometer
AVHRR	Advanced very high resolution radiometer
BTD	brightness temperature difference
DU	Dobson unit
ERS	European remote sensing
ESO	Earth system observatory
GOME-2	Global Ozone Monitoring Experiment 2
IASI	Infrared atmospheric sounding interferometer
IFOV	instantaneous field of view
InSAR	Interferometric synthetic aperture radar
MIR	Middle-infrared
MODIS	Moderate resolution imaging spectroradiometer
NDVI	Normalized Difference Vegetation Index
OMI	Ozone mapping instrument
PS	Permanent scatterer
SWIR	Short wave infrared
TOMS	Total ozone mapping spectrometer
TIR	Thermal infrared
VAAC	Volcanic Ash Advisory Centers
VNIR	Visible and near-infrared

8.1 INTRODUCTION

There are about 1500 active and potentially active volcanoes on Earth, of which an average of 70 erupt in any given year (Siebert et al., 2010). This sub-areal volcanism (i.e., that which is not hidden from us beneath the surface of the ocean) is the most obvious and dramatic manifestation of energy loss from Earth's interior. The impact of volcanism on the Earth and humanity is profound; volcanism provided us with an atmosphere that it has modified ever since (Condie, 2005), and may have influenced the evolution of life on Earth through the role eruptions have played in mass extinctions (Rampino, 2010). Volcanism has been a source of myth (Sigurdsson, 1999), a subject for art and literature (Sigurdsson and Lopes-Gaultier, 2000) and has yielded resources that society can exploit (Wohletz and Heiken, 1992).

This chapter will provide an overview of how satellite remote sensing can provide quantitative information before, during, and after a volcanic eruption. After an overview of the physical processes that take place at active volcanoes that we might be interested in measuring, the manner in which remote sensing can be used to provide quantitative information about them will be discussed, along with some contextual information about how these measurements can be made in situ. Although by no means exhaustive, a list of citations for significant papers will be provided so that the reader can trace the development of remote sensing as a tool for studying volcanism on Earth over the last several decades.

8.2 WHAT DO VOLCANOES DO THAT WE MIGHT BE INTERESTED IN MEASURING?

Before describing how remote sensing can (and cannot) be used to aid in the study of Earth's volcanoes, it will be helpful to describe what it is that an erupting volcano, or one that is about to erupt, does, that scientists might be interested in measuring (Figure 8.1), in order to place the remote sensing measurements in context.

Consider a dormant volcano that is about to reawaken. Melting of the mantle takes place at depths of between ~20 and ~150 km, depending on tectonic setting. The melt is buoyant and begins to rise from the source region. As it approaches the surface it must displace crustal rocks. This results in seismicity, as the ground shakes as the surrounding country rocks are fractured, and as gases and fluids associated with the rising magma pass through cracks and fractures. The displacement of rocks is accompanied by the deformation of the surface of the volcano—the volcano swells and shrinks as magma arrives from the depths and moves around beneath and within the edifice.

In addition to shaking and deforming the ground, this transfer of mass to shallow crustal levels also transports thermal energy and volatiles toward the surface. At depth, under pressure, species such as water, carbon dioxide, and sulfur dioxide (SO_2), are soluble in magmas. But as the magma rises, and the pressure decreases, these volatiles come out of solution promoting the growth of a vapor phase within the melt. The most insoluble volatiles (such as the noble elements and carbon dioxide) begin to exsolve at a depth of about 80 km, while at depths of about 10 km confining pressures are reduced to the extent that water enters the gas phase. Other volatiles, such as SO_2 and HF, enter the mix when the magma reaches to within about 1 km of the surface. While some of these gases are common in the atmosphere (e.g., H_2O and CO_2) others are not (e.g., SO_2), and act as almost incontrovertible proof that fresh magma has been injected beneath the volcano. Changes in the amount and type of gases exhaled by a volcano provide insights into the mass of magma involved, and the nature of its movement between the shallow magma reservoir and the surface, via the conduit.

The presence of a shallow magma body heats meteoric water, resulting in convective cycling of water within the permeable zone between the heat source and the surface. This results in relatively low temperature manifestations of volcanism at the surface (e.g., up to a few hundreds of °C), such as hot crater lakes, mud pools, geysers, and fumaroles (i.e., steam vents). Chemical reactions between this circulating groundwater and the magma also endow these fluids with chemical elements, which can be precipitated at the surface as the fluids cool. This gives rise to changes in surface composition, including sulfur, silica, and calcium deposition, as well as acid alteration of surface rocks to clays.

In many cases the story ends here, and the ascending magma stalls at several kilometers' depth in a crustal magma chamber—a failed eruption. But if magma arrives at the surface an eruption begins. Eruptions take one of two end-member forms. In an effusive eruption magma that arrives at the surface is erupted as lava (a contiguous body of silicate melt within which is suspended crystals and gas bubbles). Most commonly the lava extends downhill from the eruptive vent under the influence of gravity as one or more lava flows, which tend to be long, narrow, and thin, such as those commonly observed at Kīlauea volcano, Hawai'i, or Mount Etna, Sicily. Although dramatic, lava flows rarely pose a threat to human life as the speeds at which they travel downslope are generally little more than a fast walking speed. However, they do often have a substantial economic cost, due to the destruction of infrastructure.

In cases where the lava is more viscous, gas rich, and erupted at a lower volumetric flux rate (as happens in convergent tectonic environments, exemplified by the "Ring of Fire" that surrounds the Pacific Ocean) the lava does not flow downhill significantly, but rather piles up atop the conduit, forming a relatively small, but thick, lava dome. Active lava domes are associated with the most famously destructive volcanic eruptions (e.g., Mount St. Helens, Washington, 1980, and Montagne Pelee, Martinique,1902). This is a result of the higher gas content and viscosity of the lava and the

Remote Sensing of Volcanoes

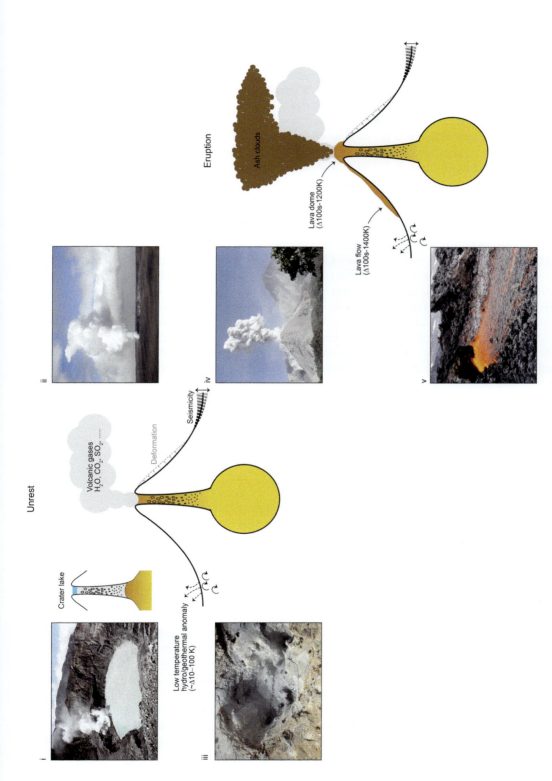

FIGURE 8.1 Cartoons illustrating the processes that occur before and during a volcanic eruption, which scientists find desirable to measure. (i) Hot, acidic, crater lake at Poás, Costa Rica; (ii) hydrothermal mud-pot in Iceland; (iii) gas plume at Kīlauea volcano, Hawai'i; (iv) lava dome and ash plume at Santa Maria volcano, Guatemala; and (v) lava flow at Mount Etna, Sicily.

fact that the domes can pile up to the extent that they become gravitationally unstable, collapse, and spawn fast-moving pyroclastic flows down the flanks of the volcano. Such pyroclastic flows were responsible for the deaths of ~28,000 people during the May 1902 eruption of Mount Pelee.

Finally, with regard to effusive eruptions, persistent lava lakes are much rarer. Here, hot, gas rich, low-viscosity lava circulating in the conduit via convection can maintain a permanent lava "lake" at the top of the conduit, within a summit crater. Such lava lakes can persist for decades, perhaps even hundreds of years. Although they pose little or no threat to life, they are of interest to volcanologists partly because of their rarity, and partly because they are sites where circulation of magma and volatiles between the surface and the shallow chamber can be directly studied. Whether it results in a lava flow, lava dome, or lava lake, an effusive eruption is characterized by the emplacement of lava, with temperatures as high as ~1150 °C, onto the surface of the volcano.

At the other extreme an eruption can be explosive rather than effusive. In an explosive eruption ascending magma experiences a reduction in pressure as it approaches the surface, and rapid expansion of the gases contained in growing bubbles within it tears the stiffening magma apart, resulting in a mixture of solids (pyroclasts, or fragmented magma) suspended in gas. This mixture accelerates rapidly up the conduit before being ejected at great speed into the atmosphere. Although nature provides a spectrum of exit velocities, masses, mass flux rates, and plume heights, it is the volcanic plumes generated in large explosive eruptions, such as the eruption of Mount Pinatubo, Philippines, in 1991, or Mount St. Helens in 1980, that have the most far-reaching effects. In these eruptions the energy involved pulverizes the fragments into very small ash-sized particles, forming volcanic ash clouds. During the 1991 Pinatubo eruption the cloud reached heights of 40 km, and covered an area of 230,000 km^2. Collapse of these plumes (once they have entrained enough cold air to become denser than the surrounding atmosphere) generates dangerous pyroclastic flows. The ash can be hazardous to human health, as it may be fine enough to be respirable. Ash clouds are also a hazard to global aviation.

For the interested reader, introductions to volcanology are provided by Parfitt and Wilson (2008) and Francis and Oppenheimer (2004). A more advanced treatment of all aspects of the science of volcanology can be found in the volume edited by Sigurdsson et al. (2015).

8.3 WHY USE SATELLITE REMOTE SENSING TO STUDY ACTIVE VOLCANISM?

As subjects for study, active volcanism leverages the advantages afforded by an orbital perspective. Erupting volcanoes are temporally dynamic; active lava flows expand, volcanoes deform, volumetric flux rates vary; eruptions themselves last from days, to weeks, to months. For such targets, the acquisition of data of uniform quality, at repetitive time intervals, over extended periods of time, from the safety of space, is obviously beneficial. A synoptic view is of importance, given the size that some targets can attain; lava flows can easily attain lengths of several kilometers, while volcanic plumes ash clouds are tens of kilometers in diameter. Furthermore, these ash clouds can expand to cover large areas as they disperse; the ash cloud produced during the climactic eruption of June 15, 1991, at Mount Pinatubo extended more than 1000 km from the vent in less than 11 hours (Holasek et al., 1996). The aerosol cloud produced by the 1982 eruption of El Chichon, Mexico, circled the globe within two weeks. Such large eruption products cannot be adequately sampled from the ground, or with a narrow field of view. Remote sensing of volcanism also, as will become apparent, leverages data acquired from sub-micron to centimeter wavelengths, in the ultraviolet (for the analysis of volcanic gas plumes), through the shortwave and mid-wave infrared (for retrieving the temperatures of active lavas) and the long-wave infrared (for detecting and tracking volcanic ash clouds) to the microwave (for quantifying volcano deformation).

The relative importance of remote sensing in providing information about active volcanoes varies depending on which volcano is the subject. Very well-monitored volcanoes (such as Kīlauea volcano in Hawai'i) do not rely on satellite remote sensing to provide information about volcanic

Remote Sensing of Volcanoes

processes, given the comprehensive in situ monitoring systems that exist (including seismometers, permanent global positioning system arrays, tilt meters, ground-based gas sensing spectrometers, thermal cameras, infrasound arrays). At this class of volcano satellite remote sensing may provide only supplemental data. Alternatively, satellite remote sensing may be the only source of information regarding eruptions at poorly monitored volcanoes, such as the large number in the remote, and unpopulated, Aleutian island chain. Between these endmembers, remote sensing plays a greater or lesser role in providing information about the volcanic processes described in the previous section. Most of Earth's active or potentially active volcanoes benefit from no in situ monitoring at all (Pritchard et al., 2022), making satellite remote sensing the only means for obtaining information about them.

8.4 REMOTE SENSING OF VOLCANO DEFORMATION

Volcanoes inflate as magma is intruded from beneath and deflate as magma is erupted onto the surface or redistributed within the edifice. In situ measurements of ground deformation have been shown to reliably document these movements of magma, at many volcanoes around the world (Figure 8.2).

Measurements of volcano deformation were traditionally made using conventional field surveying techniques, including precise leveling and trilateration (e.g., Murray et al., 1995). Tilt measurements (using the wet tilt technique, Figure 8.2, or dry tilt) have also been a mainstay of deformation monitoring at volcanoes around the world. In recent years global positioning system have somewhat supplanted these methods.

Although able to resolve very subtle changes in volcano topography, these methods suffer from the fact that they provide a very incomplete spatial sample. Volcanoes are large (Mount Etna in Sicily covers an area of approximately 1200 km^2), and leveling lines, surveying benchmarks sample at a relative handful of point locations, requiring interpolation between disparate measurements to yield a volcano-wide deformation field. In addition, with the exception of the permanently recording GPS networks installed at some volcanoes, temporal sampling also leaves much to be desired, as these surveying methods require boots on the ground, with temporal sampling limited by the expense that entails. This combination of factors meant that although the deformation behavior of some volcanoes was very well constrained (e.g., Mount Etna, Sicily, and Kīlauea, Hawai'i), many (most) others, either because of remoteness, lack of funds, or lack of interest, were unmonitored.

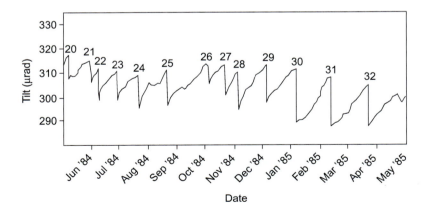

FIGURE 8.2 Tilt measurements obtained on the rim of the caldera at the summit of Kilauea volcano, Hawai'i. Increases in tilt angle correspond to periods of volcano inflation. Decreases correspond to deflation during an eruption. Numbers denote eruptive episodes.

From http://volcanoes.usgs.gov/activity/methods/deformation/tilt/kilauea.php, courtesy of the USGS.

8.4.1 Quantifying the Surface Deformation Field

The field of volcano deformation monitoring has been revolutionized by the technique of interferometric synthetic aperture radar (InSAR, sometimes referred to as IFSAR). Figure 8.3 illustrates the technique.

For those interested in more details, Burgmann et al. (2000) and Richards (2007) provide reviews of the technique, and Lu and Dzurisin (2014) provide a comprehensive overview of how InSAR has been applied to volcanoes in the Aleutian Island arc. Side-looking spaceborne SARs sample the amplitude and phase of radar echoes from ground resolution elements. Amplitude images are familiar to many. Phase information less so, but it is the crucial component of the InSAR technique. The electromagnetic waves that the SAR emits travel to the target and be scattered back to the antenna. The distance is an integer number of complete wave cycles, plus a fraction of a cycle—this fraction is the phase, which combined yield the range from the antenna to each point on the ground within the image. Figure 8.3 shows two amplitude and phase images acquired by the first European Remote Sensing satellite (ERS-1) SAR at Mount Peulik volcano, Alaska. Here, spatial resolution is about 30 m. The images encode phase as colors varying between 0° and 360° (alternatively expressed as 0 to 2π radians). One cycle of interferometric phase is called a "fringe."

An idealized description of the processing follows, before a description of how non-ideal conditions influence the applicability of the technique. Volcano deformation occurs over time, so a single image contains no deformation information. InSAR uses two images of the same target separated in time (Δt) and computes the phase for each pixel. If no deformation has occurred during the time interval there will be no phase change ($\Delta \varphi = \varphi_1 - \varphi_2 = 0$). However, if deformation has occurred then the same point on the ground will have moved to or away from the satellite, and the $\Delta \varphi$ will not be 0. First, the two SAR images are co-registered. This is necessary in order that the phase values retrieved from the first and second images that are to be differenced relate to the same physical area on the ground. Once this has been achieved the phase difference that has accumulated between observation time t_1 and t_2, for each point on the surface of the volcano can be determined, and the interferogram generated (i.e., $\varphi_1 - \varphi_2 = \Delta \varphi$; Figure 8.3, v).

If the two images were acquired by the same sensor at exactly the same position in space, then $\Delta \varphi$ would be the result of deformation alone. However, spacecraft orbital and attitude control does not allow for imaging from the exact same position, so in practice the images will be acquired from *about* the same position. This difference in position of the satellites at times t_1 and t_2 is referred to as the baseline and should be as small as possible for deformation monitoring. The limit depends on the sensor used; for ERS-1 a theoretical maximum of 1100 m has been calculated, but in practice a value of closer to 600 m was found to be more realistic. Imaging the same target from two different positions in space means that $\Delta \varphi$ is not purely the result of deformation. The change of viewing angle introduces a parallax effect, shifting the apparent position of ground points within the images, and resulting in a phase difference for each ground target due to the topography ($\varphi_{topography}$). Even in the absence of any topography a flat (or curved) earth would still yield a gradual phase change across the image ($\varphi_{flat-earth}$) because of the gradient in look angle across the scene. Both the phase difference due to topography and the phase difference caused by the difference in satellite positions over a flat earth must be removed from $\Delta \varphi$ to isolate the deformation component.

The flat-earth phase (which is manifest as strong linear fringes running across the image; Figure 8.3 vi) can be predicted (i.e., based on knowledge of the satellites' geometry with respect to a reference ellipsoid) and removed using a geometry-based approach (e.g., Zebker et al., 1994). This is referred to as flattening the interferogram ($\Delta \varphi - \varphi_{flat-earth}$, Figure 8.3 vii), and is performed before removal of the topographic phase. Removal of the topographic phase information can be achieved by either using a digital elevation model and solving for the phase difference that that topography would contribute to the observed phase (e.g., generate a synthetic interferogram from the DEM; Massonnet et al., 1993), or using a third radar image to generate a DEM from the radar data itself. In this case, the first pair of images (which contain the deformation signal) are spaced with a sufficient

Remote Sensing of Volcanoes

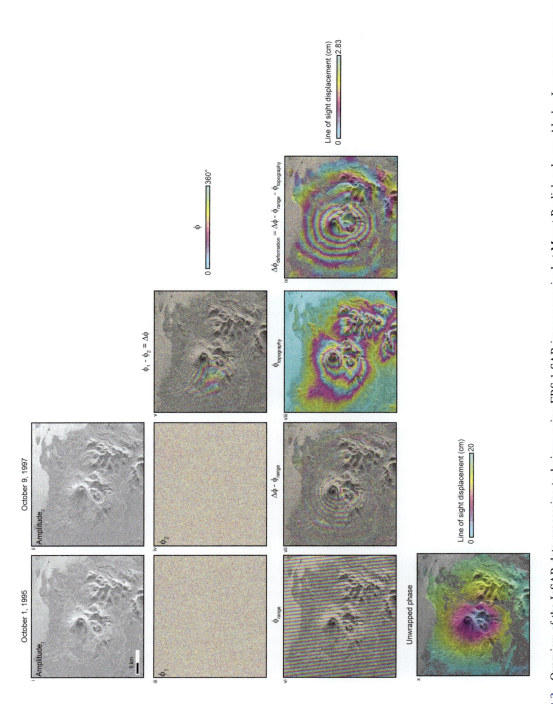

FIGURE 8.3 Overview of the InSAR data processing technique, using ERS-1 SAR images acquired at Mount Peulick volcano, Alaska. Images were processed by Zhong Lu, Southern Methodist University, Texas, and the author expresses his gratitude to Zhong for taking the time to prepare the figures.

time interval (Δt) to capture the deformation signal of interest. The third image must be captured very close in time to either of these two images, such that it can be assumed that no deformation has had chance to occur, and the only phase change this pair records is that of the topography caused by the non-zero baseline (e.g., Gabriel et al., 1989).

The result (Figure 8.3 ix, $\varphi_{deformation} = \Delta\varphi - \varphi_{flat-earth} - \varphi_{topography}$) is the phase change due to deformation of the volcano surface between time t_1 and t_2. As phase is assumed to be proportional to distance at the scale of the wavelength, in these images one cycle/fringe corresponds to a distance of 2.8 cm, the wavelength of the C-band ERS-1 SAR being 5.6 cm (ground movement of $\lambda/2$ corresponds to a whole wavelength change in the round-trip range, and is thus equal to one fringe). At this point the actual amount of deformation at any point in the image is not known, as $\varphi_{deformation}$ merely cycles, or wraps, between 0 and 2π. The interferogram depicted in Figure 8.3 ix is said to be wrapped. As previously stated, the actual range between the antenna and the satellite for each point in the image is given by $n \times 2\pi$ (where n is the complete number of wave cycles between target and antenna—the integer ambiguity) *plus* the small fraction of interferometric phase measured, $\Delta\varphi$. In order to work out the unambiguous phase for each pixel in the image a value of n has to be added to, or subtracted from, each $\varphi_{deformation}$ measurement. This process, determining the correct value of n for each pixel, is called phase unwrapping. A generalized method is to choose a seed pixel in areas of low noise and then add values of 2π to its phase value in expanding contours away from the seed, whilst minimizing the change between adjacent points. There are many phase unwrapping algorithms described in the literature (e.g., Zebker and Lu, 1998; Yu et al., 2019).

Phase unwrapping is an important part of the process, required to obtain a quantitative deformation field for the study area, detect subtle deformations, and also if the InSAR data are to be used to retrieve the properties of the sub-surface source that caused the deformation, via numerical modeling. It is common that (1) the interferogram is unwrapped, (2) the unwrapped interferogram is then used for deformation source modeling, and (3) the observed interferogram, the modeled interferogram, and residual interferogram (see next) are then plotted in a "wrapped" fashion for publication and interpretation. Figure 8.3 x shows the unwrapped interferogram for this example. As such, the relative deformation recorded in a wrapped interferogram such as the one depicted in Figure 8.3 ix can be determined by (1) choosing a point in the far-field where no fringes occur (and hence there was no deformation), and then (2) counting the number of fringes inwards from that point and multiplying by $\lambda/2$. In this example, approximately 18 cm of deformation occurred at the summit of the volcano over the two-year period sampled by the interferogram. Whether this is subsidence (movement away from the satellite) or inflation (movement toward) depends on the direction of the change in phase (or order of the colors as shown in the color scale); when counting in from the edge of the fringe pattern toward the center of the deformation feature, if the interferometric phase decreases toward the center (i.e., trends from 2π to 0) then the distance between the antenna and the ground has become shorter toward the center, so the ground must have moved toward the satellite (i.e., inflation). The opposite is true for ground subsidence. In this case the colors cycle from cyan-yellow-magenta as the center of the fringe pattern is approached—the volcano experienced a period of net inflation in the two-year period encompassed by this interferogram.

The precision with which volcano deformation can be quantified is usually quoted at the sub-cm or mm scale, and the ability to quantify deformation at this scale over the entire volcanic edifice has revolutionized ground deformation monitoring. This mm sensitivity depends on several factors. The phase change due to deformation must be isolated from the interferometric phase measured, which is also composed of phase shifts due to several other factors. Although expressed differently by different authors these components can be summarized as factors that introduce a change in phase due to uncertainty in the spacecraft orbit, errors in the DEM used to estimate topographic phase, the changing atmosphere, and the scattering properties of the surface (e.g., Massonnet and Feigl, 1998). Uncertainties in orbit (baseline uncertainties) impact the success of the flat-earth correction. For ENVISAT SAR (part of the ERS series), this uncertainty has been estimated to be equivalent to a few tenths of a mm per km across the image (Wang et al., 2009), but this depends on which SAR is used, as this error can be eliminated if spacecraft orbit is known perfectly (and they are known more, or less, perfectly for different spacecraft). As previously discussed, isolating the deformation signal requires that the topographic phase be calculated, often using a DEM. The DEMs derived from the Shuttle Radar Topography Mission have a relative vertical accuracy of less than 10 m. It may seem counterintuitive that a DEM with ~10 m uncertainties can be used as part of a process that derives ground displacements on the mm scale, but the sensitivity of the phase measurement to deformation is several thousands of times greater than to the sensitivity to the topography (equations 15 and 16 in Zebker et al., 2001). This explains why deformation can be estimated to sub-cm precision even if the DEM used to correct for topographic phase is inaccurate on the order of meters.

The at-satellite phase can also be influence by the atmosphere (mainly water vapor concentration), which alters the propagation speed of the signal. Zebker et al. (1997) estimate that a 20% change in relative humidity can translate into 10 cm of perceived deformation. A common solution to this is to compute several interferograms of the same target spanning different time periods and then stack the interferograms, driving down the error due to the atmosphere by one over the square root of the number of interferograms (Zebker et al., 1997), although this does reduce the temporal resolution of the deformation measurement. Alternatively, meteorological measurements can also be used to estimate the magnitude of the atmospheric phase contribution (Delacourt et al., 1998), as can GPS measurements, which are also affected by the atmosphere (Webley et al., 2002), or other remote sensing data sets that allow atmospheric water vapor to be estimated independently (e.g., Li et al., 2005). Addressing the impact of our fickle atmosphere on deformation retrievals remains a very active field of research (e.g., Yu et al., 2018; Xiao et al., 2021).

Finally, changes in the scattering properties of the surface (or changes in the apparent backscatter from that surface as viewing geometry changes) contribute to the interferometric phase (Zebker and Villasenor, 1992), often referred to as decorrelative noise. In short, the phase measurement from any given parcel of ground in unpredictable, being the sum of all interactions of the radar wave with all sub-pixel-sized scattering components, which means that the useful phase information that is proportional to target range cannot be extracted. But, if in the time between two SAR image acquisitions the target does not alter its scattering properties (at the scale of the radar wavelength), but merely moves toward or away from the antenna, then the difference in phase between these two observations times does tell us about this change in range, because the aforementioned random component of the phase cancels out. Coherence is a measure of how well correlated the phase measurements are between the two images. If they are perfectly correlated (high coherence), then the previous assumption, that the random component of the measured phase cancels out, holds. If not, then differencing the two phase images (Figure 8.3) does not leave the phase change due to propagation delay, and deformation may not be quantifiable.

There are several sources of decorrelation, apparent on interferograms as areas without fringes (i.e., noisy areas, Figure 8.3). One source is poor co-registration of the image pair. Decorrelation takes place when the nature of the surface scatterers within each pixel changes substantially between SAR images acquisitions. This could be because a new lava flow has been emplaced over a previously lava-free area, ash has fallen on a previously ash-free surface, or snow has fallen (a common problem on volcanoes). Because such changes tend to be cumulative over time, long time periods between SAR image acquisition lead to increased incidence of decorrelation. Decorrelation can also occur because of overly large acquisition baselines. The signal received from the same sub-pixel arrangement of surface scatterers will change as viewing geometry changes (i.e., baseline increases) as the apparent relative positions of the scatterers changes. As such, decorrelation can be ameliorated by using smaller temporal and spatial baselines (although obviously enough time must elapse for deformation to take place).

Given that decorrelation is wavelength dependent, different SAR instruments suffer to different extents. For example, as the size of leaves are of the order of the C-band wavelength (5.7 cm), interferograms derived from C-band SAR (such as ERS-1, ERS-2, ENVISAT, and the Canadian Radarsat series), suffer from low coherence over vegetated surfaces, as the arrangement of the leaves of trees between acquisitions varies with the wind, or season. As vegetation is not uncommon on volcanoes, this has been a problem. Longer wavelength SARs (such as the Japanese ALOS L-band SAR; wavelength of 24 cm) are not affected by scatterers as small as leaves; they are affected by the tree trunks, but these do not change (as much) between SAR acquisitions. Consequently, interferograms derived from such L-band data exhibit greater coherence spatially than those derived from C-band data. NISAR (https://nisar.jpl.nasa.gov/) a joint USA–Indian SAR mission due to be launched in 2024, will carry an L-band RADAR.

The technique has been widely applied to quantify volcano deformation since Massonnet et al. (1995) were the first to use ERS-1 SAR to calculate 11 cm of deflation at Mount Etna, coincident with the 1991–1993 eruption, during which almost 500 million cubic meters of lava were erupted onto the surface. (It should be noted that a single interferogram resolves motion of the ground in the spacecraft's line-of-sight direction and does not return true vertical or horizontal ground motions, as in situ surveying methods do. However, by computing interferograms from SAR images acquired from different look angles [e.g., Fialko et al., 2001] then the true 3D displacement field can be computed.) An obvious question is whether InSAR has detected deformation prior to eruptions, to which the answer is yes. Kizimen volcano, Kamchatka, eruption in 2010 for the first time since the late 1920s. Inflation of 6 cm over a nearly two-year period prior to this eruption was detected using orbital InSAR, which also allowed the source of the deformation (propagation of a near vertical dike) to be constrained from the satellite data (see later in this chapter; Ji et al., 2013). Although most dramatic deformation usually occur at mafic volcanoes (because the erupted volumes are larger) substantial pre-eruptive deformation has also been observed at silicic volcanoes using InSAR. Jay et al. (2014) describe how three cycles of inflation preceded the 2011 explosive eruption of Cordon Caulle, Chile. When deformation is not followed by an eruption, it is still a useful indicator of volcanic unrest. Eighty-four mm of subsidence were detected at Campi Flegrei (near Naples Italy), a so-called super-volcano because of its potential to generate large explosive eruptions, between 1993 and 1996 (Avallone et al., 1999). During 2003, Mauna Loa volcano, Hawai'i, received a great deal of media attention because of a renewed period of inflation (the volcano has not erupted since 1984). Although InSAR detected the inflation (Amelung et al., 2007), no eruption ensued. Another cycle of inflation was measured using InSAR beginning in 2014 (Varugu and Amelung, 2021): in this case Mauna Loa did erupt, in late 2022.

Perhaps less common is the situation when an eruption occurs with no discernable deformation signature, as observed by Moran et al. (2006). Here two eruptions at Shishaldin volcano, Alaska, in 1996 and 1999 were accompanied by no apparent ground motion. These authors speculate that the reason for this was that magma rises so fast at this volcano, and from such depth, that significant deformation of the edifice does not occur on a resolvable timescale. At Lascar volcano, Chile, Pritchard and Simons (2002) advanced similar reasons as to why no deformation was observed at that volcano during a period when several eruptions took place involving an amount of magma that should have produced resolvable deformation: either the magma source was too deep or the InSAR pair spanned a period of time before and after an inflation-deflation cycle over which time deformation was relatively short-lived and perfectly elastic. Sigmundsson et al. (1999) describe how although substantial deformation was measured by InSAR after the onset of the 1998 eruption of Piton de al Fournaise (Reunion Island) there was no deformation prior to the eruption, upon which the eruption could have been predicted. Again, very deep storage of the magma prior to eruption was advanced as the reason.

8.4.2 QUANTIFYING SUBSURFACE MAGMA BODIES

The deformation data that InSAR provides is very valuable in itself, and the technique has also been used to quantify thermal contraction of cooling lava flows (Stevens et al., 2001), lava flow volumes (Lu et al., 2002), and volcanic mass wasting (Ebmeier et al., 2010). However, its role in constraining the nature of the sub-surface magma bodies responsible for the observed deformation is probably of greater importance. For a magma source at given depth, and geometry, deforming at a given rate, the change in surface topography with distance from the source can be predicted. A common source is a spherical point source of deformation, or a "Mogi source" (Mogi, 1958). Using numerical minimization techniques, the depth and volume change of a Mogi source can be stipulated, and the surface deformation field that combination would produce can be computed and compared with that obtained via the InSAR technique. The source parameters (location and volume change) are then iterated until the residuals between the observed and the predicted deformation fields are minimized, providing a best-fit estimate of the location of the magma chamber, and the rate at which it is growing or shrinking as magma is added or removed, very important pieces of information for those wishing to assess the likelihood of future eruptions at volcanoes, by identifying which volcanoes are deforming and which are not. The properties of the magma chamber, as derived from InSAR, also provides crucial constraints for modeling other volcanological processes and behaviors (e.g., Anderson and Segall, 2013). Although the Mogi point source of dilation is commonly used as the basis for this sort of geophysical inversion, other deformation models that reflect deformations resulting from intrusion of magma as dikes and sills, non-spherical magma chambers, or deformation due to faulting can also be used, if appropriate to the volcano in question (e.g., Okada, 1985; Davis, 1986; Amelung et al., 2000). The "best-fit" of a model deformation field to observation to obtain source parameters is non-unique, resulting in uncertainty in the source parameters retrieved (in addition to that introduced by the issues raised earlier, e.g., Gong et al., 2016), and attention has been paid in recent years to determining the statistical robustness of these source parameter inversions (e.g., Bagnardi and Hooper, 2018).

The issue of decorrelation has been mentioned before, as a factor that reduces the ability to resolve deformation across the entire surface of the volcano, for most volcanoes on Earth. The kind of long temporal baselines required to resolve subtle ground motions are precluded in most cases by the issue of temporal decorrelation. Although many SAR image pairs are often acquired of a target volcano, only a subset meet the critical baseline requirements, the remainder being subject to degradation via baseline decorrelation. Ferretti et al. (2001) proposed a technique that allows deformation measurements to be made even under these non-ideal conditions, called the permanent scatterer (PS) technique. The technique identifies individual pixels in co-registered SAR images that display high coherence (correlation) over long periods of time. As such (because the phase scattering characteristics are dominated by one stable point scatterer) long temporal baselines can be employed, as well as large spatial baselines because of the high degree of correlation. These permanent scatter pixels in effect behave as a natural GPS network (Ferretti et al., 2001) allowing volcano deformation to be quantified even when no traditional InSAR fringes can be retrieved, as the stability of their scattering characteristics means that the phase information relating to changes in antenna-target range can still be retrieved for these scattered ground locations. Figure 8.4 shows

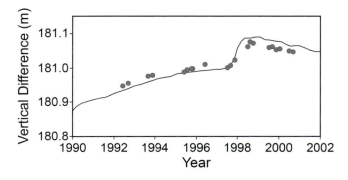

FIGURE 8.4 Comparison of ground deformation retrieved from permanent scatterer InSAR with in situ measurements. Solid line shows height changes derived from electronic distance measurements; circles are the PS InSAR data. Adapted from Hooper et al. (2004).

a comparison of ground deformation derived from the InSAR permanent scatter technique with motions derived from traditional ground surveying.

Although initially applied on a volcano-by-volcano basis, increased availability of data (as more InSAR-capable missions have been launched), combined with the advent of more widely available and user-friendly InSAR processing software (and scientists who can use that software), and an ever-lengthening archive of SAR data, has allowed the application of the technique to be expanded spatially and temporally. Local and regional analyses of volcanic deformation have been conducted (e.g., Pritchard and Simons, 2004; Lu and Dzuirsin, 2014). A recent paper by Biggs et al. (2014) described a global analysis of deformation at 198 volcanoes over an 18-year period finding that, of these, 54 exhibited deformation, while 25 of these went on to erupt. Analysis of 540 volcanoes over a shorter three-year period allowed Biggs et al. to place this deformation-eruption/no eruption relationship in the context of wider petro-tectonic factors.

8.4.3 Quantifying Volcano Topography

Finally, although this section has focused on volcano deformation, it should be noted that volcano topography itself is an important product that the InSAR technique yields (see Rowland, 1996). Knowledge of volcano topography derived from InSAR is vital for modeling the downslope propagation of volcanic flows (lava or pyroclasts). As might be imagined, large spatial baselines (to maximize the stereoscopic effect that yields the topographic phase information) and short temporal baselines (to minimize phase changes due to topography) are preferable in this instance. (In fact, some restrict the use of the term InSAR to be solely associated with derivation of topography, the term DInSAR reserved for deformation studies.) The near-global DEM data provided by NASA's Shuttle Radar Topography Mission has been used to quantify aspects of volcano morphology (e.g., Wright et al., 2006; Grosse et al., 2014). The global DEM derived from the Terra ASTER (Advanced Spaceborne Thermal Emission and Reflection Radiometer) archive is a more complete and up-to-date record of volcano morphological information (Abrams et al., 2020).

8.5 REMOTE SENSING OF VOLCANIC DEGASSING

The importance of gas measurements stems from the insights that they can yield into a wide range of volcanic phenomena. Gas measurements provide constraints on the composition, quantity, and origin of magmatic volatiles (e.g., Symonds et al., 1994), the hazards posed by an active volcano, insofar as fluxes of magmatic volatiles document changes in magma supply from depth (Roberge et al., 2009), and the deleterious effects, on both humans and landscapes, of the gases emitted (Williams-Jones and Rymer, 2000; Tortini et al., 2017).

Volcanic gas emissions comprise a relatively restricted range of elements as molecular species. Although variable from volcano to volcano, H_2O, CO_2, and SO_2 are universally the most abundant volatiles emitted, in something like the proportions listed earlier (see Table 8.3 in Symonds et al., 1994, for much more data). Direct sampling, either using a Giggenbach bottle (flask partially filled with an absorbing solution) or filter packs close to gas vents (the resulting samples subsequently analyzed using traditional wet chemistry techniques, chromatography, or mass spectrometry), has been used for a long time. Most of the previously listed species can be measured using these techniques with high precision, although in situ gas measurements suffer from poor temporal resolution, spatial resolution, and an element of danger to the person doing the sampling. Direct sampling by flying through the gas plume (e.g., Gerlach et al., 1999) somewhat overcomes the safety issue without overcoming the spatio-temporal sampling issues (and adding cost). This has led volcanologists to look toward remote sensing as a source of data on volcanic gas emissions.

The list of species amenable to characterization from low Earth orbit, using recently and currently operational sensors, is more restricted. The issue is the path length to space and the effect this has on isolating signal from noise (i.e., detecting a relatively small amount of volcanic gas in a long atmospheric path that may well contain that same gas naturally), the often coarse spatial resolution of the remote sensing instruments available, and/or the inability of these instruments to resolve the fine spectral resolution wavelength-to-wavelength contrasts that denotes the characteristic transitions of the relevant molecules, particularly given the masking effect that other species (such as H_2O) can have. Nevertheless, progress has been made.

From low earth orbit, only volcanic emissions of SO_2 have consistently, and successfully, been mapped. SO_2 is a magmatic gas that is not present naturally in the atmosphere (although it is produced by burning fossil fuels), making it a relatively easy target. It is emitted by most volcanoes in prodigious amounts. The SO_2 molecule also has absorption features at wavelengths, in the ultraviolet and the long-wave thermal infrared, that are relatively free of obscuration by other gases, and at which orbital remote sensing instruments commonly make measurements.

8.5.1 Quantifying Volcanic Sulfur Dioxide Emissions in the Ultraviolet

Kruger (1983) was the first to demonstrate that volcanic SO_2 could be measured from orbit, using data acquired by the Nimbus 7 Total Ozone Mapping Spectrometer (TOMS), during the 1982 eruption of El Chichon, Mexico. Figure 8.5 shows the extent of the SO_2 cloud on April 8, 1982, four days after the eruption occurred. TOMS-like volcanic SO_2 retrievals are based on the assumption that the reflectivity of the atmosphere in the ultraviolet is the sum of backscattering of UV light by air molecules, aerosols, and clouds, and the ground, modulated by absorption by ozone and sulfur dioxide present in the column (although TOMS was designed for mapping ozone, O_3 and SO_2 have similar absorption spectra in the UV; Figure 8.5).

Dave and Mateer (1967) described the basis for ozone mapping from orbit, something that volcanologists have managed to exploit for SO_2 mapping. The amount of ultraviolet light available to measure at the top of the atmosphere is assumed to depend on the backscatter from the atmosphere (air molecules, aerosols, and clouds) and the ground, modulated by absorption by ozone in the atmospheric column, given the O_3 molecule's absorption cross-section at these wavelengths. Clearly for the emergent signal to contain any information about O_3 concentrations (or SO_2), the effective scattering layer (i.e., the layer of the atmosphere from which most light is reflected back to the spacecraft) must be below the height in the atmosphere at which the O_3 (or SO_2) resides (if it is above, then the light never has a chance to interact with the target molecule).

By using radiative transfer modeling to predict the diffuse reflectance from the top of the atmosphere as a function of UV wavelength (for assumed solar zenith angles, O_3 profiles, and the reflectivity from the physical layer of the model atmosphere above which scattering and absorption takes place) a measurement of the actual reflectance at one wavelength could be used to infer the total amount of O_3 in the column (i.e., interpolating between the reflectance predicted for each model O_3 distribution and that observed for an a priori unknown real O_3 distribution). Two measurements made at different UV wavelengths (one where O_3 attenuation is strong, the other weak) was found to reduce the error. Here the key parameter, N_λ is given by:

$$N_\lambda = -100\log_{10}(I_\lambda/F_\lambda) \qquad (8.1)$$

where I_λ is the intensity of light reflected in the nadir direction (the observation), F_λ is the exoatmospheric solar flux normally incident, and λ corresponds to the wavelength at which the measurement is made. Figure 8.5 denotes the wavelengths at which Nimbus 7 TOMS and its successors acquired data. The model assumes no SO_2 in the atmosphere. But given the overlap in the spectra of SO_2 and O_3, presence of volcanic SO_2 would lead to an overestimate of total O_3; the issue for volcanologists is how to discriminate between attenuation by O_3 (always present) and attenuation by SO_2 (sometimes present; Krueger et al., 1995).

Several different algorithms that have been used to process TOMS-like UV data for volcanic SO_2 retrievals. For example, the original work of Krueger (1983) analyzed regions within the volcanic

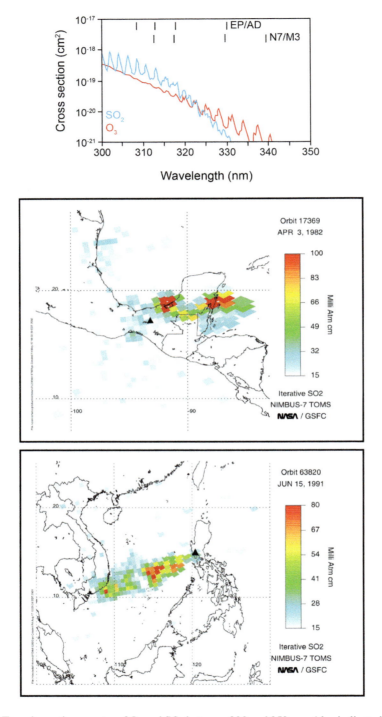

FIGURE 8.5 Top: absorption spectra of O_3 and SO_2 between 300 and 350 nm. Also indicated are the approximate wavelengths sampled by the Nimbus 7 and Meteor-3 (N and M3) and Earth Probe and ADEOS (EP and AD) TOMS instruments. OMI provides contiguous sampling across this region at 0.4 nm resolution. Middle: TOMS-derived estimate of the SO_2 placed into the 1982 eruption of El Chichon, Mexico. Bottom: the same data for the 1991 eruption of Mount Pinatubo, Philippines. The bottom two figures are courtesy of NASA/GSFC (http://so2.gsfc.nasa.gov).

SO$_2$ plume, as well as plume-free (and SO$_2$-free) regions adjacent to it. The apparent "excess ozone," SO$_2$, in the plume region was estimated by determining that typical for the local atmosphere (determined from the plume-free regions), compared to the plume region. Krueger et al. (1995) used a least squares minimization approach based on multiple TOMS wavebands. The absorption by SO$_2$ and O$_3$ is retrieved via:

$$N_\lambda = -[a + b(\lambda-\lambda_0) + s(\alpha_{1\lambda}\omega_1 + \alpha_{2\lambda}\omega_2)] \qquad (8.2)$$

Where a and b are coefficient that describe the relative radiance of a scattering atmosphere, s is optical path, ω_1 and ω_2 are the total column amount of O$_3$ and SO$_2$ from the ground to the top of the atmosphere, α_1 and α_2 are the O$_3$ and SO$_2$ absorption coefficients, and N_λ is given by equation (8.1), modified to include a term to account for changes in solar zenith angle. Here, the four unknowns (α_1, α_2, ω_1, ω_2) were determined by using four values for N_λ obtained at four TOMS wavelengths. Papers by Krotkov et al. (2006) and Yang et al. (2007) describe iterations that have allowed for improved retrievals of volcanic SO$_2$ using UV reflectance measurements.

These new algorithms have been developed as new sensors have been launched, and SO$_2$ detection limits have improved as a result of increased spatial resolution and changes in wavelengths sampled (see Table 8.1 in Carn et al., 2003). Nimbus 7 TOMS and Meteor-3 TOMS (operational 1978–1994) had spatial resolution of 50 and 63 km at nadir, with SO$_2$ detection limits of ~12,000–17,000 tonnes. The launch of Earth Probe TOMS (operational 1996–2005) and ADEOS TOMS (operational 1996–1997) provided data with a nadir resolution of 24–42 km, but an increased sensitivity down to ~1000–4000 tonnes of SO$_2$. Still, only large eruptions that injected large quantities of SO$_2$ into the upper troposphere and lower stratosphere could be quantified. The launch of the Ozone Mapping Instrument (OMI) on board NASA's Aura platform in 2004 led to a substantial improvement in our ability to detect volcanic SO$_2$ in the atmosphere, as a result of its smaller footprint (13 × 24 km at nadir), better spectral resolution and improved detector noise characteristics. This combination of factors has lowered the sensitivity limit to < 100 tonnes (using equivalent metrics) and allowed volcanologists to monitor non-explosive passive degassing of Earth volcanoes, as well as low intensity eruptive degassing that remains confined to the lower troposphere (see Carn et al., 2003).

Although the TOMS series of sensors (and now OMI) have proven to be the workhorses for characterizing volcanic SO$_2$ emissions from space, other sensors have also been used, including the Scanning Imaging Absorption Spectrometer for Atmospheric Chartography (SCHIAMACHY) launched in 2002 on board ENVISAT (see Loyola et al., 2008), and the Global Ozone Monitoring Experiment 2 (GOME-2) launched on MetOp-A in 2006 (see Rix et al., 2012). The Ozone Profiler and Mapping Suite (OMPS, SUOMI NPP) can also be used for quantifying SO$_2$ in the atmosphere (Yang et al., 2013). Once more, alternative algorithms have been developed for the data that these sensors provide. Although all of the algorithms cannot be described in this chapter (see cited papers for details) it is worth noting that regardless of sensor and algorithm, the unit universally used to quantify SO$_2$ in the remote sensing of volcanic SO$_2$ literature is the Dobson Unit (DU), which is a vertical column abundance (1 DU = 1 milli-atm-cm = 2.691016 molecules/cm^2, = 0.0285 grams of SO$_2$ per m^2), or the extinction per centimeter thickness of that gas under standard temperature and pressure. The total mass of SO$_2$ in a volcanic plume, such as that depicted in Figure 8.5, can then be easily estimated based on the instantaneous field of view of the sensor in question and the number of pixels comprising the plume, from which mass time series can be constructed using multi-temporal observations. However, field volcanologists commonly report (or ultimately desire) an estimate of the *flux* of SO$_2$ being emitted from a volcano at a given time and how this changes (i.e., in kg s^{-1}, or t d^{-1}). Converting the satellite-derived SO$_2$ masses (sometimes referred to as "burden") to fluxes is not straight forward, and Theys et al. (2013) give a detailed overview of this issue and attempts to reconcile these quantities. For example, a simple approach is to assume that the flux of SO$_2$ from a volcano (M T⁻¹) is given by the mass of SO_2 determined from a UV satellite observation (M) divided by the residence time of SO_2 in the atmosphere (T).

Given that volcanic degassing is a temporally dynamic phenomenon, the question of temporal resolution at which satellite data are acquired is of importance. The aforementioned SCHIAMACHY sensor provides complete global coverage once every six days; GOME-2 matches the temporal revisit of OMI (~24 hours), but at much coarser spatial resolution (40 × 80 km). The early TOMS sensors did not provide complete global coverage in a 24-hour period because of variations in their mission orbits. However, OMI and OMPS allow true global monitoring of volcanic SO_2 emissions (down to the quoted sensitivity limits) for most of Earth's volcanoes on a daily basis, and have been used to quantify global volcanic SO_2 emissions on decadal time scales (Carn et al., 2017).

8.5.2 Quantifying Volcanic Sulfur Dioxide Emissions in the Thermal Infrared

SO_2 also exhibits absorption features at 7.3 µm and 8.6 µm, which means that thermal infrared remote sensing also can be used to quantify its abundance in volcanic plumes (Realmuto et al., 1994; Watson et al., 2004). (For those who are interested, the paper by Guo et al. (2004) provides a direct comparison of the infrared and UV approaches for measuring volcanic SO_2, using the 1991 eruption of Mount Pinatubo as an example). Figure 8.6 shows the absorption features in question, compared with the spectral response functions of the Terra MODIS (Moderate Resolution Imaging Spectroradiometer) and ASTER (Advanced Spaceborne Thermal Emission and Reflection Radiometer) thermal infrared (TIR) channels.

ASTER data only provide information regarding the less intense 8.6 µm feature, which will be discussed first. Adjacent is an ASTER TIR false color composite image of Miyake-jima volcano, Japan. The absorption of ground radiance by SO_2 in channel 11 reduces the amount of radiance reaching the sensor in this wavelength interval, causing the plume to appear in hues of yellow in this band triplet. Realmuto et al. (1994) describe a technique for using thermal data of this kind to estimate the column abundance (g m⁻²) of SO_2 in volcanic plumes, by modeling how much SO_2 must be present to produce the observed depression in at-satellite radiance in the 8.6 µm channel. The plume appears yellow because radiance from the scene is not substantially attenuated at wavelengths sampled by ASTER bands 14 and 13.

The at-satellite spectral radiance, $L_s(\lambda)$, for a vertical path through the atmosphere, can be written as (Realmuto, 2000):

$$L_s(\lambda) = \{\varepsilon_\lambda L(\lambda, T_0) + [1-\varepsilon_\lambda]L_d(\lambda)\} \times \tau_\lambda + L_u(\lambda) \quad (8.3)$$

where ε_λ is spectral emissivity of the ground, $L(\lambda, T_0)$ is the spectral radiance from the ground surface radiating at temperature T_0, τ_λ is the spectral transmissivity of the atmosphere, and L_d and L_u represent the down-welling and upwelling radiance produced by the atmosphere, respectively. The amount of radiance leaving the ground is a function of the radiance emitted by the ground ($\varepsilon_\lambda L(\lambda,T_0)$) plus the down-welling radiance reflected by the ground ($[1-\varepsilon_\lambda]Ld(_\lambda)$). Upon passage from the ground to the sensor through the atmosphere, some of this radiance is attenuated (τ_λ) and some additional radiance is contributed ($L_u(\lambda)$).

In summary, the algorithm uses a radiative transfer model (e.g., MODTRAN, Berk et al., 1989) to predict how spectral radiance from a ground surface of temperature T_0 and emissivity ε_λ is attenuated as it passes through a model atmosphere from ground to sensor. The concentration of SO_2 at a particular altitude in the model atmosphere is varied until the predicted top-of-the-atmosphere radiance matches the observed at-satellite radiance for a range of TIR wavelengths centered on, and adjacent to, the 8.6 µm absorption feature. SO_2 column abundance is the product of SO_2 concentration and plume thickness. Integrating this for each plume pixel and multiplying by wind speed yields SO_2 flux (e.g., kg s⁻¹; Realmuto et al., 1994). The temperature and emissivity of the ground

Remote Sensing of Volcanoes

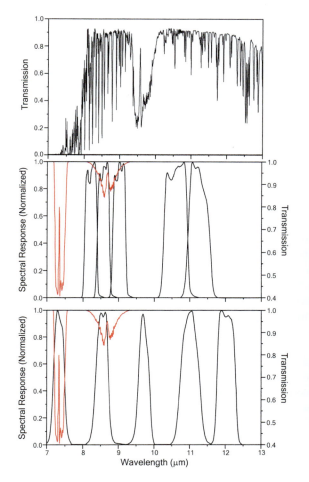
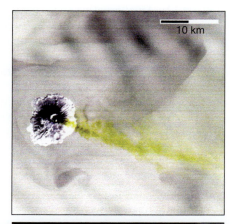

FIGURE 8.6 Left, The transmission of a model atmosphere (top) and SO_2 (red curve). Superimposed on the lower two panels are the spectral response of the Terra ASTER TIR subsystem (bands 10, 11, 12, 13, and 14 from left to right) and a selection of the MODIS emissive channels (bands 28, 29, 30, 31, and 32, left to right) sensors. Right: Terra ASTER false color composite (bands 14, 13, and 12, in red, green, blue) of Miyaka-jima volcano, Japan. The sulfur dioxide plume appears yellow in this band triplet. Below is an ASTER near-infrared and visible wavelength false color composite (bands 3N, 2, and 1, in red, green and blue). In reflected light the plume appears white, as it is predominately composed of water.

beneath the volcanic plume can be obtained from image data. Although the height and thickness of the plume (which determines the layer of the plane–parallel model atmosphere in which SO_2 concentration will be varied) and its speed (which is used to compute the mass flux) could be constrained using stereoscopic or photo-clinometric techniques, values for these parameters are often assumed, or obtained from in situ observations.

This technique has been used successfully to estimate SO_2 fluxes at a range of volcanoes using data acquired from space (e.g., Urai, 2004; Pugnaghi et al., 2006), and has the advantage of being able to image volcanic plumes by day or night (the UV approaches clearly cannot work without sunlight). The high spatial resolution of ASTER allows even low-intensity passive degassing to be detected and quantified. Comparisons between in situ measurements and ASTER-derived SO_2 abundances have shown an encouraging level of agreement (e.g., Henney et al., 2012). An alternative approach to retrieving SO$_2$ path concentrations that does not use the Realmuto et al. (1994) radiative-transfer-based approach is described by Campion et al. (2010). MODIS data can also be used as input to the algorithm (Watson et al., 2004; Novak et al., 2008), although the coarser spatial resolution (1 km, rather than 90 m) means that the volcanic emissions (e.g., plume size and SO$_2$ concentration) must be larger (see Realmuto, 2000 for a sensitivity analysis).

The technique is limited by uncertainties regarding plume geometry and speed, atmospheric water vapor content, and the fact that the transmissivity of the atmosphere at 8.6 µm is also a function of absorption and scattering by other constituents of volcanic clouds, such as ash, sulfate aerosols, and ice particles (Watson et al., 2004). (The presence of ash mixed with SO$_2$ also confounds unambiguous SO$_2$ detection in the UV.) Furthermore, using ASTER and MODIS, the technique has not been operationally implemented to document global volcanic SO$_2$ fluxes in the manner seen with the previously described UV approaches. ASTER's low duty cycle (it is not a mapping mission) means that temporal coverage for most of Earth's volcanoes can be poor. Of course, differences in IFOV, spectral resolution, and precision (i.e., NEDT) of the at-satellite radiance measurement all impact the fidelity of these gas retrievals. Corradini et al. (2021) collated data from geostationary orbit (MSG-SEVIRI) and low Earth orbit (Aqua/Terra-MODIS, Aqua-AIRS, Sentinel-5p-TROPOMI, NPP/NOAA20-VIIRS, MetopA/MetopB-IASI) during the December 2018 eruption of Mount Etna, reducing all satellite measurements to the common metric of kT of SO$_2$ per day. The good agreement between results obtained from these very different data sets shows that synergistic use of data acquired by multiple satellites can be relied upon to give an internally consistent picture of volcanic degassing history.

It has also been demonstrated that thermal infrared data can also be used to quantify volcanic SO$_2$ using the 7.3 µm absorption feature (e.g., Prata et al., 2003; Prata and Bernardo, 2007). Although the atmospheric column as a whole is opaque at this wavelength (due to water absorption), the water vapor itself emits energy toward the sensor which can be absorbed by SO$_2$ in the path (i.e., high altitude SO$_2$ clouds). Sensors such as MODIS, TOVS-HIRS (TIROS Operational Atmospheric Sounder, High Resolution Infrared Radiation Sounder), and the Aqua AIRS instrument (Atmospheric Infrared Sounder) all allow this absorption to be measured. In the case of HIRS (Prata et al., 2003), by making measurements of at-satellite radiance over an SO$_2$ cloud in wavebands adjacent to this absorption feature (6.7 µm and 11.1 µm) the radiance from an SO$_2$-free atmosphere at 7.3 µm (L$_{7.3\text{pred}}$) can be computed as radiance will vary linearly between these three wavelengths. Measured departures from this predicted radiance (L$_{7.3\text{obs}}$) can then be related to the SO$_2$ content of the cloud:

$$L_{7.3\text{obs}} - L_{7.3\text{pred}} = (1 - \tau_{SO_2}) \times (L_{\text{cloud, T}} - L_{7.3\text{pred}}) \tag{8.4}$$

where τ_{SO_2} is the transmission of SO$_2$ and L$_{\text{cloud, T}}$ is the spectral radiance from the cloud at temperature T.

As sulfate aerosols and ash particles do not strongly affect extinction at this wavelength, the 7.3 µm region allows absorption by SO$_2$ to be measured more directly than is possible at 8.6 µm. However, successful detection of volcanic SO$_2$ at this wavelength is limited to plumes with altitudes greater than ~3 km (i.e., above the majority of the H$_2$O, Watson et al., 2004), for MODIS and HIRS. AIRS has the advantage that, as an imaging spectrometer, it has sufficient spectral resolution to isolate the SO$_2$ absorption feature, and measure between the water absorption features, allowing SO$_2$ to be detected at lower altitudes. This must be offset against the fact that the poor spatial resolution of these sensors (HIRS resolution is ~19 km at nadir, increasing to ~32 km × 63 km at the edge of the scan; AIRS resolution is ~15 km^2 at nadir, increasing to 18 km × 40 km at the edge of the scan) equates to lower sensitivity. The IASI (Infrared Atmospheric Sounding Interferometer) sensor, flown on board MetOp A and B, has similarly poor spatial resolution (~12 km at nadir) but also allows volcanic sulfur dioxide to be observed (e.g., Taylor et al., 2018).

8.5.3 OTHER GASES

Although SO_2 is the gas that volcanologists have most easily been able to measure from space, it is not necessarily the most desirable. Carbon dioxide is typically emitted in greater quantities than SO_2, exsolves at greater depth, and is relatively inert in shallow hydrothermal systems and the atmosphere (Burton et al., 2000), reducing its susceptibility to the scrubbing suffered by acid gases such as SO_2 (Edmonds, 2008). This means that the flux of CO_2 from a volcano at the surface is more a direct function of that exsolved at depth than SO_2, and given that this depth is greater an increase in CO_2 output can herald an increase in volcanic unrest before the subsequent shallow degassing of SO_2 (i.e., CO_2 may provide early warning of an eruption).

It is, however, not easy to detect volcanic CO_2 from space. Unlike SO_2 (which exists in non-volcanic regions at the ppbv to pptv level; Seinfeld and Pandis, 1997) the atmosphere contains almost 400 ppmv of CO_2, against which volcanic CO_2 fluxes must be identified, and then quantified. And the volcanic signal can be low. Gerlach et al. (2002) found volcanic CO_2 concentrations (volcanic = total—background) of up to 700 ppmv at the summit of Kilauea volcano, Hawai'i, (which is a very large CO_2 source) decreasing to approximately 300 ppmv within a horizontal distance of ~300 m, and to < 100 ppmv within a lateral distance of ~1500 m.

The ability to detect volcanic CO_2 from space has long been a goal for volcanologists, given its relative inertness in shallow hydrothermal systems and the fact that it degasses at much greater depth that SO_2, thus potentially providing an earlier warning of rising magma. Schwandner et al. (2017) reported successfully detecting CO_2 emissions from Yasur volcano using NASA's Orbiting Carbon Observatory-2 (OCO-2), as did Johnson et al. (2020), this time at Kīlauea volcano. Although these measurements were made at low spatial resolution, efforts are ongoing to reduce sensitivity limits.

8.6 REMOTE SENSING OF GEOTHERMAL AND HYDROTHERMAL ACTIVITY

The arrival of magma at shallow depths in the crust results in a range of interactions that manifest themselves as geothermal and hydrothermal phenomena that can be quantified from orbit, as these systems manifest themselves at the surface in the form of fumaroles, mud pots, geysers, and crater lakes, which exhibit either anomalous temperatures, surface compositions, or both. These phenomena yield insights into the nature of the magma body and can also act as a precursor warning on impending eruptions. The famous 1902 eruption of Mount Pelée was preceded by three years of increasing fumarole temperatures (Chrétien and Brousse, 1989), with similar examples cited by Francis (1979).

Hydrothermal and geothermal activity tends to manifest itself as relatively low temperature thermal anomalies, where low temperature means elevations in temperature up the order of a few hundred degrees kelvin above ambient. Planck's blackbody radiation law relates the spectral radiance emitted by a surface to its kinetic temperature, and so a space-based measurement of the spectral radiance emitted by such a volcanic target, at an appropriate wavelength, can be inverted to retrieve its temperature. Of course, it is not just the temperature of the target that is of relevance but also its spatial abundance at the sub-pixel scale. Although the vapor vented through fumaroles can approach magmatic temperatures, the fumaroles themselves (fumaroles are cracks in rocks heated by passage of magmatic gases and steam from heated meteoric water) are very small compared to the size of the instantaneous field of view of remote sensing instruments. For example, at Vulcano, Aeolian Islands, Harris and Stevenson (1997) estimated that the fumaroles themselves occupied only 2.5% of the 16,000 m^2 area covered by the fumarole field. For a Landsat TM (Thematic Mapper) pixel this amounts to only 22 m^2 out of 900 m^2. Thus, there is a strong dilution effect whereby such targets, whilst hot enough to "glow" at visible wavelengths at the resolution of the human eye, are so small that they do not emit sufficient spectral radiance, at the instantaneous field of view (IFOV) scale, to be apparent at visible and near-infrared (VNIR), short wave infrared (SWIR) or middle-infrared (MIR) wavelengths when observed from space. Because of this, and the fact that many of these volcanic features are water-related, remote sensing of hydrothermal and geothermal phenomena has focused largely on the use of long-wave infrared (8–14 µm) data.

Figure 8.7 illustrates some of the issues. On the left are a series of Planck curves indicative of the amount of spectral radiance that a fumarole field might be expected to emit. The blue curve shows emittance for a surface at 300 K that fills the sensor IFOV. The red curve shows the emittance from a surface at 400 K that also fills the IFOV, some kind of geothermal feature (perhaps a large fumarole field, or a very hot acidic crater lake). Clearly they are separable in the thermal infrared and the middle infrared, but not sufficiently hot to radiate at all in the shortwave infrared. However, it is usual that the emitted spectral radiance is a mixture of that from ground radiating at ambient temperature (e.g., 300 K) and the geothermal feature of interest, as the volcanic target is often sub-pixel in size. The green curve shows the emittance from a pixel that contains fumaroles at 400 K (occupying only 1% of the IFOV, e.g., Harris and Stevenson, 1997) surrounded by ambient temperature ground at 300 K (i.e., occupying the remainder of the pixel). Clearly an image pixel containing this volcanic anomaly would barely be distinguishable from an adjacent pixel containing no such anomaly, making is very hard to detect and harder to quantify the excess volcanogenic radiance. When the size of the volcanic feature increases to cover 10% of the image pixel (orange curve) the anomalous pixel becomes somewhat distinct from the background (blue curve) more so at about 4 µm (the mid-infrared) than the long-wave infrared.

Although fumaroles can reach higher temperatures (e.g., Harris and Maciejewski, 2000 measured some in the range of 800 K) these very hot fumaroles occupy an even smaller fraction of the total radiating area. The black curve illustrates the emitted radiance from a pixel containing fumaroles at 800 K occupying only 0.1% of the IFOV. A pixel containing such a volcanic target would be indistinct from its neighbors in the TIR but would radiate some energy in the MIR and SWIR. However, the preceding discussion has assumed that the observations are made at night. During the day, reflected sunlight will mask thermal anomalies in the SWIR and MIR. The gray curve shows the reflected sunlight from a 15% reflector. In summary, low temperature hydrothermal and geothermal phenomena tend to radiate predominantly in the TIR. If they are sufficiently hot to radiate at shorter wavelengths, they are generally of such small sub-pixel size that the amount of energy radiated is in fact small at these wavelengths, and easily masked by reflected sunlight during the day. It should also be noted that solar heating of surfaces can also mask volcanogenic sources of heat during the day and into the early part of the evening. These issues are obviously exacerbated as the spatial resolution of the data worsens (e.g., as one moves from Landsat TM-class resolution to MODIS or VIIRS-class resolution).

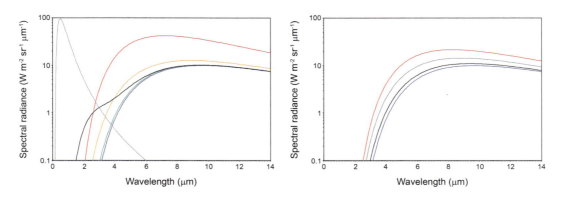

FIGURE 8.7 Simulated spectral radiance emitted from surfaces analogous to low temperature hydrothermal and geothermal volcanic phenomena.

Despite the challenges, relatively high spatial resolution sensors (including the Landsat TM and its successor, the Enhanced Thematic Mapper Plus, ETM+) and Terra ASTER have been used to successfully quantify such low-temperature volcanism. Gaonac'h et al. (1994) used Landsat TM to estimate heat fluxes at Vulcano volcano, measuring excess radiant flux of up to ~50 W m^2 in the crater region, using nighttime data. Harris and Stevenson (1997) arrived at similar values using daytime data but noted the existence of solar heating anomalies of similar magnitude to the anomaly associated with the fumarole field itself. In both cases no emittance was observed in the SWIR (Figure 8.7). Furthermore, Harris and Stevenson (1997) found that at 1 km spatial resolution of the AVHRR (Advanced Very High Resolution Radiometer) sensor, the temperature of the fumarole field was only 1 K higher than adjacent area, as a result of the dilution effect previously discussed. To be somewhat contrary, Kaneko and Wooster (1999) report thermal emission at SWIR wavelength from fumaroles at Unzen volcano, Japan, (albeit based on analysis of nighttime Landsat TM images), finding positive correlations between this thermal emission and SO_2 flux, and magma discharge rate. Patrick et al. (2004) used TIR data acquired by the Landsat 7 ETM+ to document the areal expansion rate and surface heat flux from flows erupted from mud volcanoes with temperatures in the range of 10–40°C. Yellowstone National Park plays host to the full spectrum of hydrothermal and geothermal phenomena, (including hot springs, geysers, fumaroles, and mud pots) and Vaughan et al. (2012) describe how ASTER TIR (and to a lesser extent MODIS TIR data) can be used to quantify geothermal heat fluxes at the scale of the park from all such sources, as well as monitor how these fluxes change over time. Of course, detecting small signals requires that noise be suppressed. By leveraging multi-year time series of low spatial resolution MODIS data Girona et al. (2021) were able to identify subtle, edifice-wide thermal anomalies before eruptions at several volcanoes.

Although substantial attention has been paid to measuring the thermal characteristics of low temperature volcanism, this type of volcanic activity also gives rise to signatures that can be detected via either reflectance or emittance spectroscopy. The interaction of hydrothermal fluids alters surface rock compositions; minerals are also precipitated on the surface as warm, element-charged hydrothermal fluids equilibrate to surface temperatures and (and to a lesser extent pressures). Broadly speaking, carbonates, sulfates, clays, and silica-rich compositions result. For commercial reasons, much effort has been focused on detecting these targets from orbit, and a large body of literature exists describing these exploration-centric activities (see Sabins, 1999 for a review, and Pour et al., 2013, for a recent example). Rather less has been reported on remote spectroscopic analysis for the purposes of studying active volcanic processes. Hellman and Ramsey (2004) investigated the use of ASTER data for studying active and fossil hot-spring deposits at Yellowstone. As hydrothermal alteration weakens rocks, mapping alteration on volcanoes can identify zones that are prone to catastrophic collapse and landslides. Crowley et al. (2003) describe the use of hyperspectral reflectance measurements acquired by NASA's Earth Observing-1 Hyperion sensor to do this. In a very interesting recent paper Bogue et al. (2023) showed how Landsat-derived maps of NDVI (Normalized Difference Vegetation Index) could be used to enhanced photosynthesis in surface vegetation as a result of increased exhalation of volcanic CO_2 from the soil, perhaps several years before obvious hydrothermal activity became established.

In addition to the aforementioned manifestations of hydrothermal and geothermal volcanism are volcanic crater lakes. There are more than 100 volcanic crater lakes on Earth (Delmelle and Bernard, 2000; Rouwet et al., 2015). The volume, temperature, and bulk composition (which controls water color) of the water in the lake are controlled by the influx and efflux of water, enthalpy, and chemical elements. The arrival of a fresh batch of magma beneath a crater lake will perturb the lake system, and the water (which acts as a calorimeter and chemical condenser for volcanic heat and chemical elements that rise from the underlying magmatic system) will respond.

The literature contains many examples of instances when changes in either crater lake temperature or color have preceded eruptions. For example, lake temperatures increased by 10°C in the three months preceding the 1965 eruption of Taal volcano, Philippines, an eruption that resulted in over

200 deaths (Moore et al., 1966). A similar 10°C rise (this time over a period of four months) preceded the 1990 eruption of Kelut volcano, Indonesia (Badrudin, 1994). Such temperature increases are clearly within the range that can be observed from space.

Changes in lake color have also been observed to precede eruptions. Coloration results from organic and inorganic materials dissolved and/or suspended in the water. At Ruapehu, New Zealand, a shift in color from blue-green to gray has been shown to indicate an increase in hydrothermal flux, the gray color due to the mobilization of bottom sediments by enhanced sub-aqueous fumarolic flow (Christenson, 1994). Dramatic color changes from gray-green to yellow-green at Poás, Costa Rica, have been attributed to increased fluxes of SO_2, which oxidizes dissolved iron, changing the spectral absorptance of the water (Delmelle and Bernard, 2000). Substantial volume changes (i.e., disappearance of a lake, something that would be relatively easy to detect from orbit) have been observed to precede eruptions at Poás in Costa Rica (Brown et al., 1989), as a result of increased magmatic energy input from depth.

The temperature and color of crater lakes has been measured from space (Figure 8.8). With regard to temperature, Oppenheimer (1996) presents a detailed heat budget analysis of several crater lakes based on temperatures derived from Landsat TM TIR data, noting the need to correct for the skin effect in order to force better agreement between the remote sensing data and in situ measurements, which typically report bulk water temperatures. Trunk and Bernard (2008) present an extensive analysis of ASTER-derived crater lake temperatures at four volcanoes and show how ASTER allows lake temperatures to be estimated to within 1.5°C of in situ measurements. Obviously, higher resolution data are preferable, as the mixing effect previously discussed makes the same lake temperature anomaly harder to detect in lower resolution data. In Figure 8.7b, the blue curve shows emitted spectral radiance from a 300 K blackbody, while the red curve shows the spectral radiance from a 350 K blackbody (i.e., the maximum crater lake temperature reported by Trunk and Bernard, 2008). The black curve shows the spectral radiance from a pixel containing a lake at 350 K covering 10% of a 1 km pixel (e.g., MODIS or AVHRR), assuming that 90% of the pixel radiates at the ambient background temperature of 300 K, approximately equal to the size of the lake at Ruapehu (dia. 400 m). The gray curve shows the same, but for a lake that covers 40% of the 1 km pixel (e.g., indicative of the dia. 700 km lake at Kawah Ijen, Indonesia). Regarding color, Figure 8.8 shows that the color of volcanic crater lakes can be retrieved provided true color data are acquired. Oppenheimer (1997) provides some examples of the use of Landsat TM to determine lake water color. More recently, Murphy et al. (2018) used Google Earth Engine

 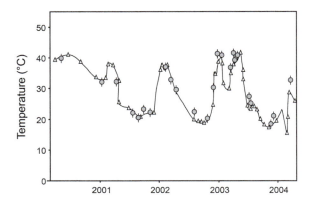

FIGURE 8.8 Left. Landsat ETM+ true color composites of the crater lakes at Maly Semyachik, Russia (left) and Irazu (Costa Rica, right) with corresponding photographs beneath. Right. ASTER-derived crater lake temperatures at Mount Ruapehu, New Zealand. Open triangles are in situ data, filled circles are ASTER measurements. Adapted from Trunk and Bernard (2008).

to quantify changes in the color and temperature of three Indonesian crater lakes over a 20-year period.

8.7 REMOTE SENSING OF ACTIVE LAVAS—EFFUSIVE ERUPTIONS

Once magma breaches the surface an eruption begins. Effusive eruptions involve the eruption of lava, which flows under the influence of gravity from the vent. At basaltic volcanoes (such as Mount Etna, Sicily) this lava may be as hot as 1150°C, and is sufficiently fluid to flow downhill, forming lava flows that are longer than they are wide, and wider than they are thick. At felsic volcanoes (such as Mount St. Helens, Washington, USA), the lava is erupted at temperatures perhaps a few hundreds of degrees lower. This combined with its different chemistry means that felsic lava tends to be rheologically stiffer and piles up over the vent forming lava domes, which have higher thickness to area ratios.

In either case the lava cools exponentially when exposed to the atmosphere. This causes rheological changes that influence the ability of the lava to expand from the vent. Ultimately the areal expansion of a lava flow is halted because either these rheological changes cause the lava to become sufficiently stiff that it will not flow any further, or the supply of lava from the vent shuts off. In the case of lava domes, growth can cease because of either of these two factors, or because the dome is disrupted in an explosive eruption, with a new dome taking its place (felsic lavas tend to have higher volatile contents, and higher viscosity, than mafic lavas, promoting generation of gas overpressures that may be released explosively).

The role of remote sensing in analyzing effusive eruption can be broadly divided into two fields. First, much attention has been diverted to developing autonomous systems that use orbiting satellites to detect the onset and cessation of effusive eruptions, based on their heat signatures observed from space. These studies have largely focused on the use of low spatial resolution data (such as AVHRR, MODIS, and even the geostationary missions, such as the Geostationary Operational Environmental Satellite series, GOES) because of the high temporal resolution at which such data are acquired, essential for timely detection and documentation of temporally dynamic eruptions that occur all over the globe. Secondly, there is a substantial body of literature targeted at analyzing the detailed thermo-physical characteristics of lava flows and domes. These studies have tended to use higher spatial and spectral resolution data sets, such as Landsat TM, Terra ASTER, and to a lesser extent EO-1 Hyperion. It will become apparent that there is overlap between these two general fields. The book by Harris (2013) should be consulted for further details.

8.7.1 Detecting the Thermal Signature of Erupting Volcanoes

It is obvious that earth orbiting satellites should be used to detect the heat signatures of erupting volcanoes, and many papers have been published describing approaches for doing this. The algorithms themselves have almost everything in common with those developed by remote sensors interested in detecting and mapping global wildfires (e.g., Prins and Menzel, 1992). Given that the temporal revisit of Landsat-class instruments (~16 days, daytime acquisitions only) would be of no use for detecting eruptions that may last a matter of days, particularly as sensors such as TM, ETM+, and ASTER do not have 100% duty cycles. Low spatial resolution but high temporal resolution sensors such as AVHRR, MODIS, and GOES have been used extensively. These provide global coverage at a temporal frequency of between 15 minutes to 12 hours, with day and night imaging.

Although the spatial resolution of these sensors is relatively coarse (about 1–4 km^2), this is not a barrier to detection from orbit given that high temperature of the targets, provided data are acquired at suitable wavelengths (Figure 8.9).

To detect an active lava body requires that the image pixel(s) that contain it are distinguishable from adjacent image pixels that do not contain active lava. In Figure 8.9, the blue curve shows the spectral radiance emitted from a blackbody at 300 K, representative of this ambient background.

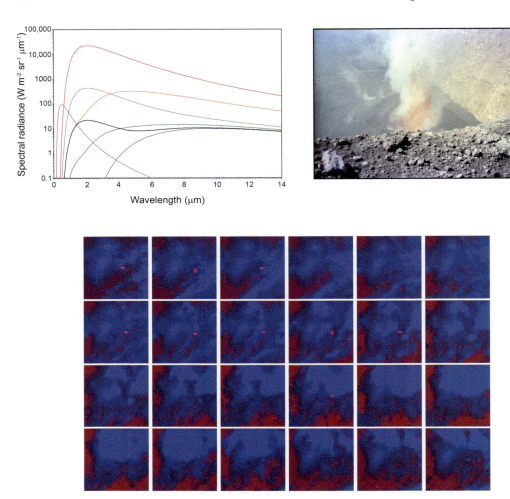

FIGURE 8.9 Left: Spectral radiance emitted from hypothetical surfaces analogous to high-temperature lavas. See text for details. Right: photograph of an active vent at Mount Etna, Sicily, September 1997. Vent is 80 m in diameter. Bottom: six hours of GOES data acquired at Kilauea volcano, Hawai'i. Images have been processed to highlight high temperature material as magenta, making the lava flows from the volcano visible in the center of each image, cloud permitting.

The red curve shows the spectral radiance emitted by a 1400 K blackbody, equivalent to the eruption temperature of an active lava flow. Clearly, they are distinguishable at all wavelengths. Note that given the high temperature of active lava, the wavelength of peak emission is in the SWIR, in contrast to the low temperature hydrothermal phenomena described in the previous section for which only TIR data were useful. Note also the change in scale in ordinate between Figures 8.7 and 8.9; active lavas emit prodigious amounts of radiance at all wavelengths.

However, lava at eruption temperature never fills a 1–4 km² IFOV. Rather, the surface of an active lava is covered by a cool crust, within which cracks expose much hotter material from the flow interior (Figure 8.9), the total spectral radiance being an areally weighted mixture of that from these components. Some active lava flows are sufficiently large that they fill several 1 km IFOVs. The orange illustrates the spectral radiance from such a pixel that contains an active lava flow, where the flow fills the pixel, and has a crust temperature of 600 K, within which cracks (occupying only 0.1% of the flow surface) expose lava at 1400 K. Such a curve is indicative of either a large lava body imaged at coarse spatial resolution data, or a smaller lava body imaged at higher spatial resolution. However, in the present context (detection of volcanic thermal signatures using low spatial resolution data) the target of interest is often (perhaps usually) smaller than the pixel. The green curve shows the spectral radiance from an IFOV where the lava body still has a crust temperature of 600 K (and hot cracks still occupying 0.1% of the lava surface), but in this case the lava body only fills 3% of the IFOV (e.g., a 100 m diameter lava dome). The gray curve shows another example, this time for a vent such as the one depicted in Figure 8.9 which, at 80 m diameter, would occupy 2% of a 1 km^2 IFOV. In this latter case there is no cool crust—the 2% of the IFOV is occupied by material at 1400 K, with the remainder at ambient temperature (300 K). Finally, the black curve depicts the situation if that vent is made even smaller, occupying only 0.1% of the IFOV (i.e., a 30 m diameter vent radiating at a temperature of 1400 K). The magenta curve shows the reflected sunlight from a 15% reflector.

This graph conveys most of that which is required to understand the detection of active lavas from space using low spatial resolution satellite data (or indeed any data). Firstly, and contrary to the situation for the low temperature geo/hydrothermal phenomena described previously, pixels containing active lava are most easy to detect if one looks in the MIR (~4 µm) and the SWIR (1.2–2.2 µm), as order of magnitude separation in the amount of emitted spectral radiance between target and background is apparent at these wavelengths. Secondly, coarse spatial resolution is no barrier to detection, as even vastly sub-pixel sized active lava bodies emit enough radiance at SWIR and MIR wavelengths to influence the gross at-satellite signal, making pixels that contain these targets easily discriminable from the background. Thirdly, the wavelength of peak emission for some of the hypothetical lava bodies depicted in this chart lies in the SWIR. At night, the SWIR signal is purely volcanogenic, with adjacent ground being too cool to radiate. Wooster and Rothery (1997) exploited this fact and used the 1.6 µm channel of the ERS-1 Along-Track Scanning Radiometer (ATSR-1) to document the SWIR spectral radiance emitted by the lava dome at Lascar volcano, Chile, a feature vastly smaller than the 1 km^2 size of the ATSR IFOV. However, as shown in Figure 8.9, during the day the SWIR signal from small active lava bodies is masked by reflected sunlight, rendering them undetectable. As a result, algorithms for the automated detection of volcanic hot spots have largely relied on the 4 µm atmospheric window. Inspection of Figure 8.9 reveals that all of the hypothetical lavas depicted are most easily distinguishable from the background (including solar heating and any reflected sunlight during the day) at this wavelength.

Several algorithms have been developed that exploit these principles, using low spatial but high temporal resolution data sources to monitor volcanoes in near real-time at both the regional (e.g., Harris et al., 2000; Dehn et al., 2000) and global scales (e.g., Wright et al., 2002). Early work focused on the use of AVHRR (e.g., Harris et al., 1997a) and ATSR (e.g., Wooster and Rothery, 1997). Harris et al. (1997b) showed how even very low resolution (4 km) data acquired by the geostationary GOES satellite could be used to detect and track variations in thermal emission (but not map) from active lava flows at Kilauea volcano, in this case leveraging the geostationary vantage point to achieve 15-minute temporal frequency (Figure 8.9).

The details of the algorithms are many and varied, and for the sake of brevity all cannot be recounted here (for details of several techniques spanning the spectrum of approaches that have been adopted, see Higgins and Harris, 1997; Dehn et al., 2000; Wright et al., 2002; Pergola, 2004; Steffke and Harris, 2011; Coppola et al., 2016; Gouhier et al., 2016). All interrogate large volumes of image data and seek to distinguish pixels that contain volcanic thermal anomalies from adjacent pixels that do not, by resolving the kinds of thermal emittance signatures depicted in Figure 8.9. Figure 8.10 shows, schematically, how a typical hot spot detection algorithm might work (in this example, the kind of spatio-spectral-contextual algorithm employed by Higgins and Harris, 1997).

All algorithms involve comparing the radiant properties of a pixel against some kind of detection threshold (either absolute, based on the behavior of that pixel's neighboring pixels, or based on time-series analysis of the thermal history of a particular volcano). As such, near real-time volcano

1. A kernal is passed over the image. It is assumed that each pixel in the image could potentially contain a volcanic "hot-spot". A pixel is reclassified as an actual volcanic hot-spot if its spectral radiance/temperature characteristics are found to differ significantly from its neighbors (background pixels).

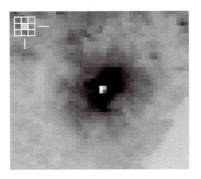

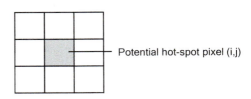

Potential hot-spot pixel (i,j)

2. The brightness temperature for each pixel in the image is computed at 4 and 12 μm(T_4 and T_{12}), in this example, ATSR-1 bands 2 and 4, respectively). Pixels that do not contain active lava will exhibit similar brightness temperatures. Pixels that are entirely filled with active lava will also have $T_4 \approx T_{12}$, but in this case the temperatures will be much higher than adjacent pixels in the kernal. Such a situation may arise when imaging an areally extensive lava flow. Pixels that contain sub-pixel-sized volcanic features are characterized by $T_4 \gg T_{12}$ (below). In this example, the elevated T_4 of the pixels in the center of the image is the result of small amounts of active lava present in Mount Etna's summit craters, vastly sub-pixel in scale, but sufficiently radiant at 3.7 μm to cause T_4 for these pixels to be significantly higher than adjacent pixels that do not contain lava. ΔT ($T_4 - T_{12}$) describes this difference and is effective at distinguishing pixels that contain small volcanic features (such as lava lakes, lava domes, and volcanic vents) from adjacent pixels that do not (background pixels).

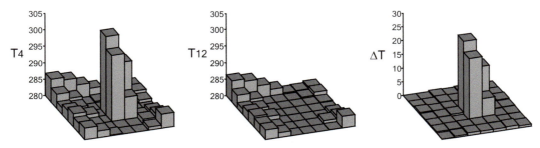

3. Comparing both T4 and ΔT for each pixel within the image with its immediate neighbors allows pixels containing active lava to be distinguished from those that do not. Hot-spot pixels are characterized by T_4 and ΔT which are more than n standard deviations (s) above the mean ($T_{4,bm}$; ΔT_{bm}) of the background pixels.

T4,b	T4,b	T4,b
T4,b	T4,(i,j)	T4,b
T4,b	T4,b	T4,b

ΔTb	ΔTb	ΔTb
ΔTb	ΔT(i,j)	ΔTb
ΔTb	ΔTb	ΔTb

Potential hot-spot pixel reclassified as an actual hot-spot pixel when:

$T_{4,(i,j)} > T_{4,bm} + n\sigma T_{4,bm}$

and

$\Delta T_{(i,j)} > \Delta T_{bm} + n\sigma \Delta T_{bm}$

FIGURE 8.10 Overview of a volcanic hot spot detection algorithm.

monitoring algorithms must balance the desire to set thresholds low enough to detect the smallest and coolest hot spots possible against the need to minimize false positives, which erode confidence in algorithm performance. Substantial progress has been made in this arena, to the point that truly autonomous, global monitoring for volcanic thermal anomalies has been achieved, using data from the Terra and Aqua MODIS sensors (http://modis.higp.hawaii.edu; see Wright et al., 2002). At this website, the details of all volcanic thermal hot spots detected MODIS around the globe are made available within about 1.5 hours of satellite overpass.

8.7.2 Quantifying the Thermo-Physical Characteristics of Active Lava Bodies

In addition to simple detection, these thermal data can also be used to analyze the nature of the volcanic activity responsible. Although initially this focused on the analysis of high-resolution Landsat-class data (e.g., Rothery et al., 1988), it quickly became apparent that volcanologically useful data could also be extracted from low spatial resolution data (e.g., Harris et al., 1997a). Determination of lava temperature was the early focus of this work, with the paper by Rothery et al. (1988) one of the most influential, in which they describe how SWIR data acquired by the Landsat TM can be used to constraint he sub-pixel thermal structure of lava flow surfaces (i.e., the temperature of the crust, the temperature of the hot cracks, and the area of the flow surface that the hot cracks cover) using a modified sub-pixel temperature un-mixing approach based on the Dozier (1981) method. This work was subsequently expanded by others (e.g., Oppenheimer, 1991). Although in principle this method works, it was in fact limited by the fact that (1) the Landsat TM only had two useable SWIR wavebands, forcing the assumption of an unrealistically simple thermal mixture model, and (2) the SWIR channels of TM readily saturated over highly radiant lava flow surfaces. Wright et al. (2010) show how the use of hyperspectral data acquired across the full 0.4–2.5 μm range alleviated these restrictions, allowing the full sub-pixel temperature distribution of active lavas to be retrieved from space, to the point that lava eruption temperature (a proxy for lava chemistry) can also be identified (Wright et al., 2011).

A particularly significant development was the realization that thermal measurements of active lava flows could be used to derive the volumetric flux of lava at the vent. This flux, the lava effusion rate, is of particular importance in volcanology as it is, after lava composition, the single most important variable that determines the final length that a lava flow can attain (Walker, 1973). Harris et al. (1997a) showed how AVHRR data acquired during the 1991–1993 eruption of Mount Etna could be used to estimate the lava effusion rate 27 times during the eruption. Not only were the individual AVHRR-derived estimates comparable with in situ estimates, but upon integration of the effusion rates (m^3 s^{-1}) over time (s) the final volume of the lava flow estimated from the AVHRR-derived fluxes was within the bounds identified from GPS surveying. In short the method exploits a simple physics-based proportionality between lava effusion rate and active flow area, which had been proposed earlier by Pieri and Baloga (1986), in which a higher effusion rate produces a larger lava flow. By making measurements of the area of active lava at the instant of satellite overpass, the remote sensing method allows the antecedent effusion rate that produced that amount of lava to be estimated, via an empirically determined constant (see Wright et al., 2001; Harris and Baloga, 2009 for details).

The importance of this development has been twofold. Firstly, as mentioned, effusion rate is of paramount importance in determining how far a lava flow will advance. Wright et al. (2008) showed how satellite-derived effusion rates could be used to autonomously drive, and update, physics-based lava flow predictions. Secondly, the total amount of lava that cools to the atmosphere during an effusive eruption places constraints on how the volume of magma available at shallow depth is partitioned between the component that is erupted and the component that is intruded within the edifice. This has been used to constrain magma budgets at several volcanoes, including those hosting permanently active lava lakes, using both high and low spatial resolution data sets (e.g., Francis et al., 1993; Harris et al., 2000; Steffke et al., 2011). Bonny and Wright (2017) showed how simply by compiling such satellite-derived effusion rate time series during eruptions could allow the end of the eruption to be predicted soon after onset, simply by extrapolating the effusion rate to zero, at which point, by definition, the eruption has ended.

It has also been found that simply plotting the amount of energy radiated by erupting volcanoes contains valuable information about volcanic processes. Trends in satellite-derived thermal emission correlate with other geophysical metrics (Figure 8.11).

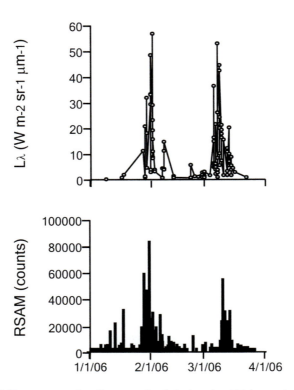

FIGURE 8.11 Top: 3.959 μm spectral radiance emitted during the 2006 eruption of Augustine volcano, Alaska, as measured by Terra and Aqua MODIS band 21. Bottom: real-time seismic amplitude measured in situ. RSAM data from Power et al. (2006).

Although many studies have sought to correlate satellite measured thermal emission measurements with volcanic processes, two studies separated by two decades perhaps illustrate the best evidence that such satellite records contain signals that can be construed precursory to eruptions taking place. Oppenheimer et al. (1993) found that large decreases in SWIR spectral radiance as measured by Landsat TM at Lascar volcano, Chile, were followed by large explosive eruptions. Here, the hot fumaroles that gave rise to the excess SWIR radiance sealed, causing gas pressures within the dome to build up to the point of explosion. More recently, Van Manen et al. (2013) report how systematic increases in emitted MIR spectral radiance at Bezymianny volcano, Russia, as measured by AVHRR, were statistically likely to be followed by explosions.

8.8 REMOTE SENSING OF VOLCANIC ASH CLOUDS—EXPLOSIVE ERUPTIONS

During an explosive volcanic eruption, the rapid rate at which volcanic ash plumes rise through the atmosphere and disperse within it makes near real-time detection of their location, and how this changes over time, essential for aviation. There have been more than 100 inadvertent encounters between aircraft and volcanic ash clouds (Webley and Mastin, 2009). Flying through an eruption cloud causes abrasion and damage to an aircraft's exterior surfaces; ingestion of ash can cause engines to stall. There have been two well-documented cases where passenger jets lost power to all four engines after inadvertently flying through ash clouds. Nine Volcanic Ash Advisory Centers (VAACs) around the world issue warnings about the location of volcanic ash clouds to the aviation community. Satellite remote sensing provides a significant amount of the information upon which these warnings are based.

Reliable detection of volcanic ash clouds depends upon the ability to effectively distinguish them from regular meteorological clouds. Prata (1989) suggested that thermal infrared satellite data could be used to distinguish volcanic ash clouds from meteorological clouds based on their transmissive properties in this wavelength region. Following Prata (1989), the at-satellite spectral radiance ($L_s(\lambda)$) when viewing the ground through a partially transparent cloud (silicate ash or H_2O) can be written as:

$$L_s(\lambda) = e^{-\tau_\lambda} L(\lambda, T_s) + (1 - e^{-\tau_\lambda}) L(\lambda, T_c) \tag{8.5}$$

where τ_λ is the cloud optical depth along the line of sight, T_c is the temperature of the cloud top, and T_s is the temperature of the ground beneath the cloud. The method assumes a plume that is not opaque to the transmission of light from the ground beneath. Spectral radiance from the ground at temperature T_s passes up and encounters the cloud or plume. As the transmissivity of H_2O is higher at ~11 μm (λ_i) than at ~12 μm (λ_j) the difference in at-satellite radiance at these two wavelengths (when expressed as a difference in brightness temperature; BTD = $T_i - T_j$)) will be positive for water clouds. Conversely, volcanic ash clouds are less transmissive at 11 μm than at 12 μm, and the equivalent brightness temperature difference will be negative. This so-called split window technique has been widely used as the basis for detecting and tracking volcanic ash clouds for more than 20 years, using images provided by any sensor that acquires data at these wavelengths (e.g., AVHRR, GOES, MODIS). Data from geostationary spacecraft allow ash clouds to be detected and tracked in near real-time (although this capability is reduced at high latitudes where the satellite zenith angle becomes large). Figure 8.12 illustrates the technique, using MODIS data acquired during the 2010 eruption of Eyjafjallajökull, Iceland. The ash cloud is clearly discriminated from other scene elements by virtue of its strongly negative BTD.

Although there are many sources of data regarding ash clouds that are used by the aviation authorities when issuing advisories (including visible wavelength satellite images and pilot reports themselves) the method of Prata (1989) remains an important means for detecting volcanic ash clouds using thermal infrared satellite data, and there are many papers in the literature that have employed the technique (e.g., Mayberry et al., 2001; Dean et al., 2004; Dubuisson et al., 2014). Although not perfect (see Prata et al., 2001), it is effective in a great many instances provided that the results are interpreted correctly, and the conditions under which, what is actually a relatively simple algorithm that should not be expected to work flawlessly, are taken into account. For example, it is known (Prata et al., 2001) that negative BTDs can result over deserts (which themselves have a silicate absorption spectrum, like the ash), clear land surfaces at night (due to surface temperature and moisture inversions), and at ash cloud edges (due to mis-registration of the two channels used to compute the BTD index), yielding false positives. It is also known that under certain circumstances ash clouds can exhibit a positive BTD (opaque ash clouds; coatings of ice on ash particles; see Watson et al., 2004). The Prata (1989) method also has the distinct advantage that it can be exploited by any remote sensing mission that acquires data in the 11 and 12 μm passbands, including those acquired from high temporal resolution geostationary instruments such as GOES, MTSAT (Multi-Functional Transport Satellite) and SEVIRI (Spinning Enhanced Visible and Infrared Imager). This allows ash clouds to be tracked with a resolution of minutes, rather than hours or days. Alternative techniques for volcanic ash detection have been published. The method of Filizzola et al. (2007) does not rely on any radiative transfer modeling of ash cloud physical properties but rather uses a change detection approach, largely equivalent to the approach described by Pergola et al. (2004).

Although this section has focused on ash detection in the thermal infrared, volcanic ash can also be detected in the UV, using TOMS (Seftor et al., 1997), based on an aerosol index computed from measurements of backscatter at two UV wavelengths. Krotkov et al. (1999) show excellent agreement between the previously described thermal infrared retrieval and this ultraviolet-derived metric

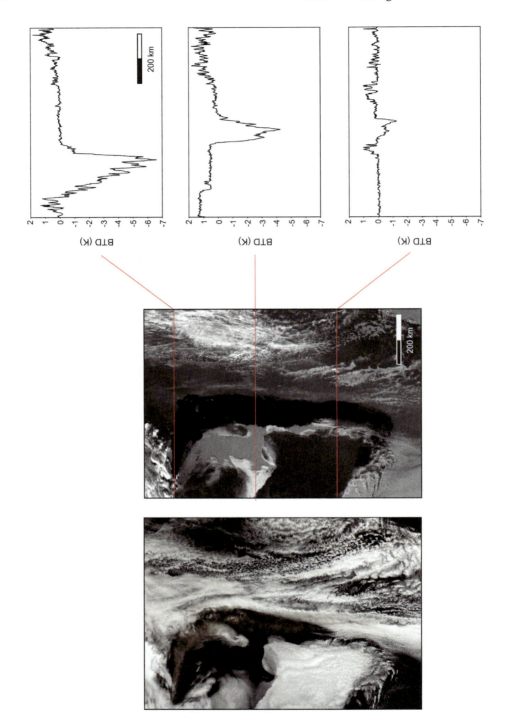

FIGURE 8.12 Left: Terra MODIS-simulated true-color composite (bands 1, 4, and 3 in red, green, and blue) of the ash cloud produced during the 2010 eruption of Eyjafjallajökull, Iceland. The ash cloud is clearly visible as the brown streak running approximately north-south. Center: MODIS band 31 (11.02 μm) minus MODIS band 32 (12.02 μm) brightness temperature difference (BTD) image. The black pixels running north-south correspond to strongly negative BTD values. Right: three BTD profiles across the image subset (red lines mark transects). The horizontal scale is the same as for the horizontal transect on the images displayed.

Remote Sensing of Volcanoes

for the same ash cloud imaged almost simultaneously by AVHRR and TOMS. Constantine et al. (2000) show less impressive convergence for a different eruption. The issue of ash (and SO_2) detection in either the ultraviolet or the thermal infrared is complicated by their frequent coexistence in the same cloud, in addition to the presence of other species (water, ice, sulfate aerosols) that also absorb light in the same spectral pass-bands (Watson et al., 2004; Figure 8.13).

Rose et al. (1995) showed how in one case the presence of ice in an ash cloud completely masked the negative BTD thermal signature described earlier. Clearly, higher spectral resolution provides a solution to this. Unfortunately, higher spectral resolution tends to come at the expense of spatial resolution (e.g., AIRS, IASI). Given the importance of rapidly (and reliably) distinguishing ash clouds from water clouds in global airspace, there is no shortage of research in this area. Instead of trying to distinguish ash clouds spectrally, Pavolonis et al. (2018) developed an algorithm to identify clouds that were growing vertically, rapidly (i.e., rising eruption plumes) from clouds that were not, in sequential, high temporal resolution LWIR images acquired from geostationary time series. With the increased availability of machine learning engines, it should be no surprise that machine

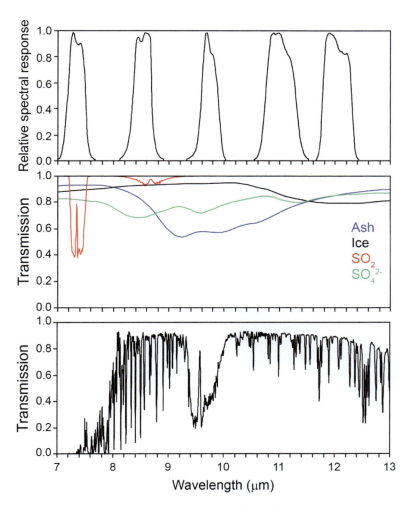

FIGURE 8.13 Top: relative spectral response of MODIS channels 28, 28, 30, 31, and 32, from left to right. Middle: transmission spectra of ice, silicate ash, SO_2 and SO_4^{2-}. Bottom: transmissivity of a model atmosphere, calculated using MODTRAN. Figure simplified from Watson et al. (2004).

learning is also being leveraged by satellite volcanologists: Bugliaro et al. (2022) show how a neural network algorithm trained on the spectral properties of atmospheric columns containing silicate ash can be used to segment pixels containing volcanic ash from adjacent pixels that do not.

Detection and tracking of ash is obviously important. But providing quantitative information about the properties of the cloud is also of value. Following from the original work of Prata (1989), Wen and Rose (1994) developed a method for using thermal infrared satellite data to retrieve the sizes and total mass of ash particles in volcanic clouds. Assuming a thin plane-parallel cloud comprising spherical ash particles overlying a homogenous surface, they write at-satellite spectral radiance as:

$$L_s(\lambda) = (1 - r_c(r_e, \tau_\lambda))L(\lambda, T_c) + t_c(r_e, \tau_\lambda)(L(\lambda, T_s) - L(\lambda, T_c))) \tag{8.6}$$

where r_c is the reflectivity of the cloud, r_e is the effective radius of the spherical ash particles, and t_c is the transmissivity of the cloud. Assuming two measurements of emitted radiance at wavelengths λ_i (~11 µm) and λ_j (~12 µm), and converting to brightness temperature, allows the possible set of contours of r_e and τ_λ to be plotted in a T_i–T_j vs. T_i brightness temperature space. Comparing real values of T_i–T_j vs. T_i values obtained from analysis of AVHRR images of volcanic clouds (where T_i and T_j are the brightness temperature in AVHRR channels 4 and 5, respectively) allows the effective radius of the ash particles comprising the cloud and its optical depth to be determined on a pixel-by-pixel basis. Knowledge of the density of the ash allows the mass of ash present to be computed from remote sensing data sets that allow a brightness temperature difference between 11 and 12 microns to be computed (e.g., Gu et al., 2005). The same basic approach was employed by Prata and Prata (2012) to determine volcanic ash concentrations during the previously described 2010 eruption of Eyjafjallajökull. In the years since the approach described earlier was published, more complicated approaches have been developed, while still aiming to achieve the same result: prediction of how surface leaving radiance is attenuated by ash clouds en route to the aperture of the telescope on an orbiting spacecraft (e.g., Prata et al., 2022).

At a more rudimentary level the planimetric shape of the ash cloud and its height can also be determined from satellite data. These are important boundary conditions for numerical models that seek to predict the dispersal of ash during an eruption, as well as providing validation data for the output of these models (e.g., Searcy et al., 1998). Ash cloud height is also an important constraint on the intensity of an eruption (i.e., the mass flux of magma from the vent) as the theoretical relationship between plume height and mass flux is well constrained (e.g., Wilson et al., 1978). Several authors have demonstrated how ash cloud height can be determined from satellite measurements of cloud-top temperature (e.g., Kienle and Shaw, 1979; Holasek et al., 1996). Here a simple measurement of cloud-top temperature can be used to estimate its height above sea level if the local lapse rate of the atmosphere is known and the cloud is in thermal equilibrium with the atmosphere (and the emissivity of the ash cloud is known or can safely be assumed). The cloud must also be opaque, to prevent the cloud top temperature being an integral over a range of depths within the cloud. In addition, the momentum of a rising ash cloud can cause it to overshoot the level of neutral buoyancy, resulting in error in height estimation.

An ash cloud casts a shadow, and measurements of shadow length cast by an ash cloud on the ground below can also be used to estimate cloud height (Glaze et al., 1989). Clearly, this technique only works during the day, but is reliable if the satellite and solar zenith and azimuth angles are known, and the plume is placed against a simple background (e.g., the ocean). If the shadow is cast upon a cloud deck at a lower altitude, then the shadow can still be used to estimate height but only in conjunction with thermal measurements of these clouds to constrain their height above sea level. The shapes of the tops of ash clouds add a level of complexity. For example, Holasek et al. (1996) found that the central region of the Pinatubo eruption cloud was ~15 km higher than the cloud edge, meaning multiple measurements were required, each with a different datum. Heights can also be retrieved from stereoscopic image pairs (see Prata and Turner, 1997). With regards to the shape of ash clouds, Glaze et al. (1999) present a method for determining their surface morphology using photoclinometric analysis of AVHRR images.

8.9 CONCLUSION

The purpose of this chapter is to introduce how satellite remote sensing can be used to better understand volcanism. As such, very little in the preceding sections has changed since the previous edition was written (2014), as the underlying principles of how we document volcano deformation using microwave data, quantify the concentration of volcanic gases using long-wave infrared and ultraviolet spectroscopy, and measure the cooling rates of lava flows using shortwave infrared radiometry, has not changed. No new, or fundamentally different, physical measurements, per se, have been developed. What has changed in the last decade, and will continue to change in the immediate future, is how these firmly established physical principles are employed in the face of (1) dramatic increases in the volume of data made available through the proliferation of new observatory-class missions launched by international space agencies, and the figurative explosion in remote sensing data acquired from constellations of microsatellites and CubeSats, and (2) rapid advances in cloud computing (and computational capabilities in general) that will allow this mountain of data to be transformed, in very near real time, into analysis-ready data and information.

Federal space agencies (e.g., NASA, ESA) have committed to launching series of Earth observation satellites, many of which acquire data that can be used to study active volcanism, using the general principles described in this chapter. Deployment of NASA's Earth System Observatory (ESO, https://science.nasa.gov/earth-science/missions/earth-system-observatory/) will see the launch of the NISAR mission in 2024 (a L-band SAR for which volcano deformation is a mission focus; Rosen and Kumar, 2021) along with the Surface Biology and Geology mission (two spacecraft, one carrying a visible and shortwave infrared hyperspectral imager, the other carrying a mid-wave and long-wave infrared thermal imager; Thompson et al., 2022) that will provide new data for studying the cooling of active lava bodies, and volcanic ash and gas emission, toward the mid-to-late part of this decade. Over the last decade NASA's investments have been matched by those of ESA's Sentinel program (a series of 10 spacecraft/payloads launched between 2014 and 2020), something that will continue in the future via ESA's Sentinel Expansion program, which will see a further six missions launch in the next decade. The Sentinel-1 mission made an enormous contribution to the study of volcanic deformation, due to the huge increase in data it provided (Biggs et al., 2022). Governments have never provided so much satellite data to the volcanological community. Of course, the amount of data provided by the private sector of relevance to volcanologists may begin to dwarf this. In 2014 about 200 CubeSats were launched; in 2023 that number had increased by an order of magnitude (Kulu, 2022). Some are designed to study active volcanoes (e.g., the recently launched HyTI—Hyperspectral Thermal Imager; Wright et al., 2023). Many constellations of microsatellites (currently there are about three dozen constellations on orbit; Swartwout, 2022) will provide information that can be used to quantify volcanic processes from space, notably the Capella (X-band SAR) and Planet (visible to shortwave infrared imaging) constellations.

The ability to analyze such large data volumes, with the short latency required to be of use in responding to often short-lived volcanic events, has also advanced. Two decades ago it was accepted that only low spatial, high temporal resolution satellite missions could be used to perform routine near real-time monitoring of global thermal anomalies, with the MODVOLC system (Wright et al., 2002) being the first. This was because high spatial resolution data (i.e., Landsat-class) were (1) not routinely acquired for all volcanoes as a result of limited duty cycles (or acquisition plans), (2) were not processed to Level 1 radiance products in a timely manner and so could not be ingested rapidly (i.e., within hours) into eruption detection algorithms, and (3) even if (1) and (2) were not true, would (at ~30–120 m per pixel) have yielded a global data volume that would have been simply unwieldy. But mission architectures now emphasize rapid conversion of L0 to L1 (and higher-level data products), something that mission ground segments can easily achieve given advances in computational capability. Importantly, cloud-computing allows vast volumes of data to be analyzed without the need to move that data from the archive to the algorithm. As a result, near real-time thermal volcano monitoring algorithms that use the high spatial resolution data now routinely acquired by the Landsat and Sentinel missions, exist (e.g., Genzano et al., 2020). Wright et al. (2008) showed how satellite-derived lava effusion rates could be used to drive and validate lava flow hazard simulations. But only conceptually. This is now a reality. Ganci et al. (2020) describe how data from a suite of sensors, with varying spatial, spectral, and temporal resolution can be fused to derive vent location, lava cooling rate, and volumetric flux, and constrain underlying topography, to drive numerical simulations of lava flow propagation, and then validate the model output. This is now done in near real-time, and is operational, not conceptual. Twenty-five years ago, volcano InSAR was the preserve of a small handful of people who wrote their own code or had access to someone's code, and could find image pairs amenable to the technique (no small problem, as only a few spacecraft actually made the relevant measurements, and the data were not freely available). The field has now evolved to the point where it has become a routine means for monitoring Earth's inventory of volcanoes (Biggs and Wright, 2020). At the time of writing, ~1.5 million interferograms of volcanoes around the world are freely available online (https://comet.nerc.ac.uk/comet-lics-portal/), where SAR data are ingested automatically, and global volcano deformation products generated automatically (Lazecký et al., 2020).

In many ways the efforts of volcanologist in developing techniques to study volcanoes from space are now being rewarded with the data and ground segment resources needed to make the most of those efforts. It is an excellent time to be a satellite volcanologist.

REFERENCES

Abrams, M., Crippen, R., & Fujisada, H. 2020. ASTER global digital elevation model (GDEM) and ASTER global water body dataset (ASTWBD). *Remote Sensing*, 12, 1156. http://doi.org/10.3390/rs12071156.

Amelung, F., Jonsson, S., Zebker, H., & Segall, P. 2000. Widespread uplift and 'trapdoor' faulting on Galapagos volcanoes observed with radar interferometry. *Nature*, 407, 993–996.

Amelung, F., Yun, S., Walter, T.R., Segall, P., & Kim, S. 2007. Stress control of deep rift intrusion at Mauna Loa volcano, Hawaii. *Science*, 316, 1026–1030.

Anderson, K., & Segal, P. 2013. Bayesian inversion of data from effusive volcanic eruptions using physics-based models: Application to Mount St. Helens 2004–2008. *Journal of Geophysical Research*, 118, 2017–2037.

Avallone, A., Zollo, A., Briole, P., Delacourt, C., & Beauducel, F. 1999. Subsidence of Campi Flegrei (Italy) detected by SAR interferometry. *Geophysical Research Letters*, 26, 2303–2306.

Badrudin, M. 1994. Kelut volcano monitoring: Hazards, mitigation and changes in water chemistry prior to the 1990 eruption. *Geochemical Journal*, 28, 233–241.

Bagnardi, M., & Hooper, A. 2018. Inversion of surface deformation data for rapid estimates of source parameters and uncertainties: A Bayesian approach. *Geochemistry, Geophysics, Geosystems*, 19, 2194–2211. https://doi.org/10.1029/2018GC007585.

Berk, A., Bernstein, L.S., & Robertson, D.C. 1989. MODTRAN: A moderate resolution model for LOWTRAN7. *Final Report, GL–TR–0122*, Airforce Geophysics Laboratory, Hanscom Airforce Base, MD.

Biggs, J., Anantrasirichai, N., Albino, F., et al. 2022. Large-scale demonstration of machine learning for the detection of volcanic deformation in Sentinel-1 satellite imagery. *Bulletin of Volcanology*, 84, 100. https://doi.org/10.1007/s00445-022-01608-x

Biggs, J., Ebmeier, S.K., Aspinall, W.P., Lu, Z., Pritchard, M.E., Sparks, R.S.J., & Mather, T. 2014. Global link between deformation and volcanic eruption quantified by satellite imagery. *Nature Communications*, 5. http://doi.org/10.1038/ncomms4471

Biggs, J., & Wright, T.J. 2020. How satellite InSAR has grown from opportunistic science to routine monitoring over the last decade. *Nature Communications*, 11, 3863. https://doi.org/10.1038/s41467-020-17587-6

Bogue, R.R., Douglas, P.M.J., Fisher, J.B., & Stix, J. 2023. Volcanic diffuse volatile emissions tracked by plant responses detectable from space. *Geochemistry, Geophysics, Geosystems*, 24, e2023GC010938. https://doi.org/10.1029/2023GC010938

Bonny, E., & Wright, R. 2017. Predicting the end of lava-flow-forming eruptions from space. *Bulletin of Volcanology*, 79. http://doi.org/10.1007/s00445-011-1134-8

Brown, G., Rymer, H., Dowden, J., Kapadia, P., Stevenson, D., Barquero, J., & Morales, L.D. 1989. Energy budget analysis for Poas crater lake: Implications for predicting volcanic activity. *Nature*, 339, 370–373.

Bugliaro, L., Piontek, D., Kox, S., Schmidl, M., Mayer, B., Müller, R., Vázquez-Navarro, M., Peters, D.M., Grainger, R.G., Gasteiger, J., & Kar, J. 2022. VADUGS: A neural network for the remote sensing of volcanic ash with MSG/SEVIRI trained with synthetic thermal satellite observations simulated with a radiative transfer model. *Natural Hazards and Earth System Science*, 22, 1029–1054. https://doi.org/10.5194/nhess-22-1029-2022

Burgmann, R., Rosen, P.A., & Fielding, E.J. 2000. Synthetic aperture radar interferometry to measure Earth's surface topography and its deformation. *Annual Reviews of Earth and Planetary Sciences*, 28, 169–209.

Burton, M.R., Oppenheimer, C., Horrocks, L.A., & Francis, P.W. 2000. Remote sensing of CO_2 and H_2O emissions from Masaya volcano, Nicaragua. *Geology*, 28, 915–918.

Campion, R., Salerno, G.G., Coheur, P., Hurtmans, D., Clarisse, L., Kazahaya, K., Burton, M., Caltabiano, T., Clerbaux, C., & Bernard, A. 2010. Measuring volcanic degassing of SO_2 in the lower troposphere with ASTER band ratios. *Journal of Volcanology and Geothermal Research*, 194, 42–54.

Carn, S.A., Fioletov, V., McLinden, C., Li, C., & Krotkov, N. 2017. A decade of global volcanic SO_2 emissions measured from space. *Scientific Reports*, 7, 44095. https://doi.org/10.1038/srep44095

Carn, S.A., Krueger, A.J., Bluth, G., Schaefer, S.J., Krotkov, N.A., Watson, I.M., & Datta, S. 2003. Volcanic eruption detection by the Total Ozone Mapping Spectrometer (TOMS) instruments: A 22-year record of sulphur dioxide and ash emissions In: C. Oppenheimer, et al. (eds) *Volcanic Degassing*. London: Geological Society of London, pp. 177–202.

Chrétien, S., & Brousse, R. 1989. Events preceeding the great eruption of 8 May, 1902 at Mount Pelée, Martinique. *Journal of Volcanology and Geothermal Research*, 38, 67–75.

Christenson, B.W. 1994. Convection and stratification in Ruapehu crater lake, New Zealand: Implications for Lake Nyos–type gas release eruptions. *Geochemical Journal*, 28, 185–197.

Condie, K.C.2005. *Earth as an Evolving Planetary System*. London: Elsevier Academic Press.

Constantine, E.K., Bluth, G.J.S., & Rose, W.I. 2000. TOMS and AVHRR observations of drifting volcanic clouds from the August 1991 eruptions of Cerro Hudson. In: P.J. Mouginis-Mark et al. (eds) *Remote Sensing of Active Volcanism*. Washington, DC: AGU Geophysical Monograph Series, pp. 45–64.

Coppola, D., Laiolo, M., Cigolini, C., Donne, D.D., & Ripepe, M. 2016. Enhanced volcanic hot-spot detection using MODIS IR data: Results from the MIROVA system. In: A. Harris, T. DeGroove, F. Garel, and S. Carn (eds) *Detecting, Modelling, and Responding to Effusive Eruptions*. London: Special Publication of the Geological Society of London, pp. 426, 181–205.

Corradini, S., Guerrieri, L., Brenot, H., Clarisse, L., Merucci, L., Pardini, F., Prata, A.J., Realmuto, V.J., Stelitano, D., & Theys, N. 2021. Tropospheric volcanic SO_2 mass and flux retrievals from satellite. The Etna December 2018 eruption. *Remote Sensing*, 13(11), 2225. https://doi.org/10.3390/rs13112225

Crowley, J.K., Hubbard, B.C., & Mars, J.C. 2003. Analysis of potential debris flow source areas on Mount Shasta, California, by using airborne and satellite remote sensing data. *Remote Sensing of Environment*, 87, 345–358.

Dave, J.V., & Mateer, C.L. 1967. A preliminary study on the possibility of estimating total atmospheric ozone from satellite measurements. *Journal of Atmospheric Sciences*, 24, 414–427.

Davis, P.M.1986. Surface deformation due to inflation of an arbitrarily oriented triaxial ellipsoidal cavity in an elastic half-space, with reference to Kilauea Volcano, Hawaii. *Journal of Geophysical Research*, 91, 7429–7438.

Dean, K., Dehn, J., Papp, K.R., Smith, S., Izbekov, P., Peterson, R., Kearney, C., & Steffke, A. 2004. Integrated satellite monitoring of the 2001 eruption of Mount Cleveland, Alaska. *Journal of Volcanology and Geothermal Research*, 135, 51–74.

Dehn, J., Dean, K., & Engle, K. 2000. Thermal monitoring of North Pacific volcanoes from space. *Geology*, 28, 755–758.

Delacourt, C., Briole, P., & Achache, J. 1998. Tropospheric corrections of SAR interferograms with strong topography: Application to Etna. *Geophysical Research Letters*, 25, 2849–2852.

Delmelle, P., & Bernard, A. 2000. Volcanic lakes. In: H. Sigurdsson et al. (eds) *Encyclopedia of Volcanoes*. London: Academic Press, pp. 877–895.

Dozier, J.1981. A method for satellite identification of surface temperature fields of subpixel resolution. *Remote Sensing of Environment*, 11, 121–129.

Dubuisson, P., Herbin, H., Minvielle, F., Compiegne, M., Thieuleux, F., Parol, F., & Pelon, J. 2014. Remote sensing of volcanic ash plumes from thermal infrared: A case study analysis from SEVERI, MODIS and IASI instruments. *Atmospheric Measurement Techniques*, 7, 359–371.

Ebmeier, S.K., Biggs, J., Mather, T.A., Wadge, G., and Amelung, F. 2010. Steady downslope movement on the western flank of Arenal volcano, Costa Rica. *Geochemistry, Geophysics and Geosystems*, 11, Q12004.

Edmonds, M. 2008. New geochemical insights into volcanic degassing. *Philosophical Transactions of the Royal Society of London*, 366, 4559–4579.

Ferretti, A., Prati, C., & Rocca, F. 2001. Permanent scatterers in SAR interferometry. *IEEE Transactions on Geoscience and Remote Sensing*, 39, 8–20.

Fialko, Y., Simons, M., & Agnew, D. 2001. The complete (3-D) surface displacement field in the epicentral area of the 1999 Mw 7.1 Hector Mine earthquake, California, from space geodetic observations. *Geophysical Research Letters*, 28, 3063–3066.

Filizzola, C., Lacava, T., Marchese, F., Pergola, N., Scaffidi, I., & Tramutoli, V. 2007. Assessing RAT (Robust AVHRR Techniques) performances for volcanic ash cloud detection and monitoring in near real-time: The 2002 eruption of Mount Etna (Italy). *Remote Sensing of Environment*, 107, 440–454.

Francis, P.W. 1979. Infra-red techniques for volcano monitoring and prediction—a review. *Journal of the Geological Society of London*, 136, 355–359.

Francis, P.W., & Oppenheimer, C. 2004. *Volcanoes*. 2nd ed. Oxford: Oxford University Press.

Francis, P.W., Oppenheimer, C., & Stevenson, D.S. 1993. Endogenous growth of persistently active volcanoes. *Nature*, 366, 544–557.

Gabriel, A.K., Goldstein, R.M., & Zebker, H.A. 1989. Mapping small elevation changes over large areas: Differential radar interferometry. *Journal of Geophysical Research*, 94, 9183–9191.

Ganci, G., Cappello, A., Bilotta, G., & Del Negro, C. 2020. How the variety of satellite remote sensing data over volcanoes can assist hazard monitoring efforts: The 2011 eruption of Nabro volcano. *Remote Sensing of Environment*, 236, 111426. https://doi.org/10.1016/j.rse.2019.111426.

Gaonac'h, H., Vandemeulebrouck, J., Stix, J., & Halbwachs, M. 1994. Thermal infrared satellite measurements of volcanic activity at Stromboli and Vulcano. *Journal of Geophysical Research*, 99, 9477–9485.

Genzano, N., Pergola, N., & Marchese, F. 2020. A Google Earth Engine tool to investigate, map and monitor volcanic thermal anomalies at global scale by means of mid-high spatial resolution satellite data. *Remote Sensing*, 12, 3232. http://doi.org/10.3390/rs12193232.

Gerlach, T.M., Doukas, M.P., McGee, K.A., & Kessler, R. 1999. Airborne detection of diffuse carbon dioxide emissions at Mammoth Mountain, California. *Geophysical Research Letters*, 26, 3661–3664.

Gerlach, T.M., McGee, K.A., Elias, T., Sutton, A.J., & Doukas, M.P. 2002. Carbon dioxide emission rate of Kilauea volcano: Implications for primary magma and summit reservoir. *Journal of Geophysical Research*, 107. http://doi.org/10.1029/2001JB000407.

Girona, T., Realmuto, V., & Lundgren, P. 2021. Large-scale thermal unrest of volcanoes for years prior to eruption. *Nature Geoscience*. https://doi.org/10.1038/s41561-021-00705-4.

Glaze, L.S., Francis, P., Self, S., & Rothery, D. 1989. The 16 September 1986 eruption of Lascar volcano, north Chile: Satellite investigations. *Bulletin of Volcanology*, 51, 149–160.

Glaze, L.S., Wilson, L., & Mouginis-Mark, P.J. 1999. Volcanic eruption plume top topography and heights determined from photoclinometric analysis of satellite data. *Journal of Geophysical Research*, 104, 2989–3001.

Gong, W., Lu, Z., & Meyer, F. 2016. Uncertainties in estimating magma source parameters from in SAR observation. In: K. Riley, P. Webley, and M. Thompson (eds) *Natural Hazard Uncertainty Assessment: Modeling and Decision Support, Geophysical Monograph 223*. Hoboken, NJ: Wiley, pp. 89–104.

Gouhier, M., Guéhenneux, Y., Labazuy, P. Cacault, P., Decriem, J., & Rivet, S. 2016. HOTVOLC: A web-based monitoring system for volcanic hot spots. In: A. Harris, T. DeGroove, F. Garel, and S. Carn (eds) *Detecting, Modelling, and Responding to Effusive Eruptions*. London: Special Publication of the Geological Society of London, pp. 426, 223–241.

Grosse, P., Euillades, P.A., Euillades, L.D., & van Wyk de Vries, B. 2014. A global database of composite volcano morphometry. *Bulletin of Volcanology*, 76, 784. http://doi.org/10.1007/s00445-013-0784-4.

Gu, Y., Rose, W.I., Schneider, D.J., Bluth, G.J.S., & Watson, I.M. 2005. Advantageous GOES IR results for ash mapping at high latitudes: Cleveland eruptions 2001. *Geophysical Research Letters*, 32, L02305. http://doi.org/10.1029/2004GL021651.

Guo, S., Bluth, G.J.S., Rose, W.I., Watson, I.M., & Prata, A.J. 2004. Re-evaluation of SO_2 release of the 15 June 1991 Pinatubo eruption using ultraviolet and infrared satellite sensors. *Geochemistry, Geophysics and Geosystems*, 5, Q04001.

Harris, A.J.L. 2013. *Thermal Remote Sensing of Active Volcanoes: A User's Manual.* Cambridge: Cambridge University Press.

Harris, A.J.L., & Baloga, S.M. 2009. Lava discharge rates from satellite measured heat flux. *Geophysical Research Letters*, 36, L19302. http://doi.org/10.1029/2009GL039717.

Harris, A.J.L., Blake, S., Rothery, D.A., & Stevens, N.F. 1997a. A chronology of the 1991 to 1993 Etna eruption using advanced very high resolution radiometer data: Implications for real–time thermal volcano monitoring. *Journal of Volcanology and Geothermal Research*, 102, 7985–8003.

Harris, A.J.L., Flynn, L.P., Rothery, D.A., Oppenheimer, C., & Sherman, S.B. 1999. Mass flux measurements at active lava lakes: Implications for magma recycling. *Journal of Geophysical Research*, 104, 7117–7136.

Harris, A.J.L., Keszthelyi, L., Flynn, L.P., Mouginis-Mark, P.J., Thornber, C., Kauahikikaua, J., Sherrod, D., Trusdell, F., Sawyer, M.W., & Flament, P. 1997b. Chronology of the episode 54 eruption at Kilauea Volcano, Hawai'i, from GOES-9 satellite data. *Geophysical Research Letters*, 24, 3181–3184.

Harris, A.J.L., & Maciejewski, A.J.H. 2000. Thermal surveys of the Vulcano Fossa fumarole field 1994–1999: Evidence for fumarole migration and sealing. *Journal of Volcanology and Geothermal Research*, 102, 119–147.

Harris, A.J.L., & Stevenson, D.S. 1997. Thermal observations of degassing open conduits and fumaroles at Stromboli and Vulcano using remotely sensed data. *Journal of Volcanology and Geothermal Research*, 76, 175–198.

Hellman, M.J., & Ramsey, M.S. 2004. Analysis of hot springs and associated deposits in Yellowstone National Park using ASTER and AVIRIS remote sensing. *Journal of Volcanology and Geothermal Research*, 135, 195–219. https://doi.org/10.1016/j.jvolgeores.2003.12.012

Henney, L.A., Rodriguez, L.A., & Watson, I.M. 2012. A comparison of SO_2 retrieval techniques using mini-UV spectrometers and ASTER imagery at Lascar volcano, Chile. *Bulletin of Volcanology*, 74, 589–594.

Higgins, J., & Harris, A.J.L. 1997. VAST: A program to locate and analyse volcanic thermal anomalies automatically from remotely sensed data. *Computers and Geoscience*, 23, 627–645.

Holasek, R., Self, S., & Woods, A.W. 1996. Satellite observations and interpretation of the 1991 Mount Pinatubo eruption plumes. *Journal of Geophysical Research*, 101, 27, 635–27, 655.

Hooper, A., Zebker, H., Segall, P., & Kampes, B. (2004). A new method for measuring deformation on volcanoes and other natural terrains using InSAR persistent scatterers. *Geophysical Research Letters*, 31, L23611. http://doi.org/10.1029/2004GL021737.

Jay, J., Costa, F., Pritchard, M., Lara, L., Singer, B., & Herrin, J., 2014. Locating magma reservoirs using InSAR and petrology before and during the 2011–2012 Cordon Caulle silicic eruption. *Earth and Planetary Science Letters*, 395, 254–266.

Ji, L., Lu, Z., Dzurisin, D., & Senyukov, S. 2013. Pre-eruption deformation caused by dike intrusion beneath Kizimen volcano, Kamchatka, Russia, observed by InSAR. *Journal of Volcanology and Geothermal Research*, 256, 87–95.

Johnson, M.S., Schwandner, F.M., Potter, C.S., Nguyen, H.M., Bell, E., Nelson, R.R., et al. 2020. Carbon dioxide emissions during the 2018 Kilauea volcano eruption estimated using OCO-2 satellite retrievals. *Geophysical Research Letters*, 47, e2020GL090507. https://doi.org/10.1029/2020GL090507

Kaneko, T., & Wooster, M.J. 1999. Landsat infrared analysis of fumarole activity at Unzen volcano: Time-series comparison with gas and magma fluxes. *Journal of Volcanology and Geothermal Research*, 89, 57–64.

Kienle, J., & Shaw, G.E. 1979. Plume dynamics, thermal energy and long distance transport of vulcanian eruption clouds from the Augustine volcano. *Journal of Volcanology and Geothermal Research*, 6, 139–164.

Krotkov, N.A., Carn, S.A., Krueger, A.J., Bhartia, P.K., & Yang, K. 2006. Band residual difference algorithm for retrieval of SO_2 from the Aura Ozone Monitoring Instrument (OMI). *IEEE Transactions on Geoscience and Remote Sensing*, 44, 1259–1266.

Krotkov, N.A., Torres, O., Seftor, C., Krueger, A.J., Kostinski, A., Rose, W.I., Bluth, G.J.S., Schneider, D., & Schaefer, S.J. 1999. Comparison of TOMS and AVHRR volcanic ash retrievals from the August 1992 eruption of Mt. Spurr. *Geophysical Research Letters*, 26, 455–458.

Krueger, A.J., Walter, L.S., Bhartia, P.K., Schnetzler, C.C., Krotkov, N.A., Spord, I., & Bluth, G.J.S. 1995. Volcanic sulfur dioxide measurements from the total ozone mapping spectrometer instruments. *Journal of Geophysical Research*, 100, 14057–14076.

Krueger, A.J.1983. Sighting of El Chichon sulfur dioxide clouds with the Nimbus 7 total ozone mapping spectrometer. *Science*, 220, 1377–1379.

Kulu, E. 2022. Nanosatellite Launch Forecasts—track record and latest prediction. *Proceedings of the AIAA/ USU Conference on Small Satellites*, Swifty Session 1, SSC22-S1-04. https://digitalcommons.usu.edu/smallsat/2022/all2022/7.

Lazecký, M., Spaans, K., González, P.J., Maghsoudi, Y., Morishita, Y., Albino, F., Elliott, J., Greenall, N., Hatton, E., Hooper, A., et al. 2020. LiCSAR: An automatic InSAR tool for measuring and monitoring tectonic and volcanic activity. *Remote Sensing*, 12(15), 2430. https://doi.org/10.3390/rs12152430.

Li, Z., Muller, J.P., Cross, P., & Fielding, E.J. 2005. Interferometric synthetic aperture radar (InSAR) atmospheric correction: GPS, Moderate Resolution Imaging Spectroradiometer (MODIS), and InSAR integration. *Journal of Geophysical Research*, 110. http://doi.org/10.1029/2004JB003446.

Loyola, D., van Geffen, J., Valks, P., Erbertseder, T., Van Roozendael, M., Thomas, W., Zimmer, W., & Wißkirchen, K. 2008. Satellite-based detection of volcanic sulphur dioxide from recent eruptions in Central and South America. *Advances in Geosciences*, 14, 35–40.

Lu, Z., & Dzurisin, D. 2014. *InSAR Imaging of Aleutian Volcanoes: Monitoring a Volcanic Arc from Space*. Heidelberg: Springer Praxis Books.

Lu, Z., Wicks, C., Dzurisin, D., Power, P., Moran, C., & Thatcher, W. 2002. Magmatic inflation at a dormant stratovolcano: 1996–1998 activity at Mount Peulik volcano, Alaska, revealed by satellite radar interferometry. *Journal of Geophysical Research*, 107. http://doi.org/10.1029/2001JB000471.

Massonnet, D., Briole, P., & Arnaud, A. 1995. Deflation of Mount Etna monitored by spaceborne radar interferometry. *Nature*, 375, 567–570.

Massonnet, D., & Feigl, K.L., 1998. Radar interferometry and its application to changes in the earth's surface. *Reviews in Geophysics*, 36, 441–500.

Massonnet, D., Rossi, M., Carmona, C., Adragna, F., Peltzer, G., Feigl, K., & Rabaute, T. 1993. The displacement field of the Landers earthquake mapped by radar interferometry. *Nature*, 364, 138–142.

Mayberry, G.C., Rose, W.I., & Bluth, G.J. 2001. Dynamics of the volcanic and meteorlogical clouds produced by the December 26 (Boxing Day) 1997 eruption of Soufriere Hills volcano, Montserrat, WI. In: T.H. Druitt and B.P. Kokelaar (eds) *The Eruption of Soufriere Hills Volcano, Montserrat, from 1995 to 1999*. London: Geological Society of London, pp. 539–556.

Mogi, K.1958. Relations between the eruptions of various volcanoes and the deformations of the ground surfaces around them. *Bulletin of the Earthquake Research Institute of Tokyo*, 36, 99–134.

Moore, J.G., Nakamura, K., & Alcarez, A. 1966. The eruption of Taal volcano, Philippines, September 28–30, 1965. *Science*, 151, 955–960.

Moran, S.C., Kwoun, O., Masterlark, T., & Lu, Z. 2006. On the absence of InSAR-detected volcano deformation spanning the 1995–1996 and 1999 eruptions of Shishaldin volcano, Alaska. *Journal of Volcanology and Geothermal Research*, 150, 119–131.

Murphy, S., Wright, R., & Rouwet, D. 2018. Color and temperature of Kelimutu's volcanic crater lakes. *Bulletin of Volcanology*, 80. https://doi.org/10.1007/s00445-017-1172-2.

Murray, J.B., Pullen, A.D., & Saunders, S. 1995. Ground deformation surveying of active volcanoes. In: W.J. McGuire et al. (eds) *Monitoring Active Volcanoes*. London: UCL Press, pp. 113–150.

Novak, M.A.M., Watson, I.M., Delgado-Granados, H., Rose, W.I., Cardenas-Gonzalez, L., & Realmuto, V.J. 2008. Volcanic emissions from Popocatepetl volcano, Mexico, quantified using moderate resolution imaging spectroradiometer (MODIS) infrared data: A case study of the December 2000–January 2001 emissions. *Journal of Volcanology and Geothermal Research*, 170, 76–85.

Okada, Y.1985. Surface deformation due to shear and tensile faults in a half space. *Bulletin of the Seismological Society of America*, 75, 1135–1154.

Oppenheimer, C. 1991. Lava flow cooling estimated from Landsat Thematic Mapper infrared data: The Lonquimay eruption (Chile, 1989). *Journal of Geophysical Research*, 96, 21865–21878.

Oppenheimer, C.1996. Crater lake heat losses estimated by remote sensing. *Geophysical Research Letters*, 23, 1793–1796.

Oppenheimer, C.1997. Remote sensing of the color and temperature of volcanic crater lakes. *International Journal of Remote Sensing*, 18, 5–37.

Oppenheimer, C., Francis, P.W., Rothery, D.A., Carlton, R.W.T., & Glaze, L.S. 1993. Infrared image analysis of volcanic thermal features: Lascar Volcano, Chile, 1984–1992. *Journal of Geophysical Research*, 98, 4269–4286.

Oshawa, S., Saito, T., Yoshikawa, S., Mawatari, H., Yamada, M., Amita, K., Takamatsu, N., Sudo, Y., & Kagiyama, T. 2010. Color change of lake water at the active crater lake of Aso volcano: Is it in response to change in water quality induced by volcanic activity? *Limnology*, 11, 207–215.

Parfitt, E.A., & Wilson, L. (2008). *Fundamentals of Physical Volcanology*. Oxford: Blackwell Publishing.

Patrick, M., Dean, K., & Dehn, J. (2004). Active mud volcanism observed with Landsat 7 ETM+. *Journal of Volcanology and Geothermal Research*, 131, 307–320.

Pavolonis, M.J., Sieglaff, J., & Cintineo, J. 2018. Automated detection of explosive volcanic eruptions using satellite-derived cloud vertical growth rates. *Earth and Space Science*, 5, 903–928. https://doi.org/10.1029/2018EA000410

Pergola, N., Marchese, F., and Tramutoli, V. 2004. Automated detection of thermal features of active volcanoes by means of infrared AVHRR records. *Remote Sensing of Environment*, 93, 311–327.

Pieri, D.C., & Baloga, S.M. 1986. Eruption rate, area, and length relationships for some Hawaiian lava flows. *Journal of Volcanology and Geothermal Research*, 30, 29–45.

Pour, A.B., Hashim, M., & Genderan, J. 2013. Detection of hydrothermal alteration zones in a tropical region using satellite remote sensing data: Bau goldfield, Sarawak, Malaysia. *Ore Geology Reviews*, 54, 181–196.

Power, J.A., Nye, C.J., Coombs, M.L., Wessels, R.L., Cervelli, P.F., Dehn, J., Wallace, K.L., Freymueller, J.T., & Doukas, M.P. 2006. The reawakening of Alaska's Augustine volcano, EOS. *Transactions of the American Geophysical Union*, 87, 373.

Prata, A.J. 1989. Infrared radiative transfer calculations for volcanic ash clouds. *Geophysical Research Letters*, 16, 1293–1296.

Prata, A.J., & Bernardo, C. 2007. Retrieval of volcanic SO_2 column abundance from Atmospheric Infrared Sounder data. *Journal of Geophysical Research*, 112. http://doi.org/10.1029/2006JD007955.

Prata, A.J., Bluth, G., Rose, W.I., Schneider, D., & Tupper, A. 2001. Comments on failures in detecting volcanic ash from a satellite-based technique. *Remote Sensing of Environment*, 78, 341–346.

Prata, A.T., Grainger, R.G., Taylor, I.A., Povey, A.C., Proud, S.R., & Poulsen, C.A. 2022. Uncertainty-bounded estimates of ash cloud properties using the ORAC algorithm: Application to the 2019 Raikoke eruption. *Atmospheric Measurement Techniques*, 15, 5985–6010. https://doi.org/10.5194/amt-15-5985-2022.

Prata, A.J., & Prata, A.T. 2012. Eyjafjallajökull volcanic ash concentrations determined using Spin Enhanced Visible and Infrared Imager measurements. *Journal of Geophysical Research*, 117, D00U23. http://doi.org/10.1029/2011JD016800.

Prata, A.J., Rose, W.I., Self, S., & O'Brien, D.M. 2003. Global, long–term sulphur dioxide measurements from TOVS data: A new tool for studying explosive volcanism and climate. In: A. Robock et al. (eds) *Volcanism and the Earth's Atmosphere*. Washington, DC: American Geophysical Union, pp. 75–92.

Prata, A.J., & Turner, P.J. 1997. Cloud top height determination from the ATSR. *Remote Sensing of Environment*, 59, 1–13.

Prins, E.M., & Menzel, W.P. 1992. Geostationary satellite detection of biomass burning in South America. *International Journal of Remote Sensing*, 13, 2783–2799.

Pritchard, M.E., Poland, M., Reath, K., Andrews, B., Bagnardi, M., Biggs, J., Carn, S., Coppola, D., Ebmeier, S.K., Furtney, M.A., Girona, T., Griswold, J., Lopez, T., Lundgren, P., Ogburn, S., Pavolonis, M., Rumpf, E., Vaughan, G., Wauthier, C., Wessels, R., Wright, R., Anderson, K.R., Bato, M.G., & Roman, A. 2022. Optimizing satellite resources for the global assessment and mitigation of volcanic hazards—Suggestions from the USGS Powell Center Volcano Remote Sensing Working Group: U.S. *Geological Survey Scientific Investigations Report 2022–5116*, 69 p. https://doi.org/10.3133/sir20225116.

Pritchard, M.E., & Simons, M. 2002. A satellite geodetic survey of large-scale deformation of volcanic centres in the central Andes. *Nature*, 418, 167–171.

Pritchard, M.E., & Simons, M. 2004. An InSAR-based survey of volcanic deformation in the southern Andes. *Geophysical Research Letters*, 31. http://doi.org/10.1029/2004GL020545.

Pugnaghi, S., Gangale, G., Corradini, S., & Buongiorno, M.F. 2006. Mt. Etna sulfur dioxide flux monitoring using ASTER–TIR data and atmospheric observations. *Journal of Volcanology and Geothermal Research*, 152, 74–90.

Rampino, M. 2010. Mass extinctions of life and catastrophic flood basalt volcanism. *Proceedings of the National Academy of Sciences of the United States of America*, 107, 6555–6556.

Realmuto, V.J. 2000. The potential use of Earth Observing System data to monitor the passive emission of sulfur dioxide from volcanoes. In: P.J. Mouginis-Mark et al. (eds) *Remote Sensing of Active Volcanism.* Washington, DC: AGU Geophysical Monograph Series, pp. 101–115.

Realmuto, V.J., Abrams, M.J., Buongiorno, M.F., & Pieri, D.C. 1994. The use of multispectral thermal infrared image data to estimate the sulfur dioxide flux from volcanoes: A case study from Mount Etna, Sicily, July 29, 1986. *Journal of Geophysical Research*, 99, 481–488.

Richards, M.A.2007. A beginner's guide to interferometric SAR concepts and signal processing. *IEEE Aerospace and Electronic Systems Magazine*, 22, 5–29.

Rix, M., Valks, P., Hao, N., Loyola, D., Schlager, H., Huntrieser, H., Flemming, J., Koehler, U., Schumann, U., & Inness, A. 2012. Volcanic SO_2, BrO and plume height estimations using GOME-2 satellite measurements during the eruption of Eyjafjallajökull in May 2010. *Journal of Geophysical Research*, 117. http://doi.org/10.1029/2011JD016718.

Roberge, J., Delgado-Grandos, H., & Wallace, P.J. 2009. Mafic magma recharge supplies high CO_2 and SO_2 gas fluxes at Popocatepetl volcano, Mexico. *Geology*, 37, 107–110.

Rose, W.I., Delene, D.J., Schneider, D.J., Bluth, G.J.S., Krueger, A.J., Sprod, I., McKee, C., Davies, H.L., & Ernst, G.J. 1995. Ice in the 1994 Rabaul eruption: Implications for volcanic hazard and atmospheric effects. *Nature*, 375, 477–479.

Rosen, P.A., & Kumar, R. 2021. *NASA-ISRO SAR (NISAR) Mission Status, 2021 IEEE Radar Conference (RadarConf21)*. Atlanta, GA, 1–6. http://doi.org/10.1109/RadarConf2147009.2021.9455211.

Rothery, D.A., Francis, P.W., & Wood, C.A. 1988. Volcano monitoring using short wavelength infrared data from satellites. *Journal of Geophysical Research*, 93, 7993–8008.

Rouwet, D., Christenson, B., Tassi, F., & Vandemeulebrouck, J. 2015. *Volcanic Lakes*. Berlin: Springer.

Rowland, S.K. 1996. Slopes, lava flow volumes and vent distributions on Volcán Fernandina, Galápagos Islands. *Journal of Geophysical Research*, 101, 27657–27672.

Sabins, F.F.1999. Remote sensing for mineral exploration. *Ore Geology Reviews*, 14, 157–183.

Schwandner, F.M., Carn, S.A., Kuze, A., Kataoka, F., Shiomi, K., Goto, N., Popp, C., Ajiro, M., Suto, H., Takeda, T., Kanekon, S., Sealing, C., & Flower, V. 2014. Can satellite-based monitoring techniques be used to quantify volcanic CO_2 emissions? *Abstract presented at 2014 General Assembly, European Geophysical Union*, Vienna, 27 April–2 May.

Schwandner, F.M., Gunson, M.R., Miller, C.E., Carn, S.A., Eldering, A., Krings, T., Verhulst, K.R., Schimel, D.S., Nguyen, H.M., Crisp, D., O'Dell, C.W., Osterman, G.B., Iraci, L.T., & Podolske, J.R. 2017. Spaceborne detection of localized carbon dioxide sources. *Science*. http://doi.org/10.1126/science.aam5782.

Searcy, C., Dean, K., & Stringer, W. 1998. PUFF: A high resolution volcanic ash tracking model. *Journal of Volcanology and Geothermal Research*, 80, 1–16.

Seftor, C.J., Hsu, N.C., Herman, J.R., Bhartia, P.K., Torres, O., Rose, W.I., Schneider, D.J., & Krotkov, N. 1997. Detection of volcanic ash clouds from Nimbus-7/total ozone mapping spectrometer. *Journal of Geophysical Research*, 102, 16749–16759.

Seinfeld, J.H., & Pandis, S.N. 1997. *Atmospheric Chemistry and Physics*. New York: John Wiley and Sons.

Siebert, L., Simkin, T., & Kimberly, P. 2010. *Volcanoes of the World* (3rd ed.). Los Angeles: University of California Press.

Sigmundsson, F., Durand, P., & Massonnet, D. 1999. Opening of an eruptive fissure and seaward displacement at Piton de la Fournaise volcano measured by RADARSAT satellite radar interferometry. *Geophysical Research Letters*, 26, 533–536.

Sigurdsson, H.1999. *Melting the Earth: The History of Ideas on Volcanic Eruptions*. Oxford: Oxford University Press.

Sigurdsson, H., Houghton, B., McNutt, S.R., Rymer, H., & Stix, J. (Eds). 2015. *Encyclopedia of Volcanoes* (2nd ed.). London: Academic Press.

Sigurdsson, H., & Lopes-Gautier, R. 2000. Volcanoes in literature and film. In: H. Sigurdsson et al. (eds) *Encyclopedeia of Volcanoes*. London: Academic Press, pp. 1339–1360.

Steffke, A.M., & Harris, A.J.L. 2011. A review of algorithms for detecting volcanic hot spots in satellite infrared data. *Bulletin of Volcanology*, 73, 1109–1137.

Steffke, A.M., Harris, A.J.L., Burton, M., Caltabiano, T., & Salerno, G. 2011. Coupled use of COSPEC and satellite measurements to define the volumetric balance during effusive eruptions at Mt. Etna, Italy. *Journal of Volcanology and Geothermal Research*, 205, 47–53.

Stevens, N.F., Wadge, G., Williams, C.A., Morley, J.G., Muller, J.-P., Murray, J.B., & Upton, M. 2001. Surface movements of emplaced lava flows measured by synthetic aperture radar interferometry. *Journal of Geophysical Research*, 106, 11292–11313. https://doi.org/10.1029/2000JB900425

Swartwout, M. 2022. Cubesats/Smallsats/Nanosats/Picosats/Rideshare(sats) in 2022: Making Sense of the Numbers. *2022 IEEE Aerospace Conference (AERO)*, Big Sky, MT, 1–10. http://doi.org/10.1109/AERO5 3065.2022.9843832.

Symonds, R.B., Rose, W.I., Bluth, G.J.S., & Gerlach, T.M. 1994. Volcanic-gas studies: Methods, results and applications. *Reviews in Mineralogy*, 30, 1–66.

Taylor, I.A., Preston, J., Carboni, E., Mather, T.A., Grainger, R.G., Theys, N., et al. 2018. Exploring the utility of IASI for monitoring volcanic SO_2 emissions. *Journal of Geophysical Research: Atmospheres*, 123, 5588–5606. https://doi.org/10.1002/2017JD027109

Theys, N., Campion, R., Clarisse, L., Brenot, H., van Gent, J., Dils, B., Corradini, S., Merucci, L., Coheur, P.-F., Van Roozendael, M., Hurtmans, D., Clerbaux, C., Tait, S., & Ferrucci, F. 2013. Volcanic SO_2 fluxes derived from satellite data: A survey using OMI, GOME-2, IASI and MODIS. *Atmospheric Chemistry and Physics*, 13. http://doi.org/10.5194/acp-13-5945-2013.

Thompson, D.R., et al. 2022. Ongoing progress toward NASA's surface biology and geology mission. *IGARSS 2022–2022 IEEE International Geoscience and Remote Sensing Symposium*, Kuala Lumpur, Malaysia, 5007–5010. http://doi.org/10.1109/IGARSS46834.2022.9884123.

Tortini, R., van Manen, S.M., Parkes, B.R.B, & Carn, S.A. 2017. The impact of persistent volcanic degassing on vegetation: A case study at Turrialba volcano, Costa Rica. *International Journal of Applied Earth Observation and Geoinformation*, 59, 92–103. https://doi.org/10.1016/j.jag.2017.03.002.

Trunk, L., & Bernard, A. 2008. Investigating crater lake warming using ASTER thermal imagery: Case studies at Ruapehu, Poás, Kawah Ijen, and Copahué Volcanoes. *Journal of Volcanology and Geothermal Research*, 178, 259–270.

Urai, M., 2004. Sulfur dioxide flux estimation from volcanoes using Advanced Spaceborne Thermal Emission and Reflection Radiometer—a case study of the Miyakejima volcano, Japan. *Journal of Volcanology and Geothermal Research*, 134, 1–13.

Van Manen, S., Blake, S., Dehn, J., & Valcic, L. 2013. Forecasting large explosions at Bezymianny Volcano using thermal satellite data. *Geological Society of London Special Publication*, 380, 187–201.

Varugu, B., & Amelung, F. 2021. Southward growth of Mauna Loa's dike-like magma body driven by topographic stress. *Scientific Reports*, 11, 9816. https://doi.org/10.1038/s41598-021-89203-6.

Vaughan, R.G., Keszthelyi, L.P., Lowenstern, J.B., Jaworowski, C., & Heasler, H. 2012. Use of ASTER and MODIS thermal infrared data to quantify heat flow and hydrothermal change at Yellowstone National Park. *Journal of Volcanology and Geothermal Research*, 233–234, 72–89.

Walker, G.P.L. 1973. Lengths of lava flows. *Philosophical Transactions of the Royal Society of London*, 273, 107–118.

Wang, H., Wright, T., & Biggs, J. 2009. Interseismic slip rate of the northwestern Xianshuihe fault from InSAR data. *Geophysical Research Letters*, 36, L03302. http://doi.org/10.1029/2008GL036560.

Watson, I.M., Realmuto, V.J., Rose, W.I., Prata, A.J., Bluth, G.J.S., Gu, Y., Bader, C.E., & Yu, T. 2004. Thermal infrared remote sensing of volcanic emissions using the moderate resolution imaging spectroradiometer. *Journal of Volcanology and Geothermal Research*, 135, 75–89.

Webley, P.W., Bingley, R.M., Dodson, A.H., Wadge, G., Waugh, S.J., & James, I.N. 2002. Atmospheric water vapor correction to InSAR surface motion measurements on mountains: Results from a dense GPS network on Mount Etna. *Physics and Chemistry of the Earth*, 27, 363–370.

Wen, S., & Rose, W.I., 1994. Retrieval of sizes and total mass of particles in volcanic ash clouds using AVHRR bands 4 and 5. *Journal of Geophysical Research*, 99, 5421–5431.

Williams-Jones, G., & Rymer, H. 2000. Hazards of volcanic gases. In: H. Sigurdsson et al. (eds) *Encyclopedia of Volcanoes*. London: Academic Press, pp. 997–1004.

Wilson, L., Sparks, R.S.J., Huang, T.C., & Watkins, N.D. 1978. The control of volcanic column heights by eruption energetics and dynamics. *Journal of Geophysical Research*, 83, 1829–1836.

Wohletz, K., & Heiken, G. 1992. *Volcanology and Geothermal Energy*. Los Angeles: University of California Press.

Wooster, M.J., & Rothery, D.A. 1997. Thermal monitoring of Lascar Volcano, Chile using infrared data from the Along Track Scanning Radiometer: A 1992–1995 time-series. *Bulletin of Volcanology*, 58, 566–579.

Wright, R., Blake, S., Rothery, D.A., & Harris, A.J.L. 2001. A simple explanation for the space–based calculation of lava eruption rates. *Earth and Planetary Science Letters,* 192, 223–233.

Wright, R., Flynn, L.P., Garbeil, H., Harris, A.J.L., & Pilger, E. 2002. Automated volcanic eruption detection using MODIS. *Remote Sensing of Environment,* 82, 135–155.

Wright, R., Garbeil, H., Baloga, S.M., & Mouginis–Mark, P.J. 2006. An analysis of Shuttle Radar Topography Mission digital elevation data for studies of volcano morphology. *Remote Sensing of Environment,* 105, 41–53.

Wright, R., Garbeil, H., & Davies, A.G. 2010. Cooling rate of some active lavas determined using an orbital imaging spectrometer. *Journal of Geophysical Research,* 115. http://doi.org/10.1029/2009JB006536.

Wright, R., Garbeil, H., & Harris, A.J.L. 2008. Using infrared satellite data to drive a thermo–rheological/stochastic lava flow emplacement model: A method for near–real–time volcanic hazard assessment. *Geophysical Research Letters,* 35, L19307. http://doi.org/10.1029/2008GL035228.

Wright, R., Glaze, L., & Baloga, S.M. 2011. Constraints on determining the composition and eruption style of terrestrial lavas from space. *Geology,* 39, 1127–1130.

Wright, R., Nunes, M.A., Lucey, P.G., Gunapala, S., Rafol, S., Ting, D., Ferrari-Wong, C., Flynn, L.P., & George, T. 2023. The HyTI mission. *Proceeding of SPIE 12729, Sensors, Systems, and Next-Generation Satellites XXVII,* 1272906. https://doi.org/10.1117/12.2679541

Xiao, R., Yu, C., Li, Z., & He, X. 2021. Statistical assessment metrics for InSAR atmospheric correction: Applications to generic atmospheric correction online service for InSAR (GACOS) in Eastern China. *International Journal of Applied Earth Observation and Geoinformation,* 96, 102289. https://doi.org/10.1016/j.jag.2020.102289.

Yang, K., Dickerson, R.R., Carn, S.A., Ge, C., & Wang, J. 2013. First observations of SO_2 from the satellite Suomi NPP OMPS: Widespread air pollution events over China. *Geophysical Research Letters,* 40, 4957–4962. http://doi.org/10.1002/grl.50952.

Yang, K., Krotkov, N.A., Krueger, A.J., Carn, S.A., Bhartua, P.K., & Levelt, P.F. 2007. Retrieval of large volcanic SO_2 columns from the Aura Ozone Monitoring Instrument: Comparison and limitations. *Journal of Geophysical Research,* 112. http://doi.org/10.1029/2007JD008825.

Yu, C., Li, Z., & Penna, N.T. 2018. Interferometric synthetic aperture radar atmospheric correction using a GPS-based iterative tropospheric decomposition model. *Remote Sensing of Environment,* 204, 109–121. https://doi.org/10.1016/j.rse.2017.10.038.

Yu, H., Lan, Y., Yuan, Z., Xu, J., & Lee, H. 2019. Phase unwrapping in InSAR: A review. *IEEE Geoscience and Remote Sensing Magazine,* 7, 40–58. http://doi.org/10.1109/MGRS.2018.2873644

Zebker, H.A., Amelung, F., & Jonsson, S. 2001. Remote sensing of volcano surface and internal processes using radar interferometry. In: J.P. Mouginis-Mark et al. (eds) *Remote Sensing of Active Volcanism.* Washington, DC: AGU Geophysical Monograph Series, pp. 179–205.

Zebker, H.A., & Lu, Y., 1998. Phase unwrapping algorithms for radar interferometry: Residue cut, least squares and synthesis algorithms. *Journal of Optical Applications,* 15, 586–598.

Zebker, H.A., Rosen, P.A., Goldstein, R.M., Gabriel, A., & Werner, C.L. 1994. On the derivation of coseismic displacement fields using differential radar interferometry: The Landers earthquake. *Journal of Geophysical Research,* 99, 19617–19634.

Zebker H.A., Rosen P.A., & Hensley S. 1997. Atmospheric effects in interferometric synthetic aperture radar surface deformation and topographic maps. *Journal of Geophysical Research,* 102, 7547–7563.

Zebker, H.A., & Villasenor, J.A. 1992. Decorrelation in interferometric radar echoes. *IEEE Transactions on Geoscience and Remote Sensing,* 30, 950–959.

Part IV

Fires

9 Satellite-Derived Nitrogen Dioxide Variations from Biomass Burning in a Subtropical Evergreen Forest, Northeast India

Krishna Prasad Vadrevu and Kristofer Lasko

ACRONYMS AND DEFINITIONS

AMF	Air mass factor
AOD	Aerosol optical depth
CV	Coefficient of variation
DJF	December, January, and February
DOAS	Differential optical absorption spectroscopy
DW	Durbin–Watson
FRP	Fire radiative power
GHG	Greenhouse gas
GOME	Global ozone monitoring experiment
OLS	Ordinary least square
OMI	Ozone monitoring instrument
SCIAMACHY	Scanning imaging absorption spectrometer for atmospheric cartography
TEMIS	Tropospheric emission monitoring internet service
VOC	Volatile organic compounds

9.1 INTRODUCTION

Biomass burning is an important source of greenhouse gas emissions and aerosols, including carbon dioxide (CO_2), methane (CH_4), carbon monoxide (CO), nitrogen oxides (NOx), ammonia (NH_3), volatile organic compounds (VOC) (Liu et al., 2024; Shi et al., 2020; Zhou et al., 2017; Andreae and Merlet, 2001). Global annual area burned for the years 1997 through 2011 vary from 301 to 377 Mha, with an average of 348 Mha (Giglio et al., 2013). Of the different regions, tropical Asia is considered a major source of biomass burning (Zhao et al., 2023; Li et al., 2019; Bray et al., 2018; Vadrevu and Justice, 2011). Important sources of biomass burning emissions in tropical Asia include deforestation (Li et al., 2020; Zhang et al., 2020; Van Der Werf et al., 2008), slash-and-burn agriculture (Prasad et al., 2000; Langner et al., 2007), agricultural residue burning (Cheewaphongphan and Garivait, 2013; Badarinath et al., 2009; Vadrevu et al., 2011, 2012), management fires (Murdiyarso and Level, 2007), peat land burning (Heil et al., 2007), and more. Present estimates suggest that globally wildfires contribute about 20% of the fossil fuel carbon emissions to the atmosphere and global fire emissions averaged over 1997–2009 amount to 2.0 Pg C/yr (Curado et al., 2024; Zhao et al., 2023; Andreae, 2019; Yin et al., 2019a; van der Werf et al.,

2010). It is estimated that carbon monoxide (CO) and nitrogen dioxide (NO_2) emissions from fires comprise approximately 30% and 15% of global total direct emissions, respectively (Schill et al., 2020; Arellano et al., 2006; Müller and Stavrakou, 2005; Jaeglé et al., 2005). Enhanced CO and NO_2 concentrations can impact tropospheric ozone formation and affect the oxidizing capacity of the atmosphere by regulating the OH lifetime (Logan et al., 1981). Aerosols released from the biomass burning can be elevated by mid-latitude wave cyclones and sometimes can travel long distances to possibly influence climate and weather patterns. Specific to climate impacts, Wang et al. (2014) has shown that Asian pollution invigorates winter cyclones over the Pacific Northwest, increasing precipitation by 7% and net cloud radiative forcing by 1.0 W m^{-2} at the top of the atmosphere and by 1.7 W m^{-2} at the Earth's surface. No single system can provide all the necessary data and to address air quality and climate impacts of GHGs; several studies infer the need to integrate both top-down and bottom-up approaches, including modeling (Andreae, 2019; Mota and Wooster, 2018; Martin et al., 2002).

The science of Earth observation, more specifically, remote detection of tropospheric gases using satellite instruments has significantly improved over the last 20 years (Burrows et al., 2011). Several greenhouse gas (GHG) concentrations, such as O_3, CO, CO_2, CH_4, HCHO, NO_2, SO_2, and BrO can be measured from nadir-looking sensors that record these gases in the lower troposphere (< 6 km) (Huang et al., 2024; Schill et al., 2020; Shi et al., 2020; Bray et al., 2018; Zhou et al., 2017; Boersma et al., 2004). Important sources of these GHGs include both fossil fuel combustion (automobile combustion, industrial production, and heating) and natural sources (terrestrial vegetation, soils, lightning, wetlands, biomass burning). Of the different natural sources, biomass burning that is prevalent in tropical regions contributes significantly to GHG emissions, including NO_2. The exposure to NO_2 pollutants can cause or worsen respiratory disease, such as emphysema and bronchitis (Schwartz, 2004).

NO_2 is an important trace gas in both the troposphere and stratosphere that exhibits high atmospheric variability. In the troposphere, NO_2 is a precursor for ozone formation. The photolysis of NO_2 in the presence of strong solar radiation releases atomic oxygen, which then combines with molecular oxygen to form ozone (Logan, 1983). In an unpolluted atmosphere, the natural sources of SO_2 and NO_2 provide a mechanism by which the pH of aerosols and rain are expected to be slightly acidic. However, large amounts of NO, NO_2, and SO_2 produced during fossil fuel and biofuel combustion can result in acid rain (Burrows et al., 2011). NO_2 is not only involved in catalytic ozone depletion in the stratosphere but also reacts with halogen oxides to form reservoir substances and thereby reduces the ozone depletion potential of Cl and Br. NO_2 is removed from the atmosphere through chemical conversion to other nitrogen contained species, nitrate aerosols, and uptake by vegetation and soils (Shaw, 1976). It also exhibits a distinct diurnal cycle (Thomas et al., 1998).

While the main sources and source regions of NO_2 are known, large uncertainties remain as to the individual source strengths and their latitudinal and seasonal variation (Richter and Burrows, 2002). NO_2 can be measured either by *in situ* chemical methods, on airborne or balloon platforms, or by remote sensing in the ultra-violet/visible and infrared spectral regions. From satellite instruments, NO_2 can be measured as a column integral from solar backscatter instruments from space, since it absorbs light in the visible portion of the electromagnetic spectrum. Tropospheric NO_2 columns are retrieved from the total columns by subtracting the stratospheric contribution, assuming zonal invariance of the stratospheric contribution. Because the satellite signal is closely related to the total amount of NO_2, there is a more direct link with area-averaged concentrations than with surface *in situ* observations, which depend strongly on local sources and local removal processes (Burrows et al., 2011). Tropospheric NO_2 columns retrieved from satellite measurements, e.g., by the Global Ozone Monitoring Experiment (GOME), GOME-II, Scanning Imaging Absorption Spectrometer for Atmospheric Cartography (SCIAMACHY), and Ozone Monitoring Instrument (OMI) have contributed to mapping spatiotemporal variations in NO_2 sources (e.g., Liu et al., 2024;

Ceamanos et al., 2023; Schill et al., 2020; Li et al., 2019; Andreae, 2019; Bray et al., 2018; Zhou et al., 2017; Burrows et al., 1999; Richter and Burrows, 2002; Martin et al., 2006; Van der A et al., 2006; Boersma et al., 2007; Stavrakou et al., 2008; Kurokawa et al., 2009; Zhao and Wang, 2009; Lin et al., 2010; Russell et al., 2011; Miyazaki et al., 2012). For example, Richter and Burrows (2002), using GOME measurements, observed enhanced tropospheric NO_2 from biomass burning above Africa during the fall of 1997. Similarly, Thomas et al. (1998) recorded a twofold increase in the vertical NO_2 content over large parts of the smoke cloud formed from rainforest biomass burning episodes in Java and Borneo during September 1997. Using the combined GOME and SCIAMACHY retrievals, van der A et al. (2006, 2008) mapped a significant increase in the NO_2 concentrations over China from 1996 to 2006 and attributed these emissions to the rapidly growing economy and the associated increase of fossil fuel consumption, industries, and traffic. Other examples of using satellite-derived NO_2 products for characterizing anthropogenic emissions can be found in several papers (Zhao et al., 2023; Pan et al., 2020; Andreae, 2019; Bray et al., 2018; Zhou et al., 2017; Beirle et al., 2003; Bertram et al., 2005; Jaeglé et al., 2005; Kim et al., 2006; Martin et al., 2006; Boersma et al., 2007; Hudman et al., 2007; Ghude et al., 2008; Mijling et al., 2009; Bucsela et al., 2010; Lin et al., 2010; Valin et al., 2011; Mebust et al., 2011; Zhou et al., 2012; David and Nair, 2013).

In contrast to these studies conducted all over the globe, evaluation of OMI and SCIAMACHY-NO_2 signals in relation to evergreen forest fires in India have not yet been attempted. Assessment of satellite observations of tropospheric NO_2 columns is needed over a range of environments to improve the validation efforts (Lamsal et al., 2010). In this study, the following questions relating to fire-NO_2 concentrations were addressed: (1) How does biomass burning due to slash-and-burn agriculture impact NO_2 concentrations for sub-tropical evergreen forests? (2) How are the MODIS-retrieved fire counts and fire radiative power (FRP) products related to NO_2 concentrations in evergreen forest burning? (3) Which satellite product, i.e., SCIAMACHY-NO_2 or OMI-NO_2, best correlates with the MODIS fire products? (4) What is the correlation strength between OMI and SCIAMACHY-NO_2 products in a biomass burning landscape? (5) How do these products correlate with MODIS-retrieved aerosol optical depth (AOD)? (6) How much variance in NO_2 concentrations can be accounted by the MODIS Terra and Aqua fire counts and the FRP together?

We answered these questions using MODIS active fires, SCIAMACHY-NO_2 (2003–2011), OMI-NO_2 (2005–2011), and MODIS Aqua aerosol optical depth (AOD) (2003–2011) datasets. In addition to the questions, we hypothesized that the correlation strength between the FRP and the NO_2 concentrations will be much stronger than the correlation strength between the number of fire counts and NO_2 concentrations because the FRP has been previously associated with the strength of the fires (Wooster et al., 2005). From these questions and hypothesis testing, the results from this study are expected to provide robust information on fire-NO_2 relationships in subtropical evergreen forests.

9.2 STUDY AREA

To test the fire-NO_2 relationships, we selected the subtropical evergreen forests of northeast India, where biomass burning is prevalent due to slash-and-burn agriculture. Northeast India refers to the easternmost region of India consisting of the contiguous seven states of Arunachal Pradesh, Assam, Manipur, Meghalaya, Mizoram, Nagaland, and Tripura, occupying ~255,083 km² (see boxed area in Figure 9.1a). Seventy percent of the region is occupied by hills. Nearly 400,000 families belonging to 100 different indigenous tribes practice slash-and-burn agriculture, locally called *jhum*. Nearly 3863 km² is affected by slash-and-burn cultivation annually (Majumder et al., 2011). *Jhum* cultivation involves the clearing of forest vegetation and burning (March–May) just before the monsoon (June), followed by mixed cropping on steep slopes of 30–40°. In the *jhum* plots, varieties of crops are grown, including cereals (*Oryza sativa, Zea mays.*), tuberous crops (*Manihot esculenta, Dioscorea* sp.), vegetable crops (*Cucurbita moschata, Solanum melongena*, etc.), and spices

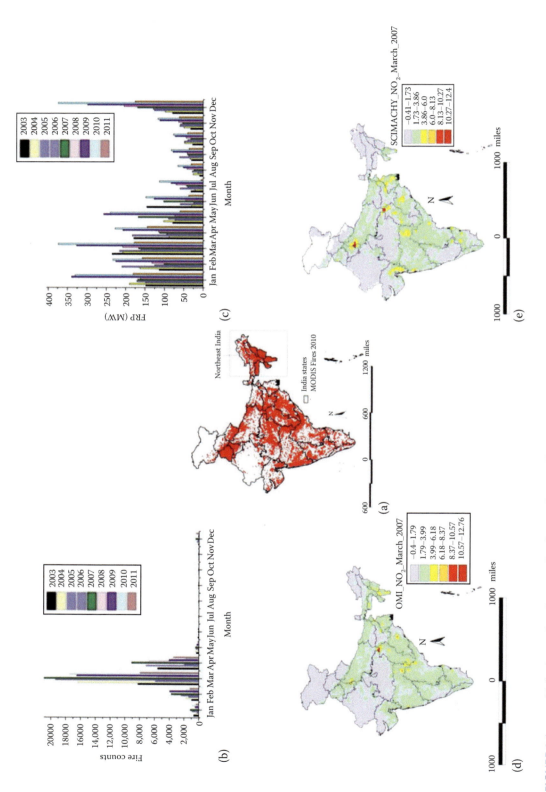

FIGURE 9.1 a–e (a) Study area location map showing northeast India (in box) with MODIS active fires; (b) Fire counts for different months and years in the northeastern states; (c) FRP (MW) for different months and years in the northeast; (d) OMI-NO_2 (x-axis values in 1015molecules cm^{-2}) for the peak month of March (2007); (e) SCIAMACHY-NO_2 (x-axis values in 1015 molecules cm^{-2}) for the peak month of March (2007), northeast India.

(*Zingiber officinale*, *Capsicum* sp. etc.). Due to an increasing population, overexploitation of forest resources and loss of soil fertility, the *jhum* cycle has reduced to 3–5 years from the more traditional 20–30 years on land that had already been occupied for slash-and-burn agriculture (Majumder et al., 2011). The forests are mainly subtropical evergreen type, interspersed by bamboo forest. Subtropical temperate climax forests of mixed broad-leaved forest with early successional species of pure pine or mixed pine were also reported from this region.

9.3 DATASETS

9.3.1 ACTIVE FIRES AND FIRE RADIATIVE POWER (FRP)

We aggregated the monthly active fires and FRP data from MODIS for the northeastern states of India. The two MODIS sun-synchronous, polar-orbiting satellites pass over the equator at approximately 10:30 a.m. (Terra) and 1:30 p.m. (Aqua) and they have a revisit time of one to two days. The data collected by the sensor is processed by the MODIS Advanced Processing System (MODAPS), using the enhanced contextual fire detection algorithm (Giglio et al., 2003) into the Collection 5 Active Fire product. For this study, we analyzed the data from 2003 to 2011. The fire data are at 1 km nominal spatial resolution at nadir; however, under ideal conditions flaming fires as small as 50m² can be detected. FRP is the rate of fire energy released per unit time, measured in megawatts (Kaufman et al., 1998). The MODIS algorithm for FRP is calculated as the relationship between the brightness temperature of fire and background pixels in the middle infrared (band center near 4 μm) (Pan et al., 2020; Pistone et al., 2019; Bray et al., 2018; Chen et al., 2017; Zhou et al., 2017). It is given as (Kaufman et al., 1998),

$$FRP = 4.34 \times 10^{-19}(T_{MIR}^8 - T_{bgM}^8) \quad (9.1)$$

where FRP is the rate of radiative energy emitted per pixel, 4.34 X 10–19 (MW km^{-2} K^8) is the constant derived from the simulations, T_{MIR} (Kelvin) is the radiative brightness temperature of the fire component, T_{bgMIR} (Kelvin) is the neighboring non-fire background component, and MIR refers to the middle infrared wavelength here, 3.96μm. In this study, we utilized the Collection 5 Terra and Aqua monthly climate modeling grid datasets (MOD14CMH/MYD14CMH) that represent cloud- and overpass-corrected fire pixels data along with the mean FRP data.

9.3.2 OMI-NO$_2$

OMI is one of four instruments on board NASA's EOS-Aura satellite, launched on July 15, 2004. OMI is a nadir-viewing imaging spectrometer (Levelt et al., 2006; Boersma et al., 2007) (Table 9.1). Aura traces a sun-synchronous, polar orbit with a period of 100 min and has a local equator crossing time of about 13:45 p.m. OMI provides measurements of both direct and atmospheric backscattered sunlight in the ultraviolet-visible range from 2.7 to 5μm useful to retrieve tropospheric NO$_2$ columns (Miyazaki et al., 2012). OMI pixels are 13 × 24 km at nadir, increasing in size to 24 × 135 km for the largest viewing angles. The instrument achieves near-daily coverage, and the retrievals are sufficient to extract global NO$_2$ concentrations on a daily basis. These capabilities of OMI make it unique as compared to GOME and SCIAMACHY retrievals, which have relatively lower spatial and temporal resolutions and less frequent global coverage (Liu et al., 2024; Li et al., 2020; Zhang et al., 2020; Bray et al., 2018).

In this study, we specifically used the Dutch OMI-NO$_2$ (DOMINO) product (version 2.0) (Boersma et al., 2011). The DOMINO is a post-processing data set based on the most complete set of OMI orbits, improved level-1b (ir)radiance data (collection 3), analyzed meteorological fields, and

TABLE 9.1

SCIAMACHY and OMI Instrument Characteristics

Instrument	Satellite	Nadir view	Global coverage	Wavebands	Spectral resolution (at 440nm)
SCIAMACHY	Envisat	30 × 60 km²	6 days	UV-SWIR: 240–314, 309–3405, 394–620, 604–805, 785–1050, 1000–1750, 1940–2040 and 2265–2380nm	0.44nm
OMI	EOS-Aura	13 × 24 km²	1 day	270-500nm	0.63nm

actual spacecraft data. The better data coverage, the improved calibration of level-1b data, and the use of analyzed rather than forecast data make the DOMINO product superior to the near real-time NO_2 data (Boersma et al., 2011). The Differential Optical Absorption Spectroscopy (DOAS) spectral analysis technique is used to determine NO_2 slant column densities, and then the stratospheric portion of the column is subtracted to yield a tropospheric slant column. The multiplicative air mass factor (AMF), is then used to convert the slant column to a vertical column based on the output from a radiative transfer model that accounts for terrain, profile, cloud, and viewing parameters. The DOMINO data contain geolocated column-integrated NO_2 concentrations, or NO_2 columns (in units of molecules cm^{-2}). DOMINO data constitute a pure level 2 product, i.e., it provides geophysical information for every ground pixel observed by the instrument, without the additional binning, averaging, or gridding typically applied for level 3 data. In addition to vertical NO_2 columns, the product contains intermediate results, such as the result of the spectral fit, fitting diagnostics, assimilated stratospheric NO_2 columns, the averaging kernel, cloud information, and error estimates. To reduce the errors resulting from cloud cover, we excluded those pixels with an effective cloud fraction exceeding 20%. Further, for the study area, there was no contamination due to row anomalies and pixels at the swath edges.

9.3.3 SCIAMACHY-NO$_2$

The SCIAMACHY instrument on board ENVISAT is an eight-channel UV/visible/NIR grating spectrometer covering the wavelength region of 220–2400 nm with a 0.2–1.5 nm spectral resolution 10:00 a.m. local time equator crossing, and a global coverage of every six days (Table 9.1). It measures trace gas constituents in nadir, limb, and occultation configuration. The UV/VIS nadir measurements of SCIAMACHY are very similar to those performed by GOME, the main difference being the better spatial resolution (30 × 30 km² to 30 × 240 km²) as compared to 40 × 320 km² for GOME. Similar to OMI data, we used the data from Tropospheric Emission Monitoring Internet Service (TEMIS). These datasets have been validated against *in situ* and aircraft measurements, and compared with regional air-quality models (e.g., Liu et al., 2024; Pan et al., 2020; Zhang et al., 2020; Yin et al., 2019a; Yin et al., 2019b; Chen et al., 2017; Schaub et al., 2006; Blond et al., 2007). As discussed by Boersma et al. (2007) and Lin et al. (2010), systematic errors in OMI (DOMINO v1) and SCIAMACHY retrievals are expected to correlate well with each other, since these retrievals are derived with a very similar algorithm.

9.3.4 MODIS Aerosol Optical Depth (AOD) Variations

We used the MODIS Collection 5.1 (MYD08_M3.051) AOD at 550 nm (Remer et al., 2005; Levy et al., 2007) level 3 monthly product for characterizing the fire-AOD variations from 2003 to 2011. The aerosol properties are derived from the inversion of the MODIS-observed reflectance using pre-computed radiative transfer look-up tables based on aerosol models (Curado et al., 2024; Zhao et al., 2023; Li et al., 2020; Li et al., 2019; Bray et al., 2018; Mota and Wooster, 2018; Wu et al., 2018; Chen et al., 2017; Zhou et al., 2017; Remer et al., 2005; Levy et al., 2007).

9.4 METHODS

9.4.1 Descriptive Statistics

Mean NO_2 concentrations from OMI (2005–2011), SCIAMACHY (2003–2011), and corresponding active fire numbers and FRP (MW) values from the MODIS datasets were extracted for the northeast India states. Both the inter-annual and the seasonal variations winter (January–February), summer or pre-monsoon (March–May), rainy season (June–September), post-monsoon season (October–December) variations in NO_2 were analyzed. We also extracted MODIS AOD values corresponding to NO_2 and fires. A yearly coefficient of variation (CV) in NO_2 was calculated to determine the variability and amplitude based on the temporal data. Time-series datasets of active fire numbers, FRP, NO_2, and AOD were plotted to assess fire and FRP-NO_2-AOD signal variations. Pearson correlation coefficients (two-tailed test of significance 0.05 level) were computed among the datasets. In addition to reporting descriptive statistics of fire counts, FRP, OMI-NO_2, SCIMACHY NO_2, and MODIS AOD values, we also used cumulative relative frequency plots to assess the NO_2 concentrations from OMI and SCIAMACHY.

9.4.2 Time-Series Regression

We used the time-series regression to assess the combined contribution of fire counts and FRP affecting the NO_2 concentrations from OMI and SCIAMACHY independently. We used fire counts and FRP as predictor variables of NO_2. When using the time-series data, Ordinary Least Square (OLS) estimates may not be reliable. A certain amount of smoothing is introduced in the time-series data, by averaging the data over months (or months from days, or quarters to years). Thus, some of the randomness inherent in the data is lost. The smoothing of data can lead to systematic patterns in the error terms, thus leading to the possibility of autocorrelation. As a result, the estimated variances of the OLS estimators are biased and tend to underestimate the true variances and standard errors which may inflate the "t" values, thus potentially leading to erroneous conclusions. As a result, the usual F and t tests are not reliable, including the estimated R^2. To account for these errors, Prais–Winsten regression was used, which utilizes the generalized least-squares method to estimate the parameters in a linear regression model in which the errors are serially correlated. The errors are assumed to follow a first-order autoregressive process. Once a model is estimated from the time-series data, the Durbin–Watson (DW) statistic (also known as the d-statistic) is calculated for diagnosing whether the residuals are serially correlated (Durbin and Winsten, 1950). DW statistic is given as,

$$DW = \sum_{i=1}^{n}(e_i - e_{i-1})^2 / \sum_{i=1}^{n} e_i^2 \qquad (9.2)$$

Residual (e_i) is the difference between the observed value and the predicted value at a certain level of X:

$$e_i = y_i - \hat{y}_i \quad (9.3)$$

The DW statistic lies in the 0–4 range, with a value near two indicating no first-order autocorrelation. Positive serial correlation is associated with DW values below two and negative serial correlation with DW values above two. DW values close to two are desirable, suggesting no autocorrelation errors.

9.5 RESULTS

9.5.1 FIRES IN NORTHEAST INDIA

Active fires retrieved from the MODIS datasets for the extent of India are shown in Figure 9.1(a). The aggregated yearly and monthly MODIS fire counts and FRP for northeast India from 2003 to 2011 are shown in Figures 9.1(b) and 9.1(c). MODIS recorded 21,417 fire counts per year from northeast India, which mostly correspond to evergreen forest fires from slash-and-burn agriculture. Of the different years, 2009 recorded the highest number of fire counts, followed by 2006 and 2010. Analysis of fire counts for monthly variations (Figure 9.1b) suggested that March had the highest fire counts, with 63% of all fires occurring during that month followed by April (21%), February (9.15%), etc.

Monthly fire counts as well as the mean FRP (MW) averaged over a nine-year time period are shown in Figures 9.2(a,b). Results suggested that fire counts in the early afternoon from Aqua were six times higher than the morning Terra detections. The higher fire detections from Aqua are justified as most of the clear felled slash is left until late afternoon for drying and then burnt by the locals (Majumder et al., 2011). The averaged FRP for Terra was 12.5–108.7 MW, whereas Aqua 12.6–104.8 MW, with the peak during March (91.02 MW averaged for Aqua and Terra) (Figure 9.2b).

9.5.2 NO_2 TEMPORAL AND SEASONAL VARIATIONS

Monthly and seasonal variations in the mean columnar NO_2 concentrations obtained by averaging OMI (2005–2011) and SCIAMACHY (2003–2011) data are shown in Figure 9.3(a, b) and the

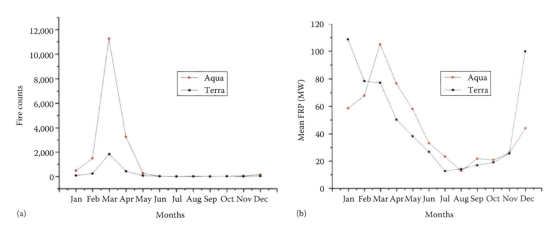

FIGURE 9.2A-B Nine-year data has been averaged to arrive at the monthly fire counts and FRP (MW). (a) Aqua and Terra fire counts for northeast India; (b) Mean FRP for northeast India.

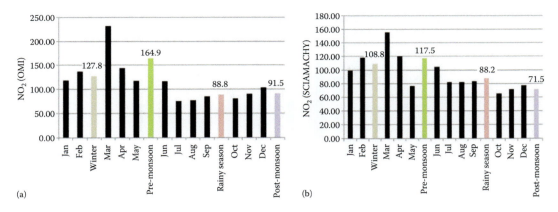

FIGURE 9.3 (a) Seasonal variations in OMI-NO$_2$ (y-axis values in 10^{13} molecules cm^{-2}) and (b) SCIAMACHY-NO$_2$ (y-axis values in 10^{13} molecules cm^{-2}) in northeast India.

peak concentration of NO$_2$ spatial patterns during March in Figure 9.1(d, e). The temporal mean of OMI data suggested a concentration of 1.4 × 10^{15} molecules cm^{-2} with 1.59 × 10^{15} molecules cm^{-2} (+1σ) and 0.70 × 10^{15} molecules cm^{-2} (-1 σ). Relatively higher concentrations of NO$_2$ were observed during March (2.3 × 10^{15} molecules cm^{-2}), April (1.4 × 10^{15} molecules cm^{-2}), February (1.18 × 10^{15} molecules cm^{-2}), and the lowest concentration was observed during July (0.75 × 10^{15} molecules cm^{-2}). From the SCIAMACHY data, the nine-year temporal mean suggested an NO$_2$ concentration of 0.95 × 10^{15} molecules cm^{-2}, with 1.27 × 10^{15} molecules cm^{-2} (+1σ) and 0.63 × 10^{15} molecules cm^{-2} (-1σ). Although with lower concentrations, SCIAMACHY-NO$_2$ also exhibited a similar trend with the highest NO$_2$ values during March (1.5 × 10^{15} molecules cm^{-2}), April (1.20 × 10^{15} molecules cm^{-2}), and February (1.18 × 10^{15} molecules cm^{-2}), however, with the least NO$_2$ during October (0.65 × 10^{15} molecules cm^{-2}). Evaluation of the seasonal patterns in OMI-NO$_2$ concentrations showed the highest concentrations during pre-monsoon, followed by winter, post-monsoon, and the lowest concentration during the rainy season (Figure 9.3a). Similar to OMI-NO$_2$, SCIAMACHY-NO$_2$ showed higher concentrations during pre-monsoon followed by winter; however, the rainy season concentrations were relatively higher than that of the post-monsoon (Figure 9.3b). Analysis for inter-annual variations in NO$_2$ (Figure 9.4) suggested that OMI showed relatively high variations during 2006, 2007, 2009, and 2010. The empirical cumulative distribution function plots with the cumulative relative frequency for OMI-NO$_2$ and SCIAMACHY-NO$_2$ data are shown in Figures 9.5(a, b). The OMI-NO$_2$ data had the maximum of (3.0 × 10^{15} molecules cm^{-2}), mean of (1.1 × 10^{15} molecules cm^{-2}), and median of (1.05 × 10^{15} molecules cm^{-2}) compared to the SCIAMACHY maximum of (1.75 × 10^{15} molecules cm^{-2}), mean of (0.95 × 10^{15} molecules cm^{-2}), and median of (0.9 × 10^{15} molecules cm^{-2}) values. The other statistics are shown in the Quantile plots, where the 90th percentile value for the OMI-NO$_2$ data was 1.63 × 10^{15} molecules cm^{-2} in contrast to the SCIAMACHY data with 1.41 × 10^{15} molecules cm^{-2}. These plots (Figures 9.5a, b) clearly suggest relatively higher concentrations of NO$_2$ captured by OMI than by the SCIAMACHY data.

9.5.3 Correlations and Time-Series Regression

A clear increase in NO$_2$ signal corresponding with the fire counts during the peak biomass burning month of March can be seen in the OMI-NO$_2$ time series data (Figure 9.6a), and with a lesser correspondence in the SCIAMACHY-NO$_2$ data (Figure 9.6b). OMI-NO$_2$ coincides more strongly with

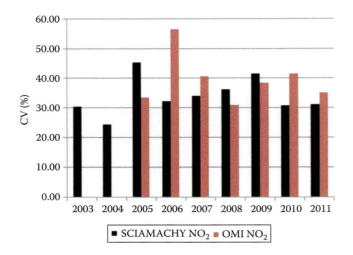

FIGURE 9.4 Coefficient of variation (CV) % in the OMI and SCIAMACHY NO$_2$ retrievals.

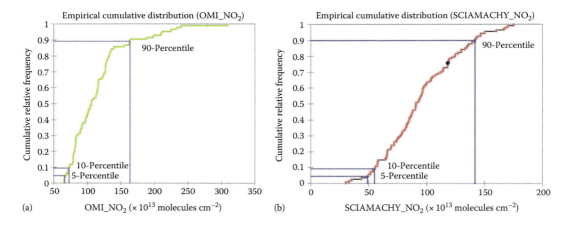

FIGURE 9.5 (a,b) Empirical cumulative distribution plots for the OMI-NO$_2$ and the SCIAMACHY-NO$_2$ obtained from long-term datasets. For the OMI NO$_2$ the value of the 5th percentile was 66.4 ($\times$ 10^{13} molecules cm^{-2}), 10th percentile was 72.54 ($\times$ 10^{13} molecules cm^{-2}), and 90th percentile was 163.6 ($\times$ 10^{13} molecules cm^{-2}); for SCIAMACHY NO$_2$ the value of the 5th percentile was 48.96 ($\times$ 10^{13} molecules cm^{-2}), 10th percentile was 55.05 ($\times$ 10^{13} molecules cm^{-2}), and 90th percentile was 141.5 ($\times$ 10^{13} molecules cm^{-2}). Clearly, the OMI-NO$_2$ showed relatively higher concentrations (90th percentile) than the SCIAMACHY-NO$_2$.

the AOD signal than SCIAMACHY-NO$_2$ (Figures 9.6c,d). Scatter plots with correlations for different datasets are shown in Figure 9.7(a–i). The results suggested a stronger correlation of fire counts with OMI - NO$_2$ ($R^2 = 0.74$) than fire counts with SCIAMACHY - NO$_2$ ($R^2 = 0.35$) (Figures 9.7a,b). In addition, the FRP and NO$_2$ correlation was weak compared to the sum of fire counts and NO$_2$ (Figures 9.7c,d). OMI-NO$_2$ showed a stronger correlation coefficient with MODIS-AOD ($R^2 = 0.54$) than SCIAMACHY - NO$_2$ and MODIS - AOD ($R^2 = 0.36$) (Figures 9.7e,f and 9.6e). The results also suggested a relatively higher correlation of the sum of fire counts with AOD ($R^2 = 0.40$) than FRP with AOD (0.21) (Figures 9.7g,h). OMI-NO$_2$ and SCIAMACHY-NO$_2$ showed correlation strength of 57% (Figures 9.7i and 9.6f).

We used a time-series regression to address the total variance in the NO_2 observations contributed by the fire counts and FRP. Results using the Prais–Winsten time-series regression for OMI-NO_2 and SCIAMACHY-NO_2 data are shown in Tables 9.2 and 9.3. For the Prais–Winsten regression, the regression coefficients, residual sum of squares, F-test, R^2, adjusted R^2, residual standard errors, t-statistic, p-values, 95% confidence intervals, "rho" and the DW statistic were all reported. The adjusted R^2 value is useful for comparing the explanatory power of models with different numbers of predictors. A model with more terms may appear to have a better fit simply because it has more terms, thus adjustment is needed to account for all the predictors. As depicted in the adjusted R^2 values, fire counts and FRP together explained 78% of NO_2 variance in the OMI data compared to 33% in the SCIAMACHY data. The t-statistic and its corresponding p-value determine whether a regression coefficient is significantly different from zero. In our case, the p-values for the estimated coefficients for the sum of fire counts and mean FRP are both 0.000 at alpha = 0.05, indicating that they are significantly related to NO_2 concentrations in the OMI data. In the SCIAMACHY-NO_2 data, FRP seems to be less significant than the fire counts (Table 9.3). The standard errors reflect the variability and are a measure of the precision with which the regression coefficients are measured. In both the OMI-NO_2 and SCIAMACHY-NO_2 models, the standard errors seem to show high precision. An F-test assesses how well the set of independent variables, as a group, explains the variation in the dependent variable. The significant F-value (Tables 9.2 and 9.3) suggests that the calculation of R^2 in the model was best fit. The "rho" parameter in the model indicates the level of autocorrelation and in the Prais–Winsten regression, it is calculated iteratively. The "rho" parameters converged at 0.24 in the OMI-NO_2 model and 0.15 in the SCIAMACHY-NO_2 model, indicating that both of these datasets were more stationary time series having a clear structure and were not too far from the mean. Finally, the DW statistic clearly indicated that the Prais–Winsten transformation was helpful with the values close to 2 in both the models, indicating that the error term is serially independent, thus no autocorrelation.

TABLE 9.2

Time Series Prais–Winsten Regression of Fire Counts and FRP with OMI NO_2

Omi_NO2	Coef	Std. Err	t	P> |t|
Sum_FC	.0081107	.0005831	13.91	0.000
Sum_FRP	.2182609	.0435501	5.01	0.000
cons	81.55099	4.757556	17.14	0.000

R-square =0.79; Adjusted R-square =0.78

TABLE 9.3

Time Series Prais–Winsten Regression of Fire Counts and FRP with SCIAMACHY NO_2

sciamachy_~2	Coef	Std. Err	t	P> |t|
Sum_FC	.0038616	.0007331	5.27	0.000
Sum_FRP	.2318535	.0969293	2.39	0.019
cons	78.53321	5.036624	15.59	0.000

R-square = 0.34; Adjusted R-square = 0.33

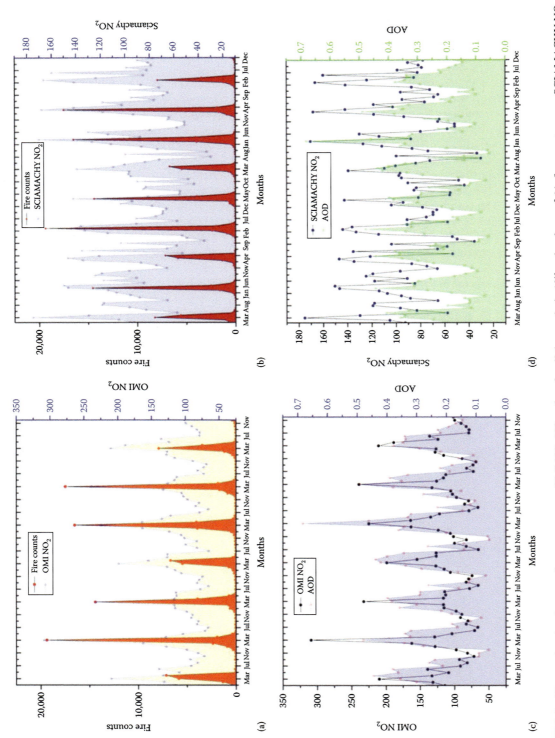

FIGURE 9.6 Time series plots between fire counts versus OMI-NO$_2$, NO$_2$ values for all the graphs in 10^{13} molecules cm^{-2} (a); fire counts versus SCIAMACHY-NO$_2$ (b); OMI-NO$_2$ and AOD (c); SCIAMACHY-NO$_2$ and AOD (d); fire counts and AOD (e); OMI-NO$_2$ and SCIAMACHY-NO$_2$ (f).

Satellite-Derived Nitrogen Dioxide Variations

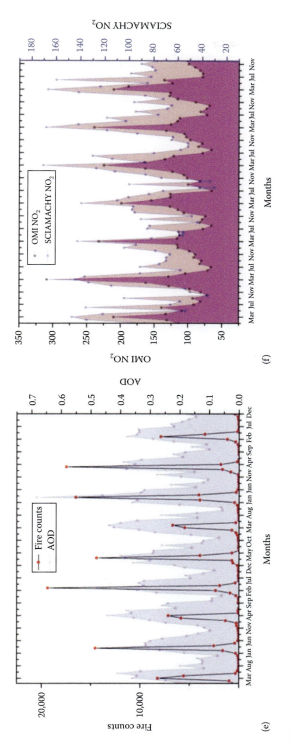

FIGURE 9.6 (Continued)

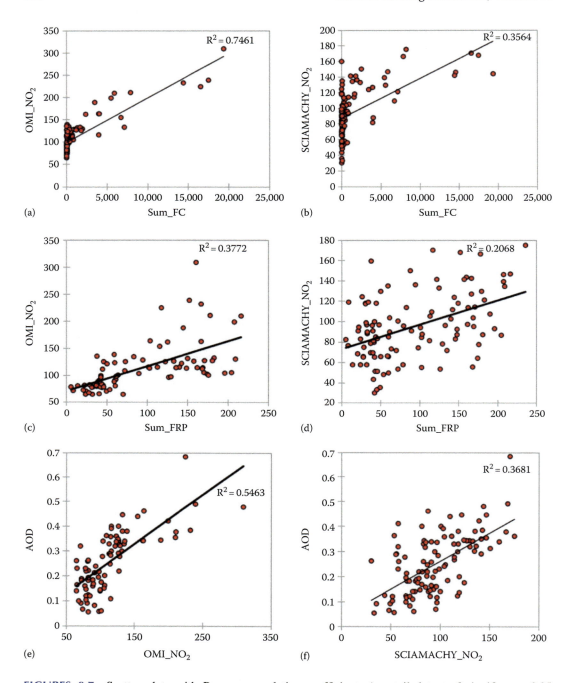

FIGURES 9.7 Scatter plots with Pearson correlation coefficients (two-tailed test of significance 0.05 level). (a) Sum of fire counts (FC) versus OMI-NO$_2$ values in 10^{13} molecules cm^{-2}; (b) sum of FC versus SCIAMACHY-NO$_2$ values in 10^{13} molecules cm^{-2}; (c) sum of fire radiative power (FRP in MW) versus OMI-NO$_2$ values in 10^{13} molecules cm^{-2}; (d) sum of FRP (MW) versus SCIAMACHY-NO$_2$ values in 10^{13} molecules cm^{-2}; (e) OMI-NO$_2$ (values in 10^{13} molecules cm^{-2}) versus aerosol optical depth (AOD); (f) SCIAMACHY-NO$_2$ (values in 10^{13} molecules cm^{-2}) versus aerosol optical depth (AOD); (g) sum of FC versus AOD; (h) sum of FRP versus AOD; (i) OMI-NO$_2$ versus SCIAMACHY NO$_2$ (values x10^{13} molecules cm^{-2}).

Satellite-Derived Nitrogen Dioxide Variations

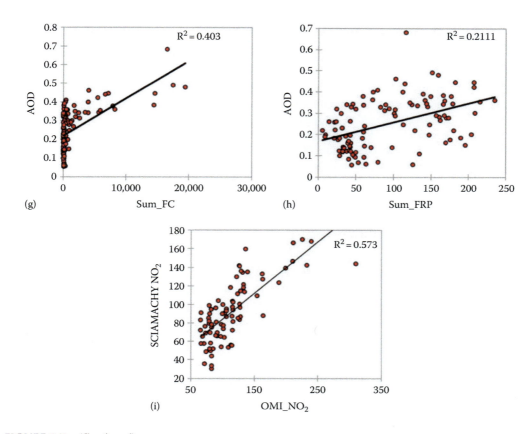

FIGURE 9.7 (Continued)

9.6 DISCUSSION

Accurate data on biomass burning emissions can help to assess land-atmosphere interactions effectively. In tropical regions, uncertainties still exist for biomass burning emissions, and thus there is a need to evaluate different approaches and datasets that are useful for quantifying the emissions. Inter-comparison of OMI and SCIAMACHY-NO$_2$ in relation to fires clearly suggested stronger correlations with the OMI than SCIAMACHY. Comparison of seasonal trends clearly showed that the NO$_2$ patterns from both OMI as well as SCIAMACHY matched well with the peak biomass burning episodes (March, April, and February). However, OMI-NO$_2$ captured 1.5 times more NO$_2$ concentrations during the peak month of March than the SCIAMACHY-NO$_2$ as observed from the time-series data (Liu et al., 2024; Ceamanos et al., 2023; Shi et al., 2020; Yin et al., 2019b; Chen et al., 2017; Zhou et al., 2017). Earlier, Martin et al. (2002), using the GOME data, reported seasonal enhancements in NO$_2$ during December, January, and February (DJF) over northern Africa; March, April, and May (MAM) in India; and from June to October in central Africa and South America. Further, comparison of NO$_2$ concentrations in the other biomass-burning regions suggests that NO$_2$ concentrations from the evergreen forest fires of Northeast India are relatively lower (1.14 x 10^{15} molecules cm^{-2}) than the other biomass burning regions. For example, Thomas et al. (1998), reported total NO$_2$ concentrations of around 2.5 x 10^{15} (plus or minus 0.8 × 10^{15} mol cm^{-2}) over southern Borneo, Indonesia, mainly due to rainforest burning. Similarly, Jaegle et al. (2005)

reported NO$_2$ values in the range of 2.5–4.0 × 10^{15} molecules cm^{-2} in the South African savanna burning areas using GOME data.

The relatively higher NO$_2$ column retrievals observed by OMI than SCIAMACHY during the evergreen forest fires are attributed to the satellite overpass time, i.e., SCIAMACHY during the morning (10:00 a.m.) and OMI during the afternoon (13:45 p.m.). This inference in our case is also supported by the relatively higher number of fire counts captured by the MODIS AQUA having later afternoon satellite overpass compared to Terra. For example, Aqua-MODIS captured six times more fires compared to Terra-MODIS in the study area. The differences in NO$_2$ from different vegetation types are attributed to the amount of biomass burnt and gas-phase oxidation of NO$_2$ with OH in different environments. In addition, comparison of urban NO$_2$ concentrations elsewhere with the biomass burning NO$_2$ concentrations clearly suggests relatively higher NO$_2$ in urban areas than biomass burning areas. For example, using the GOME NO$_2$ retrievals, Velders et al. (2001) noted the maximum NO$_2$ columns during January for the eastern United States, Western Europe, and Eastern China in the range of 33–43 × 10^{15} molecules cm^{-2} compared to the other months in the range of 14–20 × 10^{15} molecules cm^{-2}. Also, NO$_2$ from evergreen forest fires of northeast India seem to be relatively low compared to other large point sources located in India, such as from thermal power plants, steel plants, cement plants, industrial processes, fossil fuel extractions, etc., which exceed 4.0 × 10^{15} molecules cm^{-2} (Ghude et al., 2008; David and Nair, 2013). These inferences suggest that relative to the biomass burning, NO$_2$ emissions from urban sources should be given priority for mitigation efforts.

Boersma et al. (2007, 2011) noted that SCIAMACHY observes higher NO$_2$ than OMI (up to 40%) in most industrial regions of northern mid-latitudes, while it observes lower NO$_2$ than OMI (up to 35%) in tropical biomass burning regions. Further, Boersma et al. (2007) through the detailed analysis of slant columns and AMF concluded that differences between SCIAMACHY and OMI-NO$_2$ columns in source regions could not be ascribed to a retrieval artifact but to the underlying chemistry governing the NO$_2$ concentrations. The lifetime of NO$_2$ varies from one hour at the surface to days in the upper atmosphere, depending on the season, latitude, altitude, and other local atmospheric properties. The lifetime variations during different seasons reflect gas-phase oxidation of NO$_2$ with OH in the summer with an increasing role for heterogeneous chemistry in the winter (Lamsal et al., 2010; Miyazaki et al., 2012). The seasonal differences highlight the temporal variability of NO$_2$ from different sources and regions.

Although we hypothesized that FRP would be a better predictor of NO$_2$ concentrations, we found that fire counts had a better correlation. We also found that use of fire counts in conjunction with FRP increased the correlation with NO$_2$ concentrations. The use of FRP values for characterizing the fire cycle may be constrained due to various other reasons, such as smoke plumes, MODIS viewing geometry, and FRP sampling (Ichoku et al., 2008; Vadrevu et al., 2011). Considering these discrepancies, we infer that more studies are needed on using FRP as a surrogate measure for quantifying GHG concentrations and emissions. Further, results from the time-series regression clearly suggested combined use of the sum of fire counts and mean FRP as predictors of NO$_2$ concentrations. Relating to the other uncertainties, although the errors in the tropospheric NO$_2$ columns caused due to cloud fraction (~30%), surface albedo (25%), air mass factors (10%), etc. are addressed in previous studies using the DOMINO product (Boersma et al., 2011), more detailed analysis including modeling might be needed in order to address the remaining uncertainties specific to Indian monsoon climate.

9.7 CONCLUSIONS

Biomass burning is an important source of greenhouse gas emissions and aerosols (Curado et al., 2024; Huanget al., 2024; Liu et al., 2024; Ceamanos et al., 2023; Zhao et al., 2023; Li et al., 2020; Pan et al., 2020; Schill et al., 2020; Shi et al., 2020; Zhang et al., 2020; Li et al., 2019; Andreae, 2019; Pistone et al., 2019; Yin et al., 2019a; Yin et al., 2019b; Bray et al., 2018; Mota and Wooster,

2018; Wu et al., 2018; Chen et al., 2017; Zhou et al., 2017). We evaluated the nitrogen dioxide (NO_2) emissions from OMI and SCIAMACHY data from the biomass burning of subtropical evergreen forests, northeast India. We also assessed the long-term (2003 – 2011) MODIS – Aerosol Optical Depth (AOD) signal in relation to MODIS fire retrievals. Results suggested a stronger correlation of MODIS fire counts with OMI-NO_2 (R2 = 0.74) than with SCIAMACHY-NO_2 (R2 = 0.35). Also, OMI-NO_2 showed a stronger correlation with MODIS - AOD (R2 = 0.54) than SCIAMACHY-NO_2 (R2 = 0.36). Further, inter-comparison of OMI and SCIAMACHY - NO_2 showed correlation of 57%. We attribute these results to the satellite overpass time, i.e., SCIAMACHY during the morning (10:00 a.m.) and OMI during the afternoon (13:45 p.m.), ecosystem variations and oxidation reactions influencing NO_2 formation. Results from time-series modeling suggested combined use of fire counts and FRP for explaining NO_2 concentrations than either of those variables alone. Our results highlight the potential MODIS fires and OMI datasets in characterizing biomass-burning episodes from subtropical evergreen forests. We infer the need for field campaigns that characterize the relationship between satellite retrievals of emissions and ground-based measurements to resolve satellite-based uncertainties.

9.8 ACKNOWLEDGMENTS

Authors thank the MODIS fire, MODIS aerosol, SCIAMACHY, and OMI-NO_2 science teams for the datasets. This research was supported by NASA grant NNX10AU77G.

REFERENCES

Andreae, M.O. 2019. Emission of trace gases and aerosols from biomass burning—an updated assessment. *Atmospheric Chemistry and Physics*, 19, 8523–8546. https://doi.org/10.5194/acp-19-8523-2019.

Andreae, M.O., & Merlet, P. 2001. Emission of trace gases and aerosols from biomass burning. *Global Biogeochemical Cycles*, 15(4), 955–966.

Arellano, A.F., Kasibhatla, P.S., Giglio, L., van der Werf, G.R., Randerson, J.T., & Collatz, G.J. 2006. Time-dependent inversion estimates of global biomass-burning CO emissions using Measurement of Pollution in the Troposphere (MOPITT) measurements. *Journal of Geophysical Research*, 111, D09303. https://doi.org/10.1029/2005JD006613.2006.

Badarinath, K.V.S., Kharol, S.K., Sharma, A.R., & Krishna Prasad, V. 2009. Analysis of aerosol and carbon monoxide characteristics over Arabian Sea during crop residue burning period in the Indo-Gangetic Plains using multi-satellite remote sensing datasets. *Journal of Atmospheric and Solar-Terrestrial Physics*, 71(12), 1267–1276.

Beirle, S., Platt, U., Wenig, M., & Wagner, T. 2003. Weekly cycle of NO_2 by GOME measurements: A signature of anthropogenic sources. *Atmospheric Chemistry and Physics*, 3, 2225–2232.

Bertram, T.H., Heckel, A., Richter, A., Burrows, J., & Cohen, R.C. 2005. Satellite measurements of daily variations in soil NOx emissions. *Geophysical Research Letters*, 32, L24812. https://doi.org/10.1029/2005GL024640.

Blond, N., Blond, N., Boersma, K.F., Eskes, H.J., van der, R.J.A., Roozendael, M.V., Smedt, De, I., Bergametti, G., & Vautard, R. 2007. Intercomparison of SCIAMACHY nitrogen dioxide observations, in-situ measurements, and air quality modeling results over Western Europe. *Journal of Geophysical Research*, 112, D10311. https://doi.org/10.1029/2006JD007277.

Boersma, K.F., Eskes, H.J., & Brinksma, E.J. 2004. Error analysis for tropospheric NO_2 retrieval from space. *Journal of Geophysical Research*, 109, D04311. https://doi.org/10.1029/2003JD003962.

Boersma, K.F., Eskes, H.J., Dirksen, R.J., van der A, R.J., Veefkind, J.P., Stammes, P., Huijnen, V., Kleipool, Q.L., Sneep, M., Claas, J., Leitão, J., Richter, A., Zhou, Y., & Brunner, D. 2011. An improved retrieval of tropospheric NO_2 columns from the Ozone Monitoring Instrument. *Atmospheric Measurement Techniques*, 4, 1905–1928.

Boersma, K.F., Eskes, H.J., Veefkind, J.P., Brinksma, E.J., van der, R.J., Sneep, M., van den Oord, G.H. Levelt, P.F., Stammes, P., Gleason, J.F., & Bucsela, E.J. 2007. Near-real time retrieval of tropospheric NO_2 from OMI. *Atmospheric Chemistry and Physics*, 2013–2128.

Bray, C.D., Battye, W., Aneja, V.P., Tong, D.Q., Lee, P., & Tang, Y. 2018. Ammonia emissions from biomass burning in the continental United States. *Atmospheric Environment*, 187, 50–61. ISSN 1352–2310. https://doi.org/10.1016/j.atmosenv.2018.05.052. https://www.sciencedirect.com/science/article/pii/S1352231018303625

Bucsela, E.J., Pickering, K.E., Huntemann, T.L., Cohen, R.C., Perring, A., Gleason, J.F., Blakeslee, R.J., Albrecht, R.I., Holzworth, R., Cipriani, J.P., & Vargas-Navarro, D. 2010. Lightning-generated NOx seen by the ozone monitoring instrument during NASA's tropical composition, cloud and climate coupling experiment (TC4). *Journal of Geophysical Research: Atmospheres*, 115(D10).

Burrows, J.P., Platt, U., & Borrell, P. 2011. Tropospheric remote sensing from space. In: J.P. Burrows, Ulrich Platt and P. Borell (eds.). *The Remote Sensing of Tropospheric Composition from Space.Physics of Earth and Space Environments*. Heidelberg and New York: Springer.

Burrows, J.P., Weber, M., Buchwitz, M., Rozanov, V., Weißenmayer, A.L., Richter, A., DeBeek, R., Hoogen, R., Bramstedt, K., Eichmann, K.-U., Eisinger, M., & Perner, D. 1999. The Global Ozone Monitoring Experiment (GOME): Mission concept and first scientific results. *Journal of Atmospheric Science*, 56, 151–175.

Ceamanos, X., Coopman, Q., George, M., et al. 2023. Remote sensing and model analysis of biomass burning smoke transported across the Atlantic during the 2020 Western US wildfire season. *Scientific Reports*, 13, 16014. https://doi.org/10.1038/s41598-023-39312-1

Cheewaphongphan, P., & Garivait, S. 2013. Bottom up approach to estimate air pollution of rice residue open burning in Thailand. *Asia-Pacific Journal of Atmospheric Sciences*, 49(2), 139–149.

Chen, J., Li, C., Ristovski, Z., Milic, A., Gu, Y., Islam, M.S., Wang, S., Hao, J., Zhang, H., He, C., Guo, H., Fu, H., Miljevic, B., Morawska, L., Thai, P., LAM, Y.F., Pereira, G., Ding, A., Huang, X., & Dumka, U.C. 2017. A review of biomass burning: Emissions and impacts on air quality, health and climate in China. *Science of The Total Environment*, 579, 1000–1034. ISSN 0048-9697. https://doi.org/10.1016/j.scitotenv.2016.11.025. https://www.sciencedirect.com/science/article/pii/S0048969716324561

Curado, L.F.A., de Paulo, S.R., da Silva, H.J.A., et al. 2024. Effect of biomass burning emission on carbon assimilation over Brazilian Pantanal. *Theoretical and Applied Climatology*, 155, 999–1006. https://doi.org/10.1007/s00704-023-04673-0

David, L.M., & Nair, P.R. 2013. Tropospheric column O3 and NO2 over the Indian region observed by Ozone Monitoring Instrument (OMI): Seasonal changes and long-term trends. *Atmospheric Environment*, 65(2013), 25–39.

Durbin, J., & Winsten, G.S. 1950. Testing for serial correlation in least-squares regression II. *Biometrika*, 37, 409–428.

Ghude, S.D., Fadnavis, S., Beig, G., Polade, S.D, and van der, R.J.A. 2008. Detection of surface emission hot spots, trends, and seasonal cycle from satellite-retrieved NO2 over India. *Journal of Geophysical Research*, 113. http://doi.org/10.1029/2007JD009615.

Giglio, L., Descloitres, J., Justice, C.O., & Kaufman, Y. 2003. An enhanced contextual fire detection algorithm for MODIS. *Remote Sensing of Environment*, 87, 273–282.

Giglio, L., Randerson, J.T., & van der Werf, G.R. 2013. Analysis of daily, monthly, and annual burned area using the fourth-generation global fire emissions database (GFED4). *Journal of Geophysical Research: Biogeosciences*, 118(1), 317–328.

Heil, A., Langmann, B., & Aldrian, E. 2007. Indonesian peat and vegetation fire emissions: Study on factors influencing large-scale smoke haze pollution using a regional atmospheric chemistry model. *Mitigation and Adaptation Strategies for Global Change*, 12(1), 113–133.

Huang, H., Wang, S., Lau, W.K.M., Wang, S.S., & da Silva, A.M. 2004. Impact of regional climate patterns on the biomass burning emissions and transport over Peninsular Southeast Asia, 2000–2019. *Atmospheric Research*, 297, 107067. ISSN 0169-8095. https://doi.org/10.1016/j.atmosres.2023.107067. https://www.sciencedirect.com/science/article/pii/S0169809523004647

Hudman, R.C., Jacob, D.J., Turquety, S., Leibensperger, E.M., Murray, L.T., Wu, S., Gilliland, A.B., Avery, M., Bertram, T.H., Brune, W., Cohen, R.C., Dibb, J.E., Flocke, F.M., Fried, A., Holloway, J., Neuman, J.A., Orville, R., Perring, A., Ren, X., Sachse, G.W., Singh, H.B., Swanson, A., & Wooldridge, P.J. 2007. Surface and lightning sources of nitrogen oxides over the United States: Magnitudes, chemical evolution, and outflow. *Journal of Geophysical Research*, 112, D12S05. http://doi.org/10.1029/2006JD007912.

Ichoku, C., Giglio, L., Wooster, M.J., & Remer, L.A. 2008. Global characterization of biomass-burning patterns using satellite measurements of fire radiative energy. *Remote Sensing of Environment*, 112, 2950–2962.

Jaeglé, L., Steinberger, L., Martin, R.V., & Chance, K. 2005. Global partitioning of NOx sources using satellite observations: relative roles of fossil fuel combustion, biomass burning and soil emissions. *Faraday Discussion*, 130, 407–423. http://doi.org/10.1039/b502128f.

Kaufman, Y.J., Justice, C.O., Flynn, L., Kendall, J.D., Prins, E.M., Giglio, L., Ward, D., Menzel, W., & Setzer, A. 1998. Potential global fire monitoring from EOS-MODIS. *Journal of Geophysical Research*, 103, 32215–32238.

Kim, S.-W., Heckel, A., McKeen, S.A., Frost, G.J., Hsie, E.Y., Trainer, M.K., Richter, A., Burrows, J.P., Peckham, S.E., & Grell, G.A. 2006. Satellite-observed U.S. power plant NOx emission reductions and their impact on air quality. *Geophysical Research Letters*, 33, L22812. http://doi.org/10.1029/2006GL027749.

Kurokawa, J.I., Yumimoto, K., Uno, I., & Ohara, T. 2009. Adjoint inverse modeling of NOx emissions over eastern China using satellite observations of NO2 vertical column densities. *Atmospheric Environment*, 43, 1878–1887. http://doi.org/10.1016/2008.12.030.

Lamsal, L.N., Martin, R.V., van Donkelaar, A., Celarier, E.A., Bucsela, E.J., Boersma, K.F., Dirksen, R., Luo, C., & Wang, Y. 2010. Indirect validation of tropospheric nitrogen dioxide retrieved from the OMI satellite instrument: Insight into the seasonal variation of nitrogen oxides at northern midlatitudes. *Journal of Geophysical Research*, 115, D05302. http://doi.org/10.1029/2009JD013351.

Langner, A., Miettinen, J., & Siegert, F. 2007. Land cover change 2002–2005 in Borneo and the role of fire derived from MODIS imagery. *Global Change Biology*, 13(11), 2329–2340.

Levelt, P.F., van den Oord, G.H.J., Dobber, M.R., Malkki, A., Visser, H., de Vries, J., Stammes, P., Lundell, J.O.V., & Saari, H. 2006. The ozone monitoring instrument. *IEEE Transactions on Geoscience and Remote Sensing*, 44, 1093–1101.

Levy, R.C., Remer, L.A., & Dubovik, O. 2007. Global aerosol optical properties and application to Moderate Resolution Imaging Spectroradiometer aerosol retrieval over land. *Journal of Geophysical Research*, 112, D13210. http://doi.org/10.1029/2006JD007815.

Li, F., Zhang, X., & Kondragunta, S. 2020. Biomass burning in Africa: An investigation of fire radiative power missed by MODIS using the 375 m VIIRS active fire product. *Remote Sensing*, 12(10), 1561. https://doi.org/10.3390/rs12101561

Li, F., Zhang, X., Roy, D.P., & Kondragunta, S. 2019. Estimation of biomass-burning emissions by fusing the fire radiative power retrievals from polar-orbiting and geostationary satellites across the conterminous United States. *Atmospheric Environment*, 211, 274–287. ISSN 1352–2310. https://doi.org/10.1016/j.atmosenv.2019.05.017. https://www.sciencedirect.com/science/article/pii/S1352231019303164

Lin, J.T., McElroy, M.B., & Boersma, K.F. 2010. Constraint of anthropogenic NOx emissions in China from different sectors: A new methodology using multiple satellite retrievals. *Atmospheric Chemistry and Physics*, 10, 63–78. http://doi.org/10.5194/acp-10-63-2010.

Liu, Y., Chen, J., Shi, Y., Zheng, W., Shan, T., & Wang, G. 2024. Global emissions inventory from open biomass burning (GEIOBB): Utilizing fengyun–3D global fire spot monitoring data. *Earth System Science Data Discussions* [preprint]. https://doi.org/10.5194/essd-2023-527.

Logan, J.A. 1983. Nitrogen oxides in the troposphere: Global and regional budgets. *Journal of Geophysical Research*, 88, 10785–10807.

Logan, J.A., Prather, M.J., Wofsy, S.C., & McElroy, M.B. 1981. Tropospheric chemistry: A global perspective. *Journal of Geophysical Research*, 86, 7210–7254. http://doi.org/10.1029/JC086iC08p07210

Majumder, M., Shukla, A.K., & Arunachalam, A. 2011. Agricultural practices in Northeast India and options for sustainable management. In: E. Lichtfouse (ed.). *Biodiversity, Biofuels, Agroforestry and Conservation Agriculture*. Sustainable Agriculture Rev., 5. Dordrecht, Netherlands: Springer, 287–315. http://doi.org/10.1007/978-90-481-9513-8_10

Martin, R.V., Chance, K., Jacob, D.J., Kurosu, T.P., Spurr, R.J.D., Bucsela, E., Gleason, J.F., Palmer, P.I., Bey, I., Fiore, A.M., Li, Q., Yantosca, R.M., & Koelemeijer, R.B.A. 2002. An improved retrieval of tropospheric nitrogen dioxide from GOME. *Journal of Geophysical Research*, 107(D20), 4437. http://doi.org/10.1029/2001JD001027.

Martin, R.V., Sioris, C.E., Chance, K., Ryerson, T.B., Bertram, T.H., Wooldridge, P.J., Cohen, R.C., Neuman, J.A., Swanson, A., & Flocke, F.M. 2006. Evaluation of space-based constraints on global nitrogen oxide emissions with regional aircraft measurements over and downwind of eastern North America. *Journal of Geophysical Research*, 111, D15308. http://doi.org/10.1029/2005JD006680.

Mebust, A.K., Russell, A.R., Hudman, R.C., Valin, L.C., & Cohen, R.C. 2011. Characterization of wildfire NOx emissions using MODIS fire radiative power and OMI tropospheric NO2 columns. *Atmospheric Chemistry and Physics*, 11, 5839–5851.

Mijling, B., van der A.R.J., Boersma, K.F., Van Roozendael, M., De Smedt, I., & Kelder, H.M. 2009. Reductions of NO₂ detected from space during the 2008 Beijing Olympic Games. *Geophysical Research Letters*, 36, L13801. http://doi.org/10.1029/2009GL038943.

Miyazaki, K., Eskes, H.J., & Sudo, K. 2012. Global NOx emission estimates derived from an assimilation of OMI tropospheric NO2 columns. *Atmospheric Chemistry and Physics*, 12, 2263–2288.

Mota, B., & Wooster, M.J. 2018. A new top-down approach for directly estimating biomass burning emissions and fuel consumption rates and totals from geostationary satellite fire radiative power (FRP). *Remote Sensing of Environment*, 206, 45–62. ISSN 0034–4257. https://doi.org/10.1016/j.rse.2017.12.016. https://www.sciencedirect.com/science/article/pii/S0034425717305886

Müller, J.-F., & Stavrakou, T. 2005. Inversion of CO and NOx emissions using the adjoint of the IMAGES model. *Atmospheric Chemistry and Physics*, 5, 1157–1186. http://doi.org/10.5194/acp-5-1157-2005.

Murdiyarso, D., & Lebel, L. 2007. Local to global perspectives on forest and land fires in Southeast Asia. *Mitigation and Adaptation Strategies for Global Change*, 12(1), 3–11.

Palmer, P.I., Jacob, D.J., Chance, K., Martin, R.V., Spurr, R.J.D., Kurosu, T.P., Bey, I., Yantosca, R., Fiore, A., & Li, Q. 2001. Air-mass factor formulation for spectroscopic measurements from satellites: application to formaldehyde retrievals from the Global Ozone Monitoring Experiment. *Journal of Geophysical Research*, 106, 539–550.

Pan, X., Ichoku, C., Chin, M., Bian, H., Darmenov, A., Colarco, P., Ellison, L., Kucsera, T., da Silva, A., Wang, J., Oda, T., & Cui, G. 2020. Six global biomass burning emission datasets: Intercomparison and application in one global aerosol model. *Atmospheric Chemistry and Physics*, 20, 969–994. https://doi.org/10.5194/acp-20-969-2020.

Pistone, K., Redemann, J., Doherty, S., Zuidema, P., Burton, S., Cairns, B., Cochrane, S., Ferrare, R., Flynn, C., Freitag, S., Howell, S.G., Kacenelenbogen, M., LeBlanc, S., Liu, X., Schmidt, K.S., Sedlacek III, A.J., Segal-Rozenhaimer, M., Shinozuka, Y., Stamnes, S., van Diedenhoven, B., Van Harten, G., & Xu, F. 2019. Intercomparison of biomass burning aerosol optical properties from in situ and remote-sensing instruments in ORACLES-2016. *Atmospheric Chemistry and Physics*, 19, 9181–9208. https://doi.org/10.5194/acp-19-9181-2019.

Prasad, K.V., K Gupta, P., Sharma, C., Sarkar, A.K., Kant, Y., Badarinath, K.V.S., & Mitra, A.P. (2000). NO x emissions from biomass burning of shifting cultivation areas from tropical deciduous forests of India–estimates from ground-based measurements. *Atmospheric Environment*, 34(20), 3271–3280.

Remer, L.A., Kaufman, Y.J., Mattoo, S., Martins, J.V., Ichoku, C., Levy, R.C., Kleidman, R.G., Tanré, D., Chu, D.A., Li, R.R., Eck, T.F., Vermote, E., & Holben, B.N. 2005. The MODIS algorithm, products and validation. *Journal of Atmospheric Science*, 62, 947–973. http://doi.org/10.1175/JAS3385.1

Remer, L.A., Kaufman, Y.J., Tanré, D., Mattoo, S., Chu, D.A., Martins, J.V., Li, Ichoku, C., Levy, R.C., Kleidman, R.G., Eck, T.F. Vermote, E., & Holben, B.N. 2005. The MODIS aerosol algorithm, products and validation. *Journal of Atmospheric Science*, 62, 947–973.

Richter, A., & Burrows, J.P. 2002. Tropospheric NO2 from GOME measurements. *Advances in Space Research*, 29, 1673–1683.

Russell, A.R., Perring, A.E., Valin, L.C., Bucsela, E., Browne, E.C., Min, K.E., Wooldridge, P.J., & Cohen, R.C. 2011. A high spatial resolution retrieval of NO₂ column densities from OMI: Method and evaluation. *Atmospheric Chemistry and Physics*, 11, 12411–12440.

Schaub, D., Boersma, K.F., Kaiser, J.W., Weiss, A.K., Folini, D., Eskes, H.J., & Buchmann, B. 2006. Comparison of GOME tropospheric NO₂ columns with NO₂ profiles deduced from ground-based in situ measurements. *Atmospheric Chemistry and Physics*, 6, 3211–3229.

Schill, G.P., Froyd, K.D., Bian, H., et al. 2020. Widespread biomass burning smoke throughout the remote troposphere. *Nature Geoscience*, 13, 422–427. https://doi.org/10.1038/s41561-020-0586-1

Schwartz, J. 2004. Air pollution and children's health. *Pediatrics*, 113, 1037–1043.

Shaw, G.E. 1976. Nitrogen dioxide-optical absorption in the visible. *Journal of Geophysical Research*, 81, 5791–5792.

Shi, Y., Zang, S., Matsunaga, T., & Yamaguchi, Y. 2020. A multi-year and high-resolution inventory of biomass burning emissions in tropical continents from 2001–2017 based on satellite observations. *Journal of Cleaner Production*, 270, 122511. ISSN 0959–6526. https://doi.org/10.1016/j.jclepro.2020.122511. https://www.sciencedirect.com/science/article/pii/S0959652620325580

Stavrakou, T., Muller, J.F., Boersma, K.F., De Smedt, I., & van der A, R.J. 2008. Assessing the distribution and growth rates of NOx emission sources by inverting a 10-year record of NO2 satellite columns. *Geophysical Research Letters*, 35, L10801. http://doi.org/10.1029/2008GL033521.

Thomas, W., Hegels, E., Slijkhuis, S., Spurr, R., & Chance, K. 1998. Detection of biomass burning combustion products in Southeast Asia from backscatter data taken by the GOME spectrometer. *Geophysical Research Letters*, 25, 1317–1320.

Vadrevu, K.P., Ellicott, E., Badarinath, K.V.S., & Vermote, E. 2011. MODIS derived fire characteristics and aerosol optical depth variations during the agricultural residue burning season, north India. *Environmental Pollution*, 159, 1560–1569.

Vadrevu, K.P., Ellicott, E., Giglio, L., Badarinath, K.V.S., Vermote, E., Justice, C., & Lau, W.K. 2012. Vegetation fires in the Himalayan region–Aerosol load, black carbon emissions and smoke plume heights. *Atmospheric Environment*, 47, 241–251.

Vadrevu, K.P., & Justice, C.O. 2011. Vegetation fires in the Asian region: Satellite observational needs and priorities. *Global Environmental Research*, 15(1), 65–76.

Valin, L.C., Russell, A.R., Bucsela, E.J., Veefkind, J.P., & Cohen, R.C. 2011. Observation of slant column NO_2 using the super-zoom mode of AURA-OMI. *Atmospheric Measurement Techniques Discussion*, 20, 1989–2005. http://doi.org/10.5194/amtd-4-1989-2011.

van der A, R.J., Eskes, H.J., Boersma, K.F., van Noije, T.P.C., van Roozendael, M., De Smedt, I., Peters, D.H.M.U., & Meijer, E.W. 2008. Trends, seasonal variability and dominant NOx source derived from a ten year record of NO2 measured from space. *Journal of Geophysical Research*, 113, 1–12. http://doi.org/10.1029/2007JD009021.

van der A, R.J., Peters, D.H.M.U., Eskes, H., Boersma, K.F., Van Roozendael, M., De Smedt, I., & Kelder, H.M. 2006. Detection of the trend and seasonal variation in tropospheric NO_2 over China. *Journal of Geophysical Research*, 111, D12317. http://doi.org/10.1029/2005JD006594.

Van der Werf, G.R., Dempewolf, J., Trigg, S.N., Randerson, J.T., Kasibhatla, P.S., Giglio, L., & DeFries, R.S. 2008. Climate regulation of fire emissions and deforestation in equatorial Asia. *Proceedings of the National Academy of Sciences*, 105(51), 20350–20355.

van der Werf, G.R., Randerson, J.T., Giglio, L., Collatz, G.J., Mu, M., Kasibhatla, P.S., Morton, D.C., deFries, R.S., Jin, Y., & van Leeuwen, T.T. 2010. Global fire emissions and the contribution of deforestation, savanna, forest, agricultural, and peat fires (1997–2009). *Atmospheric Chemistry and Physics*, 10, 11707–11735.

Velders, G.J.M., Granier, C., Portmann, R.W., Pfeilsticker, K., Wenig, M., Wagner, T., Platt, U., Richter, A., & Burrows, J.P. 2001. Global tropospheric NO_2 column distributions: Comparing three-dimensional model calculations with GOME measurements. *Journal of Geophysical Research*, 106, D12.

Wang, Y., Renyi, Z., & Saravanan, R. 2014. Asian pollution climatically modulates mid-latitude cyclones following hierarchical modelling and observational analysis. *Nature Communications*, 5.

Wooster, M.J., Roberts, G., Perry, G.L.W., & Kaufman, Y.J. 2005. Retrieval of biomass combustion rates and totals from fire radiative power observations: FRP derivation and calibration relationships between biomass consumption and fire radiative energy release. *Journal of Geophysical Research*, 110, D24311–D24311.

Wu, J., Kong, S., Wu, F., Cheng, Y., Zheng, S., Yan, Q., Zheng, H., Yang, G., Zheng, M., Liu, D., Zhao, D., & Qi, S. 2018. Estimating the open biomass burning emissions in central and eastern China from 2003 to 2015 based on satellite observation. *Atmospheric Chemistry and Physics*, 18, 11623–11646. https://doi.org/10.5194/acp-18-11623-2018.

Yin, L., Du, P., Zhang, M., Liu, M., Xu, T., & Song, Y. 2019a. Estimation of emissions from biomass burning in China (2003–2017) based on MODIS fire radiative energy data. *Biogeosciences*, 16, 1629–1640. https://doi.org/10.5194/bg-16-1629-2019.

Yin, S., Wang, X., Zhang, X., Guo, M., Miura, M., & Xiao, Y. 2019b. Influence of biomass burning on local air pollution in mainland Southeast Asia from 2001 to 2016. *Environmental Pollution*, 254(Part A), 112949. ISSN 0269-7491. https://doi.org/10.1016/j.envpol.2019.07.117. https://www.sciencedirect.com/science/article/pii/S0269749119316896

Zhang, X., Liu, J., Han, H., Zhang, Y., Jiang, Z., Wang, H., Meng, L., Chen, Y., & Liu, Y. 2020. Satellite-observed variations and trends in carbon monoxide over Asia and their sensitivities to biomass burning. *Remote Sensing*, 12(5), 830. https://doi.org/10.3390/rs12050830

Zhao, C., & Wang, Y. 2009. Assimilated inversion of NOx emissions over east Asia using OMI NO_2 column measurements. *Geophysical Research Letters*, 36, L06805. https://doi.org/10.1029/2008GL037123.

Zhao, S., Wang, L., Shi, Y., Zeng, Z., Nath, B., & Niu, Z. 2023. Methane emissions in boreal forest fire regions: Assessment of five biomass-burning emission inventories based on carbon sensing satellites. *Remote Sensing*, 15(18), 4547. https://doi.org/10.3390/rs15184547

Zhou, W., Cohan, D.S., Pinder, R.W., Neuman, J.A., Holloway, J.S., Peischl, J., Ryerson, T.B., Nowak, J.B., Flocke, F., & Zheng, W.G. 2012. Observations and modelling of the evolution of Texas plant plumes. *Atmospheric Chemistry and Physics*, 12, 455–468.

Zhou, Y., Xing, X., Lang, J., Chen, D., Cheng, S., Wei, L., Wei, X., & Liu, C. 2017. A comprehensive biomass burning emission inventory with high spatial and temporal resolution in China. *Atmospheric Chemistry and Physics*, 17, 2839–2864. https://doi.org/10.5194/acp-17-2839-2017.

10 Remote Sensing–Based Mapping and Monitoring of Coal Fires

Anupma Prakash, Claudia Kuenzer, Santosh K. Panda, Anushree Badola, and Christine F. Waigl

ACRONYMS AND DEFINITIONS

ASTER	Advanced spaceborne thermal emission spectrometer
ATCOR	Atmospheric correction and haze reduction
AVHRR	A very high resolution radiometer
BIRD	Bi-spectral infrared detection
CBERS	China-Brazil Earth Resources Satellite
ETM	Enhanced thematic mapper
FLAASH	Fast line-of-sight atmospheric analysis of hypercubes
MODIS	Moderate resolution imaging spectrometer
MODTRAN	MODerate resolution atmospheric TRANsmission
NDVI	Normalized Difference Vegetation Index
STTBT	Spatio-temporal temperature-based thresholding
SWCM	Strength and weakness constraint method
SWIR	Shortwave infrared
TIR	Thermal infrared

10.1 INTRODUCTION

Coal fires, or fires occurring in reserves of coal, are a phenomenon of global occurrence. Almost all large coal mining operations in the world have been threatened by coal fires at some time. An excellent compilation of case studies of fires in different parts of the world is presented in Stracher et al. (2013), which includes an online database and a scalable map showing locations of several reported coal fires worldwide (Gens, 2013). This section introduces commonly used terminology for different types of coal fires, why these fires are so common, what damage they can cause, and how remote sensing can be used to investigate coal fires.

10.1.1 Terminology

Depending on the location of coal fires, several different terms have been coined. A good first review of classification of coal fires is presented by Zhang et al. (2004a). A fire in a coal mine is generally referred to as a coal mine fire. If the fire is in an in situ coal seam it is also referred to as coal seam fire. If the coal seam that is on fire is exposed on the surface it is occasionally referred to as an outcrop fire. The bigger differentiation for remote sensing–based investigation is based on whether the fire is at the surface and exposed (then known as surface coal fire) or if it is buried at depth (then known as subsurface or underground coal fire). Surface coal fires can include coal seam fires or also fires that occur in mining-related waste dumps (overburden dumps), in which case they are commonly called overburden coal fires. A further in-depth classification of coal fires was also published by Kuenzer and Stracher (2011), who presented a detailed overview on coal fire–induced geomorphologic features, and classified fires not only according to their spatial occurrence (surface, subsurface, within a seam, or in an artificial coal storage of coal waste pile), but also differentiated them according to age (paleo coal fires versus recent coal fires), genesis (natural coal fires versus human-induced coal fires), and burning behavior (steadily burning, accelerating, burning out, extinct, re-ignited). In this chapter we will restrict to using the terms surface and subsurface coal fires, the former being associated with very high-temperature fires exposed on the surface and the latter being associated with fires underground that are associated with warm surfaces, much lower in temperature than surface fires (Figure 10.1). More attributes of these coal fires are discussed later in this chapter.

Domestic and industrial coal fires are not a subject of consideration in this study. This study presents only the use of thermal infrared (TIR) and shortwave infrared (SWIR) remote sensing (used for temperature mapping) that forms the foundation of remote sensing investigations of coal fires. Related work (see Table 10.1; adapted and updated from Prakash, 2010) such as use of geophysical techniques for identification of active or previously burned areas, interferometric techniques to map fire- and mining-related subsidences, and optical images to study environmental impacts and land use/land cover change due to coal are beyond the scope of this chapter.

10.1.2 Causes

The most commonly reported cause of coal fires is spontaneous combustion, a process wherein the presence of some moisture acting as a catalyst, carbon in the coal reacts with oxygen in an

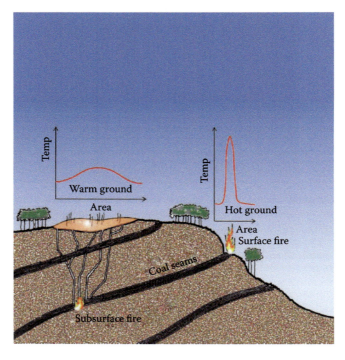

FIGURE 10.1 A conceptual diagram showing surface and subsurface fires in coal seams and the differences in temperature anomalies they cause at the land surface.

TABLE 10.1

Regions of the Electromagnetic Spectrum, Commonly Associated Wavelength Ranges, and the Main Coal Mining and Coal Fire Applications Associated with These Regions. Selected References for Each Application Area Are Also Listed

Region of Spectrum	Wavelength Range	Main Coal Mining and Coal Fire Applications	Example References
Visible	0.4–0.7 μm	Land cover/land use	Kumar and Pandey (2013)
			Kuenzer et al. (2012)
			Kuenzer et al. (2008a)
			Martha et al. (2010)
			Kuenzer et al. (2007d)
			Prakash and Gupta (1998)
		Geomorphology and faults	Kuenzer and Stracher (2011)
			Yang et al. (2008)
Near infrared (NIR)	0.7–1.3 μm	Vegetation	Kuenzer et al. (2005)
			Gupta and Prakash (1998)
			Lambin and Ehrlich (1996)
		Soil moisture	Hummel et al. (2001)
		Gas emissions	Engle et al. (2012a, 2012b)
			Van Dijk et al. (2011)
			Kuenzer et al. (2007c)
			Gangopadhyay (2007)
Shortwave infrared (SWIR) through mid-Infrared (MIR)	1.3–3.0 μm through 3.0–8.0 μm	Soil composition	Ben-Dor (2002)
		Associated forest fires	Waigl et al. (2014)
			Prakash et al. (2011)
			Whitehouse and Mulyana (2004)
		High-temperature surface coal fires	He et al. (2020)
			Kuenzer et al. (2013)
			Kuenzer et al. (2008b, 2008c, 2008d)
			Hecker et al. (2007)
			Zhang and Kuenzer (2007)
			Zhang et al. (2007)
			Tetzlaff (2004)
			Voigt et al. (2004)
			Zhang et al. (2004b)
			Prakash and Gupta (1999)
Thermal infrared (TIR)	8.0–14 μm	Underground coal fires	Du et al. (2022)
			He et al. (2020)
			Biswal and Gorai (2020)
			Martha et al. (2010)
			Kuenzer et al. (2008b, 2008c, 2008d)
			Kuenzer et al. (2007a)
			Hecker et al. (2007)
			Zhang and Kuenzer (2007)
			Zhang et al. (2007)
			Tetzlaff (2004)
			Cassells (1998)
			Prakash et al. (1997, 1999)
			Zhang et al. (1997)

(Continued)

TABLE 10.1 (*Continued*)
Regions of the Electromagnetic Spectrum, Commonly Associated Wavelength Ranges, and the Main Coal Mining and Coal Fire Applications Associated with These Regions. Selected References for Each Application Area Are Also Listed

Region of Spectrum	Wavelength Range	Main Coal Mining and Coal Fire Applications	Example References
		Fire depths	Berthelote et al. (2008)
			Wessling et al. (2008a, 2008b)
			Prakash and Berthelote (2007)
			Peng et al. (1997)
			Prakash et al. (1995b)
			Panigrahi et al. (1995)
			Mukherjee et al. (1991)
		Gas emissions	Van Dijk et al. (2011)
			Kolker et al. (2009)
Microwave	0.75 cm–1 m	Faults	Bonforte et al. (2013)
			Gens (2009)
		Subsidence	Wang et al. (2022)
			Liu et al. (2021)
			Jiang et al. (2011)
			Voigt et al. (2004)
			Prakash et al. (2001), Genderen et al. (2000)

Adapted and updated from Prakash (2010).

exothermic reaction, causing enough heat accumulation to set the coal on fire. For an excellent background on this topic, and on the propensity of different coal types to ignite by spontaneous combustion, the readers are referred to the book by Banerjee (1985). In the Jharia, India, coalfield, the first fire started in the Bhowrah colliery in 1916 due to spontaneous combustion, and nearly 100 years later this fire is still reported to be active. Other natural causes of coal fires are lightning strikes, forest fires, and fires occurring in peat lands or waste dumps—the latter being a cause for the well-known coal fire under the town of Centralia, Pennsylvania, USA. In many cases, forest fires and coal fires are closely and cyclically related with one being the cause of the other (Prakash et al., 2011; Waigl et al., 2014; Whitehouse and Mulyana, 2004).

Most other causes of coal fires are linked to mining or related human activities. These include mining-induced subsidence, short circuits, frictional heat from mining equipment, dust explosions, or negligence of safety protocols in mining operations (e.g., Chen et al., 2012; Zheng et al., 2009). Occasionally coal fires have been reported to start from people using abandoned mines as safe abodes for illegal distilling of alcohol or to hide from cold harsh weather (Glover, 2011; Prakash, 2014). Fire in an overburden dump can come in contact with a coal seam and set the coal seam aflame. Fire in one coal seam can migrate and ignite the coal in another seam above or below it. Regardless of how a fire starts, it continues to propagate as long the three essential elements (1) coal, (2) heat, and (3) oxygen are present (van Dijk et al., 2011; Kim, 2010; Stracher et al., 2010; Kuenzer et al., 2007a).

10.1.3 Hazards

Hazards from coal fires are widely documented in popular visual and printed media. This is especially true for some underground fires in the USA, India, and China, where these fires have caused substantial adverse impacts to the environment and communities. In the formerly prosperous town of Centralia (PA, USA) an underground coal seam caught fire arguably in 1962 and decades later the fire still persists. Incidences of sudden land subsidence and constant gaseous emissions from the burning coal made the living conditions dangerous, requiring relocation of all residents. According to the 2020 census data, Centralia is now an uninhabited ghost town where the population dwindled from 2,761 in 1890 to five remaining long-time residents who were granted permission in 2013 to live out their lives there. The condition in the Jharia coal mining area in India and larger coal fire areas in China are similar, but the situation is more complicated due to the high population density in the mining areas.

The obvious issue posed by fires is the loss of coal, a precious non-renewable energy resource. The coal burns down to ash, reducing its economic value to nothing. As the coal burns, mining the unburned coal around it becomes dangerous and sometimes practically impossible, causing further economic loss. Burning coal produces oxides and dioxides of carbon, nitrogen, and sulphur, methane, toxic compounds, along with steam and particulate matter (Carras et al., 2009; Engle et al., 2012a, 2012b; Gangopadhyay, 2007; Kolker et al., 2009) that contribute to atmospheric pollution. The environmental pollution due to coal fires affects not only the atmosphere but also the surrounding land and water resources (Kuenzer et al., 2007b). Locally dry patches of land devoid of healthy vegetation, referred to as barren aureoles (Gupta and Prakash, 1998), are common in some coal fire areas. Such patches show anomalous signatures in both optical and TIR images (Lambin and Ehrlich, 1996). A larger issue associated with underground coal fires is land stability. As the coal burns underground and turns to ash, the material volume decreases, leaving voids underground that ultimately cause the overlying land surface to sink in and subside (Figure 10.2). Massive mining- and fire-related subsidences are reported from many parts of the world (Genderen et al., 2000; Jiang et al., 2011; Prakash et al., 2001; Voigt et al., 2004, Kuenzer and Stracher, 2011) where buildings,

FIGURE 10.2 A sink hole in a coalfield in India, created due to sudden subsidence of land as subsurface coal fires consumed material. Heat emanating from the underground fires has baked the rocks and even turned some to ash (see gray region on one side of the sinkhole). The surface of this sink hole is approximately 3 m by 3 m in this area. Photo Credit: Anupma Prakash, 2006.

transportation networks, and mining equipment have been completely engulfed or have suffered massive damage due to subsidence and coal fires. Environmental pollution leads to serious health hazards (Finkelman, 2004; Hower et al., 2009; Stracher and Taylor, 2004) and a general socioeconomic decline in the affected areas.

Early detection and containment of a coal fire, before it becomes uncontrollable, is critical. Remote sensing offers a powerful tool for ongoing mapping and monitoring of coal fires to effectively target and manage fire-fighting efforts, which vary depending on the severity of the fire, its proximity to infrastructure and communities, the availability of resources, and funding for containment efforts. In principle, all fire-fighting efforts are targeted at cutting off access to one or more of the three factors (oxygen, heat, and coal) required to sustain a coal fire. The most common practice for putting out surface and near-surface coal fires is to simply isolate the burning area by trenching or building a retaining structure, and then douse the flames and excavate the remaining coal. For subsurface fires, it is more effective to cut off the oxygen supply by completely blanketing the area with sand and loess (e.g., Figure 10.3b). In other cases, prior to sand blanketing, the voids underground are also filled with different materials such as a sand-slurry mixture, swelling clays, fire-resistant colloidal mixtures, or by injecting water, inert gasses, or foam. In more sophisticated and effective fire-fighting efforts the entire fire areas were flushed with liquid nitrogen to bring down the temperature of coal seams so that the fire would not re-ignite (Ray and Singh, 2007).

10.1.4 Attributes of Surface and Subsurface Coal Fires

Surface coal fires in coal mining areas are much easier to locate in the field than subsurface fires. Sometimes they are associated with flames (Figure 10.3a). At other times they become visible when the surface material collapses, exposing the coal to a fresh gust of wind that fans the hot surface and makes it glow. Some surface fires are associated with dense smoke, whereas the others just smolder and emanate small amounts of smoke. Large surface fires can extend for hundreds of meters when a long stretch of a coal seam outcrop is on fire, as in some coalfields in India and China. Sometimes the coal outcrop is on fire at several different locations causing a string of surface fires. However, the most common occurrence is an individual instance of surface fire that may persistently burn for a long time or appear and disappear intermittently, due to seasonal factors or fire-fighting efforts.

Detecting subsurface fires in the field is hard (Zhang and Kuenzer, 2007) because one has to rely on indirect indicators such as the presence of heated ground, smoke-emitting vents (Figure 10.3b), dry and barren soils, local subsidence pockets, mineral occurrences in cracks (Stracher et al., 2005), and sometimes steam effusions after a rainfall event. Such surface manifestations are subtle and span a large spatial extent, making field-based identification and delineation more complex.

Remote sensing techniques have proven to be increasingly successful in recent years in more reliably delineating regions affected by both surface and subsurface fires, as described in the following section.

10.2 REMOTE SENSING OF COAL FIRES

Coal fires are high-temperature phenomena that lend well to remote sensing–based detection as fire areas emit higher amounts of energy due to elevated temperatures. Use of remote sensing technology starting from the early seventies to the present day is presented along with a discussion on the future use of this technology for coal fire studies.

10.2.1 History and Recent Evolution

It is no surprise that remote sensing–based fire detection dates as far back as the development and use of passive sensors that measure the Earth's emitted energy. Coal fires were mapped in the early

Remote Sensing–Based Mapping and Monitoring of Coal Fires

FIGURE 10.3 Field photo of coal fires in the Jharia coalfield, India. (a) A surface fire showing a burning coal seam. Flames, red hot rock, and ash are clearly visible. The width shown in this photograph is approximately 3 m. (b) A subsurface fire area that was blanketed with topsoil in efforts to fill in the cracks and contain the fire. The cracks resurfaced soon. In this photo smoke is seen emitting from these cracks. Note the huts and trees in the background for scale. Photo Credit: Anupma Prakash, 2006.

1970s using broadband TIR cameras mounted on aircrafts by Ellyett and Fleming (1974). Satellite-based TIR remote sensing lagged behind due to the coarse spatial resolution of the TIR sensors. Mansor et al. (1994) demonstrated the utility of even coarse 1.1 km spatial resolution images from A Very High Resolution Radiometer (AVHRR) TIR channel to map large regions of coal fires in the eastern part of the Jharia coalfield in India. Kuenzer et al. (2008b) and Hecker et al. (2007) also successfully used the Moderate Resolution Imaging Spectrometer (MODIS) TIR bands to map and monitor coal fires in the Jharia coalfield. However, the use of AVHRR and MODIS data is limited only to regional investigations of a few very large coal fire areas. TIR bands of the Landsat Thematic Mapper (TM) and Enhanced Thematic Mapper (ETM+), and the Advanced Spaceborne Thermal Emission Spectrometer (ASTER) are by far the most widely used satellite data for coal fire investigations in the last two decades (see Table 10.1). Data from some experimental satellite missions, such as the Bi-spectral Infrared Detection (BIRD) operated by the German Aerospace Agency (DLR), have also successfully demonstrated the potential of sensors operating in the mid-infrared wavelength ranges for mapping surface coal fires (Tetzlaff, 2004; Zhukov et al., 2006).

10.2.2 Spatial and Temporal Resolution

The spatial and temporal resolution of popular remote sensing systems used for mapping and monitoring coal fires is shown in Figure 10.4. Earth observing satellites such as Landsat and ASTER

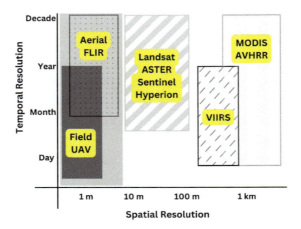

FIGURE 10.4 Spatial and temporal resolution of popular coal fire mapping and monitoring systems. Commonly used satellite sensors are shown in boxes and labeled with yellow highlights. The spatial and temporal resolutions of aerial and field surveys are not fixed, as shown in this figure, but are dependent on resource availability.

are particularly useful for coal fire studies, as their spatial resolution is commensurate with the spatial extent of coal fires that range typically from a few meters to hundreds of meters. Their temporal resolution of 16–18 days also works well for long-term monitoring of large underground coal fires. Coal fires are not as dynamic as forest fires. Though some small surface coal fires can move fast in the order of meters per day, most subsurface coal fires migrate at a much slower rate, with perceptible change on satellite images occurring over several months to years. Nonetheless, several recent studies have demonstrated the potential of low-altitude ultra high-resolution UAV-based TIR imaging in detection and very precise mapping of surface and subsurface coal fire areas (He et al., 2020).

As indicated in Figure 10.4, the higher spatial resolution airborne TIR systems are ideal for detailed mapping and investigation of coal fires on a one-time basis, or in some cases for monitoring coal fires, contingent on whether the mining agency can afford to acquire frequent airborne coverage. The coarser resolution satellite images, such as AVHRR and MODIS, have limited potential and are useful only for regional-scale studies with little practical use in decision support for targeted fire-fighting. VIIRS, Landsat, ASTER, Sentinel, and Hyperion provide a cost-effective means of monitoring coal fires.

10.2.3 Spectral Resolution

Different spectral regions have different significance for remote sensing of coal fires and related phenomena (Table 10.1). As mentioned in Section 10.1, here we focus only on the gray highlighted portion of the table that deals with characterizing the surface and subsurface coal fires. The SWIR region (1.3–3 μm) and the TIR region (8–14 μm) of the electromagnetic spectrum are by far the most important regions for remote sensing of coal fires, as they allow for quantitative estimates of temperatures of underground and surface coal fires. The most popularly used satellites for coal fire studies that have sensors operating in these spectral regions are ASTER and the Landsat series of satellites. Table 10.2 provides details of the spatial and spectral resolution of the satellite sensor that are conducive for coal fire research.

TABLE 10.2
Characteristics of Important Satellite Sensors and Their Applicability for Coal Fire Research

	Satellite/Sensors							
Resolution	**Bands**	**Landsat 4/5 TM**	**Landsat 7 ETM+**	**Landsat 8 OLI &TIRS**	**ASTER**	**VIIRS***	**Sentinel-2**	**Application in Coal Fire Studies**
Spatial resolution	VIS/NIR	30m × 30m	30m × 30m	30m × 30m	30m × 30m	750m × 750m	10m × 10m	Thermal band of Landsat 7 is most useful for coal fire studies due to its superior spatial resolution
	SWIR					750m × 750m	20m × 20m	
	TIR	120m × 120m	60m × 60m	100m × 100m	90m × 90m	750m × 750m		
	PAN		15m × 15m	15m × 15 m	15m × 15m			
Spectral resolution	VIS/NIR	1: 0.45–0.52	1: 0.45–0.52	1: 0.43–0.45	1: 0.52–0.60	M1: 0.40–0.42	1: 0.43–0.45	Primarily used for land use/land cover mapping. ASTER 3a (nadir) and 3b (backward looking) also used for elevation mapping
		2: 0.52–0.60	2: 0.52–0.60	2: 0.45–0.51	2: 0.63–0.69	M2: 0.44–0.45	2: 0.45–0.52	
		3: 0.63–0.69	3: 0.63–0.69	3: 0.53–0.59	3a: 0.78–0.86	M3: 0.48–0.50	3: 0.54–0.57	
		4: 0.76–0.90	4: 0.77–0.90	4: 0.64–0.67	3b: 0.78–0.86	M4: 0.54–0.56	4: 0.65–0.68	
				5: 0.85–0.88		M5: 0.66–0.68	5: 0.69–0.71	
						M6: 0.74–0.75	6: 0.73–0.74	
						M7: 0.85–0.88	7: 0.77–0.79	
						DNB: 0.5 –0.9	8: 0.78–0.89	
							8a: 0.85–0.87	
							9: 0.93–0.95	
	SWIR	5: 1.55–1.75	5:1.55–1.75	6: 1.57–1.65	4: 1.600–1.700	M8: 1.23–1.25	10: 1.36–1.39	Mapping and monitoring of high-temperature surface fires; sub-pixel temperature, and area analysis
		7: 2.08–2.35	7:2.09–2.35	7:2.11–2.29	5: 2.145–2.185	M9: 1.37–1.38	11: 1.56–1.65	
					6: 2.185–2.225	M10: 1.57–1.63	12: 2.10–2.28	
					7: 2.235–2.285	M11: 2.23–2.28		
					8: 2.295–2.365			
					9: 2.360–2.430			
	TIR	6: 10.40–12.50	7: 10.40–12.50	10: 10.60–11.19	10: 8.12–8.47	M12: 3.60–3.79		Mapping/monitoring subsurface fires; temperature estimation; emissivity (ASTER and VIIRS only)
				11:11.50–12.51	11: 8.47–8.82	M13: 3.99–4.14		
					12: 8.92–9.27	M14: 8.41–8.75		
					13:10.25–10.95	M15:10.23–11.25		
					14: 10.95–11.65	M16:11.40–12.32		
	PAN		8: 0.52–0.90	8: 0.50–0.68				As high spatial resolution base image

To detect coal fires, we rely on the fact that fires have elevated temperatures, and these can be derived from satellite images acquired in the TIR and SWIR regions. The principle and techniques of temperature estimation from remote sensing data are covered in detail in literature (e.g., Kuenzer and Dech, 2013; Kuenzer, 2015; Chander et al., 2009; Gupta, 2003; Quattrochi et al., 2009). Coal fire temperature estimation is based on the principle of inversing Planck's function to change an image derived spectral radiance to temperature values (e.g., Kuenzer et al., 2007a; Prakash and Gupta, 1999; Prakash et al., 1995a; Saraf et al., 1995; Zhang et al., 1997). Figure 10.5 shows the typical emission curves of blackbodies at various temperatures as determined by Planck's function. The locations of the Landsat and ASTER spectral bands are also plotted on this graph to give an idea of the temperature sensitivity ranges of these spectral bands.

The key message from Figure 10.5 and previous studies is that data from the broad band TIR region (approximately 8–14 µm) are better suited for deriving surface temperatures associated with underground fires, whereas the data from SWIR regions (1.3–3 µm) are better suited for estimating temperature of surface fires and open flames.

10.3 METHODS

Methods used for remote sensing of coal fires rely primarily on the accurate land surface temperature estimations. They can be divided into methods for (1) atmospheric correction and emissivity compensation, (2) land surface temperature estimation, (3) thresholding for fire delineation, (4) multi-source data analysis for improved detection, and (5) time-series analysis for fire monitoring.

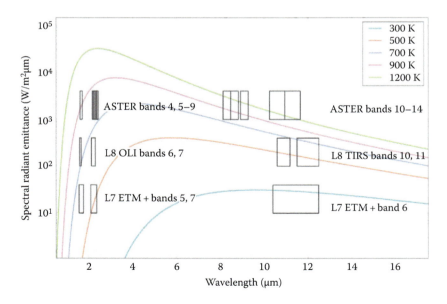

FIGURE 10.5 Positions of the SWIR and TIR bands of the Aster, Landsat 7, and Landsat 8 satellites, superimposed on Planck's emission curves for blackbodies at different temperatures. Note that only the spectral bandwidths are indicated and the saturation temperatures of individual spectral bands are not reflected in this figure. Generally, the SWIR bands are used for surface fire mapping and the TIR bands are used for subsurface fire mapping.

10.3.1 Atmospheric Correction and Land Surface Emissivity Variations

An important consideration in accurate temperature retrievals from remote sensing data is accounting for atmospheric correction and land surface emissivity variations (Gens and Cristobal, 2015).

The emitted spectral radiance measured at the sensor (L_s) is different from the spectral radiance emitted from the ground (L_g), as it includes the path radiance (L_p) or the upwelling spectral radiance emitted from the atmospheric particles. The signal reaching the sensor is also attenuated in complex ways that can be corrected by accounting for atmospheric transmissivity (τ). The ground emitted spectral radiance can be computed using equation (10.1)

$$L_{g\lambda} = \frac{L_{s\lambda} - L_{p\lambda}}{\tau_\lambda} \tag{10.1}$$

where λ is the central wavelength (in μm) of the spectral band in consideration; L_g, L_s, and L_p are all measured in $Wm^{-2}\mu m^{-1}sr^{-1}$; and τ is unitless.

Water vapor and aerosols (smoke, dust, haze—all very common in coal fire areas) are the main atmospheric components influencing the at-sensor spectral radiance, and therefore are the most important inputs in most radiative transfer models, such as MODerate resolution atmospheric TRANsmission (MODTRAN), used in atmospheric corrections of TIR data (Cristóbal et al., 2009; Jiménez-Muñoz, 2014; Schott et al., 2012). The water vapor values are typically either taken from field measurements where available, or from higher-order image products from other satellites, e.g., MODIS (MOD05 water vapor), and AVHRR (Sobrino et al., 1999). Transmissivity is estimated from general conditions of visibility reported from field sites. Commercial software packages such as Atmospheric Correction and Haze Reduction (ATCOR) and Fast Line-of-sight Atmospheric Analysis of Hypercubes (FLAASH) provide a graphical user interface to input these variables (or use pre-set standard atmospheric conditions based on general parameters) and run a MODTRAN or MODTRAN-like atmospheric correction model in the background to atmospherically correct TIR data and retrieve spectral radiance, which can be used further for temperature retrievals (see following Section 10.3.2).

Emissivity is the ratio of the spectral radiance emitted by a target at a given temperature and the spectral radiance emitted by a blackbody at the same temperature. Being a ratio, emissivity has no unit. The emissivity of a blackbody (an ideal energy absorber) is 1 and that of a white body (an ideal reflector) is 0. All natural objects typically have an emissivity value between 0 and 1. Several approaches have been taken into account in coal fire studies to assign target emissivity values. These include (1) using a constant, uniform, and high emissivity value usually ranging from 0.95 to 0.98; (2) assigning each land cover class a general published emissivity value; (3) using NDVI to determine emissivity; or (4) using image-based emissivity products, such as the ASTER emissivity product. Methods to retrieve emissivity from remote sensing images are well documented in the literature (Gillespie et al., 1998; Hook, 1992; Hulley et al., 2012; Martha et al., 2010; Schmugge et al., 1998; Sobrino et al., 2008).

10.3.2 Fire Temperature Estimation

Spectral radiance can be related to land surface temperature using the Planck's function $B_{\lambda T}$:

$$L_{\lambda T} = \varepsilon_\lambda B_{\lambda T} = \varepsilon_\lambda \frac{C_1 \lambda^{-5}}{\pi \left[\exp\left(\frac{C_2}{\lambda T}\right) - 1 \right]} \tag{10.2}$$

where $L_{\lambda T}$ is the spectral emitted radiance in $Wm^{-2}\mu m^{-1}sr^{-1}$ at wavelength λ (μm) and temperature T (K); ε_λ is spectral emissivity (dimensionless); $C_1 = 2\pi hc^2 =$ a constant value of $3.7418*10^{-16}$ W m²; C_2 = hc/k = a constant value of 1.4388*10⁻² m*K. Other values used to derive C_1 and C_2 are Planck's constant (h = 6.626076*10⁻³⁴ Js), speed of light (c= 2.998*10⁸ ms⁻¹), and Boltzmann constant (k= 1.3806*10⁻²³ JK⁻¹).

Equation (10.2_ can be inverted to directly derive land surface temperature (T).

$$T = \frac{C_2}{\lambda \ln\left[\frac{\varepsilon_\lambda C_1}{\pi \lambda^5 L} + 1\right]} \quad (10.3)$$

To find warm areas associated with subsurface coal fires TIR images acquired in the 8–14 µm are very useful. However, the temperature of the exposed coal fires and their flaming fronts may rise well over 1000°C, saturating the 8–14 µm wavelength bands, which cannot be used for temperature retrievals. This makes it necessary to rely on shorter 1.3–3 µm wavelength infrared channels for temperature retrievals.

Many surface fires are also smaller in size than the spatial resolution of the sensor that is being used to study them. In such cases dual-channel processing techniques need to be applied to retrieve subpixel area and subpixel temperature estimates of surface fires (Matson and Dozier, 1981; Prakash and Gupta, 1999; Tetzlaff, 2004). The main limitation in using Landsat SWIR images for subpixel-level investigation is the limited availability of spectral bands in the SWIR region. The stray light artifacts in Landsat-8 Band 11 made the dual-channel split-window technique impractical for retrieving surface temperature without requiring atmospheric data. To address this limitation, Cristobal et al. (2018) proposed a single-channel methodology to improve retrieval of surface temperature from Landsat-8 TIRS Band 10 using near-surface air temperature (Ta) and integrated atmospheric column water vapor (w). Their algorithm yielded overall errors on the order of 1 K and a bias of −0.5 K validated against in situ data, providing a better performance compared to other models.

10.3.3 Thresholding to Delineate Fire Area

Once the land surface temperatures have been computed for an image, it is important to set a threshold value (a temperature cut-off value) to delineate fire areas from cooler non-fire background areas. Threshold determination is an important step and researchers have approached it differently. An overview of some of the common thresholding techniques is presented by Prakash and Gens (2010), Kuenzer et al. (2013), and Raju et al. (2013).

The most common practice is to determine the threshold by trial-and-error guided by field knowledge (Gangopadhyay et al., 2005; Prakash et al., 1995a). The success of this practice stems from the fact that it works well locally for areas that are well studied in the field, as is the case with many important coal mining areas. Some researchers have simply used a fixed image standard deviation as a threshold criterion, e.g., a threshold of two standard deviation was deemed to be good for delineating fires in the Jharia coalfield by Prakash et al. (1999). However, more recent studies have shown that such low standard delineations may not work well in all conditions. Waigl et al. (2014) demonstrated that in automated processing of large scenes from high-latitude regions that show a great variation in land surface temperatures, a standard deviation of four is more suitable for delineating hot spots related to near surface coal fires. A study by Raju et al. (2013) determined that the highest spectral radiance that can be attributed to solar reflection on Landsat band-7 SWIR images can serve as a conservative threshold to segregate the surface fire pixels from non-fire pixels.

Other techniques rely on the derivation of scene-dependent relative thresholds based on the image histogram, e.g., using the change in slope of the histogram of a small image window around a fire (Rosema et al., 1999; Prakash and Vekerdy, 2004; Zhang et al., 2004a; Biswal and Gorai, 2020), or the first local minimum after the main maximum digital value during several iterations of a moving window kernel (Kuenzer et al., 2007a), as a good indicator to separate fire and non-fire areas. This latter automatic method enables detection of thermal anomalies of different temperatures within an image: an anomalous cluster of 65°C in a surrounding of 50°C heated bedrock will likewise be detected, just as a cluster of 70°C in a surrounding background of 60°C heated bedrock. This method is attributed to the fact that coal fire related anomalies are (e.g., compared to forest fires) extremely subtle, which are hard to detect against the background. A dark shale surface or a coal dust–covered surface can easily reach 60°C and more at 10:30 during local Landsat overpass time. Thus, natural surfaces heated by solar radiation easily mimic coal fire related anomalies. Simple thresholding with a temperature threshold in a full thermal satellite scene will thus not only lead to the extraction of coal fire–related anomalies but also to the extraction of heated bedrock surfaces, sun-exposed slopes, and other thermal anomalies. So trial-and-error-based thresholding only helps to delineate coal fire areas that are already well-known from field surveys. However, the main purpose of remote sensing should be to detect phenomena in hard-to-access and large areas where the coal fire locations are not known. This challenge could be addressed in China (Kuenzer et al., 2007a) and in a coal fire in the boreal forest in Alaska (Prakash et al., 2011), where unknown coal fires were first detected in remote sensing imagery and later validated in the field.

Du et al. (2022) tested the efficacy of nighttime ASTER thermal infrared images to monitor subsurface fire propagation. By applying an adaptive-edge-threshold algorithm on the nighttime images for mining areas in the Wuda coalfield, they successfully mapped subsurface coal fire areas and fire spread. They achieved higher accuracy in coal fire detection from the nighttime ASTER thermal infrared image than that of the daytime Landsat thermal infrared image due to the absence of the external influences of solar irradiance, topographic relief, and land cover.

10.3.4 Multi-Source Data for Improved Coal Fire Mapping

Some of the latest coal fire research combined land surface temperature with ground subsidence mapped from SAR data for improved subsurface coal fire detection (Liu et al., 2021; Karanam et al., 2021; Kim et al., 2021; Riyas et al., 2021; Wang et al., 2022). Liu et al. (2021) developed a novel approach to improve the detection accuracy of coal fires compounded by non–coal fire thermal anomalies and surface subsidence in mining areas. They proposed a multi-source information Strength and Weakness Constraint Method (SWCM) using temperature, normalized difference vegetation index (NDVI), and subsidence information based on Landsat 8 and Sentinel-1 data. They reported higher accuracy for coal fire identification and significantly reduced commission (70.4%) and omission (30.6%) error caused by non-coal-fire-related thermal anomalies and subsidence. Wang et al. (2022) developed a Spatio-Temporal Temperature-Based Thresholding (STTBT) algorithm to detect thermal anomalies. They combined it with the ground deformation obtained by the Polarimetric Persistent Scatterer Interferometry (PolPSI) method using a two-stage bandpass filter for the coal fire detection. For the Fukang coal fire area, China, they demonstrated that the STTBT method achieves higher accuracy of thermal anomalies detection compared with the conventional methods.

10.3.5 Time Series Analysis for Fire Monitoring

Time series analysis implies analysis of many images acquired at different times over the same study area. One purpose of carrying out a time series analysis is to determine whether the thermal anomaly detected by thresholding TIR images is transient or persistent. Transient anomaly is one that appears infrequently on a TIR image of one day or a few days, but then is not detectable on other days. This may happen due to some unusual heating locally at a particular time or due to local weather conditions, especially fluctuating wind speeds, rain events, etc. On the other hand persistent thermal anomalies tend to appear again and again at the same spot in a temporal stack of processed images. Waigl et al. (2014) classified a time-series of 40 Landsat scenes and defined a thermal anomaly as persistent if it appeared in more than one-third of the scenes in which the target pixel was present (in some cases the pixel was masked-out due to cloud cover).

The second and more important purpose of a time series analysis is to monitor the movement or progression of a coal fire over a longer period of time based on multi-temporal Earth observation data. Early efforts in coal fire monitoring were reported by Mansor et al. (1994), Saraf et al. (1995), and Prakash et al. (1999), with publications following from Zhang et al. (2004a), Hecker et al. (2007), Chatterjee et al. (2007), and Kuenzer et al. (2008a, 2008b, 2008c, 2008d), all employing multi-temporal data sets of daytime and nighttime to extract fire-related anomalies based on manual, semi-automated, and automated methods. The monitoring activities, amongst others, aimed to determine if fires shrink or increase in size, estimate how large the variability of the fire-related thermal anomaly is, and monitor if extinguished fires remain extinguished.

10.4   SELECTED RESULTS

To map the locations of surface and subsurface coal fires in a selected part of the Jharia coalfield, India, we used a recent Landsat image acquired in April 2014, and processed it using the approach outlined by Prakash et al. (1997). It is important to remember that in coal mining areas it is common to find instances where coal fires exist only at the surface, or only in the subsurface, or co-exist on the surface and subsurface due to the geologic setting and fire history. In the subsurface, one or more coal fires may occur in one seam or in different coal seams at different depths. Figure 10.6 shows the processing result.

Figure 10.6a is a false color composite generated by coding the Landsat 8 spectral band 7 (SWIR 2), band 6 (SWIR 1), and band 4 (Red) in red, green, and blue respectively. On this figure surface fires that emit significantly in the SWIR regions stand out as yellow and red pixels. Figure 10.6b is a density-sliced, color-coded TIR image (Landsat 8, band 10) of the same area. On this image red, orange and yellow represent successively warmer land surface temperatures associated with subsurface fires. The gray-scale image in the background shows variation in the background land surface temperatures, with the darkest gray tones associated with the cooler temperatures of the Damodar River. Comparing Figure 10.6a with 10.6b we clearly see a situation marked with a yellow arrow, where there is a large temperature anomaly likely due to a subsurface fire but there is no obvious surface fire present. In contrast there is a small surface fire south of the river (marked with a red arrow on Figure 10.6a), which shows no corresponding thermal anomaly on the TIR image, possibly due to the fact that the thermal signature disappears when averaged over the coarser TIR pixels and because there is no anomalous heating from a co-existing subsurface fire. The blue and green arrows show more complex situations of co-existing fires at the same location or where a subsurface fire is shifted but in proximity to the surface fires, a situation comparable to the conceptual coal fire scenario shown in Figure 10.1 in which a coal seam is on fire where it is exposed at the surface, as well as at a location where it is buried deeper down.

We also processed a VIIRS daytime image (Figure 10.7a) and a VIIRS nighttime image (Figure 10.7b) from mid-February 2023 to map the more recent locations of large active fires in the Jharia coalfield. Figure 10.7c shows the hot spots detected from the VIIRS nighttime image overlaid on a false color composite image generated by coding the Landsat 8 spectral band 7 (SWIR 2), band 6 (SWIR 1), and band 4 (Red) in red, green, and blue, respectively. This image clearly shows the sickle-shaped Jharia coalfield with coal seams in shades of brown and vegetation appearing in green. The Damodar River flows south of the coalfield.

The typical average ambient temperatures in mid-February in this region range from 12°C to 28°C, with diurnal temperature fluctuations in the order of 15 degrees. As anticipated, the temperatures on the daytime images appear higher due to the additive component of solar reflection. The high solar reflection reduces the accuracy of coal fire detection. As an example, the sand bars on the Damodar River, and some barren mounds near the western part of the coalfield show as false positives in the fire maps on the daytime images. The nighttime VIIRS image generates a superior product for fire detection and is one that can be used for routine monitoring of the larger coal fires in the Jharia coalfield.

Remote Sensing–Based Mapping and Monitoring of Coal Fires

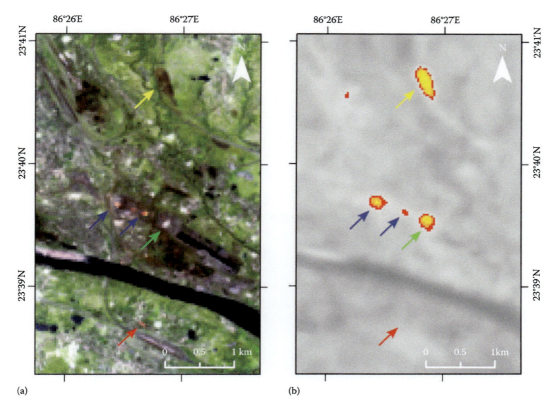

FIGURE 10.6 A comparative analysis of locations of surface and subsurface coal fires in the southeastern part of the Jharia coalfield in India, using Landsat 8 satellite images from April 2014. (a) is a false color composite generated by coding the shortwave infrared bands in red and green colors so that high-temperature surface fires show in shades of yellow and red. (b) is a density-sliced, color-coded thermal infrared image of the same date showing successively higher land surface temperatures (LST) due to subsurface fires in red (36°C to 37°C), orange (37°C to 39°C), and yellow (39°C to 46°C). Fires may occur only on the surface (red arrow), only in the subsurface (yellow arrow), or both on the surface and underground at the same location (blue arrows) or adjacent location (green arrow).

The third example (Figure 10.8) shows results of a coal fire monitoring study in the Wuda coalfield in China that was based on a combination of in situ mapping and processing a Landsat 7 image acquired on December 10, 2002. The right panel shows the fire zones number 1, 2, and 3 in the Wuda coal fire area, and the left panel shows an enlarged view of fire zone 3.

For in situ mapping the field crew traversed the terrain in a predefined grid pattern, logged the GPS track, and recorded the land surface temperature using contact thermometers and handheld radiometers. This helped to map the surface thermal expression, or footprint, of the underground fire (Kuenzer et al., 2008c). In situ mapping was carried out during 2000, 2003, and 2004, and fire boundaries from these surveys are shown as orange, blue, and yellow polygons, respectively. The boundaries of fire zones 1, 2, and 3 are clear for the 2000 survey, which was carried out by mining engineers. Fire 1–related anomalies could not be found in 2003 and 2004 field campaigns. Fire 2 was still active in 2003 but had burned out by 2004. Fire 3 had formed new high-temperature clusters in 2004.

The 2002 Landsat scene was processed using the method presented by Kuenzer et al. (2007a), where an unbiased automatic moving window approach was implemented to extract regional

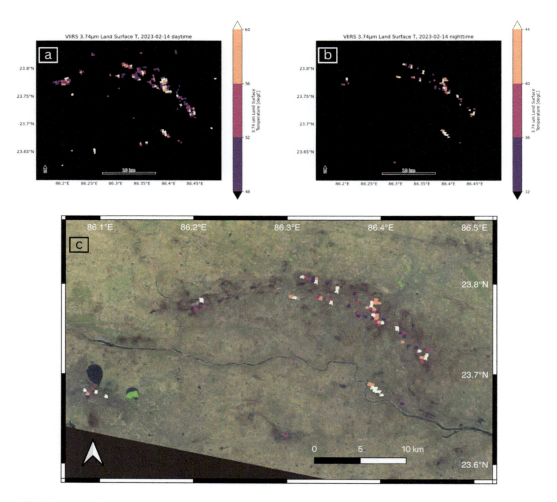

FIGURE 10.7 Coal fires in the Jharia coalfield in India as detected from February 2023 VIIRS thermal infrared daytime image (a) and VIIRS thermal infrared nighttime image (b). Purple, pink, orange, and yellow show successively higher temperatures associated with coal fires. In (c), fires detected from the VIIRS nighttime image have been overlaid on a Landsat 8 false color composite image taken two days later to provide locational context.

thermal anomalies. As opposed to using a single pre-defined threshold for the entire image to delineate fire area, this technique defined variable region-specific threshold values, picking up "relative" hot spots. This implied that a 50°C pixel in a 40°C background was treated the same as a 60°C pixel in a 50°C surrounding, and the algorithm defined both cases as a thermal anomaly. This processing technique was successful in extracting only the hottest areas of fires 1 and 3. Large parts of fire 3 and the entire area of fire 2 remained undetected.

Monitoring studies, as exemplified, are very important. So far no frequent long-term monitoring study has been published to our knowledge. The most extensive monitoring study published has been based on repeated in situ field mappings of the coal fires of Wuda, China (Kuenzer et al., 2008c), depicting that fires can move for several tenths to hundreds of meters per year, and that coal fire dynamics are quite accentuated. This publication also highlighted that remote sensing–based analyses miss many coal fire–related hot spots in a coal fire area, and can only detect the hottest and largest fires.

Remote Sensing–Based Mapping and Monitoring of Coal Fires

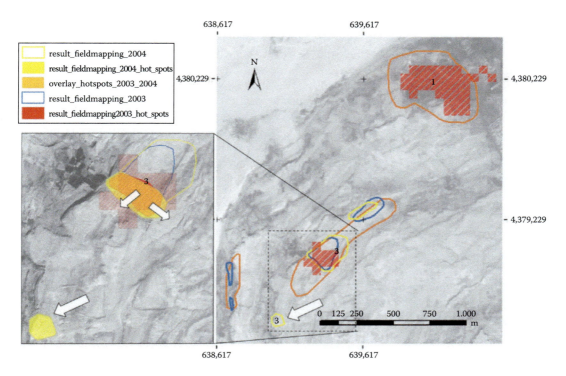

FIGURE 10.8 Coal fire monitoring for part of the Wuda coalfield, China. A panchromatic QuickBird image from October 5, 2003, is used as a background. Coal fire areas mapped in situ in 2000, 2003, and 2004 are shown as orange, blue, and yellow outlined polygons, whereas the hottest areas are depicted as solid filled polygons. The red striped areas are coal fire extents extracted from the thermal infrared band of Landsat ETM+ acquired on December 10, 2002. Only the hottest fire areas could be detected and extracted from the remote sensing images.

10.5 DISCUSSION

Efforts in remote sensing–based coal fire mapping and monitoring have steadily grown since the mid-nineties. However, compared to the research community using remote sensing for studying other high-temperature events, such as volcanic eruptions and wildfires, the coal fire research community is relatively small and transient, in part due to the sporadic nature of funding in this area. Based on the body of existing research and our experience in coal fire studies we outline the issues and limitations of current research, discuss future needs, and make recommendations for remote sensing of coal fires. The following sections are centered on some of the main issues.

10.5.1 COAL FIRE RESEARCH IS LOCAL IN NATURE

In the past decades, most scientists focused on specific coal fire localities and consequently the research publications have a very local, or at best regional, focus. A global, standardized approach to monitor coal fires does not exist to date. While it is true that each coal fire/coal fire area is different, with variability due to locality, climate, weather, bedrock, soils, type of coal, depth of the fire, mining methods, etc., it is important to establish a generally accepted, standardized approach to map and monitor the fires to generate comparable and continuous results. Just as there are established standard algorithms for atmospheric correction, or multispectral image classification, or time series analyses, it would be most useful to have a "coal-fire toolbox" with a set of standardized algorithms and tools to analyze or monitor a coal fire area based on a certain type of thermal data. More refined remote sensing tools for estimating depth of underground coal fires, fire heat flux (e.g., Fischer et al., 2011; Haselwimmer et al., 2013), amount of coal burned, and material requirement for fire containment need to be developed and tested. This will require a larger dedicated team of researchers for developing and testing algorithms to work for varying conditions and standardizing them as an operational tool-set for the global coal fire community.

Currently, the biggest growth in the number of coal fire remote sensing scientists is in China, as this country is one that is most seriously impacted by coal fires. The challenge here is that much of their research is published only in Chinese and thus not accessible to researchers outside of China. However, it is likely that future coal fire research will be driven by the fleet of young researchers there who are investigating further in this area.

10.5.2 Coal Fires Are Dynamic and Baseline Data Does Not Exist

Coal fire–related parameters (such as the occurrence of thermal anomalies, the size of thermal anomalies, the temperature of anomalies, amount of emissions, etc.) fluctuate a lot with local climate, weather, and other influences. To get a reliable estimate of these parameters and quantify the average energy released (that helps to estimate the amount of coal burned) requires analysis of a large volume of thermal data. Most published coal fire research is based on the analysis of only a few images, or at-best four thermal images in a given year. Therefore, a comprehensive baseline data for coal fires, comparable to the baseline data available for volcanoes, does not exist. This is a much required gap. It will be important to see future coal fire remote sensing researchers be involved in processing large volumes of thermal data, (e.g., all available cloud-free Landsat and ASTER TIR scenes for targeted coal fire areas), to establish and grow the baseline data on coal fires.

10.5.3 Timing of Data Acquisition Is Important

Coal fire monitoring literature indicates that there can be significant diurnal temperature variability of natural bedrock surfaces with daytime temperature easily mimicking coal fire–related anomalies. To limit the influence of solar heating, nighttime data is better suited for coal fire analyses. The nighttime thermal data of Landsat is not truly nighttime as the overpass time is around 10 p.m., and it still contains effects of uneven solar heating during the day. Additionally, as there is very limited availability of nighttime TIR images, most studies rely on daytime TIR images, but minimize the effect of solar heating by empirical methods or by using energy budget models (Deng et al., 2001; Haselwimmer and Prakash, 2013).

Pre-dawn is best for thermal anomaly extraction data. MODIS has pre-dawn acquisitions (e.g., images acquired around 4:00 a.m.), which are ideally timed but the images lack sufficient spatial resolution to be useful.

For coal fire studies summertime data should be avoided. Snow-free winter data, or data from spring or fall, are best suited, as here the contrast between the anomaly and the background temperatures will be most accentuated.

10.5.4 Spaceborne TIR Imagery Is Coarse and Discontinuous

The largest gap in coal fire research is indirectly a result of limited data continuity and changing sensor characteristics in the TIR region. While the free Landsat archive now enables optical data access for most spots on the Earth over the past 50+ years, the situation is complicated in the thermal domain. Landsat MSS did not have a thermal band, so Landsat thermal data exists only since the launch of Landsat 4 in 1982. However, the thermal bands of Landsat 4 and Landsat 5 have a coarse spatial resolution of 120 m, which only enables them to pick up thermal anomalies induced by extremely hot and large surface or subsurface fires. Even gaseous emissions at 500°C are insufficient to make a detectable signature on the coarse 120 m thermal pixel, if such emissions come from vents that are only in the order of centimeters in magnitude.

The coal fire research experienced a strong push forward with the launch of Landsat ETM+ in 2001, which had a low gain and a high gain thermal band at 60 m spatial resolution. Unfortunately, Landsat 7 ETM+ experienced a scanner problem in May 2003, and since then hardly any useful thermal data has been available. This was one reason that publications in the late 2000s focused on analyses of data acquired before May 2003, or exploited the potential of ASTER or other low-resolution thermal data, such as MODIS. Although ASTER had five thermal bands enabling emissivity retrievals, it proved to be inferior to Landsat 7 ETM+, for coal fire research due to its coarser 90 m spatial resolution. Unfortunately, ASTER also encountered the failure of its SWIR bands in 2009 and had limited data availability that posed challenges in data analysis. Thus, little coal fire–related research has been published based on thermal data newer than 2009. The launch of Landsat 8 in February 2013 gave new hope to the coal fire community. However, unlike ETM+ at 60 m resolution, this sensor now has a thermal band at only 100 m spatial resolution—a step backwards for coal fire–related research.

It is anticipated that the next years will bring out some repeated local studies in well-known coal fire areas, such as Jharia in India, Wuda in China, or selected fires in other localities based on such Landsat-8, as well as selected China-Brazil Earth Resources Satellite (CBERS) thermal data (similar resolution): however, ground-breaking developments will require a sustained availability of higher spatial resolution thermal data. The optimal sensor for coal fire–related thermal monitoring would have a band or two in the SWIR region and at the least a band in the TIR region with a spatial resolution of 45 m or better. It would have a repeat cycle of less than 15 days with a possibility of on-demand nighttime acquisitions. Furthermore, data would need to be free of charge to facilitate research and grow the community of scientists working in this important area.

In the absence of a reliable source of higher-resolution spaceborne TIR imagery, in situ field mapping along with airborne thermal campaigns are the methods of choice to extract and deliver information for a comprehensive and stakeholder-relevant assessment of coal fires in a region.

10.6 CONCLUSIONS

Coal fires occur on the surface and in the subsurface in coal mines across the globe. They present a threat to the environment and also an opportunity in remote sensing research to characterize, map, and monitor these fires with the intent to support mining operations, firefighters, and local communities. Based on a review of past research on remote sensing of coal fires, and near future plans of space agencies for new satellite launches, we conclude that:

- Remote sensing of coal fires is an area of research that has witnessed slow progress due sporadic and limited funding but has the potential to grow in the era of heightened environmental concerns.
- Coarse spatial resolution images from AVHRR and MODIS type sensors are not useful for providing meaningful fire-specific information. Most research has relied on the use of TIR and SWIR images from Landsat and ASTER, which are good for studying the larger and known fires, but often miss small individual fires.
- VIIRS, especially the nighttime VIIRS data, has proven very useful for coal fire detection and can serve as a cost-effective solution especially for monitoring large active coal fires.
- Pre-dawn cloud-free and snow-free airborne images acquired in the non-summer seasons, coupled with concurrent field investigations, are ideal for coal fire studies.

- More research in remote sensing of coal fires will help to test and improve algorithms for fire detection and characterization, which could then be standardized for wide-scale implementation and global coal fire database generation.
- There is a pressing need in coal fire research for a global spaceborne TIR mission that could acquire TIR images at better than 45 m spatial resolution with at least a repeat coverage of twice in a given month.

10.7 ACKNOWLEDGMENTS

Authors would like to thank Derek Starkenburg for their assistance in generating Figure 10.1.

REFERENCES

Banerjee, S.C., 1985, *Spontaneous Combustion of Coal and Mine Fires*. Rotterdam, Netherlands: Oxford and IBH Publisher Co., 168 p. ISBN: 9061915740, 9789061915744

Ben-Dor, E., 2002, Quantitative remote sensing of soil properties. *Advances in Agronomy*, 75, 173–231.

Berthelote, A.R., Prakash, A., and Dehn, J., 2008, An empirical function to estimate the depths of linear hot sources: Applied to the Kuhio Lava tube, Hawaii. *Bulletin of Volcanology*, 70(7), 813–824.

Biswal, S.S., and Gorai, A.K., 2020, Change detection analysis in coverage area of coal fire from 2009 to 2019 in Jhari a Coalfield using remote sensing data. *International Journal of Remote Sensing*, 41(24), 9545–9564. http://doi.org/10.1080/01431161.2020.1800128.

Bonforte, A., Federico, C., Giammanco, S., Guglielmino, F., Liuzzo, M., and Neri, M., 2013, Soil gases and SAR measurements reveal hidden faults on the sliding flank of Mt. Etna (Italy). *Journal of Volcanology and Geothermal Research*, 251, 27–40.

Carras, J.N., Day, S.J., Saghafi, A., and Williams, D.J., 2009, Greenhouse gas emissions from low-temperature oxidation and spontaneous combustion at open-cut coal mines in Australia. *International Journal of Coal Geology*, 78(2), 161–168.

Cassells, C.J.S., 1998, *Thermal Modelling of Underground Coal Fires in Northwest China*. University of Dundee, Dundee. (ITC publication 51).

Chander, G., Markham, B.L., and Helder, D.L., 2009, Summary of current radiometric calibration coefficients for Landsat MSS, TM, ETM+, and EO-1 ALI sensors. *Remote Sensing of Environment*, 113, 893–903.

Chatterjee, R.S., Shah, A., Raju, E.V.R., Lakhera, R.C., and Dadhwal, V.K., 2007, Dynamics of coal fire in Jharia Coalfield, Jharkhand, India during the 1990s as observed from space. *Current Science*, 92(1), 61–68.

Chen, H., Qi, H., Long, H., and Zhang, M., 2012, Research on 10-year tendency of China coal mine accidents and the characteristics of human factors. *Safety Science*, 50(4), 745–750.

Cristóbal, J., Jiménez-Muñoz, J.C., Prakash, A., Mattar, C., Skoković, D., and Sobrino, J.A., 2018, An improved single-channel method to retrieve land surface temperature from the Landsat-8 thermal band. *Remote Sensing*, 10, 431. https://doi.org/10.3390/rs10030431.

Cristóbal, J., Jiménez-Muñoz, J.C., Sobrino, J.A., Ninyerola, M., and Pons, X., 2009, Improvements in land surface temperature retrieval from the Landsat series thermal band using water vapour and air temperature. *Journal of Geophysical Research*, 114, D08103, 1–16.

Deng, W., Wan, Y.Q., and Zhao, R.C., 2001, Detecting coal fires with a neural network to reduce the effect of solar radiation on Landsat Thematic Mapper thermal infrared images. *International Journal of Remote Sensing*, 22, 933–944.

Du, X., Sun, D., Li, F., and Tong, J., 2022, A study on the propagation trend of underground coal fires based on night-time thermal infrared remote sensing technology. *Sustainability*, 14, 14741. https://doi.org/10.3390/su142214741

Ellyett, C.D., and Fleming, A.W., 1974, Thermal infrared imagery of the Burning Mountain coal fire. *Remote Sensing of Environment*, 3, 79–86.

Engle, M.A., Radke, L.F., Heffern, E.L., O'Keefe, J., Hower, J.C., Smeltzer, C.D., Hower, J.M., Olea, R., Eatwell, R.J., Blake, D., Emsbo-Mattingly, S.D., Stout, S.A., Queen, G., Aggen, K.L., Kolker, A., Prakash, A., Henke, K.R., Stracher, G.B., Schroeder, P.A., Román-Colón, Y., and ter Schure, A., 2012a, Gas emissions, minerals, and tars associated with three coal fires, Powder River Basin, USA. *Science of Total Environment*, 420, 146–159.

Engle, M.A., Radke, L.F., Heffern, E.L., O'Keefe, J., Smeltzer, C.D., Hower, J.C., Hower, J.M., Prakash, A., Kolker, A., Eatwell, R.J., ter Schure, A., Queen, G., Aggen, K.L., Stracher, G.B., Henke, K.R., Olea, R., and Román-Colón, Y., 2012b, Quantifying greenhouse gas emissions from coal fires using airborne and ground-based methods. *International Journal of Coal Geology*, 88(2–3), 147–151.

Finkelman, R.B., 2004, Potential health impacts of burning coal beds and waste banks. *International Journal of Coal Geology*, 59(1–2), 19–24.

Fischer, C., Li, J., Ehrler, C., and Wu, J., 2011, Radiative energy release quantification of subsurface coal fires. *Proceedings of the ISPRS Conference*. Available at http://www.isprs.org/proceedings/2011/isrse-34/211104015Final00345.pdf

Gangopadhyay, P.K., 2007, Application of remote sensing in coal-fire studies and coalfire related emissions. *Reviews in Engineering Geology*, 18, 239–248.

Gangopadhyay, P.K., Maathuis, B., and Dijk, P. van., 2005, ASTER derived emissivity and coal fire related surface temperature anomaly: A case study in Wuda, North China. *International Journal of Remote Sensing*, 24, 5555–5751.

Genderen, J.L. van, Prakash, A., Gens, R., Veen, B.S. van, Liding, C., Tao, T.X., and Feng, G., 2000, *Coal Fire Interferometry*. Netherlands Remote Sensing Board, USP-2-99-32. Delft: BCRS. Available at https://www.osti.gov/etdeweb/biblio/20171154

Gens, R., 2009, Spectral information content of remote sensing imagery. In Li Deren, et al. (eds.) *Geospatial Technology for Earth Observation*. New York: Springer, 558 p. ISBN-13: 978-1441900494.

Gens, R., 2013, Global distribution of coal and peat fires. In *Interactive World Map of Coal and Peat Fires Made to Accompany Stracher et al. (eds.) Coal and Peat Fires: A Global Perspective, Volume 2, Photographs and Multimedia Tours*. Elsevier. Available at http://booksite.elsevier.com/brochures/coalpeatfires/interactivemap.html

Gens, R., and Cristobal, J.R., 2015, Remote sensing data normalization. Chapter 5 this book.

Gillespie, A., Rokugawa, S., Matsunaga, T., Cothern, J.S., Hook, S., and Kahle, A.B., 1998, A temperature and emissivity separation algorithm for advanced spaceborne thermal emission and reflection radiometer (ASTER) images. *IEEE Transactions on Geoscience and Remote Sensing*, 36(4), 1113–1126.

Glover, S.S., 2011, *Coal Mining in Jefferson County*. Alabama: Arcadia Publishing, 128 p. ISBN: 0738582174, 9780738582177.

Gupta, R.P., 2003, *Remote Sensing Geology*, 2nd edition. Berlin: Springer, 656 p. ISBN-13: 978–3540431855.

Gupta, R.P., and Prakash, A., 1998, Reflection aureoles associated with thermal anomalies due to subsurface mine fires in the Jharia Coalfield, India. *International Journal of Remote Sensing*, 19(14), 2619–2622.

Haselwimmer, C., and Prakash, A., 2013, Chapter 22—Thermal infrared remote sensing of geothermal systems. In C. Kuenzer and S. Dech (eds.) *Thermal Infrared Remote Sensing—Sensor, Methods, Applications*. Remote Sensing and Digital Image Processing Series, Vol. 17. Dordrecht: Springer, pp. 453–474, 537 p. ISBN: 978-9400766389.

Haselwimmer, C., Prakash, A., and Holdmann, G., 2013, Quantifying the heat flux and outflow rate of hot springs using airborne thermal imagery: Case study from Pilgrim Hot Springs, Alaska. *Remote Sensing of Environment*, 136, 37–46.

He, X., Yang, X., Luo, Z., and Guan, T., 2020, Application of unmanned aerial vehicle (UAV) thermal infrared remote sensing to identify coal fires in the Huojitu coal mine in Shenmu city, China. *Scientific Reports*, 10(1), 13895.

Hecker, C., Kuenzer, C., and Zhang, J., 2007, Remote sensing based coal fire detection with low resolution MODIS data. *Reviews in Engineering Geology*, 18, 229–239.

Hook, S.J., 1992, A comparison of techniques for extracting emissivity information from thermal infrared data for geologic studies. *Remote Sensing of Environment*, 42(2), 123–135.

Hower, J.C., Henke, K., O'Keefe, J.M.K., Engle, M.A., Blake, D.R., and Stracher, G.B., 2009, The Tiptop coalmine fire, Kentucky: Preliminary investigation of the measurement of mercury and other hazardous gases from coal-fire gas vents. *International Journal of Coal Geology*, 80(1), 63–67.

Hulley, G.C., Hughes, C.G., and Hook, S.J., 2012, Quantifying uncertainties in land surface temperature and emissivity retrievals from ASTER and MODIS thermal infrared data. *Journal of Geophysical Research*, 117, D23113. http://doi.org/10.1029/2012JD018506.

Hummel, J.W., Suddutha, K.A., and Hollingerb, S.E., 2001, Soil moisture and organic matter prediction of surface and subsurface soils using an NIR soil sensor. *Computers and Electronics in Agriculture*, 32(2), 149–165.

Jiang, L., Lin, H., Ma, J., Kong, B., and Wang, Y, 2011, Potential of small-baseline SAR interferometry for monitoring land subsidence related to underground coal fires: Wuda (Northern China) case study. *Remote Sensing of Environment*, 115(2), 257–268.

Jiménez-Muñoz, J.C., Sobrino, J.A., Skoković, D., Mattar, C., and Cristóbal, J., 2014, Land surface temperature retrieval methods from Landsat-8 thermal infrared sensor data. *IEEE Geoscience and Remote Sensing Letters*, 11(10), 1840–1843.

Karanam, V., Motagh, M., Garg, S., and Jain, K., 2021, Multi-sensor remote sensing analysis of coal fire induced land subsidence in Jharia Coalfields, Jharkhand, India. *International Journal of Applied Earth Observation and Geoinformation*, 102, 102439. ISSN 1569–8432. https://doi.org/10.1016/j.jag.2021.102439.

Kim, A.J., 2010, Coal formation and the origin of coal fires. In G.B. Stracher, A. Prakash and E.V. Sokol (eds.) *Coal and Peat Fires: A Global Perspective, Volume1, Coal—Combustion and Geology*. Oxford: Elsevier.

Kim, A.J., Lin, S.-Y., Singh, R.P., Lan, C.-W., and Yun, H.-W., 2021, Underground burning of Jharia coal mine (India) and associated surface deformation using InSAR data. *International Journal of Applied Earth Observation and Geoinformation*, 103, 102524. ISSN 1569–8432. https://doi.org/10.1016/j.jag.2021.102524.

Kolker, A., Engle, M., Stracher, G., Hower, J., Prakash, A., Radke, L., ter Schure, A., and Heffern, E., 2009, Emissions from coal fires and their impact on the environment. *U.S. Geological Survey Fact Sheet*, 2009–3084, 4 p.

Kuenzer, C., 2015, Chapter X, this book

Kuenzer, C., Bachmann, M., Mueller, A., Lieckfeld, L., and Wagner, W., 2008a, Partial unmixing as a tool for single surface class detection and time series analysis. *International Journal of Remote Sensing*, 29(11), 1–23.

Kuenzer, C., and Dech, S., 2013, Theoretical background of thermal infrared remote sensing. In C. Kuenzer and S. Dech (eds.) *Thermal Infrared Remote Sensing—Sensors, Methods, Applications*. Remote Sensing and Digital Image Processing Series, Vol. 17. Dordrecht: Springer, pp. 1–26, 572 p. ISBN: 978-94-007-6638-9.

Kuenzer, C., Hecker, C., Zhang, J., Wessling, S., Kuenzer, C., and Wagner, W., 2008b, The potential of multidiurnal MODIS thermal band data for coal fire detection. *International Journal of Remote Sensing*, 29(3), 923–944.

Kuenzer, C., and Stracher, G.B., 2011, Geomorphology of coal seam fires. *Geomorphology*, 138, 209–222.

Kuenzer, C., Voigt, S., and Morth, D., 2005, Investigating land cover changes in Chinese coal mining environments using partial unmixing, *GGRS Conference: Applications in Geo-Sciences*. Göttingen, Germany, pp. 31–37. Available at https://elib.dlr.de/57851/

Kuenzer, C., Wessling, S., Zhang, J., Litschke, T., Schmidt, M., Schulz, J., Gielisch, H., and Wagner, W., 2007c, Concepts for green house gas emission estimating of underground coal seam fires. *Geophysical Research Abstracts*, Vol. 9, 11716, EGU 2007, 16–20th April 2007, Vienna.

Kuenzer, C., Zhang, J., Hirner, A., Bo, Y., Jia, Y., and Sun, Y., 2008c, Multitemporal in-situ mapping of the Wuda coal fires from 2000 to 2005—assessing coal fire dynamics. In *UNESCO Beijing, 2008, Spontaneous Coal Seam Fires: Mitigating a Global Disaster*. ERSEC Ecological Book Series, Vol. 4. Beijing: Springer, pp. 132–148, 606 p. ISBN: 978-7-302-17140-9.

Kuenzer, C., Zhang, J., Li, J., Voigt, S., Mehl, H., and Wagner, W., 2007a, Detecting unknown coal fires: Synergy of automated coal fire risk area delineation and improved thermal anomaly extraction. *International Journal of Remote Sensing*, 28(20), 4561–4585.

Kuenzer, C., Zhang, J., Sun, Y., Jia, Y., and Dech, S., 2012, Coal fires revisited: The Wuda coal field in the aftermath of extensive coal fire research and accelerating extinguishing activities. *International Journal of Coal Geology*, 102, 75–86.

Kuenzer, C., Zhang, J., Tetzlaff, A., and Dech, S., 2013, Thermal infrared remote sensing of surface and underground coal fires. In C. Kuenzer and S. Dech (eds.) *Thermal Infrared Remote Sensing—Sensors, Methods, Applications*, Vol. 17. Remote Sensing and Digital Image Processing Series. Dordrecht: Springer, pp. 429–451, 572 p. ISBN: 978-94-007-6638-9.

Kuenzer, C., Zhang, J., Tetzlaff, A., Dijk, P, van, Voigt, S., Mehl, H., and Wagner, W., 2007b, Uncontrolled coal fires and their environmental impacts: Investigating two arid mining regions in north-central China. *Applied Geography*, 27, 42–62.

Kuenzer, C., Zhang, J., Tetzlaff, A., Voigt, S., and Wagner, W., 2008d, Automated demarcation, detection and quantification of coal fires in China using remote sensing data. In *UNESCO Beijing, 2008, Spontaneous Coal Seam Fires: Mitigating a Global Disaster*. ERSEC Ecological Book Series, Vol. 4. Beijing: Springer, pp. 362–380, 602 p. ISBN: 978-7-302-17140-9.

Kuenzer, C., Zhang, J., Voigt, S., and Wagner, W., 2007d, Remotely sensed land-cover changes in the Wuda and Ruqigou-Gulaben coal mining areas China. *Reviews in Engineering Geology*, 18, 219–228.

Kumar, A., and Pandey, A.C., 2013, Evaluating impact of coal mining activity on landuse/landcover using temporal satellite images in South Karanpura coalfields and environs, Jharkhand State, India. *International Journal of Advanced Remote Sensing and GIS*, 2(1), 183–197.

Lambin, E.F., and Ehrlich, D., 1996, The surface temperature-vegetation index space for land cover and land-cover change analysis. *International Journal of Remote Sensing*, 17(3), 463–487.

Liu, J., Wang, Y., Yan, S., Zhao, F., Li, Y., Dang, L., Liu, X., Shao, Y., and Peng, B., 2021, Underground coal fire detection and monitoring based on Landsat-8 and Sentinel-1 data sets in miquan fire area, XinJiang. *Remote Sensing*, 13, 1141.

Mansor, S.B., Cracknell, A.P., and Shilin, B.V., 1994, Monitoring of underground coal fires using thermal infrared data. *International Journal of Remote Sensing*, 15, 1675–1685.

Martha, T.R., Guha, A., Kumar, K.V., Kumaraju, M.V.V., and Raju, E.V.R., 2010, Recent coal-fire and land-use status of Jharia Coalfield, India from satellite data. *International Journal of Remote Sensing*, 31(12), 3243–3262.

Matson, M., and Dozier, J., 1981, Identification of subresolution high temperature sources using a thermal IR sensor. *Photogrammetric Engineering and Remote Sensing*, 47(9), 1311–1318.

Mukherjee, T.K., Bandhopadhyay, T.K., and Pande, S.K., 1991, Detection and delineation of depth of subsurface coalmine fires based on an airborne multispectral scanner survey in part of Jharia Coalfield, India. *Photogrammetric Engineering and Remote Sensing*, 57, 1203–1207.

Panigrahi, D.C., Singh, M.K., and Singh, C., 1995, Predictions of depth of mine fire from the surface by using thermal infrared measurement. *Proceedings of the National Seminar on Mine Fires, Varanasi*, India, 24th–25th February 1995. Banaras Hindu University, Varanasi, pp. 122–134.

Peng, W.X., Genderen, J.L. van, Kang, G.F., Guan, H.Y., and Yongjie, T., 1997, Estimating the depth of underground coal fires using data integration techniques. *Terra Nova*, 9(4), 180–183.

Prakash, A., 2010, Coal fire research: Heading from remote sensing to remote measurement. *Second International Conference on Coal Fire Research*, May 19–21, Berlin, Germany.

Prakash, A., 2014, *Coal Fires*. http://www2.gi.alaska.edu/~prakash/coalfires/causes_hazards.html (last accessed April, 2014).

Prakash, A., and Berthelote, A.R., 2007, Subsurface coal mine fires: Laboratory simulation, numerical modeling and depth estimation. *Reviews in Engineering Geology*, 18, 211–218.

Prakash, A., Fielding, E.J., Gens, R., Genderen, J.L. van, and Evans, D.L., 2001, Data fusion for investigating land subsidence and coalfire hazards in a coal mining area. *International Journal of Remote Sensing*, 22(6), 921–932.

Prakash, A., and Gens, R.P., 2010, Remote sensing of coal fires. In G.B. Stracher, A. Prakash and E.V. Sokol (eds.) *Coal and Peat Fires: A Global Perspective, Volume 1, Coal—Combustion and Geology*. Oxford: Elsevier.

Prakash, A., Gens, R.P., and Vekerdy, Z., 1999, Monitoring coal fires using multi-temporal night-time thermal images in a coalfield in North-west China. *International Journal of Remote Sensing*, 20(14), 2883–2888.

Prakash, A., and Gupta, R.P., 1998, Land-use mapping and change detection in a coal mining area—a case study of the Jharia Coalfield, India. *International Journal of Remote Sensing*, 19(3), 391–410.

Prakash, A., and Gupta, R.P., 1999, Surface fires in Jharia Coalfield, India—their distribution and estimation of area and temperature from TM data. *International Journal of Remote Sensing*, 20(10), 1935–1946.

Prakash, A., Gupta, R.P., and Saraf, A.K., 1997, A Landsat TM based comparative study of surface and subsurface fires in the Jharia Coalfield, India. *International Journal of Remote Sensing*, 18(11), 2463–2469.

Prakash, A., Saraf, A.K., Gupta, R.P., Dutta, M., and Sundaram, R.M., 1995a, Surface thermal anomalies associated with underground fires in Jharia Coal Mine, India. *International Journal of Remote Sensing*, 16(12), 2105–2109.

Prakash, A., Sastry, R.G.S., Gupta, R.P., and Saraf, A.K., 1995b, Estimating the depth of buried hot feature from thermal IR remote sensing data, a conceptual approach. *International Journal of Remote Sensing*, 16(13), 2503–2510.

Prakash, A., Schaefer, K., Witte, W.K., Collins, K., Gens, R., and Goyette, M., 2011, Remote sensing—GIS based investigation of a boreal forest coal fire. *International Journal of Coal Geology*, 86(1), 79–86.

Prakash, A., and Vekerdy, Z., 2004, Design and implementation of a dedicated prototype GIS for coal fire investigations in North China: Challenges met and lessons learnt. *International Journal of Coal Geology*, 59, 107–119.

Quattrochi, D.A., Prakash, A., Evena, M., Wright, R., Hall, D.K., Anderson, M., Kustas, W.P., Allen, R.G., Pagano, T., and Coolbaugh, M.F., 2009, Thermal remote sensing: Theory, sensors, and applications. In Mark Jackson (ed.) *Manual of Remote Sensing 1.1: Earth Observing Platforms & Sensors*. Baton Rouge: ASPRS, 550 p. ISBN: 1-57083-086-4.

Raju, A., Gupta, R.P., and Prakash, A., 2013, Delineation of coalfield surface fires by thresholding Landsat TM-7 day-time image data. *Geocarto*, 28(4), 343–363.

Ray, S.K., and Singh, R.P., 2007, Recent developments and practices to control fire in undergound coal mines. *Fire Technology*, 43, 285–300.

Riyas, M.J., Syed, T.H., Kumar, H., and Kuenzer, C., 2021, Detecting and analyzing the evolution of subsidence due to coal fires in Jharia Coalfield, India using Sentinel-1 SAR data. *Remote Sensing*, 13, 1521. https://doi.org/10.3390/rs13081521

Rosema, A., Guan, H., van Genderen, J.L., Veld, H., Vekerdy, Z., Ten Katen, A.M., and Prakash, A.P., 1999, *Manual of Coal Fire Detection and Monitoring*. Report of the Project 'Development and implementation of a coal fire monitoring and fighting system in China'. Netherlands Institute of Applied Geoscience, Utrecht, NITG 99-221-C, ISBN: 90-6743-640-2, 245 p.

Saraf, A.K., Prakash, A., Sengupta, S., and Gupta, R.P., 1995, Landsat TM data for estimating ground temperature and depth of subsurface coal fire in Jharia Coal Field, India. *International Journal of Remote Sensing*, 16(12), 2111–2124.

Schmugge, T., Hook, S.J., and Coll, C., 1998, Recovering surface temperature and emissivity from thermal infrared multispectral data. *Remote Sensing of Environment*, 65(2), 121–131.

Schott, J.R., Hook, S.J., Barsi, J.A., Markham, B.L., Miller, J., Paduala, F.P., and Raqueno, N.G., 2012, Thermal infrared radiometric calibration of the entire Landsat 4, 5, and 7 archive (1982–2010). *Remote Sensing of Environment*, 122, 41–49.

Sobrino, J.A., Jiménez-Muñoz, J.C., Sòria, G., Romaguera, M., Guanter, L., Moreno, J., Plaza, A., and Martínez, P., 2008, Land surface emissivity retrieval from different VNIR and TIR sensors. *IEEE Transactions on Geoscience and Remote Sensing*, 46, 316–327.

Sobrino, J.A., Raissouni, N., Simarro, J., Nerry, F., and François, P., 1999, Atmospheric water vapor content over land surfaces derived from the AVHRR data: Application to the Iberian Peninsula. *IEEE Transactions on Geoscience and Remote Sensing*, 37, 1425–1434.

Stracher, G.B., Prakash, A., Schroeder, P., McCormack, J., Zhang, X.M., and van Dijk, P., 2005, New mineral occurrences and mineralization processes: Wuda coal-fire gas vents of Inner Mongolia. *American Mineralogist*, 90(11–12), 1729–1739.

Stracher, G.B., Prakash, A., and Sokol, E.V. (eds.), 2010, *Coal and Peat Fires: A Global Perspective, Volume 1, Coal- Combustion and Geology*. Amsterdam: Elsevier, 335 p. ISBN: 978-0444528582.

Stracher, G.B, Prakash, A., and Sokol, E.V. (eds.), 2013, *Coal and Peat Fires: A Global Perspective, Volume 2, Photographs and Multimedia Tour*. Amsterdam: Elsevier, 564 p. ISBN: 978-0444594129.

Stracher, G.B., and Taylor, T.P., 2004, Coal fires burning out of control around the world: Thermodynamic recipe for environmental catastrophe. *International Journal of Coal Geology*, 59, 7–18.

Tetzlaff, A., 2004, *Coal Fire Quantification Using ASTER, ETM, and BIRD Satellite Instrument Data*. LMU Munich Ph.D. Dissertation. http://edoc.ub.uni-muenchen.de/4398/ (Last accessed, April 2014).

van Dijk, P., Zhang, J., Jun, W., Kuenzer, C., and Wolf, K.H., 2011, Assessment of the contribution of in-situ combustion of coal to greenhouse gas emission; based on a comparison of Chinese mining information to previous remote sensing estimates. *International Journal of Coal Geology*, 86(1), 108–119.

Voigt, S., Tetzlaff, A., Zhang, J., Künzer, C., Zhukov, B., Strunz, G., Oertel, D. Roth, A., Dijk, P.M. van, and Mehl, H., 2004, Integrating satellite remote sensing techniques for detection and analysis of uncontrolled coal seam fires in North China. *International Journal of Coal Geology*, 59(1–2), 121–136.

Waigl, C., Prakash, A., Ferguson, A., and Stuefer, M., 2014, Chapter 24—Delineating coal fire hazards in high latitude coal basins: A case study from interior Alaska. In G.B. Stracher, A. Prakash and E.V. Sokol (eds.) *Coal and Peat Fires: A Global Perspective, Volume 3, Case Studies* (in press).

Wang, T., Wang, Y., Zhao, F., Feng, H., Liu, J., Zhang, L., Zhang, N., Yuan, G., and Wang, D., 2022, A spatiotemporal temperature-based thresholding algorithm for underground coal fire detection with satellite thermal infrared and radar remote sensing. *International Journal of Applied Earth Observation and Geoinformation*, 110, 102805. ISSN 1569-8432. https://doi.org/10.1016/j.jag.2022.102805.

Wessling, S., Kessels, W., Schmidt, M., and Krause, U., 2008a, Investigating dynamic underground coal fires by means of numerical simulation. *Geophysical Journal International*, 172(1), 439–454.

Wessling, S., Kuenzer, C., Kessels, W., and Wuttke, M.W., 2008b, Numerical modeling for analyzing thermal surface anomalies induced by underground coal fires. *International Journal of Coal Geology*, 74, 175–184.

Whitehouse, A.E., and Mulyana, A.A.S., 2004, Coal fires in Indonesia. *International Journal of Coal Geology*, 59, 91–97.

Yang, B., Li, J., Chen, Y., Zhang, J., and Kuenzer, C., 2008, Automated detection and extraction of surface cracks from high resolution Quickbird imagery. In *UNESCO Beijing, 2008, Spontaneous Coal Seam Fires: Mitigating a Global Disaster*. ERSEC Ecological Book Series, Vol. 4. Beijing: Springer, pp. 381–389, 602 p. ISBN: 978-7-302-17140-9.

Zhang, J., and Kuenzer, C., 2007, Thermal surface characteristics of coal fires: Results of in-situ measurements. *Journal of Applied Geophysics*, 63, 117–134.

Zhang, J., Kuenzer, C., Tetzlaff, A., Oertl, D., Zhukov, B., and Wagner, W., 2007, Thermal characteristics of coal fires 2: Results of measurements on simulated coal fires. *Journal of Applied Geophysics*, 63, 135–147.

Zhang, J., Wagner, W., Prakash, A., Mehl, H., and Voigt, S., 2004a, Detecting coal fires using remote sensing techniques. *International Journal of Remote Sensing*, 25(16), 3193–3220.

Zhang, X.M., Genderen, J.L. van, and Kroonenberg, S.B., 1997, A method to evaluate the capability of landsat-5 TM band 6 data for sub-pixel coal fire detection. *International Journal of Remote Sensing*, 18, 3279–3288.

Zhang, X.M., Zhang, J., Kuenzer, C., Voigt, S., and Wagner, W., 2004b, Capability evaluation of 3–5μm and 8–12,5μm airborne thermal data for underground coalfire detection. *International Journal of Remote Sensing*, 25(12), 2245–2258.

Zheng, Y., Feng, C., Jing, G., Qian, X., Li, X., and Liu, Z., 2009, A statistical analysis of coal mine accidents caused by coal dust explosions in China. *Journal of Loss Prevention in the Process Industries*, 22(4), 528–532.

Zhukov, B., Lorenz, E., Oertel, D., Wooster, M., and Robert, G., 2006, Spaceborne detection and characterization of fires during the bi-spectral infrared detection (BIRD) experimental small satellite mission (2001–2004). *Remote Sensing of Environment*, 100(1), 29–51.

Part V

Urban

11 Urban Growth and Climatic Mapping of Mega Cities
Multi-Sensor Approach

Hasi Bagan, Chaomin Chen, and Yoshiki Yamagata

ACRONYMS AND DEFINITIONS

ALOS	Advanced Land Observing Satellite
AVNIR-2	Advanced visible and near infrared radiometer type 2
DEM	Digital elevation model
DMSP	Defense Meteorological Satellite Program
GCP	Ground control points
GIS	Geographic information system
LiDAR	Light detection and ranging
LSTs	Land surface temperatures
OLS	Operational linescan system
PRISM	Panchromatic remote-sensing instrument for stereo mapping
SAR	Synthetic aperture radar
SRTM	Shuttle radar topography mission

11.1 INTRODUCTION

Urban areas occupy a relatively small fraction of the Earth's land area, but at present more than half of the global population lives in urban areas, and this proportion is expected to increase in the coming decades (http://esa.un.org/unup/). Urban areas contribute significantly to climate change as a result of the use of fossil fuels for electricity generation, transportation, and industry. Already the intensive burning of carbon fuels in the world's urban areas accounts for about 70% of global greenhouse gas emissions (Solecki et al., 2013). In addition, previous research suggests that a 10% increase in urban land cover in a country is associated with an increase of more than 11% in the country's total CO_2 emissions (Angel et al., 2011). Urban form and structure are key factors that determine urban climate, energy use, and emissions, and urbanization fundamentally changes the urban form and urban spatial structure, including the number of buildings, their geometry, pattern, distribution, and density (Frolking et al., 2013; Chen et al., 2022).

Along with the increasing amount of human activity that accompanies urban development, the energy balances at and near the surface are altered in urban areas, typically resulting in higher air and surface temperatures in urban areas than in surrounding rural areas (i.e., the urban heat island phenomenon) (Voogt and Oke, 2003; Masson et al., 2020). The concept of a "local climate zone" (LCZ) was introduced in 2012 by Stewart and Oke to quantify the relationship between urban morphology and the urban heat island phenomenon. The LCZ framework provides a good opportunity to link land cover types and urban form with corresponding thermal properties, which can further quantify the association between the urban internal spatial structure and the surface thermal environment. LCZ maps have become increasingly valuable in terms of urban form and urban climate, which can help support urban planning, policymaking, and climate-responsive designs. From a

DOI: 10.1201/9781003541417-16

policy perspective, climate change adaptation strategies for cities are becoming increasingly important. The identification of surface temperature differences for each LCZ class provides insight into the prioritization and targeting of such mitigation and adaptation strategies.

Monitoring and mapping of urban growth and developing effective urban planning strategies requires spatio-temporal extent and expansion trends of cities. Urban land use and land cover changes are linked to socioeconomic activities (Lambin et al., 2003; Small and Cohen, 2004; Doll et al., 2006; Avelar et al., 2009), and urbanization includes both the physical growth of a city and the movement of people to urban areas. Therefore, it is essential to combine remote sensing–derived parameters with socioeconomic parameters to analyze the spatio-temporal changes of urban growth (Bagan and Yamagata, 2012). In addition, little attention has been paid to the quantitative analysis of relationships among land cover category changes (Bagan and Yamagata, 2014).

Remote sensing and techniques have already proven useful for mapping and modeling urban growth and climate. Various available global data sets are used for measuring, analyzing, and, hence, understanding the complex processes of urbanization (Potere et al., 2009). Examples are global urban extent maps based on, e.g., NOAA Air Force Defense Meteorological Satellite Program (DMSP) Operational Linescan System (OLS) sensor nighttime lights imagery (Elvidge et al., 2007; Doll, 2008) or MODIS data (Schneider et al., 2009). The distribution of cities around the world broadly corresponds to the brightness distribution of DMSP nighttime lights (Zhang and Seto, 2011; Parés-Ramos et al., 2013). However, no single DMSP brightness threshold is valid for extracting the urban extent of all cities because small settlements that are not frequently lit are likely to be excluded (Small et al., 2005). Although coarse spatial resolution (from 250 m to 2 km) monitoring provides global and national estimates of urban growth, coarse data may be less reliable for correctly estimating the urban area of cities and often results in either overestimation or underestimation of urban areas (Potere and Schneider, 2007). Townshend and Justice (2002) argued that "a substantial proportion of the variability of land cover change has been shown to occur at resolutions below 250 m," and Giri et al. (2013) reported that land parcels managed at a local scale are often smaller than the resolution of coarse spatial resolution satellite data. Thus, coarse spatial resolution products still lack appropriate temporal, spatial, and/or thematic resolution to effectively support detailed analyses on the characteristics of cities and analyzing changes over time (Taubenböck et al., 2014).

In recent years, higher-resolution and more multisource sensor systems are available for monitoring the spatial effects of urbanization and mapping LCZs that provide spatial information content that is hundreds of times better than coarse spatial resolution data sets. Small (2003) compares 14 cities at a very high geometric level using QuickBird data with a sub-meter geometric resolution and derives parameters such as vegetation fraction; Berger et al. (2013) extract urban land cover information from high spatial resolution multi-spectral and light detection and ranging (LiDAR) data; Taubenböck et al. (2012) analyze the spatial effects of urbanization over a span of almost 40 years using Landsat data. Currently, the data utilized for LCZ classification comprise mainly optical satellite images (e.g., Landsat and Sentinel-2). The combination of multisource remote sensing and geographic information system (GIS) data is the current trend associated with diverse LCZ classification methods. For example, optical imagery (e.g., Landsat and Sentinel-2) and synthetic aperture radar (SAR) (e.g., Sentinel-1, PALSAR, and PALSAR-2) data were combined to map LCZs (La et al., 2020; Chen et al., 2021; Hou et al., 2023; Huang et al., 2023). Chen et al. (2023) combined airborne LiDAR data, Sentinel-2 imagery, and GIS vector data to develop a multiscale automated LCZ classification scheme.

This chapter has two major objectives with a goal of demonstrating the value of remote sensing in urban studies. These objectives are illustrated taking three case studies. In the first study (Section 11.3.1–11.3.3), we used grid cells to investigate spatial and temporal land cover changes in the Tokyo metropolitan area, specifically combining remote sensing data with population census data to investigate the past and present patterns and trends of urban growth. With 8.5% of its land area is urbanized (GSI, 2012), Japan is one of the world's most urbanized countries. The rapid growth of the Japanese economy has concentrated its population, industry, and other economic activities in metropolitan regions to an extreme degree (Bagan and Yamagata, 2012). Thus, urban growth represents one of the most important land cover/land use change events in recent Japanese history. In the second study (Section 11.3.4), we analyzed the relationship between LCZs and multidate land surface temperatures (LSTs) in the 23 special wards of Tokyo using grid cells. In the third study (Section 11.3.5), we investigated spatial-temporal urban LCZ changes in Shanghai, China.

11.2 GRID CELL PROCESS

Grid cells, the most common mapping units, enable the user to perform integrated data analyses from multiple sources and to achieve a balance between the feasibility of computations and the need for details. In addition, the advantage of grid cell system is that it can avoid the potential problem of changing boundaries of administrative units during the time interval of interest. Thus, grid cells with unique IDs enabled us to link multisource data (e.g., population census, nighttime lights, and LSTs) with land cover maps for spatial and temporal land cover change analysis. The grid cells enabled us to aggregate the categories for each map and to calculate their proportions in basic grid cells. To do this, it is necessary to represent the land cover maps in grid square cells.

Recently, spatial-temporal analyses of land cover changes using 1 km^2 grid cells have demonstrated that grid cells provide a new way to obtain spatial and temporal information about areas that are smaller than the municipal scale and uniform in size (Bagan and Yamagata, 2012; Qian et al., 2014) and to further develop the change dynamics analysis in order to better characterize the phenomena using limited available data. The relationships among changes in urban land use/land cover patterns can be better understood if these data are mapped onto a grid composed of square grid cells.

11.3 MULTI-SENSOR APPROACH FOR URBAN MAPPING

11.3.1 Spatial-Temporal Changes of Land Cover in Tokyo

The Tokyo Metropolitan Area includes the city and prefecture of Tokyo as well as Kanagawa, Saitama, Chiba, and parts of Ibaraki prefectures (Figure 11.1). The population of the study area reached 37.6 million in 2010, which is about 29.4% of the country's population of 128.1 million, while the area occupy less than 3% of the Japan's land surface (Statistics Bureau, Japan, 2011).

Three Landsat scenes, namely, the center, west, and east, cover the study area, as shown in Figure 11.1. The center scene accounts for about 90% of the total study area; thus, as a convention, we hereafter refer to the mosaic Landsat imagery by the date of the center scene.

We acquired Landsat MSS, TM, and ETM+ images to interpret land use and land cover changes for the study area from four separate dates (nominally 1972, 1987, 2001, and 2011). All Landsat data are processed for standard terrain corrections by the US Geological Survey. Table 11.1 provides information on the image data. Only images from April to November, the green vegetation season, and with low cloud cover were considered to maximize the vegetation information content for each monitoring date. All analyses were based on the optical and thermal infrared bands of the MSS, TM, and ETM+ sensors, while excluding panchromatic bands (Table 11.1).

Precise geometric registration to a common map reference and co-registration between individual images are crucial for ensuring the reliable detection of temporal changes of land cover. All Landsat imagery was geometrically rectified to a common map reference system (UTM map projection Zone 54 North, WGS-84 geodetic datum) using a different number (around 25–36) of ground control points (GCP) for each image. The GCPs were selected using ortho-rectified Advanced Land Observing Satellite (ALOS) RGB color images (2.5 m spatial resolution, derived from Advanced Visible and Near Infrared Radiometer type 2 (AVNIR-2) and Panchromatic Remote-sensing Instrument for Stereo Mapping (PRISM) from 2006 to 2009; data provided by NTT Data, Japan)

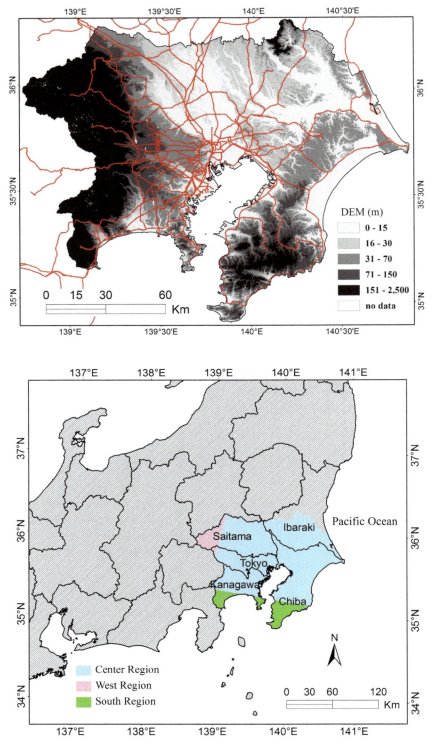

FIGURE 11.1 Left: Site location of the Tokyo metropolitan area. A mosaic of three Landsat images is also shown with the south, center, and west regions shaded in green, blue, and pink, respectively (Table 11.1). Right: The gray image is a digital elevation model (DEM) generated from Shuttle Radar Topography Mission (SRTM) data. Red lines refer to the major railway and metro lines.

TABLE 11.1
The Path/Row and Acquisition Dates for the Landsat Time-Series Scenes Used in This tudy

Sensor	Date Acquired	Path/Row	Location	Spatial (meters)	Spectral Bands
Landsat-1 MSS	1972-11-26	115/35	Center	60	2 bands visible
Landsat-1 MSS	1972-11-26	115/36	South		2 bands infrared
Landsat-1 MSS	1972-11-09	116/35	West		
Landsat-5 TM	1987-05-21	107/35	Center	30	3 bands visible
Landsat-5 TM	1993-05-21	107/36	South		3 bands infrared
Landsat-5 TM	1990-12-06	108/35	West		1 band thermal
Landsat-7 ETM+	2001-09-24	107/35	Center	30	3 bands visible
Landsat-7 ETM+	2001-09-24	107/36	South		3 bands infrared
Landsat-7 ETM+	2001-11-02	108/35	West		1 band thermal
Landsat-5 TM	2011-04-05	107/35	Center	30	3 bands visible
Landsat-5 TM	2011-04-05	107/36	South		3 bands infrared
Landsat-5 TM	2010-04-25	108/35	West		1 band thermal

and the Digital Map 2500 (Spatial Data Framework) at a scale of 1:2500 (provided by the Geospatial Information Authority of Japan [GSI]) as references. The GCPs were well dispersed throughout each scene and yielded root-mean-square errors of less than 0.7 pixels.

Tokyo is a data-rich area in which to study the dynamics of spatial and temporal land cover change patterns. In addition to ALOS data, we acquired and utilized numerous ancillary data for determining typical land cover classes and selecting ground reference sites for each Landsat recording date:

1. Vegetation maps for 1973, 1983–1986, and 1994–1998, with scales ranging from 1:25,000 to 1:200,000, which were derived from field surveys and airborne images (published by the Geospatial Information Authority of Japan);
2. Land use GIS datasets from 1976, 1987, 1997, and 2008 with a spatial resolution of 100 m (published by the Geospatial Information Authority of Japan);
3. High spatial resolution remote sensing data such as IKONOS (acquired in 2003–2007), World View-2 (acquired in 2011), and aerial photography data (acquired in 2005–2007, provided by the Geospatial Information Authority of Japan); and
4. Shuttle Radar Topography Mission (SRTM) elevation data.

Using the field investigation results, GIS datasets, visual interpretation of the remote sensing data, and with consideration of the Landsat scene acquisition dates, we designated five to nine land cover types in this experiment (Table 11.2).

The supervised subspace classification method was applied to each of the four dataset groups (Table 11.1). Subspace methods (Oja, 1983), widely used for pattern recognition and computer vision, have been applied to remote sensing data classification (Bagan et al., 2008; Bagan and Yamagata, 2010). In this method, high-dimensional input data are projected onto a low-dimensional feature space, and the different classes are then represented in their own low-dimensional subspace. Subspace methods proved to be well suited for classifying mosaic Landsat imagery from a heterogeneous and dynamic urban environment with accuracies ranging between 84.5% and 93.5% for the test datasets.

As described in Section 11.3.2, after aggregation, the final five categories in the land-cover maps were forest, urban/built-up, cropland, grassland, and water. Figure 11.2 shows the after aggregation land-cover classification maps for each of the four years. There are several major trends evident in the changes of land-cover that are consistent over the period 1972–2011. The urban/built-up area increased rapidly, and there was a marked decrease in cropland area. Although the area of forest decreased between 1972 and 1987, after 1987 there was a steady increase of forest growth.

At the same time, the overall population of the study area increased by 49.5%, from 25.2 million in 1970 to 37.6 million in 2010.

TABLE 11.2
Description of the Land Cover Classification System and Training and Test Pixel Counts for Data from 1972, 1987, 2001, and 2011 Data

	1972		1987		2001		2011	
Class	Training	Testing	Training	Testing	Training	Testing	Training	Testing
1. Forest	1478	680	2672	1375	2287	1259	2300	1157
2. Urban/built-up	875	456	2277	1156	1940	1101	2014	1164
3. Cropland	872	458	2062	1022	1949	1033	1609	903
4. Paddy	–	–	1528	823	–	–	–	–
5. Grassland	800	427	1140	624	1267	542	1080	556
6. Sparse	–	–	1170	588	1002	570	1022	562
7. Water	692	331	1582	775	1308	606	1138	618
8. Flooded paddy	–	–	–	–	–	–	797	433
9. Wheat	–	–	–	–	–	–	796	422
10. Bare	–	–	559	300	490	291	–	–
11. Snow	–	–	–	–	–	–	742	386
Total	*4717*	*2352*	*12,990*	*6663*	*10,243*	*5402*	*11,498*	*6201*

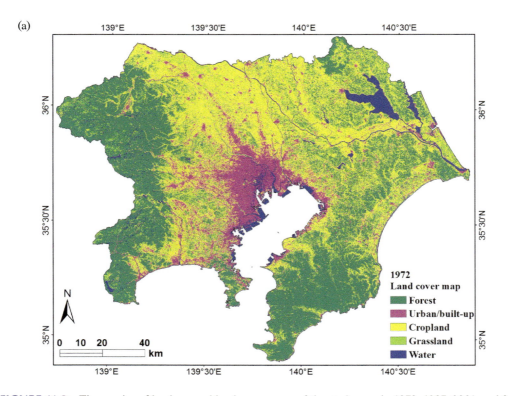

FIGURE 11.2 Time series of land use and land cover maps of the study area in 1972, 1987, 2001, and 2011.

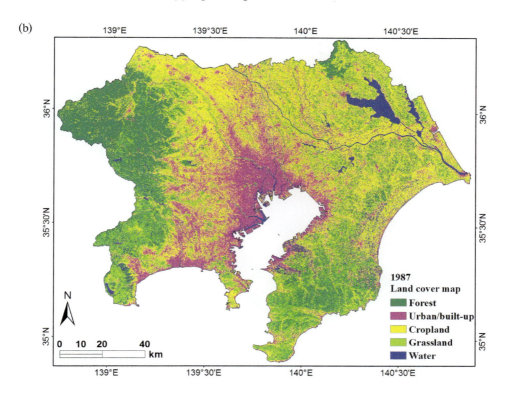

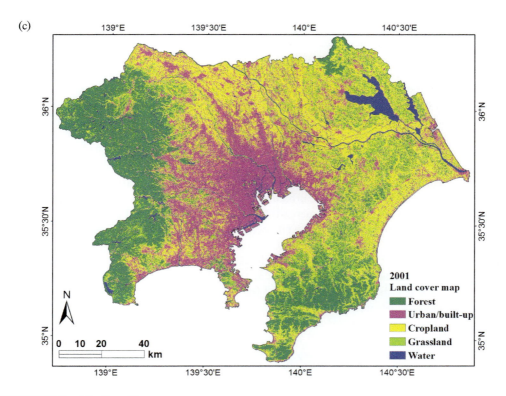

FIGURE 11.2 *(Continued)*

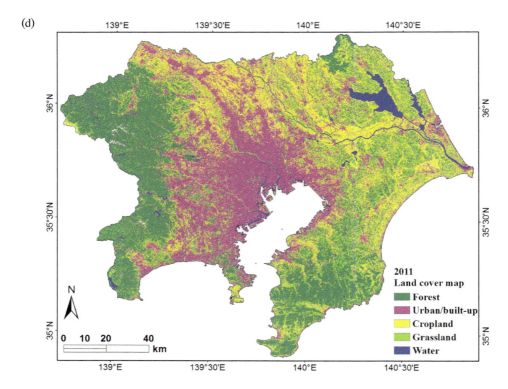

FIGURE 11.2 *(Continued)*

As explained in Section 11.3.2, we integrated the classified images with the empty basic grid cells to compute the area of each land-cover category within a cell. The generated grid cell land-cover classification maps facilitated our calculation of the percentage land-cover category changes from 1972 to 2011 at the scale of 1 km². Furthermore, since the size and location of the grid cells were exactly the same as those of the grid cells used for population census data (grid cell system census), we were able to use these to determine statistical associations between changes in land-cover classes and the variation of population densities.

Figure 11.3 shows the grid-cell-based spatial changes of urban area from 1972 to 2011. The grid cell values were calculated by subtracting the urban area of 1972 from that of 2011 in each grid cell and then dividing the changed area by the cell area. As Figure 11.3 illustrates, the urban/built-up area rapidly expanded to the surrounding suburban area where it was mainly flat or along transportation lines, whereas, in contrast, the urban/built-up area decreased in the center of the city. In fact, many high-rise (e.g., office or commercial) buildings and city-planned parks and green spaces replaced the dense low-rise buildings in the city center concurrently with the development of housing estates along railway lines in the suburbs (Okata and Murayama, 2011).

Figure 11.4 shows the grid-cell-based spatial changes of cropland area from 1972 to 2011, which were calculated in the same way as in Figure 11.3. Typically the greatest decreases in agricultural land area took place in either relatively large and flat areas or in proximity to urban regions where the demand for land for urban purposes is high (Saizen et al., 2006; Catalán et al., 2008).

To further investigate the spatial change in population density, we computed the difference in population density between 1970 and 2010 by subtracting the 1970 cell values from the 2010 values as shown in Figure 11.5. A number of trends are immediately clear: the dominant feature is a massive decentralization of population from the metropolitan core to the surrounding region. By contrast, the suburban areas surrounding these core areas of Tokyo, Kawasaki, and Yokohama

Urban Growth and Climatic Mapping of Mega Cities 345

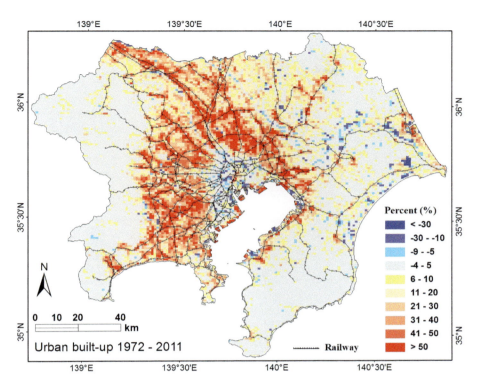

FIGURE 11.3 Percentage change of urban/built-up area in 1 km² grid cells from 1972 to 2011.

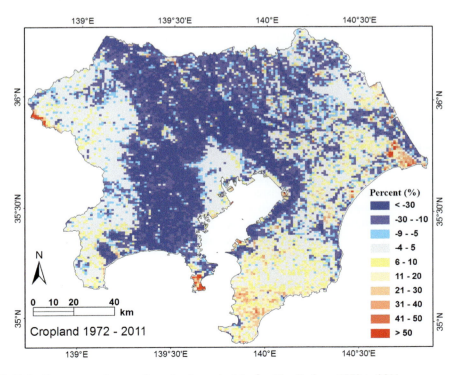

FIGURE 11.4 Percentage change of cropland area in 1 km² grid cells from 1972 to 2011.

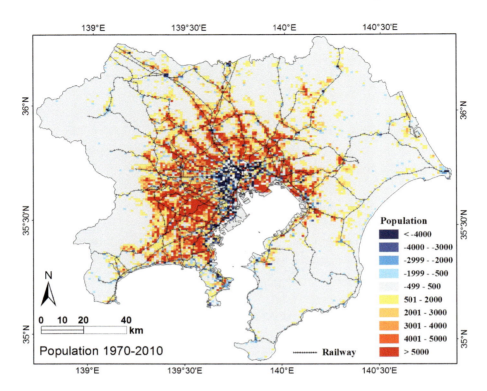

FIGURE 11.5 Change in population density between 1970 and 2010.

have seen enormous increases in population. Figure 11.5 also indicates that the western part of the study area has experienced a much faster rate of population growth along the transportation lines, and the spatial distribution of population increase trends is similar to the urban expansion trends (Figure 11.3).

11.3.2 Relationship Between Land Cover Changes and Population Census in Tokyo

Based on Section 11.3.1, we analyzed two different types of data to quantify the urban spatial extent: population census data (1 km spatial resolution; Basic Grid Square data) and land cover map derived from Landsat.

The Basic Grid Square population census data are available from the Statistics Bureau of Japan for every five-year period from 1970. Each square (area about 1 km²) is assigned an appropriate unique eight-digit ID number based on its longitude and latitude that is used as location information in the National Land Numerical Information databank (Statistics Bureau, Japan, 1973).

To investigate the relationship between the census population and the amount of land-cover category changes, we calculated the correlation coefficients of land-cover categories (i.e., forest, cropland, urban/built-up, grassland, and water) and population density changes based on the grid cells. All 15,851 grid cells in the study area were used for the statistical analysis. Table 11.3 presents a summary of the linear correlation coefficient matrix between the changes of land cover categories from 1972 to 2011 and population density changes from 1970 to 2010 based on the 15,851 samples.

As shown in Table 11.3, the linear correlation coefficient was −0.77 between urban/built-up and cropland, 0.54 between urban/built-up and population change, and −0.44 between cropland and

TABLE 11.3
Correlations (r) among the Changes of Land Cover Categories during 1972–2011 and Population Density Change during 1970–2010

R	Urban/Built-up	Cropland	Forest	Grassland	Water	Population
Urban/built-up	1					
Cropland	−0.7691	1				
Forest	0.0213	−0.3915	1			
Grassland	−0.1904	−0.0445	−0.4515	1		
Water	−0.1025	−0.1306	−0.1153	0.0431	1	
Population	0.5418	−0.4442	−0.0050	−0.0697	0.0134	1

population change. Meanwhile, forest change was negatively correlated with cropland ($r = -0.39$) and grassland ($r = -0.45$) owing to the portion of abandoned agricultural land and grassland that transformed to forest during the past four decades. The correlations between urban/built-up, cropland, and population change were statistically significant. These results are consistent with earlier findings in the Dhaka metropolitan area of Bangladesh (Dewan and Yamaguchi, 2009) and the Barcelona metropolitan region (Catalán et al., 2008).

Figure 11.6(a) shows the relationship between urban/built-up changes and cropland changes from 1972 to 2011, and Figure 11.6(b) shows the relationship between the urban/built-up changes from 1972 to 2011 and population density changes from 1970 to 2010. We found a strong, negative linear relationship between urban/built-up change and cropland change ($r = -0.77$) (Figure 11.6(a)), suggesting that a vast area of cropland has been converted to urban/built-up area during the last four decades.

The urban/built-up change has a significant positive correlation with the population density change ($r = 0.54$) (Figure 11.6(b)). However, there are areas with poor correlation caused by sparse population density in some industrial regions (e.g., the industrial region around Tokyo Bay, where large businesses have left under-utilized buildings) or dense populations in regions with overcrowded high-rise apartment buildings.

To better understand the trends of urban growth, we divided the land-cover change during 1972–2011 and the population density change during 1970–2010 into three intervals: 1972–1987 land cover versus 1970–1985 population, 1987–2001 land cover versus 1985–2000 population, and 2001–2011 land cover versus 2000–2010 population. We used the 1970, 1985, 2000, and 2010 population census data because they most closely matched the Landsat-derived land cover maps.

Results from our correlation analyses between land-cover change and population density change for the three intervals are reported in Tables 11.4. The correlations between the urban/built-up and cropland are highly negative: −0.40, −0.79, and −0.47, respectively, for the three periods. The correlations between urban/built-up and population are positive for 1972–1987 ($r = 0.36$) and 1987–2001 ($r = 0.44$). It is not until 2001–2011 that almost no correlation is found ($r = 0.06$).

Forest and grassland had strong negative correlations for 1972–1987 ($r = -0.69$) and 1987–2001 ($r = -0.79$), whereas the correlation was small for 2001–2011 ($r = -0.16$) (Table 11.4). In contrast, almost no correlation appeared in the linear relationship between cropland and grassland for 1972–1987 ($r = 0.01$), but the negative correlation dramatically increased to −0.21 and −0.40 during the last two periods, respectively. There were strong negative correlations between forest and cropland for 1972–1987 ($r = -0.50$) and 2001–2011 ($r = -0.54$), but little correlation was found for 1987–2001 ($r = 0.01$).

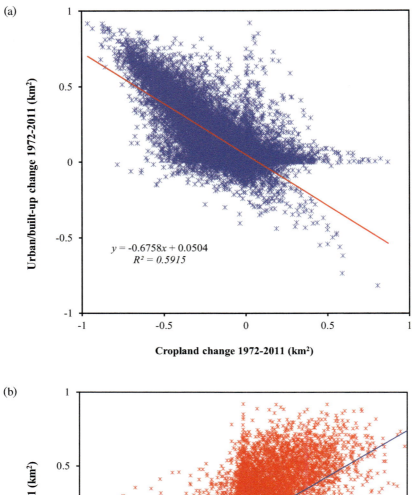
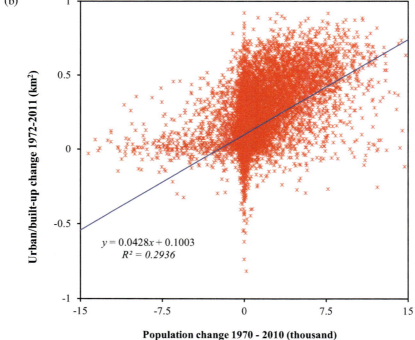

FIGURE 11.6 (a) Scatterplot of urban/built-up changes versus cropland changes for 1972–2011. The red line indicates the linear least-squares fit and *r* is the correlation coefficient. (b) The correlation between the urban/built-up changes for 1972–2011 and population density changes for 1970–2010. The blue line indicates the linear least-squares fit.

TABLE 11.4
Correlations (r) between the Changes of Land Cover Categories and Population Density Change: 1972–1987 vs. 1970–1985, 1987–2001 vs. 1985–2000, and 2001–2011 vs. 2000–2010, Respectively for the Three Periods

a. Changes of land cover categories for 1972–1987 vs. population density change for 1970–1985.

r	Urban/built-up	Cropland	Forest	Grassland	Water	Population
Urban/built-up	1					
Cropland	−0.4031	1				
Forest	−0.0009	−0.4966	1			
Grassland	−0.1555	0.0091	−0.6900	1		
Water	−0.1970	−0.1699	−0.1281	0.0080	1	
Population	0.3560	−0.1662	0.0622	−0.1591	−0.0034	1

b. Changes of land cover categories for 1987–2001 vs. population density change for 1985–2000.

r	Urban/built-up	Cropland	Forest	Grassland	Water	Population
Urban/built-up	1					
Cropland	−0.7904	1				
Forest	−0.2001	0.0109	1			
Grassland	0.1276	−0.2116	−0.7931	1		
Water	0.1583	−0.3698	−0.0453	−0.1363	1	
Population	0.4400	−0.3204	−0.0563	0.0360	0.0313	1

c. Changes of land cover categories for 2001–2011 vs. population density change for 2000–2010.

r	Urban/built-up	Cropland	Forest	Grassland	Water	Population
Urban/built-up	1.0000					
Cropland	−0.4701	1.0000				
Forest	−0.1450	−0.5429	1.0000			
Grassland	−0.0827	−0.4036	−0.1612	1.0000		
Water	−0.0271	−0.0545	−0.0426	−0.1774	1.0000	
Population	0.0562	−0.0194	−0.0252	0.0212	−0.0338	1.0000

Water change was not correlated with other land cover changes or population changes in any of the three periods, except for 1987–2001 when the relationship between water and cropland had a negative correlation of −0.37. This slightly negative relationship may reflect the fact that the 1987 Landsat image was acquired over flooded paddy fields on May 24 during the rice growing season; as a consequence, some paddy fields may have been misclassified into water classes.

11.3.3 Relationship among DMSP, Urban Area, and Population Census in Tokyo

The distribution of cities around the world broadly corresponds to the brightness distribution of DMSP nighttime lights (Small et al., 2005). Syntheses of population data, urban/built-up area, and satellite nighttime lights images can be used to identify and characterize the spatio-temporal extent and expansion trends of urban growth. Thus, the relationships among population census data, land-use data, and DMSP nighttime lights data can be better understood if these data are mapped onto a grid composed of square grid cells. Furthermore, it allows the commonly used multivariate linear regression model to be used to analyze urbanization.

The mean annual stable nighttime lights from the NOAA's DMSP/OLS nighttime lights sensors images have a 30-second spatial resolution. The cleaned up avg_vis contains the lights from cities, towns, and other sites with persistent lighting. Ephemeral events, such as fires have been discarded. Then the background noise was identified and replaced with values of zero. The DMSP digital number (DN) is an integer between 0 (no light) and 63. The DMSP/OLS are available at: http://www.ngdc.noaa.gov/dmsp/downloadV4composites.html.

For link between DMSP/OLS with census data and land cover data, we resampled DMSP maps to a 1 m spatial resolution pixel size. This process can reduce the error caused by the pixel size. Then we overlaid the resampled DMSP images on the grid cells to compute for each cell the percentage of DN values within it and stored the results in a new attribute table. Thus, the attribute table include 64 new added DMSP attributes, i.e., DN value from 0 to 63.

We compared the spatial distributions of DMSP DN values (2010), population density (2010), and percent density of urban/built-up area (2011) in Tokyo at the scale of 1 km^2 (Figure 11.7).

The result showed that the spatial distributions of population density, DMSP nighttime lights brightness, and urban land use are broadly consistent, suggesting a spatial correlation among them.

Furthermore, we examined the correlations of DMSP DN values with urban land use area and population density in Tokyo (Figure 11.8).

As the DN values of DMSP nighttime lights increase, the correlation coefficient between the DMSP DN values and population density, and that between DMSP DN values and urban land use area, also increase (Figure 11.8).

As population density showed a strong correlation with both land cover and DMSP data, it is possible to use multiple linear regression to predict population density at a 1 km spatial resolution from the combination of land cover and DMSP data.

However, as shown in Figure 11.8, the urban land use area did not reflect the vertical component (i.e., high-rise buildings), and DMSP nighttime lights may be saturated in core city areas where the population density is very high (Frolking et al., 2013). Thus, to improve the prediction accuracy of the population density in the future, the analysis should include data of urban 3D structures.

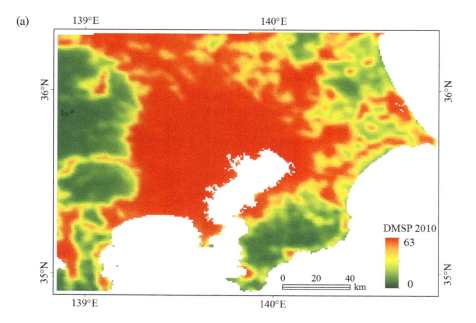

FIGURE 11.7 (a) DMSP in 2010; (b) population density in 2010 at the scale of 1 km^2; (c) proportion of urban/built-up area in 2011 at the scale of 1 km^2.

Urban Growth and Climatic Mapping of Mega Cities 351

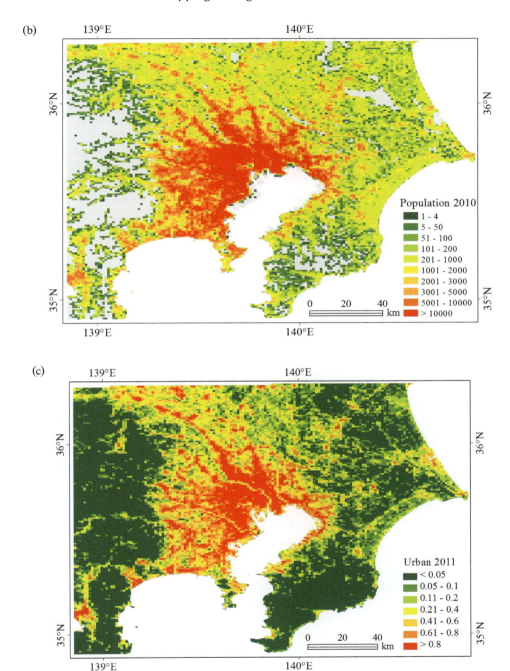

FIGURE 11.7 *(Continued)*

11.3.4 Relationship between Local Climate Zones and Land Surface Temperatures in Tokyo 23-Ku

Tokyo consists of 23 special wards (Tokyo 23-Ku), 26 cities, 5 towns, and 8 villages. The study area was selected by combining the administrative boundaries of Tokyo 23-Ku in 2020 and the actual area covered by the data.

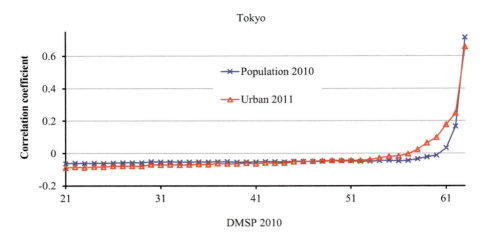

FIGURE 11.8 Correlations between DMSP DN values and urban land use area and between DMSP DN values and population density in Tokyo. There are saturation issues at high DMSP values in city areas where the population density is high.

We acquired airborne LiDAR data, Sentinel-2 imagery, and GIS vector (buildings and roads) data to classify local climate zones for the study area. Multidate thermal infrared data (Landsat-8 and ASTER data) were used to analyze the relationship between LCZs and multidate LSTs. Table 11.5 provides the details of multisource remote sensing and GIS data. Airborne LiDAR data were acquired from the Geospatial Information Authority of Japan (GSI). The building footprint polygons were derived from the Zmap-TOWNII product, a residential map database from the ZENRIN Company, Japan. The road polygons were obtained from the basic land use status survey database of the Tokyo Metropolitan Development Bureau. The road centerlines were obtained from the standard national digital road map (DRM) database of the Japan DRM Association. As depicted in Table 11.6, the standard LCZ scheme comprises two major types: built types (LCZ classes 1–10) and land cover types (LCZ classes A–G).

Based on multisource remote sensing and GIS data, we calculated seven LCZ properties, including height of roughness elements, sky view factor, aspect ratio, building surface fraction, impervious surface fraction, pervious surface fraction, and surface albedo. Then, we optimized the original thresholds provided in the LCZ definition based on the statistical distribution of the LCZ reference samples in the context of Tokyo. Next, GIS-based LCZ mapping was implemented using fuzzy logic classifiers at the 100-m grid-cell and block scales. For the reference samples, the overall accuracy of the LCZ classification at the block scale (66.88%) was lower than that at the 100-m grid-cell scale (80.34%).

As presented in Figure 11.9, at the 100-m grid-cell and block scales, the central region of Tokyo 23-Ku was identified as LCZ 1 (compact highrise), surrounded by LCZs 2 (compact mid-rise), 3 (compact low-rise), and 6 (open low-rise). LCZs 4 (open high-rise), 5 (open mid-rise), and 9 (sparsely built) were scattered within a "radial ring" LCZ pattern. LCZs 1 and 2 were clustered not only in the central area of Tokyo 23-Ku, but also in the surrounding subcity areas. At the 100-m grid-cell and block scales, LCZ 6 occupied the largest area, followed by LCZs 3 and G (water).

We examined the relationships between 100-m grid-cell LCZs and multidate LSTs (Figure 11.10). In general, there were differences in the statistical distributions of LSTs (both daytime and nighttime) among LCZ classes, regardless of the date. For the LCZ compact classes (LCZs 1–3), the mean daytime LSTs were ranked in descending order as LCZs 3, 2, and 1; the mean nighttime LSTs were ranked in descending order as LCZs 1, 2, and 3. For LCZ open classes (LCZs 4–6), the

TABLE 11.5
Summary of Multisource Remote Sensing and Geographic Information System Data Used in This Study

Data	Date (yyyy/mm/dd, Local Time)	Spatial Resolution/Scale	Usage
Airborne LiDAR data	2001–2002	4–5 points/m^2	Local climate zone classification
Sentinel-2 MSI Level 1C imagery	2015/01/01–2016/12/31	10 m, 20 m	Local climate zone classification
Landsat-8 collection-2 level 2 daytime surface temperature products	2016/03/01 10:15 (Spring) 2016/03/17 10:15 (Spring) 2016/05/04 10:15 (Spring) 2016/07/07 10:15 (Summer) 2016/10/27 10:16 (Autumn) 2016/12/30 10:16 (Winter)	100 m	Land surface temperature
ASTER level 2 AST-08 nighttime surface temperature products	2011/03/26 21:48 (Spring) 2012/12/25 21:48 (Winter) 2013/08/31 21:42 (Summer) 2013/10/09 21:48 (Autumn) 2014/12/24 21:42 (Winter) 2015/02/01 21:48 (Winter) 2017/06/23 21:42 (Summer)	90 m	Land surface temperature
Zmap-TOWNII building footprint product	2017	Polygon vector (1:2500)	Local climate zone classification
Road vector data	2016	Polygon vector (1:2500)	Local climate zone classification
	2013	Line vector (1:2500)	Division of block units

TABLE 11.6
Standard Local Climate Zone (LCZ) Scheme from Stewart and Oke (2012)

Type	Class	Description
Built types	LCZ 1	Compact high-rise
	LCZ 2	Compact mid-rise
	LCZ 3	Compact low-rise
	LCZ 4	Open high-rise
	LCZ 5	Open mid-rise
	LCZ 6	Open low-rise
	LCZ 7	Lightweight low-rise
	LCZ 8	Large low-rise
	LCZ 9	Sparsely built
	LCZ 10	Heavy industry
Land cover types	LCZ A	Dense trees
	LCZ B	Scattered trees
	LCZ C	Bush, scrub
	LCZ D	Low plants
	LCZ E	Bare rock or paved
	LCZ F	Bare soil or sand
	LCZ G	Water

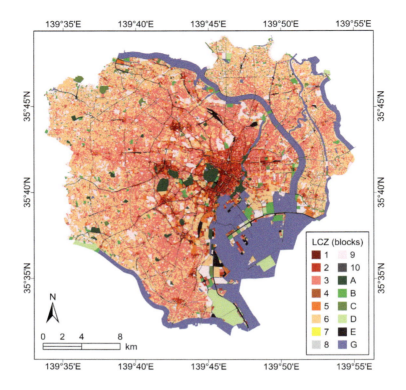

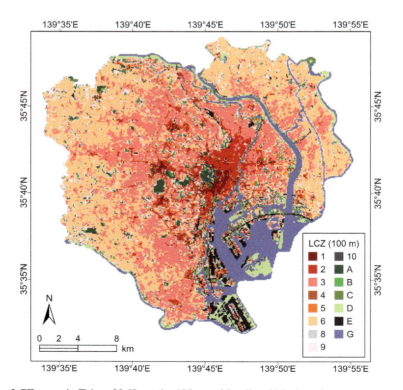

FIGURE 11.9 LCZ maps in Tokyo 23-Ku at the 100-m grid-cell and block scales.

Urban Growth and Climatic Mapping of Mega Cities

mean daytime LSTs were ranked in descending order as LCZs 6, 5, and 4; the mean nighttime LSTs were ranked in descending order as LCZs 4, 5, and 6. Regardless of the date, the mean LST (both daytime and nighttime) was higher for LCZ 3 (compact low-rise) than for LCZ 6 (open low-rise). The mean LST was lower for LCZ 1 (compact high-rise) than for LCZ 4 (open high-rise) only on December 30 (winter daytime), March 26 (spring nighttime), and February 1 (winter nighttime). The mean LST was lower for LCZ 2 (compact mid-rise) than for LCZ 5 (open mid-rise) only on March 1 (spring daytime), March 17 (spring daytime), October 27 (autumn daytime), and December 30 (winter daytime).

As shown in Figure 11.10, regardless of the date, the mean daytime LSTs of LCZ E (bare rock or paved) were highest among the land cover LCZ classes (LCZs A–G). For dates excluding December

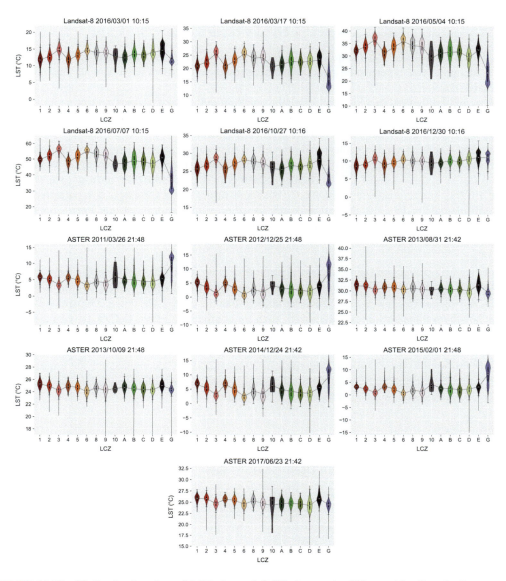

FIGURE 11.10 Violin density plots of LSTs for each LCZ class at the 100-m grid-cell scale (dates are in the format year/month/day). The white circles indicate the mean, and the horizontal white lines indicate the median.

30 (winter), the mean daytime LST of LCZ G (water) was lowest among the land cover LCZ classes. On December 30 (winter), the mean daytime LST of LCZ A (dense trees) was lowest, followed by that of LCZ C (bush, scrub). On March 1 (spring), March 17 (spring), and October 27 (autumn), the mean daytime LSTs of LCZ A (dense trees) were the second lowest among the land cover LCZ classes. The mean daytime LSTs of LCZ D (low plants) were the second lowest among the land cover LCZ classes on May 4 (spring) and July 7 (summer). In spring (March 26) and winter (December 25, December 24, and February 1), among the land cover LCZ classes, the mean nighttime LSTs of LCZ G were highest, whereas the mean nighttime LSTs of LCZs D and C were lowest. In summer (August 31 and June 23) and autumn (October 9), among the land cover LCZ classes, the mean nighttime LSTs of LCZ E were highest, whereas the mean nighttime LSTs of LCZs G and D were lowest.

11.3.5 Spatial–Temporal Analyses of Local Climate Zones in Shanghai

Shanghai is one of the most important Chinese cities in terms of economic activity. Since the 1990s, Shanghai has experienced rapid economic growth, urban land development, relocation, and deindustrialization.

We combined optical imagery and synthetic aperture radar (SAR) data (Landsat-5 and PALSAR for 2008; Sentinel-2 and PALSAR-2 for 2020) to map the LCZs in Shanghai, China (Table 11.7). Random forest models were used to perform LCZ classification.

We used a nearest neighbor interpolation method to resample the band spatial resolution to 10 m for processing. Considering LCZ types with similar characteristics in the study area, we combined LCZ 8 (large low-rise) and LCZ 10 (heavy industry) into LCZ 8&10. In addition, we also combined LCZ B (scattered trees) and LCZ C (bush and scrub) into LCZ B&C. We collected ground truth samples of the 13 LCZ classes randomly based on visual interpretations from Google Earth. We randomly split the ground truth samples into training samples and test samples to ensure separation in their spatial distributions. Table 11.8 shows the number of training and test pixels for each LCZ class.

The overall accuracy of the LCZ classification results for 2020 (80.40%) was higher than that for 2008 (75.35%). As shown in Figure 11.11, LCZ 1 (compact high-rise), LCZ 2 (compact mid-rise), and LCZ A (dense trees) were much smaller in area than the other LCZ types. The area was considerably larger for open buildings (LCZs 4–6) than for compact buildings (LCZs 1–3). In 2008, LCZ D (low plants) accounted for the largest area among all LCZ types, followed by LCZ B&C (scattered trees with bush and scrub). Among the built types, LCZ 6 (open low-rise) had the largest area in 2008, followed by LCZ 5 (open mid-rise). In 2020, LCZ B&C accounted for the largest area of all LCZ types, followed by LCZ F (bare soil or sand). Among the built types, LCZ 5 had the largest area in 2020, followed by LCZs 4 (open high-rise) and 6.

TABLE 11.7
Summary of Remote Sensing Data Used for LCZ Classification

Satellite Data	Date	Band	Spatial Resolution (m)
Landsat-5 TM C1 Level-2	25 April 2008	Band 1–5, 7	30
ALOS PALSAR RTC	28 April 2008	HH, HV	20
	15 May 2008		
	01 June 2008		
Sentinel-2 MSI L2A	28 April 2020	Band 1–8, 8a, 9, 11, 12	10, 20, 60
ALOS-2 PALSAR-2 L3.1	02 May 2020	HH, HV	10
	30 May 2020		

TABLE 11.8
The Number of Training and Test Pixels for Images Acquired in 2008 and 2020

		2008		2020	
Class	Description	Training	Test	Training	Test
LCZ 1	Compact high-rise	1264	241	4261	741
LCZ 2	Compact mid-rise	15,656	4873	3640	932
LCZ 3	Compact low-rise	53,494	14,445	11,180	4092
LCZ 4	Open high-rise	69,933	10,287	40,791	12,035
LCZ 5	Open mid-rise	211,189	70,466	59,962	19,785
LCZ 6	Open low-rise	114,805	25,423	22,547	7332
LCZ 8&10	Large low-rise and heavy industry	105,346	33,776	21,620	7700
LCZ A	Dense trees	3050	254	1463	1132
LCZ B&C	Scattered trees with bush and scrub	73,182	21,031	26,407	6134
LCZ D	Low plants	135,671	43,186	23,148	5051
LCZ E	Bare rock or paved	67,727	24,252	37,434	10,821
LCZ F	Bare soil or sand	41,915	9893	35,953	3978
LCZ G	Water	27,203	6088	23,349	4384
Total		920,435	264,215	311,755	84,117

As shown in Figure 11.12, from 2008 to 2020, the expansion of LCZ 1 was primarily attributed to the transformation of LCZs 4, 5, and 8&10. The increase in LCZ 2 was mainly derived from LCZs 3, 5, and 8&10. The increase in LCZ 3 was mainly derived from LCZs 5, 6, and 8&10. The increase in LCZ 4 was mainly derived from LCZs 5, 6, D, and E. The increase in LCZ 5 was mainly derived from LCZs 6, 8&10, D, and E. The increase in LCZ 6 was mainly derived from LCZ 5, B&C and D. The increase in LCZ 8&10 was mainly derived from LCZs 5, D, and E. The increase in LCZ B&C was mainly derived from LCZs 6, D, and E. The increase in LCZ D was mainly derived from LCZs 6, B&C, and E. The increase in LCZ F was mainly derived from LCZs 6, B&C and D. The reason for the significant reduction in LCZ D is mainly attributed to rapid urbanization and changes in vegetation types, such as the conversion of LCZ D to LCZ B&C and F. The shift from LCZs 3, 5, and 6 to LCZ 4 may be related to the renovation of old urban residential areas and the expansion of urban built-up areas.

11.4 CONCLUSIONS

The first study demonstrated that rapid expansion of the area of urban land use around Tokyo metropolitan area accompanied the massive population increases that occurred during 1972–2011. The dominant changes that took place in the study area during this period were urban expansion over a wide area along transportation systems and large areas of cropland transformed into urban/built-up area, while such changes were accompanied by residents migrating to the outlying suburban areas. The second study in the 23 special wards of Tokyo showed that the mean daytime/nighttime LSTs in most LCZs differed significantly from each other. The multidate LSTs are able to reflect the unique surface thermal characteristics of each LCZ class. The third study in Shanghai showed that the areas of open high-rise (LCZ 4) and open mid-rise (LCZ 5) buildings significantly increased from 2008 to 2020.

The grid-cell-based method presented in this chapter is an innovative approach to the study of spatial-temporal urban growth patterns since it is capable of linking disparate data from many different agencies and organizations into a single comprehensive dataset that covers a wide range of

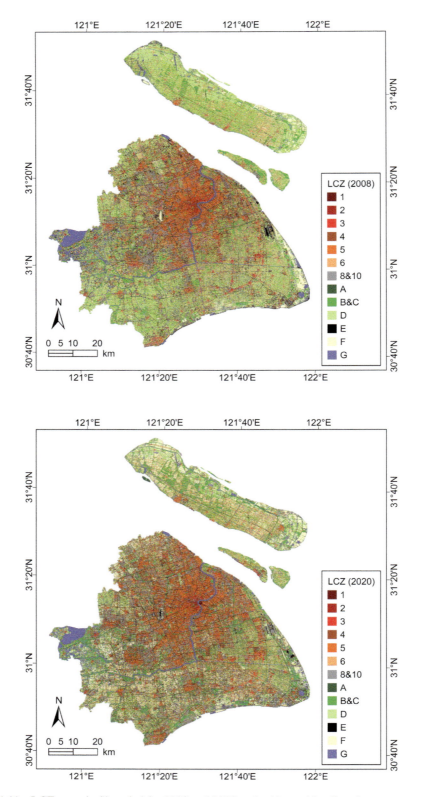

FIGURE 11.11 LCZ maps in Shanghai for 2008 and 2020 at the 10 m grid-cell scale.

Urban Growth and Climatic Mapping of Mega Cities

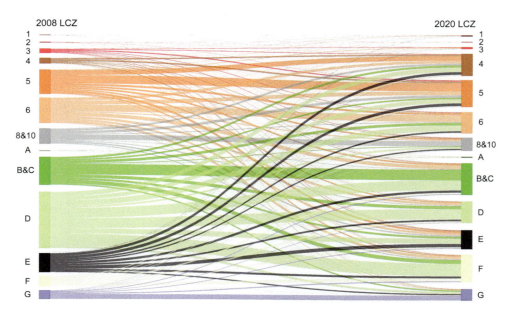

FIGURE 11.12 Sankey diagram demonstrating the transitions in the LCZs from 2008 to 2020. The width of the lines is proportional to the quantity of flow. The left side shows the outflow in 2008, and the right side shows the inflow in 2020.

spatial and temporal scales. The method as applied in the study has demonstrated the capability to improve the knowledge, understanding, and analysis of urban dynamics. The results confirm that the grid-square-based analysis allowed us to describe the spatial-temporal dynamics of the urban growth and climatic patterns in more detail.

REFERENCES

Angel, S., J. Parent, D. L. Civco, A. Blei, and D. Potere. 2011. The dimensions of global urban expansion: Estimates and projections for all countries, 2000–2050. *Progress in Planning* 75: 53–107.

Avelar, S., R. Zah, and C. Tavares-Corrêa. 2009. Linking socioeconomic classes and land over data in Lima, Peru: Assessment through the application of remote sensing and GIS. *International Journal of Applied Earth Observation and Geoinformation* 11: 7–37.

Bagan, H., and Y. Yamagata. 2010. Improved subspace classification method for multispectral remote sensing image classification. *Photogrammetric Engineering and Remote Sensing* 76(11): 1239–1251.

Bagan, H., and Y. Yamagata. 2012. Landsat analysis of urban growth: How Tokyo became the world's largest megacity during the last 40 years. *Remote Sensing of Environment* 127: 210–222.

Bagan, H., and Y. Yamagata. 2014. Land-cover change analysis in 50 global cities by using a combination of Landsat data and analysis of grid cell. *Environmental Research Letters* 9(6): 064015.

Bagan, H., Y. Yasuoka, T. Endo, X. Wang, and Z. Feng. 2008. Classification of airborne hyperspectral data based on the average learning subspace method. *IEEE Geoscience and Remote Sensing Letters* 5(3): 368–372.

Berger, C., M. Voltersen, S. Hese, I. Walde, and C. Schmullius. 2013. Robust extraction of urban land cover information from HSR multi-spectral and LiDAR data. *IEEE Journal of Selected Topics in Applied Earth Observations and Remote Sensing* 6: 2196–2211.

Catalán, B., D. Saurí, and P. Serra. 2008. Urban sprawl in the Mediterranean? Patterns of growth and change in the Barcelona Metropolitan Region 1993–2000. *Landscape and Urban Planning* 85: 174–184.

Chen, C., H. Bagan, X. Xie, Y. La, and Y. Yamagata. 2021. Combination of Sentinel-2 and PALSAR-2 for local climate zone classification: A case study of Nanchang, China. *Remote Sensing* 13(10): 1902.

Chen, C., H. Bagan, and T. Yoshida. 2023. Multiscale mapping of local climate zones in Tokyo using airborne LiDAR data, GIS vectors, and Sentinel-2 imagery. *GIScience & Remote Sensing* 60(1): 2209970.

Chen, C., H. Bagan, T. Yoshida, H. Borjigin, and J. Gao. 2022. Quantitative analysis of the building-level relationship between building form and land surface temperature using airborne LiDAR and thermal infrared data. *Urban Climate* 45: 101248.

Dewan, A. M., and Y. Yamaguchi. 2009. Using remote sensing and GIS to detect and monitor land use and land cover change in Dhaka Metropolitan of Bangladesh during 1960–2005. *Environmental Monitoring and Assessment* 150: 237–249.

Doll, C. N. H. 2008. *CIESIN Thematic Guide to Night-Time Light Remote Sensing and Its Applications*. Palisades, NY: Center for International Earth Science Information Network (CIESIN), Columbia University.

Doll, C. N. H., J. P. Muller, and J. G. Morley. 2006. Mapping regional economic activity from night-time light satellite imagery. *Ecological Economics* 57: 75–92.

Elvidge, C. D., B. T. Tuttle, P. C. Sutton, K. E. Baugh, A. T. Howard, C. Milesi, B. Bhaduri and R. Nemani. 2007. Global distribution and density of constructed impervious surfaces. *Sensors* 7: 1962–1979.

Frolking, S., T. Milliman, K. C. Seto, and M. A. Friedl. 2013. A global fingerprint of macro-scale changes in urban structure from 1999 to 2009. *Environmental Research Letters* 8(2): 024004. http://doi.org/10.1088/1748-9326/8/2/024004

Geospatial Information Authority of Japan (GSI) (2012). *Maps, Aerial Photographs and Survey Results (in Japanese)*. Available at: http://www.gsi.go.jp/tizu-kutyu.html (last accessed: 15 May 2013).

Giri, C., B. Pengra, J. Long and, and T. R. Loveland. 2013. Next generation of global land cover characterization, mapping, and monitoring. *The International Journal of Applied Earth Observation and Geoinformation* 25: 30–37.

Hou, X., X. Xie, H. Bagan, C. Chen, Q. Wang, and T. Yoshida. 2023. Exploring spatiotemporal variations in land surface temperature based on local climate zones in Shanghai from 2008 to 2020. *Remote Sensing* 15(12): 3106.

Huang, F., S. Jiang, W. Zhan, B. Bechtel, Z. Liu, M. Demuzere, Y. Huang, et al. 2023. Mapping local climate zones for cities: A large review. *Remote Sensing of Environment* 292: 113573.

La, Y., H. Bagan, and Y. Yamagata. 2020. Urban land cover mapping under the local climate zone scheme using Sentinel-2 and PALSAR-2 data. *Urban Climate* 33: 100661.

Lambin, E. F., H. J. Geist, and E. Lepers. 2003. Dynamics of land-use and land-cover change in tropical regions. *Annual Review of Environment and Resources* 28: 205–241.

Masson, V., A. Lemonsu, J. Hidalgo, and J. Voogt. 2020. Urban climates and climate change. *Annual Review of Environment and Resources* 45(1): 411–444.

Oja, E. 1983. *Subspace Methods of Pattern Recognition*. Letchworth, UK: Research Studies Press and John Wiley & Sons.

Okata, J., and A. Murayama. 2011. Tokyo's urban growth, urban form and sustainability. In A. Sorensen and J. Okata (Eds.), *Megacities: Urban form, Governance, and Sustainability* (pp. 15–41). Tokyo: Springer.

Parés-Ramos, I. K., N. L. Álvarez-Berríos, and T. M. Aide. 2013. Mapping urbanization dynamics in major cities of Colombia, Ecuador, Perú, and Bolivia using night-time satellite imagery. *Land* 2: 37–59.

Potere, D., and A. Schneider. 2007. A critical look at representations of urban areas in global maps. *GeoJournal* 69: 55–80.

Potere, D., A. Schneider, S. Angel, and D. L. Civco. 2009. Mapping urban areas on a global scale: Which of the eight maps now available is more accurate? *International Journal of Remote Sensing* 30: 6531–6558.

Qian, T., H. Bagan, T. Kinoshita, and Y. Yamagata. 2014. Spatial–temporal analyses of surface coal mining dominated land degradation in Holingol, Inner Mongolia. *IEEE Journal of Selected Topics in Applied Earth Observations*, in press.

Saizen, I., K. Mizuno, and S. Kobayashi. 2006. Effects of land-use master plans in the metropolitan fringe of Japan. *Landscape and Urban Planning* 78(4): 411–421.

Schneider, A., M. A. Friedl, and D. Potere. 2009. A new map of global urban extent from MODIS satellite data. *Environmental Research Letters* 4: 044003.

Small, C. 2003. High spatial resolution spectral mixture analysis of urban reflectance. *Remote Sensing of Environment* 88: 170–186.

Small, C., and J. E. Cohen. 2004. Continental physiography, climate, and the global distribution of human population. *Current Anthropology* 45(2): 269–277.

Small, C., F. Pozzi, and C. D. Elvidge. 2005. Spatial analysis of global urban extents from the DMSP-OLS night lights. *Remote Sensing of the Environment* 96(3–4): 277–291.

Solecki, W., K. C. Seto, and P. J. Marcotullio. 2013. It's time for an urbanization science. *Environment: Science and Policy for Sustainable Development* 55(1): 12–17.

Statistics Bureau, Japan. 1973. *Standard Grid Square and Grid Square Code Used for the Statistics*. Available at: http://www.stat.go.jp/english/data/mesh/02.htm.

Statistics Bureau, Japan. 2011. *A Guide to the Statistics Bureau, the Director-General for Policy Planning (Statistical Standards) and the Statistical Research and Training Institute* (pp. 85–87). Available at: http://www.stat.go.jp/english/info/guide/2011ver/pdf/2011ver.pdf.

Stewart, I. D., and T. R. Oke. 2012. Local climate zones for urban temperature studies. *Bulletin of the American Meteorological Society* 93(12): 1879–1900.

Taubenböck, H., T. Esch, A. Felbier, M. Wiesner, A. Roth, and S. Dech. 2012. Monitoring urbanization in mega cities from space. *Remote Sensing of Environment* 117: 162–176.

Taubenböck, H., M. Wiesner, A. Felbier, M. Marconcini, T. Esch, and S. Dech. 2014. New dimensions of urban landscapes: The spatio-temporal evolution from a polynuclei area to a mega-region based on remote sensing data. *Applied Geography* 47: 137–153.

Townshend, J. R., and C. O. Justice. 2002. Towards operational monitoring of terrestrial systems by moderate-resolution remote sensing. *Remote Sensing of Environment* 83: 351–359.

Voogt, J. A., and T. R. Oke. 2003. Thermal remote sensing of urban climates. *Remote Sensing of Environment* 86(3): 370–384.

Zhang, Q., and K. C. Seto. 2011. Mapping urbanization dynamics at regional and global scales using multi-temporal DMSP/OLS nighttime light data. *Remote Sensing of Environment* 115: 2320–2329.

12 High-Resolution Remote Sensing and Visibility Analysis Method for Smart Environment Design

Yoshiki Yamagata, Daisuke Murakami, Hajime Seya, and Takahiro Yoshida

ACRONYMS AND DEFINITIONS

ASTER Advanced spaceborne thermal emission and reflection radiometer
DEM Digital elevation model
DSM Digital surface model
EV Electric vehicle
GIS Geographic information system
LiDAR Light detection and ranging
PV Photovoltaic
TM Thematic mapper

12.1 INTRODUCTION

12.1.1 Remote Sensing and Urban Analysis

Since the first Earth observation satellite, Landsat 1, was launched in 1972, remote sensing technology, which can provide spatial information at a high frequency of updating at low cost (Donnay et al., 2001), has received considerable attention. This is also true for regional/urban studies, especially after earlier studies clarified the usefulness of remotely sensed information as a proxy for urban conditions. Forster (1983) reveals that remotely sensed images can describes urban residential quality effectively.

A limitation of the sensors developed in the early phase of remote sensing technology is that they have coarse spatial resolutions (e.g., the Landsat-1 multispectral scanner has a spatial resolution of 80 m); a fine spatial resolution is crucial for carrying out remote sensing-based urban analysis (Welch, 1982). However, around the 1990s, second-generation sensors (e.g., the Landsat thematic mapper [TM] and advanced spaceborne thermal emission and reflection radiometer [ASTER]), with spatial resolutions of 10–30 m, were developed; therefore, beginning in the 1990s, simulations based on remote sensing technology were used increasingly for urban studies. Furthermore, after the development of third-generation sensors (e.g., the sensors of the IKONOS, QuickBird, WorldView-1, GeoEye-1, and WorldView-2 satellites), whose spatial resolutions range from 0.5 m to 5 m, an increasing number of studies started using remote sensing imagery for intra-urban scale analyses (Patino and Duque, 2013). This opened new doors to analyzing detailed scale relationships between urban morphological (e.g., floor area ratio) and environmental (e.g., land surface temperature) (Chen C. et al., 2022) indicators. Currently, remote sensing is widely accepted as a technique for monitoring detailed urban environmental conditions at a lower cost than *in situ* observations, with the latter being extremely costly in most cases (Miller and Small, 2003).

12.1.2 REMOTE SENSING AND SMART ENVIRONMENT

An increasing number of recent urban studies focus on the concept of "smart city." The definition of smart cities is vague (Hollands, 2008); according to Giffinger et al. (2007), the following sub-concepts characterize smart cities: smart economy, people, governance, mobility, environment, and living. Blaschke et al. (2011) indicated that remote sensing plays a critical role in designing a smart environment, which is characterized by attractive natural conditions (climate, green space, etc.), low pollution, effective resource management, and environmental protection. Indeed, several remote-sensing studies on urban environments provided fruitful insights for improving the "smart environment."

The "smart environment" concept highlights the smart use of natural resources. In other words, natural resources must be managed sensibly, while considering their value as both market (goods traded in markets) and non-market (goods that are not traded in markets) goods.

12.1.2.1 Remote Sensing and Natural Resources as Market Goods

Energy is the principal market good derived from natural resources. Among the various natural energy sources, renewable energy sources (e.g., solar radiation and wind power) are key to sustainable development and have received considerable attention in recent years. Notably, electric vehicles (EVs) and solar photovoltaic (PV) panels are important components of sustainable development because they do not involve CO_2 emissions, provided that EVs are charged solely using renewable-energy-based electricity, e.g., electricity generated from solar PVs (Giannouli and Yianoulis, 2012; Denholm et al., 2013). Notably, using non-renewable energy for charging EVs may increase CO_2 emissions (Wu et al., 2012). Therefore, the management of solar PVs can be a determining factor for sustainable development.

Solar PV systems must be allocated based on the actual amount of solar radiation received by a region. Although carrying out *in situ* observations of solar radiation across a region can be costly, the solar radiation in a region can be estimated accurately using remote sensing imagery and light detection and ranging (LiDAR) data, which can then, be used to describe the urban fabric of a region in three dimensions (3D) (Tooke et al., 2014; Chun and Guldmann, 2014). Figure 12.1 portrays an example of the LiDAR-based estimates of rooftop solar radiation on an average day in August in a central area of Yokohama City, Japan, created using the ArcGIS Spatial Analyst extension. Thus, the high-resolution solar radiation amounts at arbitrary times can be easily estimated using LiDAR data.

12.1.2.2 Remote Sensing and Natural Resources as Non-market Goods

Natural resources are important non-market goods that improve human well-being. The value of natural resources has been studied extensively by applying the hedonic approach (Rosen, 1974), which is a typical way to evaluate the economic value of non-market goods (sometimes involving the use of a regression model). Most hedonic studies focus on the two-dimensional (2D) structure of natural resources, e.g., the accessibility to natural resources or their scale (e.g., Tyrvainen and Miettinen, 2000; Irwin, 2002; Morancho, 2003; Tajima, 2003; Kong et al., 2007; Cho et al., 2008), and remote sensing images have often been used in these studies to acquire the 2D placement data of natural resources (e.g., the actual placement of green areas in a region).

However, natural resources are perceived by the human eye as 3D objects; therefore, it may be important to consider a third dimension, that is, height (in addition to length and breadth). Fortunately,

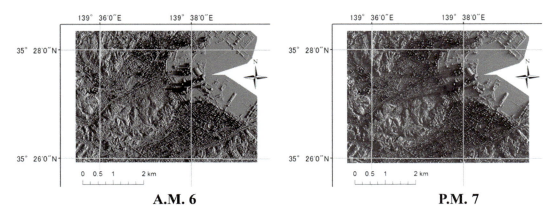

FIGURE 12.1 Example of the estimation of rooftop solar radiation for the central part of Yokohama City in August; the dark color denotes less radiation, whereas the light color denotes more radiation. These images were created as follows: (i) airborne light detection and ranging (LiDAR) data observed in February 2006, February 2007, and April 2007 were integrated to form one image. Then, (ii) a digital surface model (DSM; spatial resolution: 0.5 m) was created using the integrated LiDAR data, and (iii) solar radiation was estimated using solar radiation analysis tools available within the ArcGIS Spatial Analyst extension.

recent remotely sensed 3D data, such as LiDAR data, can accurately describe 3D objects, including buildings and natural resources (e.g., trees and mountains). As a result, an increasing number of hedonic studies have begun to realize the economic value of 3D perception or 3D view.

12.1.2.3 Remote Sensing and Natural Resources: Summary

This section focuses on energy and hedonic analyses, with respect to the management of smart environments. The application of remote sensing in energy analysis has been summarized by Liang et al. (2010). Notably, although remote sensing data are often applied in the hedonic analysis of 2D natural resources, the application of remote sensing in the hedonic analysis of 3D views remains limited.

Thus, the remainder of this chapter focuses on hedonic analysis. First, we reviewed the commonly used 3D-view evaluation approaches and then reviewed the hedonic analyses of the 3D views. Subsequently, we conducted a hedonic analysis of 3D views, including open, green, and ocean views, using the condominium and LiDAR data of Yokohama City, Japan.

12.2 VISIBILITY ANALYSIS

12.2.1 CLASSICAL VISIBILITY EVALUATION APPROACHES

Visibility was discussed earlier than the appearance of high-resolution sensors. Earlier studies used dummy variables; 1 indicated that the target object was visible and 0 indicated that it was not. McLeod (1984) used the dummy variable of river view, whereas Benson et al. (1998) used the dummy variables of ocean, lake, and mountain views. Other studies evaluated different views based on field investigations. For instance, Luttik (2000) extracted information on environmental factors and other location factors from maps, complemented by field investigations, whereas Tyrväinen and Miettinen (2000) and Lange and Schaeffer (2001) conducted field investigations to obtain window-view information; Brown and Raymond (2007) performed a mail survey.

However, both the dummy variable–based approach and the field investigation–based approach have drawbacks: the former ignores the quality of view, i.e., views are evaluated as 1 (visible) or 0 (invisible), and the latter is time-consuming and difficult to implement for large samples.

12.2.2 Viewshed and Isovist Analyses

There are two more extensively used visibility-evaluation approaches (Leduc et al., 2010): isovist analysis (Benedikt, 1979), which has been mainly discussed in architectural studies and urban studies, and viewshed analysis (Lynch, 1976), which has been mainly discussed in landscape studies.

An isovist is a space that is visible from an observation point (Benedikt, 1979). It is generally evaluated in a two-dimensional (2D) space and described by a 2D polygon. A typical isovist is a 2D horizontal slice of human perception (Yu et al., 2007). This approach is suitable for evaluating the views in urban spaces whose qualities strongly depend on the actual placement of buildings and other objects and also whose 2D structure can be described by 2D polygons. An obvious disadvantage of the isovist approach is its ignorance of the third dimension (i.e., height). A 3D isovist (e.g., Morello and Ratti, 2009) is an extended method that addresses this problem. The embodied 3D isovist (Krukar et al., 2021) is an additional extended method that incorporates the salient properties of space from the standpoint of embodiment and cognitive psychology.

Notably, viewshed analysis quantifies 3D views by examining whether each cell in a 3D raster is visible from an observation point. This approach is rarely adopted in an urban setting, because of the lack of 3D raster (known to precisely describe complex 3D urban geometry) (Yu et al., 2007; Leduc et al., 2010). However, recent high-resolution sensors, which incorporate not only the digital elevation model (DEM) but also the digital surface model (DSM), can solve this problem (e.g., ALOS World 3D); thus, an increasing number of studies that employ viewshed analyses are discussing 3D urban views (Inglis et al., 2022).

Currently, if the 3D data that describes urban geometry are available, both isovist- and viewshed-based approaches are applicable to quantify 3D views in urban spaces; high-resolution remote sensing data must have an important role in this respect. Although the standard 2D isovist was originally a vector-based approach that did not use raster data, the 3D isovist approach relied on raster data. Hence, high-resolution remote sensing data must contribute to not only viewshed analysis (originally a raster-based approach) but also to 3D isovist analyses. Such a similarity between viewshed and (3D) isovist analyses suggests that the difference between them is evolving to become increasingly obscure. Some studies have discussed them in an integrated manner (e.g., Llibera, 2003; Yu et al., 2007). For more recent reviews, refer to Inglis et al. (2022) and Wu et al. (2023).

12.2.3 Three-Dimensional (3D) View Indexes

Numerous indexes for quantifying and characterizing 3D views have been developed under the framework of isovist or viewshed analyses.

Some indexes evaluate the volume of space that is visible from a viewpoint, similar to the approach employed in a standard isovist. Table 12.1 summarizes such isovist-oriented 3D indexes, including the Sky View Factor (Ratti, 2002) and the Sky Opening (Teller, 2003), Spatial Openness (Fisher-Gewirtzman and Wagner, 2003), and view (Yu et al., 2007) indexes. The first two measures are used to perceive openness, by evaluating the amount of view of the sky. The Sky View Factor is now a well-accepted index in energy analysis (Yu et al., 2007); for example, solar radiation can be estimated using this index, and the result can be utilized to address the

TABLE 12.1
Isovist-Oriented Three-Dimensional (3D) Indexes Used in This Study: *s* Denotes the Viewpoint, and *S* Denotes the Space That Is Visible from the Viewpoint

Index	Definition	Volume of 3D view from a site *s*: *V*(s)	Search space: S (volume: *V(S)*)
Sky View Factor		Volume of the space in *S* that line of sights from *s* to the sky through	A hemisphere representing the visible area from *s*
Sky Opening Index		Volume of the area in *S* representing the sky	A fish-eye (2D) circle whose center is *s*
Spatial Openness Index		Volume of the visible space in *S*	A sphere representing the visible area from *s*
Viewsphere Index		Volume of the space in S that line of sights from s to ground objects through	A hemisphere representing the visible area from *s*

TABLE 12.2
Viewshed-Oriented Three-Dimension (3D) Indexes

Indexes	Description
Standard Viewshed	Whether each cell is visible or not from an observation point
Cumulative Viewshed	Number of observation points that each cell can see
Total Viewshed	
Iso-visi-matrix	
Visual Expose	Perceived size of cells. They are evaluated by modeling visual angle or visual span (both horizontal and vertical) of the cells.
Visual Magnitude	

solar PV allocation problem and other issues related to the urban heat island (e.g., Chun and Guldmann, 2014). The last two measures are perceived density. The Spatial Openness Index has frequently been adopted to evaluate built-up environments (e.g., Fisher-Gewirtzman and Wagner, 2003). For example, the Viewsphere Index can be used to discuss facility allocation problems. For example, Yang et al. (2007) applied this approach to address the campus allocation problem in Singapore.

Notably, some other indexes examine the visibility of each cell in a 3D raster, similar to the approach employed in viewshed analysis. This type of index has been discussed extensively, particularly after the installation of a 3D analyst toolbox, a collection of 3D spatial analysis tools (including viewshed analysis tools), in ArcGIS (ESRI Inc.) In 1998 (Ma, 2003). Table 12.2 summarizes the viewshed-oriented indexes considered in this study. The standard view examines whether each cell is visible from an observational point of view. The cumulative and total viewsheds (Wheatley, 1995; Llibera, 2003) accumulate standard views from each observation point and provide a raster of the records of the number of observation points visible in each cell. The difference between the two indexes is that the former assumes a set of selected sites as observation points, whereas the latter assumes that the observation points are distributed throughout the target region. The application of the viewshed-oriented (or raster-based) index was also proposed in the context of the isovist analysis. Note that the isovisible matrix (Morello and Ratti, 2009) is an index.

Smart Environment Design

It provides a 3D raster map whose cells are colored according to the number of cells in the rasters that are visible. From the viewpoint of the viewshed analysis, this index appeared to be identical to the total viewshed.

While the aforementioned indexes evaluate the visibility of each cell, other indexes explicitly evaluate the quality of view of each cell. For instance, visual exposure (Llibera, 2003; Domingo-Santos et al., 2011) evaluates the perceived cell size by modeling the visible angle or visible span (both horizontal and vertical) of the cell, and visual magnitude (Grêt-Regamey et al., 2007; Chamberlain and Meitner, 2013) evaluates the perceived size in a manner similar to a simpler approach. Chamberlain and Meitner (2013) demonstrate the usefulness of indexes that consider the perceived cell size.

Viewshed-oriented indexes have been adopted for various purposes, e.g., addressing the issues related to wind-turbine allocation (Benson et al., 2004) and the selection of coastal aquaculture sites (Falconer et al., 2013), carrying out negative visual impact assessment of marble quarry expansion (Mouflis et al., 2008), conducting visibility assessments of landmark buildings (Bartie et al., 2010) and from a road (Chamberlain and Meitner, 2013), and also conducting visual exposure assessment of smoking behavior (Pearson et al., 2014). Note that viewshed-oriented indexes are often used in hedonic analyses (see the next section).

Some convenient indexes that characterize views (isovist or viewshed analysis), instead of those that quantify them, have also been proposed. For example, Teller (2003) proposed several indicators for characterizing the sky view, including the mean distance to skyline and skyline regularity, which is generally defined by the standard deviation of skyline heights, eccentricity or asymmetry of the skyline, and spread, indicating the deviation of the skyline from the circle line. Llibera (2003) evaluated the visual prominence of a cell by differentiating the visibility of the cell from the visibilities of the neighborhood cells. Bartie et al. (2010) characterized the view of a ground object by mapping the following indicators: the percentage of visible cells amounting to the cells that describe the object, the largest horizontal angle of the target object, visible façade area of the object, perceived size (calculated based on the distance to the object), clearness (defined as "visible area," referring to the visible area if all other objects are removed from the scene), and skyline (defined as the percentage of the skyline of the object that is not overshadowed by taller and more distant objects).

12.2.4 HEDONIC ANALYSIS OF VIEW

Several hedonic studies examined the economic value of different views by analyzing their impact on dwelling prices, as reviewed by Bourassa et al. (2005) and Jim and Chen (2009).

The results of classical hedonic modeling studies that rely on linear regression or its extensions are summarized in Table 12.3. As shown in the table, several types of views, including the open (openness of visibility), ocean (visibility of ocean), green (visibility of open space or forest), and urban views (visibility of urban elements, such as streets or buildings), have been analyzed in previous studies; several of these studies confirm the positive economic values of the open and water views and the negative economic value of the urban view. In particular, some studies suggest a prominent positive effect of the water view (e.g., Benson et al., 1998; Yu et al., 2007; Jim and Chen, 2009). However, the economic value of the green view is unclear; it can be both positive and negative, depending on the study. This uncertainty may partly be due to the problem of data resolution, as it is difficult to obtain high-resolution remote-sensing data that precisely captures the actual placement and height of greenery (e.g., trees). Fortunately, third-generation sensors allow us to acquire data on the actual placements and heights of trees, as demonstrated in the next section.

TABLE 12.3
Results of Hedonic Studies of Views Captured after 2005 (Black: Positively Significant at Less Than 10% Level; Gray: Negatively Significant at Less Than 10% Level; Insig.: Insignificant)

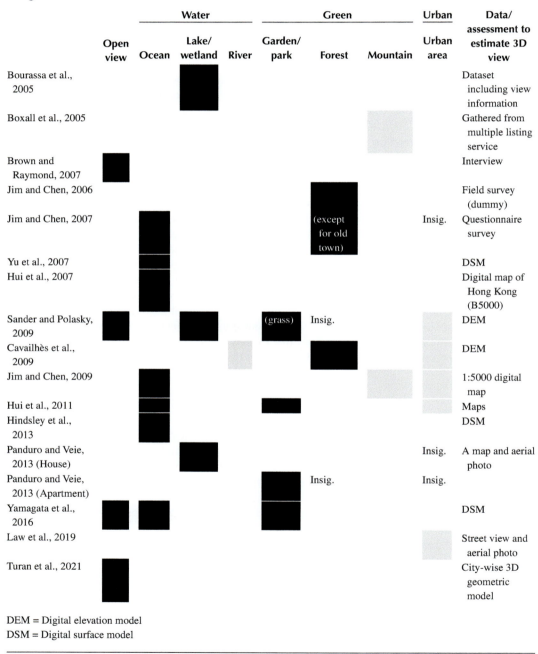

DEM = Digital elevation model
DSM = Digital surface model

Smart Environment Design

The tendency toward open data and the widespread use of machine learning have enabled the detailed evaluation of urban perceptions and their economic values. For example, Law et al. (2019) learned the visual features extracted from Google Street View and Bing aerial photos, using a convolutional neural network, and demonstrated the strong influence of urban perception on housing prices. Chen M. et al. (2022) used geotagged Flickr images to extract the features of urban scenes (using different elements, e.g., plazas, crosswalks, and palaces) and evaluated their economic values (using random forest). They suggest that the random forest model outperforms the classical hedonic model, in terms of accuracy and interpretability. Swietek and Zumwald (2023) evaluated the visual quality of individual buildings in multiple cities in Switzerland, using a large-scale 3D urban model, and mapped their economic values. They also demonstrated the higher accuracy and improved interpretability of machine-learning techniques over conventional hedonic regression models.

12.3 APPLICATION: THREE-DIMENSIONAL (3D) VIEW ANALYSIS OF YOKOHAMA CITY, JAPAN

12.3.1 Outline

This section presents the evaluation of the economic value of 3D views, including open, green, and ocean views, carried out using a hedonic analysis of apartment unit prices (Marketing Research Center Co. Ltd.). The target area was the seven central wards in Yokohama City (Figure 12.2), located approximately 30 km south of the central Tokyo area. The seven wards included 694 apartment buildings comprising 27,446 apartment units; the average prices of the apartment units are shown in Figure 12.3. Some results have been reported by Yamagata et al. (2016).

Yokohama is the second most populated city in Japan, with the population being approximately 370 million in 2014. The city opened its ports to foreign trade for the first time in 1959. Since then, Yokohama has rapidly developed as an entrance point to Japan, adopting several foreign cultures. Notably, various Western-style historic buildings and the largest Chinese town in Japan are located in Yokohama. The port-side area in Yokohama was redeveloped after the 1980s; the redeveloped area, called Minato Mirai 21, is currently a sophisticated urban area. In the modern world, Yokohama is a large city characterized by both foreign cultures and sophisticated urban facilities.

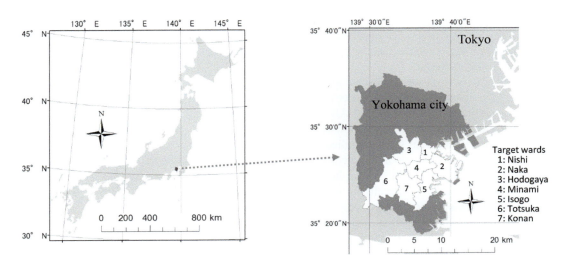

FIGURE 12.2 Map of the target area.

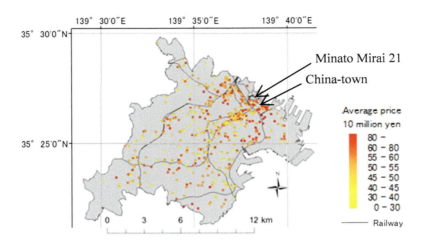

FIGURE 12.3 Map portraying the locations and average prices of the target apartments.

12.3.2 THREE-DIMENSIONAL (3D) VIEW EVALUATION

We applied the cumulative viewshed to evaluate the open, green, and ocean views. The numbers of open, green, and ocean cells visible in each apartment unit were evaluated. The details of the calculation procedure/steps are explained next:

1. The apartment data and building polygons (source: Fundamental Geospatial Data of the Geographical Survey Institute of Japan) were manually associated using Google Maps and the webpages of the apartments; the building polygons that corresponded to each apartment were identified.
2. The floor heights of each unit were identified using room numbers.
3. The 3D coordinates of each observation point were set for each unit.
4. A digital surface model (DSM) (Figure 12.4) and digital elevation model (DEM) (Figure 12.5), with spatial resolutions of 50 cm × 50 cm, were obtained using airborne LiDAR data.
5. The visibility from each observation point was evaluated using ArcGIS 3D Analyst, and the open views, defined as the number of DSM cells visible from each observation point, were also assessed.
6. The data of trees were extracted and processed, as follows:
   a. The actual placements of trees were identified by classifying aerial photos acquired simultaneously with the LiDAR data, using a likelihood maximization method.
   b. The extracted trees were spatially matched with the DSM, and the DSM cells that represented the trees were identified.
   c. The heights of trees in these cells were estimated by calculating the difference between the DSM and DEM. As the heights of trees were generally greater than 50 cm, we included only those cells in which the tree heights were > 50 cm (Figure 12.6). This process was necessary to remove noise.
7. The cells containing ocean views were identified from the geographic information system (GIS) data provided by Yokohama City.

The green and ocean views were estimated by counting the number of cells that contained trees and the ocean area, respectively.

Smart Environment Design 371

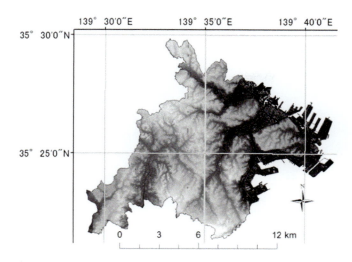

FIGURE 12.4 Digital surface model (DSM) of the study area.

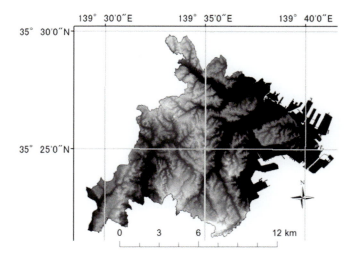

FIGURE 12.5 Digital elevation model (DEM) of the study area.

Notably, in Step 3, we had no data on the locations and directions of the windows in the apartment units. The 3D coordinates of the observation points were defined as follows [according to Yasumoto et al. (2011)]: the height was defined as [height of a floor (3 m)] × ([number of stories]−1) + [height of human eyes (1.6 m)], and the longitudes and latitudes were set at four midpoints on each side of the rectangles that contained the apartment buildings. In addition, we set the maximum visible range to 500 m, in accordance with Yu et al. (2007) and Yasumoto et al. (2011). Although the setting was somewhat subjective, the number of visible cells did not increase significantly, even if we counted the cells that were more distant than 500 m (Yasumoto et al., 2011).

The estimated open, green, and ocean views are shown in Figures 12.7, 12.8, and 12.9, respectively. The open views were relatively small in the region near Yokohama Station and the Chinatown area, wherein the buildings that could obstruct views were placed densely. While the green views portrayed similar tendencies, they exhibited a greater gap between the lower values at

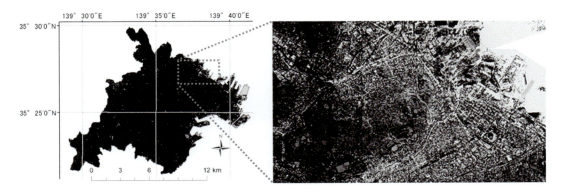

FIGURE 12.6 Extracted data for cells containing trees.

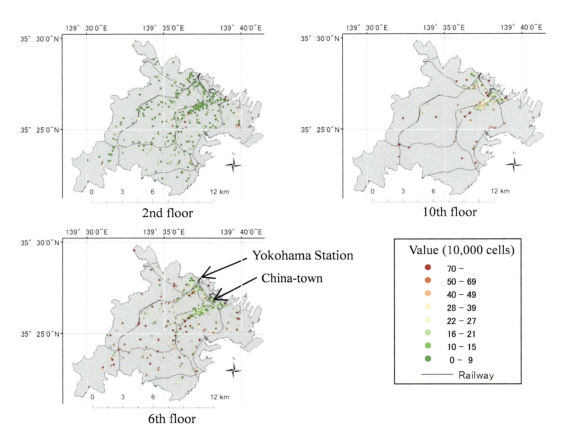

FIGURE 12.7 Analysis results for open views.

Yokohama Station and the China-town area, along with higher values in the western woody area. As most apartments were more than 500 m (i.e., the threshold distance) away from the ocean, most apartments did not have ocean views. In contrast, ocean views were prominent in the redeveloped bayside area, Minato Mirai 21.

Smart Environment Design

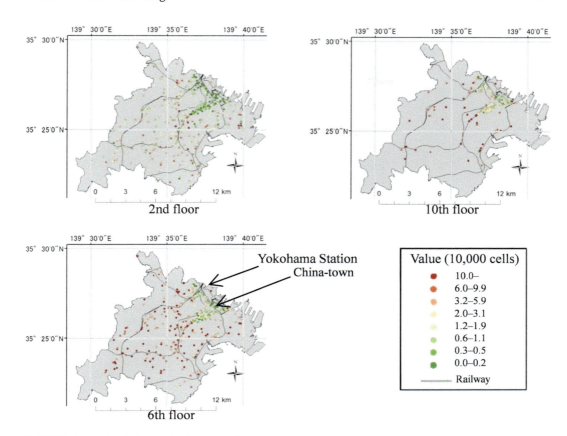

FIGURE 12.8 Analysis results for green views.

12.3.3 Hedonic Analysis Results

The economic values of the view indexes were evaluated using hedonic analysis. A multilevel model (e.g., Hox, 1998) consisting of a regression model with unit- and building-wise disturbances was applied. The explained variables were the logarithms of the condominium prices, and the explanatory variables were the open, green, and ocean views and other control variables (Table 12.4). For variable selection, a stepwise method was applied, and the control variables that were not significant at the 10% level were omitted.

Table 12.5 summarizes the estimation results. Among the control variables, the area (+), floor (+), major deviation (+), park (−), ocean (−), C1 res. (−), and semi-ind. (−) were significant at the 5% level. These observations were intuitively consistent.

The open and ocean views were positively significant at the 1% level. Thus, the significance of the open- and ocean-view designs in urban management was confirmed. However, the green view was negatively significant at the 1% level and thus, was intuitively inconsistent. As discussed by Jim and Chen (2009), in the bay-side area, green views can have a negative impact because they are less preferred than ocean views. However, in inland areas with no ocean views, green views may be preferred.

Therefore, we conducted a ward-wise hedonic analysis. The estimated significances of the view variables are plotted in Figure 12.10. Open views had a positive impact in the majority of wards;

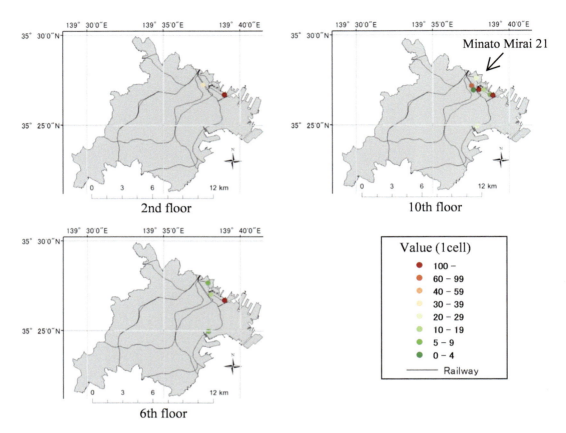

FIGURE 12.9 Analysis results for ocean views.

ocean views had a positive impact only around the Minato Mirai 21 area. Green views were insignificant (or negatively significant) in the bay-side wards, including the Nishi, Naka, and Isogo wards; green views were positively significant in the western inland wards, including the Totsuka and Konan wards. In other words, in inland areas, a green view is generally preferred. Such insights into natural resources are useful for designing smart urban environments.

12.4 CONCLUDING REMARKS

This work discusses the roles of remote sensing in designing a smart environment and explains the importance of analyzing the economic values of the 3D views of natural resources. After the techniques for 3D-view evaluations were briefly reviewed, a hedonic analysis of the open, green, and ocean views was conducted.

In this study, we demonstrate that remote sensing can serve as a useful tool for urban planning and design. A significant challenge in the approach employed in this study is the use of large amounts of social media-based remote sensing data. Notably, unlike traditional socioeconomic surveys, social media–based remote sensing data are accumulated in almost real time, providing the updated data of a region. Even though such large spatiotemporal data may contribute to the design of smart cities, it is important to develop models that can evolve to deal with such data efficiently and effectively.

TABLE 12.4
Control Variables Considered in This Study

Variables	Description	Variables	Description
Const.	Constant	Cl high	Dummy of C1 medium-to-high exclusive RD
Area	Log. of unit area [m²]	Cl exclusive	Dummy of Cl exclusive RD
Floor	Log. of floor of unit	C2 res.	Dummy of category 2 (C2) RD
SRC	Dummy of the steel reinforced concrete structure	C2 high	Dummy of C2 medium-to-high exclusive RD
WRC	Dummy of the steel wall concrete structure	C2 exclusive	Dummy of C2 exclusive RD
Dev.	Numbers of related developers	Industry	Dummy of industrial districts
Major dev.	The ratio of major developers called MAJOR 8[(1)] accounting for Dev.	Semi ind.	Dummy of semi-industrial districts
Station	Log. of the travel time to the nearest station	Commerce	Dummy of commercial districts
Green.	Log. of the number of green cells within 500m	Neigh.com.	Dummy of neighborhood commercial districts
Park	Log. of the distance to the nearest city park [km]	Cl res.	Dummy of category 1 (Cl) residential districts (RD)
Ocean	Log. of the distance to the ocean [km]	Cl low	Dummy of C1 low-rise exclusive RD
Year	Year dummies (1993 – 2007)		

[(1)] MAJOR 8 includes Sumitomo Realty & Development Co., Ltd., Tokyu Land Corporation, Mitsubishi Estate Co., Ltd., Towa Real Estate Development Co., Ltd., Daikyo Inc., Nomura Real Estate Development Co., Ldt., Mitsui Fudosan Residential Co., Ltd. and Tokyo Tatemono Co., Ltd.

[(2)] Green is calculated using the tree cells shown in Figure 6, Park and Ocean are calculated using the GIS data provided by Yokohama city, and the other variables, except for the view variables, are from the condominium dataset of Marketing Research Center Co. Ltd.

TABLE 12.5
Estimation Results of This Study

Variables	Estimate	t value	
Const.	3.28	1.87×10^2	***
Area	1.13	4.09×10^2	***
Floor	6.19×10^{-2}	6.47×10^1	***
Major dev.	6.19×10^{-2}	4.78	***
Park	-1.16×10^{-2}	−2.39	**
Ocean	-5.22×10^{-2}	−9.65	***
Cl res.	-3.39×10^{-2}	−2.04	**
Cl exclusive	4.53×10^{-2}	1.77	*
Semi ind.	-8.08×10^{-2}	−3.97	***
Open view	1.28×10^{-1}	2.21×10^1	***
Green view	-4.71×10^{-1}	-1.87×10^1	***
Ocean view	2.52×10^2	9.71	***
AIC		−72818	

[1)] *, **, and *** denote significant levels of 10%, 5%, and 1%. respectively.

[2)] Estimate of year dummies are omitted.

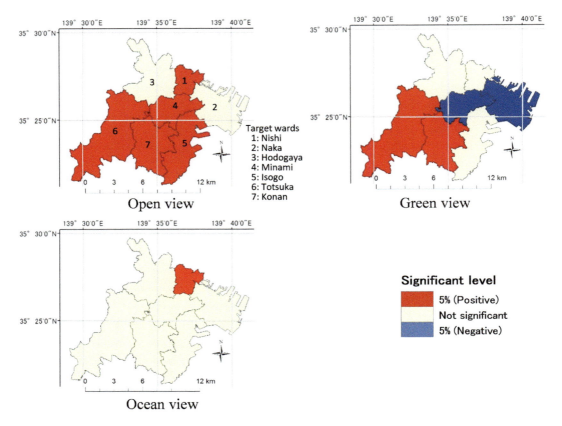

FIGURE 12.10 Significance levels of the view variables.

REFERENCES

Bartie, P., Reitsma, F., Kingham, S. & Mills, S. 2010. Advancing visibility modeling algorithms for urban environments. *Computers, Environment and Urban Systems*, 34, 518–531.

Benedikt, M.L. 1979. To take hold of space: Isovists and isovist fields. *Environment and Planning B: Planning and Design*, 6, 47–65.

Benson, E.D., Hansen, J.L., Schwartz Jr., A.L. & Smersh, G.T. 1998. Pricing residential amenities: The value of a view. *Journal of Real Estate Finance and Economics*, 16, 55–73.

Benson, J.F., Scott, K.E., Anderson, C., Macfarlane, R., Dunsford, H. & Turner, K. 2004. Landscape capacity study for onshore wind energy development in the Western Isles. *Scottish Natural Heritage Commissioned Report*, 42.

Blaschke, T., Hay, G.J., Weng, Q. & Resch, B. 2011. Collective sensing: Integrating geospatial technologies to understand urban systems-An overview. *Remote Sensing*, 3, 1743–1776.

Bourassa, S.C., Hoesli, M. & Sun, J. 2005. The price of aesthetic externalities. *Journal of Real Estate Literature*, 13, 165–188.

Brown, G. & Raymond, C. 2007. The relationship between place attachment and landscape values: Toward mapping place attachment. *Applied Geography*, 27, 89–111.

Chamberlain, B.C. & Meitner, M.J. 2013. A route-based visibility analysis for landscape management. *Landscape and Urban Planning*, 111, 13–24.

Chen, C., Hasi Bagan, H., Yoshida, T., Borjigin, H. & Gao, J. 2022. Quantitative analysis of the building-level relationship between building form and land surface temperature using airborne LiDAR and thermal infrared data. *Urban Climate*, 45, 101248.

Chen, M., Liu, Y., Arribas-Bel, D. & Singleton, A. 2022. Assessing the value of user-generated images of urban surroundings for house price estimation. *Landscape and Urban Planning*, 226, 104486.

Cho, S.-H., Poudyal, N.C. & Roberts, R.K. 2008. Spatial analysis of the amenity value of green space. *Ecological Economics*, 66, 2–3.
Chun, B. & Guldmann, J.M. 2014. Spatial statistical analysis and simulation of the urban heat island in high-density central cities. *Landscape and Urban Planning*, 125, 76–88.
Denholm, P., Kuss, M. & Margolis, R.M. 2013. Co-benefits of large-scale plug-in hybrid electric vehicle and solar PV deployment. *Journal of Power Sources*, 236, 350–356.
Domingo-Santos, J.M., de Villarán R.F. & Rapp-Arrarás, I. 2011. The visual exposure in forest and rural landscapes: An algorithm and a GIS tool. *Landscape and Urban Planning*, 101, 52–58.
Donnay, J.-P., Barnsley, M.J. & Longley, P.A. 2001. *Remote Sensing and Urban Analysis*. Taylor & Francis, London.
Falconer, L., Hunter, D.-C., Telfer, T.C. & Ross, L.G. 2013. Visual, seascape and landscape analysis to support coastal aquaculture site selection. *Land Use Policy*, 34, 1–10.
Fisher-Gewirtzman, D. & Wagner, I.A. 2003. Spatial openness as a practical metric for evaluating built-up environments. *Environment and Planning B: Planning and Design*, 30, 37–49.
Forster, B. 1983. Some urban measurements from Landsat data. *Photogrammetric Engineering and Remote Sensing*, 49, 1693–1707.
Giannouli, M. & Yianoulis, P. 2012. Study on the incorporation of photovoltaic systems as an auxiliary power source for hybrid and electric vehicles. *Solar Energy*, 86, 441–451.
Giffinger, R., Fertner, C., Kramar, H., Kalasek, R, Pichler-Milanovic, N. & Meijers, E. 2007. *Smart Cities-Ranking of European Medium-Sized Cities*. Vienna University of Technology, Vienna.
Grêt-Regamey, A., Bishop, I.D. & Bebi, P. 2007. Predicting the scenic beauty value of mapped landscape changes in a mountainous region through the use of GIS. *Environment and Planning B: Planning and Design*, 34, 50–67.
Hollands, R.G. 2008. Will the real smart city please stand up? *City*, 12, 303–320.
Hox, J.J. 1998. Multilevel modeling: When and why. In Balderjahn, I., Mathar, R. & Schader, M. (Eds.), *Classification, Data Analysis, and Data Highways*. Springer, New York.
Inglis, N.C., Vukomanovic, J., Costanza, J. & Singh, K.K. 2022. From viewsheds to viewscapes: Trends in landscape visibility and visual quality research. *Landscape and Urban Planning*, 224, 104424.
Irwin, E.G. 2002. The effects of open space on residential property values. *Land Economics*, 78, 465–480.
Jim, C.Y. & Chen, W.Y. 2009. Value of scenic views: Hedonic assessment of private housing in Hong Kong. *Landscape and Urban Planning*, 91, 226–234.
Kong, F., Yin, H. & Nakagoshi, N. 2007. Using GIS and landscape metrics in the hedonic price modeling of the amenity value of urban green space: A case study in Jinan City, China. *Landscape and Urban Planning*, 79, 240–252.
Krukar, J., Manivannan, C., Bhatt, M. & Schultz, C. 2021. Embodied 3D isovists: A method to model the visual perception of space. *Environment and Planning B: Urban Analytics and City Science*, 48(8), 2307–2325.
Lange, E. & Schaeffer, P. 2001. A comment on the market value of a room with a view. *Landscape and Urban Planning*, 55, 113–120.
Law, S., Paige, B. & Russell, C. 2019. Take a look around: Using street view and satellite images to estimate house prices. *ACM Transactions on Intelligent Systems and Technology*, 10(5), 1–19.
Leduc, T., Miguet, F., Tourre, V. & Woloszyn, P. 2010. Towards a spatial semantics to analyze the visual dynamics of the pedestrian mobility in the urban fabric. In Painho, M., Santos, M.Y. & Pundt, H. (Eds.), *Geospatial Thinking*, Springer, Berlin Heidelberg.
Liang, S., Wang, K., Zhang, X. & Wild, M. 2010. Review on estimation of land surface radiation and energy budgets from ground measurement, remote sensing and model simulations. *IEEE Journal of Selected Topics in Applied Earth Observations and Remote Sensing*, 3, 225–240.
Llibera, M. 2003. Extending GIS-based visual analysis the concept of visualscapes. *International Journal of Geographical Information Science*, 17, 25–48.
Luttik, J. 2000. The value of trees, water and open space as reflected by house prices in the Netherlands. *Landscape Urban Planning*, 48, 161–167.
Lynch, K. 1976. *Managing the Sense of Regions*. MIT Press, Cambridge, MA.
Ma, J. 2003. From professional to people's software- Tracing the development of 3D GIS software at ESRI. In Buhnann E. (Ed.), *Trends in Landscape Modeling: Proceedings at Anhalt University of Applied Science 2003*. Wichman, Heidelberg.
McLeod, P.B. 1984. The demand for local amenity: A hedonic price analysis. *Environment and Planning A*, 16, 389–400.

Miller, R.B. & Small, C. 2003. Cities from space: Potential applications of remote sensing in urban environmental research and policy. *Environmental Science and Policy*, 6, 129–137.

Morancho, A.B. 2003. A hedonic valuation of urban green areas. *Landscape and Urban Planning*, 66, 35–41.

Morello, E. & Ratti, C. 2009. A digital image of the city: 3D isovists in Lynch's urban analysis. *Environment and Planning B: Planning and Design*, 36, 837–853.

Mouflis, G.-D., Gitas, I.-Z., Iliadou, S. & Mitri, G. 2008. Assessment of the visual impact of marble quarry expansion (1984–2000) on the landscape of Thasos Island, NE Greece. *Landscape and Urban Planning*, 86, 92–102.

Patino, J.E. & Duque, J.C. 2013. A review of regional science applications of satellite remote sensing in urban settings. *Computers, Environment and Urban Systems*, 37, 1–17.

Pearson, A., Nutsford, D. & Thomson, G. 2014. Measuring visual exposure to smoking behaviours: A viewshed analysis of smoking at outdoor bars and cafés across a capital city's downtown area. *BMC Public Health*, 14, 300.

Ratti, C. 2002. *Urban Analysis for Environmental Prediction*. PhD thesis, University of Cambridge, Cambridge.

Rosen, S. 1974. Hedonic prices and implicit markets: Product differentiation in pure competition. *Journal of Political Economy*, 82, 34–55.

Swietek, A.R. & Zumwald, M. 2023. Visual capital: Evaluating building-level visual landscape quality at scale. *Landscape and Urban Planning*, 240, 104880.

Tajima, K. 2003. New estimates of the demand for urban green space: Implications for valuing the environmental benefits of Boston's Big Dig Project. *Journal of Urban Affairs*, 25, 641–655.

Teller, J. 2003. A spherical metric for the field-oriented analysis of complex urban open spaces. *Environment and Planning B: Planning and Design*, 30, 339–356.

Tooke, T.R., Coops, N.C. & Webster, J. 2014. Predicting building ages from LiDAR data with random forests for building energy modeling. *Energy and Buildings*, 68, 603–610.

Tyrvainen, L. & Miettinen, A. 2000. Property prices and urban forest amenities. *Journal of Environmental Economics and Management*, 39, 205–223.

Welch, R. 1982. Spatial resolution requirements for urban studies. *International Journal of Remote Sensing*, 3, 139–146.

Wheatley, D. 1995. Cumulative viewshed analysis: A GIS-based method for investigating intervisibility and its archaeological application. In Lock, G. & Stancic, Z. (Eds.), *Archaeology and Geographic Information Systems: A European Perspective*. Taylor and Francis, London.

Wu, Y., Yang, Z., Lin, B., Liu, H., Wang, R., Zhou, B. & Hao, J. 2012. Energy consumption and CO_2 emission impacts of vehicle electrification in three developed regions of China. *Energy Policy*, 48, 537–550.

Wu, Z., Wang, Y., Gan, W., Zou, Y., Dong, W., Zhou, S. & Wang, M. 2023. A survey of the landscape visibility analysis tools and technical improvements. *International Journal of Environmental Research and Public Health*, 20(3), 1788.

Yamagata, Y., Murakami, D., Yoshida, T., Seya, H. & Kuroda, S. 2016. Value of urban views in a bay city: Hedonic analysis with the spatial multilevel additive regression (SMAR) model. *Landscape and Urban Planning*, 151, 89–102.

Yang, P.P.-J., Putra, S.Y. & Li, W. 2007. Viewsphere: A GIS-based 3D visibility analysis for urban design evaluation. *Environment and Planning B, Planning and Design*, 34, 971–992.

Yasumoto, S., Jones, A., Nakaya, T. & Yano, K. 2011. The use of a virtual city model for assessing equality in access to views. *Computers, Environment and Urban Systems*, 35, 464–473.

Yu, S.-M., Han, S.-S. & Chai, C-H. 2007. Modeling the value of view in high-rise apartments: A 3D GIS approach. *Environment and Planning B, Planning and Design*, 34, 139–153.

Part VI

Nightlights

13 Nighttime Light Remote Sensing—Monitoring Human Societies from Outer Space

Qingling Zhang, Noam Levin, Christos Chalkias, Husi Letu, and Di Liu

ACRONYMS AND DEFINITIONS

AVHRR	Advanced very high-resolution radiometer
BRWL	Bright rich white light
CBAS	Center of big data for sustainable development goals
CIESIN	Center for International Earth Science Information Network
CUMULOS	CubeSat multispectral observing system
DMSP	Defense Meteorological Satellite Program
DNB	Day/night band
EnMAP	Environmental Mapping and Analysis Program
GDP	Gross domestic product
HPS	High-pressure sodium
HSC	High sensitivity camera
HSTC	High sensitivity technological camera
ISS	International Space Station
LED	Light emitting diodes
MISR	Multi-angle imaging spectroradiometer
NCEP	National Centers for Environmental Prediction
NGDC	National Geophysical Data Center
NPOESS	National Polar-orbiting Operation Environmental Satellite System
NLP	Nighttime light pollution
NTL	Nighttime light
OLS	Operational linescan system
SDGSAT-1	Sustainable Development Science Satellite 1
TIR	Thermal infrared
VANUI	Vegetation Adjusted Normalized Urban Index
VIIRS	Visible/infrared imager/radiometer suite
VNIR	Visible-near infrared

13.1 INTRODUCTION

The land surface of our home planet Earth is a finite resource that is central to human welfare and the functioning of Earth systems. Human population growth is one of the major global-scale forcing that underlie the most recent global-scale state shift in Earth's biosphere (Barnosky et al. 2012). Due to industrialization, economic growth, technology advances, and population explosion, human activities worldwide are transforming the terrestrial environment at unparalleled rates and scales. Prime grasslands and forests have been converted to croplands and pastures, which cover about 40% of the global land surface (Foley et al. 2005), to support the rising need for food by more than seven billion people (UNFPA 2011). Among anthropogenic activities, urbanization is the most irreversible and human-dominated form of land use, modifying land cover, hydrological systems, biogeochemistry, climate, and biodiversity worldwide (Grimm et al. 2008). Worldwide, urban expansion is one of the primary drivers of habitat loss and species extinction (Hahs et al. 2009). Urban areas also affect their local climates through the modification of surface albedo, evapotranspiration, and increased aerosols and anthropogenic heat sources, thereby creating elevated urban temperatures (Arnfield 2003) and changes in regional precipitation patterns (Marshall Shepherd et al. 2002; Rosenfeld 2000; Seto and Shepherd 2009). In many developing countries, urban expansion takes place on prime agricultural lands (del Mar López et al. 2001; Seto et al. 2000). 60–80% of final energy is consumed in urban areas worldwide (Johansson et al. 2012). However, cities also show themselves as a potential solution to climate change through efficient resource use. Compact urban development coupled with high residential and employment densities can reduce energy consumption, vehicle miles traveled, and carbon dioxide emissions (Gomez-Ibanez et al. 2009). Per capita energy use and greenhouse emissions are often lower in cities than national averages (Dodman 2009). Furthermore, increasing urban albedo could offset greenhouse gas emissions (Akbari et al. 2009).

At the same time, human societies are vulnerable to climate change. Low-lying human settlements in the coastal zones and on islands are increasingly threatened by sea level rise (Nicholls and Cazenave 2010) and extreme weather events, such as extremely powerful hurricanes (Abramson and Redlener 2012; Brinkley 2006). People living in mountains are at risk of landslides (Galli and Guzzetti 2007) and flash floods (e.g., the 2013 North India floods, http://www.foxnews.com/world/2013/07/15/india-says-5748-missing-in-floods-now-presumed-dead/, retrieved on February 27, 2014). The two and half billion people living in dry lands worldwide are vulnerable to desertification and land degradation (Reynolds et al. 2007).

Monitoring and understanding the human dimensions of global change, including economic, political, cultural, and socio-technical systems, and their interactions with the environmental systems is thus vital to the understanding of global environmental change and its consequences (Stern et al. 1991). However, monitoring human societies at the global scale can be very challenging. Social scientists have to rely on census and field survey to collection essential variables about humans, including government policies, land-tenure rules, GDP, population, population density, energy consumption, and carbon emissions, etc. Census and field survey can be very labor and cost intensive, and in many cases are just not impossible. Remote sensing data has long been used to monitor the global environment and its change due to its capability of taking measurements without contacting the objects of interest in a repeatable and relatively cheap way. However, conventional remote sensing techniques are limited to detecting only visible human artifacts such as buildings, crop fields, and roads, which are of less interest to many social scientists than the abstract variables that explain their appearance and transformation (National Research Council et al. 1998). The variables of greatest interest to these social scientists, such as land use (e.g., residential, commercial, industrial), people's movement and focal locations of social interactions, population characteristics (e.g., size, poverty) as well as government policies, are not readily measurable from space. Thus, there has been a deep doubt that remote sensing can measure anything considered important in social sciences.

We are currently living in an era of global change in hydrology, climatology, and biology that significantly differs from previous episodes of global change in the extent to which it is human in origin (Stern et al. 1991). A complex of social, political, economic, technological, and cultural variables are believed to be driving forces that influence the human activities that proximately cause global change (Stern et al. 1991). Understanding the linkages among all the driving forces is a major scientific challenge that will require developing new interdisciplinary teams, including social scientists and physical scientists.

Nighttime Light Remote Sensing

This chapter focuses on a unique type of remote sensing technology: observing nighttime lights from outer space to monitor human societies. Contrasting to conventional environmental remote sensing, nighttime light remote sensing aims at measuring human activities from outer space. The Operational Linescan System on board the Defense Meteorological Satellite Program (DMSP/OLS) has been collecting routine nighttime light observations globally since early 1970s and have been proved as a close proxy to human activities at night (Croft 1978). Nighttime light remote sensing provides the great potential to fill the gap between social sciences and remote sensing. Furthermore, nighttime light analysis has its own environmental value: assessing the impacts of anthropogenic activities on the environmental systems.

Given the previous discussions, this chapter will focus on the three areas of nighttime light remote sensing. First, a review of the history of nighttime light remote sensing and its evolution is provided. Second, major applications of nighttime light remote sensing related to social and natural sciences are illustrated. Third, challenges in future directions of nighttime lights remote sensing are discussed and a number of suggestions are made.

13.1.1 THE RATIONALE THAT UNDERLIES NIGHTTIME LIGHT (NTL) REMOTE SENSING

Diurnality is a common phenomenon in the animal kingdom. Many animals' daily activities strictly depend on sunlight, active during the day and sleeping at night. Human beings used to be a member of the diurnal club. During its early history, human activity at night was also greatly confined until they learned how to use fire to light up their living spaces. Technical advances in artificial lighting have given people greater flexibilities as to where, when, and how long their activities can take place. The use of street lighting was first recorded in the city of Antioch from the 4th century (Luckiesh 1920), later in the Arab Empire from the 9th–10th centuries, especially in Cordova, and then in London from 1417 when Henry Barton (Lovatt and O'Connor 1995), the mayor, ordered "lanterns with lights to be hanged out on the winter evenings between Hallowtide and Candlemasse." At that time candles were used for street lighting. After the invention of gas light by William Murdoch in 1792, cities in Britain began to light their streets using gas. After Edison pioneered electric use, light bulbs were developed for streetlights as well. On March 31, 1880, Wabash, Indiana, became the first electrically lighted city in the world (http://www.cityofwabash.com/city-information/history/electric-light-article/). Open Arc lamps were used in the late 19th and early 20th century by many large cities for street lighting. Some 20 years later, incandescent light bulbs were introduced. In the late 1930s the fluorescent lamp first became common, followed by mercury vapor streetlight assembly in 1948. Sodium vapor streetlights were put into service around 1970. In recent years, new types of streetlights have been introduced, including metal halide, ceramic discharge metal halide lamp, induction lamp, compact fluorescent lamp, and light emitting diodes (LED). Over the past centuries, lighting technologies have become more and more efficient, allowing to use more light per capita at lower costs (Fouquet and Pearson 2006; Tsao and Waide 2010). The vivid nightscape dominated by artificial illumination of human settlements is the other side of our blue marble earth (known as *black marble* (Kohrs et al. 2014; Román et al. 2018) and can be seen from space by astronauts at night.

Artificial illumination allows human beings break through its natural diurnality so their social, cultural, and economic activities can extend into the night, creating the so-called nighttime economy (Lovatt and O'Connor 1995). At the first locations with gas and electricity lighting, the light was there to illuminate the goods on display and the people attracted to them (Benjamin and Baudelaire 1973; Geist 1983). Since then, whether it is for a nightclub, a bar, or a cinema, lighting becomes necessary. In modern days, lighting is first for extending our activity times, convenience, and safety. Night light is thus an important and reliable indicator of human activities immediately at night and indirectly during daytime. Artificial illumination of buildings, transportation corridors, parking lots, and other elements of the built environment has become the hallmark of many contemporary urban settlements and urban activity. Our world is now a highly domesticated nature, with no place on our planet is truly free of human beings' footprints (Kareiva et al. 2007). This can be well illustrated by night light satellite images (Figure 13.1). It is a very obvious fact that night on our planet is also highly domesticated. At the beginning of the millennium more than two third of the world's population (99% of the US and the European Union population) and almost 20% of world terrain were under skies influenced by anthropogenic lights (Cinzano et al. 2001).

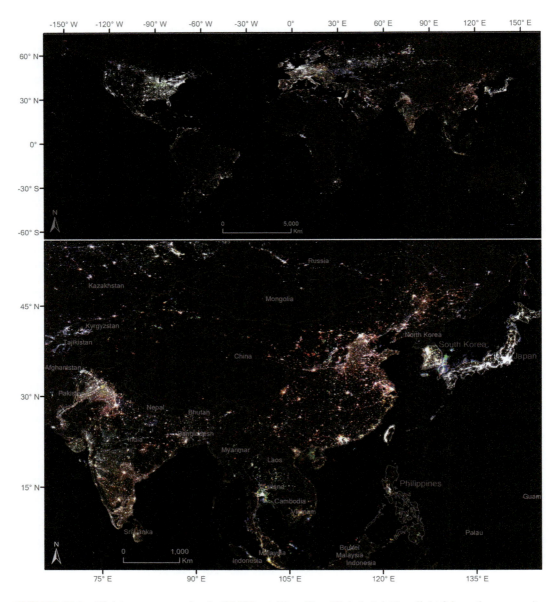

FIGURE 13.1 Nightscape as seen by the DMSP satellites. Top: Global nighttime light false color composite from DMSP/OLS average stable visible images (Red: F162009, Green: F152000, Blue: F101992); Bottom: Close up to Asia. Urban intensity is indicated by the color intensity and change in urban intensity is indicated by the color tune. Red and Yellow suggest recent growth, while Blue means decline and White means stable dense urban.

13.1.2 History of Nighttime Light Remote Sensing

Experiments in nighttime aerial photos were already done toward the end of World War I over East Anglia (Kingslake 1942), and during World War II (starting in 1940) nighttime aerial photos were often acquired using flash-bombs for reconnaissance purposes (Katz 1948; Schulman and Rader 2012). However, it was not until the age of spaceborne imagery that the potential of this data source can contribute to our understanding of human activity has been realized. The Operational Linescan System (OLS) onboard the US Air Force Defense Meteorological Satellite Program (DMSP) was originally designed to detect clouds illuminated by moonlight for military bombing operations at nights. The first DMSP satellite with night light capabilities was launched in 1971 (Levin et al. 2020), and the first one carrying OLS began flying in 1976, soon after the pioneering program Sensor Aerospace Vehicle Electronomics Package (1970–1976) ended. Since then, a series of DMSP satellites have been launched into orbit and the latest four (F15–F18) are still in operation today. In most years there were two or more satellites operating in space to form a constellation: one passing at dawn and the other at dusk. The DMSP program was declassified in 1972. A digital archive for the DMSP/OLS data was established in mid-1992 at the NOAA National Geophysical Data Center (NGDC) and made available to public access through internet. Digital DMSP/OLS data from 1972 to 1992 were not archived and all scientific access to them was through films archived at the National Snow and Ice Data Center, University of Colorado.

OLS is an oscillating scan radiometer with a swath of about 3000 km. With 14 orbits per day, each OLS is capable of providing global nighttime coverage every 24 hours. For cloud-detection purposes, OLS has two spectral bands: visible-near infrared (VNIR: 0.5–0.9 um) and thermal infrared (TIR: 10.5–12.5 um). Since moonlight is a weak illumination source compared with daytime solar illumination, the OLS VNIR band was designed to allow detecting radiances down to 10^{-9} W/sr•μm, which is more than five orders of magnitude lower than comparable bands of other daytime sensors, such as the NOAA AVHRR (Advanced Very High-Resolution Radiometer) and the Landsat Thematic Mapper (TM). OLS can acquire data in two spatial resolution modes: "fine" and "smoothed." In the "fine" mode, OLS captures full resolution data with a nominal spatial resolution of 0.5 km. Onboard averaging of 5 by 5 pixels of fine data produces "smoothed" data with a nominal spatial resolution of 2.7 km. Most of the data received by the NOAA-NGDC is in the smooth spatial resolution mode.

Shortly after DMSP's declassification the scientific research community found out that due to its low-light imaging capability, OLS can also detect nocturnal artificial lighting in clear night conditions without moonlight. Croft (1978) described the detection of cities, fires, fishing boats, and gas flares from the DMSP program, showing the potential of NTL data as an indicator of human activity. Since then, studies have shown strong relationships between NTL data and key socioeconomic variables such as urban population estimates (Amaral et al. 2006; Balk et al. 2006; Elvidge et al. 1997a; Sutton et al. 2001), population density (Sutton et al. 2003; Zhuo et al. 2009), economic activity (Doll et al. 2006), energy use, carbon emissions (Doll et al. 2000), impervious surfaces (Elvidge et al. 2007b), and sub-national estimates of gross domestic product (GDP) (Sutton et al. 2007). As such, they are often used as proxies for urban settlements and have been increasingly used to map aggregate measures of urban areas such as total area extent (Small et al. 2011) and regional and global urbanization dynamics (Zhang and Seto 2011).

However, it was clear that there was a pressing need in the economic and policy community to map urban areas at finer spatial scales than those offered by the DMSP/OLS imagery. Various sensors have been launched since 2010 that enable higher spatial resolution mapping of night lights from space. Such detailed spatial resolution nighttime light imagery can help to bridge the gap between social sciences (interested in explaining social processes) and remote sensing (offering the means to map spatial patterns and processes) (Rindfuss and Stern 1998). The High Sensitivity Camera (HSC) on board the Aquarius/SAC-D, launched in June 2011, offered nighttime light images at a spatial resolution of 200–300 m (Sen et al. 2006). Since the launch of the National Polar-orbiting Operation Environmental Satellite System (NPOESS) Preparatory Project Satellite

(NPP), which was later renamed Suomi NPP, on October 28, 2011, the onboard Visible/Infrared Imager/Radiometer Suite (VIIRS) has been collecting high-quality nighttime images at a spatial resolution of 750 m in the Day/Night Band (DNB) spanning the visible and infrared regions (Lee et al. 2006; Miller et al. 2006; Miller et al. 2005; Miller et al. 2012). With its dynamic radiometric range and advanced onboard calibration facilities, VIIRS takes continuous and consistent measurements of daytime Earth surface reflectivity and nighttime lights free of saturation, becoming the second and next generation sensor to provide daily global nighttime light observations from outer space. Primary evaluation shows that the VIIRS DNB is performing pretty well on-orbit (Liao et al. 2013) and its capability to estimate GDP is much better than DMSP/OLS at various scales evaluated (Li et al. 2013d; Shi et al. 2014).

Other new sensors that can be used for mapping night lights include astronaut photography from the International Space Station (ISS, spatial resolution 10–100) (Anderson et al. 2010; Dawson et al. 2012; Doll 2008), dedicated airborne campaigns (Elvidge and Green 2005; Tardà et al. 2011), and the High Sensitivity Technological Camera (HSTC) on board SAC-C launched in 2000 offering nighttime light images at a spatial resolution of 300 m (Colomb et al. 2003). The most accessible of these sources for medium spatial resolution images of night lights are the astronaut photographs, which were also the first to offer color images of nighttime lights from space. Available since mission 6 of the ISS, using a handheld camera Kodak DSC 760 camera (and other models later on), there are now thousands of images showing night lights of hundreds of cities around the world (Doll 2008), available for downloading through the NASA Gateway for Astronaut's Photography (http://eol.jsc.nasa.gov/). These images have been successful not only to capture public attention to light pollution through some visually stunning videos (as in Michael König's collection: http://vimeo.com/32001208), but can also be used to analyze spatial inequality in streetlights between cities and across borders with night color imagery (Levin and Duke 2012). ISS photos have also been useful to explain the nesting of sea turtles which prefer dark beaches (Mazor et al. 2013). However, ISS night photography suffers from some shortcomings, including imprecise geolocation, inconsistency in the spatial resolution and in the clarity of the image, and as the images were captured by relatively simple digital cameras, initially it was not possible be calibrate them quantitatively into photometric units (Doll 2008). Work led by Alejandro Sanchez de Miguel at the Complutense University of Madrid has made many of these ISS images easier to find, with many of them georeferenced, and a procedure was developed to calibrate them (Sanchez de Miguel et al. 2019, 2021b).

Imaging spectrometry of nighttime lights may enable a nearly full characterization of lighting type (Kruse and Elvidge 2011), as has been recently also demonstrated using the EnMAP (Environmental Mapping and Analysis Program) high-resolution imaging spaceborne spectrometer (Bachmann and Storch 2023). However this may also be partly accomplished using a multispectral sensor (Elvidge et al. 2010), and lamp classes were successfully identified from aerial color night light images (Hale et al. 2013). The potential use of information from high spatial resolution nighttime light sensors has been summarized in a proposal submitted to NASA, to launch a nighttime light dedicated sensor to be termed Nightsat, with a suggested spatial resolution between 50–100 m (Elvidge et al. 2007c; Hipskinda et al. 2011).

The recent launching of new sensors now allows us to study nighttime lights at higher spatial/spectral resolution than was impossible before, and thus will enable us to gain new insights on earth observation. The advent of high spatial resolution nighttime light satellite capabilities began with Israeli EROS-B satellite, launched in 2006 (Levin et al. 2014). But it only made nighttime acquisitions, publicly available in 2013. In 2017, the Chinese JL1–3B (Jilin-1) satellite became the first commercial satellite to offer multispectral (red, green, and blue) nighttime light images at a resolution of 0.92 m (Zheng et al. 2018), allowing to better distinguish nighttime light patterns used by different population sectors (Guk and Levin 2020). The Sustainable Development Science Satellite 1 (SDGSAT-1) (Guo et al. 2023), which was developed and operated by the International Research Center of Big Data for Sustainable Development Goals (CBAS), is the world's first scientific satellite dedicated to serving the United Nations 2030 Agenda for Sustainable Development (2030 Agenda), and has been used to assess humanitarian crisis situations, such as the 2023 Turkey earthquake (Levin 2023a). The Glimmer Imager aboard SDGSAT-1 was innovatively designed to have 3 RGB bands plus a panchromatic band. The spatial resolution is 10 m for the panchromatic band and 40 m for the RGB bands (https://www.sdgsat.ac.cn/satellite/describe). Operating in "Thermal Infrared + Multispectral" during the daytime, "Thermal Infrared + Glimmer" at night, and single sensor observing modes, SDGSAT-1 can collect three types of data of Earth surface in 24 hours. These advanced satellites with high spatial resolution enable the detailed study of urban land use and the potential identification and classification of lighting sources, and allow to better understand the correspondence between ground-based and spaceborne measurements of night lights (Levin 2023b). Figure 13.2 shows some examples of color nighttime images of Jerusalem collected by these sensors.

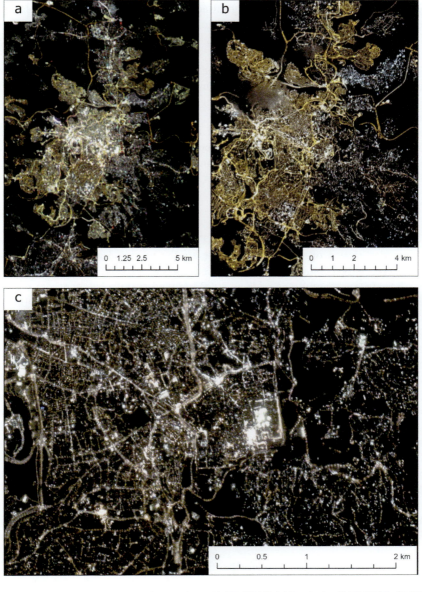

FIGURE 13.2 Color nighttime images of Jerusalem: (a) SDGSAT-1 (40 m), April 27, 2023; (b) ISS astronaut photograph (10 m), May 22, 2021; (c) Jilin-1 (1 m), April 6, 2023.

The current transformation in space-based remote sensing revolves around the use of small satellite missions. Planet Labs was the pioneer in providing global daily multispectral coverage of the entire Earth at a high spatial resolution of 3 m, using a constellation of approximately 150 nanosatellites (Strauss 2017). In coming years, researchers may benefit from similar CubeSats offering nighttime capabilities (such as NITEsat in Walczak et al. 2017). Various CubeSats have been launched in recent years, such as the CubeSat multispectral observing system (CUMULOS) and the multispectral AeroCube, demonstrating the capabilities these new sensors provide for nighttime imaging (Pack and Hardy 2016; Pack et al. 2017, 2018, 2019). Luojia-1 and Qimingxing-1(QMX-1) were built by Wuhan University, China, and were launched in 2018 and 2022, providing nighttime images at spatial resolutions of 130 m and 40 m (Jiang et al. 2018; Li et al. 2018). Recent studies have shown that Luojia-1 images are capable to accurately map urban extent and to monitor the construction of infrastructure at a moderate spatial resolution.

Moreover, the methodology for acquiring an aerial high spatial resolution night lights image has been demonstrated over Berlin (1 m) by Küechly et al. (2012). This has been followed by an even finer (10 cm) aerial high spatial resolution night lights orthophoto acquired over Birmingham, United Kingdom (Hale et al. 2013).

Linking the gap between ground based and spaceborne measurements of night lights, drones are starting to being used for calibration and validation, using vertical images (Li et al. 2020; Massetti et al. 2022) as well as panoramic images providing side views at various angles (Kong et al. 2019; Karpińska and Kunz 2023).

Additional research using higher spatial and temporal resolution night lights imagery will undoubtedly expand and "demonstrate the real potential and utility of satellite remote sensing for regional science in urban environments" (Patino and Duque 2013).

The full list of available nighttime light sensors is summarized in Table 13.1.

13.1.3 THE DMSP/OLS DATA ARCHIVE

Given the long historical archive of DMSP/OLS NTL data and the increasing interest of using them, we give a short description of the process to generate the version 4 annual composites, based on the DMSP/OLS user guide (https://eogdata.mines.edu/products/dmsp/#docs). OLS has a wide view over the Earth surface—scanning as wide as 3000 km land surface below across the sub-satellite track (orbital swath) in every single pass. The nighttime light data collected by the DMSP/OLS measures light on Earth's surface such as those generated by human settlements, gas flares, fires, and illuminated marine vessels. The current image processing practiced by NOAA is to generate annual global composites through extracting best quality observations only. There are a number of sources that can contaminate NTL observations, including clouds, solar illumination, lunar illumination, aurora in the northern hemisphere, and lightning, etc. Clouds are first excluded with the help of the OLS thermal band and NCEP (National Centers for Environmental Prediction, NOAA) surface temperature grids. Sunlit and moonlit data are excluded based on solar elevation angle and moon phase. Lighting from aurora is screened and excluded in the northern hemisphere on an orbit-by-orbit manner using visual inspection. Finally, only data from the center half of each 3000 km wide OLS swath goes into the final composites to ensure high-quality observations in terms of geolocation, pixel size and the consistency of radiometry. The version 4 annual composites contain three products in a nominal 30 arc second resolution (about 1 km):

1. Total number of cloud-free observations. Data quality in areas with low number of cloud-free observations is often reduced. In some years there are areas with zero cloud-free observations.
2. Average visible band value with no further filtering. This product is generated by averaging the visible band digital number of all cloud-free observations and valid data range from 0–63. Areas with zero cloud-free observations are flagged by the value 255.

TABLE 13.1
Comparison of Available Spaceborne Sensors for Night Lights Mapping, Sorted by Spatial Resolution

Sensor	Spatial Resolution (m)	Operational Years	Temporal Resolution	Products	Radiometric Range	Spectral Bands	Main References
DMSP/OLS	3000	Declassified in 1972, digital archive available from 1992 onwards	Global coverage can be obtained every 24 hours	Stable lights, Radiance calibrated, Average DN	10^{-6} to 10^{-9} watts/cm^2/sr/μm 6 bit	Panchromatic 500–900 nm	(Doll 2008; Elvidge et al. 1997b)
Fengyun-3E MEdium Resolution Spectral Imager–low light (MERSI-LL)	1000 (at nadir)	Launched in 2021		Annual and monthly low lights (at a spatial resolution of 0.02°)	$3 \times 10{-5}$ to 90 W/m2/sr 12 bit	Panchromatic 500–900 nm	(Yu et al. 2023; Zhang et al., in press)
NPP-VIIRS	740	Launched in Oct 2011	Nightly images can be downloaded conflict	Near real time from http://ngdc.noaa.gov/eog/viirs/download_dbs.html In development, including fire and power outage detection. Cloud-free composites available in radiance units of nano-Watts/(cm2*sr), at annual, monthly, and nightly temporal resolutions	14 bit	Panchromatic 505–890 nm	(Miller et al. 2012)
SAC-C HSTC	300	Launched in Nov 2000	Sporadic	N\A	8 bit	Panchromatic 450–850 nm	(Colomb et al. 2003)
SAC-D HSC	200–300	Launched in June 2011	Sporadic	N\A	10 bit	Panchromatic 450–900 nm	(Sen et al. 2006)
Astronauts' photographs on board the International Space Station (ISS)	10–200	From 2003 onwards (since mission ISS006)	Photos taken irregularly	Photos can be searched and downloaded from: http://eol.jsc.nasa.gov/	8 bit	RGB	(Doll 2008; Levin and Duke 2012b)

(*Continued*)

TABLE 13.1 (Continued)
Comparison of Available Spaceborne Sensors for Night Lights Mapping, Sorted by Spatial Resolution

Sensor	Spatial Resolution (m)	Operational Years	Temporal Resolution	Products	Radiometric Range	Spectral Bands	Main References
CUMULOS	150	Experimental Cubesat, 2018	Sporadic	N/A	N/A	Panchromatic	Pack et al. (2018), 2019
LuoJia1–01	130	Launched June 2018	15 day revisit time	Freely available	DN values with lab calibration	Panchromatic, 460–980 nm	(Li et al. 2018)
SDGSAT-1	40 multispectral, 10 panchromatic	Launched 2021	Sporadic	Freely available for registered users via https://data.sdgsat.ac.cn	DN values (12 bit), can be calibrated to radiance using provided coefficients	RGB Panchromatic	(Guo et al. 2023)
QMX-1	10–40	Launched 2022	Sporadic	N/A	DN values with lab calibration	RGB Panchromatic	(Zhong et al. 2023)
Jilin-1 (JL1–3B)	0.9	Launched January 2017;	Commercial satellite, acquires images on demand	N/A	8 bit	RGB	(Zheng et al. 2018)
EROS-B	0.7	Night lights images offered since mid-2013	Commercial satellite, acquires images on demand	N/A	16 bit	Panchromatic	(Levin 2014)

3. Stable average visible band value after cleaning. Ephemeral events, such as wildfires, have been screened and discarded. The final product contains lights from cities, towns, and other sites with persistent lighting year-round. This is the commonly used DMSP/OLS NTL product.

The similar procedure is also carried out to generate a 2012 annual mosaic from VIIRS DNB imagery (Baugh et al. 2013) and is also available for download from the website of NOAA NGDC. The generation of DMSP products ended in 2013, given the superior radiometric, temporal, and spatial quality of the VIIRS/DNB products that have since replaced those of the DMSP/OLS (Elvidge et al. 2013).

13.1.4 THE VIIRS/DNB DATA ARCHIVE

The Visible Infrared Imaging Radiometer Suite (VIIRS) sensor, situated on Suomi National Polar-orbiting Partnership (NPP) satellite was launched in October 2011. VIIRS introduced a specialized panchromatic sensor known as the Day and Night Band (DNB), specifically designed to measure nighttime lights, a task previously undertaken by DMSP/OLS sensor. The VIIRS/DNB marked a significant advancement over its predecessor, the DMSP/OLS sensor, in several key aspects (Elvidge et al. 2013):

1. VIIRS/DNB data became available on a daily basis, providing a higher frequency of images to researchers and the public at no cost.
2. The spatial resolution improved substantially, reaching 740 m, a significant enhancement compared to the approximately 3 km resolution offered by DMSP/OLS.
3. VIIRS/DNB provided radiometrically calibrated data that was sensitive to lower light levels, preventing saturation even in densely urbanized areas.
4. VIIRS/DNB exhibited reduced overglow, a phenomenon where excessive light spills over into neighboring areas, further enhancing the accuracy of nighttime light data.

This improvement led to a transition in global nighttime lights product generation from DMSP to VIIRS data in 2012. VIIRS/DNB data paved the way for the creation of global monthly and nightly composites of night lights, available from April 2012 onward. These datasets have significantly advanced our understanding of various phenomena, such as seasonal variations in nighttime brightness (Levin 2017) and the detection of adverse effects resulting from military conflicts (Li et al. 2017). Unlike its predecessor DMSP/OLS, raw VIIRS/DNB imagery is freely accessible for nightly download, enabling researchers to gain insights into the anisotropic characteristics of artificial lights (Li et al. 2019).

Recently, NASA introduced the Black Marble nighttime lights product suite (VNP46 series) at a remarkable spatial resolution of 500 meters. These products (VNP46A1 and VNP46A2), as detailed by Román et al. (2018), offer cloud-free, atmospheric-, terrain-, vegetation-, snow-, lunar-, and stray light-corrected radiances. These corrections enhance the accuracy of estimating daily nighttime lights (NTL), allowing for precise tracking of populations affected by conflicts, assessing damages to the electricity grid following disasters, and identifying specific events and locations where people congregate (Román and Stokes 2015). This suite of products represents a significant leap in our ability to monitor and understand nighttime illumination patterns on a global scale.

13.2 MAJOR APPLICATIONS OF NIGHTTIME LIGHTS

Since Croft (1978) first identified the potential of DMSP/OLS nighttime light imagery as an indicator of human activities, numerous applications have been reported in the literature to use them. A thorough review of all of them is not practical here due to limit of space (audiences are advised to the Center for International Earth Science Information Network [CIESIN] thematic guide to nighttime light remote sensing [Doll 2008] for a detailed review of major applications before 2008) and to the more recent review by Levin et al. [2020]). We place our current effort on new trends of applications emerging in recent years. Here we summarize them into several categories and focus on introducing the most representative cases within each category. The majority of nighttime light remote sensing applications in the past have used annual DMSP/OLS composites (Bennett and Smith 2017) and in the past decade have opted to the better-quality, original and modified VIIRS/NDB composites at annual, monthly or nightly temporal resolutions. They can be further grouped into urban related applications, such as urban extent, urban population, urban energy use, urban carbon emissions, urban population, urban GDP, conflict, and in-use stocks, and environmental-related applications, such as fisheries, gas flare, wildfires, and light pollution impacts on fauna, flora, and human health.

13.2.1 Urban Extent and Socioeconomic Variables

As one looks at the nighttime light images (Figure 13.1), the first impression one immediately gets is that nighttime light intensity varies with urban built-up intensity. It is often true that cities are well lit while rural areas and remote areas are not, which indicates that it is possible to find a threshold to delineate urban extents from DMSP/OLS NTL. Numerous efforts have been tried to find that optimal threshold (Imhoff et al. 1997; Small et al. 2005a). However, the conclusion is that there is no single optimal threshold that can accurately delineate both large cities and small cities simultaneously. A larger threshold might be good for delineating large cities but will underestimate small cities. A smaller threshold can bring back small cities but tends to overestimate the extents of large cities. This dilemma mainly comes from the overglow effect in DMSP/OLS NTL, but may also be related to the type of lighting as well as differing street lighting standards in different countries (Small 2005). Due to the fact that cities are brightly lit during the night, urban areas can be easily identified in nighttime light remote sensing data. Indeed, one of the first uses of NTL data from DMSP/OLS was to delineate urban extents, and DMSP/OLS data is one of the earliest datasets available for mapping our urbanizing planet (Zhu et al. 2019), and have been shown as useful for tracking electrification rates also in rural areas. At the other end of the spectrum, a recent study found that 19% of the world's overall settlement footprint had no detectable artificial radiance (McCallum et al. 2022).

Empirical analyses find strong correlations between NTL and socioeconomic variables, such as population, gross domestic products, carbon emissions, and energy use as mentioned earlier. Aggregated urban extent and later stable NTL intensity are used to build regression models. These models can then be used to disaggregate population (Cox et al. 2022; Zhuo et al. 2009), carbon emission (Liu et al. 2022; Oda and Maksyutov 2011), GDP (Chen and Nordhaus 2010; Farzanegan and Fischer 2021; Man et al. 2021), and in-use stock (Rauch 2009), which are often reported by administrative units, to a higher resolution or at the pixel level. However, it is found that the overglow and saturation problems cause biases in the derived models, and there might be a discrepancy between lighting and living when mapping the distribution of population at the pixel level. For example, industrial zones, airports, and commercial strips can have high NTL intensity but with few residents living there. This leads to efforts trying to figure out the relationship between lighting and land use land cover with finer resolution nighttime light images (Hale et al. 2013).

13.2.2 Nighttime Light Pollution (NLP)

There is a dark side to nighttime light (Hölker et al. 2010a). Light pollution, the disturbance of the natural dark night sky due to night light emissions, has been identified as one factor of environmental pollution. Night Light Pollution described as "one of the most rapidly increasing alterations to the natural environment"; a problem whereby "mankind is proceeding to envelop itself in a luminous fog" (Cinzano et al. 2001). Recently light pollution is considered not only as an alteration of the night sky for an observer but as a real environmental pollution (Ayudyanti and Hidayati 2021; Cinzano and Falchi 2013; Hu et al. 2018).

The rapid growth of NLP—especially over areas of intense human activity—not only affects humans' ability to perceive the universe (Marin 2009), but also has significant impacts on all kinds of environmental, health, and socioeconomic factors related to the disturbance of the nocturnal environment (Navara and Nelson 2007). Astronomers are among the most concerned due to direct impacts on astronomical observations (Falchi and Cinzano 2000; IDSA 1996), but environmentalists and medical professionals are also concerned about direct and indirect impacts on wildlife, as well as reductions in the overall "quality of life" for people exposed to significant nighttime light emissions (Gaston et al. 2013). Over the past decade, the impacts of NLP on nighttime visibility, socioeconomic processes, the environment, and human health have been reported (Navara and Nelson 2007; Figure 13.3). Thus, there are various methods and applications for measuring and mapping nighttime light pollution (Mander et al. 2023).

13.2.2.1 Night Vision

In the document of Proclamation of 2009 as International Year of Astronomy (presented in 2005 at the 33rd Session of the UNESCO General Conference), the sky observation is presented as follows: "Humankind has always observed the sky either to interpret it or to understand the physical laws that govern the universe. This interest in astronomy has had profound implications for science, philosophy, religion, culture and our general conception of the universe." Thus, the sky is a significant component of our common universal heritage. The consequences of artificial night lighting to night vision and sky observation are profound. Many contemporary mega cities produce a glow in the night sky that can be seen from more than 150 km away. At the beginning of the millennium more than two third of the world's population (99% of the US and the European Union population) and almost 20% of world terrain was under light polluted skies (Cinzano et al. 2001). The newer version of the world atlas of artificial night sky brightness provides updated statistics at the country level (Falchi et al. 2016). Furthermore, light pollution is considered an important driver for the loss of aesthetic values such as the visibility of the Milky Way (Smith 2008), which are also of cultural importance for indigenous populations globally (Hamacher et al. 2020).

13.2.2.2 Human Health—Medicine

Recently, many researchers argue that the light at night may lead to circadian desynchrony (among others, Pauley 2004; Salgado-Delgado et al. 2011). Accordingly, this desynchrony leads individuals

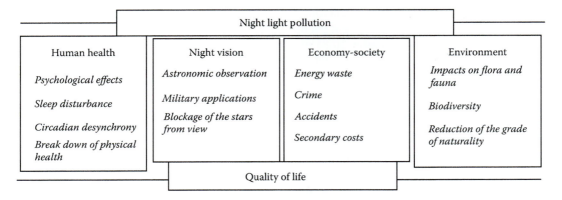

FIGURE 13.3 The main consequences of night light pollution.

to the loss of internal temporal order and to a wide psychological disorder including impulsivity, mania and depression. Moreover, many other health problems such as sleep and metabolic disorders are associated with internal desynchronization and potentially with the NLP. Exposure to NLP suppresses the production of the pineal hormone melatonin, and since melatonin is an anti-carcinogenic agent, lower levels in blood may encourage the tumorigenesis (Bullough et al. 2006; Kloog et al. 2009). Exposure to artificial light is believed to play a role in certain diseases, such as certain types of cancers, obesity, and depression. This influence stems from the disruption of the circadian rhythm and disturbances in sleep patterns. Additionally, artificial light exposure can suppress the production of melatonin, a hormone closely linked to the surrounding environmental light levels. Other research studies have explored ALAN data derived from DSMP in connection with various health conditions, including hormone-dependent cancers (Bauer et al. 2013; Hurley et al. 2014; Rybnikova et al. 2015, 2017; Portnov et al. 2016; James et al. 2017; Kim et al. 2017; Rybnikova et al. 2018), obesity (Rybnikova et al. 2016; Rybnikova and Portnov 2016; Koo et al. 2016), and sleep quality (Koo et al. 2016). Conducted in diverse regions and among different population groups, these studies offer mutually reinforcing evidence of significant correlations between ALAN and a wide range of adverse health conditions.

13.2.2.3 Environment

The ecologic effects of NLP have been well documented. Light pollution has been shown to affect both plants and animals. The duration of darkness controls the metabolism as well as the growth of plants. Thus NLP can disrupt plants by distorting their natural day–night cycle (Longcore and Rich 2004). Research on many wildlife animals shows that light pollution can alter their habitats, orientation, foraging areas, and even their physiology, not only in urban centers but in rural areas as well (Salmon 2003). Gaston et al. (2013) proposed a framework that focuses on the ways in which NLP influences biological systems by making the distinction between light as a resource and light as an information source. In their research they argue that NLP has downstream effects on the structure and function of biological organizations from cell to ecosystem. They also underline that even low levels of nighttime light pollution can have significant ecological impacts. Nowadays, NLP is treated as a biodiversity threat (Hölker et al. 2010b).

Artificial lighting also disturbs the "tranquility" (grade of naturality) of an area. This kind of pollution is directly correlated to the presence of human activities and for this reason is considered of high interest. Tranquility maps are a valuable tool for the classification of parts of the countryside, as well as for the classification of areas that are relatively undisturbed by noise and visual intrusion, areas representative of "unspoilt" countryside. Tranquility can be defined as "the sense of peace, quiet and natural pureness of the countryside." While tranquility disturbance is profound in modern urban areas, suburban and rural areas also face the same problems due to urban growth, intense cultivation activities, transportation network expansion, etc. Thus, research has been addressed in order to assess the potential use of night time remotely sensed images for modeling light pollution, as well as to estimate the grade of light pollution in suburban areas (Chalkias et al. 2006). Due to the coronavirus pandemic, (Ściężor 2021) discovered that upon switching off street lighting, the zenith surface brightness of the clear night sky decreased by 15–39%. This reduction was directly proportional to the population of the city. Although this decrease in radiance was relatively small, it led to a significant 40% reduction in the surface brightness of the night sky, regardless of atmospheric conditions.

13.2.2.4 Economy—Society

The increasing prevalence of exposure to light at night has significant social, behavioral, and economic consequences—among others—that are only recently becoming apparent (Navara and Nelson 2007). At the same time there is an urgent need for light pollution policies given the dramatic increase in artificial light at night. This increase varies (from 0 to 20% per year—mean value: 6%) depending on geographic region (Hölker et al. 2010a). Light pollution generates significant costs related to not only the profound waste of energy but to negative impacts on wildlife, health, and astronomy (Ayudyanti and Hidayati 2021). According to Gallaway et al. (2010) the amount of this loss in the United States is approximately $7 billion annually. Current scientific models of light pollution are purely population based. Fractional logit models in their research showed that both population and GDP are significant explanatory variables of NLP.

Major applications using nighttime light time series are summarized in Table 13.2.

TABLE 13.2
Summary of Major Applications Using Nighttime Lights

Application	Strengths/Limitations	Data Source	References (selected)
Urban extent	Quick/simple estimations at global scale; many differences according to the size of city and type of lighting; overglow effect.	DMSP/OLS; Suomi NPP/VIIRS DNB	(Imhoff et al. 1997; Li and Chen 2018; Shi et al. 2014; Small et al. 2005b)
Population/population density	Efficient representations of the spatial heterogeneity; low accuracy of the final outputs; differences across NLE data from various satellites.	DMSP/OLS; Suomi NPP/VIIRS DNB; ISS	(Cox et al. 2022; Levin and Duke 2012; Levin and Zhang 2017; Zhuo et al. 2009)
GDP	Beneficial in countries with poor statistical systems; limited use to provide accurate indicators of economic activity, or for assessing regional GDP.	DMSP/OLS; Suomi NPP/VIIRS DNB	(Bickenbach et al. 2013; Chen and Nordhaus 2010; Elvidge et al. 2007c; Man et al. 2021; Sutton et al. 2007)
Carbon emissions	Strong correlation only in developed countries; coarse estimations; stable lights are less appropriate than radiance calibrated products due to pixel saturation.	DMSP/OLS; Suomi NPP/VIIRS DNB	(Ghosh et al. 2010; Letu et al. 2014; Liu et al. 2022; Oda and Maksyutov 2011)
Energy consumption	Suitable for global studies; limited availability of energy data time series in order to evaluate statistical analysis; underestimations caused by saturated pixels; advantage of VIIRS over DMSP due to improved spatial resolution and the availability of data in physical units (W cm-2 sr-1).	DMSP/OLS; Suomi NPP/VIIRS DNB	(Amaral et al. 2005; Coscieme et al. 2013)
In-use stock	Useful estimations of the in-use stock deposits dynamics; evaluation of the in-use stock in areas where statistical data are incomplete; not suitable for precise estimations; improvements in methodology expected by using the new VIIRS sensor.	DMSP/OLS	(Hattori et al. 2014; Rauch 2009; Takahashi et al. 2010)
Light pollution	Suitable for modeling light pollution as well as estimating light pollution in suburban areas; whereas DMSP/OLS data is limited by its coarse resolution, fine spatial resolution airborne and spaceborne sensors are now becoming available.	DMSP/OLS; emerging fine spatial resolution sensors (airborne, EROS-B)	(Ayudyanti and Hidayati 2021; Chalkias et al. 2006; Falchi et al. 2011; Hale et al. 2013; Hu et al. 2018; Kuechly et al. 2012; Kyba et al. 2011; Levin et al. 2014; Liu et al. 2013; Miller 2006; Navara and Nelson 2007; Olsen et al. 2013; Ściężor 2021; Simpson and Hanna 2010)

13.3 STUDY OF URBAN DYNAMICS WITH NTL TIME SERIES

13.3.1 Mapping Urban Extent Dynamics

The long historical archive of DMSP/OLS NTL data has the high potential in characterizing urbanization dynamics at regional and global scales. However, such a potential was not well explored until very recently due to differences in satellite orbits (dawn pass versus dusk pass) and sensor degradation, which causes NTL data collected by sensors onboard different DMSP satellites to be significantly different, even when there are no real changes occurred on the ground. The version 4 DMSP/OLS NTL annual composites include data from five satellites: F10, F12, F14, F15, and F16 (Figure 13.3). In theory, a time series of NTL data could capture the dynamics of urbanization despite possible errors from sensor differences. This would require the signal of change to be larger than the error signal and also large enough to render the error signal (noise) unimportant.

We can derive five stylized facts from Figure 13.4:

1. The time series are very noisy. The shift between satellites brings in very large biases.
2. The biases show up in all countries in a similar way, which indicates systematic processes may have been in effect.
3. However, within the span of each satellite (except for F15, which has two distinct segments) there exists a clear urban development pattern in each country.

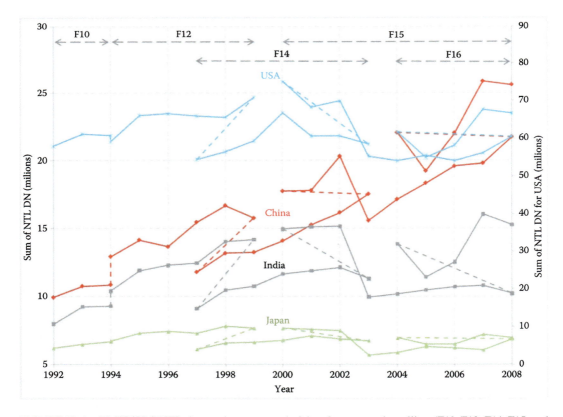

FIGURE 13.4 DMSP/OLS NTL time series composed of data from several satellites (F10, F12, F14, F15, and F16). Dotted line segments in the time series indicate satellite shifting. (Adapted from Zhang and Seto 2011).

4. The overlapping sections of F12 and F14 as well as between F14 and F15 show very similar patterns, although there are differences in magnitude.
5. Each satellite can capture an urban development trend, i.e., the noise is not large enough to overwhelm the signal within the span of each satellite.

These facts strongly suggest that the signal of change captured by the DMSP/OLS sensors is larger than the error signals and can be used as an indicator of urban change. This is further confirmed by validation results using population data at the country level as well as those using land cover data at state and regional levels in the United States (Zhang and Seto 2011). Zhang and Seto (2011) first explored the potential of utilizing the time series NTL to map urbanization dynamics at regional and global scales. Using time series constructed with 17 NTL images from four satellites, F10 (1992–1994), F12 (1995, 1996), F14 (1997–1999), and F15 (2000–2008), they successfully identified five different types of urbanization dynamics in various regions (Figure 13.5 shows results in the Pearl River Delta region) with an iterative unsupervised classification procedure.

A number of applications have emerged to utilize multitemporal DMSP/OLS NTL annual composites for characterizing urbanization dynamics in different regions, such as China (Liu et al. 2012; Ma et al. 2012), northeast China (Yi et al. 2014), South America (Álvarez-Berríos et al. 2013), India (Pandey et al. 2013), the conterminous United States (Zhang et al. 2014), and Hanoi, Vietnam (Castrence et al. 2014). Recently, Bennie et al. (2014a) examined the trends of light pollution across Europe from 1995 to 2010 using time series of DMSP/OLS NTL.

Liu et al. (2012) reported that urban growth patterns from 1992 to 2008 revealed by DMSP/OLS nighttime light time series in three cities of China are very similar to those revealed by Landsat TM/ETM+ time series, with an average overall accuracy of 82.74%. However, mainly due to its relatively coarse spatial resolution and low radiometric resolution, urban expansion detected with DMSP/OLS are much larger than that with Landsat TM/ETM+. During that period, urban areas expanded 107,593 ha in Beijing, 52,838 ha in Chengdu, and 41,004 ha in Zhengzhou, as detected with the NTL time series. However, those numbers are only 49,733 ha in Beijing, 48,583 ha in Chengdu, and 25,992 ha in Zhengzhou, as detected with Landsat TM/ETM+ time series.

In contrast, VIIRS DNB presents significant improvements over DMSP/OLS, including higher resolution, a broader dynamic range, 14-bit quantization, and in-flight calibrations (Elvidge et al. 2013) Consequently, VIIRS DNB data are expected to be extensively employed in scientific applications, particularly in urban studies, replacing DMSP/OLS to overcome the limitations of the previous generation (Cao et al. 2022; Li et al. 2021; Zhang et al. 2017).

13.3.2 Detecting Socioeconomic Changes

Characterizing urbanization dynamics can help understanding the relationship between urban changes and socioeconomic changes. Jiang et al. (2012) modeled urban expansion and cultivated land conversion for hot spot counties in China. Frolking et al. examined macro-scale changes in urban structure in a number of global cities from 1999 to 2009 through the combination of DMSP/OLS NTL and NASA's SeaWinds microwave scatterometer data (Frolking et al. 2013). Research also finds that armed conflict events have significant impacts on nighttime light through an analysis of the relationship between armed conflicts and DMSP/OLS nighttime light variation from 1992 to 2010 in 159 countries (Li et al. 2013a). Shortland et al. (2013) discover that NTL provides striking illustrations of economic decline and recovery and clearly shows the contrast between the stable regions of northern Somalia and the chaos and anarchy of Southern Somalia. Similarly decadal economic decline in Zimbabwe was successfully detected using DMSP/OLS NTL time series (Li et al. 2013c), and the impact of armed conflicts in the Middle East following the Arab Spring were identified using monthly time series of nighttime lights (Levin et al. 2018). In summary, the significance of integrating night light data into economic analyses lies in several aspects: (1) It allows the estimation of GDP at a more detailed spatial resolution than what official statistics provide. (2)

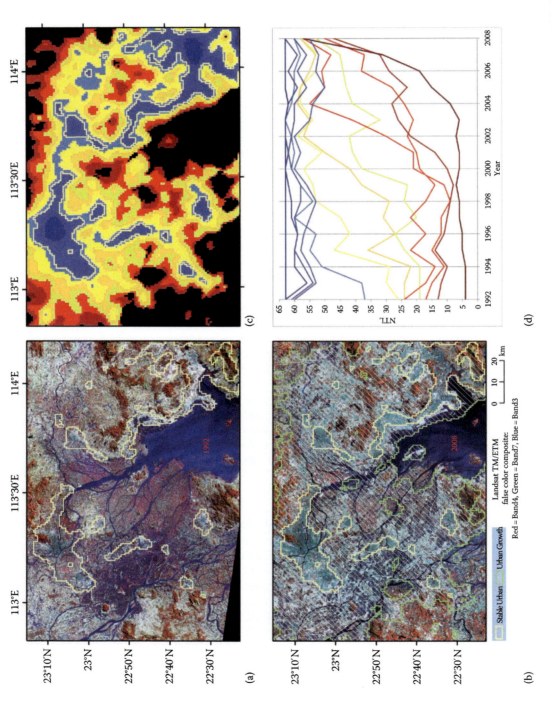

FIGURE 13.5 Urban dynamics in the Pearl River Delta region, China from 1992 to 2008. A) Stable urban areas in 1992; B) Urban growth areas in 2008; C) Multi-temporal urbanization dynamics as detected from NTL; D) Temporal NTL profiles of different types of urbanization dynamics. (Adapted from Zhang and Seto 2011).

It enables the assessment of GDP changes over time, offering a high temporal frequency, as demonstrated in studies such as Bennie et al. (2014a, 2014b). (3) Night lights data facilitates GDP estimation in regions where reporting is inadequate or nonexistent, as highlighted in the research conducted by Henderson et al. in 2012.

13.3.3 Tracking Social Events with High Temporal Frequency NTL Time Series

The majority of change detection applications mentioned earlier depend on time series constructed from the DMSP/OLS NTL annual composites. However, considering that DMSP/OLS can cover the entire globe every day, its temporal resolution is greatly reduced after the construction of annual NTL composites. Bharti et al. (2011) utilized time series constructed from daily DMSP/OLS imagery and found that high temporal frequency DMSP/OLS NTL data offers a very powerful tool to measure seasonal variation in population density for studying measles epidemics. Measles epidemics in West Africa cause a significant proportion of vaccine-preventable childhood mortality. Epidemics are strongly seasonal, but the drivers of these fluctuations are poorly understood, which limits the predictability of outbreaks and the dynamic response to immunization (Pavlačka et al. 2023; Stokes and Roman 2022). Spatiotemporal changes in population density as measured by NTL are useful to explain measles seasonality (Zhao et al. 2018). With dynamic epidemic models, measures of population density are essential for predicting epidemic progression at the city level and for improving intervention strategies. The ability to measure fine-scale changes in population density with high temporal frequency nighttime light data has implications for public health, crisis management, and socioeconomic development.

Such a potential can be further expanded by the VIIRS Day and Night Band (DNB), from which well-calibrated and normalized time series with daily coverage can be relatively easier to obtain. A preliminary analysis of the 2012 Ramadan, the ninth month of the Islamic calendar, in Cairo, Egypt, illustrates the improved potential of VIIRS DNB (Figure 13.6). Muslims worldwide observe Ramadan as a month of fasting. While fasting from dawn until sunset during Ramadan, Muslim people have food and drink after sunset and before sunrise every day, which means increased activities during night compared with non-Ramadan periods. Although still quite noisy, the unfiltered VIIRS DNB time series shows significant increase in nighttime light intensity during Ramadan in Cairo to reflect increased human activities during that special period (Figure 13.5). In recent years, various studies have shown that major religious and cultural events associated with increased night lights can be detected using spaceborne imagery of night lights (Román and Stokes 2015; Liu et al. 2019; Ramirez et al. 2023).

Major applications using nighttime lights time series are summarized in Table 13.3.

13.4 CHALLENGES IN REMOTE SENSING OF NIGHT LIGHTS

The DMSP/OLS instrument was initially designed to observe moonlit clouds (Elvidge et al. 1997c). Due to its low-light imaging capability, the instrument can also detect artificial lighting at night in clear conditions without moonlight. The NTL data collected by the DMSP/OLS uniquely measures light on Earth's surface such as that generated by human settlements, gas flares, fires, and illuminated marine vessels. Although NTL does not directly measure human settlements or urban land cover, the data are strongly correlated with many characteristics of urban settlements, including carbon emissions (Doll et al. 2000) and economic activity (Sutton et al. 2007). However, a number of issues, including sensor degradation, limited radiometric range, and satellite orbit difference (dawn pass vs. dusk pass), limit the utility of NTL data for many applications. Chief among these

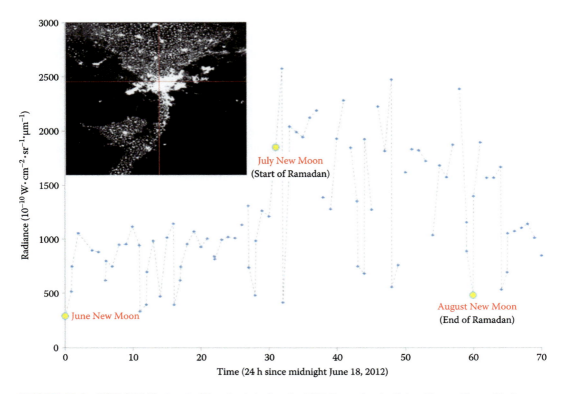

FIGURE 13.6 VIIRS DNB signals (blue dots) during the 2012 Ramadan in Cairo, Egypt. The red hair cross indicates the location where the VIIRS DNB time series was extracted.

is the well-documented saturation of data values in core urban areas (Elvidge et al. 2007a). Data values for the core of urban centers—where there are many lights—are much brighter entities than moonlit clouds. However, due to the limited radiometric range of DMSP/OLS, data values in these regions tend to be truncated. This problem of saturation is a significant challenge in using NTL for characterizations of inter-urban areas.

VIIRS/DNB imagery has become the most commonly used nightlight remote sensing imagery at present. In comparison to DMSP/OLS data, this data does not exhibit the phenomenon of "saturation." However, due to the high sensitivity of this sensor, it often results in a significant amount of "noise" in the data, leading to abnormal pixel values for fishing boat lights and highly reflective ice and snow. Additionally, the VIIRS/DNB sensor has improved the image resolution, broader dynamic range, finer quantization, and in-orbit radiometric calibration, significantly enhancing its capability to detect artificial nighttime illumination from cities, ships, and natural gas flares in the absence of moonlight. Nevertheless, there are also certain limitations in the band configuration of VIIRS/DNB, particularly in its ability to identify new forms of lighting dominated by blue light, such as Light Emitting Diodes (LEDs).

In recent years, there have been some new sensors introduced into the field as well. These new sensors, more or less, also face certain issues. The Luojia-1 nighttime remote sensing sensor has a relatively short lifespan, and the data it captures are primarily limited to the Chinese region. Currently, its data updates have been suspended. Astronauts using digital cameras on board the International Space Station (ISS) capture nighttime Earth images, but these lack radiometric calibration, have no fixed revisit cycle, and exhibit low coverage in areas with weak light sources. The Jilin-1 nighttime remote sensing data have a relatively small coverage area, and the sensor has certain defects in acquiring continuous data over a short period of time.

TABLE 13.3

Summary of Major Applications Using Nighttime Lights Time Series

Application	Strengths/Limitations	Data Source	References (selected)
Urban extent dynamics	Simple and fast for regional and global urban growth patterns detection; accuracy is limited due to its coarse spatial resolution and low radiometric resolution.	DMSP/OLS VIIRS/DNB SDGSAT, Jilin 1, ISS	(Bennie et al. 2014a; Li and Chen 2018; Liu et al. 2012; Ma et al. 2012; Ściężor 2021; Small and Elvidge 2013; Zhang et al. 2014; Zhang and Seto 2011; Zhang et al. 2017)
Armed conflicts	Night lights variation provides a good indicator for armed conflicts; can be confused with natural disasters.	DMSP/OLS VIIRS/DNB SDGSAT, Jilin 1, ISS	(Li et al. 2013a; Li and Li 2014; Li et al. 2017; Levin et al. 2018)
Natural disasters	Night lights can assist in identifying areas impacted by natural disasters and the rates of recovery.	DMSP/OLS VIIRS/DNB SDGSAT, Jilin 1, ISS	Román et al. 2019; Levin and Phinn 2022; Levin 2023a
Economic decline and recovery	Valuable in regions with poor economic data; reliability is limited by its coarse resolution and inter-satellite variation.	DMSP/OLS VIIRS/DNB SDGSAT, Jilin 1, ISS	(Li et al. 2013c; Shortland et al. 2013)
Seasonal population density variation	Useful for understanding the relationship between population density and disease transmission; high temporal resolution data is not readily available.	DMSP/OLS VIIRS/DNB SDGSAT, Jilin 1, ISS	(Bharti et al. 2011; Zhao et al. 2018)
Cultural celebrations	Simple and fast for tracking social events.	DMSP/OLS VIIRS/DNB SDGSAT, Jilin 1, ISS	Román and Stokes 2015; Liu et al. 2019;

In order to better address future geographical challenges, we advocate the development of more advanced nighttime remote sensing sensors.

13.4.1 Correcting Saturation in DMSP/OLS NTL Imagery

Various methods have been proposed to correct or reduce NTL saturation in DMSP/OLS imagery. The correction algorithms vary in the spatial scale at which they are implemented, their complexity, and their success. In general, the correction methods fall into two categories: those that utilize only NTL data and those that use other satellite data to correct the NTL data. Of those that only use NTL data for the correction, Ziskin et al. (2010) generate non-saturated NTL data by adding additional data taken at a "low gain" setting in dense urban areas to the operational NTL data that are taken at a "high gain" setting. It is the operational "high gain" setting of DMSP/OLS that causes bright urban centers to be saturated whereas the "low gain" setting allows bright urban centers to be captured with finer details. In terms of accuracy and data quality, Ziskin's method, utilizing the dynamic gain settings, is ideal. However, this method as currently applied with DMSP/OLS is very labor and cost intensive, and therefore the corrected NTL data are only available for a very limited number of years (Elvidge et al. 1999; Ziskin et al. 2010), and the method is unlikely to be used to correct the entire historical NTL archive. Letu et al. (2010) applied cubic regression models to correct saturation within an administrative unit. However, this method cannot be applied at the pixel scale. Letu et al. (2012) later proposed a method to correct NTL saturation at the pixel scale based on 1999 non-saturated NTL data. A major assumption of this approach is that the nighttime light intensity in saturated areas did not change from 1996 to 1999. Based on this assumption, they built a linear regression model based on data from the non-saturated (in the 1996–1997 stable light image) regions and applied the derived model to correct saturation in NTL for the 1996–1997 image.

The major drawback of this method comes from the assumption on which it is based. While that assumption may hold in countries or regions with high urbanization levels (e.g., Japan, USA), it is not valid in many developing countries such as India and China where rapid urban expansion is the norm rather than the exception. Thus, their method has limited applicability for correcting NTL saturation globally.

It has been shown that vegetation abundance is closely and inversely correlated with impervious surfaces, a characteristic of many urban areas (Bauer et al. 2004; Pozzi and Small 2005; Small 2001; Weng et al. 2006; Weng et al. 2004). Several studies have attempted to combine information from NDVI and NTL to greatly enhance urban features (Cao et al. 2009; Lu et al. 2008), to examine the impact of urbanization on net primary productivity (Milesi et al. 2003), and to develop more accurate maps of urban areas (Lu et al. 2008; Schneider et al. 2003). This second category of NTL correction methods is based on the rationale that key urban features are inversely correlated with vegetation health and abundance. The Human Settlement Index (HSI) normalizes the NTL with MODIS NDVI, measures of vegetation health and quantity (Rouse Jr et al. 1974; Tucker 1979), to mitigate the saturation effects (Lu et al. 2008).

Zhang et al. (2013) proposed a simple method, Vegetation Adjusted Normalized Urban Index (VANUI), to correct NTL saturation with MODIS NDVI:

$$VANUI = (1 - NDVI) * NTL \qquad (13.1)$$

where NDVI is NDVI derived from MODIS. As negative MODIS NDVI values are usually associated with water and glaciers, NDVI values here are constrained to the range of non-negative values between 0 and 1. The parenthetical expression 1 − NDVI inverses the shape of NDVI transects. This simple calculation assigns larger non-vegetative weights (as evidenced by 1 − NDVI) to core urban areas than to peri-urban areas, which results in an increased variability of data values within urban cores. Therefore, we expect core urban areas will have positive VANUI values close to 1, while non-urban, non-illuminated areas—usually those with an abundance of vegetation—will have low VANUI values close to 0. We also expect peri-urban areas and regions affected by overglow to exhibit lower VANUI values than the city core. Figure 13.7 shows VANUI for select urban regions around the world, chosen to represent a wide variation in city size, geography, and economic base. For all urban areas examined, the normalized NTL values were saturated in the urban cores and within urban NTL variations were not detectable. In Contrast, VANUI captures the fine spatial details in and around urban areas. Following the development of the VANUI, follow-up studies have suggested the use other spectral indices such as Enhanced Vegetation Index or the Normalized Difference Built-up Index to reduce the saturation problem of DMSP/OLS data (Zhuo et al. 2015; Zhang et al. 2022).

13.4.2 Intercalibrating the DMSP/OLS NTL Time Series

The DMSP/OLS sensors on the different DMSP satellites were not calibrated, generating inconsistent values when comparing the data between different DMSP satellites and between years for the same satellite. The challenge to achieve successful radiometric calibration of remote sensing imagery obtained at different times is to find invariant ground targets that can be used as references for reliable comparison over time. Elvidge et al. (2009) chose Sicily, Italy, as the reference site and derive second-order regression models between the reference image F121999 and other images individually with all data points in Sicily. These models were then applied to calibrate the entire time series from 1992 to 2008. This method successfully reduced differences caused by satellite shift to some degree. However, it is questionable that models derived in Sicily can be generalized

Nighttime Light Remote Sensing

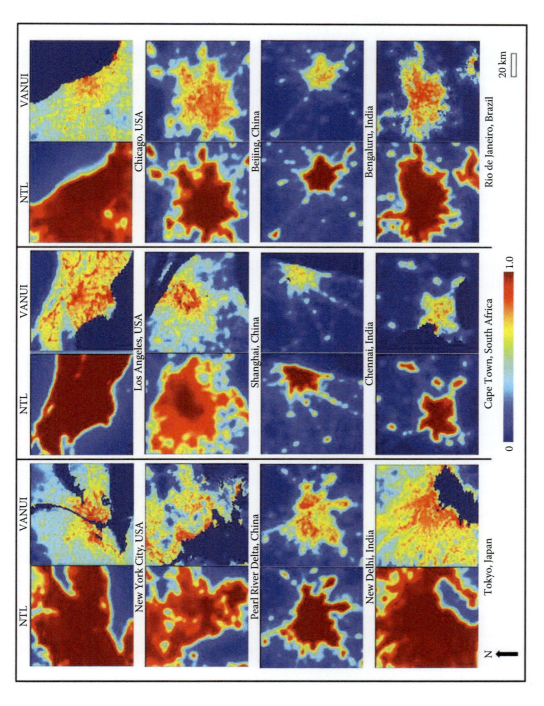

FIGURE 13.7 2006 normalized NTL with saturation and VANUI for select cities and urban regions around the world. Note: NTL and VANUI are normalized and unitless and are both scaled to 0–1.0. (Adapted from Zhang et al. 2013).

to cover the entire globe, since noises introduced by various sources might not be geographically consistent. For this reason, regional urbanization dynamic researchers derived their models that can suit closely to their specific regions by choosing local reference sites (Liu et al. 2011; Liu et al. 2012; Nagendra et al. 2013; Pandey et al. 2013). In an attempt to produce more generalized models for the entire globe, Wu et al. (2013) extended the Elvidge et al. (2009) method by selecting more reference sites, including Mauritius, Puerto Rico, and Okinawa, Japan. Despite that the method developed by Wu et al. (2013) achieves improvement, the way they identified invariant regions was not essentially different than that applied by Elvidge et al. (2009) and also suffered from the limitation of subjectivity. Li et al. (2013b) designed an automatic method to find invariant pixels in Beijing, China. They assumed that there exist stable pixels in the region and run a regression with all pixels without any screening. Based on the resulting regression model, they look for outliers, which are considered pixels that experienced change, and discard them iteratively. A final model can be built with only stable pixels identified. This automatic method can minimize bias introduced by subjective selection of invariant regions and has the potential to be extended to the entire globe. However, since the region of Beijing experienced dramatic change in the past decades, this method might lead to overcorrection to the NTL time series. Furthermore, the iterative procedure to identify stable pixels is very computation intensive thus cannot be directly implemented at the global scale, considering the gigantic number of pixels.

13.4.3 Intercalibrating DMSP/OLS and VIIRS/DNB for Generating Continuous Time Series

Since the launch of VIIRS/DNB, it has replaced DMSP/OLS as the satellite providing global nighttime lights images at higher spatial resolution, sensitive to lower nighttime lights and not suffering from saturation (Elvidge et al. 2013, 2017). However, to examine changes in nighttime lights since the early 1990s to the present, there is a need to intercalibrate those two sensors, bridging their differences in spatial resolution and radiometric sensitivity. Many papers have been published on this cross-sensor calibration in recent years for various applications, with one of the first papers aiming to evaluate the impact of the war in Syria on the country's major cities (Li et al. 2017), using a power function model to simulate DMSP/OLS data from VIIRS/DNB. Various methods have been proposed to achieve this cross-sensor calibration, such as the removal of seasonality from VIIRS/DNB and the application of a geographically weighted regression model (Zheng et al. 2019), the application of a Biphasic Dose Response model (Ma et al. 2020), or by using stable pixels and removing outlier pixels (Dong et al. 2021).

13.4.4 Quantifying the Impacts of the Transitions to LED Lighting on Light Pollution

The world is in the midst of a "lighting revolution" due to the development of Light Emitting Diode (LED) technology (Pust et al. 2015). LED lighting is more efficient, offering energy savings (Charles 2009; Pagden et al. 2020), greater flexibility in controlling the timing, color and directionality of the lighting (Cole and Driscoll 2013). However, LED lighting can also increase ecological light pollution (Pawson and Bader 2014), given that LED lights typically have peak emissions also in the blue light and that their lower costs might enhance a greater use of light at night (Tsao et al. 2010). The global spectral shift due to adoption of bright rich white light (BRWL) LEDs is a major challenge for astronomers, both because the blue component of white light produces more skyglow, and because many current ground and space-based sensors are not sensitive to blue light (Levin et al. 2020), and therefore we currently do not monitor this transition. Sánchez de Miguel (2015) using images from the International Space Station (ISS), and Zheng et al. (2018) using images from the new Jilin-1 satellite, were able to distinguish high-pressure sodium (HPS) lamps from white LEDs. However, there are fundamental limitations for multispectral sensors to distinguish between similar color light sources, like fluorescents/compact fluorescents and LEDs of same color temperature or HPS lamps and PC-Amber LEDs (Sánchez de Miguel et al. 2019). Ground-based measurements can provide more frequent measurements than currently available space based sensors and can fill in the blind spot of the lack of sensitivity to blue light from LEDs (Kyba et al. 2017). Several studies have used calibrated RGB cameras to track lighting remodeling from vapor lamps to LEDs (Kolláth et al. 2016; Barentine et al. 2018). Modeling work done by Bará et al. (2019b) indicates that for certain transition scenarios (from High Pressure Sodium, HPS, to LED), the VIIRS sensor may detect reduction in artificial zenithal sky brightness, even if sky brightness in reality increases, due to the loss of the HPS line in the near-infrared, and the inability of the VIIRS to detect blue light. Stokes and Seto (2019) detected a decrease in nighttime lights as measured by VIIRS/DNB in city centers in the United States and India, and attributed this to the transition to LED lighting, which emits light in the blue wavelength, which is not measured by VIIRS/DNB. Sánchez de Miguel et al. (2021a) provided the first estimate of the potential increase in radiance due to the transition to LED lighting, suggesting that it might be even greater by 270–400% in some areas. Using composites of calibrated multispectral ISS images for Europe, Sánchez de Miguel et al. (2022) identified a spectral shift towards whiter lighting in Europe following the transition to LED lights, indicating an increase in the risk to ecosystems from light pollution. Further work in this direction, combining multispectral ground photometers (e.g., LANcube) and spaceborne sensors (such as Jilin-1 and SDGSAT-1) is needed to better evaluate the impact of the transition to LED lighting on light pollution (Levin 2023b).

13.5 OUTLOOK—FINE RESOLUTION NIGHTTIME LIGHT REMOTE SENSING

The choice of the appropriate sensor to use is a function of the spatial extent of the area that needs to be covered, the temporal dynamics of the studied phenomena, its spatial heterogeneity and the purpose for which this mapping will be used. The same general principles apply for mapping night lights. While DMSP/OLS and VIIRS are the sensors of choice for mapping global patterns of night lights and trends over time, medium spatial resolution sensors are required to examine urban patterns. Using ISS imagery, the development of new built-up areas and of lighting infrastructure can be mapped at spatial resolutions equivalent to those available from Landsat-type satellites. One of the most rapid developing regions in the world is that of the United Arab Emirates, where oil and gas revenues fuel the economy, and where the emirates try to position their countries on the global stage (Bagaeen 2007; Figure 13.8—urban changes in Abu-Dhabi between 2003 and 2013). When finer details are of interest, e.g., to examine lighting at the street level or even for mapping individual streetlights, fine spatial resolution images (covering small areas) are required. While in the past such fine spatial resolution was only available from dedicated aerial campaigns, since mid-2013 ImageSat is offering high spatial resolution (< 1 m) panchromatic night lights images from its EROS-B satellite (Levin et al. 2014), and in 2017 the Chinese Chang-Guang Satellite Technology launched their JL1–3B satellite offering multi-spectral night lights images at less than 1 m resolution (Zheng et al. 2018). Medium and fine spatial resolution nighttime images can also assist to calibrate and validate models that aim at predicting light pollution. Such models use GIS layers of streetlights (location and type of streetlights), the topographic layout of terrain, buildings heights (and vegetation if available) and shading algorithms so that we can better understand the extent of artificial light pollution (Bennie et al. 2014b; Chalkias et al. 2006; Gaston et al. 2012; Teikari 2007).

13.5.1 Applications of Fine Spatial Resolution Night Lights

One of the basic needs for moderate and fine spatial resolution night light images is to form a better understanding of the spatial distribution of point sources emitting night lights and of the land use associated with light pollution. The contribution of street lights to directed light pollution has been estimated using nighttime aerial photos in Berlin by Kuechly et al. (2012) to be 31.6% of in Berlin, whereas Hale et al. (2013) estimated using aerial photos for Birmingham (UK) that street lights constituted about 38% of lit area. Based on these two studies, land use types that contribute most of the emitted night lights are manufacturing, commercial, and housing. However, the contribution of streetlights to light pollution can vary widely between and within cities; recent studies have used dimming experiments to try and assess the actual contribution of streetlights (Barentine et al. 2020; Kyba et al. 2021). Capturing nighttime aerial photos of the University of Arizona campus at night, Kim and Hong (2013) suggested that high-reflective materials used in the built environment increase light pollution, even when full cutoff lights are used. However, additional studies from different cities around the world are needed to generalize such conclusions.

Once the spatial distribution of night lights is mapped, it can contribute to different aspects of urban planning. Lit areas were found to be safer for traffic (Jackett and Frith 2013), and are often perceived as safer areas from crime (Loewen et al. 1993). Thus, by mapping and planning streetlights, planners can have an impact on the spatial distribution of crime and of accidents, which is known not to be randomly distributed in the city (Weisburd and Amram 2014). Another important application using night lights mapping for urban planning, is related to the monitoring of energy consumption in the city, so that a city's carbon footprint can be reduced. Quantifying urban carbon footprint is not an easy task (Sovacool and Brown 2010), and night lights mapping can contribute to this end. Urban planning can also use night lights mapping to examine issues of spatial equality, examining the types and amount of street lights in different neighborhoods, so as to determine whether all citizens are given similar standards of streetlights (Coulter 1983).

Last but not least, light pollution has been found to have negative impacts not only for astronomers that need dark sky, but also on humans, animals, and plant species, due to the alteration of natural light regimes affecting our circadian clock (Chepesiuk 2009; Longcore and Rich 2004; Pauley 2004). In order to reduce our exposure to light pollution, mapping it is a basic requirement.

13.5.2 CALIBRATING WITH GROUND MEASUREMENTS

While fine spatial resolution mapping of night lights is expected to benefit various applications, there are certain research gaps that need to be overcome in order to transform this data to be more quantitative. While in traditional optical remote sensing satellite images are atmospherically corrected to derive their reflectance values, it is not so clear which units should be used in night light imagery. The DMSP/OLS imagery products are distributed as stable lights or average lights x percent (DN values between 0 and 63). Often these products are used to calculate the total lit area or the total lights; however, these data are not in luminance units. Photometry is the measurement of the intensity of electromagnetic radiation in photometric units, like lumen, lux, etc., or magnitudes. Radiometry is the measurement of optical radiation, with some of the many typical units encountered are watts/m^2 and photons/sr. The main difference between photometry and radiometry is that photometry is limited to the visible spectra as defined by the response of the human eye (Teikari 2007).

In recent years there have been some attempts to calibrate fine spatial resolution images to photometric units. Hale et al. (2013) used ground measurements of incident lux along linear transects to calibrate their aerial night light images into illuminance units. Another approach for field mapping of night lights that can be used for calibrating aerial or spaceborne night light imagery is using ground networks of instruments such as the Sky Quality Meter (SQM, manufactured by Unihedron, measuring the brightness of the night sky in magnitudes per square arc second; http://www.unihedron.com/projects/darksky/); however, ground networks aimed at monitoring light pollution are fairly recent (den Outer et al. 2011; Pun and So 2012; Pun et al. 2014). An additional network of sky brightness photometers which has been developed in recent years is that of the panchromatic Telescope Encoder and Sky Sensor (TESS-W) developed by the European Union's Stars4All project, produced by (Bará et al. 2019). In an interesting study using Extech EasyView 30 light meters to

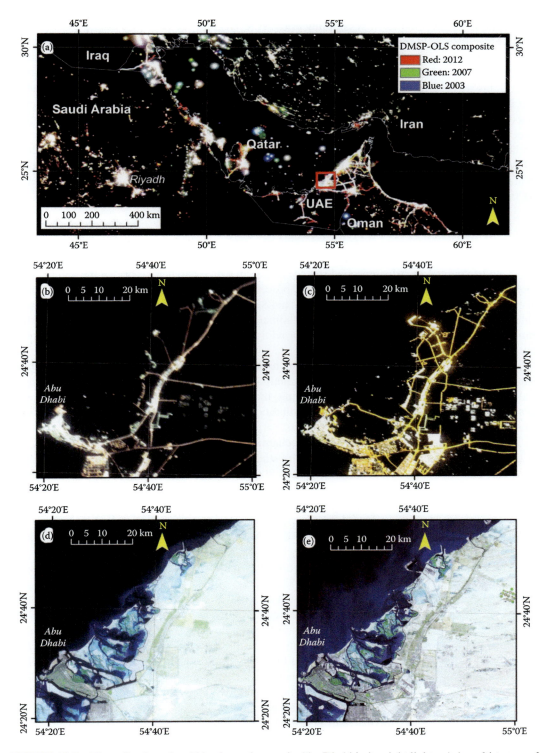

FIGURE 13.8 The reflection of rapid land use changes in Abu Dhabi in its night lights: a) A multi-temporal composite of DMSP/OLS stable lights images around the Persian Gulf (R = 2012, G = 2007, B = 2003); b) ISS photo of Abu-Dhabi, 4/2/2003, spatial resolution of 90 m; c) ISS photo of Abu-Dhabi, 11/12/2013, spatial resolution of 120 m; d) Landsat 7 false color composite of Abu-Dhabi, 28/5/2003; e) Landsat 8 false color composite of Abu-Dhabi, 9/12/2013.

map night brightness along a 10 m sampling grid on the Virginia Tech campus, brightness was measured twice: First with the light meter pointing upward to catch direct light from the light fixtures at 30 cm from the ground, then with the light meter pointing down to measure reflected light (Kim 2012). Thus, in addition to the inconsistency in the photometric units used for calibrating aerial night lights images, there is a gap with regards to how one should measure light on the ground so that it best corresponds with what an airborne or a spaceborne captures. Recent studies have shown that emissions of artificial night lights are anisotropic (i.e., they vary in their horizontal and vertical directions) (Li et al. 2019, 2022; Wang et al. 2021). Therefore, to better understand the sources of artificial lights as measured from space, ground based measurements should also be conducted in different directions, as has been done either with SQM photometers pointed in different directions (Katz and Levin 2016), or with the multidirectional photometer known as LANcube (Aubé et al. 2020; Levin 2023b).

13.6 SUMMARY AND FUTURE DIRECTIONS

Nighttime light remote sensing beginning with DMSP/OLS ushered in a range of important applications and has proved its value in social science research, which can help to bridge the gap between social sciences (interested in explaining social processes) and remote sensing (offering the means to map spatial patterns and processes) (Rindfuss and Stern 1998). Nighttime light remote sensing provides a powerful tool to social scientists for estimating population size, economic activity, carbon emissions, and additional human activities, providing a bridge between physical scientists, social scientists, and the remote sensing scientists.

Although almost as old as daytime optical remote sensing, such as Landsat, nighttime light remote sensing is still in its primary—mainly qualitative—stage. Many applications are still empirical. To advance nighttime light remote sensing, there still exist various challenges, with regards to sensor availability, imagery acquisition, calibration between sensors and with ground measurements, the development of operational products.

There is a lack of understanding of the mechanism behind nighttime light remote sensing, due to the lack of studies at the ground level. The studies by researchers in the light pollution field (Chalkias et al. 2006; Gaston et al. 2012; Teikari 2007) aim to understand light emitted from lighting sources and how they are transferred and then make predictions of light pollution through modeling. This kind of work is highly required to better interpret nighttime light remote sensing images, just as radiative transfer models to daytime optical remote sensing. Understanding nighttime light transfer from lighting sources through the air to the sensor is critical for designing future nighttime light remote sensing sensors.

Another important issue is to generate high-quality data from existing nighttime light sensors, such as DMSP/OLS and Suomi NPP/VIIRS. The release of the digital archive of DMSP/OLS and the annual composites have made remote sensing data easy to use among social scientists that often might have little training in remote sensing image processing and may feel uncomfortable with intense image processing. Numerous studies have benefited from this great remote sensing resource. However, during the compositing process, a lot of valuable information is discarded. The selection of high-quality observations is subjective. The high temporal frequency of DMSP/OLS and VIIRS DNB are valuable for intra-annual change analysis. But daily images of scientific quality are not readily available. New algorithms to generate better quality images from the long historical DMSP/OLS archives and the relatively new VIIRS data are highly desired.

Another point to make is that both DMSP/OLS and Suomi NPP/VIIRS are not dedicated to observing city lights at night, thus bringing numerous obstacles to nighttime light remote sensing. Sensors that are designed with a specific purpose of lighting sources on the ground are highly required in the future.

Currently, both DMSP/OLS and VIIRS DNB are single-band sensors (the thermos infrared band is only for cloud screening and provides no information about nighttime lights). Multi-spectral band

Nighttime Light Remote Sensing

sensors are needed in order to discriminate different types of lighting techniques, which can convey very useful information.

Furthermore, both DMSP/OLS and Suomi NPP/VIIRS DNB are wide-view sensors, with swath widths greater than 3000 km, which means they can accumulate angular observations varying in a large range. Angular observations sometimes are not preferred, because they often cause variation across geography, thus making mosaicking a big challenge. However, angular information is proved carrying valuable structural information and ironically is critical to normalize observations to the standard viewing-illuminating geometry, as seen in MODIS (Schaaf et al. 2002) and MISR (Multi-angle Imaging Spectroradiometer) (Diner et al. 1998). Due to the variation in street layout and building height, nighttime light is also expected to vary accordingly (Kyba et al. 2013). Angular observations from both DMSP/OLS and Suomi NPP/VIIRS DNB thus will release structural information about urban areas. Such information still remains unmined up to date. Future research is highly required to extract this invaluable information from both DMSP/OLS and Suomi NPP/VIIRS DNB.

In addition, one recent trend in optical remote sensing is to increase observation frequencies to meet the urgent need for effective monitoring of ephemeral events or phenomena on Earth from space. Increasing the number of satellites can surely help to increase the temporal resolution of remote sensing observations, but there are still challenges at the night side, when there is no sunlight available to illuminate the Earth's surface. Such a situation can be even worse in the polar regions, where sunlight is not available for almost half the year (Liu et al. 2023). With the advent of low light–detecting technologies, nightlight remote sensing makes it possible to detect artificial lights from space, forming a convenient and powerful tool to characterize and understand human beings' altered diurnality. Although moonlight has been an important factor that affects many nocturnal animals and plants (Liu et al. 2021), mainstream nightlight remote sensing image processes either try their best to totally avoid moonlight, or to remove the moonlight component from observations with tuned algorithms.

Combining nighttime light remote sensing with other remote sensing modalities, such as hyperspectral or radar data, can provide a more comprehensive view of the Earth's surface at night. Integrating multiple data sources can improve the detection and characterization of various phenomena, including urban heat islands, vegetation condition, and water bodies. This multimodal approach can provide a more holistic understanding of nocturnal processes and improve the accuracy of remote sensing applications.

Finally, we may suggest several promising research avenues in the field of fine (< 10 m) and medium (10–100 m) spatial resolution night light remote sensing:

1. Calibration—protocols and experience need to be gained about how to perform ground measurement of downward emitted light and upward reflected light that can be used to calibrate aerial and spaceborne images of night lights, and which units are best to use.
2. Mapping—the next step after the 2D mapping of nightscapes, can be their mapping in three dimensions (e.g., also light escaping from windows), similar to 3D thermal studies, e.g., using video cameras (as in Chudnovsky et al. 2004), or using drones and hemispheric photography to quantify vertical variation in light pollution (e.g., Degen et al. 2022; Karpińska and Kunz 2022).
3. So far almost all remote sensing studies using time series of nighttime lights relied on annual, monthly, or nightly images. It is time to explore the temporal dynamics at higher resolutions, e.g., exploring the hourly changes in night lights through the night (see example of Dobler et al. 2021).
4. Once the technical and methodological aspects of fine and medium spatial resolution imagery of night lights are solved, thematic issues can be explored. Two examples include the correspondence between lit areas and hot areas (measured using a thermal sensor), and the correspondence between night lights and human activity patterns (e.g., using traffic counts).

5. With the recent development of low light–detection technology, the night light remote sensing community started to realize that moonlight can be a very useful illumination source for detecting weather and climate parameters at night, instead of being treated as a noise source for city light detection. Miller et al. (2013) made a detailed insight into many potential applications for nocturnal low-light visible satellite observations and presented a long list of key variables that could be obtained under moonlight using VIIRS/DNB from space. They found that sometimes nighttime moonlight remote sensing even showed advantages over daytime sunlight remote sensing. These include the detection of snow cover (Levin et al. 2017; Huang et al. 2022; Chen et al. 2023; Liu et al. 2023), rainfall distributions across arid/semi-arid surfaces, the ability to peer through optically thin clouds to reveal sea ice, and the detection of oceanic currents, etc. Although these studies demonstrated a comprehensive potential for nighttime low-light measurements, quantitative assessment is still needed.

REFERENCES

Abramson, D.M., & Redlener, I. (2012). Hurricane Sandy: Lessons learned, again. *Disaster Medicine and Public Health Preparedness*, 6, 328–329.

Akbari, H., Menon, S., & Rosenfeld, A. (2009). Global cooling: Increasing world-wide urban albedos to offset CO2. *Climatic Change*, 94, 275–286.

Álvarez-Berríos, N.L., Parés-Ramos, I.K., & Aide, T.M. (2013). Contrasting patterns of urban expansion in Colombia, Ecuador, Peru, and Bolivia between 1992 and 2009. *Ambio*, 42, 29–40.

Amaral, S., Câmara, G., Monteiro, A.M.V., Quintanilha, J.A., & Elvidge, C.D. (2005). Estimating population and energy consumption in Brazilian Amazonia using DMSP night-time satellite data. *Computers, Environment and Urban Systems*, 29, 179–195.

Amaral, S., Monteiro, A., Camara, G., & Quintanilha, J. (2006). DMSP/OLS night-time light imagery for urban population estimates in the Brazilian Amazon. *International Journal of Remote Sensing*, 27, 855–870.

Anderson, S.J., Tuttle, B.T., Powell, R.L., & Sutton, P.C. (2010). Characterizing relationships between population density and nighttime imagery for Denver, Colorado: Issues of scale and representation. *International Journal of Remote Sensing*, 31, 5733–5746.

Arnfield, A.J. (2003). Two decades of urban climate research: A review of turbulence, exchanges of energy and water, and the urban heat island. *International Journal of Climatology*, 23, 1–26.

Aubé, M., Marseille, C., Farkouh, A., Dufour, A., Simoneau, A., Zamorano, J., . . . & Tapia, C. (2020). Mapping the melatonin suppression, star light and induced photosynthesis indices with the LANcube. *Remote Sensing*, 12(23), 3954.

Ayudyanti, A.G., & Hidayati, I.N. (2021). Impact of Optical Aerosol Depth (AOD) on light pollution level: A spatio-temporal analysis. In *IOP conference series: Earth and environmental science* (p. 012037). IOP Publishing.

Bachmann, M., & Storch, T. (2023). First nighttime light spectra by satellite—By EnMAP. *Remote Sensing*, 15(16), 4025.

Bagaeen, S. (2007). Brand Dubai: The instant city; or the instantly recognizable city. *International Planning Studies*, 12, 173–197.

Balk, D., Deichmann, U., Yetman, G., Pozzi, F., Hay, S., & Nelson, A. (2006). Determining global population distribution: Methods, applications and data. *Advances in Parasitology*, 62, 119–156.

Bará, S., Rigueiro, I., & Lima, R.C. (2019b). Monitoring transition: Expected night sky brightness trends in different photometric bands. *Journal of Quantitative Spectroscopy and Radiative Transfer*, 239, 106644.

Bará, S., Tapia, C.E., & Zamorano, J. (2019). Absolute radiometric calibration of TESS-W and SQM night sky brightness sensors. *Sensors*, 19(6), 1336.

Barentine, J.C., Kundracik, F., Kocifaj, M., Sanders, J.C., Esquerdo, G.A., Dalton, A.M., . . . & Kyba, C.C. (2020). Recovering the city street lighting fraction from skyglow measurements in a large-scale municipal dimming experiment. *Journal of Quantitative Spectroscopy and Radiative Transfer*, 253, 107120.

Barentine, J.C., Walker, C.E., Kocifaj, M., Kundracik, F., Juan, A., Kanemoto, J., & Monrad, C.K. (2018). Skyglow changes over Tucson, Arizona, resulting from a municipal LED street lighting conversion. *Journal of Quantitative Spectroscopy and Radiative Transfer*, 212, 10–23.

Barnosky, A.D., Hadly, E.A., Bascompte, J., Berlow, E.L., Brown, J.H., Fortelius, M., Getz, W.M., Harte, J., Hastings, A., & Marquet, P.A. (2012). Approaching a state shift in Earth/'s biosphere. *Nature*, *486*, 52–58.

Benjamin, W., & Baudelaire, C. (1973). *A Lyric Poet in the Era of High Capitalism*. NLB.

Bauer, M.E., Heinert, N.J., Doyle, J.K., & Yuan, F. (2004). Impervious surface mapping and change monitoring using Landsat remote sensing. In *ASPRS Annual Conference Proceedings, May 23–28,2004*. Denver, CO.

Bauer, S.E., Wagner, S.E., Burch, J., Bayakly, R., & Vena, J.E. (2013). A case-referent study: light at night and breast cancer risk in Georgia. *International Journal of Health Geographics*, *12*, 1–10.

Baugh, K., Hsu, F.-C., Elvidge, C.D., & Zhizhin, M. (2013). Nighttime lights compositing using the VIIRS day-night band: Preliminary results. *Proceedings of the Asia-Pacific Advanced Network*, *35*, 70–86.

Bennett, M.M., & Smith, L.C. (2017). Advances in using multitemporal night-time lights satellite imagery to detect, estimate, and monitor socioeconomic dynamics. *Remote Sensing of Environment*, *192*, 176–197.

Bennie, J., Davies, T.W., Duffy, J.P., Inger, R., & Gaston, K.J. (2014a). Contrasting trends in light pollution across Europe based on satellite observed night time lights. *Scientific Reports*, *4*.

Bennie, J., Davies, T.W., Inger, R., & Gaston, K.J. (2014b). Mapping artificial lightscapes for ecological studies. *Methods in Ecology and Evolution*, *5*(6), 534–540.

Bharti, N., Tatem, A.J., Ferrari, M.J., Grais, R.F., Djibo, A., & Grenfell, B.T. (2011). Explaining seasonal fluctuations of measles in Niger using nighttime lights imagery. *Science*, *334*, 1424–1427.

Bickenbach, F., Bode, E., Lange, M., & Nunnenkamp, P. (2013). Night lights and regional GDP (No. 1888). In *Kiel Working Paper*. chrome-extension://efaidnbmnnnibpcajpcglclefindmkaj/https://www.econstor.eu/bitstream/10419/88764/1/775763411.pdf

Brinkley, D. (2006). *The Great Deluge: Hurricane Katrina, New Orleans, and the Mississippi Gulf Coast*. William Morrow.

Bullough, J.D., Rea, M.S., & Figueiro, M.G. (2006). Of mice and women: Light as a circadian stimulus in breast cancer research. *Cancer Causes & Control*, *17*, 375–383.

Cao, C., Zhang, B., Xia, F., & Bai, Y. (2022). Exploring VIIRS night light long-term time series with CNN/SI for urban change detection and aerosol monitoring. *Remote Sensing*, *14*.

Cao, X., Chen, J., Imura, H., & Higashi, O. (2009). A SVM-based method to extract urban areas from DMSP-OLS and SPOT VGT data. *Remote Sensing of Environment*, *113*, 2205–2209.

Castrence, M., Nong, D.H., Tran, C.C., Young, L., & Fox, J. (2014). Mapping urban transitions using multitemporal Landsat and DMSP-OLS night-time lights imagery of the red river delta in Vietnam. *Land*, *3*, 148–166.

Chalkias, C., Petrakis, M., Psiloglou, B., & Lianou, M. (2006). Modelling of light pollution in suburban areas using remotely sensed imagery and GIS. *Journal of Environmental Management*, *79*, 57–63.

Charles, D. (2009). Leaping the efficiency gap. *Science*, *325*, 804–811.

Chen, B., Zhang, X., Ren, M., Chen, X., & Cheng, J. (2023). Snow cover mapping based on SNPP-VIIRS day/night band: A case study in Xinjiang, China. *Remote Sensing*, *15*(12), 3004.

Chen, X., & Nordhaus, W.D. (2010). The value of luminosity data as a proxy for economic statistics. *NBER Working Paper*, *16317*.

Chepesiuk, R. (2009). Missing the dark: Health effects of light pollution. *Environmental Health Perspectives*, *117*, A20.

Chudnovsky, A., Ben-Dor, E., & Saaroni, H. (2004). Diurnal thermal behavior of selected urban objects using remote sensing measurements. *Energy and Buildings*, *36*, 1063–1074.

Cinzano, P., & Falchi, F. (2013). Quantifying light pollution. *Journal of Quantitative Spectroscopy and Radiative Transfer*, *139*, 13–20.

Cinzano, P., Falchi, F., & Elvidge, C.D. (2001). Naked-eye star visibility and limiting magnitude mapped from DMSP-OLS satellite data. *Monthly Notices of the Royal Astronomical Society*, *323*, 34–46.

Cole, M., & Driscoll, T. (2013). The lighting revolution: If we were experts before, we're novices now. *IEEE Transactions on Industry Applications*, *50*(2), 1509–1520.

Colomb, R., Alonso, C., & Nollmann, I. (2003). SAC-C mission and the international AM constellation for Earth observation. *Acta Astronautica*, *52*, 995–1005.

Coscieme, L., Pulselli, F.M., Bastianoni, S., Elvidge, C.D., Anderson, S., & Sutton, P.C. (2013). A thermodynamic geography: Night-time satellite imagery as a proxy measure of energy. *Ambio*, 1–11.

Coulter, P.B. (1983). Inferring the distributional effects of bureaucratic decision rules. *Policy Studies Journal*, *12*, 347–355.

Cox, D., de Miguel, A.S., Bennie, J., Dzurjak, S., & Gaston, K. (2022). Majority of artificially lit Earth surface associated with the non-urban population. *Science of the Total Environment, 841*, 156782.

Croft, T. (1978). Nighttime images of the earth from space. *Scientific American, 239*, 86–98.

Dawson, M., Evans, C., Stefanov, W., Wilkinson, M.J., Willis, K., & Runco, S. (2012). Human settlements in the South-Central US, viewed at night from the International Space Station. In *Geological Society of America-South Central Meeting* (No. JSC-CN-25552).

Degen, T., Kolláth, Z., & Degen, J. (2022). X, Y, and Z: A bird's eye view on light pollution. *Ecology and Evolution*, 12(12), e9608.

del Mar López, T., Aide, T.M., & Thomlinson, J.R. (2001). Urban expansion and the loss of prime agricultural lands in Puerto Rico. *Ambio: A Journal of the Human environment, 30*, 49–54.

den Outer, P., Lolkema, D., Haaima, M., Hoff, R.V.D., Spoelstra, H., & Schmidt, W. (2011). Intercomparisons of nine sky brightness detectors. *Sensors, 11*, 9603–9612.

Diner, D.J., Beckert, J.C., Reilly, T.H., Bruegge, C.J., Conel, J.E., Kahn, R.A., Martonchik, J.V., Ackerman, T.P., Davies, R., & Gerstl, S.A. (1998). Multi-angle Imaging SpectroRadiometer (MISR) instrument description and experiment overview. *Geoscience and Remote Sensing, IEEE Transactions on, 36*, 1072–1087.

Dobler, G., Bianco, F.B., Sharma, M.S., Karpf, A., Baur, J., Ghandehari, M., ... & Koonin, S.E. (2021). The urban observatory: A multi-modal imaging platform for the study of dynamics in complex urban systems. *Remote Sensing, 13*(8), 1426.

Dodman, D. (2009). Blaming cities for climate change? An analysis of urban greenhouse gas emissions inventories. *Environment and Urbanization, 21*, 185–201.

Doll, C.N. (2008). CIESIN thematic guide to night-time light remote sensing and its applications. *Center for International Earth Science Information Network of Columbia University*. Palisades, NY.

Doll, C.N., Muller, J., & Elvidge, C. (2000). Night-time imagery as a tool for global mapping of socioeconomic parameters and greenhouse gas emissions. *Ambio, 29*, 157–162.

Doll, C.N., Muller, J., & Morley, J. (2006). Mapping regional economic activity from night-time light satellite imagery. *Ecological Economics, 57*, 75–92.

Dong, K., Li, X., Cao, H., & Tong, Z. (2021). Intercalibration between night-time DMSP/OLS radiance calibrated images and NPP/VIIRS images using stable pixels. *IEEE Journal of Selected Topics in Applied Earth Observations and Remote Sensing, 14*, 8838–8848.

Elvidge, C.D., Baugh, K., Dietz, J., Bland, T., Sutton, P., & Kroehl, H. (1999). Radiance calibration of DMSP-OLS low-light imaging data of human settlements. *Remote Sensing of Environment, 68*, 77–88.

Elvidge, C.D., Baugh, K.E., Kihn, E.A., Kroehl, H.W., & Davis, E.R. (1997b). Mapping city lights with nighttime data from the DMSP operational linescan system. *Photogrammetric Engineering and Remote Sensing, 63*, 727–734.

Elvidge, C.D., Baugh, K., Kihn, E., Kroehl, H., Davis, E., & Davis, C. (1997a). Relation between satellite observed visible-near infrared emissions, population, economic activity and electric power consumption. *International Journal of Remote Sensing, 18*, 1373–1379.

Elvidge, C.D., Baugh, K.E., Kihn, E.A., Kroehl, H.W., Davis, E.R., & Davis, C.W. (1997c). Relation between satellite observed visible-near infrared emissions, population, economic activity and electric power consumption. *International Journal of Remote Sensing, 18*, 1373–1379.

Elvidge, C.D., Baugh, K.E., Zhizhin, M., & Hsu, F.C. (2013). Why VIIRS data are superior to DMSP for mapping nighttime lights. *Proceedings of the Asia-Pacific Advanced Network, 35*, 62.

Elvidge, C.D., Baugh, K., Zhizhin, M., Hsu, F.C., & Ghosh, T. (2017). VIIRS night-time lights. *International Journal of Remote Sensing, 38*(21), 5860–5879.

Elvidge, C.D., Cinzano, P., Pettit, D., Arvesen, J., Sutton, P., Small, C., Nemani, R., Longcore, T., Rich, C., & Safran, J. (2007c). The Nightsat mission concept. *International Journal of Remote Sensing, 28*, 2645–2670.

Elvidge, C.D., & Green, R.O. (2005). High-and low-altitude AVIRIS observations of nocturnal lighting. In *13th JPL Airborne Earth Science Workshop, Pasadena, California, May 24–27, 2005*. Pasadena, CA: Jet Propulsion Laboratory, National Aeronautics and Space Administration.

Elvidge, C.D., Keith, D.M., Tuttle, B.T., & Baugh, K.E. (2010). Spectral identification of lighting type and character. *Sensors, 10*, 3961–3988.

Elvidge, C.D., Safran, J., Tuttle, B., Sutton, P., Cinzano, P., Pettit, D., Arvesen, J., & Small, C. (2007a). Potential for global mapping of development via a nightsat mission. *GeoJournal, 69*, 45–53.

Elvidge, C.D., Tuttle, B., Sutton, P., Baugh, K., Howard, A., Milesi, C., Bhaduri, B., & Nemani, R. (2007b). Global distribution and density of constructed impervious surfaces. *Sensors*, *7*, 1962–1979.

Elvidge, C.D., Ziskin, D., Baugh, K.E., Tuttle, B.T., Ghosh, T., Pack, D.W., Erwin, E.H., & Zhizhin, M. (2009). A fifteen year record of global natural gas flaring derived from satellite data. *Energies*, *2*, 595–622.

Falchi, F., & Cinzano, P. (2000). Measuring and modeling light pollution. Cinzano P. ed., *Memorie della Societa Astronomica Italiana*, *71*, 139.

Falchi, F., Cinzano, P., Duriscoe, D., Kyba, C.C., Elvidge, C.D., Baugh, K., . . . & Furgoni, R. (2016). The new world atlas of artificial night sky brightness. *Science Advances*, *2*(6), e1600377.

Falchi, F., Cinzano, P., Elvidge, C.D., Keith, D.M., & Haim, A. (2011). Limiting the impact of light pollution on human health, environment and stellar visibility. *Journal of Environmental Management*, *92*, 2714–2722.

Farzanegan, M.R., & Fischer, S. (2021). Lifting of international sanctions and the shadow economy in Iran—a view from outer space. *Remote Sensing*, *13*, 4620.

Foley, J.A., DeFries, R., Asner, G.P., Barford, C., Bonan, G., Carpenter, S.R., Chapin, F.S., Coe, M.T., Daily, G.C., & Gibbs, H.K. (2005). Global consequences of land use. *Science*, *309*, 570–574.

Fouquet, R., & Pearson, P.J. (2006). Seven centuries of energy services: The price and use of light in the United Kingdom (1300–2000). *The Energy Journal*, *27*(1).

Frolking, S., Milliman, T., Seto, K.C., & Friedl, M.A. (2013). A global fingerprint of macro-scale changes in urban structure from 1999 to 2009. *Environmental Research Letters*, *8*, 024004.

Gallaway, T., Olsen, R.N., & Mitchell, D.M. (2010). The economics of global light pollution. *Ecological Economics*, *69*, 658–665.

Galli, M., & Guzzetti, F. (2007). Landslide vulnerability criteria: A case study from Umbria, Central Italy. *Environmental Management*, *40*, 649–665.

Gaston, K.J., Bennie, J., Davies, T.W., & Hopkins, J. (2013). The ecological impacts of nighttime light pollution: A mechanistic appraisal. *Biological Reviews*, *88*, 912–927.

Gaston, K.J., Davies, T.W., Bennie, J., & Hopkins, J. (2012). REVIEW: Reducing the ecological consequences of night-time light pollution: Options and developments. *Journal of Applied Ecology*, *49*, 1256–1266.

Geist, J.F. (1983). *Arcades: The History of a Building Type*. MIT Press.

Ghosh, T., Elvidge, C.D., Sutton, P.C., Baugh, K.E., Ziskin, D., & Tuttle, B.T. (2010). Creating a global grid of distributed fossil fuel CO2 emissions from nighttime satellite imagery. *Energies*, *3*, 1895–1913.

Gomez-Ibanez, D.J., Boarnet, M.G., Brake, D.R., Cervero, R.B., Cotugno, A., Downs, A., Hanson, S., Kockelman, K.M., Mokhtarian, P.L., & Pendall, R.J. (2009). Driving and the built environment: The effects of compact development on motorized travel, energy use, and CO2 emissions. In *Oak Ridge National Laboratory (ORNL)*. https://www.researchgate.net/publication/236448413_Driving_and_the_Built_Environment_The_Effects_of_Compact_Development_on_Motorized_Travel_Energy_Use_and_CO2_Emissions

Grimm, N.B., Faeth, S.H., Golubiewski, N.E., Redman, C.L., Wu, J., Bai, X., & Briggs, J.M. (2008). Global change and the ecology of cities. *Science*, *319*, 756–760.

Guk, E., & Levin, N. (2020). Analyzing spatial variability in night-time lights using a high spatial resolution color Jilin-1 image–Jerusalem as a case study. *ISPRS Journal of Photogrammetry and Remote Sensing*, *163*, 121–136.

Guo, B., Hu, D., & Zheng, Q. (2023). Potentiality of SDGSAT-1 Glimmer imagery to investigate the spatial variability in nighttime lights. *International Journal of Applied Earth Observation and Geoinformation*, *119*, 103313.

Hahs, A.K., McDonnell, M.J., McCarthy, M.A., Vesk, P.A., Corlett, R.T., Norton, B.A., Clemants, S.E., Duncan, R.P., Thompson, K., & Schwartz, M.W. (2009). A global synthesis of plant extinction rates in urban areas. *Ecology Letters*, *12*, 1165–1173.

Hale, J.D., Davies, G., Fairbrass, A.J., Matthews, T.J., Rogers, C.D., & Sadler, J.P. (2013). Mapping lightscapes: Spatial patterning of artificial lighting in an urban landscape. *PLoS ONE*, *8*, e61460.

Hamacher, D.W., De Napoli, K., & Mott, B. (2020). Whitening the Sky: Light pollution as a form of cultural genocide. *arXiv preprint arXiv:2001.11527*.

Hattori, R., Horie, S., Hsu, F.-C., Elvidge, C.D., & Matsuno, Y. (2014). Estimation of in-use steel stock for civil engineering and building using nighttime light images. *Resources, Conservation and Recycling*, *83*, 1–5.

Henderson, J.V., Storeygard, A., & Weil, D.N. (2012). Measuring economic growth from outer space. *American Economic Review*, *102*(2), 994–1028.

Hipskinda, S., Elvidgeb, C., Gurneyc, K., Imhoffd, M., Bounouad, L., Sheffnera, E., Nemania, R., Pettite, D., & Fischerf, M. (2011). Global night-time lights for observing human activity. In *Proceeding of the 34th international symposium on remote sensing of environment* (Vol. 373, pp. 1–5).

Hölker, F., Moss, T., Griefahn, B., Kloas, W., Voigt, C.C., Henckel, D., Hänel, A., Kappeler, P.M., Völker, S., & Schwope, A. (2010a). The dark side of light: A transdisciplinary research agenda for light pollution policy. *Ecology & Society*, 15.

Hölker, F., Wolter, C., Perkin, E.K., & Tockner, K. (2010b). Light pollution as a biodiversity threat. *Trends in Ecology & Evolution*, 25, 681–682.

Hu, Z., Hu, H., & Huang, Y. (2018). Association between nighttime artificial light pollution and sea turtle nest density along Florida coast: A geospatial study using VIIRS remote sensing data. *Environmental Pollution*, 239, 30–42.

Huang, Y., Song, Z., Yang, H., Yu, B., Liu, H., Che, T., . . . & Xu, J. (2022). Snow cover detection in mid-latitude mountainous and polar regions using nighttime light data. *Remote Sensing of Environment*, 268, 112766.

Hurley, S., Goldberg, D., Nelson, D., Hertz, A., Horn-Ross, P.L., Bernstein, L., & Reynolds, P. (2014). Light at night and breast cancer risk among California teachers. *Epidemiology* (Cambridge, Mass.), 25(5), 697.

IDSA. (1996). Astronomy's problem with light pollution. *ISDA Information Sheet, No 1, May 1996*. Tuscon: International Dark-Sky Association.

Imhoff, M.L., Lawrence, W.T., Stutzer, D.C., & Elvidge, C.D. (1997). A technique for using composite DMSP/OLS "city lights" satellite data to map urban area. *Remote Sensing of Environment*, 61, 361–370.

Jackett, M., & Frith, W. (2013). Quantifying the impact of road lighting on road safety—A New Zealand Study. *IATSS Research*, 36, 139–145.

James, P., Bertrand, K.A., Hart, J.E., Schernhammer, E.S., Tamimi, R.M., & Laden, F. (2017). Outdoor light at night and breast cancer incidence in the Nurses' Health Study II. *Environmental Health Perspectives*, 125(8), 087010.

Jiang, L., Deng, X., & Seto, K.C. (2012). Multi-level modeling of urban expansion and cultivated land conversion for urban hotspot counties in China. *Landscape and Urban Planning*, 108, 131–139.

Jiang, W., He, G., Long, T., Guo, H., Yin, R., Leng, W., Liu, H., & Wang, G. (2018). Potentiality of using Luojia 1–01 nighttime light imagery to investigate artificial light pollution. *Sensors*, 18(9), 2900.

Johansson, T.B., Patwardhan, A.P., Nakićenović, N., & Gomez-Echeverri, L. (Eds.). (2012). *Global Energy Assessment: Toward a Sustainable Future*. Cambridge: Cambridge University Press.

Kareiva, P., Watts, S., McDonald, R., & Boucher, T. (2007). Domesticated nature: Shaping landscapes and ecosystems for human welfare. *Science*, 316, 1866–1869.

Karpińska, D., & Kunz, M. (2022). Vertical variability of night sky brightness in urbanised areas. *Quaestiones Geographicae*, 42(1), 5–14.

Karpińska, D., & Kunz, M. (2023). Relationship between the surface brightness of the night sky and meteorological conditions. *Journal of Quantitative Spectroscopy and Radiative Transfer*, 306, 108621.

Katz, A.H. (1948). Aerial photographic equipment and applications to reconnaissance. *JOSA*, 38, 604–605.

Katz, Y., & Levin, N. (2016). Quantifying urban light pollution—A comparison between field measurements and EROS-B imagery. *Remote Sensing of Environment*, 177, 65–77.

Kim, K.Y., Lee, E., Kim, Y.J., & Kim, J. (2017). The association between artificial light at night and prostate cancer in Gwangju City and South Jeolla Province of South Korea. *Chronobiology International*, 34(2), 203–211.

Kim, M. (2012). Modeling nightscapes of designed spaces–case studies of the University of Arizona and Virginia Tech Campuses. *13th International Conference on Information Technology in Landscape Architecture Proceedings* (pp. 455–463). chrome-extension://efaidnbmnnnibpcajpcglclefindmkaj/https://www.researchgate.net/profile/Mintai-Kim/publication/265879890_Modeling_Nightscapes_of_Designed_Spaces_-_Case_Studies_of_the_University_of_Arizona_and_Virginia_Tech_Campuses/links/558d34d408ae591c19da55ae/Modeling-Nightscapes-of-Designed-Spaces-Case-Studies-of-the-University-of-Arizona-and-Virginia-Tech-Campuses.pdf

Kim, M., & Hong, S.-H. (2013). Relationship between the reflected brightness of artificial lighting and land-use types: A case study of the University of Arizona campus. *Landscape and Ecological Engineering*, 1–7.

Kingslake, R. (1942). Lenses for aerial photography. *Journal of the Optical Society of America A*, 32, 129–134.

Kloog, I., Haim, A., Stevens, R.G., & Portnov, B.A. (2009). Global co-distribution of light at night (LAN) and cancers of prostate, colon, and lung in men. *Chronobiology International*, 26, 108–125.

Kohrs, R.A., Lazzara, M.A., Robaidek, J.O., Santek, D.A., & Knuth, S.L. (2014). Global satellite composites—20 years of evolution. *Atmospheric Research*, *135*, 8–34.

Kolláth, Z., Dömény, A., Kolláth, K., & Nagy, B. (2016). Qualifying lighting remodelling in a Hungarian city based on light pollution effects. *Journal of Quantitative Spectroscopy and Radiative Transfer*, *181*, 46–51.

Kong, W., Cheng, J., Liu, X., Zhang, F., & Fei, T. (2019). Incorporating nocturnal UAV side-view images with VIIRS data for accurate population estimation: A test at the urban administrative district scale. *International Journal of Remote Sensing*, *40*(22), 8528–8546.

Koo, Y.S., Song, J.Y., Joo, E.Y., Lee, H.J., Lee, E., Lee, S.K., & Jung, K.Y. (2016). Outdoor artificial light at night, obesity, and sleep health: Cross-sectional analysis in the KoGES study. *Chronobiology International*, *33*(3), 301–314.

Kruse, F.A., & Elvidge, C.D. (2011). Characterizing urban light sources using imaging spectrometry. In *Urban Remote Sensing Event (JURSE), 2011 Joint* (pp. 149–152). IEEE.

Kuechly, H.U., Kyba, C., Ruhtz, T., Lindemann, C., Wolter, C., Fischer, J., & Hölker, F. (2012). Aerial survey and spatial analysis of sources of light pollution in Berlin, Germany. *Remote Sensing of Environment*, *126*, 39–50.

Kyba, C.C., Kuester, T., De Miguel, A.S., Baugh, K., Jechow, A., Hölker, F., ... & Guanter, L. (2017). Artificially lit surface of Earth at night increasing in radiance and extent. *Science Advances*, *3*(11), e1701528.

Kyba, C.C., Ruby, A., Kuechly, H.U., Kinzey, B., Miller, N., Sanders, J., ... & Espey, B. (2021). Direct measurement of the contribution of street lighting to satellite observations of nighttime light emissions from urban areas. *Lighting Research & Technology*, *53*(3), 189–211.

Kyba, C.C., Ruhtz, T., Fischer, J., & Hölker, F. (2011). Cloud coverage acts as an amplifier for ecological light pollution in urban ecosystems. *PLoS ONE*, *6*, e17307.

Kyba, C.C., Ruhtza, T., Lindemanna, C., Fischera, J., & Hölkerb, F. (2013). Two camera system for measurement of urban uplight angular distribution. In *Radiation Processes in the Atmosphere and Ocean (IRS2012): Proceedings of the International Radiation Symposium (IRC/IAMAS)* (pp. 568–571). AIP Publishing.

Lee, T.E., Miller, S.D., Turk, F.J., Schueler, C., Julian, R., Deyo, S., Dills, P., & Wang, S. (2006). The NPOESS VIIRS day/night visible sensor. *Bulletin of the American Meteorological Society*, *87*.

Letu, H., Hara, M., Tana, G., & Nishio, F. (2012). A saturated light correction method for DMSP/OLS nighttime satellite imagery. *Geoscience and Remote Sensing, IEEE Transactions on*, *50*, 389–396.

Letu, H., Hara, M., Yagi, H., Naoki, K., Tana, G., Nishio, F., & Shuhei, O. (2010). Estimating energy consumption from night-time DMPS/OLS imagery after correcting for saturation effects. *International Journal of Remote Sensing*, *31*, 4443–4458.

Letu, H., Nakajima, T.Y., & Nishio, F. (2014). Regional-scale estimation of electric power and power plant CO2 emissions using DMSP/OLS nighttime satellite data. *Environmental Science & Technology Letters*, *1*(5), 259–265.

Levin, N. (2017). The impact of seasonal changes on observed nighttime brightness from 2014 to 2015 monthly VIIRS DNB composites. *Remote Sensing of Environment*, *193*, 150–164.

Levin, N. (2023a). Using night lights from space to assess areas impacted by the 2023 Turkey Earthquake. *Remote Sensing*, *15*(8), 2120.

Levin, N. (2023b). Quantifying the variability of ground light sources and their relationships with spaceborne observations of night lights using multidirectional and multispectral measurements. *Sensors*, *23*(19), 8237.

Levin, N., Ali, S., & Crandall, D. (2018). Utilizing remote sensing and big data to quantify conflict intensity: The Arab Spring as a case study. *Applied Geography*, *94*, 1–17.

Levin, N., & Duke, Y. (2012). High spatial resolution night-time light images for demographic and socio-economic studies. *Remote Sensing of Environment*, *119*, 1–10.

Levin, N., Johansen, K., Hacker, J.M., & Phinn, S. (2014). A new source for high spatial resolution night time images—The EROS-B commercial satellite. *Remote Sensing of Environment*, *149*, 1–12.

Levin, N., Kyba, C.C., Zhang, Q., de Miguel, A.S., Román, M.O., Li, X., ... & Elvidge, C.D. (2020). Remote sensing of night lights: A review and an outlook for the future. *Remote Sensing of Environment*, *237*, 111443.

Levin, N., & Phinn, S. (2022). Assessing the 2022 flood impacts in Queensland combining daytime and nighttime optical and imaging radar data. *Remote Sensing*, *14*(19), 5009.

Levin, N., & Zhang, Q. (2017). A global analysis of factors controlling VIIRS nighttime light levels from densely populated areas. *Remote Sensing of Environment, 190*, 366–382.

Li, F., Li, E., Zhang, C., Samat, A., Liu, W., Li, C., & Atkinson, P.M. (2021). Estimating artificial impervious surface percentage in Asia by fusing multi-temporal MODIS and VIIRS nighttime light data. *Remote Sensing, 13*.

Li, K., & Chen, Y. (2018). A genetic algorithm-based urban cluster automatic threshold method by combining VIIRS DNB, NDVI, and NDBI to monitor urbanization. *Remote Sensing, 10*.

Li, X., Chen, F.R., & Chen, X.L. (2013a). Satellite-observed nighttime light variation as evidence for global armed conflicts. *IEEE Journal of Selected Topics in Applied Earth Observations and Remote Sensing, 6*, 2302–2315.

Li, X., Chen, X., Zhao, Y., Xu, J., Chen, F., & Li, H. (2013b). Automatic intercalibration of night-time light imagery using robust regression. *Remote Sensing Letters, 4*, 45–54.

Li, X., Ge, L., & Chen, X. (2013c). Detecting Zimbabwe's decadal economic decline using nighttime light imagery. *Remote Sensing, 5*, 4551–4570.

Li, X., Levin, N., Xie, J., & Li, D. (2020). Monitoring hourly night-time light by an unmanned aerial vehicle and its implications to satellite remote sensing. *Remote Sensing of Environment, 247*, 111942.

Li, X., & Li, D. (2014). Can night-time light images play a role in evaluating the Syrian Crisis? *International Journal of Remote Sensing, 35*, 6648–6661.

Li, X., Li, D., Xu, H., & Wu, C. (2017). Intercalibration between DMSP/OLS and VIIRS night-time light images to evaluate city light dynamics of Syria's major human settlement during Syrian Civil War. *International Journal of Remote Sensing, 38*(21), 5934–5951.

Li, X., Ma, R., Zhang, Q., Li, D., Liu, S., He, T., & Zhao, L. (2019). Anisotropic characteristic of artificial light at night–Systematic investigation with VIIRS DNB multi-temporal observations. *Remote Sensing of Environment, 233*, 111357.

Li, X., Shang, X., Zhang, Q., Li, D., Chen, F., Jia, M., & Wang, Y. (2022). Using radiant intensity to characterize the anisotropy of satellite-derived city light at night. *Remote Sensing of Environment, 271*, 112920.

Li, X., Xu, H., Chen, X., & Li, C. (2013d). Potential of NPP-VIIRS nighttime light imagery for modeling the regional economy of China. *Remote Sensing, 5*, 3057–3081.

Li, X., Zhao, L., Li, D., & Xu, H. (2018). Mapping urban extent using Luojia 1–01 nighttime light imagery. *Sensors, 18*(11), 3665.

Liao, L., Weiss, S., Mills, S., & Hauss, B. (2013). Suomi NPP VIIRS day-night band on-orbit performance. *Journal of Geophysical Research: Atmospheres, 118*, 12, 705–712, 718.

Liu, D., Shen, Y., Wang, Y., Wang, Z., Mo, Z., & Zhang, Q. (2023). Monitoring the spatiotemporal dynamics of arctic winter snow/ice with moonlight remote sensing: Systematic evaluation in Svalbard. *Remote Sensing, 15*(5), 1255.

Liu, D., Zhang, Q., Wang, J., Wang, Y., Shen, Y., & Shuai, Y. (2021). The potential of moonlight remote sensing: A systematic assessment with multi-source nightlight remote sensing data. *Remote Sensing, 13*(22), 4639.

Liu, J., Peng, X.C., Zhong, Q.J., Lin, K., Feng, H.L., Hong, H.J., Dong, J.H., Fang, Q.L., & Wu, Y.Y. (2013). Case study of investigation and evaluation on the urban light pollution in Macau. *Applied Mechanics and Materials, 295*, 678–687.

Liu, S., Li, X., Levin, N., & Jendryke, M. (2019). Tracing cultural festival patterns using time-series of VIIRS monthly products. *Remote Sensing Letters, 10*(12), 1172–1181.

Liu, W., Luo, Z., & Xiao, D. (2022). Age structure and carbon emission with climate-Extended STIRPAT model-a cross-country analysis. *Frontiers in Environmental Science, 9*, 667.

Liu, Z., He, C., & Yang, Y. (2011). Mapping urban areas by performing systematic correction for DMSP/OLS Nighttime Lights Time Series in China from 1992 to 2008. In *Geoscience and Remote Sensing Symposium (IGARSS), 2011 IEEE International* (pp. 1858–1861). IEEE.

Liu, Z., He, C., Zhang, Q., Huang, Q., & Yang, Y. (2012). Extracting the dynamics of urban expansion in China using DMSP-OLS nighttime light data from 1992 to 2008. *Landscape and Urban Planning, 106*, 62–72.

Loewen, L.J., Steel, G.D., & Suedfeld, P. (1993). Perceived safety from crime in the urban environment. *Journal of Environmental Psychology, 13*, 323–331.

Longcore, T., & Rich, C. (2004). Ecological light pollution. *Frontiers in Ecology and the Environment, 2*, 191–198.

Lovatt, A., & O'Connor, J. (1995). Cities and the night-time economy. *Planning Practice and Research, 10*, 127–134.

Lu, D., Tian, H., Zhou, G., & Ge, H. (2008). Regional mapping of human settlements in southeastern China with multisensor remotely sensed data. *Remote Sensing of Environment, 112*, 3668–3679.

Luckiesh, M. (1920). *Artificial Light: Its Influence Upon Civilization*. University of London Press.

Ma, J., Guo, J., Ahmad, S., Li, Z., & Hong, J. (2020). Constructing a new inter-calibration method for DMSP-OLS and NPP-VIIRS nighttime light. *Remote Sensing, 12*(6), 937.

Ma, T., Zhou, C., Pei, T., Haynie, S., & Fan, J. (2012). Quantitative estimation of urbanization dynamics using time series of DMSP/OLS nighttime light data: A comparative case study from China's cities. *Remote Sensing of Environment, 124*, 99–107.

Man, D.C., Tsubasa, H., & Fukui, H. (2021). Normalization of VIIRS DNB images for improved estimation of socioeconomic indicators. *International Journal of Digital Earth, 14*, 540–554.

Mander, S., Alam, F., Lovreglio, R., & Ooi, M. (2023). How to measure light pollution—A systematic review of methods and applications. *Sustainable Cities and Society, 92*, 104465.

Marin, C. (2009). StarLight: A common heritage. *Proceedings of the International Astronomical Union, 5*, 449–456.

Marshall Shepherd, J., Pierce, H., & Negri, A.J. (2002). Rainfall modification by major urban areas: Observations from spaceborne rain radar on the TRMM satellite. *Journal of Applied Meteorology, 41*.

Massetti, L., Paterni, M., & Merlino, S. (2022). Monitoring light pollution with an unmanned aerial vehicle: A case study comparing RGB images and night ground brightness. *Remote Sensing, 14*(9), 2052.

Mazor, T., Levin, N., Possingham, H.P., Levy, Y., Rocchini, D., Richardson, A.J., & Kark, S. (2013). Can satellite-based night lights be used for conservation? The case of nesting sea turtles in the Mediterranean. *Biological Conservation, 159*, 63–72.

McCallum, I., Kyba, C.C.M., Bayas, J.C.L., Moltchanova, E., Cooper, M., Cuaresma, J.C., . . . & Fritz, S. (2022). Estimating global economic well-being with unlit settlements. *Nature Communications, 13*(1), 2459.

Milesi, C., Elvidge, C.D., Nemani, R.R., & Running, S.W. (2003). Assessing the impact of urban land development on net primary productivity in the southeastern United States. *Remote Sensing of Environment, 86*, 401–410.

Miller, M.W. (2006). Apparent effects of light pollution on singing behavior of American robins. *The Condor, 108*, 130–139.

Miller, S.D., Hawkins, J.D., Kent, J., Turk, F.J., Lee, T.F., Kuciauskas, A.P., Richardson, K., Wade, R., & Hoffman, C. (2006). NexSat: Previewing NPOESS/VIIRS imagery capabilities. *Bulletin of the American Meteorological Society, 87*.

Miller, S.D., Lee, T.F., Turk, F.J., Kuciauskas, A.P., & Hawkins, J.D. (2005). Shedding new light on nocturnal monitoring of the environment with the VIIRS day/night band. In *Optics & Photonics 2005* (pp. 58900W-58900W-58909). International Society for Optics and Photonics.

Miller, S.D., Mills, S.P., Elvidge, C.D., Lindsey, D.T., Lee, T.F., & Hawkins, J.D. (2012). Suomi satellite brings to light a unique frontier of nighttime environmental sensing capabilities. *Proceedings of the National Academy of Sciences, 109*, 15706–15711.

Miller, S.D., Straka, W., Mills, S., Elvidge, C., Lee, T., Solbrig, J., Walther, A., Heidinger, A., & Weiss, S. (2013). Illuminating the capabilities of the Suomi National Polar-Orbiting partnership (NPP) visible infrared imaging radiometer suite (VIIRS) day/night band. *Remote Sensing, 5*, 6717–6766.

Nagendra, H., Lucas, R., Honrado, J.P., Jongman, R.H., Tarantino, C., Adamo, M., & Mairota, P. (2013). Remote sensing for conservation monitoring: Assessing protected areas, habitat extent, habitat condition, species diversity, and threats. *Ecological Indicators, 33*, 45–59.

National Research Council, Board on Environmental Change, & Committee on the Human Dimensions of Global Change. (1998). *People and Pixels: Linking Remote Sensing and Social Science*. National Academies Press.

Navara, K.J., & Nelson, R.J. (2007). The dark side of light at night: Physiological, epidemiological, and ecological consequences. *Journal of pineal research, 43*, 215–224.

Nicholls, R.J., & Cazenave, A. (2010). Sea-level rise and its impact on coastal zones. *Science, 328*, 1517–1520.

Oda, T., & Maksyutov, S. (2011). A very high-resolution (1 km× 1 km) global fossil fuel CO_2 emission inventory derived using a point source database and satellite observations of nighttime lights. *Atmospheric Chemistry and Physics, 11*, 543–556.

Olsen, R.N., Gallaway, T., & Mitchell, D. (2013). Modelling US light pollution. *Journal of Environmental Planning and Management*, 1–21.

Pack, D., Coffman, C., Rowen, D.W., Santiago, J.R., Kinum, G., & Russell, R.W. (2018). Earth remote sensing results from the CUbesat MULtispectral Observing System, CUMULOS. *Moon*, *19*, 20.

Pack, D.W., Coffman, C.M., & Santiago, J.R. (2019). *A Year in Space for the Cubesat Multispectral Observing System: CUMULOS*. https://digitalcommons.usu.edu/smallsat/2019/all2019/148/

Pack, D.W., Coffman, C.M., Santiago, J.R., & Russell, R. (2022). *Earth Remote Sensing Results from the CUbesat MULtispectral Observing System*. CUMULOS. Authorea Preprints.

Pack, D.W., & Hardy, B.S. (2016). *CubeSat Nighttime Lights*. chrome-extension://efaidnbmnnnibpcajpcglclefindmkaj/http://mstl.atl.calpoly.edu/~workshop/archive/2016/Summer/Day%201/Session%204/7_DeePack.pdf

Pack, D.W., Hardy, B.S., & Longcore, T. (2017). *Studying the Earth at Night from CubeSats*. chrome-extension://efaidnbmnnnibpcajpcglclefindmkaj/https://core.ac.uk/download/pdf/84292767.pdf

Pagden, M., Ngahane, K., & Amin, M.S.R. (2020). Changing the colour of night on urban streets-LED vs. part-night lighting system. *Socio-Economic Planning Sciences*, *69*, 100692.

Pandey, B., Joshi, P., & Seto, K.C. (2013). Monitoring urbanization dynamics in India using DMSP/OLS night time lights and SPOT-VGT data. *International Journal of Applied Earth Observation and Geoinformation*, *23*, 49–61.

Patino, J.E., & Duque, J.C. (2013). A review of regional science applications of satellite remote sensing in urban settings. *Computers, Environment and Urban Systems*, *37*, 1–17.

Pauley, S.M. (2004). Lighting for the human circadian clock: Recent research indicates that lighting has become a public health issue. *Medical Hypotheses*, *63*, 588–596.

Pavlačka, D., Vyvlečka, P., Barvíř, R., Rypl, O., & Burian, J. (2023). Influence of COVID-19 on night-time lights in Czechia. *Journal of Maps*, *19*.

Pawson, S.M., & Bader, M.F. (2014). LED lighting increases the ecological impact of light pollution irrespective of color temperature. *Ecological Applications*, *24*(7), 1561–1568.

Portnov, B.A., Stevens, R.G., Samociuk, H., Wakefield, D., & Gregorio, D.I. (2016). Light at night and breast cancer incidence in Connecticut: An ecological study of age group effects. *Science of the Total Environment*, *572*, 1020–1024.

Pozzi, F., & Small, C. (2005). Analysis of urban land cover and population density in the United States. *Photogrammetric Engineering and Remote Sensing*, *71*, 719–726.

Pun, C.S.J., & So, C.W. (2012). Night-sky brightness monitoring in Hong Kong. *Environmental Monitoring and Assessment*, *184*, 2537–2557.

Pun, C.S.J., So, C.W., Leung, W.Y., & Wong, C.F. (2014). Contributions of artificial lighting sources on light pollution in Hong Kong measured through a night sky brightness monitoring network. *Journal of Quantitative Spectroscopy and Radiative Transfer*, *139*, 90–108.

Pust, P., Schmidt, P.J., & Schnick, W. (2015). A revolution in lighting. *Nature Materials*, *14*(5), 454.

Ramírez, F., Cordón, Y., García, D., Rodríguez, A., Coll, M., Davis, L.S., . . . & Carrasco, J.L. (2023). Large-scale human celebrations increase global light pollution. *People and Nature*, *5*(5), 1552–1560.

Rauch, J.N. (2009). Global mapping of Al, Cu, Fe, and Zn in-use stocks and in-ground resources. *Proceedings of the National Academy of Sciences*, *106*, 18920–18925.

Reynolds, J.F., Smith, D.M.S., Lambin, E.F., Turner, B., Mortimore, M., Batterbury, S.P., Downing, T.E., Dowlatabadi, H., Fernández, R.J., & Herrick, J.E. (2007). Global desertification: Building a science for dryland development. *Science*, *316*, 847–851.

Rindfuss, R.R., & Stern, P.C. (1998). Linking remote sensing and social science: The need and the challenges. *People and Pixels: Linking Remote Sensing and Social Science*, 1–27.

Román, M.O., & Stokes, E.C. (2015). Holidays in lights: Tracking cultural patterns in demand for energy services. *Earth's Future*, *3*(6), 182–205.

Román, M.O., Stokes, E.C., Shrestha, R., Wang, Z., Schultz, L., Carlo, E.A.S., . . . & Enenkel, M. (2019). Satellite-based assessment of electricity restoration efforts in Puerto Rico after Hurricane Maria. *PLoS ONE*, *14*(6), e0218883.

Román, M.O., Wang, Z., Sun, Q., Kalb, V., Miller, S.D., Molthan, A., . . . & Masuoka, E.J. (2018). NASA's Black Marble nighttime lights product suite. *Remote Sensing of Environment*, *210*, 113–143.

Rosenfeld, D. (2000). Suppression of rain and snow by urban and industrial air pollution. *Science*, *287*, 1793–1796.

Rouse Jr, J., Haas, R., Schell, J., & Deering, D. (1974). Monitoring vegetation systems in the Great Plains with ERTS. *NASA Special Publication*, *351*, 309.

Rybnikova, N., Haim, A., & Portnov, B.A. (2015). Artificial light at night (ALAN) and breast cancer incidence worldwide: A revisit of earlier findings with analysis of current trends. *Chronobiology International*, 32(6), 757–773.

Rybnikova, N.A., Haim, A., & Portnov, B.A. (2016). Does artificial light-at-night exposure contribute to the worldwide obesity pandemic? *International Journal of Obesity*, 40(5), 815–823.

Rybnikova, N.A., Haim, A., & Portnov, B.A. (2017). Is prostate cancer incidence worldwide linked to artificial light at night exposures? Review of earlier findings and analysis of current trends. *Archives of Environmental & Occupational Health*, 72(2), 111–122.

Rybnikova, N.A., & Portnov, B.A. (2016). Artificial light at night and obesity: Does the spread of wireless information and communication technology play a role? *International Journal of Sustainable Lighting*, 18, 16–20.

Rybnikova, N.A., & Portnov, B.A. (2018). Population-level study links short-wavelength nighttime illumination with breast cancer incidence in a major metropolitan area. *Chronobiology International*, 35(9), 1198–1208.

Rybnikova, N., Stevens, R.G., Gregorio, D.I., Samociuk, H., & Portnov, B.A. (2018). Kernel density analysis reveals a halo pattern of breast cancer incidence in Connecticut. *Spatial and Spatio-Temporal Epidemiology*, 26, 143–151.

Salgado-Delgado, R., Tapia Osorio, A., Saderi, N., & Escobar, C. (2011). Disruption of circadian rhythms: A crucial factor in the etiology of depression. *Depression Research and Treatment*, 2011.

Salmon, M. (2003). Artificial night lighting and sea turtles. *Biologist*, 50, 163–168.

Sánchez de Miguel, A. (2015). *Variación espacial, temporal y espectral de la contaminación lumínica y sus fuentes: Metodología y resultados* (Doctoral dissertation, Universidad Complutense de Madrid).

Sánchez de Miguel, A., Bennie, J., Rosenfeld, E., Dzurjak, S., & Gaston, K.J. (2021a). First estimation of global trends in nocturnal power emissions reveals acceleration of light pollution. *Remote Sensing*, 13(16), 3311.

Sánchez de Miguel, A., Bennie, J., Rosenfeld, E., Dzurjak, S., & Gaston, K.J. (2022). Environmental risks from artificial nighttime lighting widespread and increasing across Europe. *Science Advances*, 8(37), eabl6891.

Sánchez de Miguel, A., Kyba, C.C., Aubé, M., Zamorano, J., Cardiel, N., Tapia, C., . . . & Gaston, K.J. (2019). Colour remote sensing of the impact of artificial light at night (I): The potential of the International Space Station and other DSLR-based platforms. *Remote Sensing of Environment*, 224, 92–103.

Sánchez de Miguel, A., Zamorano, J., Aubé, M., Bennie, J., Gallego, J., Ocana, F., . . . & Gaston, K.J. (2021b). Colour remote sensing of the impact of artificial light at night (II): Calibration of DSLR-based images from the International Space Station. *Remote Sensing of Environment*, 264, 112611.

Schaaf, C.B., Gao, F., Strahler, A.H., Lucht, W., Li, X., Tsang, T., Strugnell, N.C., Zhang, X., Jin, Y., & Muller, J.-P. (2002). First operational BRDF, albedo nadir reflectance products from MODIS. *Remote Sensing of Environment*, 83, 135–148.

Schneider, A., Friedl, M.A., & Woodcock, C.E. (2003). Mapping urban areas by fusing multiple sources of coarse resolution remotely sensed data. In *Geoscience and Remote Sensing Symposium, 2003. IGARSS'03. Proceedings. 2003 IEEE International* (pp. 2623–2625). IEEE.

Ściężor, T. (2021). Effect of street lighting on the urban and rural night-time radiance and the brightness of the night sky. *Remote Sensing*, 13.

Sen, A., Kim, Y., Caruso, D., Lagerloef, G., Colomb, R., & Le Vine, D. (2006). Aquarius/SAC-D mission overview. In *Remote Sensing* (pp. 63610I-63610I-63610). International Society for Optics and Photonics.

Seto, K.C., Kaufmann, R.K., & Woodcock, C.E. (2000). Landsat reveals China's farmland reserves, but they're vanishing fast. *Nature*, 406, 121–121.

Seto, K.C., & Shepherd, J.M. (2009). Global urban land-use trends and climate impacts. *Current Opinion in Environmental Sustainability*, 1, 89–95.

Shi, K., Yu, B., Huang, Y., Hu, Y., Yin, B., Chen, Z., Chen, L., & Wu, J. (2014). Evaluating the ability of NPP-VIIRS nighttime light data to estimate the gross domestic product and the electric power consumption of china at multiple scales: A comparison with DMSP-OLS data. *Remote Sensing*, 6, 1705–1724.

Shortland, A., Christopoulou, K., & Makatsoris, C. (2013). War and famine, peace and light? The economic dynamics of conflict in Somalia 1993–2009. *Journal of Peace Research*, 50, 545–561.

Simpson, S.N., & Hanna, B.G. (2010). Willingness to pay for a clear night sky: Use of the contingent valuation method. *Applied Economics Letters*, 17(11), 1095–1103.

Small, C. (2001). Estimation of urban vegetation abundance by spectral mixture analysis. *International Journal of Remote Sensing, 22*, 1305–1334.

Small, C. (2005). A global analysis of urban reflectance. *International Journal of Remote Sensing, 26*, 661–682.

Small, C., & Elvidge, C.D. (2013). Night on Earth: Mapping decadal changes of anthropogenic night light in Asia. *International Journal of Applied Earth Observation and Geoinformation, 22*, 40–52.

Small, C., Elvidge, C.D., Balk, D., & Montgomery, M. (2011). Spatial scaling of stable night lights. *Remote Sensing of Environment, 115*, 269–280.

Small, C., Pozzi, F., & Elvidge, C.D. (2005a). Spatial analysis of global urban extent from DMSP-OLS night lights. *Remote sensing of environment, 96*, 277–291.

Small, C., Pozzi, F., & Elvidge, C.D. (2005b). Spatial analysis of global urban extent from DMSP-OLS night lights. *Remote Sensing of Environment, 96*, 277–291.

Smith, M. (2008). Time to turn off the lights. *Nature, 457*, 27–27.

Sovacool, B.K., & Brown, M.A. (2010). Twelve metropolitan carbon footprints: A preliminary comparative global assessment. *Energy Policy, 38*, 4856–4869.

Stern, P.C., Young, O.R., & Druckman, D. (1991). *Global Environmental Change: Understanding the Human Dimensions*. National Academies Press.

Stokes, E.C., & Roman, M.O. (2022). Tracking COVID-19 urban activity changes in the Middle East from nighttime lights. *Scientific Reports, 12*, 8096.

Stokes, E.C., & Seto, K.C. (2019). Characterizing urban infrastructural transitions for the Sustainable Development Goals using multi-temporal land, population, and nighttime light data. *Remote Sensing of Environment, 234*, 111430.

Strauss, M. (2017). Planet Earth to get a daily selfie. *Science* 782–783.

Sutton, P.C., Elvidge, C.D., & Ghosh, T. (2007). Estimation of gross domestic product at sub-national scales using nighttime satellite imagery. *International Journal of Ecological Economics & Statistics, 8*, 5–21.

Sutton, P.C., Elvidge, C.D., & Obremski, T. (2003). Building and evaluating models to estimate ambient population density. *Photogrammetric Engineering and Remote Sensing, 69*(5), 545–554.

Sutton, P.C., Roberts, D., Elvidge, C., & Baugh, K. (2001). Census from Heaven: An estimate of the global human population using night-time satellite imagery. *International Journal of Remote Sensing, 22*, 3061–3076.

Takahashi, K.I., Terakado, R., Nakamura, J., Adachi, Y., Elvidge, C.D., & Matsuno, Y. (2010). In-use stock analysis using satellite nighttime light observation data. *Resources, Conservation and Recycling, 55*, 196–200.

Tardà, A., Palà, V., Arbiol, R., Pérez, F., Viñas, O., Pipia, L., & Martínez, L. (2011). Detección de la iluminación exterior urbana nocturna con el sensor aerotransportado CASI 550. *Proceedings of the International Geomatic Week, Barcelona, Spain, March, 2011*.

Teikari, P. (2007). Light pollution: Definition, legislation, measurement, modeling and environmental effects. *Universitat politécninca de Catalunya. Barcelona, Catalunya, 10*.

Tsao, J.Y., Saunders, H.D., Creighton, J.R., Coltrin, M.E., & Simmons, J.A. (2010). Solid-state lighting: An energy-economics perspective. *Journal of Physics D: Applied Physics, 43*(35), 354001.

Tsao, J.Y., & Waide, P. (2010). The world's appetite for light: Empirical data and trends spanning three centuries and six continents. *Leukos, 6*(4), 259–281.

Tucker, C.J. (1979). Red and photographic infrared linear combinations for monitoring vegetation. *Remote Sensing of Environment, 8*, 127–150.

UNFPA (2011). *State of World Population 2011: People and Possibilities in a World of 7 Billion*. The United Nations Population Fund.

Walczak, K., Gyuk, G., Kruger, A., Byers, E., & Huerta, S. (2017). Nitesat: A high resolution, full-color, light pollution imaging satellite mission. *International Journal of Sustainable Lighting, 19*(1), 48–55.

Wang, Z., Román, M.O., Kalb, V.L., Miller, S.D., Zhang, J., & Shrestha, R.M. (2021). Quantifying uncertainties in nighttime light retrievals from Suomi-NPP and NOAA-20 VIIRS Day/Night Band data. *Remote Sensing of Environment, 263*, 112557.

Weisburd, D., & Amram, S. (2014). The law of concentrations of crime at place: The case of Tel Aviv-Jaffa. *Police Practice and Research*, 1–14.

Weng, Q., Lu, D., & Liang, B. (2006). Urban surface biophysical descriptors and land surface temperature variations. *Photogrammetric Engineering & Remote Sensing, 72*, 1275–1286.

Weng, Q., Lu, D., & Schubring, J. (2004). Estimation of land surface temperature–vegetation abundance relationship for urban heat island studies. *Remote Sensing of Environment, 89*, 467–483.

Wu, J., He, S., Peng, J., Li, W., & Zhong, X. (2013). Intercalibration of DMSP-OLS night-time light data by the invariant region method. *International Journal of Remote Sensing, 34*, 7356–7368.

Yi, K., Tani, H., Li, Q., Zhang, J., Guo, M., Bao, Y., Wang, X., & Li, J. (2014). Mapping and evaluating the urbanization process in Northeast China using DMSP/OLS nighttime light data. *Sensors, 14*, 3207–3226.

Yu, T., Chen, L., Xu, N., Xu, H., Hu, X., & Zhang, X. (2023). Fengyun-3E low light observation and nighttime lights product. *IEEE Transactions on Geoscience and Remote Sensing, 61*, 3292236.

Zhang, Q., He, C., & Liu, Z. (2014). Studying urban development and change in the contiguous United States using two scaled measures derived from nighttime lights data and population census. *GIScience & Remote Sensing*, 1–20.

Zhang, Q., Schaaf, C., & Seto, K.C. (2013). The Vegetation Adjusted NTL Urban Index: A new approach to reduce saturation and increase variation in nighttime luminosity. *Remote Sensing of Environment, 129*, 32–41.

Zhang, Q., & Seto, K.C. (2011). Mapping urbanization dynamics at regional and global scales using multi-temporal DMSP/OLS nighttime light data. *Remote Sensing of Environment, 115*, 2320–2329.

Zhang, Q., Wang, P., Chen, H., Huang, Q., Jiang, H., Zhang, Z., Zhang, Y., Luo, X., & Sun, S. (2017). A novel method for urban area extraction from VIIRS DNB and MODIS NDVI data: A case study of Chinese cities. *International Journal of Remote Sensing, 38*, 6094–6109.

Zhang, Q., Zheng, Z., Wu, Z., Cao, Z., & Luo, R. (2022). Using multi-source geospatial information to reduce the saturation problem of DMSP/OLS nighttime light data. *Remote Sensing, 14*(14), 3264.

Zhao, X., Li, D., Li, X., Zhao, L., & Wu, C. (2018). Spatial and seasonal patterns of night-time lights in global ocean derived from VIIRS DNB images. *International Journal of Remote Sensing, 39*, 8151–8181.

Zheng, Q., Weng, Q., Huang, L., Wang, K., Deng, J., Jiang, R., Ye, Z. and Gan, M. (2018). A new source of multi-spectral high spatial resolution night-time light imagery—JL1–3B. *Remote Sensing of Environment, 215*, 300–312.

Zheng, Q., Weng, Q., & Wang, K. (2019). Developing a new cross-sensor calibration model for DMSP-OLS and Suomi-NPP VIIRS night-light imageries. *ISPRS Journal of Photogrammetry and Remote Sensing, 153*, 36–47.

Zhong, Q., Xiao, R., Cao, H., Li, X., & Wu, J. (2023). Evaluation of Qimingxing-1 nighttime light image. *Geomatics and Information Science of Wuhan University, 48*(8), 1273–1285.

Zhu, Z., Zhou, Y., Seto, K.C., Stokes, E.C., Deng, C., Pickett, S.T., & Taubenböck, H. (2019). Understanding an urbanizing planet: Strategic directions for remote sensing. *Remote Sensing of Environment, 228*, 164–182.

Zhuo, L., Ichinose, T., Zheng, J., Chen, J., Shi, P., & Li, X. (2009). Modelling the population density of China at the pixel level based on DMSP/OLS non-radiance-calibrated night-time light images. *International Journal of Remote Sensing, 30*, 1003–1018.

Zhuo, L., Zheng, J., Zhang, X., Li, J., & Liu, L. (2015). An improved method of night-time light saturation reduction based on EVI. *International Journal of Remote Sensing, 36*(16), 4114–4130.

Ziskin, D., Baugh, K., Hsu, F.C., Ghosh, T., & Elvidge, C. (2010). Methods used for the 2006 radiance lights. *Proceedings of the Asia-Pacific Advanced Network, 30*(0), 131–131.

Part VII

Summary and Synthesis for Volume VI

14 Summary Chapter for Remote Sensing Handbook, Volume VI

Droughts, Disasters, Pollution, and Urban Mapping

Prasad S. Thenkabail

ACRONYMS AND DEFINITIONS

ACF	Action Contre la Faim
ADFC	Agricultural drought frequency change
ALEXI	Atmosphere-land exchange inverse
ASIS	Agricultural Stress Index System
CPC	Climate Prediction Center
DSI	Drought Severity Index
ESI	Evaporative Stress Index
FAO	Food and Agriculture Organization
FEWSNET	Famine early warning system network
FSA	Farm Service Agency
GIIDI	Geographically Independent Integrated Drought Index
GVH	Global vegetation health
JRC	Joint Research Centre
LST	Land surface temperature
LULC	Land use/land cover
MODIS	Moderate-resolution imaging spectroradiometer
NDVI	Normalized difference vegetation index
NMME	North American multimodal ensemble
OBDI	Objective Blend Drought Index
OVDI	Optimized Vegetation Drought Index
PASG	Percent annual seasonal greenness
PDSI	Palmer Drought Severity Index
RDI	Reclamation Drought Index
SDCI	Scaled Drought Condition Index
SDI	Synthesized Drought Index
SOSA	Start of season anomaly
SPI	Standardized Precipitation Index
SMADI	Soil Moisture Agricultural Drought Index
SMCI	Soil Moisture Condition Index
SMOS	Soil moisture and ocean salinity
SPEI	Standardized Precipitation Evapotranspiration Index

DOI: 10.1201/9781003541417-21

SWSI	Surface Water Supply Index
TCI	Temperature Condition Index
TRMM	Tropical rainfall measuring mission
USDA	US Department of Agriculture
USDM	US drought monitor
VCI	Vegetation Condition Index
VegDRI	Vegetation Drought Response Index
VH	Vegetation health
VHI	Vegetation Health Index
WDI	Water Deficit index

This chapter provides a brief summary of all 13 chapters appearing in Volume VI of the *Remote Sensing Handbook*. The volume has a focus on water resources, disasters, and urban remote sensing. The chapters are broadly classified into: (1) droughts and drylands, (2) disasters, (3) volcanoes, (4) fires, (5) urban, and (6) night lights. Under each of these topics one or more chapters provide comprehensive coverage. For example, there are six chapters under droughts and drylands. The summary in this chapter provides a "window view" of what exists in each chapter as well as inter-linkages to various chapters. You can read the summary chapter in three ways: (1) before reading the chapters to get an overview, (2) after reading all the chapters to re-cap and refresh major highlights, or (3) do both (1) and (2) to capture the totality of all the chapters. The chapter summaries also help the reader establish inter-linkages that exist between chapters, in a nutshell.

14.1 AGRICULTURAL DROUGHTS USING VEGETATION HEALTH METHODS

Droughts are broadly classified as: (1) meteorological, (2) hydrological, and (3) agricultural. Meteorological drought occurs when there is prolonged period of precipitation deficiency,

---

Chapter 14: Summary Chapter for
Remote Sensing Handbook (Second Edition, Six Volumes): Volume VI

Volume VI: Droughts, Disasters, Pollution, and Urban Mapping

Chapter 1 and 2: Agricultural Droughts using Remote Sensing-derived Vegetation Health, biophysical Quantities,

Chapter 3 and 4: Drought Monitoring Advances, U. S. Drought Monitor, and New Remote Sensing Drought Indices

Chapters 5: Remote Sensing of Drylands

Chapters 6 : Disaster Assessment and Management using Remote Sensing

Chapters 7: Study of Humanitarian Disasters using Remote Sensing

Chapters 8: Remote Sensing of Volcanoes

Chapter 9: Biomass Burning and Greenhouse gas Emission Assessments from Remote Sensing

Chapter 10: Coal Fires, Emissions, Pollution, and Disasters from Remote Sensing

Chapter 11 and 12: Urban Studies and Smart Cities using Remote Sensing

Chapter 13: Nightlights Modeling and Mapping using Remote Sensing

FIGURE 14.0 Overview of the chapters in Volume VI of the *Remote Sensing Handbook* (Second Edition).

hydrological droughts occur when there are below normal water levels in lakes, reservoirs, and rivers, and agricultural drought occurs when there is insufficient soil moisture for healthy growth of croplands and rangelands. Food security is heavily dependent on healthy crop productivity, which in turn is feasible only with sufficient water availability. Extreme conditions of agricultural droughts (e.g., Figure 14.1) will result in full or partial crop failure and/or substantial decreases in crop yields, which in turn will lead to food insecurity, price rise, and may even lead to famine. Conventional drought monitoring is performed using meteorological, hydrological, and soil moisture data and includes such indices as Standardized Precipitation Index (SPI), Palmer Drought Severity Index (PDSI), Crop Moisture Index, Palmer Hydrological Drought Index, Surface Water Supply Index (SWSI), Reclamation Drought Index (RDI), deciles, and percent of normal. The US Drought Monitor integrates several of these indices along with ancillary data to provide weekly operational drought maps. In contrast, remote sensing offers consistent global coverage, and better spatial representation for drought studies. In its simplest form, drought can be monitored using remote sensing by simple deviation of NDVI from its long-term mean.

One of the most widely used, attractive, simple, and yet powerful remote sensing measure of drought is the vegetation health (VH) method of NOAA/NESDIS developed by Dr. Felix Kogan and reported in Chapter 1 by Dr. Felix Kogan and Wei Guo. They identify vegetation health–based drought products to include moisture and thermal stress, drought start/end, intensity, duration, magnitude, area, season, origination, and impacts. Vegetation Condition Index (VCI), which is a proxy for moisture condition; Temperature Condition Index (TCI), which is a proxy for thermal condition; and Vegetation Health Index (VHI), which is a proxy for the combination of moisture and thermal conditions (Zeng et al., 2023; Kloos et al., 2021; Wu et al., 2020). These indices are represented in scale of 0 (extreme vegetation stress) to 100 (optimal conditions of vegetation health). VHI is somewhat limited in energy-limited environments like high elevations and high latitudes (see Chapters 1–3 of this volume).

Emphasis in Chapter 1 is on Global Vegetation Health (GVH) computation using 4 km resolution NOAA AVHRR GAC data (Pinzon et al., 2023; Pinzon et al., 2014; Tucker et al., 20025), now available from 1981 to the present, through three indices: VCI, TCI, and VHI. GVH maps provide sophisticated and powerful, yet simple to visualize drought conditions, severity, and progression maps and data. This is especially useful in assessing such measures as global food production and famine early warning. An exciting prospect is the continuity of this 4 km, 34-year drought record through much higher resolution of 375 m from VIIRS sensor onboard NPOESS NPP. They demonstrate VH by studying droughts in the warmer world of the 21st century (this chapter, Pinzon et al., 2023; Pinzon et al., 2014; Tucker et al., 20025). Even though there generally accepted temperature "pause" over the last 15 years (0.04°C temperature rise during 1998 and 2013 when compared with 0.18°C increase in the 1990s), climate (e.g., precipitation, temperature) variability over space and time have caused severe droughts in many parts of the world (e.g., 2012 Midwestern agricultural drought in the USA) (Chapters 1–3 of this volume).

More recently, a global long-term (1981–2021), high-resolution (4 km) improved VHI dataset integrating climate, vegetation, and soil moisture was developed by Zeng et al. (2023). The PDSI best detects natural agricultural drought, and the VHI best detects actual agricultural drought. Wu et al. (2020) proposed two agricultural drought mitigation evaluation indices: (1) the agricultural drought frequency change (ADFC) and (2) agricultural drought area change (ADAC), which are calculated by combining PDSI and VHI. The TCI and VHI correlate strongly with soil moisture and agricultural yield anomalies, and hence both indices have the potential to detect agricultural droughts (Kloos et al., 2021).

14.2 AGRICULTURAL DROUGHT MONITORING USING BIOPHYSICAL PARAMETERS

Chapter 2 discusses the use of biophysical indicators derived from optical satellite remote sensing in current agricultural drought monitoring systems. Most operational systems exploit the qualitative

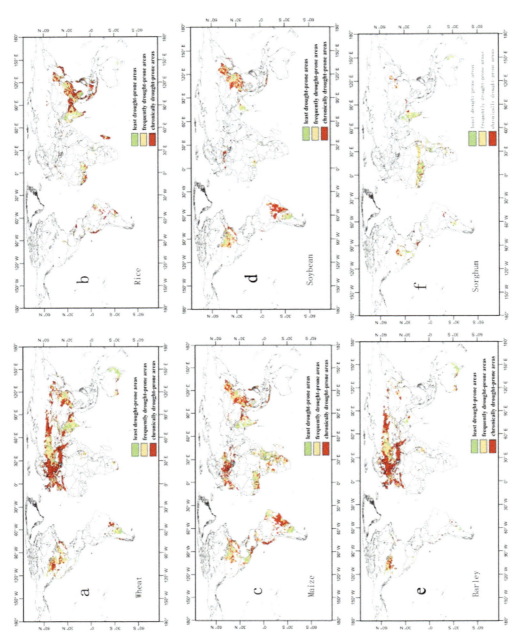

FIGURE 14.1 The spatial distribution of the frequency of severe drought events for main crops during the 1950–2008 period (a, b, c, d, e, and f show the spatial distribution of the frequency of severe drought events in wheat, rice, maize, soybean, and barley crop-planting regions, respectively) determined based on Standardized Precipitation Evapotranspiration Index (SPEI). Global chronically drought-prone areas have increased significantly, from 16.19% in 1902–1949 to 41.09% in 1950–2008

Source: Wang et al., 2014.

approach (e.g., Figure 14.2) of looking at a pixel or an area and determining the deviation of vegetation condition for the same pixel or area with respect to its long-term mean. The deviation from long-term mean of NDVI, VCI, TCI, VHI, and other vegetation indices all constitute various forms of this approach. Typically, these studies analyze one of these vegetation indexes and look for changes over time for a given pixel or area. Such a pragmatic approach is widely used and works reasonably well. Some of these methods are also discussed in Chapter 2.

However, agriculture productivity depends on a complex set of conditions (Jiao et al., 2021) that include features such as changes in:

- Crop type or cropping pattern over years;
- Agronomic inputs (e.g., high-input agriculture in previously low-input agriculture, irrigation on previously non-irrigated lands), leading to changes in biomass and yield;
- Phenology (e.g., late crop planting will result in a different phonological cycle); and
- Cultivar (e.g., different cultivar leading to different biomass and yield).

All these conditions lead to changes in biophysical indicators at a given time of the year or season relative to the same time during a previous year or season. As a result, the observed changes cannot be attributed univocally to drought conditions. For many of these factors little information is available at regional and national scale and biophysical indicators derived mainly from satellite image are the main data source.

Chapter 2 focuses on three recent approaches for monitoring agricultural drought at regional scale, implemented by different international organizations: the Food and Agriculture Organization (FAO) of the United Nations, the non-governmental organization Action Contre la Faim (ACF), and the Joint Research Centre (JRC) of the European Commission. The Agricultural Stress Index System (ASIS) of FAO (Rojas, 2020) aims at detecting agricultural areas with a high likelihood of drought at the global level using optical and thermal channels from METOP-AVHRR instrument. ASIS determines the likelihood of drought occurrence based on three distinctive drought features: intensity, duration, and spatial extent. The ACF approach is currently being developed to target pastures in the Sahel by considering three key elements: biomass production, water availability to animals, and livestock movements (Hanadé Houmma et al., 2023; Van Hoolst et al., 2016). SPOT-VEGETATION data are used to map pasture biomass production (with a light-use efficiency model) and surface water bodies.

The JRC system is based on a statistical approach aiming to provide early estimations of the probability of experiencing a critical biomass production deficit (Rojas, 2020; Hanadé Houmma et al., 2023; Van Hoolst et al., 2016). The method is built upon the similarity between the current seasonal FAPAR profile and the past ones and provide the analyst with an information that is adjusted for its reliability, as estimated form past forecasts.

Other novel and promising modeling frameworks for drought monitoring are then mentioned. These include data mining techniques, combined use of biophysical parameters and precipitation data, and explicit modeling of the spatiotemporal evolution of extreme events.

Eventually, current critical needs in agricultural drought monitoring with remote sensing are discussed: ensuring the continuity of long-term time series from different satellite missions, developing reliable and operational un-mixing techniques, achieving a better balance between timeliness and noise reduction by the use of simplified vegetation growth models and data assimilation techniques, and finally profiting from consolidated techniques such as thermal remote sensing and emerging ones such as the analysis of fluorescence and the photochemical reflectance index.

A newly proposed a Soil Moisture Agricultural Drought Index (SMADI) that integrates Soil Moisture and Ocean Salinity (SMOS) soil moisture and MODIS NDVI and LST identified 80% of global drought events (Sánchez et al., 2018). In a global and regional assessment of droughts using MODIS data Ghazaryan et al. (2020) found evaporative stress index (ESI), which is based on the ratio of actual to potential evapotranspiration, anomalies had higher correlations with maize and wheat yield anomalies than other indices, indicating that prolonged periods of low ESI during the growing season are highly correlated with reduced crop yields. They also found droughts identified by ESI and LST were more intense than NDVI-based results. Overall, multi-sensor-based drought monitoring efforts are the best. They involve extracting different types of information from different sensors for different purposes. For example, (1) precipitation data is gathered from satellites such as TRMM and PERSIANN-CCS\CDR; (2) land surface temperature (LST) data from Landsat, MODIS, and AVHRR; (3) soil moisture data from SMAP, SSM/I, and AMSR-E; (4) groundwater data from GRACE and GRACE-FO; (5) snow data from MODIS, Landsat, AMSR-E, SSM/I, and AMST2; (6) evapotranspiration from MODIS, Landsat, GLDAS, and METRIC; and (7) vegetation and biomass data from suite of satellites (Jiao et al., 2021).

14.3 DROUGHT MONITORING ADVANCES AND TOOLKITS USING REMOTE SENSING

Chapter 3 by Dr. Brian D. Wardlow et al. provides a number of innovative advances in using remote sensing in drought studies. Prior to data, approaches, and methods discussed in this chapter drought studies using remote sensing was primarily based on either looking at NDVI deviation from long-term mean and/or using VCI, TCI, and VHI discussed in Chapter 1 and Chapter 2. In contrast, Chapter 3 by Dr. Brian D. Wardlow et al. provides six remote sensing-based advances in drought studies. These are:

1. US Drought Monitor (USDM) approach (Leeper et al., 2022). The USDM approach utilizes multiple conventional, non-remote sensing indices such as PDSI, SPI, and SWSI along with guidance from experts such as climatologists, agricultural scientists, and water resources managers. In addition, more recently, USDM utilizes remote sensing–based VCI, TCI, and VHI using data from such sensors as AVHRR and MODIS. USDM provides regular drought maps that depict drought intensities over large area extents, such as a county or multiple counties. It is somewhat limited in depicting localized drought conditions. Nevertheless, the USDM approach is also attracting considerable international attention, with a similar approach used in the US Department of Agriculture (USDA) Farm Service Agency (FSA) and Famine Early Warning System Network (FEWSNET).
2. Vegetation drought response index (VegDRI) (Tadesse et al., 2017) identifies vegetation stress by integrating remote sensing, climatic, and biophysical data. Remote sensing data derived either from AVHRR or MODIS, component makes use of (1) percent annual seasonal greenness (PASG), (2) start of season anomaly (SOSA), and (3) out of season (OS) to represent NDVI during non-growing crop season. Climate part includes PDSI and SPI. Biophysical part takes dominant land use/land cover (LULC) type for each pixel. The VegDRI model integrates these remote sensing, climatic, and biophysical variables. VegDRI and USDM provide comparable results. However, the higher spatial resolution of VegDRI helps in better understanding of local (e.g., sub-county level) droughts. The USDM now routinely consult VegDRI to produce more refined USDM maps.
3. Vegetation Outlook (VegOut) (Bayissa et al., 2019) is akin to VegDRI in that it also integrates and uses remote sensing, climatic, and biophysical data. However, VegOut also uses seven Oceanic/Atmospheric indices that make it distinct from VegDRI.
4. Evaporative stress index (ESI) (Qiu et al., 2021) measures standardized anomalies in a ratio of actual to reference evapotranspiration (ET_a/ET_{ref}) retrieved using the remote sensing–based Atmosphere-Land Exchange Inverse (ALEXI) two-source surface energy

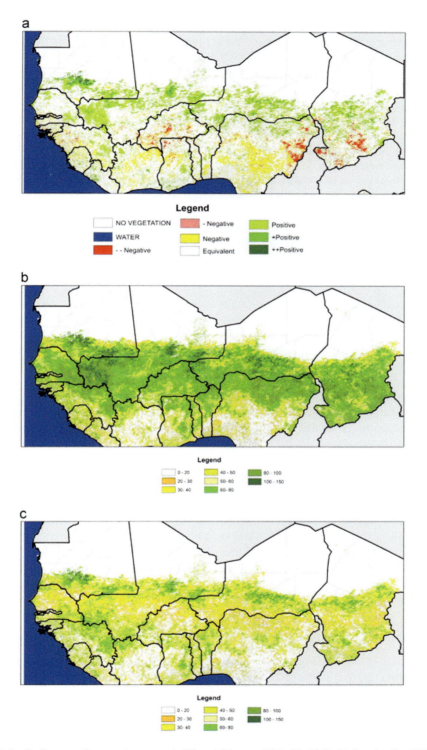

FIGURE 14.2 Indicators of vegetation status in West Africa as of July 31, 2013. (a) Standardized Normalized Difference Vegetation Index (sNDVI). (b) Normalized Growth Index (Indice de Croissance Normalize, ICN). (c) Vegetation Condition Index (VCI).

Source: AGRHYMET monthly bulletin, August 2013; in Traore et al., 2014.

balance model. ALEXI (Akasheh et al., 2023) requires at least two measurements of land surface temperature (LST) measured during morning and obtained from thermal images (e.g., 10-km resolution TIR data from GOES sounder instrument). Higher-resolution applications require field or subfield scale sampling, and is achieved through disaggregated ALEXI (DisALEXI) (Knipper et al., 2023).

5. Microwave-based surface soil moisture retrievals from sensors such as SSM/I and AMSR-E. Surface soil moisture retrievals are better achieved using the low-frequency L band (1–2 GHz), but also with high frequency (2–20 GHz) microwave observations. Since the microwave can only penetrate the topsoil (< 5 cm) and not the plan rooting depth (30–150 cm), microwave-based soil moisture retrievals to agricultural drought have generally been confined to their use in land data assimilation systems.
6. GRACE microgravity measurements are used to retrieve satellite-based soil moisture and groundwater (Fatolazadeh et al., 2022). However, GRACE data is very coarse (150,000 km^2) and is available over an area only in about 2–4 months' time frame. These spatial and temporal resolution limitations allow for only broad understanding of drought situation. However, since no other satellite offers a reasonably good understanding of the groundwater fluctuations, GRACE offers a good understanding at regional, national, and river basin scales.

The best approaches and methods for drought assessments and monitoring require comprehensive analysis framework that integrate remote sensing data with in situ data, climate data and other data, such as socioeconomic data. For example, studies have demonstrated the value of the climate forecast from North American Multimodal Ensemble (NMME) has been among the most salient progress in climate prediction and its application for drought prediction whereas the US Drought Monitor (USDM) has played a critical role in drought monitoring with different drought categories to characterize drought severity, which has been employed to aid decision-making by a wealth of users, such as natural resource managers and authorities (Hao et al., 2017). There are many other standard toolkits exist for drought studies at various scales and levels. For example (Abatzoglou et al., 2017), (1) the National Centers for Environmental Information (NCEI) Climate at a Glance tools provide drought and climate information coarse spatial scales (e.g., climate divisions), which are valuable for large-scale regional assessments but may provide insufficient spatial detail for local decision-making; (2) the PRISM Climate Group (PRISM is described in the next section) provides maps of hydroclimate anomalies from 1980 to the present, but does not present drought indices; and (3) a heavily used tool for drought declaration and decision-making is the USDM, which relies on a blend of quantitative metrics and local impact information to produce weekly maps of drought severity (Abatzoglou et al., 2017). Graw et al. (2019) developed EvIDENZ (Earth Observation–based Information Product for Drought Risk Reduction on the National Level) project, which takes advantage from the integration of RS information and socioeconomic data.

14.4 NEW REMOTE SENSING-BASED DROUGHT INDICES

Can we monitor drought using remote sensing data alone? This is a question Chapter 4 by Jinyoung Rhee et al. tries to answer. State-of-the-art drought monitoring systems like the USDM uses remote sensing, non-remote sensing, as well as some subjective expert opinion to come with drought maps. Chapter 4 lists the six key indicators used by the USDM: (1) Palmer Drought Severity Index (PCMI), (2) Percent of Normal Precipitation, (3) Standardized Precipitation Index (SPI; Alahacoon and Edirisinghe, 2022), (4) US Geological Survey (USGS) Daily Stream Flow Percentiles (Zarei et al., 2023; Wei et al., 2021), (5) model-based CPC Soil Moisture Model Percentiles, and (6) satellite-based VHI. Several ancillary data are used in the USDM, such as the PCMI, the Keetch–Bryam Drought Index, as well as reservoir and lake levels, groundwater levels, and soil moisture field observations. For the western United States, Snowpack Telemetry

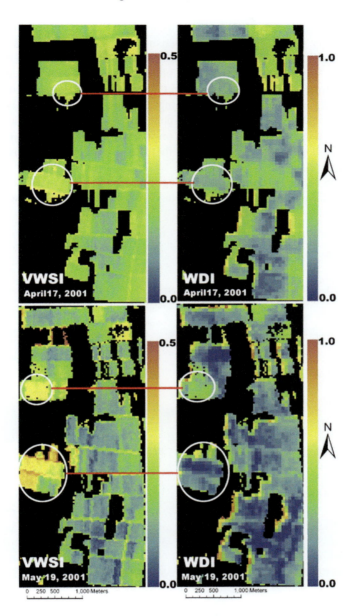

FIGURE 14.3 Advances in drought indices. Modern-day remote sensing data from different sensors can be used to derive various drought indices that address specific issues of crops and other vegetation. For example, this figure shows estimating crop water stress with Enhanced Thematic mapper Plus (ETM+) near infrared (NIR), and short-wave infrared (SWIR) data water deficit index (WDI) and vegetation water stress index (VWSI). VWSI describes the canopy water stress rather than water content, and it is of interest particularly for agricultural drought monitoring.

Source: Ghulam et al., 2008.

is also used. Further USDM used subjective judgment of experts. Other drought monitoring systems such as the Objective Blend Drought Index (OBDI) of the Climate Prediction Center (CPC) and the CPC Seasonal Drought Outlook also use remote sensing and non-remote sensing data, methods, and approaches.

Chapter 4 by Jinyoung Rhee et al. focuses on discussing and developing new advanced drought indices that are purely dependent on remote sensing. Their own (Rhee et al., 2010) Scaled Drought Condition Index (SDCI) uses only remote sensed data that include MODIS Land Surface Temperature (LST), NDVI, and Tropical Rainfall Measuring Mission (TRMM) precipitation through linear combination. This index was found useful for agricultural drought monitoring in both arid and humid regions. Another remote sensing data driven drought index is Drought Severity Index (DSI) that uses MODIS evapotranspiration, potential evapotranspiration, and NDVI products (e.g., Mu et al., 2013). Generally, remote sensing only derived drought indices do as well or even better than drought indices that either use both remote sensing or non-remote sensing data or just the non-remote sensing data.

Figure 14.4, for example, illustrates monitoring meteorological drought in semi-arid regions using multi-sensor microwave remote sensing data. With data from multiple sensors that include

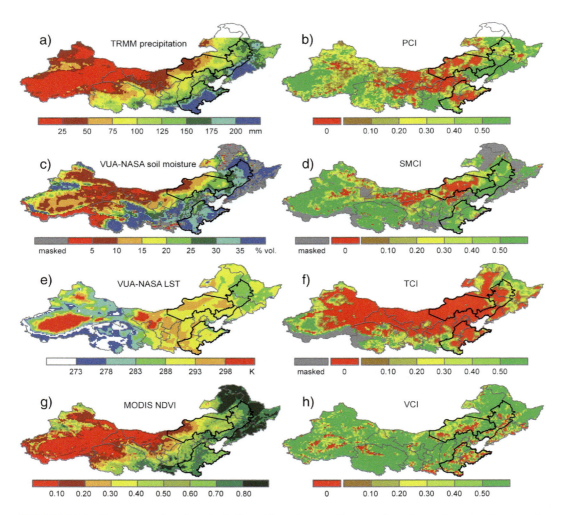

FIGURE 14.4 Remote sensing drought indices. These images illustrate detecting and monitoring drought timely at regional scale using remote sensing data and indices: TRMM Precipitation Condition Index (PCI), the Soil Moisture Condition Index (SMCI) and the Temperature Condition Index (TCI) based on microwave remote sensed TRMM precipitation, AMSR-E soil moisture, and land surface temperature retrievals from 2003 to 2010. The remotely sensed variables were linearly scaled from 0 to 1 for each pixel based on absolute minimum and maximum values for each variable over time in order to discriminate the weather-related

FIGURE 14.4 (*Continued*) component from the ecosystem component, as done for VCI using NDVI (see Chapter 1 by Kogan et al.). After normalization, the scaled value changed from 0 to 1, corresponding to the precipitation changes from extremely low to optimal. Remote sensing drought indices studied here are: TRMM Precipitation Condition Index (PCI); Soil Moisture Condition Index (SMCI); Temperature (Land Surface Temperature) Condition Index (TCI); Vegetation Condition Index (VCI); TRMM Precipitation and Soil Moisture Condition Index (PSMCI); TRMM Precipitation and Temperature Condition Index (PTCI); Soil Moisture and Temperature Condition Index (SMTCI); and Microwave Integrated Drought Index (MIDI). These drought indices are defined as: PCI = (TRMM$_i$ − TRMM$_{min}$)/(TRMM$_{max}$ − TRMM$_{min}$); SMCI = (SM$_i$ − SM$_{min}$)/(SM$_{max}$ − SM$_{min}$); TCI = (LST$_{max}$ − LST$_i$)/(LST$_{max}$ − LST$_{min}$); VCI = (NDVI$_i$ − NDVI$_{min}$)/(NDVI$_{max}$ − NDVI$_{min}$); PSMCI = α * PCI + (1 − α) * SMCI; PTCI = α * PCI + (1 − α) * TCI; SMTCI = α * SMCI + (1 − α) * TCI; MIDI = α * PCI + β * SMCI + (1 − α − β)TCI. Where, α and β represented the weight of single index while constituting the integrated drought indices; I = current time period (e.g., SMCI of this week), min = long-term minimum (e.g., SMCI minimum of this week over, say, the last ten years), max = long-term maximum (e.g., SMCI maximum of this week over, say, the last 10 years). The image shows drought detected by remote sensing indices for July of 2010 over northern China. (a), (c), (e), and (g) were 2003–2010 mean TRMM precipitation, VUA-NASA soil moisture, VUA-NASA land surface temperature, and MODIS NDVI for July; while (b), (d), (f), and (h) were drought detected by PCI, SMCI, TCI, and VCI, respectively, for July of 2010. Soil moisture and SMCI of forest were masked out in (c) and (d), while areas of SMCI/TCI in (d)/(f) corresponding to zero soil moisture/land surface temperature in July of 2010 were masked out too.

Source: Zhang and Jia, 2013.

optical, radar, microwave, hyperspectral, LiDAR, thermal, hyperspatial and as well as other modern sensors, data of which are now becoming more frequently available, there are opportunities to develop newer more specific and advanced drought indices that are based purely on remote sensing data and indices (e.g., Figure 14.4).

There is increasing trend in using drought indices based on: (1) multiple sensors and (2) integrating remote sensing indices with non-remote sensing indices. For example, Jiao et al. (2019) developed the Geographically Independent Integrated Drought Index (GIIDI), which combined TCI from the Moderate-resolution Imaging Spectroradiometer (MODIS), the Vegetation Condition Index (VCI) developed using the Vegetation Index based on Universal Pattern Decomposition method (VIUPD), the Soil Moisture Condition Index (SMCI) derived from the Advanced Microwave Scanning Radiometer–Earth Observation System (AMSR-E), and the Precipitation Condition Index (PCI) derived from the Tropical Rainfall Measuring Mission (TRMM). The GIIDI provided superior performance relative conventional drought indices across diverse temporal and spatial scales, and hence has the ability to monitor drought across a range of biomes and climates. Alahacoon and Edirisinghe (2022) reviewed and found that there are 111 drought indices, of which 44 belong to the traditional indices and 67 belong to remote sensing. They determined that the PDSI, SPI, and NDVI indices were the most popular in drought monitoring and were used equally for all parts of the world. The most popular drought indices include the Palmer Drought Severity Index (PDSI), the Standardized Precipitation Index (SPI), Vegetation Condition Index (VCI), the Temperature Condition Index (TCI) derived from Moderate Resolution Imaging Spectroradiometer (MODIS) data, the Precipitation Condition Index (PCI) derived from Tropical Rainfall Measurement Mission (TRMM) data, and the TCI and Soil Moisture Condition Index (SMCI) derived from Advanced Microwave Scanning Radiometer for the Earth Observing System (AMSR-E) data, as well as combined drought indices, including the Microwave Integrated Drought Index (MIDI), Optimized Vegetation Drought Index (OVDI), Optimized Meteorological Drought Index (OMDI), Scale Drought Conditions Index (SDCI), and Synthesized Drought Index (SDI), were analyzed and compared to evaluate their applicability (Zarei et al., 2023; Wei et al., 2021). Indeed, more specific drought indices can be developed that address specific issues of crops or other vegetation (e.g., Figure 14.3, 14.4). In Figure 14.3 a water deficit index (WDI) is compared to vegetation water stress index (VWSI) developed using Enhanced Thematic mapper Plus (ETM+)

near infrared (NIR) and short-wave infrared (SWIR) bands for wheat crop. Figure 14.4 show various other drought indices illustrated for regional-level studies. The integrated agricultural drought index (IDI), which describes the relationship between agricultural drought conditions and multiple meteo-hydrological variables such as precipitation, land surface temperature (LST), normalized difference vegetation index (NDVI), soil water capacity, and elevation are considered robust and effectively monitor drought onset, duration, extent, and intensity (Liu et al., 2020).

14.5 REMOTE SENSING OF DRYLANDS

Drylands cover ~40% of the global land area and support ~1/3rd of the world's population (Andela et al., 2013) with food and fiber or by providing grazing resources through rangelands. (e.g., Figure 14.5a,b). Agriculture in drylands is mostly rainfed and thus highly susceptible to droughts. At present, ~25% of the drylands are affected by drought every year (e.g., Figure 14.5c) and it is expected that these areas will increase to as much as 50% (see Chapters 1–4) due to climate change. Besides this direct provision of ecosystem services, rangelands are also rich in biodiversity, but may be prone to wildfires, depending on vegetation composition and climatic conditions. Even though many research initiatives aimed at assessing the status of drylands, the estimated rate of land degradation diverges greatly.

Global drylands cover extended areas (Figure 14.5) and are often hard to access and study; hence, remote sensing offers an ideal platform to observe, characterize, and study drylands. Further, image acquisitions over drylands are relatively easier than over tropical areas due to fewer cloudy days in a year. Chapter 5 by Dr. Marion Stellmes et al. provides us an overview of dryland studies using remote sensing. As the variety of studies is huge the authors focused on vegetation-related studies based on medium- to coarse-scale satellite imagery that provide regular information of the Earth's surface since the 1980s. The overview the authors provide covers the following major themes:

1. Assessing the extent of land degradation and desertification as well as the condition of ecosystems considering climate variability.
2. Monitoring conversions and modifications in dryland land cover based on biological productivity over long time periods.
3. Integrating remote sensing–derived results with the human dimension to identify drivers of land degradation.

In reviewing the previous studies, Chapter 5 by Dr. Marion Stellmes et al. discusses various satellites and sensors used in dryland studies, methods and approaches adopted, and the use of various vegetation indices. Many qualitative and quantitative measures of drylands studies are based on vegetation indices related to greenness, fraction of absorbed photosynthetic active radiation (faPAR), net primary productivity (NPP) or on techniques like for instance Spectral Mixture Analysis (SMA). The chapter also gives an overview of techniques used to assess conversions and modifications in drylands and evaluates these. These comprise a large variety of change detection methods based on multi-temporal classifications, annual time series approaches of vegetation cover or phenological dynamics. Moreover, short outlooks of new techniques and sensors are given that, e.g., comprise the combination of medium- and coarse-scale resolution data such as STARFM or future satellite based hyperspectral sensor platforms that will allow for the operational derivation of enhanced land degradation indicators such as topsoil organic matter or pigment and water content. The chapter also includes a critical review of uncertainties in land degradation assessment by remote sensing.

Resolution of imagery is a key factor in capturing vegetation greening in the drylands. For example, coarse spatial resolution remote sensing data such as AVHRR and MODIS miss the greening area compared with the Landsat data (Zhang et al., 2023). Sub-meter to 5 m very high spatial-resolution imagery (VHRI), for example, detect, map, and count individual trees in drylands such as African Sudan savannas, Sahel, and other drylands (Guirado et al., 2020). Specific narrowbands in hyperspectral data can help model and map specific quantities such as dryland biomass, plant moisture, and land cover. More complex vegetations such as species type are often mapped by data fusion. For example, dryland woody-cover species are mapped by fusing hyperspectral data with LiDAR (Norton et al., 2022). Future studies of drylands using remote sensing can be improved substantially, by (Zhang et al., 2023; Norton et al., 2022; Smith et al., 2019): (1) multisensor approach (e.g., optical, radar, LiDAR, hyperspatial, hyperspectral) to characterize, model, and map dryland quantities; (2) utilizing near-continuous observations (e.g., from geostationary satellites or constellations of polar-orbiting satellites) to capture changes in drylands; and (3) setting up ground sensors to collect data on a continuous basis to help in model development and product testing and validation.

14.6 DISASTER RISK MANAGEMENT THROUGH REMOTE SENSING

Disasters come in many forms and are of both natural and human origin. Natural disasters include cyclones, tsunamis, floods, droughts, volcanoes, earthquakes, fire, and landslides. Costs associated with disaster events have risen substantially since about 1995, but also vary strongly between years, currently typically ranging between US$100 and US$200 billion per year (Figure 14.6) (Mavrouli et al., 2023). Casualty numbers have been even more variable, with statistics skewed toward particularly devastating individual events that have claimed millions of lives, in particular flood and drought events in the 1920s and 30s (Mateos et al., 2023). Nearly, 95% of all disasters also take place in the economically developing world (Mateos et al., 2023; Mavrouli et al., 2023).

Remote sensing is widely used to study all phases of disasters: (1) hazard and risk assessment, (2) prevention and mitigation, (3) early warning, 4d) seen-event monitoring, (5) post-disaster response and damage assessment, and (6) reconstruction and rehabilitation (e.g., Figure 14.6). In Chapter 6, by Dr. Norman Kerle, different eras of disaster risk management with remote sensing are identified:

1. Early disaster damage mapping starting with the pioneering effort by George R. Lawrence to study the severe 1906 San Francisco earthquake, using an airborne kite-based camera.
2. The early satellite era driven by military interests, in particular the Corona, Argon, and Lanyard missions in the 1960s, imagery from which was only de-classified between 1995–2002.
3. The early civilian satellite era with the launch of TIROS in 1960, NOAA's VHRR in 1970, and Landsat in 1972.
4. New advanced era of satellite Earth observation beginning in the early 1990s with the launch of numerous advanced satellites, both by governments (e.g., Landsat, SPOT, IRS, various SAR systems from such as Radarsat to very recent Sentinels), and commercial platforms (e.g., IKONOS, QuickBird, GeoEye).

Disaster risk management (DRM) concepts and approaches, and the many uses of remote sensing data, are the centerpiece of Chapter 6 by Dr. Norman Kerle. These are broadly identified as follows:

1. Hazard assessment and risk identification. This involves identifying hazard elements at risk (EaR) that are highly diverse and range from the obvious, such as buildings or infrastructure, to features less frequently considered, such as places of cultural significance or heritage sites, national parks, sites of biological diversity, and so on.
2. Assessment of vulnerability, which is defined as the capacity for loss, and which can include aspects such as physical, social, and environmental, i.e., the damage that EaR will sustain in each hazard event.
3. Monitoring and early warning to detect potential hazardous situations once hazards are well understood through (1) and (2).

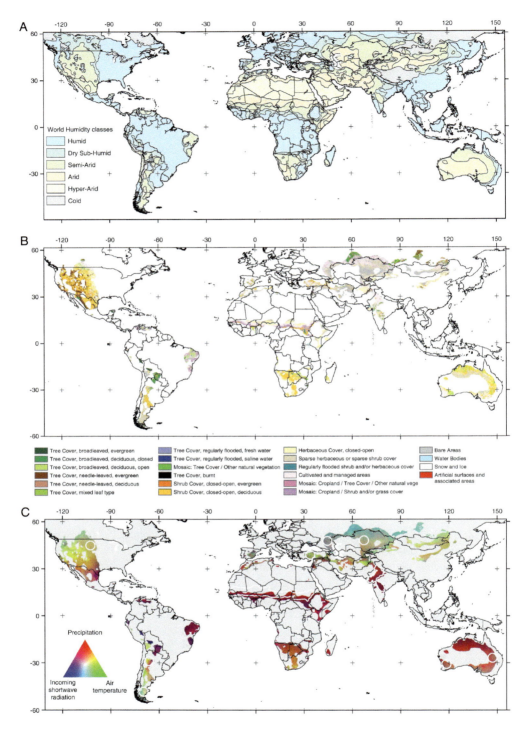

FIGURE 14.5 (A) Humidity classes as defined in "The world atlas of desertification" (UNEP, 1997). (B) Global land cover (GLC-2000) for the semi-arid areas (Bie et al., 2020). (C) Geographic distribution of potential climatic constraints to plant growth (Fensholt et al., 2012). Note: for the semi-arid areas overlaid by areas (white circles and lines) selected for analysis of seasonality by Fensholt et al. in their paper.

Source: Fensholt et al., 2012.

Summary Chapter for Remote Sensing Handbook, Volume VI

The efforts will help in hazard identification and mitigation, adaption to present hazards, and early warning for threatening events. Yet, in many cases hazards cannot be avoided, thus necessitating:

4. Post-disaster response and damage assessment.

Currently, remote sensing provides an ideal platform to study, assess, and respond to each of the above (1) to (4). First, for most parts of the world freely available, web-enabled baseline satellite image archives now exist. Second, numerous new satellite platforms in various spatial and spectral domains are being launched or developed. These satellites acquire data frequently, if tasked appropriately providing data of any land area on earth on any given day, by one satellite or another, and/or have the capacity to respond at short notice to acquire images over an area of interest. Third, advances in analysis methods, computing, and processing power of modern-day remote sensing makes it feasible to respond to disasters rapidly and accurately. Finally, there have been increasing global efforts, such as the International Charter "Space and Major Disaster" or the Global Earth Observation System of Systems (GEOSS), to make coordinated use of the growing set of remote sensing tools and instruments.

Chapter 6 does not deal with technological disasters, industrial disasters, or the complex humanitarian crisis situations associated with political or ethnic conflicts, nor with disasters such as famine that are of natural origin, but that predominantly result from political or societal failures.

The modern disaster management is fed by disruptive technologies such as the Internet of Things (IoT), image processing, artificial intelligence (AI), big data and smartphone applications (Munawar et al., 2022). All of this, of course, needs big-data analytics, machine learning, deep learning, artificial intelligence, and cloud commuting (Zhu and Zhang, 2022). Remotely sensed data fusion with data from multiple satellite sensors, enhances the credibility and reliability of disaster mapping and modeling for events such as wildfire (Sakellariou et al., 2022), floods (Zhu and Zhang, 2022), landslides (Casagli et al., 2023), and tropical cyclones (Hoque et al., 2017), to mention a few.

FIGURE 14.6 Concept of Sentinel Asia Step 3 with the goal to expand activities to cover all the disaster management cycle—mitigation/preparedness phase, response phase, and recovery phase—utilizing various and many satellites such as earth observation, communication, and navigation satellite, under further collaboration for operation and human networking by Sentinel Asia Joint Project Team.

Source: Kaku and Held, 2013.

14.7 HUMANITARIAN DISASTERS AND REMOTE SENSING

Humanitarian disasters can be both natural as well as human caused, and often causes are interrelated. Cyclones, tornadoes, earthquakes, floods, droughts, and any number of other events (e.g., internal and external conflicts, wars, and spread of rare diseases) can lead to great humanitarian disasters. Remote sensing provides neutral and unbiased data on most of the disasters. The ability of spaceborne remote sensing to repeatedly and consistently monitor disasters makes it the most appealing tool for disaster assessment and management, comprising short-term catastrophes with a clear peak (e.g., flood events) to long-term, protracted crises with severe impact on society and the environment on the longer run. Chapter 7, written by Dr. Stefan Lang and other experts in the field, provides an excellent introduction to all these topics. The authors show us how a wide array of remote sensing data, methods, and approaches can support humanitarian action and disaster mitigation. They provide case studies of monitoring displaced population and associated impact on the environment, detecting indications of destruction during conflicts and monitoring illegal mining and logging activities in relation to conflict situations. It is important to note that humanitarian disasters are often jointly coordinated by agencies such as the UN Office for the Coordination of Humanitarian Affairs (OCHA), United Nations High Commissioner for Refugees (UNHCR), World Food Programme (WFP), United Nations office for Disaster Risk reduction (UNISDR), United Nations Office for Outer Space Affairs (UNOOSA), United States Agency for International Development (USAID), and other National and International agencies, often through multi-national, multi-agency efforts. For example, the United Nations Platform for Space-based Information for Disaster Management and Emergency Response (UN-SPIDER) put together a large collection on remote sensing data gallery post the 2004 great Indian Ocean Tsunami.

Disasters require quick response at every level. The chapter shows the issues and importance of using OpenStreetMap (OSM) (Tzavella et al., 2024) where online volunteers help prepare maps with real time updates of events taking the typhoon in the Philippines in 2013 as an example. Whereas very high spatial resolution (e.g., sub-meter to 5 m) imagery such as QuickBird or WorldView is of great importance in assessing disasters (e.g., damaged buildings, bridges, refugee settlements), all types of remote sensing data can be useful in various types of disasters. Figure 14.7 shows the extracting of damages caused by the 2008 Ms 8.0 Wenchuan earthquake from the German TerraSAR-X (with X-band wavelength 31 mm, frequency 9.6 GHz, spatial resolution up to 1 m) data. Other slow progression disaster events such as droughts are monitored over large areas using NDVI or EVI from coarse resolution MODIS or moderate resolution Landsat. Flood events often need radar imagery to look through clouds since floods occur during heavy rainfall and cloud cover, ground penetrating radar, and thermal imagery to track people trapped in collapsed buildings can be very useful.

In emergencies and extreme disasters, remote sensing either from satellites or drones, often becomes the only source of data. Also, in such situation's rapid analysis of data from disparate sources requires machine learning and cloud computing for quick assessment and action. The Earth observation (EO)–based information services is sought in many emergencies such as refugee, IDP (internally displaced people) camps (Lang et al., 2018), human health (Greenough and Nelson, 2019), fire (Hassan et al., 2022), and earthquakes (Williams et al., 2018). Deep learning techniques (e.g., convolutional neural network [CNN]) can recognize ground objects from satellite images but rely on numerous labels for training for each specific task, which can be solved by fusing multiple freely accessible crowdsourced geographic data (Chen et al., 2019).

14.8 REMOTE SENSING OF VOLCANOES

In Chapter 8, Dr. Robert Wright provides an overview of volcanoes and the role of remote sensing in studying them. There are around 1500 active or potentially active volcanoes around the world of which roughly about 5% erupt every year.

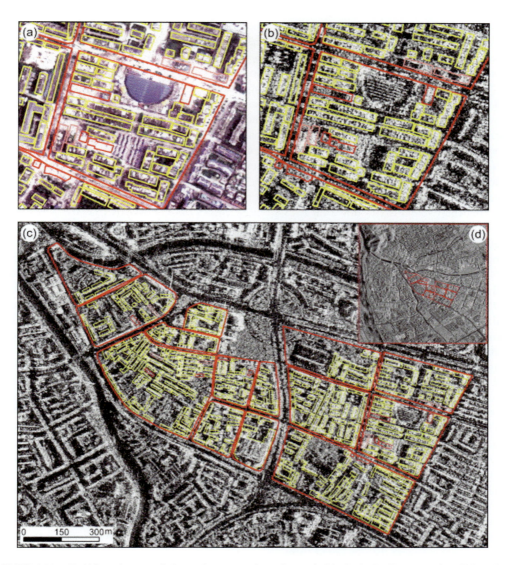

FIGURE 14.7 Building damages information extraction of sample blocks in Dujiangyan city, China, from aerial photograph and TerraSAR-X (with X-band wavelength 31 mm, frequency 9.6 GHz, spatial resolution up to 1 m) image, the GIS data of blocks and buildings are extracted from a pre-earthquake QuickBird image acquired on July 22, 2005: (a) aerial photograph of Hehuachi residential area acquired on May 18, 2008; (b) Backscattering coefficient image of TerraSAR-X of Hehuachi residential area acquired on May 15, 2008; (c) building damage information of sample city blocks extracted from TerraSAR-X image; and (d) index map showing the location of study blocks in Dujiangyan city. (Dong et al., 2011).

Chapter 8 shows us how remote sensing is used to study various volcanic processes. These include:

1. Topographic deformation (e.g., Figure 14.8) associated with sub-surface magma movements.
2. Measurement of the gases and ash released by volcanoes.
3. Geothermal and hydrothermal manifestations of active volcanism.
4. Lava flows and lava domes produced during effusive volcanic eruptions.

The phenomena associated with active volcanism are many and no single wavelength region provides data at which all (e.g., high temperature lavas, low temperature hydrothermal activity, passive emissions of gas, explosive eruptions of ash, deformation of the volcanoes themselves) can be studied adequately. Furthermore, changes in these processes occur on both short and long timescales (e.g., daily to interannual), as well as over large spatial extents (e.g., meters to hundreds of kilometers) (e.g., Figure 14.8). By leveraging data acquired by many sensors, at wavelengths from visible to microwave wavelengths, from both low earth and geostationary orbits, scientists have developed techniques for quantifying much about active volcanic processes from space, taking advantage of the synoptic, repeated coverage offered by the remote sensing technique.

No single mission has been launched to specifically study volcanism, so volcanologists have exploited a wide range of sensors for this purpose. Interferometric synthetic aperture radar (InSAR) data, which revolutionized the study of volcano geodesy, uses synthetic aperture radar data sets, including the C-band SARs flown on ERS-1, ERS-2, ENVISAT, RADARSAT, the X-band SAR on board TerraSAR-X, and the L-band SAR carried on board the Advanced Land Observing Satellite "DAICHI" (ALOS). Studies of volcanic gases have relied heavily on the Total Ozone Mapping Spectrometer (TOMS) series of sensors, the Ozone Mapping Instrument (OMI) on-board NASA's Aura, the Scanning Imaging Absorption Spectrometer for Atmospheric Chartography (SCHIAMACHY) sensor onboard ENVISAT, and the Global Ozone Monitoring Experiment 2 (GOME-2) onboard MetOp-A, in addition to the Terra ASTER and Aqua AIRS instruments, using a mixture of ultraviolet and long-wave infrared measurements, acquired at a range of spatial resolutions. Sensors such as MODIS, AVHRR, and GOES have been the workhorses of volcanic ash detection, due to their ability to discriminate silicate ash clouds from water clouds based on their contrasting long-wave infrared transmissivities, in a timely manner. The Landsat Thematic Mapper (and its successors), the Terra ASTER (Advanced Spaceborne Thermal Emission and Reflection radiometer), and the EO-1 Hyperion have been widely used for measuring the thermal properties of high-temperature lavas. Low spatial (but high temporal resolution) sensors such as MODIS, AVHRR, and GOES have been utilized to detect the onset of volcanic eruptions around the globe in near real-time, based on their thermal emission signatures. Although low temperature targets utilize the longwave infrared (8–14 µm), studies of active lavas also exploit the middle (3–5 µm) and shortwave (1.2–2.2 µm) infrared.

The volcano characteristics pre-eruption, during eruptions and post-eruptions require variety of remote sensing data. Satellite monitoring of volcanic activity typically includes four primary observations (Poland et al., 2020): (1) deformation and surface change, (2) gas emissions, (3) thermal anomalies, and (4) ash plumes. Ganci et al. (2020) showed these volcano activities are studied using distinct remote sensing data gathered from different portions of the electromagnetic spectrum: (1) time-averaged discharge rates (TADRs) from coarse-resolution imagery (e.g., MODIS, SEVIRI), (2) time-varying evolution of lava flow emplacement from moderate to fine resolution multispectral imagery (e.g., EO-ALI, Landsat, Sentinel-2, ASTER), (3) lava flow thickness variation from the topographic monitoring by using stereo or tri-stereo optical data (e.g., Pléiades, PlanetScope, ASTER). Coppola et al. (2020) developed a comprehensive assessment of thermal remote sensing for global volcano monitoring: experiences from the Middle Infrared Observation of Volcanic Activity (MIROVA) system, an automatic volcano hot spot detection system, based on the analysis of MODIS (Moderate Resolution Imaging Spectroradiometer) data.

14.9 BIOMASS BURNING AND GREENHOUSE GAS EMISSIONS STUDIED USING REMOTE SENSING

Biomass burning contributes to anywhere from 2290–2714 Tg C/year (Ito and Penner, 2004, Mieville et al., 2010) when compared with 8,180 Tg C/year by fossil fuel combustion, cement production, and gas flaring (Barker et al., 2007). Chief sources of biomass burning are savannas (42.1%),

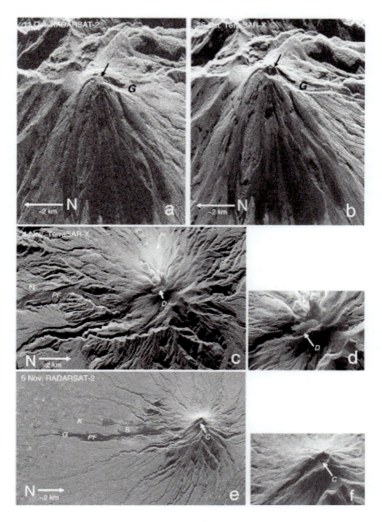

FIGURE 14.8 Synthetic Aperture Radar (SAR) images of the summit of Merapi volcano before and after the times of the October 26, 2009, explosive eruption and the November 4 explosive eruptions. For clarity, images are oriented with respect to line of sight of the radar. Arrows indicate north (N) direction and approximate scale. G (Kali Gendol), K (Kali Kuning), Kj (Kinahrejo). a. RADARSAT image, October 11, 2009. Arrow indicates the 2006 lava dome. b. TerraSAR-X image, October 26, showing new summit crater (arrow) produced by explosive eruption of October 26. c. TerraSAR-X image, November 4, 2010, was showing large (~ 5×10^6 m^3) lava dome (D). Pyroclastic flow deposits (PF) from the October 26 eruption appear dark in the radar images. d. Enlargement of the summit area of image a. e. RADARSAT image of November 5, 2010, showing pyroclastic flow deposits (PF, dark gray) and surge deposits (S, light gray). These deposits formed earlier during the main phase of the November 4–5 explosive eruption. An enlarged, elongate crater, produced by the November 4–5 eruption is also evident at the summit. f. Enlargement of the summit area of image c.

Source: Surono, 2012.

agricultural wastes (23.1%), tropical forests (14.5%), fuelwood (16.2%), temperate and boreal forests (3.3%), and charcoal (1%) (Andreae, 1991; Andreae and Merlet, 2001). Biomass emissions releases wide array of greenhouse gases such as carbon dioxide, carbon monoxide, methane, nitric oxide, ammonia, sulfur, and hydrogen (Andreae, 1991; Andreae and Merlet, 2001). These trace gases in atmosphere are best quantified through their absorption characteristics in specific wavelengths, with wavebands in visible, near infrared, and shortwave infrared. Over the years, a number of satellite sensors have been used to study trace gases. These include, Geostationary Operational Environmental Satellites (GOES) fire radiative power (FRP), Global Ozone Monitoring Experiment (GOME) sensor on board ERS-2, MODIS (e.g., MODIS burned area product), fire counts from the Tropical Rainfall Measuring Mission (TRMM)—Visible and Infrared Scanner (VIRS), and European Remote Sensing Satellites (ERS) Along Track Scanning Radiometer (ATSR) sensors.

In Chapter 9, Dr. Krishna Prasad Vadrevu and Dr. Kristofer Lasko show us how nitrogen dioxide (NO_2) emissions in troposphere are measured using Ozone Monitoring Instrument (OMI) on board Earth Observing Systems (EOS) Aura satellite, SCanning Imaging Absorption Spectrometer for Atmospheric Cartography (SCIAMACHY), and MODIS—Aerosol Optical Depth (AOD) signal varies in relation to MODIS fire retrievals. They demonstrate how one of the trace gases, NO_2, is measured, modeled, and mapped from biomass burning of subtropical evergreen forests of northeast India. The concept to measure other trace gases using sensor data is similar. Specific to south Asia, atmospheric burning the related trace gases in atmosphere is maximum in months of March, April, and February and minimum during the month of July (e.g., Figure 14.9). This is because of post-harvest fire activity and also the slashing and burning of forests during the summer, before the start of the rainy season in June in the tropics. Chapter 9 clearly shows that OMI performed better than SCIAMACHY by: (1) detecting higher NO_2 concentrations, (2) providing stronger correlation of MODIS fire counts, and (3) establishing stronger correlation with MODIS aerosol optical depth. The study highlights satellite remote sensing of NO_2 from biomass burning sources.

The TROPOspheric Monitoring Instrument (TROPOMI) carried on board ESAs GMES Sentinel 5 Precursor mission, launched on October 13, 2017, allows researchers to measure a large range of gases, including the most important trace gases, aerosols, and clouds (www.tropomi.eu/). TROPOMI is a nadir viewing spectrometer with bands in the ultraviolet, the visible, the near infrared and the shortwave infrared (www.tropomi.eu/). There are also airborne systems such as the Autonomous Modular Sensor-Wildfire (AMS) airborne multispectral imaging system.

The Paris Agreement calls on parties to undertake ambitious efforts to combat climate change by engaging in appropriate policies and measures as put forward through nationally determined contributions (NDCs), to strengthen transparency when reporting their greenhouse gas (GHG) emissions and to increase their mitigation contributions to climate action from 2020 (Prosperi et al., 2020). They use MODIS collection 6 data to estimate GHG emissions from burned areas. Liu and Popescu (2022) demonstrated clear advantages of integrating multiple satellite sensor data for biomass burning emissions. They combined the Ice, Cloud, and land Elevation Satellite-2 (ICESat-2) Land and vegetation height product (ATL08) with Landsat 8 data, and Sentinel-1 data to estimate biomass-burning emissions. Using MODIS global BA product Deshpande et al. (2023) showed GHG from agricultural residue burning has increased by a staggering 75% since 2011 in India. Similar studies using MODIS-based biomass-burning estimates have been published for tropical continents by Shi et al. (2020) and many others for other areas.

14.10 COAL FIRES STUDIES USING REMOTE SENSING

Coal contributes to ~1/3rd of today's global energy needs, meets ~40% of world's electricity needs with consumption of about ~8000 million metric tons per year (Klinlampu et al., 2023). Some even expect it to be the main energy source by 2025 beating petroleum (Belaïd et al., 2023). But this paradigm may shift and change if coals reputation of being "dirty fuel" continues. Many drawbacks of coal energy include greenhouse gas (GHG) emissions, acid rains, coal fires, and pollution of river waters and air. Yet, the importance of coal to global energy needs can't be disputed, possibly for another 50–100 years, by which time technologies may find answers for cleaner fuels to meet the global energy demand.

Coal fires are a recurring phenomenon and are caused by both natural and anthropogenic causes. One of the major contributors to GHG comes through CO_2, and overwhelmingly from burning coal.

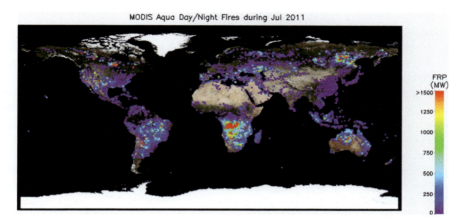

FIGURE 14.9 Fire detection from Aqua MODIS for July 2011 overlaid on a composited surface reflectance map (also from MODIS), showing the fire radiative power (FRP) value ranges for the individual fire pixels. Compared to the map scale, the fire pixels are indicated with relatively large dots to enhance visualization, causing substantial fire-pixel overlap in certain regions.

Source: Ichoku et al., 2012.

The other GHG gases released by coal include CO, CH_4, SO_x, and NO_x. Coal takes millions of years to form, and burning it will waste a non-renewable resource for ever. As a result, the study of coal fires acquires great importance. Stracher et al. (2010) provide a comprehensive global perspective of coal fires.

Remote sensing has played a key role in coal fire detection, characterization, mapping, and monitoring from beginning of satellite sensor era of the early 1960s. Chapter 10 by Dr. Anupma Prakash and Dr. Claudia Kuenzer presents remote sensing data, approaches, and methods used for coal fire detection, characterization, mapping, and monitoring. The chapter begins with an overview of the key regions of the electromagnetic spectrum used in coal mining as well as coal fire applications. These include: (1) visible (0.4–0.7 µm) spectrum for land cover/land use, geomorphological faults, (2) NIR (0.7–1.3 µm) spectrum for vegetation, soil moisture, and gas emissions; (3) SWIR (1.3–3.0 µm) and MIR (3.0–8.0 µm) spectrum for soil composition, associated forest fires, high temperature surface coal fires; (4) TIR (8.0–14 µm) spectrum for underground coal fires, fire depths, gas emissions; and (5) microwave (0.75–1.0 µm) spectrum for faults and subsidence. The TIR, MIR, and SWIR bands are most important in studying surface and sub-surface fires. Whereas TIR is most widely used, other wavebands such as MIR (e.g., Figure 14.10) are also widely used. Often, making use of daytime and nighttime images and looking at diurnal temperature anomaly. Satellite sensors such as NOAA AVHRR, Terra/Aqua MODIS, Landsat-8 OLI, BIRD (e.g., Figure 14.10), and ASTER carry wavebands in visible, NIR, SWIR, MIR, and TIR and can be ideal platforms for study of coal mining as well as coal fires. In addition, ECOSTRESS has five thermal bands in 8–12.5 µm range that are invaluable. However, limited spatial resolution of some of these sensors (e.g., AVHRR, MODIS) can be a problem.

Some of the key messages in the coal fire study using remote sensing, gathered from Chapter 10 by Dr. Anupma Prakash and Dr. Claudia Kuenzer are:

1. Surface fires are best studied using SWIR and MIR images where high temperatures of fires are easily detected.
2. Subsurface coal fires are best studied using nighttime TIR images where feasible. However, given the paucity of nighttime TIR images, one will have to, mostly, depend on daytime TIR images. Sub-surface fires are hard to map as they are detected using indirect indicators

such as heated ground and smoke from cracks/vents. Also, TIR thresholds of sub-surface fires can vary based on the region of study and there is uncertainty in this aspect, especially when other surfaces (e.g., shale, coal dust) can have similar TIR derived temperatures. But a continuous watch on coal mining areas and/or TIR images of nighttime, will help detect these indirect indicators.
3. Remote sensing is a powerful tool for early detection, mapping, and continual monitoring of coal fires due to its repeated coverage of an area. Coal fires are often concentrated in an area and their movement is slow. However, they may burn for very long time (e.g., at times even decades as in Centralia, PA, USA). So, remote sensing is a great tool for their continual monitoring.
4. Key coal fire monitoring factors using remote sensing are to: (a) understand which time of the day or night (or both) and/or which season is the best to acquire images; and (b) establish fire areal extent, duration, variability, and extinction.
5. Scope for further studies in global coal fires is substantial. For example, a global standardized approach to coal fire mapping and monitoring using remote sensing does not exist, mainly as a result of too few people working in the area.
6. Further, the need for automated or semi-automated methods and approaches is required.

An efficient investigation approach for coal fire detection over large areas can be done effectively by through integration of Landsat-8 thermal infrared band 10 data and radar interferometry data derived from Sentinel-1 C-band images (Karanam et al., 2021; Yan et al., 2020). Surface and sub-surface fires are detected using Landsat thermal data and land subsidence using PS-InSAR data (Karanam et al., 2021).

14.11 URBAN REMOTE SENSING

Global urban areal extent estimates vary widely from anywhere between 27 million hectares (vector maps), 65.8 million hectares (MODIS 500 m estimates), 72.7 million hectares (MODIS 1 km), 30.8 million hectares (global land cover 2000 or GLC2000), 32.2 million hectares (GlobCover), and 350 million hectares or 2.7% of the terrestrial earth area (Global Rural-Urban Mapping Project, or GRUMP) (see Pérez-Hoyos et al., 2017). Such wide variations in estimates are not surprising given the various definitions, datasets, approaches, and methods used in determining the areal extent. But, what is clear is that urban areas occupy a very small portion (< 3% by any estimate) of the land cover of the total terrestrial area (14.894 billion ha) and is also very small compared to other land covers such as the tree covered areas (27.7%), bare soils (15.2%), grasslands (13%), croplands (12.5%), snow and glaciers (9.7%), shrub-covered areas (9.5%), and sparse vegetation (7.7%) (Source: FAO's Global Land Cover-SHARE of year 2014—Beta-Release 1.0) (Pérez-Hoyos et al., 2017). However, urban areas are comparable to other land cover are: inland water bodies (2.6%), herbaceous vegetation (1.3%), and mangroves (0.1%). The FAO map shows artificial surfaces (includes all urban) as < 0.6%. However, global cities are sources as much as 70% of all GHG emissions as pointed out by Dr. Hasi Bagan and Dr. Yoshiki Yamagata in Chapter 11. Cities are also cause of much of deforestation as a result of various demands of urban dwellers (e.g., housing, furniture, paper) resulting in CO_2 release into the atmosphere. In the last decades, many great cities have grown (e.g., Figure 14.11). Remote sensing offers the best opportunity to map and assess the urban sprawl (e.g., Figure 14.11).

Wide arrays of remote sensing data are increasingly used to study many urban issues (see Taubenböck and Esch, 2011). These include urban: (1) areal extent, (2) structures (e.g., buildings, roads), (3) land use, (4) temperature, (5) human population, (6) peri-urban agriculture, (7) gardens and trees, (8) golf courses, (8) lawns, (9) hydrology, (10) sewage and drainage systems, (11) planning, and (12) CO_2 emissions. Based on the characteristics of remote sensing data, its use in urban studies is determined. For example, very high spatial resolution data can be used to detect and map individual buildings and road networks; thermal data in study of urban heat islands, multispectral data

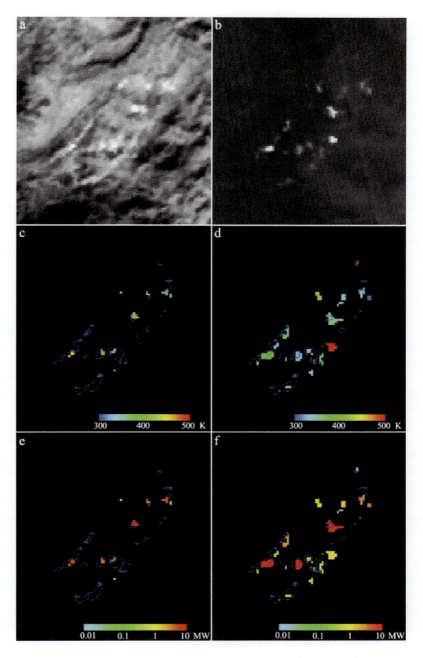

FIGURE 14.10 Coal fire detection and quantitative characterization based on the German Aerospace Center (DLR's) small satellite Bi-spectral Infrared Detection (BIRD), an experimental fire remote sensing satellite launched in October 2001, has three nadir looking bands at the wavelength of 0.84–0.90 μm [near infrared (NIR)], 3.4–4.2 μm [middle infrared (MIR)] and 8.5–9.3 μm [thermal infrared (TIR)] with a pixel spacing of 185 m. BIRD images obtained on September 21, 2002, at daytime (left) and on January 16, 2003, at night (right). (a) Daytime MIR image; (b) nighttime MIR image; (c) effective fire temperature of detected hot spots at daytime (~11 a.m. local time); (d) effective fire temperature of detected hot spots at night (~11:00 p.m. local time); (e) radiative energy release of detected hot spots at daytime; (f) radiative energy release of detected hot spots at night. Blue contours and crosses show location of coal seam fire verified on ground in September 2002, while purple triangles show location of industrial chimneys.

Source: Voigt et al., 2004.

in determining urban growth, and so on. In Chapter 11, Dr. Hasi Bagan and Dr. Yoshiki Yamagata showed us that the:

1. Use of multi-temporal Landsat data in urban studies. Ability to map major land use/land cover with emphasis on urban sprawl and its change over time and space.
2. Landsat remote sensing–derived urban/built-up changes for 1972–2011 had significant: (a) positive correlation with population density of 1970–2010 and (b) significant negative correlation with croplands during 1972–2011.
3. Defense Meteorological Satellite Program (DMSP) Operational Linescan System (OLS) nighttime lights a sensors image (~1 km) are highly correlated with population density, but does have problems of saturation at very high population densities. DMSP data is in 0 to 63 digital numbers, and the relationship begins to saturate over 55.
4. How human activity and growth of cities related to new economic activity can be studied routinely and accurately using multi-date, multispectral imagery of ~30 m spatial resolution or better.
5. Advance methods of image classification for urban areas such as subspace methods and grid cell processing.

What is clear from their study is the presence of large archives of Landsat and other remote sensing data over large cities like Tokyo. This in itself allows anyone to look through the archive and study various aspects of a city's growth. In many parts of the world rapid urbanization has occurred over the last one to two decades. So changes from the past will be massive and the availability of digital archive of remote sensing data from sensors such as Landsat, IRS, SPOT VGT, as well as very high sub-meter to few meter resolution commercial imagery makes decadal urban studies feasible, consistent, and powerful. Recent advances on Object-based image analysis (OBIA) (Weng and Quattrochi, 2006) are crucial for more refined and powerful approaches of urban studies using remote sensing.

Landsat, Sentinel, IRS, SPOT, and many other remote sensing data can be used to extract many urban features, such as urban built-up areas, and urban land cover/land use. Several other features like road networks and power lines can be extracted by sub-meter to 5 m very high-resolution imagery (VHRI), such as from Planet Labs Doves and super Doves, IKONOS, QuickBird, GeoEye and many such imagery. Data integration is the key. For example, Ma et al. (2020) integrated Landsat-8 remote sensing images used for extracting urban built-up areas by supervised neural network classifications and Geographic Information System tools, while cellular signaling data from China Unicom Inc. was used to depict human activity areas generated by spatial clustering methods. Urban heat islands (UHI) are best studied using land surface temperature (LST) data derived from Landsat or other thermal bands. The diverse urban land change characteristics of 30 global megacities can be derived from monthly nighttime light time series from VIIRS data (Zheng et al., 2021). Landsat, Sentinel and similar data also play critical role in mapping urban green space patterns and dynamics (Kuang and Dou, 2020). The COVID-19 epidemic lockdown of cities directly decreased urban socioeconomic activities, which can be captured using remotely sensed nighttime light (NTL) data (Xu et al., 2021).

14.12 REMOTE SENSING IN DESIGN OF SMART CITIES

Sensors are becoming ubiquitous, and there are many expectations that the next level of connectivity will be through smart sensors. Earlier eras of connectivity are broadly classified into age of: (1) emails and web portals, and (2) networks and social media. Now, the next in connectivity is likely to be through smart sensors. Plants can have sensors when to irrigate or fertilize automatically; vehicles can have sensors for self-driving; homes can have various types of sensors to manage the

Summary Chapter for Remote Sensing Handbook, Volume VI **449**

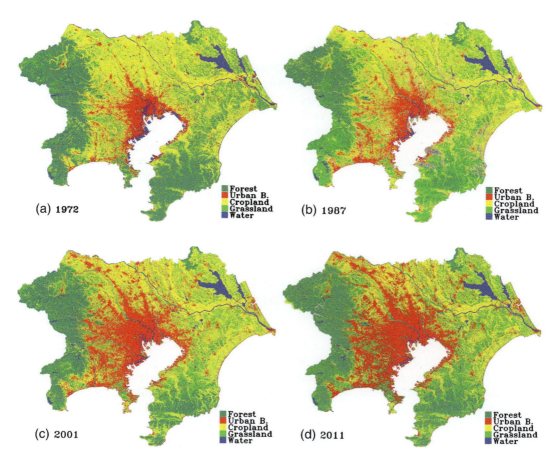

FIGURE 14.11 Time series of land use and land cover maps of the Tokyo and surroundings using Landsat images of: 1972, 1987, 2001, and 2011.

Source: Bagan and Yamagata, 2012.

house, including robots driven by sensors to vacuum or clean dishes; cities can have sensors from various platforms (ground-based, airborne such as UAVs, and spaceborne satellites) gathering data for infrastructure management, CO_2 pollution, aerosols, traffic control, and city planning. Further, megacities of today are major contributors of anthropogenic induced aerosols and CO_2 pollution. Aerosol optical thicknesses are studies using spaceborne sensors such as MODIS. LiDAR technologies are used for 3D modeling of buildings and trees. A wide array of satellite-borne sensors are routinely used in studies such as urban sprawl and urban heat islands. In addition to remote sensing data, GIS data such as street maps are mapped and updated using GPS through mechanisms like crowdsourcing. Locations of sites mapped, tracked, and updated using GPS. Even, the delivery of packages may happen through UAVs. This collection of rich sets of sensors driven management and services of cities that includes infrastructure (e.g., roads, buildings, sewage, drainage, electricity), surveillance, delivery of services, biodiversity, and habitat preservation (e.g., Figure 14.12), traffic management, and other services like health care is called Smart Cities (see Hancke et al., 2013). Smart cities require us to ensure preserving the richness of habitats (e.g., Figure 14.12), ensuring efficient movement of traffic, avoidance of pollution, efficient use of resources, and a host of other requirements. This is especially important in the 21st century when there is a swift increase in urban population around the world.

Chapter 12 by Dr. Yoshiki Yamagata et al. provides a case study of Yokohama and how the city is managed as a smart city using high-resolution remote sensing and GIS (Li et al., 2020). A wide array of remote sensing data is used to gather information on CO_2 emissions, heat emissions from buildings and transportation (road), green space ratio, and building density using various remotely sensed data. In Chapter 12, Dr. Yoshiki Yamagata et al. demonstrate the use of remote sensing for urban planning through studies of: (1) land use and (2) view. Specifically, they demonstrate:

1. LiDAR 3D modeling will help viewshed analysis to determine which buildings have full sunlight and which have less, impacting solar radiation for solar photovoltaic panels (PVs) on roof top. LiDAR is also used to determine the economic view value from buildings (e.g., ocean view versus street view). Tree heights in cities is another application of LiDAR remote sensing.
2. Solar radiation estimation on the roof top is crucial for electricity generation from PVs on roof top. They use high-resolution imagery to detect rooftop area and use LiDAR data to determine building heights.
3. DEM data, along with remote sensing data, adds value in many of these applications.
4. Determining the cost of land through remotely sensed derived information (e.g., PVs are installed in low-rent suburban regions).

Smart cities are the future and become critical with nearly 90% of the world's population going to live in cities by 2100 as per United Nations estimates (O'Sullivan, 2023; Hoornweg and Pope, 2017). The goal of smart cities is to make life of habitants convenient, safe, healthy, and inclusive. Technologies like sensor systems, wireless sensor networks (WSNs), artificial intelligence (AI), the Internet of Things (IoT), machine learning (ML), deep learning (DL), and big-data analytics have contributed immensely to the realization of this dream. Wireless sensor networks (WSNs) shape the Internet of Things (IoT) or Internet of Everything (IOE) in present-day smart city management (Ullah et al., 2020). The smart cities require extensive sensor network to monitor events and people continuously and help respond to events and emergencies rapidly. Sensors are placed everywhere these days cities (e.g., malls, airports, roads), and data is gathered from various remote sensing platforms (e.g., drones, aircrafts, spacecraft) and in many spatial resolutions and spectral wavebands. Although the use of these sensors is diverse, their application can be categorized in six different groups: energy, health, mobility, security, water, and waste management (Ramírez-Moreno et al., 2021). Smart cities must have an integrated geospatial database (Li et al., 2020) of various data layers from which any information can be derived, decisions made through instant spatial analysis, and actions taken to address all issues.

14.13 NIGHTTIME LIGHT REMOTE SENSING

Nighttime light remote sensing provides social scientists a powerful yet convenient tool to monitor human societies from space. Specifically, applications of nighttime light remote sensing include: (1) urban extents; (2) population studies; (3) adverse effects of light at night on human health, ecology, and astronomy; (4) economic growth center and poverty studies (e.g., Figure 14.13); (5) greenhouse gas (GHG) emissions; and (6) disaster management. Intensity of lights can be used to gather information on population concentration and the spread of light indicates the area extent of the cities, towns, and even villages. Nightlight data from satellites are used in economic studies (e.g., where there are street networks, markets, and growth in general) and poverty assessment especially in data scarce regions of the world. Cities contribute nearly 70% of global GHG emissions. Night light remote sensing helps in assessment of GHG emissions from specific locations

Summary Chapter for Remote Sensing Handbook, Volume VI

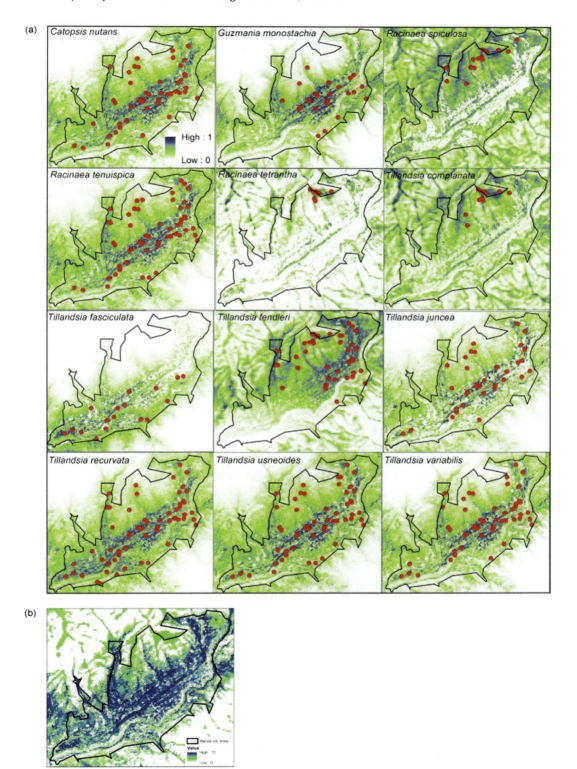

FIGURE 14.12 Smart cities need to be designed with great efficiency in terms of its planning, energy use, and a number of other factors. In this figure increasing urbanization on biodiversity in tropical regions is

FIGURE 14.12 (*Continued*) illustrated for: (a) Occurrences (dots) and modeled distributions of Bromeliaceae species in the city of Mérida (polygon). (b) Modeled species diversity. The study combined a rapid species assessment approach with environmental niche modeling based on high-resolution Advanced Spaceborne Thermal Emission and Reflection Radiometer (ASTER) satellite imagery to predict species distributions of Bromeliaceae in the city of Mérida, Venezuela.

Source: Judith et al., 2013.

of cities. For example, even though population dense cities are known to contribute less to GHG emissions, the benefit goes away when cities have extensive suburbs (Jones and Kammen, 2014). This is because dense settlements mean people walk to locations and drive less, use space and energy more efficiently and less consumptively per person, and a denser population also means lesser land disturbance elsewhere.

The chapter presents how nighttime light remote sensing is carried out by low-light imaging capabilities. For example, visible and NIR radiance down to 10^{-9} watts/cm^2·sr·μm at night, which is more than four times fainter than the minimal detectable VNIR radiances from satellite sensors optimized for daytime observation of reflected solar radiance (Elvidge et al., 1997). This chapter shows us, with clear illustrations of a wide array of applications, how night light remote sensing was pioneered by a series of Defense Meteorological Satellite Program (DMSP) Operational Linescan System (OLS) data for nearly four decades since the early 1970s. Following the DMSP program, NASA and NOAA launched in 2011 the Suomi National Polar Partnership (SNPP) satellite carrying the first Visible Infrared Imaging Radiometer Suite (VIIRS) instrument. The DMSP/OLS night light data is gathered, globally, data daily in the visible/near-infrared (VNIR, 400–1100 nm) and thermal infrared (10500–12600 nm) wavebands in 3000 m spatial resolution and six bits radiometric resolution. In contrast, VIIRS data is in 740 m spatial resolution, also daily and globally in 14 bit with a VNIR band (505–890 nm). VIIRS offers a substantial number of improvements over the OLS in terms of spatial resolution, dynamic range, quantization, calibrations, and the availability of spectral bands suitable for discrimination of thermal sources of light emissions (Elvidge et al., 2013). Chapter 26 provides an overview of additional sensors offering medium and high spatial resolutions, used in night light remote sensing. However, global studies are best carried out for the present (2011–present) using NOAA Suomi VIIRS and for the past (1992–2012) using DMSP/OLS. Indeed, numerous applications of nighttime remote sensing data such as from DMSP/OLS have appeared in the literature over the years (e.g., Figure 14.13). This number is expected to increase in the near future.

In a review of the 50 years of nightly global low-light imaging satellite observations Elvidge et al. (2022) determined that the Earth Observation Group (EOG) at NOAA has produced 65 annual global nighttime light products and over 650 monthly products. The night light data has many applications. For example, Zhao et al. (2022) generated a global dataset of annual urban extents (1992–2020) from harmonized nighttime lights. Chen et al. (2022) showed the potential of nighttime light remote sensing data to evaluate the development of digital economy. Gu et al. (2022) developed a GDP forecasting model for China's provinces using nighttime light remote sensing data. Wu et al. (2022) explored the effect of urban sprawl on carbon dioxide emissions. The recent proliferation of nighttime light (NTL) sensors, algorithms, and products create new opportunities to understand contemporary urbanization and the associated socioeconomic and environmental changes (Zheng et al., 2023).

14.14 ACKNOWLEDGMENTS

I would like to thank the lead authors and co-authors of each of the chapters for providing their insights and edits of my chapter summaries. Any use of trade, firm, or product names is for descriptive purposes only and does not imply endorsement by the US government.

Summary Chapter for Remote Sensing Handbook, Volume VI

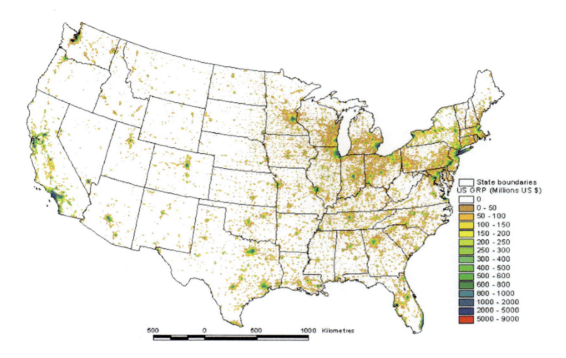

FIGURE 14.13 Map of estimated economic activity based on Defense Meteorological Satellite Program's Operational Linescan System (DMSP-OLS) radiance-calibrated nighttime lights for the United States.

Source: Doll et al., 2006.

REFERENCES

Abatzoglou, J.T., McEvoy, D.J., and Redmond, K.T. 2017. The West wide drought tracker: Drought monitoring at fine scales. *The American Meteorological Society*, 1815–1820. https://doi.org/10.1175/BAMS-D-16-0193.1

Akasheh, S.Z., Neale, C.M.U., Anderson, M.C., Hain, C.R., Roberti, D.R., Souza, V.A., Goncalves, I.Z., and Schull, M.A. 2023. Regional daily evapotranspiration estimation using remote sensing data and atmospheric-land exchange inverse energy model in Brazil. *Proc. SPIE12727, Remote Sensing for Agriculture, Ecosystems, and Hydrology*, XXV, 127270X. https://doi.org/10.1117/12.2680247.

Alahacoon, N., and Edirisinghe, M. 2022. A comprehensive assessment of remote sensing and traditional based drought monitoring indices at global and regional scale, Geomatics. *Natural Hazards and Risk*, 13(1), 762–799. http://doi.org/10.1080/19475705.2022.2044394

Andela, N., Liu, Y.Y., van Dijk, A.I.J.M., de Jeu, R.A.M., and McVicar, T.R. 2013. Global changes in dryland vegetation dynamics (1988–2008) assessed by satellite remote sensing: Comparing a new passive microwave vegetation density record with reflective greenness data. *Biogeosciences*, 10, 6657–6676. http://doi.org/10.5194/bg-10-6657-2013

Andreae, M.O. 1991. *In Global Biomass Burning: Atmospheric Climatic and Biospheric Implications*. Ed. J.S.E. Levine. The MIT Press, Cambridge, MA.

Andreae, M.O., and Merlet, P. 2001. Emissions of trace gases and aerosols from biomass burning. *Global Biogeochemical Cycles*, 15, 955–966.

Bagan, H., and Yamagata, Y. 2012. Landsat analysis of urban growth: How Tokyo became the world's largest megacity during the last 40 years. *Remote Sensing of Environment*, 127, 210–222. ISSN 0034-4257. http://doi.org/10.1016/j.rse.2012.09.011

Barker, T., Bashmakov, I., Bernstein, L., Bogner, J.E., Bosch, P.R., Dave, R., Davidson, O.R., Fisher, B.S., Gupta, S., Halsnæs, K., Heij, G.J., Kahn Ribeiro, S., Kobayashi, S., Levine, M.D., Martino, D.L., Masera, O., Metz, B., Meyer, L.A., Nabuurs, G.J., Najam, A., Nakicenovic, N., Rogner, H.H., Roy, J., Sathaye, J., Schock, R., Shukla, P., Sims, R.E.H., Smith, P., Tirpak, D.A., Urge-Vorsatz, D., and Zhou, D. 2007.

Technical summary. In: Metz, B., Davidson, O.R., Bosch, P.R., Dave, R., and Meyer, L.A. (Eds.), *Climate Change 2007:Mitigation. Contribution of Working Group III to the Fourth Assessment Report of the Intergovernmental Panel on Climate Change.* Cambridge University Press, Cambridge and New York.

Bayissa, Y., Tadesse, T., and Demisse, G. 2019. Building a high-resolution vegetation outlook model to monitor agricultural drought for the Upper Blue Nile Basin, Ethiopia. *Remote Sensing*, 11, 371. https://doi.org/10.3390/rs11040371

Belaïd, F., Al-Sarihi, A., and Al-Mestneer, R. 2023. Balancing climate mitigation and energy security goals amid converging global energy crises: The role of green investments. *Renewable Energy*, 205, 534–542. ISSN 0960–1481. https://doi.org/10.1016/j.renene.2023.01.083 https://www.sciencedirect.com/science/article/pii/S0960148123000927

Bie, Q., Luo, J., and Lu, G. 2020. Accuracy performance of three 10-m global land cover products around 2020 in an Arid Region of Northwestern China. *IEEE Access*, 11, 133215–133228, 2023. http://doi.org/10.1109/ACCESS.2023.3336733

Casagli, N., Intrieri, E., Tofani, V. et al. 2023. Landslide detection, monitoring and prediction with remote-sensing techniques. *Nature Reviews Earth & Environment*, 4, 51–64. https://doi.org/10.1038/s43017-022-00373-x

Chen, J., Zhou, Y., Zipf, A., and Fan, H. 2019. Deep learning from multiple crowds: A case study of humanitarian mapping. *IEEE Transactions on Geoscience and Remote Sensing*, 57(3), 1713–1722. http://doi.org/10.1109/TGRS.2018.2868748

Chen, Z., Wei, Y., Shi, K., Zhao, Z., Wang, C., Wu, B., Qiu, B., and Yu, B. 2022. The potential of nighttime light remote sensing data to evaluate the development of digital economy: A case study of China at the city level. *Computers, Environment and Urban Systems*, 92, 101749. ISSN 0198–9715. https://doi.org/10.1016/j.compenvurbsys.2021.101749. https://www.sciencedirect.com/science/article/pii/S0198971521001563

Coppola, D., et al. 2020. Thermal remote sensing for global volcano monitoring: Experiences from the MIROVA system. *Frontiers in Earth Science*, 7. http://doi.org/10.3389/feart.2019.00362

Deshpande, M.V., Kumar, N., Pillai, D., Krishna, V.V., and Jain, M. 2023. Greenhouse gas emissions from agricultural residue burning have increased by 75 % since 2011 across India. *Science of the Total Environment*, 904, 166944. ISSN 0048–9697. https://doi.org/10.1016/j.scitotenv.2023.166944. https://www.sciencedirect.com/science/article/pii/S0048969723055699

Doll, C.N.H., Muller, J.P., and Morley, J.G. 2006. Mapping regional economic activity from night-time light satellite imagery. *Ecological Economics*, 57(1), 75–92. ISSN 0921–8009. http://doi.org/10.1016/j.ecolecon.2005.03.007

Dong, Y., Li, Q., Dou, A., and Wang, X. 2011. Extracting damages caused by the 2008 Ms 8.0 Wenchuan earthquake from SAR remote sensing data. *Journal of Asian Earth Sciences*, 40(4), 907–914. ISSN 1367–9120. http://doi.org/10.1016/j.jseaes.2010.07.009

Elvidge, C.D., Baugh, K., Ghosh, T., Zhizhin, M., Feng, Chi Hsu, Sparks, T., Bazilian, M., Sutton, P.C., Houngbedji, K., & Goldblatt, R. 2022. Fifty years of nightly global low-light imaging satellite observations. *Frontiers in Remote Sensing*, 3, 16. http://doi.org/10.3389/frsen.2022.919937

Elvidge, C.D., Baugh, K.E., Kihn, E.A., Kroehl, H.W., and Davi, E.R. 1997. Mapping city lights with nighttime data from the DMSP operational linescan system. *Photogrammetric Engineering and Remote Sensing*, 63(6), 727–734.

Elvidge, C.D., Zhizhin, M., Hsu, F.-C., and Baugh, K. 2013. What is so great about nighttime VIIRS data for the detection and characterization of combustion sources? *Proceedings of the Asia-Pacific Advanced Network*, 35, 33–48. http://doi.org/10.7125/APAN.35.5

Fatolazadeh, F., Eshagh, M., Goïta, K., and Wang, S.A. 2022. New spatiotemporal estimator to downscale GRACE gravity models for terrestrial and groundwater storage variations estimation. *Remote Sensing*, 14, 5991. https://doi.org/10.3390/rs14235991

Fensholt, R., Langanke, T., Rasmussen, K., Reenberg, A., Prince, S.D., Tucker, C., Scholes, R.J., Le Q.B., Bondeau, A., Eastman, R., Epstein, H., Gaughan, A.E., Hellden, U., Mbow, C., Olsson, L., Paruelo, J., Schweitzer, C., Seaquist, J., and Wessels, K. 2012. Greenness in semi-arid areas across the globe 1981–2007—an Earth Observing Satellite based analysis of trends and drivers. *Remote Sensing of Environment*, 121, 144–158. ISSN 0034–4257. http://doi.org/10.1016/j.rse.2012.01.017

Ganci, G., Cappello, A., Bilotta, G., & Negro, C.D. 2020. How the variety of satellite remote sensing data over volcanoes can assist hazard monitoring efforts: The 2011 eruption of Nabro volcano. *Remote Sensing of Environment*, 236, 111426. ISSN 0034–4257. https://doi.org/10.1016/j.rse.2019.111426. https://www.sciencedirect.com/science/article/pii/S0034425719304456

Ghazaryan, G., König, S., Rezaei, E.E., Siebert, S., and Dubovyk, O. 2020. Analysis of drought impact on croplands from global to regional scale: A remote sensing approach. *Remote Sensing*, 12(24), 4030. https://doi.org/10.3390/rs12244030

Ghulam, A., Li, Z.L., Qin, Q., Yimit, H., and Wang, J. 2008. Estimating crop water stress with ETM+ NIR and SWIR data. *Agricultural and Forest Meteorology*, 148(11), 1679–1695. ISSN 0168–1923. http://doi.org/10.1016/j.agrformet.2008.05.020

Graw, V., Dubovyk, O., Duguru, M., Heid, P., Ghazaryan, G., de León, J.C.V., Post, J., Szarzynski, J., Tsegai, D., and Walz, Y. 2019. Chapter 9—Assessment, monitoring, and early warning of droughts: The potential for satellite remote sensing and beyond. In: Mapedza, E., Tsegai, D., Bruntrup, M., and Mcleman, R. (Eds.), *Current Directions in Water Scarcity Research*, Volume 2. Elsevier, 115–131. ISSN 2542–7946. ISBN: 9780128148204. https://doi.org/10.1016/B978-0-12-814820-4.00009-2. https://www.sciencedirect.com/science/article/pii/B9780128148204000092

Greenough, P.G., and Nelson, E.L. 2019. Beyond mapping: A case for geospatial analytics in humanitarian health. *Conflict and Health*, 13, 50. https://doi.org/10.1186/s13031-019-0234-9

Gu, Y., Shao, Z., Huang, X., and Cai, B. 2022. GDP forecasting model for China's provinces using nighttime light remote sensing data. *Remote Sensing*, 14(15), 3671. https://doi.org/10.3390/rs14153671

Guirado, E., Alcaraz-Segura, D., Cabello, J., Puertas-Ruíz, S., Herrera, F., and Tabik, S. 2020. Tree cover estimation in global drylands from space using deep learning. *Remote Sensing*, 12(3), 343. https://doi.org/10.3390/rs12030343

Hanadé Houmma, I., Gadal, S., El Mansouri, L., Garba, M., Gbetkom, P.G., Mamane Barkawi, M.B., and Hadria, R. 2023. A new multivariate agricultural drought composite index based on random forest algorithm and remote sensing data developed for Sahelian agrosystems. *Geomatics,Natural Hazards and Risk*, 14(1). https://doi.org/10.1080/19475705.2023.2223384

Hancke, G.P., Silva, B.C., and Hancke Jr., G.P. 2013. The role of advanced sensing in smart cities. *Sensors*, 13(1), 393–425.

Hao, Z., Xia, Y., Luo, L., Singh, V.P., Ouyang, W., and Hao, F. 2017. Toward a categorical drought prediction system based on U.S. Drought Monitor (USDM) and climate forecast. *Journal of Hydrology*, 551, 300–305. ISSN 0022–1694. https://doi.org/10.1016/j.jhydrol.2017.06.005. https://www.sciencedirect.com/science/article/pii/S0022169417304043

Hassan, M.M., Hasan, I., Southworth, J., and Loboda, T. 2022. Mapping fire-impacted refugee camps using the integration of field data and remote sensing approaches. *International Journal of Applied Earth Observation and Geoinformation*, 115, 103120. ISSN 1569–8432. https://doi.org/10.1016/j.jag.2022.103120. https://www.sciencedirect.com/science/article/pii/S1569843222003089

Hoornweg, D., and Pope, K. 2017. Population predictions for the world's largest cities in the 21st century. *Environment and Urbanization*, 29(1), 195–216. https://doi.org/10.1177/0956247816663557

Hoque, M.A., Phinn, S., Roelfsema, C., and Childs, I. 2017. Tropical cyclone disaster management using remote sensing and spatial analysis: A review. *International Journal of Disaster Risk Reduction*, 22, 345–354. ISSN 2212–4209. https://doi.org/10.1016/j.ijdrr.2017.02.008. https://www.sciencedirect.com/science/article/pii/S2212420916304794

Ichoku, C., Kahn, R., and Chin, M. 2012. Satellite contributions to the quantitative characterization of biomass burning for climate modeling. *Atmospheric Research*, 111, 1–28. ISSN 0169–8095. http://doi.org/10.1016/j.atmosres.2012.03.007

Jachowski, N.R.A., Quak, M.S.Y., Friess, D.A., Duangnamon, D., Webb, E.L., and Ziegler, A.D. 2013. Mangrove biomass estimation in Southwest Thailand using machine learning. *Applied Geography*, 45, 311–321. ISSN 0143–6228. http://doi.org/10.1016/j.apgeog.2013.09.024

Jiao, W., Tian, C., Chang, Q., Novick, K.A., and Wang, L. 2019. A new multi-sensor integrated index for drought monitoring. *Agricultural and Forest Meteorology*, 268, 74–85. ISSN 0168–1923. https://doi.org/10.1016/j.agrformet.2019.01.008. https://www.sciencedirect.com/science/article/pii/S0168192319300085

Jiao, W., Wang, L., and McCabe, M.F. 2021. Multi-sensor remote sensing for drought characterization: Current status, opportunities and a roadmap for the future. *Remote Sensing of Environment*, 256, 112313. ISSN 0034–4257. https://doi.org/10.1016/j.rse.2021.112313. https://www.sciencedirect.com/science/article/pii/S0034425721000316

Jones, C., and Kammen, D.K. 2014. Spatial distribution of U.S. household carbon footprints reveals suburbanization undermines greenhouse gas benefits of urban population density. *Journal Environmental Science & Technology (ES&T)*, 48(2), 895–902. http://doi.org/10.1021/es4034364

Judith, C., Schneider, J.V., Schmidt, M., Ortega, R., Gaviria, J., and Zizka, G. 2013. Using high-resolution remote sensing data for habitat suitability models of Bromeliaceae in the city of Mérida, Venezuela. *Landscape and Urban Planning*, 120, 107–118. ISSN 0169–2046. http://doi.org/10.1016/j.landurbplan.2013.08.012

Kaku, K., and Held, A. 2013. Sentinel Asia: A space-based disaster management support system in the Asia-Pacific region. *International Journal of Disaster Risk Reduction*, 6, 1–17. ISSN 2212–4209. http://doi.org/10.1016/j.ijdrr.2013.08.004

Karanam, V., Motagh, M., Garg, S., and Jain, K. 2021. Multi-sensor remote sensing analysis of coal fire induced land subsidence in Jharia Coalfields, Jharkhand, India. *International Journal of Applied Earth Observation and Geoinformation*, 102, 102439. ISSN 1569–8432. https://doi.org/10.1016/j.jag.2021.102439. https://www.sciencedirect.com/science/article/pii/S030324342100146X

Klinlampu, C., Chimprang, N., and Sirisrisakulchai, J. 2023. The sufficient level of growth in renewable energy generation for coal demand reduction. *Energy Reports*, 9(Supplement 10), 843–849. ISSN 2352–4847. https://doi.org/10.1016/j.egyr.2023.05.203. https://www.sciencedirect.com/science/article/pii/S2352484723008843

Kloos, S., Yuan, Y., Castelli, M., and Menzel, A. 2021. Agricultural drought detection with MODIS based vegetation health indices in Southeast Germany. *Remote Sensing*, 13(19), 3907. https://doi.org/10.3390/rs13193907

Knipper, K., Anderson, M., Bambach, N., Kustas, W., Gao, F., Zahn, E., Hain, C., McElrone, A., Belfiore, O.R., Castro, S., et al. 2023. Evaluation of partitioned evaporation and transpiration estimates within the DisALEXI modeling framework over irrigated crops in California. *Remote Sensing*, 15, 68. https://doi.org/10.3390/rs15010068

Kuang, W., and Dou, Y. 2020. Investigating the patterns and dynamics of urban green space in China's 70 major cities using satellite remote sensing. *Remote Sensing*, 12(12), 1929. https://doi.org/10.3390/rs12121929

Lang, S., Füreder, P., and Rogenhofer, E. 2018. Earth observation for humanitarian operations. In: Al-Ekabi, C., and Ferretti, S. (Eds.), *Yearbook on Space Policy 2016. Yearbook on Space Policy*. Springer, Cham. https://doi.org/10.1007/978-3-319-72465-2_10

Leeper, R.D., Bilotta, R., Petersen, B., Stiles, C.J., Helm, R., Fuchs, B., Prat, O.P., Palecki, M., and Ansari, S. 2022. Characterizing U. S. drought over the past 20 years using U. S. drought monitor. *International Journal of Climatology*, 42(12), 6616–6630. https://doi.org/10.1002/joc.7653

Li, W., Batty, M., and Goodchild, M.F. 2020. Real-time GIS for smart cities. *International Journal of Geographical Information Science*, 34(2), 311–324. http://doi.org/10.1080/13658816.2019.1673397

Liu, X., Zhu, X., Zhang, Q., Yang, T., Pan, Y., and Sun, P. 2020. A remote sensing and artificial neural network-based integrated agricultural drought index: Index development and applications. *CATENA*, 186, 104394. ISSN 0341–8162. https://doi.org/10.1016/j.catena.2019.104394. https://www.sciencedirect.com/science/article/pii/S0341816219305363

Ma, Q., Gong, Z., Kang, J., Tao, R., and Dang, A. 2020. Measuring functional urban shrinkage with multi-source geospatial big data: A case study of the Beijing-Tianjin-Hebei Megaregion. *Remote Sensing*, 12(16): 2513. https://doi.org/10.3390/rs12162513

Mateos, R.M., Sarro, R., Díez-Herrero, A., Reyes-Carmona, C., López-Vinielles, J., Ezquerro, P., Martínez-Corbella, M., Bru, G., Luque, J.A., Barra, A., et al. 2023. Assessment of the socio-economic impacts of extreme weather events on the coast of Southwest Europe during the period 2009–2020. *Applied Sciences*, 13, 2640. https://doi.org/10.3390/app13042640

Mavrouli, M., Mavroulis, S., Lekkas, E., and Tsakris, A. 2023. The impact of earthquakes on public health: A narrative review of infectious diseases in the post-disaster period aiming to disaster risk reduction. *Microorganisms*, 11, 419. https://doi.org/10.3390/microorganisms11020419

Mekonnen, M.M., and Hoekstra, A.Y. 2014. Water footprint benchmarks for crop production: A first global assessment. *Ecological Indicators*, 46, 214–223. ISSN 1470–160X. http://doi.org/10.1016/j.ecolind.2014.06.013

Mieville, A., Granier, C., Liousse, C., Guillaume, B., Mouillot, F., Lamarque, J.F., Grégoire, J.M., and Pétron, G. 2010. Emission of gases and particles from biomass burning during the 20th century using satellite data and a historical reconstruction. *Atmospheric Environment*, 44, 1469–1477.

Mu, Q., Zhao, M., Kimball, J., McDowell, N., and Running, S. 2013. A remotely sensed global terrestrial drought severity index. *Bulletin of the American Meteorological Society*, 94, 83–98.

Munawar, H.S., Mojtahedi, M., Hammad, A.W.A., Kouzani, A., and Parvez Mahmud, M.A. 2022. Disruptive technologies as a solution for disaster risk management: A review. *Science of the Total Environment*, 806(Part 3), 151351. ISSN 0048–9697. https://doi.org/10.1016/j.scitotenv.2021.151351. https://www.sciencedirect.com/science/article/pii/S0048969721064299

Norton, C.L., Hartfield, K., Holifield Collins, C.D., van Leeuwen, W.J.D., and Metz, L.J. 2022. Multi-temporal LiDAR and hyperspectral data fusion for classification of semi-arid woody cover species. *Remote Sensing*, 14(12), 2896. https://doi.org/10.3390/rs14122896

O'Sullivan, J.N. 2023. Demographic delusions: World population growth is exceeding most projections and jeopardising scenarios for sustainable futures. *World*, 4, 545–568. https://doi.org/10.3390/world4030034

Pérez-Hoyos, A., Rembold, F., Kerdiles, H., and Gallego, J. 2017. Comparison of global land cover datasets for cropland monitoring. *Remote Sensing*, 9, 1118. https://doi.org/10.3390/rs9111118

Pinzon, J.E., Pak, E.W., Tucker, C.J., Bhatt, U.S., Frost, G.V., and Macander, M.J. 2023. *Global Vegetation Greenness(NDVI) from AVHRR GIMMS-3G+, 1981–2022*. ORNL DAAC, Oak Ridge, TN. https://doi.org/10.3334/ORNLDAAC/2187

Pinzon, J.E., and Tucker, C.J. 2014. A non-stationary 1981–2012 AVHRR NDVI3g time series. *Remote Sensing*, 6, 6929–6960. https://doi.org/10.3390/rs6086929

Poland, M.P., Lopez, T., Wright, R. et al. 2020. Forecasting, detecting, and tracking volcanic eruptions from space. *Remote Sensing in Earth Systems Sciences*, 3, 55–94. https://doi.org/10.1007/s41976-020-00034-x

Prosperi, P., Bloise, M., Tubiello, F.N. et al. 2020. New estimates of greenhouse gas emissions from biomass burning and peat fires using MODIS Collection 6 burned areas. *Climatic Change*, 161, 415–432. https://doi.org/10.1007/s10584-020-02654-0

Qiu, L., Chen, Y., Wu, Y., Xue, Q., Shi, Z., Lei, X., Liao, W., Zhao, F., and Wang, W. 2021. The water availability on the Chinese loess plateau since the implementation of the grain for green project as indicated by the evaporative stress index. *Remote Sensing*, 13, 3302. https://doi.org/10.3390/rs13163302

Ramírez-Moreno, M.A., Keshtkar, S., Padilla-Reyes, D.A., Ramos-López, E., García-Martínez, M., Hernández-Luna, M.C., Mogro, A.E., Mahlknecht, J., Huertas, J.I., Peimbert-García, R.E., et al. 2021. Sensors for sustainable smart cities: A review. *Applied Sciences*, 11(17), 8198. https://doi.org/10.3390/app11178198

Rojas, O. 2020. Agricultural extreme drought assessment at global level using the FAO-Agricultural Stress Index System (ASIS). *Weather and Climate Extremes*, 27, 100184. ISSN 2212–0947. https://doi.org/10.1016/j.wace.2018.09.001. https://www.sciencedirect.com/science/article/pii/S2212094718300999

Sakellariou, S., Sfougaris, A., Christopoulou, O., and Tampekis, S. 2022. Integrated wildfire risk assessment of natural and anthropogenic ecosystems based on simulation modeling and remotely sensed data fusion. *International Journal of Disaster Risk Reduction*, 78, 103129. ISSN 2212–4209. https://doi.org/10.1016/j.ijdrr.2022.103129. https://www.sciencedirect.com/science/article/pii/S221242092200348X

Sánchez, N., González-Zamora, A., artínez-Fernández, J., Piles, M., and Pablos, M. 2018. Integrated remote sensing approach to global agricultural drought monitoring. *Agricultural and Forest Meteorology*, 259, 141–153. ISSN 0168–1923. https://doi.org/10.1016/j.agrformet.2018.04.022. https://www.sciencedirect.com/science/article/pii/S016819231830145X

Shi, Y., Zang, S., Matsunaga, T., and Yamaguchi, Y. 2020. A multi-year and high-resolution inventory of biomass burning emissions in tropical continents from 2001–2017 based on satellite observations. *Journal of Cleaner Production*, 270, 122511. ISSN 0959–6526. https://doi.org/10.1016/j.jclepro.2020.122511. https://www.sciencedirect.com/science/article/pii/S0959652620325580

Smith, W.K., Dannenberg, M.P., Yan, D., Herrmann, S., Barnes, M.L., Barron-Gafford, G.A., Biederman, J.A., Ferrenberg, S., Fox, A.M., Hudson, A., Knowles, J.F., MacBean, N., Moore, D.J.P., Nagler, P.L., Reed, S.C., Rutherford, W.A., Scott, R.L., Wang, X., and Yang, J. 2019. Remote sensing of dryland ecosystem structure and function: Progress, challenges, and opportunities. *Remote Sensing of Environment*, 233, 111401. ISSN 0034–4257. https://doi.org/10.1016/j.rse.2019.111401. https://www.sciencedirect.com/science/article/pii/S0034425719304201

Sorooshian, S., AghaKouchak, A., Arkin, P., Eylander, J., Foufoula-Georgiou, E., Harmon, R., and Hendrickx, J. 2011. Advanced concepts on remote sensing of precipitation at multiple scales. *Bulletin of the American Meteorological Society*, 92(10), 1353–1357. http://doi.org/10.1175/2011BAMS3158.1

Spalding, M., Blasco, F., Field, C. 1997. *World Mangrove Atlas*. International Society for Mangrove Ecosystems, Okinawa, Japan, p. 178.

Surono, J.P., Pallister, J., Boichu, M., Buongiorno, M.F., Budisantoso, A., Costa, F., Andreastuti, S., Prata, F., Schneider, D., Clarisse, L., Humaida, H., Sumarti, S., Bignami, C., Griswold, J., Carn, S., Oppenheimer, C., and Lavigne, F. 2012. The 2010 explosive eruption of Java's Merapi volcano—A '100-year' event. *Journal of Volcanology and Geothermal Research*, 241–242, 121–135. ISSN 0377–0273. http://doi.org/10.1016/j.jvolgeores.2012.06.018.

Tadesse, T., Champagne, C., Wardlow, B.D., Hadwen, T.A., Brown, J.F., Demisse, G.B., . . . Davidson, A.M. 2017. Building the vegetation drought response index for Canada (VegDRI-Canada) to monitor agricultural drought: First results. *GIScience & Remote Sensing*, 54(2), 230–257. https://doi.org/10.1080/15481603.2017.1286728

Taubenböck, H., & Esch, T. 2011. Remote sensing—An effective data source for urban monitoring. *Earthzine IEEE Magazine*.

Thenkabail, P.S., Gumma, M.K., Teluguntla, P., and Mohammed, I.A. 2014. Hyperspectral remote sensing of vegetation and agricultural crops. Highlight article. *Photogrammetric Engineering and Remote Sensing*, 80(4), 697–709.

Thenkabail, P.S., Nolte, C., and Lyon, J.G. 2000a. Remote sensing and GIS modeling for selection of benchmark research area in the inland valley agroecosystems of West and Central Africa. *Photogrammetric Engineering and Remote Sensing, Africa Applications Special Issue*, 66(6), 755–768.

Thenkabail, P.S., Smith, R.B., and De-Pauw, E. 2000. Hyperspectral vegetation indices for determining agricultural crop characteristics. *RemoteSensing of Environment*, 71, 158–182.

Traore, S.B., Ali, A., Tinni, S.H., Samake, M., Garba, S., Maigari, I., Alhassane, A., Samba, A., Diao, M.B., Atta, S., Dieye, P.O., Nacro, H.B., and Bouafou, K.G.M. 2014. AGRHYMET: A drought monitoring and capacity building center in the West Africa Region. *Weather and Climate Extremes*, 3, 22–30. ISSN 2212–0947. http://dx.doi.org/10.1016/j.wace.2014.03.008

Tucker, C.J., Pinzon, J.E., Brown, M.E., Slayback, D.A., Pak, E.W., Mahoney, R., Vermote, E.F., and Saleous, N.E. 2005. An extended AVHRR 8-km NDVI dataset compatible with MODIS and SPOT vegetation NDVI data. *International Journal of Remote Sensing*, 26, 4485–4498. https://doi.org/10.1080/01431160500168686

Tzavella, K., Skopeliti, A., and Fekete, A. 2024. Volunteered geographic information use in crisis, emergency and disaster management: A scoping review and a web atlas. *Geo-Spatial Information Science*, 27(2), 423–454. https://doi.org/10.1080/10095020.2022.2139642

Ullah, Z., Al-Turjman, F., Mostarda, L., and Gagliardi, R. 2020. Applications of artificial intelligence and machine learning in smart cities. *Computer Communications*, 154, 313–323. ISSN 0140–3664. https://doi.org/10.1016/j.comcom.2020.02.069. https://www.sciencedirect.com/science/article/pii/S014036641932082 1

Van Hoolst, R., Eerens, H., Haesen, D., Royer, A., Bydekerke, L., Rojas, O., . . . Racionzer, P. 2016. FAO's AVHRR-based Agricultural Stress Index System (ASIS) for global drought monitoring. *International Journal of Remote Sensing*, 37(2), 418–439. https://doi.org/10.1080/01431161.2015.1126378

Voigt, S., Tetzlaff, A., Zhang, J., Künzer, C., Zhukov, B., Strunz, G., Oertel, D., Roth, A., Dijk, P.V., and Mehl, H. 2004. Integrating satellite remote sensing techniques for detection and analysis of uncontrolled coal seam fires in North China. *International Journal of Coal Geology*, 59(1–2), 121–136. ISSN 0166–5162. http://dx.doi.org/10.1016/j.coal.2003.12.013

Wang, Q., Wu, J., Lei, T., He, B., Wu, Z., Liu, M., Mo, X., Geng, G., Li, X., Zhou, H., and Liu, D. 2014. Temporal-spatial characteristics of severe drought events and their impact on agriculture on a global scale. *Quaternary International*. Available online 5 July 2014. ISSN 1040–6182. http://doi.org/10.1016/j.quaint.2014.06.021.

Wei, W., Zhang, J., Zhou, L., et al. 2021. Comparative evaluation of drought indices for monitoring drought based on remote sensing data. *Environmental Science and Pollution Research*, 28, 20408–20425. https://doi.org/10.1007/s11356-020-12120-0

Williams, J.G., Rosser, N.J., Kincey, M.E., Benjamin, J., Oven, K.J., Densmore, A.L., Milledge, D.G., Robinson, T.R., Jordan, C.A., and Dijkstra, T.A. 2018. Satellite-based emergency mapping using optical imagery: Experience and reflections from the 2015 Nepal earthquakes. *Natural Hazards and Earth System Sciences*, 18, 185–205. https://doi.org/10.5194/nhess-18-185-2018.

Wu, B., Ma, Z., and Yan, N. 2020. Agricultural drought mitigating indices derived from the changes in drought characteristics. *Remote Sensing of Environment*, 244, 111813. ISSN 0034–4257. https://doi.org/10.1016/j.rse.2020.111813. https://www.sciencedirect.com/science/article/pii/S0034425720301838

Wu, Y., Li, C., Shi, K., Liu, S., and Chang, Z. 2022. Exploring the effect of urban sprawl on carbon dioxide emissions: An urban sprawl model analysis from remotely sensed nighttime light data. *Environmental Impact Assessment Review*, 93, 106731. ISSN 0195–9255. https://doi.org/10.1016/j.eiar.2021.106731. https://www.sciencedirect.com/science/article/pii/S0195925521001815

Xu, G., Xiu, T., Li, X., Liang, X., and Jiao, L. 2021. Lockdown induced night-time light dynamics during the COVID-19 epidemic in global megacities. *International Journal of Applied Earth Observation and Geoinformation*, 102, 102421. ISSN 1569–8432. https://doi.org/10.1016/j.jag.2021.102421. https://www.sciencedirect.com/science/article/pii/S0303243421001288

Yan, S., Shi, K., Li, Y., et al. 2020. Integration of satellite remote sensing data in underground coal fire detection: A case study of the Fukang region, Xinjiang, China. *Frontiers in Earth Science*, 14, 1–12. https://doi.org/10.1007/s11707-019-0757-9

Zarei, A.R., Mokarram, M., and Mahmoudi, M.R. 2023. Comparison of the capability of the meteorological and remote sensing drought indices. *Water Resources Management*, 37, 769–796. https://doi.org/10.1007/s11269-022-03403-x

Zeng, J., Zhou, T., Qu, Y., et al. 2023. An improved global vegetation health index dataset in detecting vegetation drought. *Scientific Data*, 10, 338. https://doi.org/10.1038/s41597-023-02255-3

Zhang, B., Wu, Y., Lei, L., Li, J., Liu, L., Chen, D., and Wang, J. 2013. Monitoring changes of snow cover, lake and vegetation phenology in Nam Co Lake Basin (Tibetan Plateau) using remote SENSING (2000–2009). *Journal of Great Lakes Research*, 39(2), 224–233. ISSN 0380–1330. http://doi.org/10.1016/j.jglr.2013.03.009.

Zhang, J., Zhang, Y., Cong, N. Tan, L., Zhao, G., Zheng, Z., and Gao, J. 2023. Coarse spatial resolution remote sensing data with AVHRR and MODIS miss the greening area compared with the Landsat data in Chinese drylands. *Frontiers in Plant Science Section Functional Plant Ecology*, 14. https://doi.org/10.3389/fpls.2023.1129665

Zhao, M., Cheng, C., Zhou, Y., Li, X., Shen, S., and Song, C. 2022. A global dataset of annual urban extents (1992–2020) from harmonized nighttime lights. *Earth System Science Data*, 14, 517–534. https://doi.org/10.5194/essd-14-517-2022.

Zheng, Q., Seto, K.C., Zhou, Y., You, S., and Weng, Q. 2023. Nighttime light remote sensing for urban applications: Progress, challenges, and prospects. *ISPRS Journal of Photogrammetry and Remote Sensing*, 202, 125–141. ISSN 0924–2716. https://doi.org/10.1016/j.isprsjprs.2023.05.028. https://www.sciencedirect.com/science/article/pii/S0924271623001521

Zheng, Q., Weng, Q., and Wang, K. 2021. Characterizing urban land changes of 30 global megacities using nighttime light time series stacks. *ISPRS Journal of Photogrammetry and Remote Sensing*, 173, 10–23. ISSN 0924–2716. https://doi.org/10.1016/j.isprsjprs.2021.01.002. https://www.sciencedirect.com/science/article/pii/S0924271621000022

Zhu, Z., and Zhang, Y. 2022. Flood disaster risk assessment based on random forest algorithm. *Neural Computing and Applications*, 34, 3443–3455. https://doi.org/10.1007/s00521-021-05757-6

Index

Note: Page numbers in *italics* indicate a figure and page numbers in **bold** indicate a table on the corresponding page.

A

Action Contre la Faim (ACF) approach, 24–25, 429
    livestock movements, integration with, 27–28
    pasture monitoring, 26, *26*, 27
    surface water monitoring, 26–27, *28*
Active Fire, 127, 291
active lavas, 246, 265
    thermal signature detection, erupting volcanoes, 265–269, *266*, *268*
    thermo-physical characteristics quantification, active lava bodies, 269–270, *270*
active volcanism, 246–247, 442
active volcanoes, 243, 246, 275
advanced microwave scanning radiometer-earth observation system (AMSR-E), 91, 94, 435
advanced spaceborne thermal emission spectrometer (ASTER), 315, 326, 327, 445
advanced very high resolution radiometer (AVHRR), 61, 75
    brightness temperature, 274–275
    coal fire areas, 315, 385
    crop yields, 170
    droughts, 158
    hydrometeorological hazards, 157
aerosol optical depth (AOD), 289, 293, *300*, 444
Agenzia Spaziale Italiana (ASI), 117
agricultural drought, 65
    applications, 68
    microwave surface soil moisture retrievals, 66–68
    monitoring systems, *see* agricultural drought monitoring systems
agricultural drought frequency change (ADFC), 427
agricultural drought monitoring systems, 17, 33, 67–68
    agricultural production, negative effects, 17–18
    biophysical parameters, 427–430, *428*
    definition, agricultural drought, 17
    food security, 18
    indices, 18–19
    low spatial resolution geostationary satellites, properties of, 20, **21**
    maximum value composite procedure (MVC), 20
    new satellite sensors, 32
    operational methods, 20–21, **31**
        Action Contre la Faim (ACF) approach targeting pastures, 24–28, *26–28*
        FAO agriculture stress index system (ASIS) approach, 24, *25*
        joint research centre (JRC) approach, early detection of biomass production deficit hot spots, 28–29
        NDVI-based vegetation anomaly indicators, 21–24, *22*
    rainfall estimate datasets, 19, **19**
    recent methodological approaches, 29–32
    vegetation health methods, 426–427
    vegetation health status, 19
    vegetation indices (VIs), 19–20
    water balance approach, 18–19
agricultural stress index system (ASIS), 24, *25*, 429

Airborne LiDAR data, 165, 338, 352
air mass factor (AMF), 292, 302
ALOS-PALSAR, 117, **218**
artificial illumination/lighting, 383, 394
artificial light at night (ALAN), 394, 395
atmosphere-land exchange inverse (ALEXI)
    energy balance model, 58, 61
    land-surface temperature, 58, 61, 432, *432*
    TSEB model, regional application, 60
atmospheric correction and haze reduction (ATCOR), 232, 319
atmospheric infrared sounder (AIRS), 74, 260
automated feature extraction, *221*, 221–222, *222*
autonomous modular sensor-wildfire (AMS), 444

B

biomass burning
    active fires, 291, 294, *294*
    aerosols, 288, 303
    correlations, 295–301
    descriptive statistics, 293
    Earth observation, 288
    fire-NO_2 relationships, testing, 289–291, *290*
    fire radiative power (FRP), 291
    greenhouse gas emissions, 287–288, 302–303, 442–444, *443*
    NO_2, 288–289, *see also* NO_2
    OMI-NO_2, 291–292, **292**
    SCIAMACHY-NO_2, 292
    time-series regression, 293–294, 295–301, *296*, **297**, *298–301*
    wildfires, 287
bi-spectral infrared detection (BIRD), 315, 445, *447*

C

catchment LSM (CLSM), 69–71
classification and regression tree (CART), 30, 90
climate change, 17, 111, 137, 162, 337, 382
climate hazards group infrared precipitation with station data (CHIRPS), 134, *135*
climate prediction center (CPC), 75, 433
climatic research unit timeseries (CRU TS), 134
coal fires, 309
    cause of, *310*, 310–312
    changing sensor characteristics, 326–327
    classification of, 309–310
    electromagnetic spectrum, regions of, **311–312**
    hazards, 313–314
    methods used, remote sensing
        atmospheric correction and emissivity compensation, 319
        land surface temperature estimation, 319–320
        multi-source data analysis for improved detection, 321
        thresholding for fire delineation, 320–321
        time-series analysis for fire monitoring, 321–322
    monitoring literature, 326

461

related parameters, 326
remote sensing, 444–446, *445*, *447*
   evolution, 315
   history, 314–315
   important satellite sensors, characteristics of, **317**
   spatial and temporal resolution, 315–316, *316*
   spectral resolution, 316–318
research, 325–326
results, 322–324, *323–325*
space agencies, future plans, 327–328
surface coal fires, 314
combined drought index (CDI), 23, 31
convention to combat desertification (CCD), 112
convolutional neural networks (CNN), 118, 167, 440
Copernicus emergency management service (CEMS), 174, 186, 205
Copernicus hyperspectral imaging mission for the environment (CHIME), 117, 118
crop moisture index (CMI), 19, 42, 90, 427
crop-specific NDVI (CNDVI) method, 23
CubeSats, 176, 275, 388
cumulative value of FAPAR (CFAPAR), 28–29, *30*

D

damage mapping, *171*, 172–174, 188
data assimilation systems, 32, 67–68, 69–74
data-mining techniques, 30, 429
defense meteorological satellite program (DMSP), 338, 349–350, 385, 399, 448, 452
   data, 350
   DN values and urban land, 350
   imagery, 401–402, *403*, 408–409
   nighttime lights, 338, 349–350
   time series, *396*, 396–397, 400
defense meteorological satellite program's operational linescan system (DMSP-OLS), 383, 388, 391, *453*
defense navigation satellite system (DNSS), 159
deformation, 248, 251–254
   measurements, 251, 253
   monitoring, 247–248
   signal, 250–251
differential optical absorption spectroscopy (DOAS), 292
digital elevation models (DEMs), 157, 365, 370, *371*
DigitalGlobe, 185, 206, 226
digital number (DN), 350, 448
digital surface model (DSM), 365, 365, 370, *371*
disaster monitoring constellation (DMC), 176, **181**
disaster risk management (DRM), 155, 437–439
   conceptualization, 156
   definition of, 155
   domain developments
      early disaster mapping, *156*, 156–157
      satellite era, 157–159, *158*
   gaps and limitations
      methodological gaps, 187
      military interests, 186–187
      standards and suitable legislation, lack of, 187–188
   operational remote sensing
      elements at risk, mapping of, 166–168, *167*
      hazard assessment, EO data, 162, 188
      key terms, 159, **160–161**
      monitoring and early warning, potentially hazardous situations, 169–171
      post-disaster recovery, 176
      post-disaster response and damage assessment, *171*, 171–175, *173–175*
      utility, natural hazard types, **163–164**
      vulnerability assessment, 168–169
   trends and developments, 188
      better data analysis methods, 185
      new platforms, 176–179, *178*, **180–183**
      new sensors, 179, **184**
      organization, 185–186
drought, 59–72, 74–76
   assessments, 68, 75
   conditions, 62, 74–75
   definition, 40, 88
   extreme climate, 39
   frequency and magnitude, 39
   hydro-meteorological variables, 88
   indicators, 62
   information, 63, 74
   intensity, 7, 68
   measuring
      remote sensing approach, 4–5
      traditional approach, 4
   monitoring, **89**
   origination, 7
   remote sensing in United States, 39
   remote sensing techniques, 88–89, *89*
   timing, 7
   21st century, 6–7, *7*
   types of, 88
   in the United States, 39, *40*, *see also* drought monitoring in United States
   vegetation health–based products, 7–11, *8–10*
   vegetation health (VH) method, 5, *6*, 13
   weather disaster, 3–4
drought monitoring, 63, 69–70, 72, 75
   applications, 67
   approaches, 90–91
   capabilities, 74
   evaporative stress index (ESI), 429
   GRACE, 432
   microwave-based surface soil moisture, 432
   remote sensing-based drought indices, 432–436, *433–434*
   tools, 74
   US drought monitor (USDM), 430
   vegetation drought response index (VegDRI), 430
   vegetation outlook (VegOut), 430
   weather satellites, 3–4, 12
drought monitoring and prediction
   data assimilation for, 92
   machine learning, 94, 102
   multi-sensor data combination issues, 94
drought monitoring in United States
   new monitoring tools
      evaporative stress index (ESI), 58–63
      forest drought index (FORDRI), 50–57, *51*, *54*, *55*, *57*
      microwave-based surface soil moisture retrievals, 63–68
      satellite gravimetry-based soil moisture, 68–74
      vegetation drought response index (VegDRI), 45–50, *46*, *50*
   satellite-based earth observing systems, 45

Index

station-based approaches, 40–42
traditional approaches, 40, 42, 44–45
USDM, 42–44, *43*
drought severity index (DSI), 434
drought/wetness indicators, 72, 74
drylands, 110
    aridity index, *110*
    assessing land condition, 119
        biological productivity of ecosystems, 119–120, *120*
        climate and its variability, 120–122, *121*
    climate change, 111, 137–138
    degradation and desertification, 112–113
    earth observation platforms, 115, *115*
        additional assessment data, 118
        coarse spatial scale satellite sensors, 116–117
        medium spatial resolution sensors, 115–116
        obtaining earth observation time series, 117
    ecosystems, conservation of, 138
    four broad biomes, 110
    integrated concepts, assessing land degradation, 128
        local scale, 128–130, *129*
        regional to global scale, 130–132, *131*
    land use, 111–112
    monitoring of land use/land cover, 122–124, *123*
        local-scale studies, 124, *125*
        regional- to global-scale studies, 125–128, *127*
    remote sensing, 436–437
    remote sensing, degradation processes, 113–114
        suitable indicators, 114–115
    remote sensing data uncertainties
        archives and their analysis, 133
        observation period, 133–135, *134–135*
        spatial scale, 136, *136*
    scientific perception, land degradation, 113
    supporting services, **111**
    time series analysis techniques, 118–119
    uncertainties and limits, 132
        land degradation, delineation of, 132
        remote sensing data, 133–136
    water scarcity, 110
drylands development paradigm (DDP), 128
Durbin–Watson (DW) statistic, 293–294
Dutch OMI-NO$_2$ (DOMINO), 291–292

E

Earth observation (EO), 157, 202, 235
Earth observation group (EOG), 452
Earth resources technology satellites (ERTS-1), 158–159
economic values, 363–364, 367, 369, 373–374
    of non-market goods, 363
elements at risk (EaR), 155, 166, 168
El Niño-Southern Oscillation (ENSO), 124
emitted spectral radiance, 262, 264, 267, 319
enhanced thematic mapper plus (ETM+), 115, 315, 432, *433*, 435
enhanced vegetation index (EVI), 114
environmental mapping and analysis program (EnMAP), 117, 386
environmental pollution, 313–314
European centre for medium-range weather forecasts (ECMWF), 26
European commission joint research center (EC-JRC), 137
European space agency (ESA), 205

evaporative stress index (ESI), 57, 57–58
    algorithm, 60–61
    applications of, 62–63, *63*
    drought conditions, 60–61, 91, 429, 430
    examples of, 61–62, *62*
    input data, 61
    regional application, TSEB model, 60
    two-source energy balance (TSEB) mode, 58–59, *59*
EvIDENZ (Earth Observation–based Information Product for Drought Risk Reduction on the National Level), 432

F

famine early warning system network (FEWSNET), 22, 430
farm service agency (FSA), 43, 430
fire radiative power (FRP), 291, 294, 302
fire temperature estimation, 319–320
flash drought, 60, 67–68, 74, 91
food and agriculture organization (FAO), 9, 22, 24, 429, 446
food security, 29, 208, 427
    drought, 17–18
forest drought index (FORDRI), 50–57, *51*, *54*, *55*, *57*
fraction of absorbed photosynthetically active radiation (FAPAR), 20, 31–32, 114
frequency magnitude analysis, 162

G

GeoEye image, *171*, *174*, 185
geographically independent integrated drought index (GIIDI), 435
geographic information system (GIS) tools, 159, 341, 352
global ozone monitoring experiment (GOME), 288–289, 301
global positioning system (GPS), 159, 247
gravity recovery and climate experiment (GRACE), 45
    GWS anomalies, 53
    hydrologic variables, 91
    microgravity measurements, 432
    natural seasonal variations, 68
    terrestrial water storage, 69–74
greenhouse gas (GHG) emissions
    biomass burning, 287–288, 442–444, *443*
    coal energy, 444, 445

H

hedonic analyses, 364, 367–369, 373–374
humanitarian-development-peace (HDP), 211
humanitarian disasters, 200, 202, 207, 219–220, 440
    case of Abyei, 227–232, *228–231*
    crisis-related earth-observable indicators, 208
        crisis monitoring, *209*, 209–210
        early warning tool, 208–209
        mid- to long-term impact, 210–211
    Earth observation capacities
        disaster and crisis initiatives, **215–216**
        EO data, usage of, 211
        nano-/microsatellites, 218–219
        optical sensors, 212–217, **214**, **216–217**
        radar sensors, 217–218, **218**
        sensors, 212
        unmanned aerial vehicles, 219

IDP camp Zam Zam, Sudan, 226–227
image analysis techniques
    automated feature extraction, 221–222, *222*
    ground reference information, 223
    image classification, 221–222, *222*
    population monitoring, 219
    spatial analysis and modeling, 222, *223*
    visual image interpretation, 219–221, *220*
logging and mining activities, Democratic Republic of the Congo (DRC), 232–235, *234*
population dynamics monitoring, refugee camp Dagahaley, Kenya, 224–226, *225*
regional conflicts, 200–202
satellite remote sensing, role of
    indication based on time series, 204
    information needs, 207–208
    objective imaging device, 202
    space policy and regulations, 205–206
    very high resolution (VHR) imagery, 202, 203
    vs. field mapping, 204, *205*
humanitarian OpenStreetMap team (HOT), 185, 203, 203, 206, 206
human settlement index (HSI), 402
HydroGenerator, 26–27

I

instantaneous field of view (IFOV), 260–262, 266–267
interferograms, 248, 250–252, 276
interferometric synthetic aperture radar (InSAR), 165, 248, *249*, 250, 252–254, 276
internally displaced persons (IDPs), 200, 200, *201*, 226, 232, 232
    camp Zam Zam, Sudan, 226–227
international network of crisis mappers (ICCM), 206, **215**
international space station (ISS), 386, 400, 404
internet of things (IoT), 439, 450
isovist analyses, 365–366
isovist-oriented three-dimensional (3D) indexes, 366

J

Jharia coalfield, India, 312, 320, *323*, *324*
joint research centre (JRC), 20, 28–29, 429
    early detection of biomass production deficit hot spots, 28–29

L

land degradation neutrality (LDN), 113, 131–132
Landsat Thematic Mapper (TM), 115, 315, 362, 385
land surface temperature (LST)
    daytime, 352, 355, 432
    drought, 92, 429–430, *430*
    grid cells, 339, 351, 352
    information, 58
    local applications, 60–61
    mean daytime/nighttime, 355
    multidate, 352, 357
    retrievals, 60
light detection and ranging (LiDAR), 118, 157
    airborne data, 162, 165
        3D modeling, 449, 450
        data, 363–364, 370

light emitting diode (LED) lighting, 404–405
linking relief, rehabilitation and development (LRRD) concept, 211
local climate zone (LCZ), 337, 339, 351–352, 355–357
low Earth orbit, 254–255

M

machine learning, 94
Madden-Julian oscillation (MJO), 94
medium resolution imaging spectrometer (MERIS), 116
medium spatial resolution sensors, *115*, 115–116
microwave-based surface soil moisture retrievals, 63–65, *63–65*, 432
    algorithm development and implementation, 65–66
    applications of, 66–68
    development and implementation, 65–66
    input data, 65
microwave integrated drought index (MIDI), 92, 435
middle-infrared (MIR) wavelengths, 262, 445
moderate resolution imaging spectrometer (MODIS), 44, 315, 319, 326, 338, 429–431

N

national oceanic and atmospheric administration (NOAA), 41, *41*, 62
    AVHRR, 4, 42, 116, 125, 130, 133
    GVI, 5
    NESDIS, 61, 427
    NGDC, 385, 391
    satellites, 157, 158
national polar-orbiting operation environmental satellite system (NPOESS), 13, 385, 427
natural resources
    market goods, 363, 364, 374
    mid- to long-term monitoring, 211
    non-market goods, 363–364
    regional conflicts, 201–202
near infrared (NIR), 4, 44, 226, 436, 444
near real-time (NRT), 18, 20, 29, 267, 275–276
net primary productivity (NPP), 110, 114, 120, 436
nighttime light pollution (NLP), 392–393, 394
    economy, 394–395
    environment, 394
    human health, 393–394
    main consequences of, *393*
    night vision, 393
nighttime light remote sensing, 381–383, 450–452, *453*
    applications, 391–392, **395**, **401**
        nighttime light pollution, *see* nighttime light pollution (NLP)
        socioeconomic variables, 392
        urban extent, 392
    challenges in, 399–401, 408
    combining with other modalities, 409
    DMSP/OLS
        data archive, 388, 391
        intercalibrating, time series, 402, 404
        LED lighting, transition to, 404–405
        NTL imagery, correcting saturation, 401–402, *403*
        VIIRS/DNB and, intercalibrating, 402, 404
    fine and medium spatial resolution, 409–410
    fine resolution nighttime lights, 405

Index

applications of, 405–406
  ground measurements, 406–408, *407*
future directions, 408–410
history of, 385–388, *387*
rationale underlying, 383–384, *384*
sensors, **389–390**
urban dynamics
  mapping, *396*, 396–397
  social events tracking, 399
  socioeconomic changes, 397–399, *398*
urbanization, 381
VIIRS/DNB data archive, 391, 400, 404–405
NO_2, 288–289
  fire-NO_2 relationships, testing, 289–291, *290*
  OMI-NO_2, 291–292, **292**, *300*
  SCIAMACHY-NO_2, 292, 297, *300*
  temporal and seasonal variations, 294–295, *295*
normalized difference infrared index (NDII), 91
normalized difference vegetation index (NDVI), 4–5, 19–20, 44–45, 90, 114
  based vegetation anomaly indicators, 20, 21–24, *22*
  emissivity, determining, 319
  MODIS-based, 52, 402
normalized difference water index (NDWI), 45, 92, 95, 99, 101
normalized multi-band drought index (NMDI), 45

O

object-based image analysis (OBIA), 166–167, *167*, 169, 213, 221, 233
objective blend drought index (OBDI), 90, 433
OpenStreetMap (OSM), 203, 440
operational drought monitoring, 66, 68, 70, 74, 76
operational linescan system (OLS), 338, 383, 385, 399, 448, 452
optical sensors, 212–213, **214**, 217
  direct mapping, 212
  disaster and crisis initiatives, **215–216**
  key characteristics and application-relevant features, **216–217**
ozone monitoring instrument (OMI), 257–258, 288–289, 444
  OMI-NO_2, 291–292, **292**, 295

P

palmer drought severity index (PDSI), 4, 11, 41–42, 89, 427, 432
  drought monitoring, 435
  limitation of, 42
  modified, *41*
  self-calibrated, 47, 48, 90
pasture monitoring, 26, *26*–27
percent annual seasonal greenness (PASG), 46–47, 90, 430
phased array type L-band synthetic aperture radar (PALSAR), 91, 179, 338, 356
phase unwrapping, 250
photochemical reflectance index (PRI), 32–33, 429
population dynamics monitoring, refugee camp Dagahaley, Kenya, 224–226, *225*
population monitoring, 219
potential evapotranspiration (PET), 52, 92, 98, 110, 430

Prais-Winsten time-series regression, 293–294, 297, **297**
precipitation condition index (PCI), 435

R

radar sensors, 217–218, **218**
rain use efficiency (RUE), 122, 126, 130, 136
rangeland and pasture productivity (RaPP) map, 124
reclamation drought index (RDI), 19, 427
regional drought monitoring, 87–90
  development of the SDCI
    advanced SDCI, 98–99, 105
    components, 95
    humid regions, 95
    validation of, 95–98, *96–98*
    vs. USDM, 95–97, *96–97*, **99**
    linearly combined SDCI, Korean peninsula, 99–100, *100–101*, 104
    nonlinearly combined SDCI, USA, 101–104, **102**, *103*, **104**
  recent trends
    customized monitoring approaches, 90–91
    data assimilation, 92
    machine learning, 94, 102
    multi-sensor data combination, 94
    new drought indices, 91–92, **93**
  type of, 88–89, **89**
residual trend analysis (RESTREND) method, 120–121, 126, 133

S

satellite gravimetry-based soil moisture and groundwater, 68–69
  application, 70–74, *71*, *73*
  development and implementation, 69–70
  input data
    GRACE/FO terrestrial water storage, 69
    meteorological data, 69
scaled drought condition index (SDCI), 90, *96–98*, 97, 104–105, 434, 435
scaled drought condition index (SDCI), development of
  advanced SDCI, 98–99, 105
  components, 95
  humid regions, 95
  linearly combined, Korean peninsula, 99–100, *100–101*, 104
  nonlinearly combined, USA, 101–104, **102**, *103*, **104**
  validation of, 95–98, *96–98*
  vs. USDM, 95–97, *96–97*, **99**
Scanning Imaging Absorption Spectrometer for Atmospheric Cartography (SCIAMACHY), 257–258, 288–289, 295, 442, 444
  SCIAMACHY-NO_2, 292, 297
self-calibrated PDSI (SC-PDSI), 47, 48, 90
Sentinel Asia program, 186
Sentinel Asia Step 3, 437, *439*
Shanghai, multi-sensor approach
  LCZ maps, 356, **357**, *358*, *359*
  spatial–temporal analyses, 356–357, **357**
shortwave infrared (SWIR), 262–263, 266–267, 310, **311**
Shuttle Radar Topography Mission, 186, 251, 254, 341
Sky View Factor, 352, 365, **366**
smart environment, 363–364, 374

smart environment design
    hedonic analysis, 367–369, **368**
    natural resources
        market goods, 363
        non-market goods, 363–364
    remote sensing, 362–363
    three-dimensional (3D) view analysis of Yokohama City, Japan, 369–374, *371–373*
    three-dimensional (3D) view indexes, 365–367
    urban analysis, 362–363
    viewshed-oriented indexes, 367
    visibility analysis
        classical approaches, 364–365
        isovist analyses, 365, **366**
        viewshed analyses, 365
socioeconomic variables, 392
soil moisture active passive (SMAP), 45, 66–68, 430
Soil Moisture Agricultural Drought Index (SMADI), 429
soil moisture and ocean salinity (SMOS), 66, 91, 429
soil moisture condition index (SMCI), 435
soil moisture drought/agricultural drought, 88, **89**
Soil Wetness Deficit Index (SWDI), 92
solar radiation estimation, 363, 365, 450, *450*
Spatial and Temporal Adaptive Reactance Fusion Model (STARFM), 117, 136, 436
Spatial Multi Criteria Evaluation (SMCE), 168
Spatial Openness Index, 365, 366, **366**
Spatio-Temporal Temperature-Based Thresholding (STTBT) algorithm, 321
spectral mixture analysis (SMA), 114, 436, *436*
SPOT (Satellite Pour l'Observation de la Terre) satellites, 116, 159, 166, 448
SPOT-VEGETATION, 19, 26, 429
Standardized Precipitation and Evapotranspiration Index (SPEI), 42, 52, 91, 99
Standardized Precipitation Evapotranspiration Index (SPEI), *41*, 42
Standardized Precipitation Index (SPI), 19, 42, 47, 52, 90, 427
start of season anomaly (SOSA), 47, 90, 430
Strength and Weakness Constraint Method (SWCM), 321
structure from motion (SfM) technique, 178, 185
subsurface coal fires, 314, 316, 322, 445, 445–446
sulfur dioxide (SO$_2$)
    thermal infrared, 258–260
    ultraviolet, 255–258, *256*
surface coal fires, 309–310, 314
surface water monitoring, 26–27, *28*
Surface Water Supply Index (SWSI), 19, 42, 427, 430
Surrey Satellite Technology Limited (SSTL), 176, **181**
Sustainable Development Science Satellite 1 (SDGSAT-1), 386–387, *387*
synthetic aperture radar (SAR), 65, 159, 338, 356

T

temperature condition index (TCI), 42, 427, *434*, 435
TerraSAR-X, 440, *441*, *443*
terrestrial water storage (TWS), 68–72, 75–76
thermal anomalies, 261–263, 267–268
thermal emission, 263, 267
thermal infrared (TIR), 45
    data process, 60–61
    methods of, 67
    remote sensing, 310, 314–315
    satellite data, 271, 274
    satellite image, 45
    sulfur dioxide emissions, 258–260, *259*
thermal signatures, 265, 267, 273
three-dimensional (3D) view analysis of Yokohama City, Japan
    evaluation, 370–372, *371–373*
    hedonic analysis results, 373–374, *374*
    map, *369*
time series analysis techniques, 118–119
time-series regression, in biomass burning
    methods of, 293
    results of, 295–296, **297**, *298–301*
Tokyo, multi-sensor approach
    DMSP, 349–351, *350*
    land cover changes *vs.* population census, 346–349, **347**, *348*, **349**
    LCZ maps, 352, **353**, *354*, 355
    local climate zones *vs.* land surface temperatures, 351–356, *352*, **353**, *354*, 355
    population census, 349–351, *350–351*
    spatial-temporal changes, 339–346, 340, **341–342**, *342–346*
    urban area, 349–351, *350–351*
total ozone mapping spectrometer (TOMS), 159, 442
transfer function analysis (TFA) method, 126
tropical rainfall measurement mission (TRMM), 92, 95, 434, *434*, 435
tsunami, 154, 169, 176
two-source energy balance (TSEB) land-surface model, 58, 60

U

ultraviolet light, 255–258, *256*
UNITAR's operational satellite application program (UNOSAT), 206
United Nations convention to combat desertification (UNCCD), 112, 114
United Nations environment programme (UNEP), 137
United Nations office for outer space affairs (UNOOSA), 206
United Nations platform for space-based information for disaster management and emergency response (UN-SPIDER), 440
United States drought monitor (USDM), 89, 90, 96, *96–97*
United States geological survey (USGS), 116
unmanned aerial vehicles (UAVs), 177, *178*, 203, 219
urban extents, 392, **395**, 450
urban growth and climatic mapping, 337–339
    climate change, 337
    grid cell process, 339
    monitoring and mapping, 338
    multisource sensor systems, 338
    Shanghai, multi-sensor approach
        LCZ maps, **356**, **357**, *358*, *359*
        spatial–temporal analyses, 356–357, **357**
    Tokyo, multi-sensor approach
        DMSP, 349–351, *350*
        land cover changes *vs.* population census, 346–349, **347**, *348*, **349**

LCZ maps, 352, **353**, *354*, *355*
local climate zones *vs.* land surface temperatures, 351–356, *352*, **353**, *354*, *355*
population census, 349–351, *350–351*
spatial-temporal changes, 339–346, 340, **341–342**, *342–346*
urban area, 349–351, *350–351*
urban heat islands (UHI), 337, 448
urban remote sensing, 446–448, *447*
US drought monitor (USDM), 42–44, *43*, 72, 74, 430
US geological survey (USGS), 208
US Land Remote Sensing Policy Act (1992), 159

V

vapor pressure deficit (VPD), 53, 74
vegetation adjusted normalized urban index (VANUI), 402, *403*
vegetation condition index (VCI)
agricultural drought monitoring systems, *431*
in malaria cases, 11
normalized difference vegetation index, 5, 23, 42, 44, 427, 435, *435*
vegetation drought response index (VegDRI), 45–46
applications of, 49–50
classification scheme, 47
examples of, 49, *50*
methodology of, *46*, 48
production system, 48, 49
remote sensing-based advances, 430
vegetation health index (VHI)
agricultural droughts, 427
agriculture stress index system, 24, *25*
color-coded map of, 5, *6*
normalized difference vegetation index, 42
satellite-based, 90
and temperature condition index, 9, 11
vegetation health (VH) method, 4, 5, 12
agricultural droughts using, 426–427
of NOAA/NESDIS, 427
vegetation index based on universal pattern decomposition method (VIUPD), 435
vegetation indices (VIs), 19–20
vegetation outlook (VegOut), 430
vegetation stress, 61–62, 67
vegetation water stress index (VWSI), 435
very high-resolution imagery (VHRI), 436, 438, 448
data, 206, 219
satellite imagery, 220, 227, 233

very high resolution radiometer (VHRR), 157
viewshed analysis, 365, 365–367
viewshed-oriented indexes, 366–367
viewshed-oriented three-dimension (3D) indexes, 366
VIIRS day and night band (DNB), 386, 399
during 2012 Ramadan, 399, *400*
for continuous time series generation, 404
data archives, 391
DMSP/OLS, Suomi NPP and, 409
VIIRS daytime/nighttime image, 322, *324*, 327
visible and shortwave infrared drought index (VSDI), 92
visible infrared imager radiometer suite (VIIRS), 13, 44, 386, 391, 452
visual image interpretation, 219
volcanic ash advisory centers (VAACs), 270
volcanic SO_2, 255–260
volcano(es), 243
active volcanism, 246–247
ash clouds, 270–271, 270–275
crater lakes, 263–264
deformation, 247–248, 254, 275
InSAR data, *249*
magma bodies, 252–254
surface field, 248–252, *249*
topography, 254
degassing, 254, 258
effusive eruptions, 265
hot spot detection algorithm, *268*
lava bodies, 269–270, *270*
thermal data, 269–270, *270*
thermal signature, 265–269, *266*, *268*
gas emissions, 254
geothermal and hydrothermal activity, 261–265, *262*, *264*
measurements, 244–246, *245*, *247*
MODIS channels, *272*, *273*
plumes, 246, 257–259
SO_2
thermal infrared, 258–260
ultraviolet, 255–258, *256*
topography, 247, 254–255
volunteer and technical communities (V&TC), 202, 203, 206
volunteered geographic information (VGI), 185, 202, 203

W

water deficit index (WDI), *433*, 435
wetland water area index (WWAI), 92
WorldView-2 images, 224, 226